T0332042

Cell Boundaries

Cell Boundaries

How Membranes and Their Proteins Work

Stephen H. White
Gunnar von Heijne
Donald M. Engelman

CRC Press
Taylor & Francis Group
Boca Raton London New York

CRC Press is an imprint of the
Taylor & Francis Group, an **informa** business

A GARLAND SCIENCE BOOK

Cover image: Wassily Kandinsky "Circles in a Circle" (1923). The Philadelphia Museum of Art: The Louise and Walter Arensberg Collection, 1950, 1950-134-104.

First edition published 2022
by CRC Press
6000 Broken Sound Parkway NW, Suite 300, Boca Raton, FL 33487-2742

and by CRC Press
2 Park Square, Milton Park, Abingdon, Oxon, OX14 4RN

ISBN: 978-0-815-34216-8 (hbk)
ISBN: 978-0-367-35716-0 (pbk)
ISBN: 978-0-429-34132-8 (ebk)

DOI: 10.1201/9780429341328

Typeset in Utopia Std
by Deanta Global Publishing Services, Chennai, India

Printed in the UK by Severn, Gloucester on responsibly sourced paper

Instructors can access the Figure Slides at the Instructor Hub, upon registering (https://routledgetextbooks.com/textbooks/instructor_downloads/).

Contents

List of Boxes

Preface

This is a book about membranes and proteins. Membranes, because they're central to all life; proteins, because—well, because the three of us have all had life-long love affairs with membrane proteins. And love sometimes makes you take on totally crazy projects, like writing a textbook ….

So here we are, 10+ years after we first sat down to hash out a list of chapters that we thought should go into a book on membranes and proteins. Our main ambition has been to bridge between the worlds of physical chemistry/biophysics, on the one hand, and molecular cell biology, on the other, without losing ourselves completely in either one. Meaning that most readers will find the treatment of their own favorite subject both inadequate and superficial, while other areas have been given far too much weight—in fact, this is precisely what we've hoped to achieve, so please let us know!

To help everyone get off to a good start, the book opens with Chapter 0: The "E" words. Or, spelt out: Energy, Enthalpy, and Entropy. Thermodynamics, in short, the favorite topic of every serious biochemist. For everyone else, Steve came up with this little pitch: "Thermodynamics can be a daunting subject and is intimidating at first to almost everyone, including the authors of this text. But persevere and keep in mind that it took some very smart people, such as Helmholtz, 150 years to develop the science, which is essential for gaining a deep understanding of biology." You don't want to miss Chapter 0.

Chapters 1–4 deal with the basics of biological membranes, lipid bilayers, and the interactions of peptides and proteins with lipid bilayers. As we have this old-men's idea that a bit of history never hurts, we try to trace the origins of some of the more important concepts as we go along.

We then shift focus to look at membranes from the point of view of molecular cell biology. How are proteins moved across cellular membranes and between membrane compartments? How are membrane proteins synthesized and inserted into different types of membranes? How do proteins help shape the cell's membranes? And can we relate some of these phenomena back to the underlying physics of protein–lipid interactions? These are questions we ask in Chapters 5–7.

Certainly, the most spectacular progress in the field of membranes and proteins in the last decade has been seen in structural biology, mainly x-ray crystallography and electron cryo-microscopy. Fortunately, we are slow writers, or this book would have been finished well before the current explosion of high-resolution membrane protein structures—on the other hand, our job would have been a lot easier then... Anyway, the final Chapters 8–15 include a primer on structure determination techniques and present the most important classes of membrane proteins: channels, transporters, the big protein complexes involved in bioenergetics, and cell-surface receptors involved in signaling. Rather than trying to be encyclopedic, our aim has been to extract some basic principles of how these different kinds of proteins actually work.

Well, that's it, folks. Thirteen long years in the making. Enjoy, as they say in our favorite Thai place.

Oh, except that there's a chapter on Membrane Protein Bioinformatics hidden in there somewhere. See if you can find it.

Steve, Gunnar, and Don

Acknowledgements

We are pleased to acknowledge and thank Bob Rogers of Garland Science Press who recognized the need for a "membrane book" and had confidence that our team was the one to do it. Once initiated, we were guided and encouraged by the Garland Science editors and associates, including Denise Schanck, Summers Scholl, Michael Roberts, Monica Toledo, and Kelly O'Connor. We are deeply grateful for their help and patience. We are especially pleased to recognize and thank Nigel Orme for his renderings of the book's figures. Finally, we are indebted to Jordan Wearing and his staff at Taylor & Francis Group for managing the production of the book.

The knowledge required to write a book such as this ranges beyond what our normal brains can encompass. The process had to, necessarily, make use of the critical views of many colleagues whose perceptions augmented and corrected ours. Their generosity in engaging with us in the project has reduced the number of erroneous views (but almost certainly has not eliminated them), amplified whatever insights we have managed to provide, and expanded the content and clarity of the visual materials. Their critical input and new discoveries have allowed us to learn as we progressed through the material, which continued to evolve as we worked, often requiring backtracking. In many cases, we have chosen to include original figures from the literature, and authors and journals have allowed us to use them, often without compensation. For each of these contributions, we applaud our community both individually and collectively.

We are grateful, indeed, to these colleagues: Åke Wieslander, Alexey Ladokhin, Anne K. Kenworthy, Anne Vidaver, Barbara Pearse, Ben de Kruijff, Bertil Hille, Charles Brooks, Chris Miller, Daniel DiMaio, David Drew, David Eisenberg, Douglas Rees, Doug Tobias, Eaton Lattman, Eric Gouaux, Erik Lindahl, Frederick Richards, Giuseppi Zaccai, Hartmut Michel, Jackie Dooley, Jaime Requena, James Bowie, James Rothman, Janos Lanyi, Jean-Luc Popot, Jeff Abramson, Jere Segrest, John Deacon, John Hunt, Kai Simons, Karen Fleming, Kim Sharp, Lucy Forrest, Lukas Tamm, Magnus Andersson, Mark Bretscher, Mark Gerstein, Mark Lemmon, Martin Caffrey, Mary Luckey, Melanie Cocco, Michael Edidin, Michael Sheetz, Nathan Joh, Olaf Andersen, Ole Mouritsen, Oleg Andreev, Olga Boudker, Paula Booth, Peter Brzezinksi, Peter Glazer, Peter Walter, Peter Moore, Pietro De Camilli, Richard Chamberlain, Richard Henderson, Robert Eisenberg, Robert Lefkowitz, Robert Stroud, Rod MacKinnon, Ron Kaback, Thomas Heimburg, Thomas McIntosh, Thomas Steitz, Tobias Baumgart, Werner Kühlbrandt, William Catterall, William F. Degrado, Yana Reshetnyak, Yong Xiong, and, finally, Robert Zelyk of the Bavarian Arms Museum.

PyMol (Schrödinger GmbH) was used for molecular graphics.

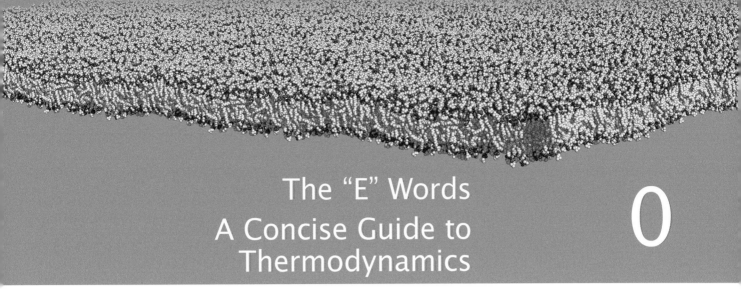

The "E" Words
A Concise Guide to Thermodynamics

0

Biological systems must obey the laws of thermodynamics. This became obvious to Hermann Helmholtz, a 19th-century German physician, who invented the ophthalmoscope (1851) and revolutionized the physiology of vision. His studies of heat production during muscle contraction led him into thermodynamics and ultimately, the enunciation of the law of conservation of energy. Helmholtz discredited the idea of "vitalism"—the belief in a life force distinct from purely chemical or physical ones—by showing that all matter, including living creatures, must obey the same physical laws. Helmholtz's work revealed the deep connection between biology and thermodynamics; understanding one elucidates the other.

The science of thermodynamics is based mainly upon two fundamental principles derived from empirical observations: the First and Second Laws of Thermodynamics. Because absolutely no exceptions to these principles have ever been observed, the principles are referred to as Laws. This introduction to *Cell Boundaries* is about these laws and the meaning and use of the "E" words that are an essential part of thermodynamics: energy, entropy, and enthalpy. Thermodynamics can be a daunting subject and is intimidating at first to almost everyone, including the authors of this text. But persevere and keep in mind that it took some very smart people, such as Helmholtz, 150 years to develop the science, which is essential for gaining a deep understanding of biology.

Cell Boundaries emphasizes quantitative approaches to research on cell membranes. Although a deep knowledge of thermodynamics is not required, some basic knowledge will help you appreciate the crucial role of thermodynamic thinking in biology. We present compactly in this Guide the essential ideas, concepts, and formulas that we believe to be useful for students of biology. If you are already familiar with thermodynamics, we hope the Guide will be a useful review and reference. If you are not yet comfortable with thermodynamics, we hope the Guide will provide some clarity and show the utility of thermodynamics in biochemistry and biophysics. If this is your first introduction to thermodynamics, we suggest that you proceed slowly section by section and give yourself time to reflect on the concepts in the course of exploring *Cell Boundaries*.

0.1 WORK, HEAT, AND ENERGY

The "E" words allow us to describe common physical experience quantitatively and with mathematical rigor. For example, heat a block of metal in a flame, and the temperature will rise; place it in an ice bath, some of the ice will melt, and the temperature of the block will decrease. Eventually, at equilibrium, the temperature of the block will equal the temperature of the bath. This simple experiment describes our experience qualitatively. How can we describe it quantitatively?

DOI: 10.1201/9780429341328-1

1

The first step is to define some terms that can be linked mathematically. The flame provides heat; the ice takes heat away by melting. Heat is a fundamental thermodynamic variable, which we represent by the symbol q.

We keep track of the general state of the block by measuring its temperature. What are we measuring, exactly? Although this may not be immediately obvious, we are measuring the internal energy of the block, which is directly proportional to temperature. We sense intuitively that a hotter block must have more energy. This sounds useful, so let's give **energy** its own symbol, **E.** It's our first "E" word.

The temperature of a block of metal can be raised in many different ways. One way is simply to bore a hole in it with a drill. Whenever we do that, we find that the temperature of the block rises due to the friction between drill and block. From the fatigue of our muscles during drilling, we sense that we are doing work. Let's represent work by the symbol w. The First Law concerns the conservation of energy; it allows one to describe mathematical relationships between q, E, and w.

0.2 FIRST LAW OF THERMODYNAMICS

A relationship between heat and work was surmised in the late 18th century by Count Rumford (Sir Benjamin Thompson) (**Figure 0.1**). While watching brass cannons being bored with iron drills turned by the power of horses at the Munich arsenal, he observed that the temperature of cannon barrels increased due to the work of the horses. He realized that the heat arising from friction between the drill and the barrel must have been generated by the horses' work. This heat could be removed by cooling the cannon with water. (It is said that he amazed the citizens of Munich by using the hot cannon to boil water!) The Count's discovery of the mechanical equivalence of heat helped establish the First Law. It was not until the 19th century, however, that the mechanical equivalence of heat was determined quantitatively by James Joule (**Figure 0.2**).

The First Law was first stated most clearly by Rudolph Clausius in the middle of the 19th century:

> Work is produced by the agency of heat, a quantity of heat is consumed which is proportional to the work done; and conversely, by the expenditure of an equal quantity of work an equal quantity of heat is produced.

In Joule's experiment, the amount of water and the chemical properties of the water were unchanged; only the temperature changed. How much the temperature changes for a given amount of work depends upon the physical properties of the material. To account for inherent properties of materials, Clausius connected the work–heat interchange by introducing the internal energy E

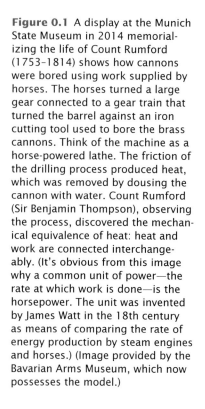

Figure 0.1 A display at the Munich State Museum in 2014 memorializing the life of Count Rumford (1753–1814) shows how cannons were bored using work supplied by horses. The horses turned a large gear connected to a gear train that turned the barrel against an iron cutting tool used to bore the brass cannons. Think of the machine as a horse-powered lathe. The friction of the drilling process produced heat, which was removed by dousing the cannon with water. Count Rumford (Sir Benjamin Thompson), observing the process, discovered the mechanical equivalence of heat: heat and work are connected interchangeably. (It's obvious from this image why a common unit of power—the rate at which work is done—is the horsepower. The unit was invented by James Watt in the 18th century as means of comparing the rate of energy production by steam engines and horses.) (Image provided by the Bavarian Arms Museum, which now possesses the model.)

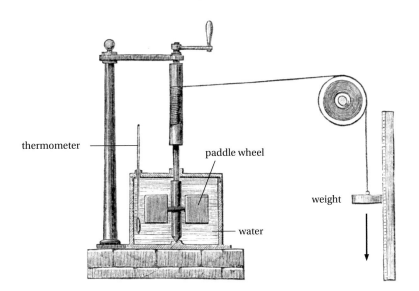

Figure 0.2 James Joule established a quantitative relationship between heat and work using this apparatus. The weight with mass m fell under the influence of gravity to turn the paddle wheel. Friction between the wheel and water raised the water temperature, demonstrating that the energy produced by the falling weight was converted to heat.

of the material. This critical step allowed Clausius to restate the First Law as follows:

> In a thermodynamic process involving a closed system, the change in the internal energy is equal to the difference between the heat accumulated by the system and the work done by it.

By a closed system, we mean one that does not exchange mass with its surroundings. More restrictively, an isolated system is a closed system that additionally cannot exchange heat or work with its surroundings. The introduction of the concept of internal energy was profoundly important, because it led to a connection, for example, between the kinetic theory of gases and thermodynamics.

0.2.1 The First Law Accounts for the Observations of Count Rumford

The idea of internal energy helps us understand more clearly—and quantitatively—Count Rumford's observations. At the start of the day, the temperature of the undisturbed cannon barrel was T_1, which we associate with the internal energy E_1. At the end of the day after the horses had done work (w) on a cannon, the temperature and energy had increased to T_2 and E_2, respectively. During the boring, however, water had been used to cool the cannon, taking away some heat q. The difference in energy ΔE between the beginning of the day (E_1) and the end of the day (E_2) required by the First Law is

$$E_2 - E_1 = \Delta E = w + q \qquad (0.1)$$

By established international convention, work or heat flowing into a system—in this case, the cannon—is defined as positive. Thus, in Count Rumford's experiment, w is positive and q negative. Thinking in terms of the mathematics of calculus, invented by Isaac Newton and Gottfried Leibnitz during the 17th century, the total energy difference ΔE is the integrated sum of infinitesimally small changes in energy dE given by

$$\Delta E = \int_1^2 dE = E_2 - E_1 \qquad (0.2)$$

Because we are primarily interested in the properties and behavior of substances, such as the bronze of Count Rumford's cannons, we distinguish thermodynamic variables that are characteristic of the substance from those that are not. The internal energy is a characteristic of the cannon, but q and w are not. Regardless of the exact way in which the work was done or the heat was taken away, at the end of the day, the temperature had changed by ΔT and internal energy by ΔE. E is what we call a state variable, because it describes the state of the system. The heat lost and the work done, on the other hand, are not state variables, because they are not inherent characteristics of the material. For infinitesimal changes in E in response to small changes in δq and δw in heat and work, we write the First Law as

$$dE = \delta q + \delta w \tag{0.3}$$

The small change in E is written as a differential dE, meaning that it is a state variable that can be treated as a differential familiar from calculus, whereas the changes in q and w are indicated by the symbol δ, because they are not state variables.

0.2.2 The Thermodynamics of Ideal Gases Reveal Quantitative Insights into the First Law

Horses and cannons are cumbersome to think about. Consider instead a simpler and more tractable system: a hollow cylinder in which, using a sliding piston, we can compress or expand an ideal gas consisting of particles that thermally collide in the absence of attractive or repulsive interactions such as those between charged particles. Ideal gases obey the equation $PV = nRT$, where P is pressure, V volume, and n the number of moles of gas, T the absolute temperature, and R the gas constant (defined later). At temperatures and pressures comfortable to humans, most gases, such as nitrogen and oxygen, behave nearly ideally.

The system consists of a cylinder of cross-sectional area A (containing a fixed mass of gas m) that is closed on one end and has a moveable piston at the other to control the cylinder's volume V (**Figure 0.3**). Let's also equip the cylinder with a thermometer for measuring the temperature T of the gas. If we apply a force F to the piston, we exert a pressure $P = F/A$ on the gas. If we move the piston a distance dy to change the volume by $dV = Ady$, the work done is force times distance: $\delta w = Fdy$ or equivalently, $\delta w = PdV$. Actually, because there is friction between the piston and the cylinder wall, some of the work will be lost due to friction, which

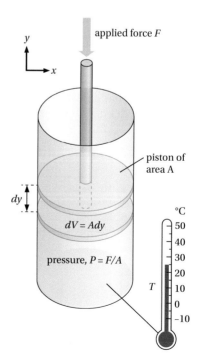

Figure 0.3 A hollow cylinder with a moveable piston provides a simple conceptual system for thinking about the First Law in the context of compression of an ideal gas. Work is done by a force that moves the piston over a distance *dy*, and the temperature rises.

is an irreversible process. For all that follows, let's assume frictionless machines. Although such machines do not exist, we can imagine that we could reduce the friction to zero by careful machining of surfaces and by moving the piston in small steps very slowly (frictional heat loss is proportional to velocity).

In the absence of friction, then, work can be done reversibly, and we write it as $\delta w_{rev} = PdV$. This type of work is called PV or mechanical work. This helps us understand the connection between energy and temperature, because for ideal gases, PV is proportional to T. In modern Standard International (SI) units, energy is measured in joules, but in older literature, centimeter-gram-seconds (CGS) units are used, and energy is measured in ergs. If we are doing only mechanical work, then the First Law requires that

$$dE = \delta q - PdV \tag{0.4}$$

If the cylinder is heated while holding the volume constant ($dV = 0$), all the heat goes toward increasing the internal energy of the gas: $dE = \delta q$. Integrating, $\Delta E = q_V$, where the V subscript means that an amount of heat q_V flowed into the cylinder at constant volume.

Suppose we pack around the cylinder a perfect insulating material (i.e., one that will not conduct heat) so that heat cannot enter or leave our cylinder from the outside environment. This is an isolated system. In that case, $\delta q = 0$ in Equation 0.4, and all the work done on the system during an infinitesimal compression increases the energy of the gas by $dE = -PdV$. The change in the energy of the system for a compression from V_1 to V_2 is $\Delta E = E_2 - E_1 = -P(V_2 - V_1)$.

0.2.3 Enthalpy Is the Most Convenient Measure of Heat in the Laboratory

Most biochemical laboratory experiments are carried out at constant atmospheric pressure. They could in principle be carried out at constant volume, but that is difficult to achieve without the use of finicky equipment. It is easy to produce a constant pressure in our cylinder by simply placing a weight of fixed mass M (kilograms) on the piston to produce a force (newtons) $F = Mg$, where g is the acceleration due to gravity (9.8 m s^{-2}).

$$\Delta E = E_2 - E_1 = q_P - w = q_P - P(V_2 - V_1) \tag{0.5}$$

The heat transfer at constant pressure (q_P) is easily found:

$$E_2 + PV_2 - (E_1 + PV_1) = q_P \tag{0.6}$$

The heat flow causes changes in internal energy and mechanical work. By doing our lab experiments at constant pressure, we see that PV work is done. But we really don't care about the work due to expansion or contraction of our beaker or flask, which is unrelated to the state of the substance contained in the beaker or flask. Because constant-pressure heat transfer is readily measured in calorimeters, we define a new thermodynamic quantity called **enthalpy (H)**, our second "E" word:

$$H \equiv E + PV \tag{0.7}$$

This is convenient, because Equation 0.6 then becomes

$$\Delta H = q_P \tag{0.8}$$

This shows that the "E" word enthalpy is simply the heat that was transferred at constant pressure. Heat is measured in joules in SI units and calories in CGS units. One calorie is the amount of heat required to raise the temperature of 1 gram of water by 1 degree Celsius. If SI units are used, 1 joule = 0.239 calories.

0.3 SECOND LAW OF THERMODYNAMICS

The First Law is an accounting principle that allows us to keep track of energy, heat, and work. But, it is not sufficient for a complete description of the behavior of matter; something is missing. Never in human history have we observed the temperature of a hot body (a cannon, for example) to increase when brought into contact with a cooler body (the water bath). Heat always flows from the warm body to the cooler one, causing the temperature of the warm body to decrease and the temperature of the cooler one to increase. How do we account for that 100%-certain observation? The Second Law of Thermodynamics was introduced to account for it. The challenge to early thermodynamicists was to construct the Law in a way that permitted quantitative predictions of the direction of chemical reactions. Clausius solved the problem in 1855 by introducing a new state variable called entropy, generally indicated by the symbol S.

Entropy (S), our third "E" word, was required to have two properties. First, it had to be a thermodynamic state variable. Second, it had to indicate the correct direction of heat flow when, for example, a hot body is immersed in cold water. Let the hot body have a temperature T_h and the cold water a temperature T_c. Clausius sought a variable whose increase indicated the direction that heat would flow. The simplest variable one can imagine is

$$dS \equiv \delta q_{rev}/T \qquad (0.9)$$

That is, the entropy increase of a system is equal to the heat the system absorbs divided by the absolute temperature. This easily accounts for the direction of the heat flow. Let's see how this works. Because $T_c < T_h$, $\delta q_{rev}/T_c$ is greater than $\delta q_{rev}/T_h$. That is, for a given reversible heat flow δq_{rev}, the entropy of the cold body increases more than the entropy of the hot body. Because we know empirically that heat only flows from the hotter to the colder body, we surmise that the heat must flow in the direction that increases entropy the most.

After the bodies come into thermal equilibrium, an amount of heat q_{rev} will have flowed between them until they have the same temperature T. The change in entropy of the system is simply $\Delta S = S_2 - S_1 = q_{rev}/T$. The First Law, Equation 0.4, can be rewritten to include entropy via Equation 0.9:

$$dE = TdS - PdV \qquad (0.10)$$

For a system at equilibrium, the volume and internal energy are constant, which means that

$$dS = 0 \qquad (0.11)$$

That is, at equilibrium, entropy is constant. But, there is more. In the language of calculus, whenever a differential is zero, the variable involved must be at an extremum, i.e., a maximum or a minimum. But, is S a maximum or a minimum at equilibrium? The following example will illustrate that it will be at a maximum, which gives thermodynamic definition of equilibrium: entropy is a maximum for constant E and V. For a system not at equilibrium, it will move toward equilibrium by maximizing entropy. The fate of the entropy thus determines the direction of chemical reactions.

0.3.1 ΔS Is Always Positive for Spontaneous Reactions at Constant Temperature

To place entropy in a more familiar context, consider two flasks of volume V_1 and V_2 connected by a closed stopcock (**Figure 0.4**). At the beginning, let flask 2 have no gas molecules and flask 1 have n moles of an ideal gas, so that $pV = nRT$, where T is the absolute temperature in Kelvins and R is the gas constant

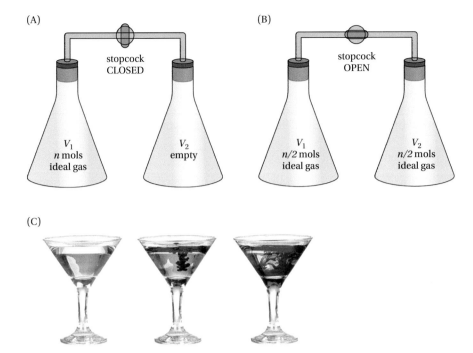

(A)

stopcock
CLOSED

V_1
n mols
ideal gas

V_2
empty

(B)

stopcock
OPEN

V_1
$n/2$ mols
ideal gas

V_2
$n/2$ mols
ideal gas

(C)

Figure 0.4 Two flasks connected by a stopcock. (A) The flasks have equal volumes V_1 and V_2, as shown, and are connected by a closed stopcock. V_1 contains n moles of an ideal gas, while V_2 is empty (vacuum). (B) The stopcock is opened slowly to allow gas to flow slowly from V_1 to V_2 so that the n moles of gas become distributed uniformly in the volume $V_1 + V_2$. (C) Diffusion of a droplet of red dye placed in a glass of liquid. The dye, driven by thermal motion, spreads until eventually, it is distributed uniformly throughout the liquid. The gas does not spontaneously return from the second flask to the first, nor does the dye concentrate itself back into a droplet, because the entropy of the system has increased in each case.

(8.314 Joule K^{-1} mol^{-1} or in CGS units, R = 1.986 cal K^{-1} mol^{-1}). Place the connected flasks into a constant-temperature bath of temperature T. Open the valve in a manner that allows the gas in flask 1 to leak slowly into flask 2, causing the system volume to change from the initial volume $V_i = V_1$ to the final volume $V_f = V_1 + V_2$. Because the initial state and the final state are at the same temperature, $dE = 0$ and $TdS = PdV$ (from Equation 0.10). Taking advantage of the ideal gas law relationship, we can write:

$$T\int_{S_i}^{S_i} dS = \int_{V_i}^{V_t} PdV = nRT\int_{V_i}^{V_t} \frac{dV}{V} = nRT\ln\left(\frac{V_f}{V_i}\right) \tag{0.12}$$

so that

$$S_f - S_i = \Delta S = nR\ln\left(\frac{V_f}{V_i}\right) \tag{0.13}$$

Because $V_f > V_i$, $\Delta S > 0$. Neat! No one has ever observed gas distributed in $V_1 + V_2$ to contract spontaneously into V_1 at constant temperature. Entropy allows us to state this fact mathematically.

Instead of volumes, let's think in terms of the concentration n/V of the gas. Before the valve is opened, the concentration is $C_i = n/V_i$, whereas after the valve is opened, the concentration is $C_f = n/V_f$. With these definitions, Equation 0.13 can be rewritten as

$$S_f - S_i = \Delta S = nRl\,n\left(\frac{n/C_f}{n/C_i}\right) = nR\ln\left(\frac{C_i}{C_f}\right) \tag{0.13a}$$

Because $\Delta S > 0$ and $V_f > V_i$, it must be true that $C_i > C_f$. Entropy thus acts to dilute the gas or minimize its concentration. A familiar example of this is the diffusion of a drop of dye placed in a glass of liquid (**Figure 0.4C**). As in the irreversible expansion of the gas into the two flasks, entropy acts to maximize the volume available to the dye molecules, which is equivalent to reducing the concentration.

An additional insight into thermodynamics is gained if we place our ideal gas in a cylinder with a piston that can be used to compress the gas back to the initial state isothermally (Figure 0.3). We start at the same final volume V_f, as in the isothermal expansion experiment, and then very slowly (reversibly) compress the gas back to volume V_i. We are now doing work w on the gas at constant temperature so that the work done is equivalent to heat (Equation 0.1). With $q = w$, the work done is given by Equation 0.12, but with the integration being from V_f to V_i, which reverses the signs. In short, the work decreases the entropy of the gas, meaning that the net entropy change for the cyclic process of expansion followed by compression is 0. This is an important general rule:

$$\Delta S_{rev} = 0 \tag{0.14}$$

In fact, this is the thermodynamic definition of reversibility. One can suppose, though, that there was a bit of friction between cylinder and piston. This would cause irreversible loss of heat q_{irr} from the system during compression. The entropy associated with this irreversible loss is $\Delta S_{irr} = q_{irr}/T$, which must be positive. The total cyclic entropy change of the system must now include the entropy gain from friction:

$$\Delta S_{rev} + \Delta S_{irr} = \Delta S_{total} > 0 \tag{0.15}$$

Therefore, for any process,

$$\Delta S_{total} \geq 0 \tag{0.16}$$

0.4 THE SECOND LAW AND STATISTICAL THERMODYNAMICS

Another way of looking at entropy comes from connecting thermodynamics to the kinetic theory of gases and statistics. This great step forward was made by Ludwig Boltzmann in the late 19th century. Consider again our two flasks (Figure 0.4) containing an ideal gas. Divide the volume of each flask into a large number n of very tiny sub-volumes (cells), each of volume v, so that the volume V of a flask can be represented by $V = nv$. Add one gas molecule at a time to the volume. For the first molecule, there are n ways of arranging the molecule in the volume. Put a second molecule into the volume. There are now n^2 ways of arranging the two molecules in the volume. This approach can be expanded to say that for N molecules, there are $W = n^N$ ways of arranging them. For our two-flask system, the ways of arranging the molecules in the initial and final volumes will be $W_i = (V_i/v)^N$ and $W_f = (V_f/v)^N$. From these equations, it follows that $V_i = vW_i^{1/N}$ and $V_f = vW_f^{1/N}$. We now use Equation 0.13:

$$\Delta S = R\ln\left(\frac{V_f}{V_i}\right) = R\ln\left(\frac{W_f^{1/N}}{W_i^{1/N}}\right) = \frac{R}{N}\ln\frac{W_f}{W_i} \tag{0.17}$$

The so-called Boltzmann constant is defined as $k = R/N$ (k is 1.381×10^{-23} Joule K^{-1} in SI or 1.381×10^{-16} erg K^{-1} in CGS units), where N is Avogadro's number (6.022×10^{23}). With this definition, Equation 0.17 can be rewritten as

$$\Delta S = k\ln\frac{W_f}{W_i} = k\ln W_f - k\ln W_i \tag{0.18}$$

Boltzmann consequently defined the entropy as

$$S = k\ln W \tag{0.19}$$

Figure 0.5 The tombstone of Ludwig Boltzmann with Equation 0.19 inscribed on it. (From Wikifoundry.com.)

This result is so profound and important that it is inscribed on Boltzmann's tombstone (**Figure 0.5**). The expansion of the gas into the larger volume V_f from the initial volume V_i means, from Equation 0.18, that $S_f > S_i$ or that $\Delta S > 0$. We can thus say that a system moves toward a state that offers the largest number of ways of arranging the molecules.

0.5 GIBBS ENERGY

The First and Second Laws form the foundation of thermodynamics. After the foundation was built by the early pioneers, the science became driven by the need to answer fundamental practical questions about chemical reactions and equilibrium. How much useful work can be extracted from a process? Will a process occur spontaneously? What is the condition for two systems to be in equilibrium with one another? The Gibbs free energy, or just Gibbs energy, developed by Josiah Willard Gibbs in the 19th century, allows these questions to be answered. We have learned that entropy dictates the direction of a reaction. This provides the basis for computing the useful work available from a chemical process.

Imagine a beaker filled with chemical reactant, such as an aqueous suspension of membrane vesicles and a membrane-active peptide (**Figure 0.6**). While we are primarily interested in what goes on in the solution, thermodynamically, we must take the beaker into account because it can exchange heat with the solution. To simplify things, put the beaker in an insulated jacket to keep it from exchanging heat with the rest of the universe (an isolated system). We are thus concerned with the entropy of the stuff in the beaker (call it ΔS_{stuff}) and the entropy of the beaker itself (call it ΔS_{ext}; ext = external); the sum of the two entropies will be ΔS_{total}. From the Second Law (Equation 0.16), $\Delta S_{total} \geq 0$. During the reaction, an amount of heat $-\Delta q$ is transferred to the beaker (minus sign because heat is leaving the solution). The entropy ΔS_{ext} gained by the beaker is therefore $-\Delta q/T$, so that the total entropy change of the system is

$$\Delta S_{stuff} + \Delta S_{ext} = \Delta S_{stuff} - \Delta q / T \geq 0$$

Multiplying the right-hand equation by T yields

$$T \Delta S_{stuff} - \Delta q \geq 0$$

If the process occurs at constant pressure, $\Delta q = \Delta q_p = \Delta H$. Multiplying through by -1 yields

$$\Delta H - T \Delta S_{stuff} \leq 0 \tag{0.20}$$

This extremely useful expression forms the thermodynamic bedrock of biology and chemistry. It is so useful, in fact, that we define a new thermodynamic energy term called the Gibbs energy:

$$\Delta G = \Delta H - T \Delta S \tag{0.21}$$

The Gibbs energy is defined more generally and formally as

$$G \equiv H - TS \tag{0.22}$$

which leads to Equation 0.21 for changes taking place at constant temperature. Mindful of Equation 0.20, we see that a reaction with $\Delta G < 0$ proceeds spontaneously, while one with $\Delta G > 0$ will not. The system is at equilibrium if $\Delta G = 0$. You will encounter Gibbs energy many times in various forms in *Cell Boundaries*.

A problem of particular concern to us in this book is the movement of a solute dissolved in water, call it X, through a membrane separating two chambers labeled 1 and 2 (**Figure 0.7**). Let the concentration of X (moles/volume),

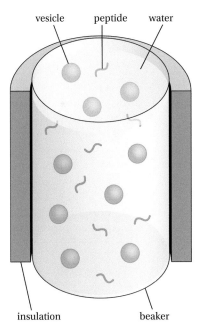

Figure 0.6 A water-filled beaker containing a solution of vesicles and interacting peptides. The beaker and its contents are insulated, making them an isolated system. The front part of the insulation is not shown in order to reveal the beaker. The favorable interaction of peptides with vesicles can produce heat, raising the temperature of beaker, water, vesicles, and peptides.

Figure 0.7 Two water-filled chambers separated by a permeable membrane that allows the solute X to diffuse between the two chambers. [X] is shorthand indicating the concentration of the solute (moles/volume); the subscript indicates the side.

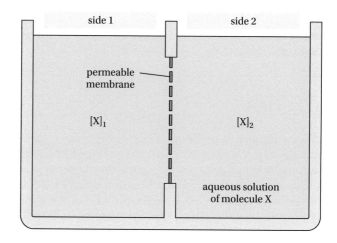

indicated by [X], on the two sides be $[X]_1$ and $[X]_2$. If the solution is ideal, i.e., there are no interactions between the solute molecules and the solute–water interactions are independent of concentration, the problem is similar to describing the movement of gas molecules between two flasks; it's an entropy problem. But, the problem is a bit more complicated because two types of molecules are present: X and water. Typical solutions that one encounters in biology are quite dilute compared with the amount of water present. The concentration of pure water is 55.3 moles liter^{-1}, which is far greater than a physiological saline solution of about 0.3 moles liter^{-1}. We can thus think of the solute molecules as flitting around in the water much as molecules do in a gas.

We learned earlier that one way of thinking about a droplet of dye diffusing throughout a glass of liquid (Figure 0.4C) is that entropy acts to minimize the overall concentration of the dye. Equation 0.13a suggests that $W_1 \propto 1/[X]_1$ and $W_2 \propto 1/[X]_2$ in Equation 0.17. Consequently,

$$\Delta S = R \ln \frac{W_1}{W_2} = R \ln \frac{[X]_2}{[X]_1} \tag{0.23}$$

We are assuming that our solutions are ideal, which means that there is no enthalpy change associated with changes in concentration (called an athermal mixture). In the Gibbs equation (0.21), $\Delta H = 0$, making the Gibbs energy

$$\Delta G = -RT \ln \frac{[X]_2}{[X]_1} \tag{0.24}$$

If the solute concentrations are the same on both sides of the membrane, $\Delta G = 0$, and the system is at equilibrium. If $\Delta G \neq 0$, then there will be a net exchange of X and water across the membrane until the concentrations are equal.

0.6 CHEMICAL POTENTIAL

Another way of describing free energy changes with solute concentration is by the use of the chemical potential, which is a concept introduced by Gilbert N. Lewis in the early 20th century. To see the point of chemical potentials, expand Equation 0.24):

$$\Delta G = -RT \ln[X]_2 + RT \ln[X]_1 \tag{0.25}$$

As noted earlier, if $\Delta G \neq 0$, then X and water will move spontaneously across the membrane until equilibrium is achieved. Chemical potential provides a convenient way of deciding the directions of the solute movements. Notice that the Gibbs energy as written in Equation 0.25 is preceded by the gas constant R, which refers to 1 mole of gas or solute. This means that as written in Equation 0.25), we can think of the Gibbs energy as the free energy per mole of solute or gas. The Gibbs energy per mole is called the chemical potential, μ. In general, we write for solute X

$$\mu = \mu_0 + RT \ln[X] \qquad (0.26)$$

where μ_0 is the chemical potential of X in a so-called reference state. Comparing Equations 0.25 and 0.26, we see that $\mu_0 = -RT\ln[X]_2$. That is, the reference state against which Gibbs energies at other concentrations are compared is the concentration $[X]_2$. Many different reference states can be used. Biochemists generally use a 1 molar solution, while physical chemists use infinite dilution. When using chemical potentials, it is important to remember that the relative Gibbs energy will depend upon the reference state. This means that before comparing the free energies published in one paper with the free energies published in another, you must be sure that the reference states are the same. As we will discuss in Chapter 3, the free energies of transfer of peptides into lipid membranes are of particular interest for understanding lipid–protein interactions. However, the free energies of partitioning depend upon the reference state used. **Figure 0.8** shows how to convert between partitioning free energy measurements that use different reference states.

The virtue of the chemical potential is that changes in compositional free energy are related to a common reference state. We can recast the problem of the free energies of solutions. For example, consider two solutions with concentrations $[X]_1$ and $[X]_2$. We are interested in the difference in chemical potential (i.e., the free energy of transfer per mole of solute) between solution 1 and solution 2, which share a common reference state μ_0. We could thus restate Equation 0.25 as

$$\Delta\mu = \mu_1 - \mu_2 = RT \ln \frac{[X]_1}{[X]_2} \qquad (0.27)$$

One advantage of this formulation is that at equilibrium, when the two concentrations are equal, $\Delta\mu = 0$. Another way of saying this is that at equilibrium,

Figure 0.8 Various systems are used for partition coefficients, association constants, and transfer free energies, so free energy data from different laboratories must be compared with caution. The simplest and most rigorous approach is to treat the association of peptides with membranes as a partitioning rather than a binding-site problem and to use mole-fraction partition coefficients for calculating standard state transfer free energies, ΔG^0 (see Box 3.4). As an example, the mole-fraction concentration x_{bil} of the peptide P bound to lipid vesicles is given by $[P]_{bil}/([L] + [P]_{bil})$, where $[L]$ is the bulk molar concentration of lipid (that is, moles per liter of water) and $[P]_{bil}$ is the is molar concentration of peptide bound to lipid vesicles. Similarly, x_w is given by $[P]_w/([W] + [P]_w)$, where $[P]_w$ is the molar concentration of peptide in the water phase not bound to vesicles. This figure summarizes different systems encountered frequently in the literature and relates them to the mole-fraction system. Conversions of molar and association free energies to mole-fraction standard transfer free energies involve only additive terms. Therefore, differential free energy terms, $\Delta\Delta G_{assoc}$ and $\Delta\Delta G^0_c$, will be identical to $\Delta\Delta G^0_x$. All equations assume partitioning from the water (w) phase to the bilayer (bil) phase. Abbreviations: $[L]$, molar concentration of lipid; $[P]$, molar concentration of free peptide; $[PL]$, molar concentration of peptide bound to lipid; $[W]$, molar concentration of water (55.3 moles liter^{-1}). The molecular volumes of lipid and water are v_{lipid} and v_{water}, respectively. For the typical phospholipid, $v_{lipid} \approx 1300$ Å^3, $v_{water} \approx 30$Å^3.

MOLE-FRACTION PARTITION COEFFICIENT

standard state: infinate dilution

$$K_x = X_{bil}/X_w, \quad \Delta G^0_x = -RT \ln K_x$$

MOLAR PARTITION COEFFICIENT

standard state: 1 molar

$$K_c = X_{bil}/X_w, \quad \Delta G^0_x = -RT \ln K_c$$

$$K_c = K_x (v_{water}/v_{lipid})$$

$$\Delta G^0_x = \Delta G^0_c + RT \ln(v_{water}/v_{lipid}) = \Delta G^0_c - 2.2 \text{ kcal mol}^{-1}$$

ASSOCIATION CONSTANT

traditional reference state: 1 molar

for L + P ↔ LP: $K_{assoc} = [PL]/[P][L]$, $\Delta G_{assoc} = -RT \ln K_{assoc}$

$$K_{assoc} = K_x/[W]$$

$$\Delta G^0_x = \Delta G_{assoc} - RT \ln([W]) = \Delta G_{assoc} - 2.38 \text{ kcal mol}^{-1}$$

the chemical potential of X is identical in all phases that are in contact with one another. The phases could be two mutually insoluble liquid phases (e.g., water and oil), a gaseous phase in equilibrium with a water phase, or even a surface phase (e.g., a monolayer) and its underlying bulk phase (e.g., water). No matter how many phases are in contact with one another, the chemical potential must be identical in all of them at equilibrium. This is the basis for the Gibbs phase rule, which we discuss later.

0.7 BOLTZMANN PRINCIPLE

There is another useful way of writing Equation 0.24. Divide both sides by RT, and then exponentiate both sides:

$$\frac{[X]_2}{[X]_1} = e^{-\Delta G/RT} \tag{0.28}$$

In general, given two states separated by an energy ΔG, the probability p of finding a molecule in a particular state is proportional to $\exp(-\Delta G/RT)$. To see this, eliminate the volume in the concentrations and think in terms of the number of moles or molecules in two states, 1 and 2. We can then write Equation 0.28 as $n_2/n_1 = \exp(-\Delta G/RT)$. The probability of being in, say, state 2 is $p_2 = n_2/(n_1 + n_2)$. But, $n_2 = n_1 \exp(-\Delta G/RT)$, which yields

$$p_2 = \frac{1}{1 + e^{\Delta G/RT}} \tag{0.29}$$

This is the equation for the S-shaped curve that is used frequently to describe the probability of ion channel opening (**Figure 0.9**). Although generally referred to as the Boltzmann curve in the physiological literature, its proper name is the Richards curve derived from the Boltzmann principle. As ΔG becomes large and positive, $p_2 \to 0$; as ΔG becomes large and negative, $p_2 \to 1$.

0.8 THE NERNST EQUATION

The Nernst equation, which describes the equilibrium distribution of ions across charged membranes, is one the most useful consequences of the chemical potential concept. The equation is important because it provides the

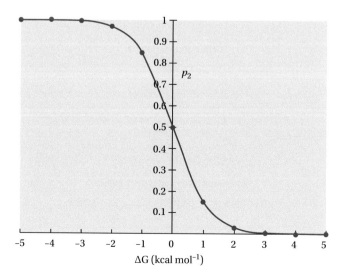

Figure 0.9 Boltzmann probability of being in state p_2 as a function of the free energy difference between states p_1 and p_2 described by Equation 0.29). $RT = 0.582$ kcal mol^{-1} at 20 °C, which is a useful number to remember. Notice that $p_2 = 0.5$ for $\Delta G = 0$, meaning that the probabilities of the two states are equal.

foundation for understanding energetics of cellular energy production, storage, and use. It is named after the German physical chemist Walther Nernst, who won the Nobel Prize in Chemistry in 1920 for his work on thermodynamics. Given two compartments separated by a charged ion permeable membrane, the equation describes the relation between ion concentrations and the electrical potential across the membrane at equilibrium (**Figure 0.10**). The question is: what transmembrane electrical potential V_m will be generated by the concentration gradient of a particular ion X under equilibrium conditions? The Nernst equation provides the answer. The electrical potential is given by

$$V_m = \frac{RT}{zF} \ln \frac{[X]_o}{[X]_i} \qquad (0.30a)$$

Or

$$V_m = 2.3 \frac{RT}{zF} \log_{10} \frac{[X]_o}{[X]_i} \qquad (0.30b)$$

where o and i mean outside and inside a cell, respectively. The factor 2.3 accounts for the change from the natural to base-10 logarithm, V_m is the electrical potential difference, R is the universal gas constant, T is the absolute temperature, z is the ion's valence ($z = 1$ for a proton or sodium ion, -1 for a chloride ion, etc.), and F is the Faraday constant (the total charge of 1 mole of monovalent ions, which is 96,485 coulombs mol^{-1}). The term $RT/F = 25$ mV at 20 °C or 27 mV at 37 °C (human body temperature). As an example, a tenfold concentration gradient between inside and outside for a monovalent ion generates a potential of 58 mV at 20 °C. Because the hydrocarbon thickness (insulating layer) of cell membranes is about 30 Å, this corresponds to an electric field of about 2×10^5 volts cm^{-1}!

Equation 0.30 can be derived as follows. Consider the situation where an ion X is present on both sides of a membrane that is permeable to X. We want to calculate the free energy change associated with moving a charge Q across the membrane. At equilibrium, the change in free energy should be zero; i.e., the chemical potential of the ion should be the same on both sides of the membrane. Chemical potentials, as we have learned, are always measured relative to some standard state; let's choose a 1 M solution of the ion as the standard state. The chemical potential of an ion at concentration [X] is then given by

$$\mu \equiv G([X]) - G(1M) = RT \ln[X] - RT \ln[1] = RT \ln[X]$$

where G is the Gibbs free energy. Therefore, the chemical potentials on either side of the membrane in the absence of a membrane potential are

$$\mu_o = RT \ln[X]_o \quad \text{and} \quad \mu_i = RT \ln[X]_i$$

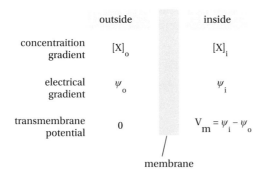

	outside	inside
concentration gradient	$[X]_o$	$[X]_i$
electrical gradient	ψ_o	ψ_i
transmembrane potential	0	$V_m = \psi_i - \psi_o$

membrane

Figure 0.10 Schematic for ions distributed across a permeable membrane in the presence of a transmembrane potential. The Nernst equation describes the condition of equilibrium for the transmembrane distribution of ions when there is a transmembrane potential of V_m volts.

At equilibrium, $\mu_o = \mu_i$, so that $[X]_o = [X]_i$, as expected. The net flux of ions across the membrane will be 0 in this case (see **Box 1.1**). As discussed in the Appendix on Electrostatics, we know that if we move a charge Q across a potential difference V_m, the work required is $\Delta w = (\Psi_i - \Psi_o)Q = V_m Q$. Because we used molar concentrations, we must use moles in the expression for work. The charge carried by 1 mole of a monovalent ion is Faraday's constant, F. Thus, for a mole of ions, $\Delta w = zFV_m$. For the situation shown in the figure, the system is in equilibrium if $\mu_o = \mu_i + \Delta w$, that is, when

$$RT \ln\left[X\right]_o = RT \ln\left[X\right]_i + zFV_m$$

Solving this equation for V_m yields the Nernst equation. In short, whatever free energy is gained/lost under the influence of the concentration gradient must be lost/gained under the influence of the electric field at equilibrium.

0.9 GENERALIZED EQUATIONS OF THERMODYNAMICS

Earlier, we expressed various thermodynamic variables as differentials, e.g., dS and dE. Almost all differentials in thermodynamics are so-called exact differentials. These differentials have special properties, illustrated here. In the following derivation, the object is to express the Gibbs energy in terms of the state variables T and P. Consider the definition of Gibbs free energy, $G = H - TS$. The differential of G is

$$dG = dH - TdS - SdT \tag{0.31}$$

The enthalpy H is defined as $H = E + PV$. Its differential is

$$dH = dE + PdV + VdP \tag{0.32}$$

Substitution of Equation 0.32 into Equation 0.31 yields

$$dG = dE + PdV + VdP - TdS - SdT \tag{0.33}$$

The first term can be replaced using the First Law, Equation 0.10, so that Equation 0.33 becomes

$$dG = TdS - PdV + PdV + VdP - TdS - SdT$$

which reduces to

$$dG = VdP - SdT \tag{0.34}$$

The significance of this equation is that dG is now defined in terms of state-variable differentials dP and dT. Note that for T and P constant, $dG = 0$. The coefficients V and S can reveal new thermodynamic relationships, because a property of exact differentials is that the coefficients are partial derivatives of the function $G(P,T)$:

$$dG = \left(\frac{\partial G}{\partial P}\right)_T dP + \left(\frac{\partial G}{\partial T}\right)_P dT \tag{0.35}$$

This yields new relationships:

$$\left(\frac{\partial G}{\partial P}\right)_T = V \text{ and } \left(\frac{\partial G}{\partial T}\right)_P = S \tag{0.36}$$

The subscripts indicate variables that are held constant.

An important property of Equation 0.35 is that it allows the Gibbs energy to be expanded to include other variables because any variable that contributes to the total energy of the system must be included in the differential to satisfy the First Law. For example, the chemical potential describes how the Gibbs energy depends on chemical composition described by the number of moles N_i of each of component i. Recognizing that G then also depends on N_i, we can expand Equation 0.19:

$$dG = VdP - SdT + \sum_i \mu_i dN \qquad (0.37)$$

where

$$\left(\frac{\partial G}{\partial N_i} \right)_{P,T,i \neq j} = \mu_i \qquad (0.38)$$

The subscript $i \neq j$ means that the concentrations of all other components are kept constant. Equation 0.38 serves as the formal definition of chemical potential.

If an interface is present that has interfacial free energy (tension) γ, then another term must be added to account for the fact that changes in interfacial area will affect the total free energy of the system, because the work to expand the interface by dA is $\delta w = \gamma dA$. In that case, Equation 0.37 becomes

$$dG = VdP - SdT + \sum_i \mu_i dN_i + \gamma dA \qquad (0.39)$$

which means that

$$\left(\frac{\partial G}{\partial A} \right)_{P,T,N_i} = \gamma \qquad (0.40)$$

This equation serves as the definition of the interfacial free energy, which we discuss in Chapter 1.

0.10 PERSPECTIVE

Over a period of several hundred years, Count Rumford, Clausius, Helmholtz, Gibbs, and many others discovered the principles of thermodynamics and developed a mathematical framework for practical applications. Gibbs' discovery of the concept of free energy provided an important foundation for modern biological research. It is interesting to contemplate that the laws of thermodynamics underly all phenomena in the universe, including life, and have never been observed to be violated. The steady expansion of the universe can be thought of as the universe obeying the Second Law. Life has evolved in the face of ever-increasing entropy through evolution of molecules and processes that consume energy to form exquisitely organized structures. Living organisms and their cells are never at equilibrium; they consume energy, do useful work, build structures, and along the way, produce heat.

Foundations of Membrane Structure

1

Why are cell membranes and their lipids and proteins worth knowing about? Simply put, membranes enable life; they organize cells into protected compartments, control the flow of nutrients and information between compartments, generate and store energy, and define cells structurally and phylogenetically. These functions make understanding membranes a key to understanding cell biology. From the point of view of physics and physical chemistry, the extreme thinness and chemical heterogeneity of cell membranes open new vistas—and challenges—at the nano scale.

The biology and physics of membranes are not easily separated; together, they have laid the foundations for understanding the structural principles of membrane function. What were the key discoveries that built the foundation? This chapter is devoted to answering this question. We will see that many of the early insights into membranes and cells came from biophysical studies based solidly on thermodynamics. Indeed, thermodynamics helps us appreciate in a deeper way the elegance of life, whose very existence must be compatible with the laws of thermodynamics. We have therefore provided a primer on thermodynamics at the beginning of this book (Chapter 0). As an aid to our discussion of foundations, we begin with a brief overview of cell structure.

1.1 MEMBRANES DEFINE CELL ANATOMY

All living creatures are divided into two broad groups: prokaryotes and eukaryotes (Greek: pro = before, eu = true, and karyon = kernel, meaning nucleus). For both groups, the cellular interior (cytoplasm) is isolated from the external environment by the plasma membrane, which protects the cell biochemically by controlling the movement of all chemical substances between an often fickle external environment and the cytoplasm. This control, exercised by membrane proteins, allows biochemical processes essential for life to proceed in an organized fashion. Eukaryotes (**Figure 1.1A**), which include all multi-cellular organisms, are distinguished from prokaryotes (e.g., bacteria, **Figure 1.1B**) by having a double membrane (nuclear membrane) that isolates the genetic material (DNA) and its regulatory machinery from the cytoplasm (except during cell division).

1.1.1 Prokaryotes Have a Minimum Complement of Membranes

Prokaryotes, further subdivided into eubacteria and archaea, generally have no internal membrane-delimited compartments but only a plasma membrane

DOI: 10.1201/9780429341328-2

(A)

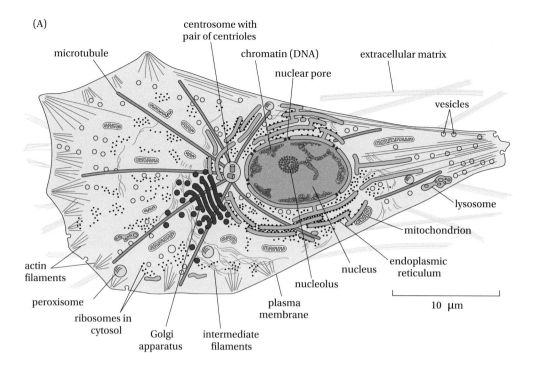

(B)

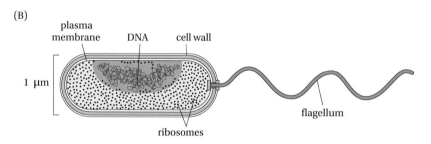

Figure 1.1 Anatomies of two types of biological cells. (A) Eukaryotic cell. (B) Prokaryotic cell. All cells have two features in common: a surface (plasma) membrane, which separates the cell interior (cytoplasm) from the external world, and DNA (deoxyribonucleic acid), which carries the inheritable genetic information. In eukaryotes, the DNA is separated from the cytoplasm by a double membrane (nuclear envelope) except during cell division. Prokaryotes do not have a nuclear membrane; the DNA resides directly in contact with the cytoplasm. A particularly important difference between prokaryotes and eukaryotes is the presence of numerous membrane-enclosed compartments in eukaryotes, such as mitochondria and the Golgi apparatus. (From Alberts B, Johnson A, Lewis JH et al. [2014] *Molecular Biology of the Cell*, 6th Edition. Garland Science, New York. With permission from W. W. Norton.)

that surrounds the cytoplasm (Figure 1.1B). In many prokaryotes, the plasma membrane is protected by a semi-rigid cell wall of variable composition and architecture. There are five different types of walls, ranging from a single layer of protein or glycoprotein (the S layer) to more complex structures comprised of an S layer plus an additional layer of chondroitin-like molecules or polysaccharides. Eubacteria are protected by a tough peptidoglycan layer, which is a composite mesh-like material built from linear polysaccharide chains that are crosslinked by short peptides. For a long time, microbiologists classified eubacteria as either Gram-positive or Gram-negative based on whether their surfaces stain blue or pink in a staining procedure invented by H.C. Gram in the late 19th century. This classification is still with us, although the staining procedure has now largely been superseded by phylogenetic analysis based on the sequence of the ribosomal 16S ribonucleic acid (RNA) molecule. In most Gram-positive bacteria (such as *Streptococcus aureus*), the cell wall is composed of a thick layer of peptidoglycan. Some simple Gram-positive bacteria (such as *Mycoplasma genitalium*) lack the peptidoglycan and only have a single lipid membrane to protect their cytoplasmic compartment.

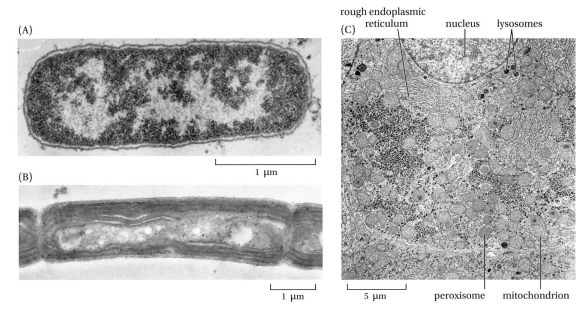

Figure 1.2 Electron micrographs of cells and their membranes. (A) The periplasmic space and the protective cell wall of *Escherichia coli* are visible in this electron micrograph. The lighter area of the cytoplasm is the cell's DNA. (B) Intracellular membranes of the photosynthetic bacterium *Phormidium laminosum* that are evolutionary harbingers of eukaryotic intracellular membranes. In this case, the intracellular membranes are the sites of photosynthesis. (C) The interiors of eukaryotic cells, in this case from human liver, are filled with membranous structures such as mitochondria that produce ATP, peroxisomes that break down fatty acids, and the endoplasmic reticulum from which membrane proteins originate. (A, From Alberts B, Johnson A, Lewis JH et al. [2014] *Molecular Biology of the Cell*, 6th Edition. Garland Science, New York. With permission from W. W. Norton. B, From Alberts B, Gray D, Hopkin K et al. [2013] *Essential Cell Biology*, 4th Edition. Garland Science, New York. With permission from W. W. Norton. C, From Alberts B, Johnson A, Lewis JH et al. [2014] Molecular *Biology of the Cell*, 6th Edition. Garland Science, New York. With permission from W. W. Norton.)

Gram-negative bacteria (such as *Escherichia coli*) have two membranes, the plasma (or inner) membrane and an outer membrane that contains polysaccharides as well as lipids and proteins. This second membrane allows some control over the immediate environment, thus providing protection to the inner membrane and cytoplasm. The space between the two membranes defines the periplasm, where important biochemical processes necessary for survival occur. Some bacteria also have specialized membrane-bounded intracytoplasmic compartments for special activities such as photosynthesis (thylakoids) or nitrogen fixation (respiratory membranes) (**Figure 1.2B**). Like Gram-negative bacteria, plant cells (which are eukaryotes) are protected by a tough cell wall. Unlike in bacteria, however, this cell wall is formed from cellulose and polysaccharides. It is not generally considered to be a cell membrane, because it lacks the characteristic thin lipid matrix of the cell wall of Gram-negative bacteria and archaea.

We often draw cells as containing a dilute, water-rich cytoplasm surrounded by a lipid bilayer membrane in which a few dispersed proteins float around, but that is highly inaccurate. The cell interior is actually a concentrated solution of macromolecules and small metabolites, and the cell membrane is stuffed full of proteins. Suggestive images, such as **Figure 1.3**, are good to keep in mind when thinking about what goes on inside a cell, and especially when comparing biochemical data obtained in dilute solutions in the test tube with data obtained from studies *in vivo*.

1.1.2 Eukaryotic Cells Have Many Compartments

The number of specialized intracellular compartments is greatly expanded in eukaryotic cells in order to sequester critical biochemical processes to specialized regions of the cell with distinctive chemical characteristics such as pH, ionic composition, and ATP/ADP ratio (**Figure 1.4**). These specialized compartments,

Figure 1.3 An artist's scale rendition of a cross-section of an *E. coli* cell. The outer and inner membranes are shown in green, ribosomes in purple, and DNA in yellow. This illustration makes clear that cell interiors are much more crowded and regional than a simple aqueous solution. Illustration by David S. Goodsell. (With permission from the Scripps Research Institute.)

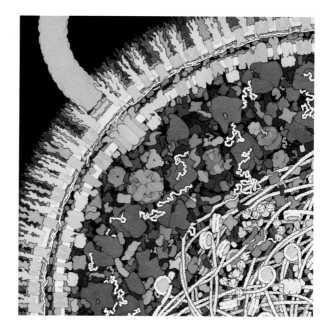

Figure 1.4 Schematic picture of an animal epithelial cell, such as those lining the gut. The cell is polarized, i.e., the membrane on the apical side is insulated by a protein barrier—a tight junction—from the membrane covering the basolateral side that faces adjacent cells and connective tissue, permitting functional differences. The apical side faces the lumen of the gut. (From Alberts B, Johnson A, Lewis JH et al. [2014] *Molecular Biology of the Cell*, 6th Edition. Garland Science, New York. With permission from W. W. Norton.)

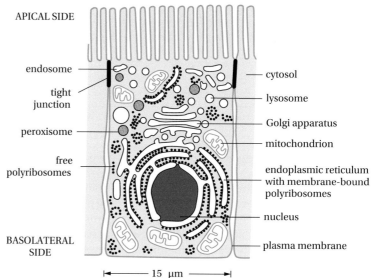

referred to as organelles, greatly increase the membrane surface area available for organizing functional membrane-associated protein complexes. The compartments are interconnected by various transport processes that allow proteins, lipids, and other molecules to move between compartments in a highly regulated fashion. Historically, the biochemical and structural characterization of intracellular organelles was pioneered by Albert Claude, George Palade, and Christian de Duve using the ultracentrifuge to separate different subcellular fractions and the electron microscope to visualize the cell architecture. They shared the 1974 Nobel Prize for physiology or medicine.

Some of the cell's compartments are connected biosynthetically. The nucleus houses the cell's DNA and is surrounded by a double membrane that is continuous with the endoplasmic reticulum (ER). Proteins destined for secretion from the cell and most of the cell's integral membrane proteins are synthesized by ER-bound ribosomes. Most proteins are then transported from the ER along the secretory pathway, first through the Golgi apparatus to the trans-Golgi network, and onwards to the plasma membrane. Proteins and other cargo can also be transported backwards from the plasma membrane to intracellular endosomes and then to lysosomes, where they can be degraded by digestive enzymes.

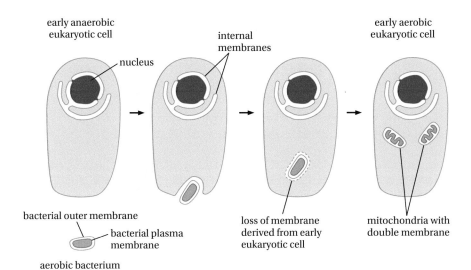

early anaerobic
eukaryotic cell

nucleus

internal
membranes

early aerobic
eukaryotic cell

bacterial outer membrane

bacterial plasma
membrane

loss of membrane
derived from early
eukaryotic cell

mitochondria with
double membrane

aerobic bacterium

Figure 1.5 Scenario for the evolution of mitochondria or chloroplasts from an invasion by a free-living bacterium. The incorporation of bacteria to form a primordial eukaryotic cell is an example of endosymbiosis, with the inclusion of aerobic bacteria giving a capacity for oxidative metabolism and the acquisition of photosynthetic bacteria giving the ability to carry out photosynthesis. (Alberts B, Johnson A, Lewis JH et al. [2014] *Molecular Biology of the Cell*, 6th Edition. Garland Science, New York. With permission from W. W. Norton.)

Other intracellular organelles include lipid droplets that are composed of a hydrophobic core of stored lipids surrounded by a protein-rich phospholipid monolayer, mitochondria containing the enzymes of the respiratory chain that converts nutrient energy into ATP, and peroxisomes that protect the cell from hydrogen peroxide as well as being involved in lipid biosynthesis. Plant cells also contain chloroplasts, the organelle that houses the photosynthetic apparatus.

Mitochondria are surrounded by an outer and an inner membrane. Chloroplasts also have both an outer and an inner membrane, plus an internal thylakoid membrane system. Both mitochondria and chloroplasts are thought to have arisen through endosymbiosis, an evolutionary process through which a eukaryotic cell "engulfs" a prokaryotic cell or the prokaryotic cell invades a eukaryotic cell (as is seen in some modern infections), and the two organisms mutually benefit (**Figure 1.5**). After the initial endosymbiotic event, large parts of the genome of the prokaryotic cells moved into the nuclear genome of the eukaryotic cell, resulting in an intracellular organelle with only a minimal genome left as a witness to its prokaryotic ancestry. There is evidence that the mitochondrial ancestors may have come from a Gram-negative Rickettsiae-like bacterium. Chloroplasts likely arose from photosynthetic cyanobacteria.

1.1.3 Compartments Are Shaped by Proteins

The shapes and dynamical behaviors of compartments result from an intimate interplay between lipids and proteins. In many instances, membranes must be locally bent or curved into tubular or vesicular structures. This is done by membrane-interacting proteins of various kinds that can force lipid bilayers to bend, as we discuss later (Chapter 7). The plasma membrane in eukaryotic cells is attached to an underlying cytoskeleton composed of filamentous structures such as microtubules, actin filaments, and intermediate filaments (**Figure 1.6**). The cytoskeleton serves to control the shape of the cell and to organize and move cytoplasmic components. It can be rapidly disassembled and reassembled in response to various internal and external stimuli.

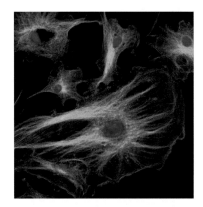

Figure 1.6 The cytoskeleton of a eukaryotic cell. Microtubules are in green and actin filaments in red. The nucleus is in blue. The cell in this picture has been chemically fixed (stabilized), and the microtubules and actin filaments have then been visualized using fluorescent antibodies that bind specifically to each kind of structure. Note the radial organization of the microtubules, the conveyor belts of the cell, and the concentration of actin, which helps shape and move a cell and its internal components, near the cell periphery.

1.2 PLASMA MEMBRANES ARE COMPOSED OF SURFACE-ACTIVE LIPIDS

Regardless of phylogenetic, structural, or biochemical status, all cell membranes share a common architecture. While they have irregular surfaces, the main structure is about 5 nm (50 Å) thick, comprised of lipids organized into

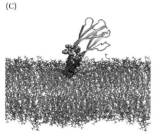

Figure 1.7 The modern understanding of protein–lipid interactions in membranes recreated by molecular dynamics simulations of isolated membrane proteins in lipid bilayers. The three panels summarize three basic structural motifs of membrane proteins: α-helical, β-barrel, and peripheral. These are the only three motifs ever observed for membrane proteins. (A) The α-helical transmembrane protein SecYEG that aids in the insertion of transmembrane proteins. The colored cylinders are schematic representations of transmembrane helices. (B) The β-barrel protein OmpT found in the outer membrane of *E. coli*. This motif is found exclusively in bacterial outer membranes, mitochondria, and chloroplasts. (C) A peripheral membrane protein domain, the C2 domain, shared by a number of monotopic membrane proteins. They are called peripheral because they interact with the bilayer surface. (B, From Baaden M, Sansom M.S.P. [2004] *Biophys J* 87: 2942–2953. With permission of Elsevier. C, Adapted from Jaud S, Tobias TJ, Falke JJ, White SH [2007] Self-induced docking site of a deeply embedded peripheral membrane protein. *Biophys J* 92: 517–524.)

a bilayer arrangement with proteins either passing through the bilayer (transmembrane proteins) or associated with the surface of the membrane (peripheral membrane proteins). The modern atomic-level view of membranes and membrane proteins (**Figure 1.7**) emerged slowly by fits and starts over a period of almost 100 years.

1.2.1 The Existence of Plasma Membranes Was Inferred from the Osmotic Properties of Plant Cells

Anton van Leeuwenhoek, a 17th-century Dutch linen merchant, was the first person to observe bacteria, various free-swimming organisms in pond water, and plant cells using a light microscope. Although these organisms have characteristic shapes suggestive of structural boundaries, the existence of a plasma membrane surrounding the cell cytoplasm was a matter of debate until well into the 1930s. Studies of plant cells by Hugo De Vries and especially Wilhelm Pfeffer during the 18th and 19th centuries were crucial for demonstrating the existence of plasma membranes. Indeed, it was Pfeffer who gave the plasma membrane its name. (De Vries, on the other hand, who rediscovered the work of Gregor Mendel, invented the term *gene* for describing hereditary traits.)

Plant cells have lysosomes of exceptional size (vacuoles, **Figure 1.8**) that can occupy much of the cell volume. De Vries discovered that isolated plant vacuoles (tonoplasts or protoplasts) could be caused to swell or shrink in response to changes in the concentration of salts and sugars in the surrounding aqueous medium, suggesting that the protoplasts were permeable to water but much less permeable to the dissolved solutes (**Box 1.1**). High solute concentrations are equivalent to low water concentrations. If the water concentration is lower outside the protoplast, there will be a net diffusion of water out of the protoplast, causing it to shrink. If the water concentration is higher (low solute concentrations), there will be a net diffusion of water into the protoplast, causing it to swell. These observations led Pfeffer to study the osmotic properties of plant protoplasts. That protoplasts respond volumetrically to osmotic challenges does not necessarily mean that they are surrounded by a semi-permeable membrane, because simple gels of large macromolecules (colloids) will also respond to osmotic challenges simply because the water activity is reduced inside the gel due to the volume occupied by the macromolecules. Pfeffer's crucial observation was that protoplasts showed differential permeability to organic dyes that was lost when the protoplast surface was disrupted. He thus proposed that the protoplasm of cells is surrounded by a thin layer of some kind, which he christened the plasma membrane. These observations led Jacobus van't Hoff to develop, during the 19th century, the physical theory of solutions, which earned him the inaugural Nobel Prize in chemistry in 1901.

Despite the simplicity and elegance of the membrane concept, it was widely disputed for many years, because the osmotic properties of colloids seemed to preclude the necessity for a membrane. In fact, there remains to this day a small band of scientists who deny the existence of plasma membranes. Most biologists, however, accepted the existence of the plasma membrane after Janet Plowe's report in 1931 on the micromanipulation of protoplasts. Her observations on the response of protoplasts to manipulation with glass needles revealed clearly the existence of a surface membrane, referred to by her as the "plasmalemma." Today, similar experiments can be performed using laser-tweezer methodology (see Chapter 2).

1.2.2 The Lipidic Nature of Membranes Was Inferred from the Osmotic Behavior of Cells

What is the chemical nature of the plasma membrane? The first clue came from extensions of Pfeffer's osmotic studies by Charles Ernest Overton in the late 19th century. His research was driven by his curiosity about what physical property of

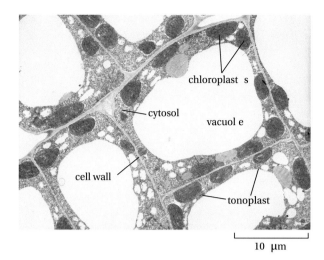

Figure 1.8 Electron micrograph of plant cells that reveals fluid-filled vacuoles, chloroplasts, and the cell wall. The membrane-bound vacuoles can be isolated and used to study the permeability of the vacuolar membrane. (From Alberts B, Johnson A, Lewis JH et al. [2007] *Molecular Biology of the Cell*, 5th Edition. Garland Science, New York. With permission from W. W. Norton.)

10 µm

BOX 1.1 MEMBRANE PERMEABILITY

The movement of uncharged molecules across membranes can be described quantitatively using the mathematical relationships shown here. The Second Law of Thermodynamics demands net movements to equalize chemical potentials (see Chapter 0). Describing the movement of charged molecules is more complicated due to the existence of transmembrane potentials.

Although the figure shows how to describe the movement (flux) of an uncharged solute across the membrane, the same ideas can be applied to the movement of water, which will generally be opposite to the movement of the solutes. As solute concentration goes up, water concentration goes down. For a high solute concentration inside the cell and a low concentration outside, the water concentration will be lower on the inside and higher on the outside, causing a net flux of water into the cell in order to equalize the water chemical potentials.

The permeability P_X of the membrane to the solute X describes how readily X can cross the membrane. P_X is a measure of the interaction of the solute with the membrane, so it depends upon the molecular structure of X and the membrane. What are the units of P_X? They can be established by appreciating that the units must be the same on both sides of the equation. Φ_{net} has units of mol cm^{-2} s^{-1} and the concentrations are in mol cm^{-3}, which means that the unit of permeability is cm s^{-1}. The unit of permeability is

water + uncharged solute X of concentration [X]

solute flux across membrane: Φ (moles/cm^2/second or mol cm^{-2} sec^{-1})

membrane permeability to X: P_X

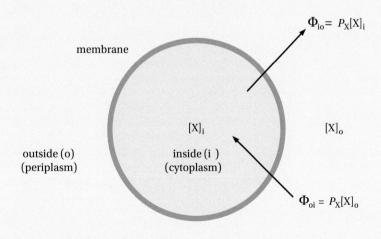

$$\Phi_{net} = \Phi_{oi} - \Phi_{io} = P_X([X]_o - [X]_i)$$

$$\Phi_{net} = 0 \text{ when } [X]_o = [X]_i$$

Figure 1 Calculation of the net passive flux of an uncharged solute across a cell membrane.

(Continued)

BOX 1.1   MEMBRANE PERMEABILITY (CONTINUED)

thus the same as for velocity, but it is not really a velocity, although it is vaguely velocity-like because it measures the ease with which a solute can cross the membrane.

The situation for solute movement into and out of cells is far more complicated than in this simple example, because there are many different solutes, some charged and some uncharged; the cell has a transmembrane potential (generally negative inside with respect to outside, except for acidophiles that live in highly acidic environments); and ions and other molecules can be transported against their electrochemical gradients by membrane transporters that consume cellular metabolic energy. The consumption of energy means that cells are not equilibrium systems, which means that one must be very careful about how to use equilibrium thermodynamics to describe them (**Figure 1**).

solutes was necessary for osmosis. His first experiments were quite simple. He examined the response of the root hairs of water poppies (*Hydrocaris morsusranae*) to osmotic challenges using various solutes. In a 7.0% sucrose solution, the root hair cells showed no signs of plasmolysis (i.e., the shrinkage of protoplast so as to pull it away from the cell wall); the 7.0% solution is isotonic with respect to the protoplast interior. In (hypertonic) 7.5% sucrose, plasmolysis occurred rapidly but was quickly reversed when the root was returned to 7.0% sucrose. He then attempted to make the 7.0% solution hypertonic by adding other solutes in addition to sucrose. The addition of alcohols, for example, caused no plasmolysis, which he interpreted to mean that the plasma membrane was as permeable to alcohols as to water. After examining a wide range of organic solutes (**Figure 1.9**), he concluded that there must be a plasma membrane whose "dissolving properties" match those of "fatty oil." If the fatty oil was made up of triglycerides or other fatty acid esters, the addition of weak solutions of sodium hydroxide would convert fatty esters to soaps through hydrolysis (saponification). Because no saponification of cells was observed, Overton surmised that the fatty layer must be composed of more than just simple fatty acids. In a remarkably prescient inductive leap, he supposed that cholesterol and perhaps lecithins must be the principal substance of the plasma membrane. This conclusion led to the need to understand the partitioning of non-polar molecules between water and the oil-like phase. The driving force for this partitioning, the hydrophobic effect (**Box 1.2**), was not understood fully until the late 1970s.

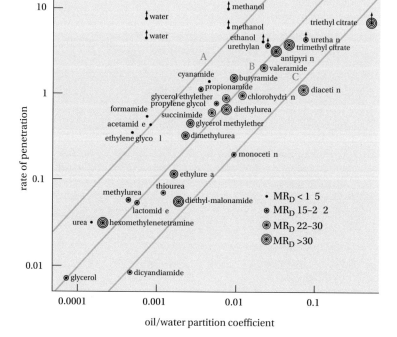

Figure 1.9 Permeability of membranes to various solutes increases with solute solubility in a non-polar phase, typically triolein-rich olive oil or octanol, and decreases with increasing molecular size. Experimental results such as these provided the first evidence that lipids form the permeability barrier of membranes. (From Collander R [2006] *Physiol Plant* 2: 300–311. With permission from John Wiley and Sons.)

BOX 1.2 THE HYDROPHOBIC EFFECT

Charles Ernest Overton concluded in 1899 that membranes had a "fatty oil" interior. This led to extensive physico-chemical studies of the partitioning of molecules between water and oil during the ensuing 75 years. The partitioning of oil-like molecules (alkanes, non-polar amino acid side chains, and anesthetics) between membranes and water is driven by the hydrophobic (HΦ) effect. The simplest way to understand the HΦ effect is to measure the solubility of hydrocarbons (e.g., alkanes) in water, which leads to values for the free energies of transfer from the bulk hydrocarbon phase to water (see Chapter 2). The solubilities of hydrocarbons in water are quite small but nevertheless measurable by gas chromatography. The thermodynamic analysis is summarized in **Figure 1A**. Notice that chemical potentials are used for the calculation of free energies of transfer and that hydrocarbon-solute chemical potentials in the two phases must be equal at equilibrium (see Chapter 0). Mole-fraction concentration units are used; the ratio X_{hc}/X_w is the mole-fraction partition coefficient K. The amount of water dissolved in the hydrocarbon phase is so small that X_{hc} can be taken as 1. The measured free energy of transfer values (hydrocarbon-to-water) are all positive (unfavorable) and increase linearly with the accessible surface area of the hydrocarbon molecule (**Figure 1B**), which is determined computationally by rolling a mathematical sphere with the radius of a water molecule (1.4 Å) over the surface. This surface area dependence of the free energy of transfer explains why hydrophobic molecules aggregate in various ways in water; aggregation minimizes the exposure of hydrocarbon surface to water. More information is provided in Chapter 2.

The classic experiments summarized in Figure 1B led to the conclusion that the energetic cost of transferring a pure hydrocarbon to water is about 21 cal mol^{-1} Å^{-2}. This number, called the solvation parameter σ, has been measured many times for many different systems; values of 20–25 cal mol^{-1} Å^{-2} are inevitably found, provided that concentrations are measured in mole-fraction units. The best modern number for hydrocarbon moieties of peptides and proteins using these units is 23 cal mol^{-1} Å^{-2}.

Near room temperature, the positive free energy of transfer is due almost entirely to an entropy decrease, which leads to the widespread idea that transfer of non-polar molecules between oil and water is entropy driven. At other temperatures, however, there is an enthalpic contribution due to the heat capacity of waters at the surface of the hydrocarbon molecule. What is happening? The hydrogen bonds between water molecules are disrupted by the hydrocarbon. Because they cannot hydrogen bond to the hydrocarbon, the waters become more ordered at the surface of the hydrocarbon (**Figure 1C**), causing the entropy of those waters to decrease (i.e., the degrees of free-dom decrease). To overcome this ordering to the greatest extent possible, the hydrogen-bond strengths increase. This increases the enthalpy of the interactions, which has heat capacity consequences.

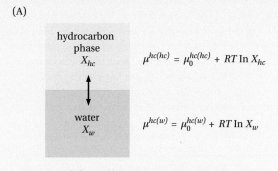

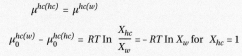

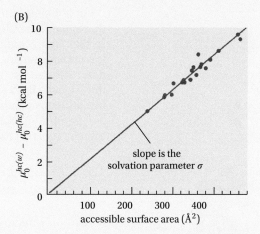

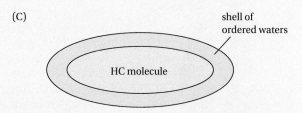

Figure 1 The partitioning of hydrophobic molecules between water and a hydrocarbon phase. (A) The hydrocarbon phase, for example *n*-decane, is immiscible with water and is less dense. (B) The free energy transfer of a hydrophobic molecule between water and hydrocarbon is a linear function of the accessible surface area. (C) Because non-polar molecules cannot form hydrogen bonds with water molecules, they form a monomolecular surface around the hydrophobic HC molecule in a manner that optimizes their hydrogen bonding with each other. This optimization results in more ordered waters, which lowers their entropy. (B, From Reynolds JA, Gilbert DB, Tanford C [1974] *PNAS* 71: 2925-2927.)

So-called Flory-Huggins-corrected volume-fraction units are sometimes used for computing free energies of partitioning because volume-fraction units account for

(Continued)

the great disparity in molecular volumes of the solvents of the two phases. Why does this matter? It matters because of the way entropy is calculated. Recall that in the statistical thermodynamics theory of entropy (see Chapter 0), the number of ways of arranging gas molecules in a volume V is given by $W = (V/v)^N$, where N is the number of gas molecules in the volume V divided into an arbitrary number n of cells of volume v. This is okay for molecules in a gas, because the average number of molecules per unit volume is small. But for liquid solvents, the number density is high, implying that the molecular volumes of the solvents must be accounted for. The Flory-Huggins correction accounts for the discrepancy in solute volumes, which can be quite

large. For example, water has a molecular volume of about 18 $Å^3$, whereas phospholipids in lipid bilayers have molecular volumes of about 1300 $Å^3$. The result is that the solvation parameter can be twice the value obtained using mole-fraction units. No consensus has ever been reached on the necessity for using Flory-Huggins-corrected volume-fraction units. It is important to realize that there is a certain arbitrariness in free energy, because unlike enthalpy, which can be measured directly, entropy is always a computed quantity; no method exists for measuring it directly. The practical solution is to be consistent in all calculations, and the simplest approach is to use mole-fraction units.

But if the membrane is oil-like, how does water, which is insoluble in oil, get through the membrane? Part of the answer is that the concentration of water is so high in any aqueous solution (about 55 moles liter^{-1}) that the low solubility is compensated for by the very strong driving force of the high water concentration. This has been demonstrated repeatedly using phospholipid vesicles and thin (black) lipid films formed across apertures in a solid septum separating two aqueous compartments (below). The other part of the answer is that there are membrane proteins (aquaporins) in some cell membranes that form selective channels for the rapid passage of water.

1.2.3 Membrane Phospholipids Are Surface Active and Form Monolayers at the Air/Water Interface

Benjamin Franklin, writing to his friend William Brownrigg, described the "calming" effect of a droplet of oil placed on a pond at London's Clapham Common (Writings of Benjamin Franklin: London, 1757–1775): "It spreads instantly many feet round, becoming so thin as to produce the prismatic Colours, for a considerable Space, and beyond them so much thinner as to be invisible except in its Effect of smoothing the Waves at a much greater Distance." These observations, which were quite sophisticated for their time, introduced the idea that some substances are surface active, i.e., have an affinity for, in Franklin's case, the air/water interface.

Franklin noted that his droplet was taken from his dinner table's oil cruet, suggesting that Franklin's droplet was probably olive oil, whose principal component is triolein, composed primarily of oleic acid esters. Like many biologically important lipids—such as phospholipids and cholesterol—olive oil is amphiphilic in that it has two chemically distinct, spatially separated regions, each with a distinct affinity for polar and non-polar environments. The phospholipids of vertebrate cell membranes (**Figure 1.10**) are good examples. The phosphoglycerol "headgroup," dominated by a negatively charged phosphate, is soluble in water, whereas the pair of alkyl (fatty acid) chains, ester-linked to the headgroup, are soluble in *n*-alkanes, which are not miscible with water. As a result, phospholipids prefer interfaces between non-polar and polar environments, such as oil/water or air/water interfaces, so that the alkyl chains associate with the non-polar environment and the headgroups with the polar environment. That is, phospholipids are surface active; at air/water or oil/water interfaces they readily form monomolecular films (monolayers). Phospholipids are so surface active that they can generate their own interfaces by spontaneously forming aggregates in order to minimize their Gibbs free energies (Chapter 0). The free energy reduction comes from association of the alkyl chains to minimize their exposure to water (Chapter 2).

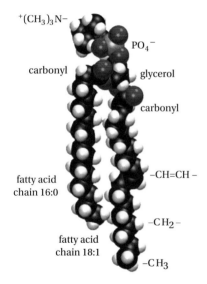

Figure 1.10 Space-filling model of a phosphatidylcholine molecule, which is a major constituent of eukaryotic plasma membranes. The molecule is amphiphilic because its phosphate and choline moieties are water soluble, while its fatty acid chains are oil soluble. This spatial separation of affinities (the usual shorthand being dual affinity or amphiphilic) for water and oil renders the molecules surface active, meaning that their free energy is lowest at oil/water or air/water interfaces.

Agnes Pockels was the first person to examine monolayers quantitatively at air/water interfaces. In 1891, she published an article in *Nature* in which she reported the ability of oleic acid to lower the surface tension of water. Pure water has a very high surface tension γ_0 (surface free energy), because the waters at the air/water interface have fewer hydrogen-bonding neighbors (**Box 1.3**). The interaction of the oleic acid carboxyl group with the water hydroxyl groups is very favorable, and the exposure of the alkyl chains to air is not nearly as unfavorable as the exposure of water to air. Thus, the oleic acid lowers the interfacial free energy (surface tension) of the water. Pockels' discovery used an early form of what came to be known as a surface balance to study this effect. She observed that as the amount of oleic acid on the surface was increased, the air/

BOX 1.3   SURFACE TENSION CONCEPTS

A surface energy imbalance exists at the interface between unlike materials that are in contact. This imbalance gives rise to a surface or interfacial energy (energy per unit area) that expresses itself as surface tension (force per unit length, which is dimensionally equivalent to energy per unit area). The simplest example of this phenomenon is a droplet of water in air (**Figure 1A**). Water molecules at the surface, unlike those in the interior, cannot satisfy all of their hydrogen bonds, which leaves them in a high-energy state (**Figure 1B**). To minimize the surface energy, the droplet assumes a spherical shape, because the sphere encloses more volume for a given surface area than any other shape. Surface-active molecules, such as fatty acids, phospholipids, and detergents, adsorbed at the interface lower the

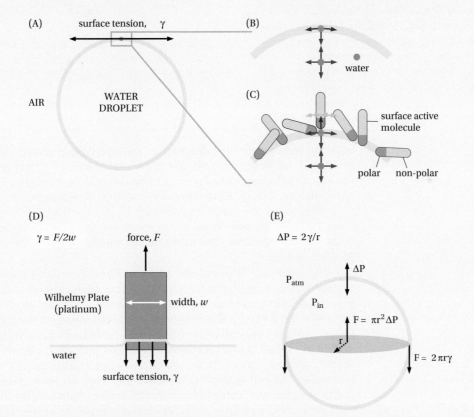

Figure 1 The origin of the surface tension of a water droplet (panel A). The waters on the surface of the droplet cannot form hydrogen bonds with the surrounding air molecules, causing them to have a higher free energy than the fully hydrogen-bonded waters in the bulk of the droplet. To minimize the energetic cost, the droplet adopts a spherical shape to lower the free energy (panel B). Surface-active detergent-like molecules on the surface of a droplet can hydrogen bond with the underlying waters via their polar groups, which lowers the free energy system (panel C). The surface tension of a liquid can be measured by dipping a thin platinum sheet into the liquid and measuring how much force is required to withdraw it (panel D). Soap bubbles have a slightly greater pressure inside compared with the outside air pressure in order to compensate for the effect of the bubble's surface tension (panel E).

(Continued)

BOX 1.3 SURFACE TENSION CONCEPTS (CONTINUED)

surface free energy (surface tension), because the water molecules can now satisfy some of their interaction needs, although imperfectly (**Figure 1C**). These ideas apply to soap bubbles as well as to water droplets. Soap bubbles are thin spherical shells of water stabilized by adsorbed surfactants, such as laundry detergents (**Figure 1.19**), that dramatically lower the interfacial tension.

The surface tension of a liquid interface can be determined in several ways. One of the earliest methods is the Wilhelmy balance (Ludwig Wilhelmy, 1812–1864). A platinum plate, roughened to improve wetting, is withdrawn through the surface, and the force exerted is measured by a balance (**Figure 1D**). The surface tension of water measured in this way is 72.8 dynes cm^{-1}. The surface tension depends upon the molecular interactions in the liquid. The values for several different liquids at the air interface are 21.8 (octane), 22.8 (ethanol), and 487 (mercury).

Curved surfaces, such as water droplets and soap bubbles, cause a pressure gradient to arise at the interface that is described by the Law of Laplace, which expresses itself most simply for spherical surfaces, such as soap bubbles (**Figure 1E**). The bubble's surface tension γ causes the surface to contract slightly to minimize its area, which increases the pressure P_{in} inside the bubble to above atmospheric pressure P_{atm} (think of an inflated balloon). The force across the film due to the pressure difference ΔP must exactly balance the opposing force exerted by the surface tension. The equation describing this balance can be derived by thinking of the force you would have to exert on a parachute's shroud lines to steady a hemispherical canopy against the force generated by a breeze. The canopy expands because the pressure on the up-wind side of the chute is higher than on the down-wind side. For our soap bubble, the force acting on a hemisphere can be calculated from the cross-sectional area of the bubble at its diameter, because Pascal's law demands that the force of hydrostatic pressure be transmitted equally in all directions throughout an isotropic fluid, which in this case, is the air trapped in the bubble. The force exerted on the hemisphere due to hydrostatic pressure is given by $F_{hp} = \pi r^2 \Delta P$, where πr^2 is the cross-sectional area at the diameter (Figure 1E). This force is balanced by surface tension acting along the circumference of the bubble to produce an opposing force $F_{st} = 2\pi r \gamma$. The relation between the pressure rise in the bubble, surface tension, and radius is obtained by demanding that $F_{hp} = F_{st}$. This leads to the following equation, which reveals that the pressure difference increases as radius decreases:

$$\Delta P = \frac{2\gamma}{r}$$

A consequence of Laplace's law is that the vapor pressure of water in very tiny droplets increases. However, the effect does not become significant until r is 1–10 μm.

Figure 1.

---

water surface tension (γ) decreased. The decrease is usually kept track of in modern times by surface pressure π, defined as $\pi = \gamma_0 - \gamma$. In this formulation, π increases as the amount of oleic acid on the surface increases. For pure water, $\gamma_0 = 72.8$ dynes cm^{-1} (7.28 newtons m^{-1}) at 20 °C. Said another way, surface pressure describes how much a surfactant lowers the air/water surface tension. Phospholipids typically have a maximum surface pressure 35–40 dynes cm^{-1}, meaning that they lower the water surface tension of the air/water interface by 33–38 dynes cm^{-1}.

Irving Langmuir adopted the ideas of Pockels to invent the first modern surface balance (1917; **Figure 1.11A**). Rather than varying surface tension by placing different amounts of oleic acid at the air/water interface as Pockels had done, he placed a fixed amount of the fatty acid on the surface and varied the surface concentration using a moveable barrier to control the area (A) occupied by the monolayer molecules. This approach allowed him to construct constant-temperature π–A curves (π–A isotherms) for surface-active molecules. His systematic studies of a wide range of chemical compounds led to the concept of amphiphilicity. Furthermore, from studies of the so-called collapse pressure of fatty acids (the surface pressure at which further reductions in A cause no increase in π), he determined that the minimum cross-sectional area of saturated fatty acids at collapse was about 20 Å^2 regardless of the number of methylene (-CH$_2$-) groups in the chain. This area, we now know, corresponds to the cross-sectional area of alkyl chains in the crystalline state. The key experimental idea to emerge from Langmuir's work was that one could measure the cross-sectional areas of molecules using π–A isotherms. More importantly, he helped start us down the road of thinking at the molecular level.

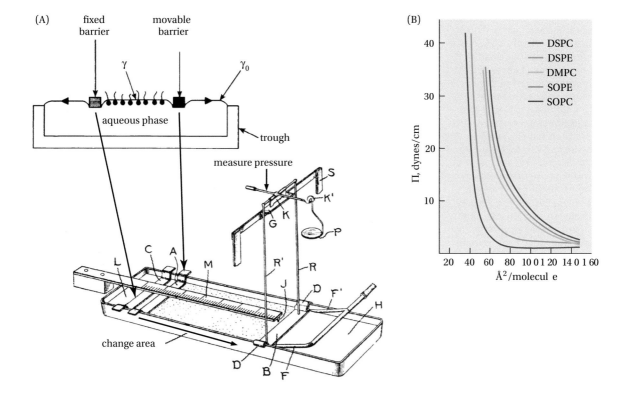

Figure 1.11 Measuring the physical properties of amphiphilic molecules at the air/water interface. (A) The surface balance designed by Irving Langmuir for studying monolayers of amphipathic molecules at the air–water interface. For a fixed number of molecules applied using a volatile solvent, the area per molecule is set by moving the barriers A and C along the trough of fixed width while measuring the surface pressure. The total area is calculated using the ruler M. The molecules on the surface exert a pressure on the free-floating barrier D, which is kept at a fixed position by placing weights in the pan P attached by a lever arm to the pivoted support S. As the area per lipid is decreased, the pressure against the barrier increases. In this way, so-called pressure–area isotherms can be generated. (B) Isotherms (i.e., pressure–area curves collected at constant temperature) for several different phospholipids. The isotherms depend upon both polar headgroup (e.g., charged headgroups repel one another) and fatty acid structure (fatty acid chains with *cis* double bonds are harder to compress than those without). Abbreviations: DSPC, distearoylphosphatidylcholine; DSPE, distearoylphosphatidylethanolamine; DMPC, dimyristoylphosphatidylcholine; SOPE, stearoyloleoylphosphatidylethanolamine; SOPC, stearoyloleoylphosphatidylcholine. (A, From Chapman D, Jones MN [1995] *Micelles, Monolayers, and Biomembranes*. With permission of Wiley-Liss. From Langmuir I [1917] *JACS* 38: 2221–2295. B, From Gaines GL [1966] *Insoluble Monolayers at Liquid-gas Interfaces*. With permission of Wiley.)

Isotherms for phospholipids with different headgroups and fatty alkyl chains are shown in **Figure 1.11B**. Note that for phospholipids without double bonds in the alkyl chains, the minimum area/lipid is about 40 Å^2, which makes sense, because there are two alkyl chains per phospholipid. When each chain of the phospholipid contains a cis double bond, the area/lipid is about 60 Å^2. This is because the double bond increases alkyl-chain disorder and prevents the monolayer equivalent of crystallization. This too makes sense because oleic acid is a liquid at room temperature.

1.2.4 Red Blood Cells Are Covered by a Layer of Lipids Two Molecules Thick

The seminal step toward describing the structural principles of the plasma membrane was made possible by Langmuir's monolayer studies, especially the idea that there is a relation between monolayer surface pressure and the number of molecules occupying a particular area at the air/water interface. In 1925, Evert Gorter and Francois Grendel extracted the lipids from erythrocytes (red blood cells: RBC) from various animals using organic solvents and determined

Figure 1.12 Architectural themes for membranes inspired by the surface balance experiments of Gorter and Grendel, which suggested that the RBC membrane is covered by a bilayer of phospholipid molecules. (A) Two early schematic representations of a lipid bilayer. Gorter and Grendel did not show cartoons such as these in their paper, however. (B) A scheme introduced by Danielli and Davson to describe the cell membrane. This was the first idea about how proteins might be arranged in membranes. Based on measurements of the surface tension of oil droplets in mackerel eggs, Danielli and Davson reasoned that proteins must be adsorbed to lipid monolayers bounding a lipid layer of uncertain thickness to lower surface tension. (B, From Danielli JF, Davson H [1935] *Journal of Cellular and Comparative Physiology* 5: 495–508. With permission from John Wiley and Sons)

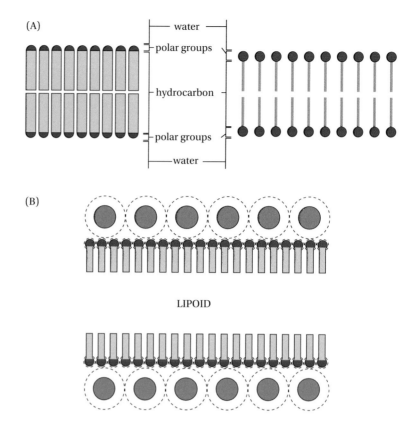

the area occupied by a specific amount of lipid at the air/water interface at a particular surface pressure. The work had to be done carefully and quantitatively. For a particular volume of blood, they first counted the number of RBCs and then estimated the surface areas of single RBCs, which have no membranes other than the plasma membrane. Next, they extracted the lipids using acetone. Evaporation of the acetone allowed an estimate of the mass of lipid in the volume of blood. Finally, they determined the area occupied by a particular mass of lipid at the air/water interface at the collapse pressure. Their measurements revealed that each RBC had enough lipid to coat the RBC surface approximately twice. That is, the RBC is covered by a lipid bilayer. Despite the care with which they executed their experiments, we now know that Gorter and Grendel made two compensating mistakes due to the methodologies available at the time: acetone is not a good solvent for phospholipids, which caused them to underestimate the amount of lipid, and shrinkage of RBC surface area determined from dried blood smears caused them to underestimate RBC area.

Gorter and Grendel did not draw in their paper the now iconic cartoons of the lipid bilayer (**Figure 1.12**). These cartoons were introduced in 1943 by James F. Danielli and Hugh Davson and reiterated in 1959 by J. David Robertson to summarize the findings of Gorter and Grendel—and to symbolize the amphiphilic character of phospholipids.

The first step toward these icons was actually made by Danielli and Davson (1935) based upon studies of the surface tension of oil droplets in mackerel eggs and theoretical considerations of the permeability of thin "lipoid" layers. They summarized their ideas in the cartoon shown in **Figure 1.12B**. They could not estimate the thickness of the membranes they studied, so they just said that there must be a lipoid layer bounded by surface-active lipids. They concluded that the low surface tension of the oil droplets could only be explained by the presence of an adsorbed layer of protein. Their seminal idea was that cell membranes must contain proteins as well as lipid. This caused membrane researchers to begin to turn their attention toward the disposition of proteins in membranes.

1.3 LIPID BILAYERS ARE THE FABRIC OF CELL MEMBRANES

1.3.1 Membranes and Lipid Dispersions Have a Trilaminar Structure When Viewed by Electron Microscopy (EM)

Vertebrate nerve cells are able to conduct nervous impulses at high velocities (up to 120 m s^{-1}) because they are surrounded by insulting layers of myelin produced by neighboring Schwann cells (**Figure 1.13A**). The layers, which wrap tightly around the nerve axon, are the plasma membranes of the Schwann cells. These layers are readily visualized in the electron microscope (**Figure 1.13B**). Because the layers are constructed from membranes, myelin was the main subject for membrane structure studies in the 1950s and 1960s, particularly because the regular repeating structure allowed them to be studied by x-ray diffraction. Fresh, untreated myelin has a repeat period of about 150 Å even though casual observation suggests that the layers observed by EM are about 75 Å thick. Both answers are correct. The layers are formed by back-to-back Schwann cell plasma membranes that are not quite symmetric, causing the x-ray Bragg spacing of 150 Å to be determined by two back-to-back membranes (**Box 1.4**).

Under EM examination, membranes of all cells and organelles from any species always appear as "railroad tracks" having a light band bounded by two dark bands, with each band appearing to have a thickness of about 25 Å (**Figure 1.14**). These railroad tracks thus have the features one might expect from the Gorter–Grendel bilayer. Because cell membranes prepared for transmission EM were fixed and stained with electron-dense osmium tetroxide (OsO$_4$), the meaning and interpretation of these bands was debated for many years, particularly regarding the disposition of membrane proteins. The situation was further complicated by the use of heavy-metal stains, such as lead and uranium, which are the actual densities seen in the electron microscope. In addition, osmium interacts with lipid double bonds, introducing yet another complication. Following the idea of Davson and Danielli that the lipoid layer must have proteins adsorbed at the surface, the dark bands were often assumed to be due to proteins. But because some of the thickness must be attributed to the heavy-metal stain, the dark bands appeared to be too thin to be globular proteins. Furthermore, there was a widespread, and largely unjustified, belief that proteins at membrane interfaces must always denature into extended polypeptide chains.

Given that all membranes look similar in transmission EM images, Robertson (1959) proposed the unit-membrane hypothesis (**Figure 1.15**) in which membranes were bilayers coated with extended polypeptide chains. All of these machinations seem quaint today, but recall that the first three-dimensional

Figure 1.13 The structure of myelin, which is an electrical insulating layer surrounding vertebrate nerve cell axons to enhance the action potential conduction velocity. The myelin, produced by Schwann cells as extensions of their cell membranes, wraps jelly roll–like around the axon (panel A). The regularly stacked layers of membranes, visualized by electron microscopy in panel B, allow structural studies using x-ray diffraction. The Bragg spacing of myelin is typically about 150 Å (15 nm), corresponding to the thickness of two back-to-back myelin membranes, each about 75 Å thick. (A, From Alberts B, Johnson A, Lewis JH et al. [2007] *Molecular Biology of the Cell*, 5th Edition. Garland Science, New York. With permission from W. W. Norton. B, From Stassart RM, Möbius W, Nave KA et al. [2018] *Front Neurosci* 12: 467. With permission from Frontiers in Neuroscience.)

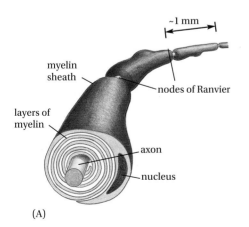

(A)

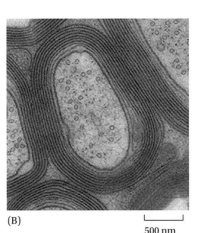

(B)   500 nm

BOX 1.4 BRAGG DIFFRACTION FROM BILAYERS

Atoms in an ordered array (crystal lattice) struck by incident x-rays re-radiate the energy as spherical waves. Viewed from some particular point in space, these waves add together. The fundamental scattering unit in diffraction experiments is the unit cell. In certain directions of space, waves scattered collectively from the sample's unit cells interfere constructively and in others, destructively. The waves interfere constructively when in-phase at integral multiples of 1 wavelength λ of the incident rays. The wavelength λ is typically 1.54 Å for x-rays produced by copper anode x-ray generators. Neutrons can also be used for diffraction measurements. They obey the same diffraction principles but scatter from the nuclei of atoms rather than from atoms' electron clouds as x-rays do.

The interference phenomenon can be described by considering either the individual atoms in the lattice (von Laue formalism due to Max von Laue, 1879–1960) or reflections from layers of atoms or lattice planes (Bragg formalism due to Sir William Bragg, 1862–1942). The Bragg condition for constructive interference from regular layers of atoms separated by the distance d is shown in **Figure 1A**. Constructive interference will occur at angles Θ_h for which the Bragg condition $2d\sin\Theta_h = h\lambda$ ($h = 1, 2, 3, \ldots$) is satisfied.

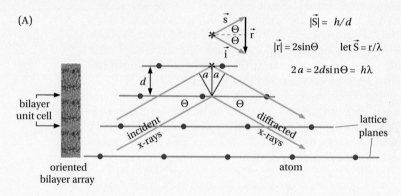

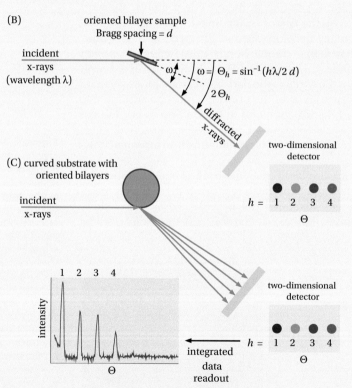

Figure 1 Bragg diffraction of x-rays or neutrons from stacks of bilayers oriented, typically, on a smooth glass or quartz plate. Panel A summarizes the principles of Bragg diffraction. Panel B shows that as the plate is rotated in the beam, Bragg peaks are observed as specific angles that satisfy the Bragg condition. Panel C shows an alternative approach to rotating a flat plate in the beam to sample the Bragg condition. Because the surface is curved, all of the Bragg peaks are sampled simultaneously.

(Continued)

BOX 1.4 BRAGG DIFFRACTION FROM BILAYERS (CONTINUED)

By defining what is called the reciprocal lattice vector $\vec{S}$ as shown in Figure 1A, the Bragg condition can also be written $|\vec{S}| = h/d$. In the figure, the vector $\vec{i}$ is a vector of amplitude 1 (unit vector) in the direction of the incident x-ray or neutron beam, and $\vec{S}$ is a unit vector in the direction of the scattered (diffracted) beam. The reciprocal lattice vector is at the heart of the Laue diffraction formalism. The value of the reciprocal lattice vector is that it is perpendicular to the reflecting planes, and its amplitude $S = 2\sin\theta/\lambda$ depends upon the angle θ. Oriented bilayer arrays (left-hand figure of Figure 1A) are well suited for the Bragg formalism. Bilayers are well ordered only along the bilayer normal. In other directions, the thermal disorder

is too great for Bragg diffraction to be seen. Bilayer diffraction is consequently often referred to as one-dimensional diffraction. The diffraction pattern for a highly ordered array of bilayers is shown in Figure 1A and B. To learn more, see Box 2.3.

The x-ray or neutron beam establishes the direction in the laboratory reference frame. Thus, as the sample is rotated around the ω axis (**Figure 1B**), diffraction maxima will be observed at angles $2\Theta_h$. A simpler approach (**Figure 1C**), and the one frequently used in bilayer diffraction experiments, is to use a curved substrate so that all diffraction peaks are observed simultaneously (Figure 1).

(A)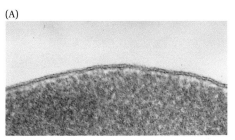

(B) cell membrane   intercellular space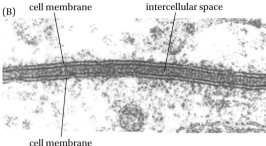

cell membrane

(C)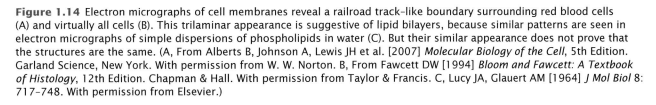

Figure 1.14 Electron micrographs of cell membranes reveal a railroad track–like boundary surrounding red blood cells (A) and virtually all cells (B). This trilaminar appearance is suggestive of lipid bilayers, because similar patterns are seen in electron micrographs of simple dispersions of phospholipids in water (C). But their similar appearance does not prove that the structures are the same. (A, From Alberts B, Johnson A, Lewis JH et al. [2007] *Molecular Biology of the Cell*, 5th Edition. Garland Science, New York. With permission from W. W. Norton. B, From Fawcett DW [1994] *Bloom and Fawcett: A Textbook of Histology*, 12th Edition. Chapman & Hall. With permission from Taylor & Francis. C, Lucy JA, Glauert AM [1964] *J Mol Biol* 8: 717–748. With permission from Elsevier.)

(3-D) structure of a protein (myoglobin) was not reported until 1960, and methods for isolating membrane proteins by sodium dodecyl sulfate (SDS; Chapter 2) and separating them on gel slabs did not come into common practice until the 1970s (**Figure 1.16**). Finally, molecular biology was at a very early stage of development. Although the structure of DNA was discovered in 1953 by James Watson and Francis Crick, the Central Dogma of Molecular Biology (i.e., the amino acid sequence of proteins is encoded in the DNA of the genome) was not enunciated by Crick until 1956, and the genetic code was not solved until the mid-1960s (**Box 1.5**). Thus, in 1959, no one really knew about the diversity and structure of membrane proteins, or any other protein for that matter during the period when Robertson was struggling to understand membrane structure. So, it was not unreasonable to assume a single conformational state for membrane proteins.

The lasting result that emerged from the years of EM studies was that simple dispersions of phospholipids in water stained with osmium tetroxide looked pretty much the same as cell membranes in myelin (Figure 1.14). But, there was a problem. Although the Gorter–Grendel model made perfect thermodynamic sense, and electron micrographs of lipids and membranes did not dispute it, there was no direct structural evidence proving the existence of lipid bilayers. The central question of membrane structure studies during the late 1960s and early 1970s became whether the trilaminar EM images of lipid dispersions really represented lipid bilayers. During this same time period, rapid progress was being made in molecular biology and protein biochemistry. Of particular importance was the widespread use of detergents for solubilizing membrane proteins. SDS solubilization and polyacrylamide gel electrophoresis revealed large numbers of proteins in RBC and other membranes (Figure 1.16).

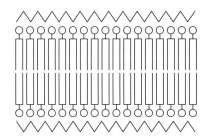

Figure 1.15 The unit-membrane model of Robertson assumed that all membranes had the same structure, because they appeared similar in electron micrographs. He proposed that layers of proteins in extended conformations formed β-sheets on the surfaces of lipid bilayers.

BOX 1.5   THE CENTRAL DOGMA OF MOLECULAR BIOLOGY

Scientific ideas proceed from hypothesis to theory; a hypothesis that resists all attempts to disprove it despite many efforts becomes a theory. The most fundamental theory of molecular biology is that the information about the development and structure of an organism is stored as sequences of deoxyribonucleic acids (DNA) in organisms' genomes. For proteins of every kind, stretches of DNA (genes) encode the primary amino acid sequence of the

protein as sequences of DNA triplets (codons), each triplet corresponding to an amino acid. All genes of organisms have names that follow naming rules, which are summarized at the end of this box. To retrieve the information, the gene is transcribed from DNA to ribonucleic acids (RNA) by a complex enzyme called RNA polymerase. This enzyme produces messenger RNA (mRNA), which is translated into the protein sequences by ribosomes. The genetic code defines the amino acids corresponding to mRNA codons. This transfer of information from DNA to RNA to protein is the most fundamental theory of biology, at least on Earth (Box Figure 1).

The theory was enunciated by Francis Crick in 1956 and named by him the Central Dogma (but also the

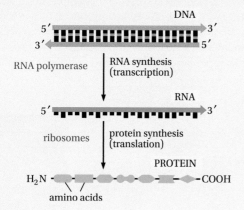

Figure 1 The Central Dogma describes the transfer of information from genomic DNA to messenger RNA to proteins, which Francis Crick sometimes called the "Doctrine of the Triad." Crick defined the Dogma as meaning: "once information has got into a protein it can't get out again." That is, there is only a one-way transfer of information. (From Alberts B, Johnson A, Lewis JH et al. [2007] *Molecular Biology of the Cell*, 5th Edition. Garland Science, New York. With permission from W. W. Norton.)

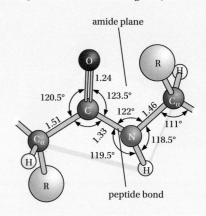

Figure 3 The peptide bond.

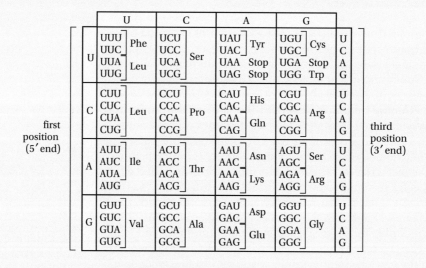

	U	C	A	G	
U	UUU ⎱ Phe UUC ⎰ UUA ⎱ Leu UUG ⎰	UCU ⎱ UCC ⎱ Ser UCA ⎰ UCG ⎰	UAU ⎱ Tyr UAC ⎰ UAA Stop UAG Stop	UGU ⎱ Cys UGC ⎰ UGA Stop UGG Trp	U C A G
C	CUU ⎱ CUC ⎱ Leu CUA ⎰ CUG ⎰	CCU ⎱ CCC ⎱ Pro CCA ⎰ CCG ⎰	CAU ⎱ His CAC ⎰ CAA ⎱ Gln CAG ⎰	CGU ⎱ CGC ⎱ Arg CGA ⎰ CGG ⎰	U C A G
A	AUU ⎱ AUC ⎱ Ile AUA ⎰ AUG ⎰	ACU ⎱ ACC ⎱ Thr ACA ⎰ ACG ⎰	AAU ⎱ Asn AAC ⎰ AAA ⎱ Lys AAG ⎰	AGU ⎱ Ser AGC ⎰ AGA ⎱ Arg AGG ⎰	U C A G
G	GUU ⎱ GUC ⎱ Val GUA ⎰ GUG ⎰	GCU ⎱ GCC ⎱ Ala GCA ⎰ GCG ⎰	GAU ⎱ Asp GAC ⎰ GAA ⎱ Glu GAG ⎰	GGU ⎱ GGC ⎱ Gly GGA ⎰ GGG ⎰	U C A G

first position (5′ end)   third position (3′ end)

amino acid names

Ala = alanine	Gln = glutamine	Leu = leucine	Ser = serine
Arg = arginine	Glu = glutamate	Lys = lysine	Thr = threonine
Asn = asparagine	Gly = glycine	Met = methionine	Trp = tryptophan
Asp = aspartate	His = histidine	Phe = phenylalanine	Tyr = tyrosine
Cys = cysteine	Ile = isoleucine	Pro = proline	Val = valine

Figure 2 The genetic code.

(Continued)

BOX 1.5   THE CENTRAL DOGMA OF MOLECULAR BIOLOGY (CONTINUED)

Doctrine of the Triad). The use of "dogma" caused some controversy because it means a belief that cannot be doubted. But in his autobiography, *What Mad Pursuit*, Crick said that he used the term to suggest "a grand hypothesis that, however plausible, had little direct experimental support." There is now so much direct experimental support that, as dogma suggests, the Doctrine of the Triad cannot be doubted.

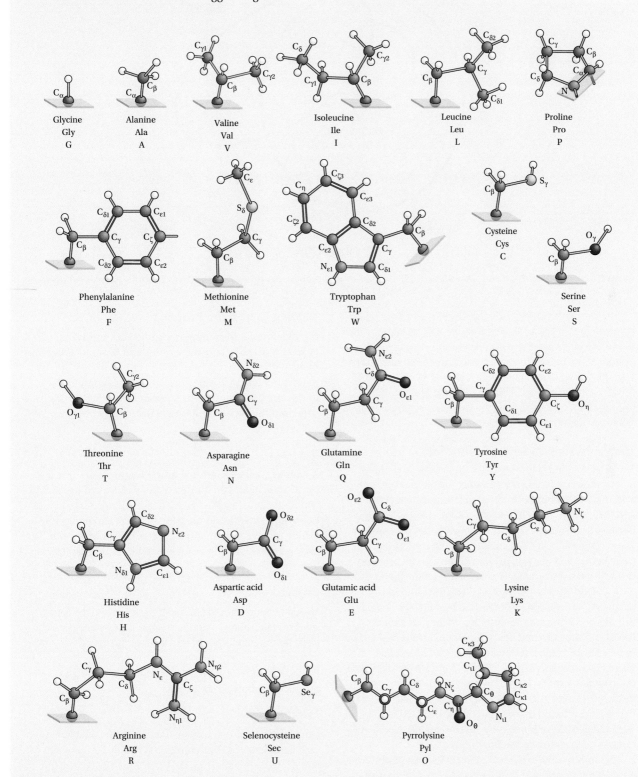

Figure 4 Structures of the 22 natural amino acids. Sec and Pyl are uncommon. (Schultz GE, Heiner R [1979] *Principles of Protein Structure*. Springer-Verlag, New York.)

(Continued)

BOX 1.5   THE CENTRAL DOGMA OF MOLECULAR BIOLOGY (CONTINUED)

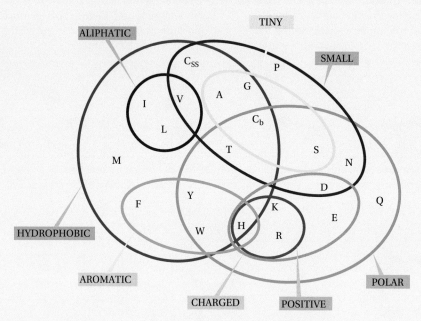

Figure 5 Physicochemical classifications of the 20 common amino acids. (M Zvelebil and J O Baum [2007] *Understanding Bioinformatics*. Garland Science, New York. With permission from Taylor & Francis.)

The common genetic code used by almost all organisms is shown in Box Figure 2. The three-letter abbreviations for the amino acids listed in the table are shown at the very bottom of the figure. The amino acids are frequently abbreviated by capitalized single letters. Having a good grasp of the physical properties of the 20 common amino acids is important for understanding the structure, synthesis, and folding of membrane proteins.

The 22 Nuclear-Encoded Amino Acids

The peptide bond is formed between the carboxyl group on one amino acid and the amino group on the next amino acid in the polypeptide. It has double-bond character and hence, does not allow rotation around the C-N bond. Because of this, the six atoms enclosed by the pink area in Figure 1 are all located in a plane, and the conformation of each amino acid in a polypeptide backbone can be summarized by the ψ and φ angles, as indicated on the figure.

There are 20 common and 2 known uncommon DNA-encoded amino acids, distinguished by their side chains (R, Box Figure 3). The atomic structures of the side chains of the 20 common amino and 2 uncommon acids are shown in Box Figure 4. Selenocysteine is encoded in archaea, bacteria, and eukarya by way of the UGA stop codon, which is recoded to express Sec (U) by a special insertion in the mRNA of the SelenoCysteine Insertion Sequence (SECIS). Pyrrolysine is found in some methanogenic archaea. Pyl is encoded in these organisms by the UAG codon, which is the "amber" stop codon in other domains.

The amino acids can be classified in various ways based on their physicochemical characteristics. A summary of some useful classifications is shown in Box Figure 5.

Naming Genes and Proteins

Biologists excel in coming up with weird, funny, or (sometimes) silly names for genes and proteins that often make it difficult to decipher the meaning of the titles of papers in biological journals. Yet, there is some order in the apparent confusion. A few of the most commonly used naming conventions are listed in the following subsections.

Bacteria

A bacterial gene is denoted by a mnemonic of three lower-case letters that indicate the pathway or process in which the gene product is involved, followed by a capital letter signifying the actual gene (e.g., *secY* in italics for a gene involved in protein secretion). The corresponding protein is named SecY (non-italics with a capital first letter).

Yeasts

The gene name should consist of three letters (the gene symbol) followed by an integer (e.g., SEC61). The three-letter gene symbol should be a mnemonic for a description of a phenotype, gene product, or gene function. The protein corresponding to a gene is often named by adding a p after the gene symbol (e.g., SEC61p).

Human

Human gene symbols are mnemonics designated by upper-case Latin letters or by a combination of upper-case letters and Arabic numerals (e.g., UGT1A1, BRCA1). For the full details, see www.genenames.org/guidelines.html.

An influential study by David E. Green (1910–1983) showed that mitochondrial membranes could be solubilized and then reconstituted by dialyzing away the detergents. This observation corresponded in time to the establishment of the Central Dogma of molecular biology that describes the connection between hereditary DNA sequences and cellular proteins (Box 1.5). Green concluded that if bilayers existed at all, they were probably not the organizing principle of membranes. Rather, he proposed that membranes were organized from protein subunits (in his case, F_0F_1 ATPase) linked together by hydrophobic interactions into sheets (**Figure 1.17**). The only role for lipids in his model was to coat certain surfaces of membrane proteins to render them hydrophilic, forcing proteins to interact at hydrophobic edges to form membrane sheets. The weakness of this idea was that all membranes would be required to have the same organizing proteins, or alternatively, all membrane proteins would have to form protein sheets.

Green used the Central Dogma to strengthen his subunit model by arguing that membrane structure must be under direct genetic control of proteins, because lipids were not represented directly by genes. Of course, this ignored the fact that the protein synthetic machinery for lipids is under genetic control. The questions that came into sharp focus through Green's controversial influence were (1) do bilayers really exist and (2) were bilayers the primary architectural element of membrane structure?

1.3.2 X-Ray Diffraction Proves the Existence of Lipid Bilayers

X-ray diffraction of lipids and membranes is so common today that it is hard to appreciate the profound influence of the direct demonstration of bilayer structure by Maurice Wilkins and Y.K. Levine in 1971. They examined diffraction patterns from hydrated egg yolk lecithin deposited on curved glass substrates (see Box 1.4) and discovered two orthogonal diffraction patterns (**Figure 1.18**) composed of arcs of intensity centered on the x-ray beam. Sitting on the axis that runs parallel to the substrate (equator), diffuse intensities corresponding to mean separations of 4.6 Å were observed. This distance and the diffuse appearance of the equatorial intensities are exactly those seen when x-rays are scattered from liquid alkanes; these intensities arise from disordered, liquid alkyl chains of the lipid. The diffraction pattern observed along the axis normal to the substrate (azimuth) has regularly spaced intensities that are close to the x-ray beam. Bilayers with these characteristics are said to be liquid-crystalline: liquid because of the acyl chain fluidity and crystalline because of the sharp lamellar diffraction lines, indicative of one-dimensional crystallinity. These lines are those expected of a lamellar structure with Bragg spacing of about 50 Å. The interpretation of these x-ray results is unambiguous: this diffraction pattern originates from stacks of lipid bilayers with alkyl chains running, on average, normal to the bilayer plane. Voilà! Bilayers exist!

Having proven the existence of bilayers, it became possible to interpret so-called powder diffraction patterns (**Figure 1.18C**) observed when phospholipids are dispersed by gentle agitation in water (**Figure 1.18B**). The bilayers in this case are completely unoriented (the liposomes are sphere-like), causing the lamellar diffraction pattern to be smeared into complete circles (imagine rotating the image in **Figure 1.18A** around the axis defined by the x-ray beam). There thus appeared to be little doubt that the EM images of lipid dispersions (Figure 1.18C) originated from heavy metal–labeled lipid bilayers.

1.3.3 Single Planar Bilayers Can Be Easily Constructed in the Laboratory

Although fundamentally definitive, the Levine and Wilkins x-ray measurements on oriented bilayers only confirmed in many minds the existence of bilayers, at least in multilamellar stacks. But cells were hypothesized to be surrounded by a

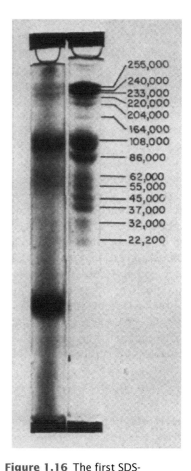

Figure 1.16 The first SDS-polyacrylamide gel of red blood cell membranes that revealed a large number of proteins with different molecular weights (right lane). SDS (sodium dodecyl sulfate; see Chapter 2) is a detergent that solubilizes the proteins. The SDS-protein dispersion is placed at the top of the gel. Application of an electrical potential along the gel causes the SDS-protein complexes to migrate down the gel. The smaller the complex, the farther it migrates in a fixed time determined by the porosity of the gel. That is, the smaller the protein, the faster it migrates. (From Lenard J. Protein and glycolipid components of human erythrocyte membranes. *Biochemistry* 9: 1129–1132. Copyright 1970 American Chemical Society.)

Figure 1.17 The subunit model of David E. Green, based upon electron micrographic studies of mitochondria (left-hand side), posited that globular proteins formed the structural core of membranes. The role of lipids was to coat selected surfaces to force the proteins to interact edge-on to form sheets (right-hand side). This idea came from studies of the reconstitution of cardiac F_1F_o ATPase and lipids solubilized in detergents. (From Fernandez Moran H, Oda T, Blair PV et al. [1964] *J Cell Biol* 22: 63–100. With permission from Rockefeller University Press.) (From Green DE, Allmann DW, Bachmann H [1967] *Arch Biochem Biophys* 119: 312–335. Used with permission from Elsevier.)

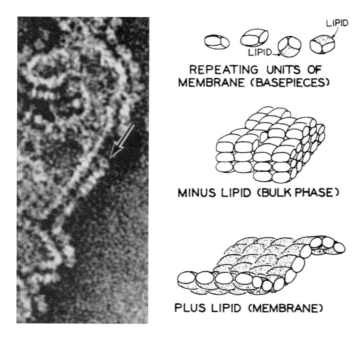

single bilayer. Was that possible? In 1962, Paul Mueller and his colleagues published a brief note in *Nature* reporting the formation of single bilayers (black lipid membranes, BLM; **Figure 1.19**) across an aperture several millimeters in diameter punched (with a hot needle!) through a polyethylene cup arranged so as to separate two aqueous compartments. The electrical properties, including a very low conductance, were those expected of a thin layer of hydrocarbon.

The creation of the BLM had a powerful positive influence on the acceptance of the bilayer membrane structural model. But Mueller took another important step that also had a profound effect on the course of membrane biophysics. He showed that a protein isolated from fermented egg whites lowered the conductance of his BLM to physiological values and was voltage dependent in a way that roughly mimicked the voltage dependence of nerve cell membranes. Today, it is common to reconstitute ion channels into BLM. In 1962, though, the existence of ion channel proteins was not widely accepted. Mueller's work was

(A)

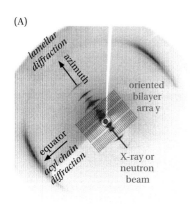

(B)

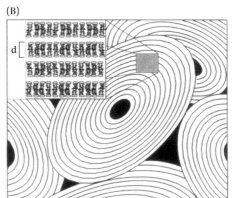

(C)

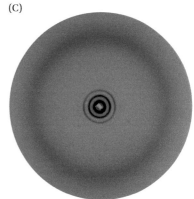

Figure 1.18 X-ray diffraction measurements of multilayers of phospholipids oriented on glass surfaces (A) or dispersed (B and C) in water. The regularly spaced narrow spots along the azimuth in panel A reveal the regular spacing of the lipid lamellae. The broader smeared spots along the equator are due to diffraction from the alkyl chains, in this case in the semi-crystalline gel state. The cartoon of a suspension of lipid liposomes shows that the bilayers are oriented in all directions of space. The diffraction pattern of such a lipid dispersion consists of concentric rings. Imagine rotating the image in panel A 360 degrees around the x-ray beam axis. Notice that the outer ring is broader in this image compared with the image in panel A. This is because the acyl chains are in a disordered fluid state. See Figure 1-20C and Box 1.4. (B, Koenig, BW, Strey HH, Gawrisch K [1997] *Biophys J* 73: 1954–1966. With permission from Elsevier.)

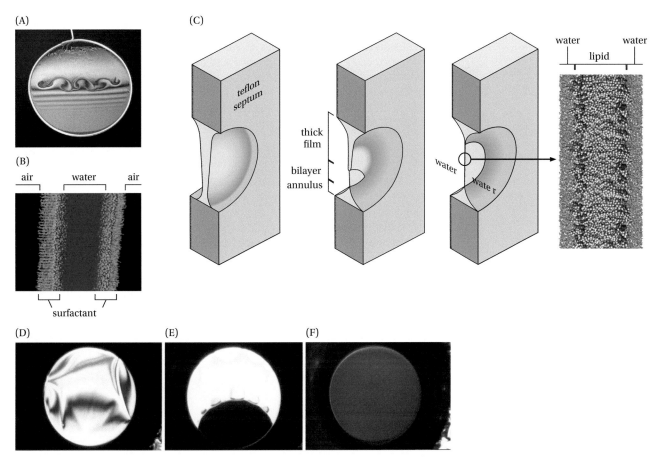

Figure 1.19 Formation of soap films and planar lipid bilayer membranes (black lipid membranes, BLM). (A) A soap film formed across a wire frame in air. (B) Soap films consist of a layer of water stabilized by surfactant monolayers. (C) BLM are formed on an aperture drilled through an inert, hydrophobic septum separating two aqueous salt solutions. A "solution" containing a surface-active lipid (e.g., phosphatidylcholine) dispersed in an alkane, typically *n*-decane, is spread across the septum, forming a thick film. Because the solution wets the hydrophobic surface, a concave (negative curvature) border is formed around the thick film. The hydrostatic pressure in the border is lower than in the flat film (Law of Laplace), which causes excess solution to flow into the border (Plateau–Gibbs border, annulus, or torus), thus causing the film to thin. When the film has a thickness on the order of a wavelength of visible light, interference between light reflected from the front and back surfaces gives rise to colored Fizeau bands (D). Further thinning to less than a wavelength of light gives rise to a uniform silver appearance. The bilayer appears spontaneously at the bottom of the film due to buoyancy (E) and spreads over the entire area (F). Panel E shows why these bilayer-thick membranes are called black films. Electrical and optical measurements prove that the black film (F) is two lipid molecules thick. Notice the similarities to the soap film in panel A that reveals black films at the top of the loop. These black films were first described by Isaac Newton. (B) From Baokina S, Monticelli L, Marrink SJ et al. [2007] *Langmuir* 23: 12617–12623. (C–F) From White SH, Petersen DC, Simon S et al. [1976] *Biophys J* 16: 481–489

thus the beginning of an eventual explosion of single-channel measurements (Chapter 11).

Extensive physicochemical measurements in the laboratory of Denis A. Haydon and others in the 1970s confirmed that BLMs (Figure 1.19) were similar to soap films. But whereas soap films are formed in the air from a thin layer of water stabilized by surfactants, BLMs are formed from surfactants, especially phospholipids, dispersed in non-polar alkanes spread across the aperture submerged in water. Soap films are stabilized against collapse mainly by Coulombic repulsion of surfactant charges across the water film; BLMs are stabilized by steric repulsion of surfactant alkyl chains acting across the hydrocarbon film. For both systems, the film-forming dispersions must make small contact angles with the supporting aperture (i.e., they must wet the surface; **Figure 1.19C**) to cause the annulus (or torus or Plateau–Gibbs border) around the film to have a negative curvature (i.e., convex as viewed from the aqueous phase). The Law of Laplace ($\Delta P \sim \gamma/R$, Box 1.3) dictates in such cases that the hydrostatic pressure difference ΔP across the curved interface with radius of curvature R must be lower than the pressure in the flat region of the film. The resulting border suction causes a flow of solvent (alkane in the case of BLM) into the torus. The film thins until it is depleted of excess solvent and the surfactant alkyl chains come into contact. The extreme thinness of the BLM causes it to reflect very little light, which makes it look black relative to its surrounding annulus (**Figure 1.19E**). Assuming a reasonable value for the dielectric constant ε of the hydrocarbon core ($\varepsilon \approx 2$), measurements of electrical capacitance were found to be consistent with BLMs being two lipid molecules thick.

1.3.4 Calorimetry and X-Ray Diffraction Demonstrate the Existence of Lipid Bilayers in Membranes

The idea of the lipid bilayer as the basic architectural element of cell membranes had gained strong inferential support by the close of the 1960s due to the invention of the BLM. Direct evidence for bilayers in membranes remained elusive, however. That quickly changed as a result of the incisive thermodynamic measurements of Joseph Steim and his colleagues. Because of the economic importance of oils and fats in the fuel and food industries, physical chemists had, by the 1950s, gained a fairly complete understanding of the structure and properties of alkanes, fatty acids, and phospholipids through x-ray and calorimetric measurements. For example, octadecane, which is solid at room temperature (23 °C), melts at 28 °C. For it to melt, it must, like ice, absorb heat from its surroundings. This latent heat of melting can be measured using a differential scanning calorimeter (**Box 1.6**) that measures the heat flow into the sample during melting.

Phospholipids dispersed in water also undergo melting transitions that can be detected by DSC. Chemically homogeneous phospholipids undergo sharp thermal phase transitions from so-called gel states to fluid (liquid-crystalline) states (**Figure 1.20A**) that depend upon the length of the alkyl chains, the degree of unsaturation, and headgroup structure, because all of these features affect how tightly the molecules can pack together. Dimyristoylphosphatidylcholine (DMPC), with two saturated 14-carbon fatty acid chains, melts at 28 °C, whereas dipalmitoylphosphatidylcholine (DPPC), with two saturated 16-carbon chains, melts at 41 °C. Dipalmitoylphosphatidylethanolamine (DPPE), however, melts at a higher temperature, about 46 °C. Distearoylphosphatidylcholine (DSPC; 18-carbon saturated chains) melts at 58 °C, while dioleoylphosphatidylcholine (DOPC; 18-carbon chains with a cis double bond at the 9,10 positions) melts at –21 °C. In all these examples, the transitions are very sharp, i.e., they occur over a very narrow temperature range. Phospholipid mixtures comprised of phospholipids with a range of acyl chain lengths and degrees of unsaturation, such as phosphatidylcholine extracted from egg yolks (egg PC, composed primarily of POPC; Figure 1.10), undergo thermal phase transitions as well, but the melting transition occurs over a much broader temperature range due to the presence of more than one lipid species.

Steim and colleagues took advantage of these thermal properties of phospholipids to demonstrate the existence of bilayers in the membranes of living cells. Their demonstration relied upon the very simplest bacteria (aphragmabacteria), such as *Mycoplasma laidlawii* (now renamed as *Acholeplasma laidlawii*), which cannot synthesize fatty acids; they are dependent solely upon fatty acids imported from the growth medium. They grew *M. laidlawii* in a medium enriched in stearic acid, which caused the bacteria to be enriched in C-18 phospholipids with relatively high melting points. Whole cells or their isolated membranes revealed two broad thermal phase transitions at around 40° and 65° (**Figure 1.20B**, curves i and ii). The lower transition was shown to be due to lipids by extracting the lipids from the bacteria using a chloroform:methanol solution (curve iv). The higher transition was shown to be due to thermal denaturation of proteins by heating isolated membranes above the transition, cooling, and then reheating. Upon reheating, the higher transition was not seen

BOX 1.6   DIFFERENTIAL SCANNING CALORIMETRY

Heat Capacity

A simple, but important, concept in thermodynamics is that of heat capacity (C), which is defined as the amount of heat required to raise the temperature of a substance by a certain amount (see Chapter 0: The "E" Words: A Concise Guide to Thermodynamics). Measurements of heat capacity by differential scanning calorimetry (DSC) have been of central importance in studies of biological lipids. The definition of heat capacity is $C = \delta q/dT$ so that $q = \int C dT$. But we know that the heat absorbed depends upon whether the process is carried out at constant volume or constant pressure. The First Law, eq. 0.4 of the Concise Guide, tells us that $\delta q = dE + PdV = CdT$. At constant volume, the heat required to raise a system from initial temperature T_i to final temperature T_f is

$$q = \int_{T_i}^{T_f} C_V dT = \Delta E \qquad (1)$$

At constant pressure,

$$q = \Delta E + \int_{V_i}^{V_f} PdV = \Delta H = \int_{T_i}^{T_f} C_p dT \qquad (2)$$

The constant volume (C_V) and pressure (C_p) heat capacities have been kept inside the integrals, because they may in some cases depend upon temperature. Heat capacities are extensive properties because they depend upon the amount of material in the system. It is therefore useful to define the heat capacity per unit mass, called specific heat capacities (c):

$$c_k = C_k / m \qquad (3)$$

where m is the mass and the subscript $k = V$ or P indicates constant-volume or constant-pressure heat capacities, respectively. The calorie is the constant-pressure specific heat capacity of water (**Figure 1**).

Differential Scanning Calorimetry

A beaker of water containing ice has two phases, the liquid water phase and the solid ice phase. By Gibbs phase rule—and experience—the temperature is fixed at the freezing point of water 273K (0 °C) until the ice melts (Chapter 2). To melt the ice, one must supply the so-called latent heat of melting. If heat is supplied to the beaker of ice water, the temperature will be able to rise only after the latent heat has been supplied. After that, the temperature will rise as more heat is supplied. Lipid bilayers behave in the same way. To move from the gel state to the fluid state, the latent heat of melting the gel-phase bilayer must be supplied. These so-called first-order thermal phase transitions can be quantitated using a differential scanning calorimeter, shown schematically in **Figure 1A**. The idea is to supply heat q_1 and q_2 steadily to raise the temperature T of the system and at the same time keep the sample temperature T_2 equal to the reference temperature T_1. If the masses and heat capacities of the reference and the sample are about the same, then

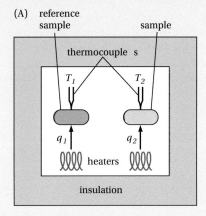

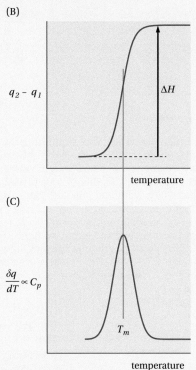

Figure 1 Principles of differential scanning calorimetry. A sample of interest that undergoes a phase transition at a particular temperature and a reference sample that does not undergo a phase transition are placed in an insulated chamber, and heat is applied to each (panel A). At the phase transition temperature of the sample, the sample absorbs additional heat (the latent heat of melting) compared with the reference sample. Plotting the differential heat flow allows the enthalpy (ΔH) of the transition to be measured (panel B). The first derivative of the differential heat flow of panel B yields the heat capacity of the sample (panel C).

the same amount of heat is added to both, and the system temperature rises steadily. But as the system temperature approaches the phase transition temperature T_m of the sample, the latent heat of melting must be supplied in order to melt the ice or gel-phase lipid. The calorimeter

(Continued)

BOX 1.6   DIFFERENTIAL SCANNING CALORIMETRY (CONTINUED)

measures the difference $q_2 - q_1$ in the amount of heat supplied to the sample and to the reference as T is increased. The curve shown in **Figure 1B** records the differential heat supplied. The total difference in heat supplied is simply the enthalpy of the melting of the ice or bilayer; normalizing for masses yields the latent heats of melting. What one usually

sees displayed on a DSC is the derivative of the curve in Figure 1B. This derivative is the heat capacity of the sample, shown in **Figure 1C**. The peak of this curve is an accurate measure of the phase transition temperature T_m provided that the scan is carried out slowly to ensure equilibrium throughout the scan.

(A)
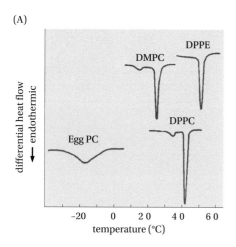

(B)
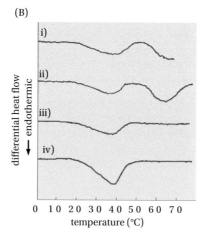

(C)
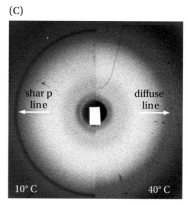

M. laidlawi i membranes

Figure 1.20 Experimental evidence supporting the lipid bilayer as the organizing principle of cell membranes. Phospholipids undergo thermal phase transitions from the gel state with well-ordered alkyl chains to a liquid-crystalline state with fluid, disordered alkyl chains as the temperature is raised. In the process, they absorb heat that can be detected by differential scanning calorimetry (DSC; Box 1.6). Bilayer lipid phase transitions depend upon lipid structure (panel A). Mycoplasma bacteria cannot synthesize fatty acids and must use fatty acids available in the growth medium. This allows the phospholipid fatty acid composition to be controlled and in turn, the thermal phase transition of the bilayers. Panel B reveals (i) thermal phase transitions in living *Mycoplasma laidlawii* (now renamed as *Acholeplasma laidlawii*) that are also seen in isolated membranes (ii—not heated, iii—reheated) and extracted lipids (iv), consistent with lipid bilayers in the cell membrane. X-ray diffraction measurements of isolated membranes (panel C) provided direct support for the existence of lipid bilayers in the membranes (compare with Figure 1.18). (A, Figure reproduced with permission from Melchior DL, Stein JM [1976] *Annu Rev Biophys Bioeng* 5: 205–38. Copyright 1976 Annual Reviews. C, From Engelman DM [1970] *J Mol Biol* 47:115–117.)

(curve iii), because the denatured protein had irreversibly coagulated (think of the egg whites of hard-boiled eggs). Careful quantitative evaluations of the data showed that most, if not all, of the lipid must be participating in the melting transition, consistent with the idea that lipid bilayers were the principal organizing feature of cell membranes. Furthermore, the absence of a major effect of the protein on the lipid transition indicated that David Green's idea that the fabric of the membrane must be protein was not correct. The molecular nature of the phase transition was established by Donald Engelman using x-ray diffraction (**Figure 1.20C**). The principal question remaining at that point was the disposition and structure of membrane proteins.

1.4 PROTEINS PENETRATE THE MEMBRANE BILAYER AND ARE MOBILE IN THE MEMBRANE PLANE

1.4.1 Freeze-Fracture EM Suggests Proteins within the Bilayer Fabric

Transmission electron micrographs of membranes (Figure 1.14) provided low-resolution images of membranes that seemed consistent with the bilayer hypothesis. But, electron micrographs raised more questions than they answered regarding the disposition of proteins in membranes, despite the efforts of many talented electron microscopists. This situation changed beginning in 1957 with the introduction of freeze-fracture/-etch EM by Russell L. Steere. This method, which requires no fixatives or stains, was used by Daniel Branton in 1966 to examine cell membranes in detail. Not only did the method provide the first clear image of proteins in membranes; it also confirmed the bilayer hypothesis quite directly (**Figure 1.21**). In his first experiments, Branton rapidly froze onion root tips stuck to a small copper plate with gum arabic by immersing them in liquid Freon. The sample was then cleaved under vacuum with a sharp blade on a pendulum that fractured the sample along the weakest surfaces, which happen to be cellular membranes. Under vacuum, the water in the sample sublimes (Box 2.2). Just as sunken objects emerge when a pond is drained, the edges and surfaces of membranes emerge as the ice recedes (**Figure 1.21A**). Platinum sputtered onto the exposed fractured surface provided a replica that could be examined in the electron microscope. Close examination of the freeze-etch images revealed that often the fracture lines ran through the middle of the membrane, as expected of bilayer membranes (**Figure 1.21B**). Split membranes revealed 85 Å particles embedded within them. At first thought to be lipid micelles, these particles were later proven to be proteins, suggesting that membrane proteins were globular and embedded within the lipid bilayer. These images sounded the death knell of the unit-membrane hypothesis.

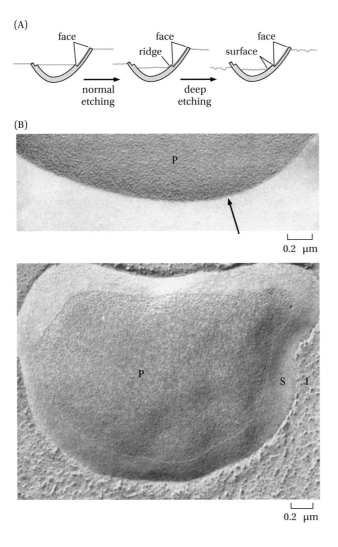

Figure 1.21 Freeze-fracture electron microscopy reveals the architecture of cell membranes. (A) Schematic representation of a cell in ice during the freeze-fracture process. After rapid freezing, the suspension of cells is placed in an evacuated chamber and fractured with a swinging blade. The fracture path follows the membrane surfaces. Controlled periods of vacuum sublimation remove different amounts of ice before a thin layer of platinum is sputtered onto the fractured surface to produce a replica, which is then examined in the electron microscope. Vacuum sublimation causes the surrounding ice to recede, thus exposing the surface of the cell. (B) Micrographs of freeze-fractured cells. The arrow in panel B (upper image) indicates the edge of the fractured membrane before deep etching (longer sublimation). Sometimes the fracture path passes through the middle of the bilayer, revealing particles (P). These particles are protein molecules that reside within the bilayer (B, lower pane). (A, From Branton D [1966] *PNAS* 55: 1048–1056. With permission from D. Branton. B, From Tillack, T W Marchesi, V T [1970] *J Cell Biol* 45: 649–653. Permission from Rockefeller University Press.)

1.4.2 Chemical Labeling of Membrane Proteins Shows That They Span Membranes

Branton's freeze-fracture EM studies revealed particles embedded in lipid bilayers, but one could not conclude from those observations alone that membrane proteins really could span the lipid bilayer. The first solid biochemical evidence demonstrating membrane-spanning proteins was provided by Mark S. Bretscher at Cambridge University. He developed a radio-labeled reagent that reacted covalently with proteins by linking to exposed amino and hydroxyl groups. He demonstrated its utility by labeling the proteins of RBC membranes. After reacting the reagent with intact or disrupted RBCs (RBC ghosts), he solubilized the membranes with the detergent SDS and separated proteins by polyacrylamide gel electrophoresis (SDS-PAGE), as in the experiment of John Lenard (Figure 1.16). After separation, the locations of the proteins on the gel could be detected by placing over the gels photographic film that is sensitive to the radioactivity of the label (**Figure 1.22**).

Bretscher made an interesting observation. If the label was applied to RBC ghosts (the membranes of ruptured RBCs), all of the membrane proteins were labeled. But if he labeled intact RBCs using his reagent, which he showed not to be membrane permeable, only two proteins could be detected on the gel. He called them protein a and protein b. If he treated the RBCs with the proteolytic enzyme pronase that breaks down proteins, then protein b disappeared (it was a peripheral membrane protein). Although protein a (MW ≈ 100 kD) also disappeared, a new protein (protein c) appeared that had a lower molecular weight (MW ≈ 70 kD); he surmised that this was a fragment of protein a that was protected from proteolysis by the membrane. Pursuing the matter further, as summarized in Figure 1.22, he found that by preparing ghosts after first "shaving" them with pronase, protein c was again labeled. He concluded that protein a must be exposed to both the extracellular solution and the cytoplasm. That is, protein a must span the membrane. To prove that protein c was a fragment of protein a, he removed proteins a and c from the gels, treated them with thermolysin (another, but milder, protease that does not completely degrade proteins to very short peptides), and examined the so-called fingerprint of peptide fragments produced. The fingerprints showed conclusively that protein c was a protein a fragment that could be labeled from both sides of the membrane. Bretscher's protein a became known as the band 3 RBC membrane protein because of its location on SDS-PAGE gels. Band 3 was eventually determined to be the chloride-bicarbonate exchanger of RBC membranes that passively

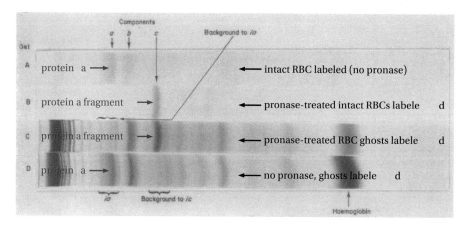

Figure 1.22 The labeling experiment of Mark Bretscher, which demonstrated that membrane proteins can span the membrane bilayer. Component *a* is the major protein component on the surface of human erythrocytes. It has been labeled from one, or both, sides of the cell membrane by labeling intact cells or cell ghosts. Fingerprint maps of labeled peptides derived from component *a*, labeled in these two ways, indicate that different parts of protein *a* reside on each side of the membrane. (Bretscher MS [1971] *J Mol Biol* 59: 351–357. With permission from Elsevier.)

exchanges bicarbonate ions for chloride ions. This exchange is crucial for pH regulation of the body, for which RBCs have a heavy responsibility.

1.4.3 Proteins in Plasma Membranes Are α-Helical

The year 1966 was a good one for advances in membrane structure, and not only because of Branton's work. To examine the unit-membrane assumption that the protein was organized as extended chains, as in β-sheets (**Figure 1.23**), A.H. Maddy and B.R. Malcolm attempted in 1966 to find β-sheet secondary structure in RBC membranes using infrared (IR) spectroscopy. Finding none, they concluded that the unit-membrane hypothesis was not tenable for RBC membranes.

But what was the secondary structure of membrane RBC membrane proteins? The alternatives were random-coil peptide chains or α-helices, but Maddy and Malcolm's measurements could not distinguish between these two secondary structure states. The issue was resolved in 1966 by John Lenard and S. Jonathan Singer, who discovered by means of optical rotary dispersion (ORD) and circular dichroism (CD) spectroscopy that the dominant secondary structure of proteins in RBC and other plasma membranes was α-helical. These IR and CD spectroscopic measurements together completely ruled out

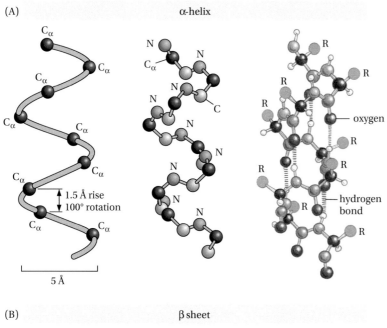

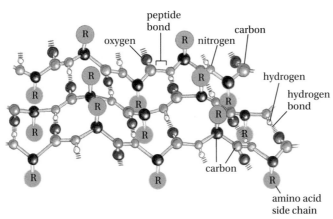

Figure 1.23 Non-covalent hydrogen bonds between the peptidebond carbonyl (C=O) oxygens and the amide (NH) hydrogens cause the formation of two kinds of regular secondary structure: α-helices (panel A) and β-sheets (panel B). The hydrogen bonds are indicated by the dashed lines between the carbonyl oxygens (red) and the amide nitrogens (blue).

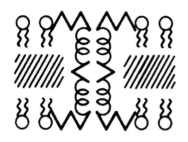

Figure 1.24 The membrane protein structure scheme proposed by Lenard and Singer. The coils represent α-helices and the jagged lines β-sheets.

Robertson's unit-membrane hypothesis. Although the CD measurements could show the main secondary structure of RBC membrane proteins to be α-helical, they could not say exactly how α-helices were situated in the membranes. The very first crystallographic structure of a protein by Sir John Kendrew and colleagues in 1960 revealed that myoglobin was constructed largely of α-helices and that the amino acids buried in the interior of the protein were largely the hydrophobic ones, such as leucine, valine, and phenylalanine (Box 1.5). With those ideas in mind, Lenard and Singer presciently proposed that membrane proteins had hydrophobic α-helices that spanned the non-polar interior of lipid bilayers (**Figure 1.24**).

1.4.4 Proteins Diffuse Freely in the Plane of Plasma Membranes

Janet Plowe's observations on the response of plant protoplasts to manipulation with glass needles showed not only that there must be a plasma membrane but that the plasma membrane responds with rapid changes in shape when it is poked or stretched with the needles. Animal cells change shape rapidly by extending and retracting pseudopods as they move. These simple observations led Larry Frye and Michael Edidin to think of the plasma membrane as being fluid-like. If that description was right, they reasoned that some membrane components, specifically proteins, must be free to move in the plane of the membrane. Elegant experiments published in 1970 showed their hypothesis to be correct (**Figure 1.25**).

Sendai virus can cause cells to fuse into a single cell. Frye and Edidin used the virus to fuse tissue-cultured mouse cells with tissue-cultured human cells. But before fusion, they prepared human and mouse antibodies that would only bind to the surface of their respective cell type. The antibodies were then each labeled with a characteristic fluorescent dye that caused mouse antibodies to fluoresce green and human antibodies red. The distribution of each type of antibody bound to membrane proteins of the cell could be examined as a function of time after the induction of fusion by simply photographing the cells through red or green filters. The boundaries of the two cell types shortly after fusion were obvious (**Figure 1.25A**). Several hours later, the antibody stains for mouse and human cells were distributed all over the cell surface, which provided clear evidence for the diffusion of proteins in the plane of the membrane. These measurements inspired many laboratories, particularly that of Richard Cone at Johns Hopkins University, to explore the lateral diffusion of lipids and proteins in model and biological membranes using a wide range of biophysical methods.

By 1972, there was little doubt that the fluid-state lipid bilayer was the organizing structure of cell membranes and that compact folded proteins were mostly deeply embedded in the bilayer and free to move within it. Missing was an appealing name and a cartoon that succinctly described membrane structure. S. Jonathan Singer and Garth L. Nicolson provided both: the fluid mosaic

Figure 1.25 Fluorescent labeling of cell membrane proteins reveals that the proteins diffuse within the membrane. Mouse cells can be fused with human cells by exposing both to Sendai virus. Fluorescent molecules were attached to antibodies for mouse and human cell-surface proteins. Mouse antibodies were labeled with a green fluorophore and human ones with a red fluorophore. After fusion, the surface distribution of the two fluorophores could be determined by observing the cells under the microscope using green and red filters. Panel A shows the antibody distributions immediately after cell fusion. Panels B and C show that the mouse and human surface proteins eventually become completely mixed after fusion. (From Frye LD, Edidin M [1970] *J Cell Sci* 7: 319–335. Used with permission from Michael Edidin.)

(A)   (B)   (C)

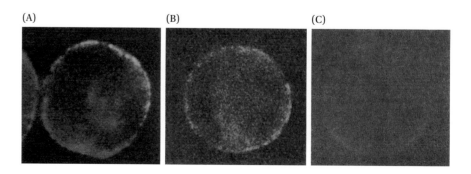

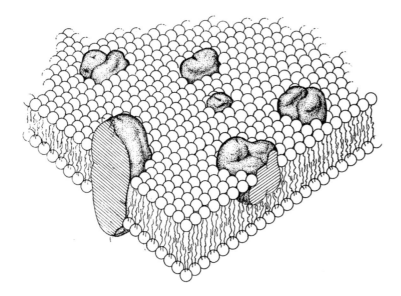

Figure 1.26 The iconic fluid mosaic "model" of Singer and Nicolson. (From Singer SJ, Nicholson GL [1972] *Science* 175: 720–731. With permission from the AAAS.)

model (**Figure 1.26**). Although not a model in the sense that it can be used for quantitative predictions or considered as an accurate structural representation, it nevertheless marks a turning point in membrane research by summarizing in one iconic image many years of research. One problem with the image, besides inaccurate protein cartoons, is the inaccurate representation of the bilayer and its lipid diversity. Another problem is the small number of proteins per unit area.

1.4.5 Electron Crystallography Reveals the First Structure of a Transmembrane Protein

The fluid mosaic model (Figure 1.26) represents membrane proteins as blobs, some say potatoes, embedded in the bilayer. But, what does a membrane protein really look like? Lenard and Singer's CD measurements suggested that they were likely to be helical and proposed that the helices might have a transmembrane orientation. The protein that provided the best chance of revealing its structure was the light-driven proton pump of bacteriorhodopsin (bR), which occurs naturally as hexagonal crystalline arrays in the membranes of *Halobacterium salinarium* (formerly *H. halobium*). As the name implies, this archaeon is an extreme halophile that thrives in saturated sodium chloride solutions. Richard Henderson and Nigel Unwin used electron diffraction to determine the 3-D structure of bR in isolated purple membranes in 1975 (**Figure 1.27**). The resolution of the structure was too low to see any features other than the seven transmembrane helices. But, the structure was a breakthrough achievement that confirmed the existence of transmembrane helices in membrane proteins, consistent with membrane proteins spanning the membrane bilayer. Ten years would pass before the structure of a membrane protein was finally determined at a high enough resolution by x-ray crystallography to allow identification of the amino acid side chains of the helices.

1.4.6 X-Ray Crystallography Reveals the Structure of a Membrane at Atomic Resolution

The report in 1985 of the structure of the photosynthetic reaction center (PSRC) of *Blastochloris viridis* (formerly *Rhodopseudomonas viridis*) at atomic resolution (**Figure 1.28**) provided the second example of an α-helical

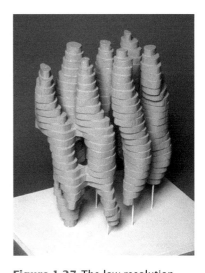

Figure 1.27 The low-resolution structure of bacteriorhodopsin (bR) in cell membranes of *Halobacterium salinarium* (formerly *H. halobium*). The protein occurs naturally in hexagonal arrays in the living organism. Isolated membranes placed in the electron beam of an electron microscope cause diffraction of electrons. The diffraction pattern was used to obtain a 3-D scattering-density map (shown as a balsa-wood model) interpreted as seven transmembrane helices. (Gift of Richard Henderson, Laboratory of Molecular Biology, Cambridge, UK.)

membrane protein structure and suggested that bundles of transmembrane α-helices were likely to be a major structural motif. The structure initiated the modern era of membrane protein structure determination and earned Harmut Michel, Johann Deisenhofer, and Robert Huber the 1988 Nobel Prize in Chemistry.

A third membrane protein structure, but at low resolution, was also reported in 1985. The protein was porin from the outer membrane of *E. coli*, and as for bR, the structure was determined by electron crystallographic measurements of two-dimensional crystals, but in this case, the crystals were obtained by reconstituting the protein into lipid bilayers. Porin is a homotrimer that forms a triple-barrel pore in the outer membrane to allow selected solutes to enter the periplasmic space of the bacteria. As long expected from spectroscopic data and analyses of its amino acid sequence, OmpF porin appeared to be composed exclusively of β-sheet secondary structure. Finally, in 1992, a high-resolution structure of OmpF, obtained from 3-D crystals by x-ray crystallography, revealed the protein in atomic detail (**Figure 1.29**). A key observation was that 16 β-strands formed a closed barrel-like structure, so that as for transmembrane α-helices, all peptide bonds of the amino acid backbone formed internal hydrogen bonds so that no peptide bonds were exposed to the lipid bilayer without a hydrogen-bonding partner.

The first membrane protein structures revealed only two motifs: α-helix bundles and β-barrels; these are the only two motifs so far observed. Furthermore, all plasma membrane proteins are helix bundles, whereas outer membrane proteins of bacteria are, with few exceptions, β-barrels. Why should there be only two motifs, and why should helix-bundle and β-barrel proteins be restricted in bacteria to the inner and outer membranes, respectively? The answers provide

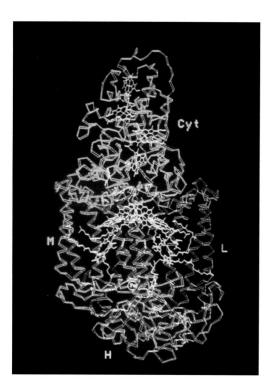

Figure 1.28 The first high-resolution structure of a membrane protein: the photosynthetic reaction center of *Blastochloris viridis* (formerly *Rhodopseudomonas viridis*). The reaction center has three transmembrane subunits (orange, blue, and violet) upon which cytochrome c is attached (light blue). This image shows only the path of the peptide backbone, but the amino acid side chains were also resolved. (Image courtesy of Hartmut Michel and Johann Deisenhofer.)

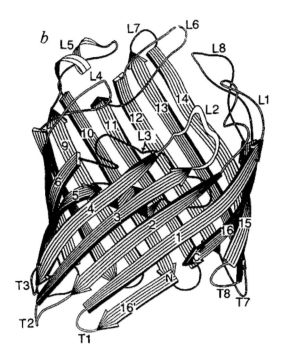

Figure 1.29 A schematic view of the crystallographic structure of OmpF, an outer membrane protein of bacteria. It is composed of β-strands arranged to form a barrel-like structure that causes all peptide bonds in the transmembrane region to be hydrogen bonded. Although the protein is naturally a trimer, this image shows only one monomer. (From Cowan SW, Schirmer T, Rummel G et al. [1992] *Nature* 358: 727–733. With permission from Springer Nature.)

entrées for understanding the general principles of membrane protein stability and biogenesis.

KEY CONCEPTS

- Membranes define cell anatomy.

- The plasma membrane separates the cell interior from the external environment.

- Prokaryotic cells generally have a single interior compartment, whereas eukaryotic cells have many compartments, called organelles, which include mitochondria and chloroplasts.

- Membranes are composed of surface-active lipids, primarily phospholipids.

- Surface-active lipids are amphiphilic: they have two chemically distinct, spatially separated regions that have distinct affinities for polar and non-polar environments.

- Amphiphilic lipids adsorb at air/water and oil/water interfaces to form lipid monolayers.

- Phospholipids dispersed in water form lipid bilayers that give characteristic diffraction patterns when placed in an x-ray beam.

- The fluid lipid bilayer is the principal organizing element of cellular membranes of all types.

- Most membrane proteins are transmembrane and diffuse freely in the plane of the membrane.

- Membrane proteins have only two structural motifs: α-helical bundles and β-barrels, the latter being found only in the outer membranes of bacteria and mitochondria.

FURTHER READING

Blaurock, A.E. (1982) Evidence of bilayer structure and of membrane interactions from X-ray diffraction analysis. *Biochim. Biophys. Acta* 650:167-207.

Bretscher, M.S. (1973) Membrane structure: Some general principles. *Science* 181:622-629.

Edidin, M. (1974) Rotational and translational diffusion in membranes. *Annu. Rev. Biophys. Bioeng.* 3:179-201.

Kondepudi, D., and Prigogine, I. (1998) *Modern Thermodynamics*, John Wiley & Sons, Chichester.

Lee, A.G. (1977) Lipid phase transitions and phase diagrams. I. Lipid phase transitions. *Biochim. Biophys. Acta* 472:237-281.

Liljas, A., Liljas, L., Piskur, J., Lindblom, G., Nissen, P., and Kjeldgaard, M. (2009) *Textbook of Structural Biology*, World Scientific, Singapore.

Pagano, R.E., Cherry, R.J., and Chapman, D. (1973) Phase transitions and heterogeneity in lipid bilayers. *Biochim. Biophys. Acta* 181:557-559.

Serdyuk, I.N., Zaccai, N.R., and Zaccai, J. (2007) *Methods in Modern Molecular Biophysics*, Cambridge University Press, New York. 1120 pp.

Small, D.M. (1977) Liquid crystals in living and dying systems. *J. Colloid Interface Sci.* 58:581-602.

Stoeckenius, W., and Engelman, D.M. (1969) Current models for the structure of biological membranes. *J. Cell Biol.* 42:613-646.

Tanford, C. (1980) *The Hydrophobic Effect: Formation of Micelles and Biological Membranes*, 2nd ed., John Wiley & Sons, New York.

KEY LITERATURE

Bar, R.S., Deamer, D.W., and Cornwell, D.G. (1966) Surface area of human erythrocyte lipids: Reinvestigation of experiments on plasma membrane. *Science* 153:1010-1012.

Branton, D. (1966) Fracture faces of frozen membranes. *Proc. Natl. Acad. Sci. U.S.A.* 55:1048-1056.

Bretscher, M.S. (1971) A major protein which spans the human erythrocyte membrane. *J. Mol. Biol.* 59:351-357.

Cherry, R.J. (1975) Protein mobility in membranes. *FEBS Lett.* 55:1-7.

Collander, R. (1949) The permeability of plant protoplasts to small molecules. *Physiologia Plantarum* 2:300-311.

Danielli, J.F. (1936) Some properties of lipoid films in relation to the structure of the plasma membrane. *Journal of Cellular and Comparative Physiology* 7:393-408.

Deisenhofer, J., Epp, O., Miki, K., Huber, R., and Michel, H. (1985) Structure of the protein subunits in the photosynthetic reaction centre of Rhodopseudomonas viridis at 3Å resolution. *Nature* 318:618-624.

Engelman, D.M. (1970) X-ray diffraction studies of phase transitions in the membrane of Mycoplasma laidlawii. *J. Mol. Biol.* 47:115-117.

Everitt, C.T., and Haydon, D.A. (1968) Electrical capacitance of a lipid membrane separating two aqueous phases. *J. Theor. Biol.* 18:371-379.

Finean, J.B. (1960) Electron microscope and x-ray diffraction studies of the effects of dehydration on the structure of nerve myelin. *Journal of Biophysical and Biochemical Cytology* 8:13-29.

Frye, L.D., and Edidin, M. (1970) The rapid intermixing of cell-surface antigens after formation of mouse-human heterokaryons. *J. Cell Sci.* 7:319-335.

Gorter, E., and Grendel, F. (1925) On bimolecular layers of lipoids on the chromocytes of the blood. *J. Exp. Med.* 41:439-443.

Green, D.E. (1966) Membranes as expressions of repeating units. *Proc. Natl. Acad. Sci. U.S.A.* 55:1295-1302.

Henderson, R., and Unwin, P.N.T. (1975) Three-dimensional model of purple membrane obtained by electron microscopy. *Nature* 257:28-32.

Langmuir, I. (1917) The constitution and fundamental properties of solids and liquids. II. Liquids. *J. Am. Chem. Soc.* 39:1848-1906.

Lenard, J. (1970) Protein components of erythrocyte membranes from different animal species. *Biochemistry* 9:5037-5040.

Lenard, J., and Singer, S.J. (1966) Protein conformation in cell membrane preparations as studied by optical rotatory dispersion and circular dichroism. *Proc. Natl. Acad. Sci. U.S.A.* 56:1828-1835.

Levine, Y.K., and Wilkins, M.H.F. (1971) Structure of oriented lipid bilayers. *Nature New Biol.* 230:69-76.

Mueller, P., Rudin, D.O., Tien, H.T., and Wescott, W.C. (1962) Reconstitution of cell membrane structure in vitro and its transformation into an excitable system. *Nature* 194:979-980.

Overton, E. (1895) Über die osmotischen Eigenschaften der lebenden Pflanzen und Tierzelle. *Vierteljahrsschrift der Naturforschenden Gesellschaft in Zürich* 40:159-201.

Plowe, J.Q. (1931) Membranes in the plant cell. *Protoplasma* 12:196-221.

Reinert, J.C., and Steim, J.M. (1970) Calorimetric detection of a membrane-lipid phase transition in living cells. *Science* 168:1580-1582.

Singer, S.J., and Nicolson, G.L. (1972) The fluid mosaic model of the structure of cell membranes. *Science* 175:720-731.

Steim, J.M., Tourtellotte, M.E., Reinert, J.C., McElhaney, R.N., and Rader, R.L. (1969) Calorimetric evidence for the liquid-crystalline state of lipids in a biomembrane. *Proc. Natl. Acad. Sci. U.S.A.* 63:104-109.

Stoeckenius, W. (1962) The molecular structure of lipid-water systems and cell membrane models studied with the electron microscope, in *The Interpretation of Ultrastructure* (Harris, R. J. C., Ed.), pp. 349-367, Academic Press, New York.

EXERCISES

1. It has been pointed out that Franklin missed an opportunity to define an upper bound for a molecular dimension when he famously spread oil on the surface of a pond in Clapham Common. If he used a teaspoon and observed that it spread to visibly still the waves over half an acre, what approximate thickness would the layer be (in nm)?

2. Similarly, the pressure-area curves in Figure 1.11B rise to collapse at certain areas per molecule. By using density measurements from a series of normal alkanes, one can obtain values for the volumes of CH_3 and CH_2 in the liquid state. Use these values to get a thickness for the hydrophobic portion of a DMPC (DiMyristoylPhosphatidylCholine) monolayer, combining the area at collapse with your volume calculation. How large an area of water would 1 ml of DMPC cover as a monolayer?

3. Again referring to the pressure-area curves, note that there are two groups, a,b and c,d,e. Why do you think this is the case? (Hint: note that the measurement is at room temperature, and then consider the melting behavior shown by calorimetry and diffraction.)

4. Several early models for membranes have a layer between two layers of lipids; see Figures 1.12 and 1.22. Why? (Check literature.) What would stabilize the thickness of such a structure? How do measures of bilayer thickness affect these views?

5. Agnes Pockels is an interesting figure. Research her personal history.

6. While alpha helices are a major theme in membrane protein structure, polypeptides can form other kinds of helices. What are they? Do you expect such helices to be found in proteins?

Lipid Bilayers

2

A key event in the emergence of the first living organisms was the evolution of lipid bilayers to separate the chemistry inside primitive cells from the environment, thus defining them as individual entities. We can only speculate about the nature of the earliest cell membranes, but given what we learned in Chapter 1, they surely must have been composed of surface-active lipids. Whatever its evolutionary path may have been, the lipid bilayer as the organizing fabric of membranes has been extraordinarily successful. Besides providing the basis of cell membrane structure, lipids participate in many other ways in life processes, such as energy storage and conversion, cellular communication, inflammatory responses, neuronal signal transmission, and carbohydrate metabolism. They are also involved in many disease states, including Alzheimer's, asthma, cancer, malaria, and rheumatoid arthritis. The role of lipids in each of these diseases might be a long article, or even a book, so we confine our focus instead to the fundamental role of lipid structure in the formation and properties of lipid bilayers.

The lipid bilayer, as we shall see, provides a formidable barrier to the movement of most water-soluble molecules. The task of selectively—and controllably—regulating the transbilayer movement of molecules and ions across membranes falls to membrane proteins. Because the proteins must function in the bilayer environment, our exploration of membrane proteins and membrane dynamics must begin with a close examination of lipids and the properties of lipid bilayers. There are many questions to answer. How does the molecular design of lipids lead to bilayer structures? How does membrane lipid composition affect membrane properties? How can we determine the structures of lipid bilayers? What characteristics of lipid bilayers do the structures reveal? What are the useful properties of bilayers that help cells live? By answering these and other questions, we will be able to appreciate in later chapters how the structures of membrane proteins meet the challenges of controlled movement of ions and other molecules across membranes.

While almost all major lipids of cells have surely been identified, many minor lipids are still being found. Importantly, minor lipids, such as sphingosine and sphingosine phosphate, are often the ones that profoundly affect biological function through signaling activities. Liquid chromatography and mass spectrometry (LC/MS) analyses are well suited to finding new minor lipids, even in the course of quantifying the major ones. There are now extensive databases dedicated to the "lipidome." Any attempt to catalog here the kinds and properties of all of these lipids—there are thousands of them—would largely replicate these databases. Consequently, we discuss only selected examples in order to focus on the principles that undergird the formation and properties of biological membranes.

DOI: 10.1201/9780429341328-3

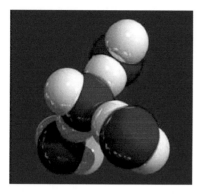

Figure 2.1 Four water molecules interacting by hydrogen bonds. The central molecule has three bonds and one unsatisfied bonding possibility. In liquid water, the hydrogen bonds form, break, and reform frequently, creating a liquid state with strongly associating H_2O units. (From Hirohito Ogasawara, www.eurekalert.org/features/doe /2004-05/dlac-uts050604.php. With permission.)

2.1 BIOLOGICAL MEMBRANE LIPIDS SPONTANEOUSLY FORM BILAYERS

The amphipathic molecular design of membrane lipids leads to spontaneous lipid bilayer formation when dispersed in water, often in the form of vesicles. These vesicles have a critically important property: they can control solution entropy by enclosing molecules at high concentration, which prevents dispersal. It is the lipid bilayer's vesicular control of entropy that has allowed life to evolve on our planet. The logical starting point for understanding the spontaneous formation of bilayer membranes is to consider the underlying molecular properties and interactions, including the hydrophobic effect, polar interactions, dynamics, and shape.

2.1.1 Hydrophobic Interactions Arise from Hydrogen Bonding in Water

Books have been written on the hydrophobic (Gr; "water fearing") effect, which was introduced in Chapter 1 (see Box 1.2) and will be expanded upon in Chapter 3. This attention is warranted, because it is the basis of so much fundamental biology, including the folding of soluble proteins as well the formation of membranes. The hydrogen-bonding ability of water lies at the heart of the hydrophobic effect.

A water molecule has a separation of charge that comes from the shift of electrons away from the two hydrogens and toward the electronegative oxygen atom, making the hydrogens into excellent hydrogen-bond donors and the oxygen into an acceptor. The orbital structure of the oxygen favors the acceptance of two hydrogen bonds, potentially allowing each water molecule to hydrogen bond to four others—two accepted by the oxygen and two donated by the hydrogens. In ice, most of these bonds are formed, but in liquid water, fewer are found on average because thermal energy disrupts them (**Figure 2.1**). The average number of hydrogen bonds per water molecule is still under study, but the cohesive properties of liquid water are obvious from looking at a raindrop or a droplet on a non-polar surface (**Figure 2.2**). The droplets tend to be spherical to minimize their surface area, and consequently, there is an energetic cost of unsatisfied hydrogen bonds at the droplet surface that gives rise to surface tension (Box 1.4). Said another way, water molecules in the droplets seek to maximize hydrogen-bond interactions with each other and pull themselves toward a spherical shape, which has the lowest surface-to-volume ratio of any geometry and hence, the lowest surface free energy.

If we think about the molecules at the surface of the raindrop or at the surface of the drop in contact with the hydrophobic surface (Figure 2.2), they are restricted in the ways that they can hydrogen bond because there are no hydrogen-bond donors or acceptors in the air or on the non-polar surface. This restriction means that the surface molecules have fewer ways to be organized to maximize hydrogen bonds and so are more ordered than molecules in bulk water. In thermodynamic terms, surface waters have lower entropy, which must be partially compensated for by a higher enthalpy than in bulk water. Enthalpic interactions arise from the reorganization of surface waters to strengthen H-bonding as much as possible. The optimal balance is achieved by forming a network of water molecules at the hydrophobic surface (**Figure 2.3**). To minimize the surface free energy further (Box 1.4), water organizes itself to minimize its surface-to-volume ratio and the area of contact with the hydrophobic surface.

Figure 2.2 Water droplet on a hydrophobic leaf surface. The mutual attraction of the water molecules via hydrogen bonding creates surface tension that causes the droplet to seek a minimum surface area, rounding up when in contact with a surface that has no hydrogen-bond donors or acceptors, such as the leaf or the air. The absence of donors or acceptors means that the water avoids the contact, and the surface is said to be hydrophobic (water fearing).

2.1.2 The Hydrophobic Effect Drives Non-Polar Molecules out of Aqueous Solution

If we apply the idea that water minimizes hydrophobic contacts, we can see at the molecular level that contact between two or more hydrophobic molecules will be favored in water because the amount of surface exposed to water will be less than if they were not in contact (Figure 2.3). Such a contact is referred to as a hydrophobic interaction, but it is really less an interaction between the non-polar molecules than the resulting improvement in the self-association of water, mainly driven by entropy at room temperature. Because this process is a surface phenomenon, the magnitude of the free energy of association can be derived from the surface area removed from contact with water, which is calculated as about 21–25 cal mol^{-1} Å^{-2} (Box 1.2). This number is called the atomic solvation parameter σ. The accessible surface area (ASA) of large non-polar amino acid side chains (e.g., Leu or Phe) in a polypeptide chain is about 100 Å^2 (Box 1.2). Assuming $\sigma = 23$ cal mol^{-1} Å^{-2}, the cost of transferring the side chain from a non-polar environment to water would be 2.3 kcal mol^{-1} (or 9.6 kJ mol^{-1} in SI units, the conversion being 4.186 kJ per kcal).

2.1.3 Amphiphiles Are Molecules That Have a Non-Polar Part Separated from a Polar Part

Amphiphiles (amphi "on both sides"; philos "loving") have a spatial segregation of hydrophilic and hydrophobic regions, often termed the "polar head" and "non-polar tail," respectively, in elongated molecules. An example is the commonly used detergent sodium dodecyl sulfate (SDS) (**Figure 2.4A**). The negative charge on the sulfate group of SDS gives strongly hydrophilic interactions with water, whereas the aliphatic methylene chain is hydrophobic. The behavior of amphiphiles in water depends on the relative polarities of the polar and non-polar groups. For example, at low concentrations, SDS is monomeric, but as the concentration of SDS is raised above the so-called critical micelle concentration (CMC), SDS molecules assemble cooperatively into micelles, which are aggregates of SDS (**Figure 2.4B**). Above the CMC, SDS exists as monomers in equilibrium with micelles (**Figure 2.4C**). Because micelle assembly is cooperative—it usually happens over a small range of concentrations—the CMC is an average. Micellar aggregates of amphiphilic monomers can take many different forms depending upon the atomic structure of the monomer. Molecules such as SDS that form aggregates containing a reasonably uniform and modest number of monomers (~60) tend to be roughly spherical. Phospholipids also form aggregates, but at extremely low concentrations and with huge numbers of phospholipids arranged as bilayers. Purists call these large phospholipid aggregates micelles because they involve cooperative assembly characterized by a CMC. Here, we reserve the name "micelle" for compact aggregates containing a relatively small number of monomers. The aggregates formed from phospholipids we refer to as liposomes (see later).

(A)   (B)

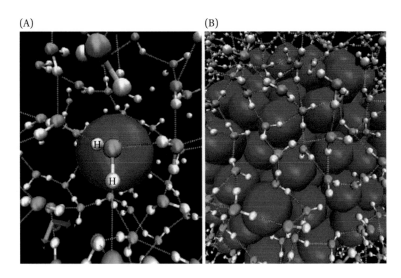

Figure 2.3 MD simulation of the hydrophobic effect. The blue and white particles represent water oxygen (O) and hydrogen (H) atoms, respectively. (A) The H-bond pattern of water around a non-polar particle (red) that is about the size of methane. (B) The H-bond pattern around a spherical cluster of methane-like particles. The water molecules next to the non-polar molecule are more ordered because they cannot make H-bonds with the surface, and are therefore restricted in their motion—an entropy effect. (From Chandler D [2005] *Nature* 437: 640–647. With permission from Springer Nature.)

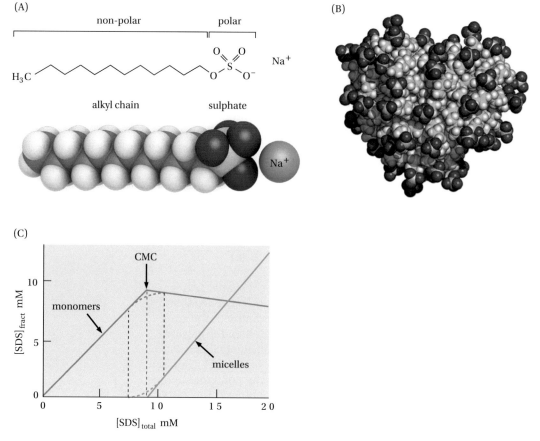

Figure 2.4 An amphiphile, like the sodium dodecyl sulfate (SDS) molecule shown in Panel A, has a polar group (sulfate in this case) and a non-polar hydrocarbon chain that are well separated from each other, allowing their distinct properties to interact with water. (B) A molecular dynamics simulation model of an SDS micelle. Notice the exposure of the hydrocarbon interior at the surface. The thermodynamic need to minimize the exposure of the non-polar hydrocarbon to water drives the formation of the micelles. (C) Self interaction of SDS molecules in aqueous solution. At low concentration, the SDS molecules are soluble as monomers, but raising the concentration leads to the cooperative formation of micelles over a narrow range of concentration, the critical micelle concentration (CMC), as indicated by the dashed red line. The transition from monomers to micelles is not a sharp one, however, as indicated by the dashed grey lines on either side of the dashed red line. Each micelle has about 60 SDS molecules. (B, From Roussel G, Michaux C, Perpète EA [2014] *J Mol Model* 20: 2469. With permission from Springer Nature. C, From Garavito RM, Ferguson-Miller S [2001] *J Biol Chem* 276: 32403–32406. Used with permission from the ASBMB.)

The basic principles driving spontaneous assembly of micelles also drive membrane formation from lipids. These principles are also the basis of detergent action and provide a set of tools for the selective disassembly of membranes by detergents. The formation of micelles relies on the favorable interaction of polar groups with water, from either hydrogen bonds or charge solvation, and the unfavorable interaction of the non-polar regions with water. The precise balance between these competing effects is determined by the atomic structure of the amphiphile, especially the relative amounts and shapes of the polar and non-polar moieties.

The CMC is a measure of the strength of amphiphile association—if the association is strong, the molecules form micelles at a low concentration; if it is weak, the CMC is high. A way to quantify the energy of association of a monomer with a micelle is to treat it as an equilibrium partitioning of the amphiphile between water and micelles. If the micelles are large enough (typically more than 50 monomers per micelle), then we can describe the free energy of transfer from micelle to water (ΔG_{mw}) using the formula described in Box 1.2 but replacing X_w with X_{CMC}, defined as the CMC of the amphiphile (mole fraction units):

$$\Delta G_{mw} = \mu_0^w - \mu_0^m = -RT \ln X_{CMC} \qquad (2.1)$$

In Equation 2.1, the superscripts w and m refer to water and micelle, respectively. SDS micelles have around 60 monomers and the CMC is about 8 mM, corresponding to $X_{cmc} = 10^{-5}$. From these numbers, we find that the energetic cost of transferring SDS from the micelle to water is about 5 kcal mol^{-1}, meaning that SDS has a strong preference for micelles. The division of properties of amphiphiles can take different forms, and the relative strengths of the hydrophilic and hydrophobic parts will vary. Because there is a direct relationship between the energy of the hydrophobic effect and molecular surface area, it is expected that longer hydrocarbon chains will be more hydrophobic and shorter ones less; this relationship is observed, with the CMC of shorter chains being higher than that of longer ones for a given polar group.

As an example, dodecylphosphocholine (DPC) (12-carbon chain) has a CMC of about 1 mM, whereas hexadecylphosphocholine (16-carbon chain) has a CMC near 0.01 mM. If the polar group has net charge, as in SDS, charge repulsion will oppose molecular association and increase the CMC. Also, an amphiphile's shape plays a role, and some amphiphiles, such as deoxycholic acid, are subtle in the division of their polar/non-polar regions (examine **Figure 2.5** and try to imagine how the amphiphilic properties arise). Deoxycholic acid has a CMC of about 5 mM.

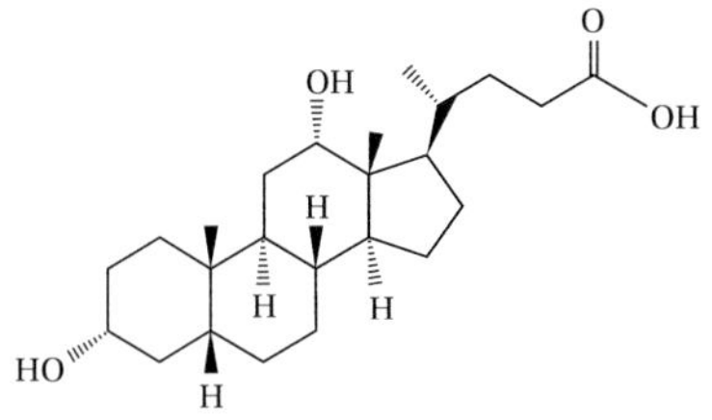

Figure 2.5 Structure of deoxycholic acid. Here, the separation of polar and non-polar parts is less obvious. Can you think how it might be an amphiphile? Would you imagine that there would be many or few molecules in a micelle?

2.1.4 Micelles Have Different Shapes and Sizes

The balance of the properties of the parts of an amphiphile results in different degrees of association and different micelle shapes. This can be understood by thinking about the mission of the micelle—it is trying to hide the hydrophobic parts from water and to cover the surface with polar groups. As the volume of the hydrophobic part becomes larger, for example by increasing chain length, it becomes more difficult to sequester using spherical geometry, so other shapes such as prolate or oblate ellipsoids are found. Micellar structures of the detergent β-octylglucoside (βOG) are illustrated in **Figure 2.6**. The kind of structure often seen in texts (panel B) makes no chemical sense, because it implies that a lot of water must be within the micelle. More realistic representations are similar to those of panels C and D. Panel C shows a model from a molecular dynamics simulation of a micelle with 20 molecules; panel D shows a micelle with 50 molecules. The deviation from sphericity gives a sense of the dynamic character of these assemblies. Packing at the center of the micelles is accomplished by the most extended of the hydrophobic chains. To get a sense of this, compare C and D in the figure—there must be a few molecules with conformations like those in B to reach and fill the micelle center. Because of this packing, micelles are

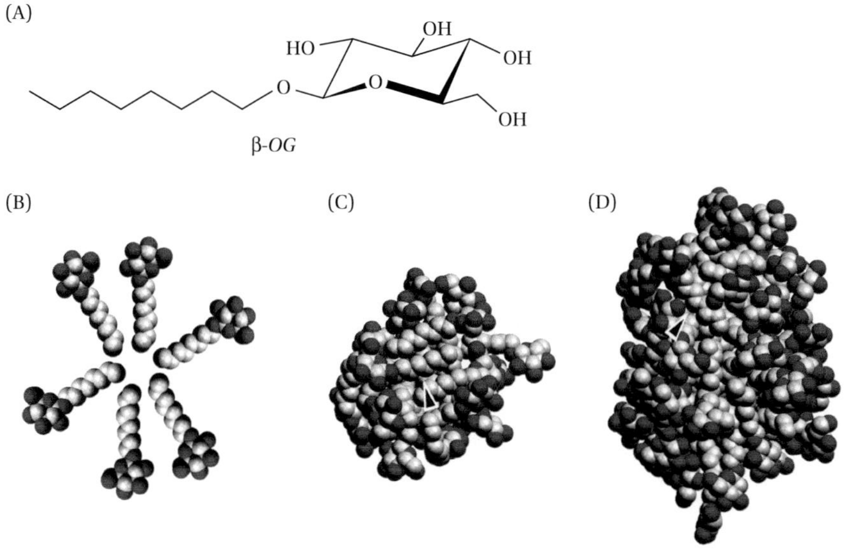

(A)

β-*OG*

(B)   (C)   (D)

Figure 2.6 Micelles of β-octyl glucoside. (A) Chemical structure. (B) Misleading schematic found in many texts. (C, D) Molecular dynamics models with different numbers of detergent molecules showing shape changes with micelle size. (From Garavito RM, Ferguson-Miller S [2001] *J Biol Chem* 276: 32403–32406. Used with permission from the ASBMB.)

BOX 2.1   WEAK AND STRONG DETERGENTS

Detergents are often used when investigating membranes because they can replace the hydrophobic interactions that stabilize lipid–lipid, lipid–protein, and protein–protein interactions by coating the hydrophobic surfaces to present polar surfaces to the aqueous milieu. Hence, detergents are often referred to as "surfactants." Solubilization of membranes results in water-soluble products that are more easily studied.

One of the parameters frequently used in describing detergents is whether they are "strong" or "weak" in disrupting membrane structure. What is meant by the words is that strong detergents solubilize membranes at detergent to lipid ratios less than 1, whereas weak detergents, with ratios greater than 1, tend to accumulate in a membrane before dissolving it. Detailed studies using isothermal titration calorimetry show a linear relationship between the partitioning of a detergent into

a membrane and its CMC. Specifically, the membrane/water partition coefficient, K, is related to the CMC as $K \times CMC \sim 1$. Strong detergents such as alkyl maltosides and tritons tend to fall below the line in Box **Figure 1**, and weak detergents, such as CHAPS and alkyl glucosides, tend to fall above the line.

Generally, when attempting to disaggregate a membrane while preserving protein structures in it, weak detergents are used. When wishing to simply dissolve the membrane as much as possible, strong detergents are used. However, the highlighted detergents, which are some that have been used successfully for membrane protein solubilization and crystallization, show that there are no hard-and-fast rules when it comes to crystallizing membrane proteins. What they do have in common, however, are CMCs greater than 10^{-3} M and bilayer partition coefficients lower than 10^3.

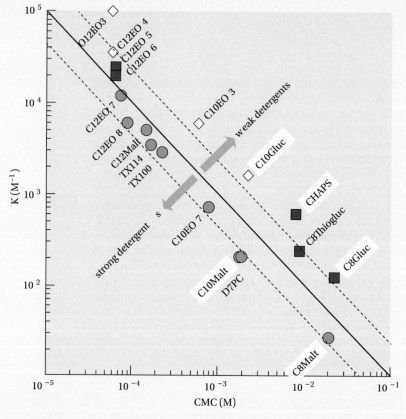

Figure 1 Properties of detergents: the binding of detergents to membranes correlates with the critical micelle concentration such that K•CMC ~1. K is the membrane/water partition coefficient. Strong detergents have K•CMC <1, and weak detergents have K•CMC >1. The detergents measured were oligo(ethylene oxide) alkyl ethers (CmEOn with m = 10/n = 3, 7 and m = 12/n = 3 ... 8); alkylglucosides (octyl, decyl); alkylmaltosides (octyl, decyl, dodecyl); diheptanoylphosphatidylcholine; Tritons (X-100, X-114); and CHAPS. Detergents commonly used in membrane protein crystallization are highlighted (see Box 9.11). The colors of the data points are not important for the present discussion. (From Heerklotz H, Seelig J [2000] *Biophys J* 78: 2435–2440. Used with permission from Elsevier.)

ordered toward their centers and disordered near their surfaces, the opposite of the ordering in the lipid bilayers found in membranes. Notice the exposure of the hydrocarbon chains to water. This exposure results in a tension that acts to minimize the surface area of the micelle.

Detergents are often used to disassemble membranes for analysis and study of their parts. A detergent is considered as "strong" if it tends to disrupt membranes when the detergent is at a low concentration relative to its CMC and to be "weak" if the detergent binds to the membrane in significant amounts before disrupting the structure (see **Box 2.1**). Thus, "mild" or "weak" detergents are favored if one wishes to preserve the folded structure of a membrane protein when removing it from a membrane.

2.1.5 Many Membrane Lipids Are Two-Chain Amphiphiles

The basic structure of membranes is organized by lipids that form bilayers, and their molecular design can be understood using the principles of amphiphilicity, dynamics, and shape that we have examined in the formation of micelles. Some of the most common membrane lipid types are surveyed in **Figure 2.7** . They vary in structural detail but often consist of a glycerol backbone to which are esterified two unbranched fatty acids and a characteristic polar group, often containing a phosphate. Typically, one of the two fatty acids is fully saturated, while the other contains one or more double bonds. In some lipids, the phosphate is absent but other polar groups are found, such as sugars or glycerol. Sphingomyelin has a different backbone that is based on sphingosine—note that one of the chains is bonded using an amide linkage, creating additional hydrogen-bonding capability in the backbone. A number of other hydrophobic molecules are found mixed with these lipids in membranes, notably cholesterol in eukaryotes and carotenoids in chloroplasts and some bacteria; other structures abound for special cases, such as the outer membranes of Gram-negative bacteria that are rich in lipopolysaccharides.

2.1.6 Membrane Lipids Form Bilayers in Water

The dialkyl lipids of Figure 2.7 have very low CMC values. For example, the measured mole fraction CMC of dipalmitoylphosphatidylcholine (DPPC) is about 8.4×10^{-12} M, corresponding to $\Delta G_{mw} = 15$ kcal mol^{-1}, which is roughly the cost of transferring 20 methylenes from a hydrocarbon phase to water. This huge free energy cost of moving a phospholipid from its micelle to water means that the aggregated state is extraordinarily stable. What form does a phospholipid micelle take? Imagine trying to form a spherical micelle from an amphiphile that has two chains instead of one, such as the lipids shown in Figure 2.7A. Because the chains are held together at the polar groups, they are limited in their relative motion and so are most ordered in this part of the molecule and are progressively disordered by thermal energy with distance from the polar end, adopting gauche conformations in the methylene chains or responding to perturbations by unsaturated bonds. Nuclear magnetic resonance (NMR) measurements (see later) show this progressive disorder down the chains quite clearly.

Recall that the micelles formed by SDS are more ordered toward the center and more disordered toward the polar end. Also, the polar parts of most micelle-forming amphiphiles are large in diameter compared with the chain, facilitating the curvature needed for a micellar shape, whereas phospholipid molecules are more cylindrical, with heads similar to the tails in cross section. One would have a hard time forming a spherical micelle given the molecular shape and dynamics of phospholipids, and indeed, they do not tend to form micelles of the sort seen for SDS and DPC. Instead, they form bilayers (**Figure 2.8**): thin sheets with a layer of lipid on each side sequestering the hydrophobic chains in the middle and coating the two surfaces with polar headgroups. The approximately cylindrical shape of the phospholipids allows the lipids to pack well. Consequently, the dynamics of the chains produce a disordered bilayer center and comparatively ordered surface regions (**Figure 2.9**). The headgroup layers can be closely packed with relatively modest unfavorable exposure of the hydrophobic interior to water.

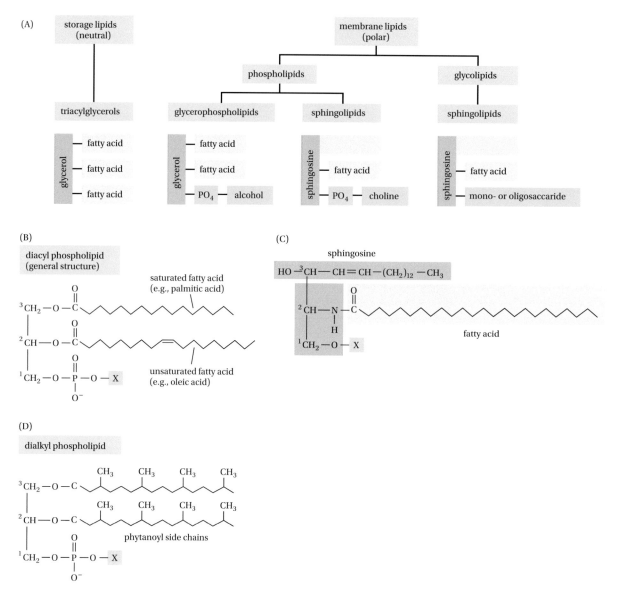

Figure 2.7 Phospholipid structures are built from modular components, typically resulting in lipids with polar headgroups and two alkyl chains. (A) Glycerol and sphingosine are the basic structural units upon which other modular components are assembled. (B) Phospholipids with two alkyl chains connected to the glycerol by ester linkages are the most common lipids of cell membranes in eukaryotes and prokaryotes. (C) Sphingosine-based phospholipids are most often found in neural tissue. Their name is derived from the Egyptian sphinx because their function seemed enigmatic when they were discovered in the 19th century. (D) In archaea, two phytanoyl fatty acid chains are linked to the glycerol backbone by ether linkages. The branched chain phytanoyl fatty acid chains have no phase transitions and provide chemical stability and fluidity over a wide range of temperatures. (E) Structures of commonly encountered glycerophospholipids. (F) Structures of commonly encountered sphingolipids. Two-chain phosphoglycero-lipids are customarily designated by two letters, representing the kinds of hydrocarbon chains (usually esterified fatty acids) followed by two letters indicating the characteristic group. If the two chains are identical, D is used to represent the "Di", as in Dipalmitoyl=DP, and the order of the letters designates the positions of the fatty acids. The most common uses are: Fatty acids: M = Myristic acid, P = Palmitic acid, S = Stearic acid, O = Oleic acid; Characteristic Groups: PC = Phosphatidylcholine, PG = Phosphatidylglycerol, PS = Phosphatidylserine, PI = Phosphatidylinositol, PA = Phosphatidicacid. Examples: DSPC = Distearoylphosphatidylcholine, POPS = Palmitoyloleoylphosphatidylserine. (Modified from Nelson, DL and Cox, MM [2000] *Lehninger Priciples of Biochemistry*, 3rd Edition.)

2.1.7 Bilayers Spontaneously Form Closed Compartments: Liposomes and Vesicles

The edge of the bilayer sheet shown in Figure 2.8 presents a problem: how to cover the hydrophobic bilayer interior and shield it from water? If we try to use the bilayer lipids to cover the edge of a bilayer sheet, we find that they cannot do it for the same reason that they do not form sphere-like micelles—they cannot

(E)

name of glycerophospholipid	name of X	formula of X	net charge (at pH 7)
phosphatidic acid	–	$-H$	–1
phosphatidylethanolamine	ethanolamine	$-CH_2-CH_2-\overset{+}{N}H_3$	0
phosphatidylcholine	choline	$-CH_2-CH_2-\overset{+}{N}(CH_3)_3$	0
phosphatidylserine	serine	$-CH_2-CH-\overset{+}{N}H_3$ $\quad\quad\; \mid$ $\quad\quad COO^-$	–1
phosphatidylglycerol	glycerol	$-CH_2-CH-CH_2-OH$ $\quad\quad\; \mid$ $\quad\quad OH$	–1
phosphatidylinositol 4,5-bisphosphate	4,5-bisphosphate	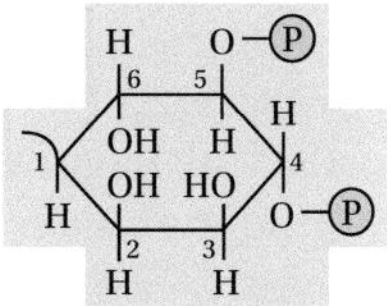 inositol ring structure	–5
cardiolipin	phosphatidylglycerol	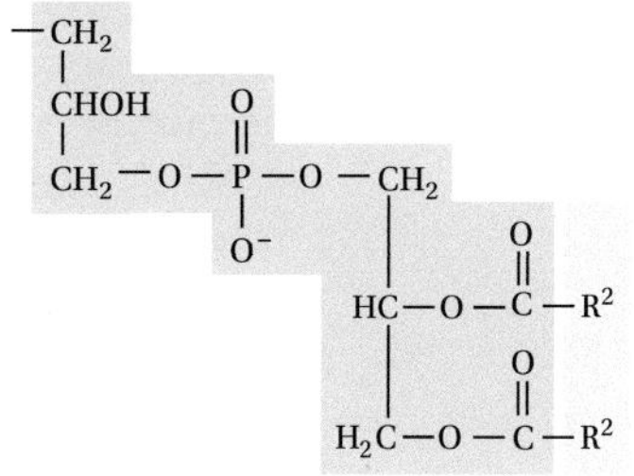 cardiolipin structure	–2

(F)

name of sphingolipid	name of X	formula of X
ceramide	–	$-H$
sphingomyelin	phosphocholine	$-\overset{\overset{\displaystyle O}{\parallel}}{\underset{\underset{\displaystyle O^-}{\mid}}{P}}-O-CH_2-CH_2-\overset{+}{N}(CH_3)_3$
neutral glycolipids glucosylcerebroside	glucose	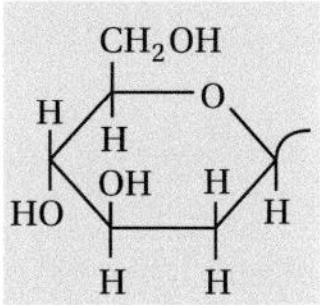 glucose ring structure
lactosylceramide	di-, tri-, or tetrasaccharide	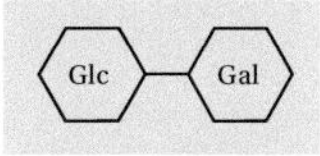 Glc–Gal
ganglioside GM2	complex oligosaccharide	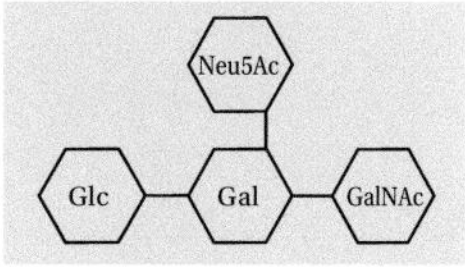 Glc–Gal–GalNAc with Neu5Ac

Figure 2.7 (Continued)

form a highly curved stable structure. Instead, the bilayer curves onto itself to form a closed bilayer surface that surrounds an interior aqueous compartment. Because the lipids cannot form highly curved surfaces, there must be a limit to how small the compartment can be. The structures take two forms. If water is added to dry phospholipids and then shaken, a milky suspension of phospholipid immediately forms. The suspension (or dispersion) is made up of closed vesicular liposomes whose walls contain many layers of bilayers (**Figure 2.10**; see also Figure 1.18B). We know that bilayers are present because of the x-ray

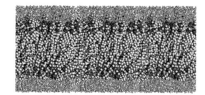

Figure 2.8 Basic structure of a lipid bilayer formed from DOPC. The polar lipid heads provide an interface with water, reducing the aqueous contact of the non-polar chains. Packing of the lipids is efficient because the heads occupy about the same area as two chains in the plane of the bilayer. Anchoring the chains to the headgroups in pairs gives more order near the interface with water than at the center of the bilayer.

diffraction pattern one obtains from them (Figure 1.18C). These spheroidal multilamellar particles of nonuniform size, whose use in biology was pioneered by Alec D. Bangham in the mid-1950s, are generally referred to as multilamellar liposomes or multilamellar dispersions.

A simpler form that phospholipids can adopt is single-bilayer (unilamellar) vesicles (**Figure 2.11**), which can be produced by passing the multilamellar liposome suspension at high pressure through a filter with very small holes. These large unilamellar vesicles (LUVs; **Figure 2.11B**), typically with diameters of 100 nm (1000 Å), are thermodynamically stable and suitable for study by a wide range of physical methods. The diameters of LUVs can be controlled by the size of the holes in the filter.

Another way of making unilamellar vesicles is by sonicating the liposome suspension. Vesicles produced this way are smaller (~25 nm) and tend to have more uniform diameters than LUVs. However, these small unilamellar vesicles (SUVs; **Figure 2.11A**) are not thermodynamically stable and will fuse over a period of days to become LUVs. The conversion of SUVs to LUVs occurs because the outer lipid monolayer is stretched and the inner layer is compressed due to high curvature. The resulting stress, induced by the energy of sonication, increases their free energy positively relative to LUVs. To lower their free energy, they slowly and spontaneously convert to LUVs. Thermodynamic measurements involving lipid bilayers should therefore use LUVs to assure thermodynamic equilibrium.

A third type of vesicle is giant unilamellar vesicles (GUVs; **Figure 2.11C**), formed by spreading a thin layer of phospholipids over the interior surface of a flask and then gently adding water. GUVs can be as large as 50 µM (50 × 10³ nm) in diameter, which is larger than most cells. Importantly, these vesicles are thermodynamically stable, which informs us that a single bilayer surrounding a biological cell is an entirely plausible proposition. Our ability to produce unilamellar vesicles easily in the laboratory supports the idea that bilayer-like structures could have formed in the early stages of life to produce compartments isolated from the surrounding aqueous phase (Figure 2.10). The bilayers of phospholipid liposomes are highly impermeable to ions and large molecules. Primordial vesicles formed from primitive lipids, such as certain combinations of fatty acids, were probably leaky, which may have allowed movement of small molecules across the bilayer without protein pumps or channels.

2.1.8 Lipids Can Form Structures Other than Bilayers

Up to this point, we have discussed two groups of lipids—detergents and bilayer lipids—as if they were completely distinct. In fact, lipid aggregates vary in form depending upon how much water is present, packing constraints, temperature, and the electrostatic properties of the headgroups. Variations in aggregate

Figure 2.9 Hydrocarbon chain conformations vary greatly in the liquid state of a bilayer interior. To emphasize the fluidity of a bilayer interior, panel A shows a gallery of chain conformations from the molecular dynamics simulation of the DOPC bilayer shown in Figure 2.8. The distribution of conformational states illustrates the importance of constraining the ends of two adjacent chains by binding them to the backbone, as seen by the close proximity of the chains near the backbone vs. the terminal methyl groups. Panel B shows a top view of the bilayer in which the alkyl chains are colored yellow, except for the terminal methyl groups, which are colored purple. Neutron diffraction experiments demonstrate that up to 20% of the methyl groups are at the membrane interface at any instant. The water exposure of the alkyl chains explains why hydrophobic peptides can interact strongly with lipid bilayers. Waters within 3 Å of the bilayer are shown in stick format and colored cyan.

(A)

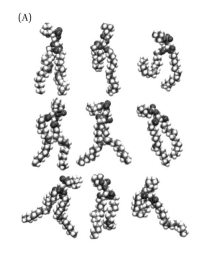

(B)

(A)   (B)

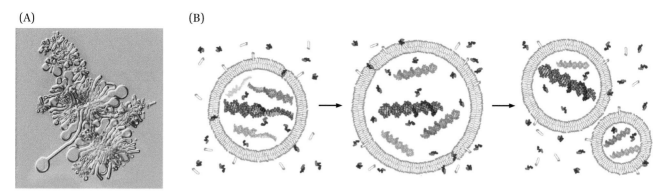

Figure 2.10 Self-assembled liposomes (A) and a hypothetical model (B) for how a protocell formed by a semi-permeable fatty acid bilayer membrane might sustain template-directed polynucleotide synthesis. Polynucleotides are shown as double helices and the building blocks as small red/blue molecules. At high concentrations, vesicles made from certain combinations of fatty acids have been shown to be thermostable and to be able to take up DNA bases from the medium at a rate sufficient to sustain template-directed DNA synthesis inside vesicles. Perhaps, the first proto-cell membrane was not made of phospholipids but of fatty acids or some such simpler molecule? (A, From Deamer DW [2008] 454: 37–38. Used with permission from Springer Nature. B, From Mansey SS et al. [2008] 454: 122–125. Used with permission from Springer Nature.)

(A)   (B)   (C)

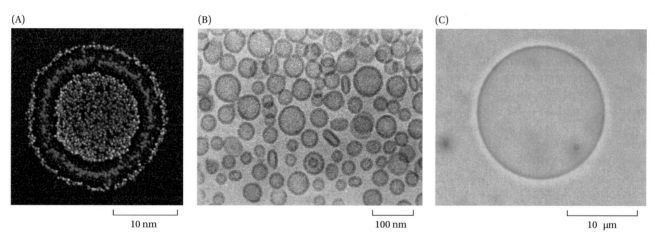

10 nm   100 nm   10 μm

Figure 2.11 Lipid vesicles. (A) Coarse-grained MD simulation of a small unilamellar vesicle (SUV) about the size of a synaptic vesicle. (B) Electron micrograph of large unilamellar vesicles (LUV), which are similar in size to small cytoplasmic organelles or microorganisms. (C) Giant unilamellar vesicle (GUV), similar to sizes of large cytoplasmic organelles or microorganisms. (A, From Marrink SJ, Mark AE [2003] *JACS* 125: 15233. Used with permission from the American Chemical Society. B, Used with permission from www.mardre.com/homepage/mic/tem/samples/colloid/pc_samples/egg_pc_liposome_dispersion_cryo _tem2.html. C, From Morscho A, Orwar O, Chiu DT et al. [1996] *PNAS* 93: 11443. Copyright (2006) National Academy of Sciences, U.S.A.)

structure are characterized by their phase behavior. Of greatest interest in biology are lamellar and hexagonal phases (**Figure 2.12**). Most dialkyl lipids (Figure 2.7) form bilayers when water is in excess. Diacylphosphatidylcholines retain a bilayer structure even at very low water contents; they form lamellar phases that produce a characteristic x-ray diffraction pattern (Figure 1.18A) if the acyl chains are in a fluid state (referred to as the L_α phase; L for lamellar). Diacylphosphatidylethanolamines, on the other hand, tend to form hexagonal II (H_{II}) phases when water content is restricted. In contrast, monoacylphosphatidylcholines, which form spherical micelles in excess water, form hexagonal I (H_I) phases at low hydration. The hexagonal-phase moniker is due to the hexagonal packing of the cylinders (**Figure 2.12A**), which produces characteristic x-ray diffraction patterns. While it is unlikely that these phases form in the excess-water environments of the cell, hexagonal lipid phases reveal the tendencies of some lipids to create curvature, and it is an intriguing idea that they may be present in the mixed bilayers of real membranes to facilitate processes like vesicle formation or fusion by lateral diffusion to the regions where shape changes are needed. Phases and phase behaviors such as those of Figure 2.12 are usually described using the Gibbs Phase Rule (in the following and **Box 2.2**).

Figure 2.12 Structural diversity of lipid phases, mostly at low water content. (A) Structures of common lipid phases (white = water, grey/red = lipid) illustrating hexagonal (H) and lamellar (L) forms. The H_I phase is associated with lipids with large headgroups, while H_{II} is associated with small headgroups. The subscript α indicates the fluid liquid crystalline state of the lamellar phase. (B) An illustration of a lipid phase diagram showing how the phases depend upon composition and temperature. The phase diagram is for glycerol monooleate, which has been used for the crystallization of membrane proteins. (C) A sketch defining the meaning of curvature tendency. (B, From Cherezov V, Clogston J, Papiz MZ et al. [2006] *J Mol Biol* 357: 1605–1618. (Used with permission from Elsevier. C, From Zimmerberg J [2000] *Traffic* 1: 366–368. Used with permission from John Wiley and Sons.)

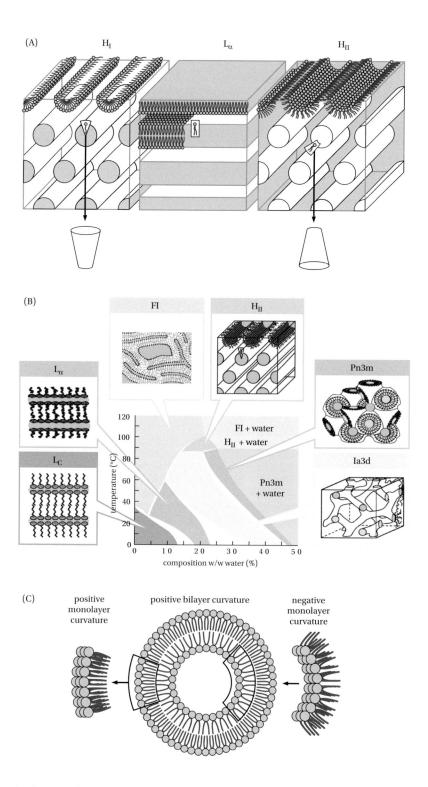

Lipids are often referred to as having a "curvature" tendency, either positive, negative, or zero (**Figure 2.12C**). Curvature refers to how a monolayer of a lipid appears in the low-hydration phase. Diacylphosphatidylcholine bilayers that remain lamellar at low hydrations have zero curvature (but they form positive curvature vesicles at high hydration). Monoacylphosphatidylcholine, i.e., lysophosphatidylcholine, forms H_I phases and has positive curvature. Diacylphosphatidylethanolamine is a negative curvature lipid because it tends to form H_{II} phases.

A common practice is to attribute the curvature tendency to the shape of the molecule (Figure 2.12A). This makes some sense for positive-curvature

BOX 2.2   GIBBS PHASE RULE

Multicomponent mixtures of chemical compounds need not be structurally and chemically homogeneous. In general, mixtures are heterogeneous, meaning that they are composed of two or more phases due to disparities in interaction energies among the components. Gibbs defined a phase as being a volume of material with uniform chemical properties bounded by a surface. The trivial example is the two-phase system of an ice cube floating in water. A phase need not be a single volume. If the ice cube is crushed, there are still only two phases, because all ice fragments have the same chemical properties. The Gibbs Phase Rule allows a systematic description of multicomponent, multiphase systems.

It is claimed by some that the 19th-century physicist Josiah Willard Gibbs made the industrial revolution possible by discovering the relationship between the number of phases, composition, and free variables in multicomponent systems. The multicomponent systems that mattered during the industrial revolution were metallic alloys. Steel, for example, is composed mainly of iron and carbon but sometimes also vanadium and chromium. The number and amounts of metallic phases within a steel alloy determine physical properties such as tensile strength and hardness. The Gibbs Phase Rule made it possible for industrial revolution metallurgists to design and characterize alloys for steel manufacture.

Why should the phase rule matter to biologists? It matters because cells are undeniably complex mixtures with myriad components. Whenever there is a large number of components with different solubilities and miscibilities, phase separations can occur. Solubility and miscibility depend upon the mutual interaction energies (U_{ij}) of the components. In a three-component (A+B+C) system, for example, we need to account for U_{AA}, U_{AB}, U_{AC}, U_{BB}, and U_{BC}. If the U_{ij} are about equal, then three components will tend to mix well. If any of the U_{ij} are very different, then the components will have favored partners, and phase separations can occur.

At the grossest level, cells are two-phase systems made up of water-insoluble lipid membranes and the cytoplasm. But at a finer level, the membrane itself has hundreds of lipid and protein components and has the possibility itself of being composed of different lipid phases with distinctive characteristics. This thinking is relevant when considering the concept of lipid rafts, for example. What, then, is the Gibbs Phase Rule?

The phase rule is mathematically simple:

$$f = C - P + n$$

where f is the number of free variables (i.e., the number of variables that can be controlled independently, generally called the "degrees of freedom"), C the number of components, P the number of phases, and n the number of intensive variables. Generally, $n = 2$ to account for the intensive variables of pressure (p) and temperature (T), but one might wish to include under some circumstances other

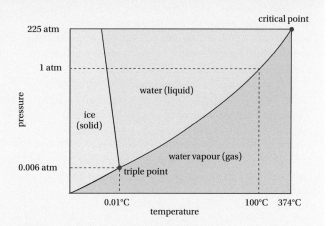

Figure 1 Phase diagram for pure water.

variables that do not depend on system size, such as electrical potential. The phase rule is easily derived, but let's first look at a traditional example of its application.

Consider a sealed rigid container partially filled with degassed water at an equilibrium temperature T. There are two phases at room temperature: water vapor and liquid. Apply the phase rule: $C = 1$, $P = 2$, and $n = 2$, so that $f = 1$. This seems odd; shouldn't one be able to vary p and T independently? No, because the vapor pressure of water depends upon T. Once T is fixed, the vapor pressure is determined; T is the single free variable.

Suppose that we now lower the temperature to produce an equilibrium mixture of ice, water, and vapor. In that case, $C = 1$, $P = 3$, and $n = 2$, which makes $f = 0$; there are no free variables. What does that statement mean? It means that there is only a single temperature and pressure at which vapor, liquid, and solid coexist. This point is the triple point shown in **Figure 1**, corresponding to $T = 273.16$ K (0.01 °C) and $p = 0.006$ atmosphere. If the temperature is lowered just a bit, the liquid phase disappears, leaving ice and vapor; there would then be only two phases, and again, $f = 1$. Raise the temperature a bit, and the solid phase disappears, leaving us at our starting point of two phases (liquid and vapor) with $f = 1$. Box Figure 1 is an example of a phase diagram. The lines on the diagram show the vapor pressures and temperatures at which the vapor phase is in equilibrium with the liquid or solid phase (or liquid and solid phases at the triple point) for a given p and T. The so-called critical point on the diagram is that p and T at which distinct liquid and vapor phases no longer exist, meaning that the properties of the liquid and gas phases approach each other as the critical temperature is approached. There is a single phase at the critical point, referred to as supercritical fluid. This brings us to the derivation of the phase rule.

When two or more phases are in equilibrium with one another, the chemical potentials of the C chemical components are equal in all P phases (see Chapter 0). In the example of water vapor in equilibrium with liquid water,

(Continued)

BOX 2.2   GIBBS PHASE RULE (CONTINUED)

the chemical potential of the water molecules is the same in both phases, meaning that there is no change in the free energy of a water molecule passing from one phase to another. The requirement for equal chemical potentials in all phases is the basis for the phase rule.

Let's measure concentrations in mole fraction units. Let the mole fraction of component k in phase i be X_k^i. To specify the system fully requires $(C-1)$ mole fraction variables for each phase P, pressure, temperature, and any other intensive variables included in n. Why $C-1$? If there were three components, for example, specifying the mole fractions of two of them automatically defines the third, because

$$\sum_{k=1}^{C} X_k^i = 1$$

The total number of variables is thus $P(C-1)+n$. Now, if the system is at equilibrium, it is restrained by the requirement that the chemical potential of component k must be the same in all P phases:

$$\mu_k^1(p,T) = \mu_k^2(p,T) = \mu_k^3(p,T) = \ldots = \mu_k^P(p,T)$$

These represent $(P-1)$ restraining equations for each of the C components. Therefore, there are a total of $C(P-1)$ restraining equations at equilibrium. The number of free variables f is thus the total number of variables minus the chemical potential restraints, yielding the phase rule:

$$f = P(C-1)+n-C(P-1) = C-P+n$$

Figure 2 shows the phase behavior of two-component mixtures of phospholipids that have different acyl chain lengths. The phase behaviors are established using a variety of techniques such as calorimetry and electron paramagnetic resonance. Let's apply the phase rule. Water is present

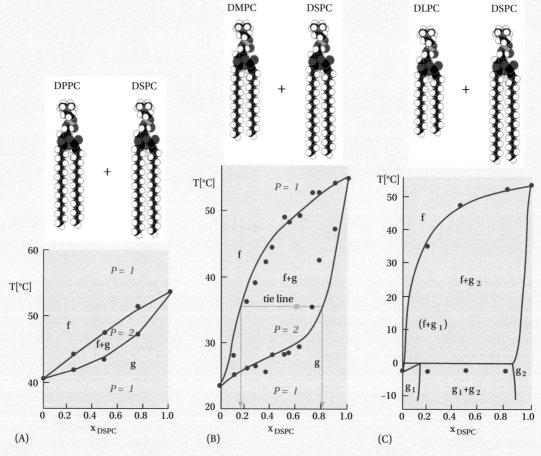

Figure 2 Phase behavior of two-component phospholipid mixtures. The pairs of lipids used to construct the phase diagrams are indicated. The diagrams show that as the mismatch between acyl chain lengths increases, the complexity of the phase behavior increases. The lipids indicated are DPPC, dipalmitoylphosphatidylcholine; DSPC, distearoylphosphatidylcholine; DMPC, dimyristoylphosphatidylcholine; DLPC, dilauroylphosphatidylcholine. Liquid phases are indicated by f and solid phases by g. The number of phases (P) in each region is indicated. The addition of a third component, such as cholesterol, causes the phase diagrams to become quite complex. (From Mouritsen OG [2005] *Life as a Matter of Fat*, Springer-Verlag, New York. With permission from O.G. Mouritsen.)

(Continued)

BOX 2.2 GIBBS PHASE RULE (CONTINUED)

in excess in all cases, and atmospheric pressure is constant. The only intensive variable is temperature, $n = 1$, and the number of components C is two. Consequently, $f = 2 - P + 1 = 3 - P$. Now, look at Box **Figure 2A**. At high temperatures, there is a single fluid (liquid-disordered) phase ($P = 1$), so that $f = 2$, meaning that we can arbitrarily vary the mole fraction of the two lipids. As the temperature is reduced, we enter a two-phase region ($P = 2$) consisting of a mixture of liquid and solid phases. In that case, $f = 1$. How does one interpret that? Look at panel B. At a particular temperature, we draw a horizontal line (green), referred to as a tie line. Chose a point on the line (green), referred to as a state point. By having chosen a particular state point ($X_{DSPC} \approx 0.6$), we have exhausted the degrees of freedom. This means that the relative amounts of the solid and liquid phases at the temperature chosen are fixed; we cannot arbitrarily choose the amounts of the two phases present. The amount of each phase is given by the intersection of the tie line with the phase boundaries. The left-hand tie-line point tells about the amount of fluid phase (X_f) and the right-hand point the amount of solid phase X_g, which must equal $1 - X_f$. One can see from panel C that the large disparity between

the chain lengths has indeed made the phase behavior quite complex.

Multicomponent, multiphase systems of current interest to biologists are model membranes with three lipid components, such as a phospholipid with unsaturated acyl chains (e.g., dioleoylphosphatidylcholine (DOPC)), a phospholipid with saturated acyl chains (e.g., distearoylphosphatidylcholine (DSPC)), and cholesterol. Such mixtures are studied as models for lipid raft formation. What kind of phase diagram do we use for three-component mixtures? At a chosen temperature, we use a so-called Gibbs triangle in which each vertex of an equilateral triangle represents one of the components; the mole fractions of the components are marked 0 to 1 along each leg of the triangle (Box **Figure 3**). To study such mixtures, GUVs are made with varying proportions of the three lipids. Phase separations can be observed by fluorescence microscopy using fluorescent dyes that partition selectively into particular phases. To see the effects of temperature, which is fixed in a particular Gibbs triangle, imagine a stack of triangles—one for each temperature. If you think about the phase behavior of a mixture of lipids extracted from a cell membrane, the phase behavior is nearly impossible to contemplate.

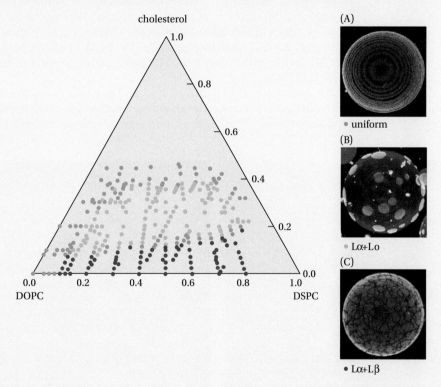

Figure 3 A Gibbs phase triangle for describing the phase behavior of three-component lipid mixtures in giant unilamellar vesicles (GUVs). On the right are images of GUVs whose regions are identified by fluorescent probes that partition selectively into the different phases. (From Zhao J, Wu J, Heberle FA et al. [1968] *Biochim Biophys Acta* 11: 2764-4776. With permission from Elsevier.)

lysophosphatidylcholine, which has a bulky headgroup and a single acyl chain, often described as ice cream cone shaped. Diacylphosphatidylethanolamine has a negative curvature tendency and consequently is thought of as having a small headgroup and bulky acyl chains, giving it the shape of a traffic cone. Nevertheless, phosphatidylethanolamine bilayers do form lamellar phases. For example, palmitoyloleoylphosphatidylethanolamine (POPE) bilayers in excess water remain lamellar up to 60 °C before forming a hexagonal phase. In contrast, POPC never undergoes a transition to a hexagonal phase.

The real situation for phosphatidylethanolamine (PE) is likely to be more subtle due to the structure and electrostatic properties of the headgroup. For example, the $-N(CH_3)_3$ group of phosphatidylcholine forms weak hydrogen bonds with neighboring lipids, whereas the $-NH_3$ group of PE forms strong hydrogen bonds. However, the geometry of the headgroup makes it difficult for all H-bonds to be satisfied simultaneously by PE. This is apparent from molecular dynamics simulations (**Figure 2.13A**), which show a nonuniform headgroup distribution in the membrane plane and consequently, nonuniform areas per lipid (**Figure 2.13B**). This is an example of geometrical frustration, in which geometrical constraints prevent simultaneous minimization of all free energies. The frustration of PE can be relieved by curvature of the interface to better optimize hydrogen-bond formation.

2.1.9 Phospholipids with Added Detergents Can Form Bilayered Micelles (Bicelles)

A disadvantage of lipid vesicles for studying membrane proteins is that their large size and consequent slow tumbling rate make them unsuitable for NMR studies of membrane proteins. This problem has been overcome by embedding proteins in bicelles, which are formed from mixtures of phospholipids and detergents. Remember that vesicles are a natural result of the energetic cost of exposing the edges of bilayers to water. The idea of the bicelle is to form flat bilayer discs by shielding the disc edges from water with short-chain detergents (**Figure 2.14**). Bicelles are typically 100–500 Å in diameter. Their advantage for NMR studies is that they have a high tumbling rate and can be oriented in magnetic fields.

A common combination of lipid and detergent is dimyristoylphosphatidylcholine (DMPC) and dihexanoylphosphatidylcholine (DHPC). Although it is a phospholipid, DHPC acts as a detergent because the acyl chains are only six

(A)   (B)

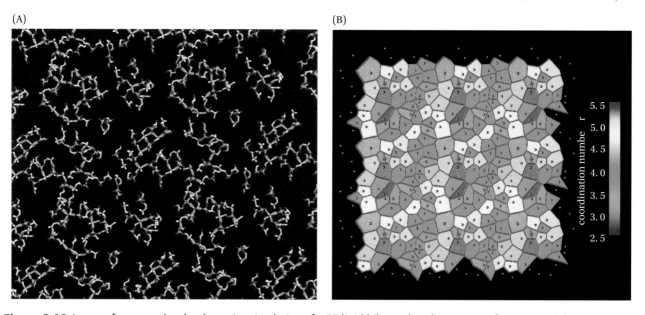

Figure 2.13 Images from a molecular dynamics simulation of a PE lipid bilayer that demonstrate frustration. (A) A snapshot showing the patchiness of the H-bonding pattern, arising from the structural restraints that prevent optimal H-bonding. H-bonds are indicated by blue-on-white dotted lines. All the PE molecules would be happier having all their H-bonds satisfied with neighbors, but the planar geometry prevents it. (B) The individual areas of the PE molecules differ as a result of differences in H-bond coordination. The color-coded patches indicate the H-bond coordination with neighbors for each PE molecule's patch. One way of relieving the frustration under the right hydration conditions is to form non-lamellar structures. (From Suits F, Pitman MC, Feller SE [2005] 122: 224714. Used with permission from AIP Publishing.)

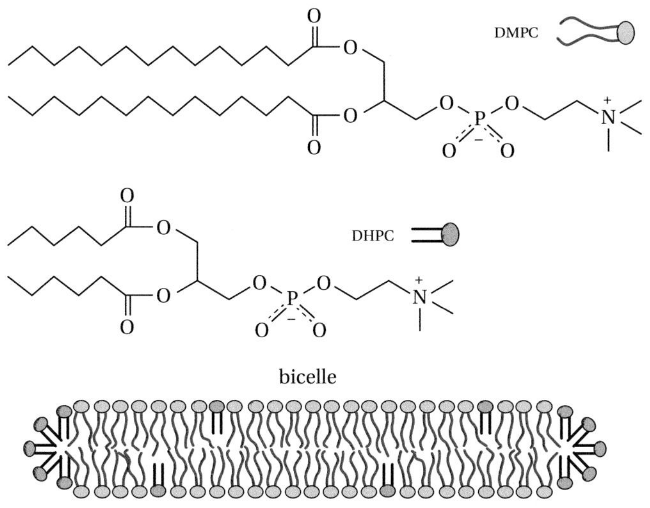

Figure 2.14 Discoidal bilayered micelles (bicelles). The DHPC detergent allows the curvature that prevents the edge of the bilayer disc from being exposed to water.

carbons long. DMPC and DHPC can form very complex phases, and consequently, there are a limited range of concentrations and mole ratios that lead to the bicelle disc structure. Single membrane proteins can be incorporated into DMPC/DHPC bicelles for NMR structural studies and for protein crystallization. Other, better means to solubilize membrane proteins have been devised, such as nanodiscs and Amphipols, which we will encounter later.

2.1.10 Lipid and Protein Composition of Membranes Varies Widely

While all cellular membranes have the same basic architecture, they can differ dramatically in protein and lipid composition (**Table 2.1**) depending on the membrane's physiological function. For example, the few proteins present in the myelin sheaths that surround nerve cells (Figure 1.13) serve primarily as structural components. Plasma membranes, on the other hand, are rich in the transport proteins required for maintaining the internal ionic environment,

These compositions give a sense of the variety of membrane compositions but must be taken as very approximate, because they will vary depending on how a membrane is isolated and washed.

Table 2.1 Compositions of Membranes

Composition of Membranes (% by weight)			
Membrane	Protein	Lipid	Carbohydrate
Myelin	20	75	5
Red blood cell ghosts	49	43	8
Plasma membranes			
Liver cells	58	42	(5–10)[a]
Ehrlich ascites	67	33	(5–10)[a]
Amoeba *Proteus*	25	32	15
Mitochondrion			
Outer membrane	52	48	(2–4)[a]
Inner membrane	76	24	(1–2)[a]
Bacteria (Gram-positive)	75	25	(10)[a]

[a]Carbohydrate is due to glycosylated proteins and lipids.
Based on data provided in Guidott G [1972] *Arch Intern Med* 129: 194–301.

Characteristic lipid compositions of a variety of membranes. Note the content of cholesterol and that each category of lipid, e.g., PC or PS, can have a variety of alkyl chains, and a variety of other lipids can be present in small amounts. Thus, the compositional complexity of diverse membranes can be very high!

Table 2.2 Approximate Lipid Compositions of Different Cell Membranes

Lipid	Liver cell plasma membrane	Red blood cell plasma membrane	Myelin	Mitochondrion (inner and outer membranes)	Endoplasmic reticulum	E. coli bacterium
Cholesterol	17	23	22	3	6	0
Phosphatidylethanolamine	7	18	15	28	17	70
Phosphatidylserine	4	7	9	2	5	trace
Phosphatidylcholine	24	17	10	44	40	0
Sphingomyelin	19	18	8	0	5	0
Glycolipids	7	3	28	trace	trace	0
Others	22	13	8	23	27	30

(From Alberts B, Johnson A, Lewis JH et al. [2007] *Molecular Biology of the Cell*, 5th Edition. Garland Science, New York. With permission from W. W. Norton.)

whereas mitochondrial membranes are rich in respiratory proteins. The differences in protein content raise the fundamental issue of how proteins are targeted to different membranes (Chapter 5). Furthermore, the lipids and proteins are not uniformly distributed in the plane of the membrane. At the grossest level, protein content varies from about 15% by mass for myelin membranes to almost 80% for the mitochondrial inner membrane; plasma membranes are typically about 50% protein.

Just as for proteins, the amounts and types of lipids also vary among membranes (**Table 2.2**). In addition, the two leaflets of the bilayer may have different lipid compositions, i.e., the lipid bilayer is asymmetric. As discussed later, different lipids impart different physicochemical properties to the bilayer. Careful studies of the lipid composition of cells suggest that fine differences in molecular structure, such as the number and position of double bonds and variations in the lipid headgroups, give rise to a very large number (>1000) of distinct lipid species. The precise biological functions are known only for a handful of lipids (mostly low-abundance lipids involved in highly regulated signaling pathways), and the reasons for the great diversity of lipids found in nature remain a biological and evolutionary riddle. But, one can imagine that the functions of many membrane proteins may depend on specific lipids in their local lipid environment. Because the main site of both lipid and protein synthesis in mammalian cells is the endoplasmic reticulum (ER), it is not surprising that that the compartments are delimited by membranes of different composition (**Box 2.3**). The formation of vesicle-like compartments is a natural tendency of phospholipids.

2.2 DIFFRACTION METHODS GIVE KEY INSIGHTS FOR UNDERSTANDING BILAYER STRUCTURE AND MEMBRANES

While many methods have contributed to our chemical understanding of bilayer structure, the most important direct structural information has come from x-ray and neutron diffraction measurements. Indeed, as we learned in Chapter 1, diffraction measurements were crucial for demonstrating that lipid bilayers form the fabric of cell membranes. The information obtainable from

BOX 2.3   LIPID SYNTHESIS IN EUKARYOTES IS IMPORTANT FOR COMPARTMENTALIZATION

The main site of lipid synthesis in mammalian cells is the ER, but some special lipids are also made in the Golgi (e.g., sphingolipids) and in mitochondria (e.g., cardiolipin). Lipids can move in different ways from their site of synthesis. The simplest route is by lateral diffusion within a continuous membrane system (e.g., from the ER to the nuclear envelope). Lipids can also be carried between compartments via vesicular transport (see later) or bound to soluble lipid transfer proteins. In the latter case, the transport reaction may take place within specialized zones of contact between the donor and acceptor organelles.

Lipids with large polar headgroups have very low rates of spontaneous flip-flop across a pure lipid bilayer. Since the synthesis of most lipids is confined to one leaflet of the bilayer, there must be ways to facilitate lipid flip-flop, both to balance the areas of the monolayers and to establish and maintain asymmetry. The ER appears to contain membrane proteins that facilitate spontaneous equilibration of lipids between the two leaflets, although some specialized lipids (e.g., those carrying oligosaccharides needed for protein glycosylation in the ER lumen) are translocated across the membrane by dedicated transport proteins. ATP-dependent lipid transport proteins are needed to do the opposite job, i.e., to generate and maintain asymmetry in lipid composition between the two halves of a bilayer in, e.g., the Golgi and the plasma membrane (**Figure 1**).

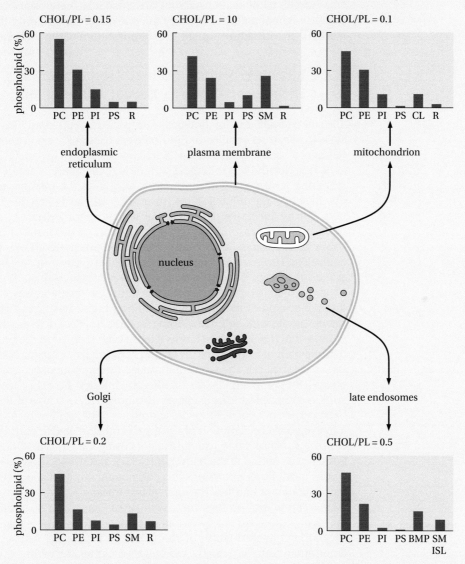

Figure 1 Lipid compositions of organelles of eukaryotic cells. Lipid compositions of the various organelle membranes are shown as bar graphs. The arrows indicate the organelles to which the compositions refer. Given the complexity of a ternary mixture (Box 2.2), complex mixing behaviors are expected. ISL stands for inositol sphingolipid. (From Van Meer G, Voelker DR, Feigenson GW [2008] 9: 112. With permission from Springer Nature.)

diffraction measurements provides a structural context for other kinds of observations, so it is useful to extend the discussion of bilayer diffraction that we began in Chapter 1. The basic principles from one-dimensional bilayer diffraction are extended to three-dimensional (3-D) structures and developed further in Chapter 9, where we discuss determination of the 3-D structures of proteins.

Figure 1.18C showed us that x-ray scattering from phospholipids dispersed in water to form unoriented stacks of bilayers have two main features: sharp lamellar diffraction peaks at small angles relative to the x-ray beam due to the stacking of bilayers, and more diffuse peaks arising from the side-to-side packing of the bilayer alkyl chains lying at large angles. When phospholipids are oriented on glass substrates (Figure 1.18A), the structural organization of bilayers is quite apparent: defining the equator as being parallel to the bilayer plane, the sharp lamellar peaks lie along the azimuth at small angles and the diffuse peaks along the equator at high angles. Remembering that in the reciprocal space of the diffraction pattern, scattering from widely separated features occurs at small angles while scattering from closely spaced features occurs at high angles, this orthogonal pattern is exactly the one expected for lipid bilayers and as we learned, provides direct qualitative confirmation of the existence of bilayers.

2.2.1 Bilayer Profiles Obtained from Lamellar Diffraction Patterns Reveal Thermal Motion

The sharp lamellar diffraction peaks from stacks of oriented bilayers make it possible to construct profiles of lipid bilayers using diffraction measurements and Fourier reconstruction (**Box 2.4**). Such profiles, shown in panels C and D of **Figure 2.15**, describe the transbilayer distribution of the bilayer's atoms in terms of their scattering power along an axis that is normal to the membrane plane. Profiles obtained by x-ray diffraction measurements are most often described using electron density (electrons per $\mathring{A}^3$) rather than scattering density , since x-rays predominantly scatter from the electrons of an atom. Because scattering density for x-rays is proportional to electron density, the two methods of representation are equivalent. However, neutrons are not sensitive to electron density; consequently, scattering density is the only reasonable way of representing the profiles. For lipids above their gel-to-liquid crystalline transition temperatures (the L_α phase), the thermal motion of the lipids is exceedingly high, causing the profiles to appear rather smooth with only a few strong features. That is to say, the thermal motion is so large that the individual atoms cannot be discerned in the profiles. A result of thermal motion is a wide range of lipid conformers inside the bilayer (Figure 2.9). But as we shall see, the scattering caused by component groups (methyls, phosphates, etc.) of the lipids can be discerned under the right experimental conditions. For example, the transbilayer distribution of the whole phosphate group can be discerned but not its individual atoms. As a rule of thumb, the width (W) of the transbilayer thermal motion of a group can be estimated from the maximum number (h_{max}) of observable structure factors using the formula $W \approx 2d/h_{max}$, where d is the repeat period of the bilayer. If $h_{max} = 8$ and $d = 50$, then the thermal motion causes transbilayer excursions of a component group of about 12 Å, corresponding to about one-fourth of the total bilayer thickness!

How can bilayer profiles such as those of Figure 2.15 be obtained? Following the ideas outlined in Box 2.4, it can be shown that the profiles can be constructed by Fourier transformation of the structure factors F_h, whose amplitudes are determined from $|F_h| = k\sqrt{I_h}$, where I_h is the intensity of the hth diffraction peak and k is an experimentally determined scale factor, which in this case places the profiles on an absolute scattering-density scale. Ignoring a lot of fussy experimental details and considering the scattering to be from a unit cell composed of two lipids and associated water, the scattering profile of a bilayer in a multilamellar stack with repeat period d is determined using Equation 2.2:

$$\rho(z) = \rho_0 + \frac{2}{d} \sum_{h=1}^{h_{max}} F_h \cos\left(\frac{2\pi hz}{d}\right) \tag{2.2}$$

BOX 2.4 RECONSTRUCTION OF BILAYER PROFILES FROM DIFFRACTION DATA

The Bilayer Unit Cell in Real and Reciprocal Space

Diffraction studies have been used to provide useful structural information about the organization and dynamics of lipid membranes. Learning to think in both the real space of the laboratory and the reciprocal space of diffraction patterns is a skill that helps with the major current methods in membrane structure: crystallography and electron cryo-microscopy (cryo-EM). This introduction may help.

As shown in Box 1.4, Bragg diffraction is observed for multilamellar stacks of bilayers oriented on glass or quartz substrates. The diffraction patterns consist, ideally, of a series of equally spaced spots on the two-dimensional (2-D) detector. The spots are diffracted images of the x-ray beam, produced by the repeated structure in the stack, which reinforces the scattering at specific angles through constructive interference of the scattered waves. Various kinds of disorder can distort the spots, but here, we consider the ideal case in which the scattered neutrons or x-rays originate from a perfect stack of bilayers. The intensities I_h of the spots are related to the sample's structure factors F_h by $|F_h| = \sqrt{I_h}$ that sample the continuous scattering of a single bilayer (panel C). One can also scatter neutrons and x-rays from single-walled lipid vesicles, which is equivalent to scattering from single, highly hydrated lipid bilayers (the continuous curve in panel C). Although diffraction patterns from bilayer stacks and vesicles are qualitatively different, they are mathematically related.

In **Figure 1**, we consider the scattering of radiation (x-rays or neutrons) from a single bilayer that defines a unit cell. In diffraction, a unit cell is the arrangement of atoms that can be translated along the x-, y-, and z-axes to reproduce the complete macroscopic crystal. For diffraction from a stack of N bilayers, the lattice is reproduced primarily by repeated translations of the single bilayer along the z-axis normal to the bilayer planes. If d_b is the width of the bilayer unit cell, then $\rho(z + nd_b) = \rho(z)$, n = 1, 2, 3 ... N − 1, which is the mathematical way of saying that the bilayer stack is periodic in z. An important point, which we ignore for now, is that the single unit cell of lipid vesicles in excess water is likely to be different in atomic detail from bilayers in stacks. The differences arise because the bilayers in the stack are less hydrated, which in turn affects the structure and dynamics of the bilayers.

Panel A shows a slice through a simulated single lipid bilayer forming the wall of a lipid vesicle. Panel B shows a neutron scattering-density profile $\rho(z)$ of the single bilayer. This profile describes how strongly neutrons are scattered from the bilayer as a function of z. A similar profile can also be constructed for the scattering of x-rays, but the details would be different because x-rays scatter from electrons, whereas neutrons scatter from nuclei (Figure 2.15). The profile of panel B is called the real-space profile, because it represents the distribution of matter in the 3-D space of the laboratory. The real-space structure and its diffraction pattern are reversibly related by Fourier transformation. The

Fourier transformation of the real-space structure of panel B yields the reciprocal-space structure shown in panel C. In reciprocal space, the bilayer structure is described uniquely by the form factor F, often called the "structure factor." (To prevent confusion, we reserve the term "structure factor" for the structure factors obtained in crystallographic experiments; see later.) Notice in panel C that the horizontal coordinate is the magnitude $S = |\vec{S}|$ of the reciprocal space vector (Box 1.4), which has units of Å^{-1}. With this inverted distance scale, widely separated features in real space are closely separated in reciprocal space.

Although panel C is the Fourier-transformed reciprocal-space representation of the bilayer, it cannot be observed directly in an experiment because detectors are sensitive only to the energy of diffracted radiation. Because energy is proportional to the square of the radiation's electric field vector, the intensity I of diffracted radiation is proportional to $|F|^2$, causing the algebraic signs of the form factor to be lost. We can readily measure the amplitude of the form factor, $|F| = \sqrt{I}$, but not its sign (phase). Panel D shows the experimentally measured intensity and illustrates the loss of phase information. The inset of panel D gives an impression of what the continuous scattering pattern from bilayer vesicles might look like if measured with a 2-D detector.

Fourier Transformation

The mathematics of Fourier transformations, developed by Jean-Baptiste Joseph Fourier in the early 19th century, is well established and described thoroughly in many textbooks. Any function that is periodic along z, such as $\rho(z)$ for a bilayer stack, can be Fourier transformed to allow it to be represented by a series of sine and cosine functions (i.e., a Fourier series). The Fourier transformation $\rho(z)$ is given by Equation 1:

$$\rho(z) = \sum_h \left\{ A_h \cos 2\pi h \frac{z}{d} + B_h \sin 2\pi h \frac{z}{d} \right\} \quad (1)$$

In this expression, h is a sequence of integers (0, 1, ... ∞), and A_h and B_h are the Fourier coefficients. Equation 1 can be simplified when the periodic function is an even function, meaning that $\rho(z) = \rho(-z)$. In the jargon of diffraction, a unit cell with this property is called centrosymmetric. The cosine function is even ($\cos(-\phi) = \cos(\phi)$), whereas the sine function is odd ($\sin(-\phi) = -\sin(\phi)$). This means that in a Fourier reconstruction of an even function describing a centrosymmetric unit cell, the sine terms will cancel, leaving only the cosine terms with amplitudes A_h. Consequently, the Fourier transform of an even function will consist of a sum of cosine terms alone, shown in Equation 2:

$$\rho(z) = \sum_h A_h \cos 2\pi h \frac{Z}{d} \quad (2)$$

(Continued)

BOX 2.4   RECONSTRUCTION OF BILAYER PROFILES FROM DIFFRACTION DATA (CONTINUED)

As we will discuss fully in Chapter 9, diffraction patterns are constructed by the superposition of waves, which is exactly what happens in Fourier transformations. The cosine functions in Equation 2 represent these waves. The amplitudes of the waves are the Fourier coefficients A_h, which are equivalent to the structure factors F_h observed in diffraction from bilayer stacks. An important question: if we know the shape of the function $\rho(z)$, can we compute the values of the coefficients A_h? Yes! Here is the formula, which is derived in basic textbooks on diffraction theory and Fourier series:

$$A_h = \frac{1}{d}\int_{-d/2}^{d/2} \rho(z)\cos 2\pi h \frac{Z}{d}\,dz \qquad (3)$$

Bilayer Form Factors and Structure Factors

Equation 3 is appropriate for crystal-like arrays such as our bilayer stacks. What do we do when we desire the form factor (panel C) of a single bilayer rather than that of a stack of bilayers? Notice in Equation 3 that the argument of the cosine function is $2\pi h(z/d)$. This expression is related to the Bragg condition $2d\sin\theta = h\lambda$. Using the reciprocal-space vector $\vec{S}$, and letting $|\vec{S}| = S$, the Bragg condition can also be written as $S = 2\sin\theta/\lambda = h/d$ (Box 1.5). S is simply a variable that describes the angle of the incident and scattered beams in a diffraction experiment. In bilayer diffraction, S happens to have non-zero values only when $S = h/d$ due to the Bragg condition. We can generalize Equation 3 by replacing h/d with S and letting S vary continuously. In that case, there are no longer discrete Fourier coefficients but rather, a continuous function, which is the form factor $F(S)$ of Equation 4:

$$F(S) = \frac{1}{d}\int_{-d/2}^{d/2} f(z)\cos(2\pi Sz)\,dz \qquad (4)$$

One can show that there is a formal relationship between Equations 3 and 4 that is imposed when the continuous function $F(S)$ becomes part of a regular periodic array. In simple terms, if we multiply $F(S)$ by a so-called interference function $G(S)$ that describes the periodicity of the stack of bilayers, then the Fourier coefficients of Equation c are obtained. That is to say,

$$A_h = F(S)G(S) \qquad (5)$$

$G(S)$ depends upon the number N of bilayers in the stack and on how much and what kind of disorder is present. If N is large (~100 or more), and the periodicity d of the stack is highly uniform, the interference function will be non-zero only when $S = h/d$, i.e., the Bragg condition. $F(S)$ then becomes a series of points $F(h/d)$, shown as dots on the form factor curve in panel C labeled h =1, ..., 4. Said another way, the interference function samples the form factor at values of $S = h/d$ to produce a set of Fourier coefficients (structure factors) that can be used to reconstruct the profile of panel B.

(A)

(B)

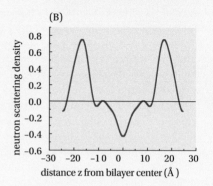

(C)

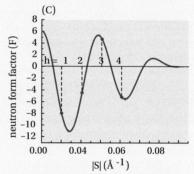

(D)

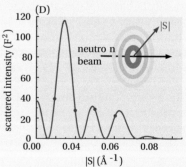

Figure 1

Notice that the F(h/d) structure factors in panel C are negative for h = 1, 2, and 4. This reminds us that for centrosymmetric structures, the signs (phases) of the structure factor are either + or −. Measured diffracted intensities I_h are proportional to $F(h/d)^2$, causing the phase information to be lost. Several methods have been developed for recovering the phases of the structure factors.

In this equation, h_{max} is the highest observable diffraction order and ρ_0 is the average scattering of the unit cell obtained by summing up the scattering lengths of all atoms in the cell. The scattering lengths, which measure how strongly atoms scatter radiation, are tabulated in various handbooks for both x-rays and neutrons.

The scattering profiles of Figure 2.15 show profiles obtained by both neutron and x-ray diffraction for oriented (DOPC) bilayers at low hydration. At first glance, the profiles are very similar; there are two strong positive peaks separated by a relatively negative peak or valley. This valley is the location of alkyl chains and especially their CH_3 end groups, which scatter less strongly than any other component group. The peaks are due to the headgroup region. Looking closely, the positions of the headgroup peaks are different for neutrons and x-rays. Why are they different? The difference occurs because x-rays scatter most strongly from the electron-rich phosphate group, whereas neutrons scatter most strongly from the hydrogen-deficient carbonyl groups. Why does hydrogen matter? It matters because as a rule, the scattering of neutrons from all nuclei is about equal except for hydrogen, which has a negative scattering length due to the paticular interaction of neutrons with the proton of the hydrogen (**Figure 2.15A**). A negative scattering length means that an H appears as a hole where other atoms appear as peaks in a density map. Deuterons, however, have a positive scattering length. If one replaces selected hydrogens with deuteriums (specific deuteration), the locations of those atoms are easily determined because they stand out strongly against the methylene background (**Figure 2.15B**). (The neutron scattering length of a CH_2 group is about zero because the negative scattering length of the hydrogens just about cancels the positive scattering length of the carbon, which is why $\rho(z)$ in **Figure 2.15D** varies around the mean value of 0.)

2.2.2 Diffraction Patterns Can Be Obtained from Single Bilayers

Although diffraction measurements are most commonly performed on multilamellar stacks of bilayers, either oriented on glass surfaces or dispersed as

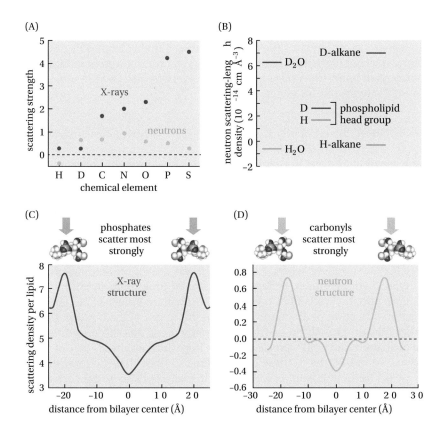

Figure 2.15 Lipid bilayer scattering-density profiles determined by x-ray and neutron diffraction emphasize different structural features. (A) X-ray and neutron scattering strengths of the chemical elements differ, because neutrons scatter from nuclei, whereas x-rays scatter from electrons. Most nuclei have about the same scattering strength, except hydrogen, which has a small negative value. X-ray scattering increases with the number of electrons, making phosphorus a stronger scatterer than carbon. (B) A comparison of neutron scattering of hydrogenated and deuterated compounds. Notice that deuteration of alkanes, a stand-in for acyl chains, has a large effect on scattering, whereas it has a smaller effect on the headgroups. (C) X-ray-determined bilayer profiles highlight the phosphate group because of its high scattering strength. (D) Neutron-determined profiles highlight the carbonyl groups because they are relatively deficient in hydrogens, which causes them to stand out relative to the acyl chains.

liposomes in water, important structural information can be obtained from single-walled vesicles as shown in Box 2.4. Scattering of x-rays or neutrons from vesicles produces not a series of sharp diffraction spots (or sharp rings for multilamellar dispersions) but rather, a series of diffuse circular patterns. These patterns represent the continuous Fourier transform of a single bilayer (Box 2.4D). If these bilayers were stacked one on top of another, the diffraction pattern would lose its diffuseness and become a series of discrete bands characteristic of multilamellar bilayer diffraction. The lattice of the bilayer stack causes the continuous form factor to be sampled at intervals of h/d, where h is the diffraction order and d is the bilayer repeat distance (Bragg spacing). The number of bilayers in the stack affects the sharpness of the lamellar diffraction peak. Generally, ten or more bilayers in a stack will lead to sharp diffraction spots.

2.2.3 Headgroup Layer Spacing Shows That Lipid Area per Molecule Is Conserved

Take a close look at the scattering intensity plot for single bilayers in Box 2.4. Notice that the major scattering feature in panel D occurs at about S = 0.03, close to the dot corresponding to h = 1 (panel C), the first-order peak in lamellar diffraction. Because S = 1/d for the first-order peak, the intensity of the first peak in panel D is related to the repeat period of the bilayer, which in turn identifies very approximately the positions of the headgroups. It can be shown that Fourier transformation of the intensity curve in panel D allows one to determine precisely the position of the phosphate groups in x-ray experiments. By using the x-ray scattering signal from bilayers in solution, one can thus measure the distance between the headgroup using the peak positions seen in Fourier transforms of the intensity curve. What the Fourier transform actually shows is the distances between features of the bilayer that are strongly correlated, namely, the phosphates of the headgroups, which mark the boundaries of the bilayer.

A series of measurements of bilayer vesicles made from lipids with esterified fatty acids having different chain lengths gives a linear plot of bilayer headgroup separation (determined from the Fourier transformation of the scattered intensity curve) versus hydrocarbon chain length, showing that the area per lipid in the bilayer is conserved over a range of bilayer thicknesses (**Figure 2.16**). This observation emphasizes the balance between the hydrophobic effect and the steric packing of the headgroups in a bilayer, and that the tendency of the chains to push apart from thermal energy is overwhelmed by the unfavorable energy of the hydrophobic effect.

2.2.4 Bilayer Scattering Can Be Seen in Natural Membranes

As we learned in Chapter 1, there was an early consensus that bilayers are the framework of membranes, but the evidence was indirect, and a number of alternative models were proposed. This situation was resolved by observing the interference of x-rays scattering from the lipid headgroups. When isolated membranes from a microorganism, *Acholeplasma laidlawii*, were concentrated in a thin-walled capillary and studied by x-ray scattering, the scattering intensity patterns showed a series of ripples at a spacing expected for lipid bilayers (**Figure 2.17**). Moreover, when the organism was grown on media with different fatty acid supplements, this simple organism took them up and incorporated them into the lipids of its membrane. Growth on longer-chain fatty acids causes the diffraction peaks to move toward the center of the pattern, showing a thicker bilayer; growth on shorter-chain fatty acids has the opposite effect. Phase changes of the lipid were also seen in these membranes (Figure 1.20). As we discussed in Chapter 1, these observations showed that the interpretation of the peaks as arising from lipid bilayer structure is likely to be correct, and when the measurements were made in the 1970s, they established that bilayers are a major structure in membranes. Showing that bilayers are a key structure

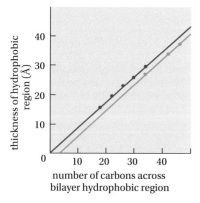

Figure 2.16 Plot of the measured thickness of the bilayer hydrocarbon region versus the number of carbons across the hydrophobic region. The linearity implies a constant area per lipid as the thickness increases. The number of carbons is taken starting at C-2 of each acyl chain. Estimated errors in the thickness measurement correspond to the size of the plotted points. The lines are least-squares fits to the data; red line indicates saturated acyl chains and blue line monounsaturated acyl chains. (From Lewis BA, Engelman DM [1983] *J Mol Biol* 166: 211–217.)

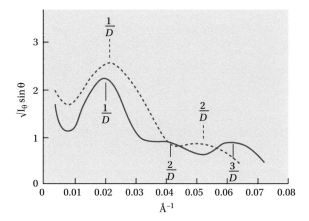

Figure 2.17 Intensities of x-rays scattered from suspensions of membranes isolated from *Mycoplasma laidlawii* have features expected for lipid bilayers and change as expected when lipid chain length is varied or the state of the lipids changes (gel vs. liquid crystal). The dashed line is for data collected at 40 °C and the solid line for data collected at 10 °C. These measurements strongly support the idea that bilayers are a predominant structural feature of membranes. (From Wilkins MHF, Blaurock AE, Engelman DM [1971] *Nature New Biol* 230: 72–76.)

in membranes motivated detailed study of bilayer properties; much effort has been devoted to this area over the last 50 years.

2.2.5 Bilayer Dynamics Can Be Inferred from Bilayer Profiles

We discussed earlier how transbilayer scattering profiles (Figure 2.15) can be obtained using both x-ray and neutron diffraction. When these two methods are used in concert to study bilayers, detailed views of the transbilayer distribution of component groups (for example, CH_3, phosphate, and carbonyl groups) can be obtained by taking advantage of the difference in neutron scattering of deuteriums compared with hydrogens. By substituting deuteriums at the double bonds of DOPC, for example, the distribution of the double bonds can be determined (**Figure 2.18**). The experiment is performed by first determining a profile of DOPC with hydrogens on the double bonds and then repeating the experiment with deuterium on the double bonds. The location of the deuteriums is then clearly revealed. The water distribution across the bilayer can be determined in a similar fashion using normal versus deuterated water. These difference-scattering experiments can also be performed with x-rays using heavy-metal labeling. Notice in Figure 2.18 that electron-dense bromine atoms substituted at the DOPC double bond also reveal the distribution of the double bonds. Heavy-atom labeling is one basis for phasing x-ray structures of proteins.

An important observation that emerges from the double-bond distribution is the width of the distribution, which directly reveals the high thermal disorder of the bilayer; the double bonds explore nearly the full width of the bilayer. We noted earlier that the average thermal width of a distribution such as this can be estimated from $W \approx 2d/h_{max}$. For $h_{max} = 8$, as in the experiments of Figure 2.18, W should be around 12 Å, which is about the width of the double-bond distribution in a single monolayer.

2.2.6 The Transbilayer Distribution of the Principal Lipid-Component Groups Provide a Detailed View of Bilayer Structure

The "structure" of a fluid bilayer is defined operationally as the time-averaged spatial distributions of the principal structural (component) groups of the lipid projected onto an axis normal to the bilayer plane. The combined use of x-ray and neutron diffraction described earlier allows one to determine these distributions, which are equivalent to probability densities that describe the probability of finding a particular structural group at a specific location (**Figure 2.19**). Complete structural images of this sort that account for all constituents of the unit cell represent fully resolved images of fluid bilayers.

Figure 2.18 Transbilayer distributions of double bonds in a DOPC lipid bilayer. (A) Two bilayer profiles determined under identical conditions except that in one case, deuteriums were placed on the double bond, as shown in the schematic. Similar results can be obtained using x-rays by adding bromines at the double bonds. (B) Difference structures obtained from the profiles in panel A show the distribution of the deuterium label. Also shown is a difference profile obtained using bromine-labeled double bonds and x-ray diffraction (dotted line). The bromine-labeled distribution is slightly broader because of the large diameter of bromine compared with hydrogen or deuterium. The double bonds explore nearly the full width of the bilayer's 30 Å-thick hydrocarbon core, illustrating the high thermal motion. (From Wiener MC, White SH [1991] *Biophys J* 60: 568–576.)

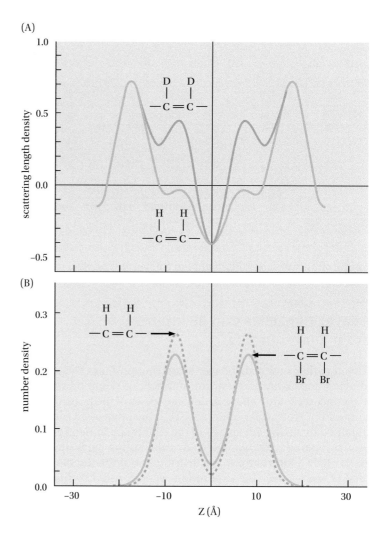

The fully resolved images are rich in information about the bilayer. The significant landmarks are noted in the profile: terminal methyl groups on the chains, location of the unsaturated bond in the oleic acid chain, glycerol backbone, choline, and phosphate groups. The distribution of water provides a gauge of the different levels of hydration across the bilayer. Note that the hydrocarbon core is about 30 Å across, as judged from the absence of water, and that carbonyl groups are good markers for the edges of the hydrocarbon core. These dimensions are key measures used in thinking about the interactions of transmembrane helices and proteins with lipid bilayers, and will be used extensively in subsequent discussions. Also, note that the layers of hydrated headgroups are each about 15 Å thick, so that the bilayer profile is about half occupied by headgroups; a significant fact, since the environment of the headgroups is very different from bulk solvent and has strong influences on the organization and function of membrane proteins.

2.2.7 Bilayer Dynamics Can Be Explored Computationally

Molecular dynamics (MD) calculations represent the chemistry and dynamics of molecules over time. The dynamics of biological molecules and assemblies are far too complex for any explicit analytical description, so a modeling strategy has been developed that takes advantage of the revolution in high-performance computing capabilities. Given the amount of experimental data required to describe the structure and dynamics of lipid bilayers, MD simulations are particularly important for studying lipid bilayers of various compositions and

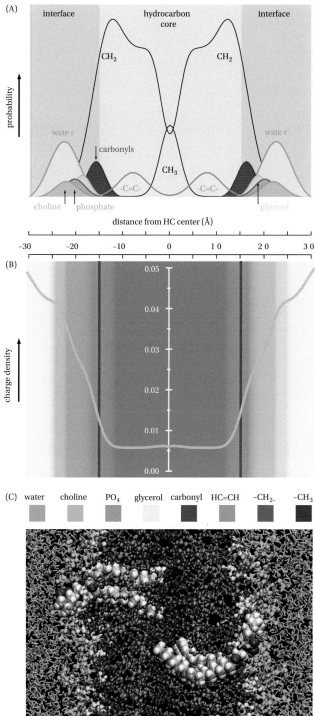

(A)

interface   hydrocarbon   interface
                      core

probability

CH₂   CH₂

water   water

carbonyls
CH₃
-C=C-   -C=C-
choline   phosphate   glycerol

distance from HC center (Å)

-30   -20   -10   0   10   20   30

(B)

charge density

0.05
0.04
0.03
0.02
0.01
0.00

(C) water choline PO₄ glycerol carbonyl HC=CH -CH₂- -CH₃

Figure 2.19 The complete structure of a DOPC bilayer determined by combining neutron and x-ray diffraction data and simulated using molecular dynamics. (A) The time-averaged distributions of the component groups (e.g., –PO₄, –CH₃, etc.) are shown projected onto a line perpendicular to the bilayer plane. The widths of the distributions are large due to the dynamics of the lipids in the fluid state. (B) The polarity gradient of a lipid bilayer. The data of panel A and the partial charges of the component groups allow the charge density of the bilayer to be plotted against transbilayer position. The polarity gradient is very steep in the ~15 Å-thick interface region. The typical diameter of an α-helix is about 10 Å, meaning that interfacial helices in membranes sit in a very steep polarity gradient. (C) A snapshot of a DOPC bilayer from a molecular dynamics simulation. (From White SH, Ladokhin, A. S., Jayasinghe, S., & Hristova, K. [2001] *J Biol Chem* 276: 32395–32398.)

their interactions with peptides and proteins. A single frame from a simulation of DOPC is shown in **Figure 2.19C**. Calculations are based on the use of a set of potential energy functions to model the interactions of atoms, called a "force field" (**Figure 2.20**). A force field typically includes elastic potentials for bond bending, torsion, and stretching, together with non-bonding coulombic potentials for charge interactions and van der Waals potentials for steric collisions, as illustrated in the figure. The atoms are set in motion at time zero by giving them kinetic energy, and their behavior is followed over time as they move within the constraints of the force field and the equations of motion. MD simulations give us our best view of the dynamic trajectories of molecules and assemblies, and

Figure 2.20 Molecular dynamics simulations use well-established principles to describe approximately the interactions of atoms in a molecule and molecules in aqueous solutions and lipid membranes. The upper row describes bonded interactions and the lower row non-bonded interactions. Because of the limits of current computers and force fields, many approximations are required. The symbol $\in$ means "element of". For the bond stretch term, for example, it means for all bonds in the molecule.

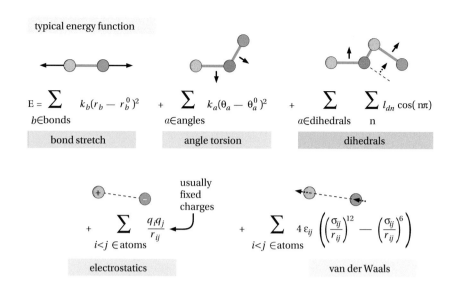

are a key part of modern structural biology. Not surprisingly, there are some limits to the accuracy of the results, mostly arising from the approximations used, the size of the system that can be examined, the time a system can be followed, and the approximations used in the force fields. Improvements are being made at a rapid pace, particularly fueled by advances in computer power. But even now, the structures of fluid bilayers surmised from MD simulations agree remarkably well with structures determined by diffraction methods.

2.2.8 NMR Order Parameters Reveal the Nature of Chain Disorder in the Hydrocarbon Core

How are the lipid alkyl chains of a bilayer configured, given their liquid-like state, which is apparent from diffraction measurements? The answer, obtained using NMR measurements, is that the chains are highly disordered in the bilayer center but become progressively more ordered near the headgroups. The NMR method that established this "order profile" is deuterium NMR, which provides order parameters when applied to multilayers of lipid bilayers in which the hydrocarbon chains have been specifically labeled with deuterium atoms. Another route to a similar view is to label specific locations in the hydrocarbon chains with deuterium and locate the labels using neutron diffraction. Both methods give useful views, but deuterium NMR is easier to implement.

Data can be obtained using NMR by placing deuteriums at each methylene of the two chains. This was originally done by a series of experiments in which methylenes were deuterated one at time, but modern NMR methods allow measurements of all order parameters in a single experiment using perdeuterated chains (**Spread 2.1**). Lipid order parameters are a measure for the orientational mobility of the C–D bond on the time scale of the NMR experiment. The order parameter is defined as

$$S = \mathrm{avg}\left[\frac{3\cos^2\theta - 1}{2}\right] \tag{2.3}$$

where θ is the (time-dependent) angle between the C–D bond vector and a reference axis, usually the bilayer normal. The brackets denote a time and ensemble average. Order parameters can be calculated from MD simulations, allowing a comparison of simulation and experiment. The order parameters for methylene groups of the two 14-carbon chains of DMPC as reported by deuterium NMR and by simulation are shown in **Figure 2.21A**. The vertical axis is the measure of order (higher values of S mean more order), and it is seen that the chains are more ordered near the backbone glycerol in the vicinity of the aqueous interface and less ordered toward the bilayer center. It is sometimes thought that the disordering

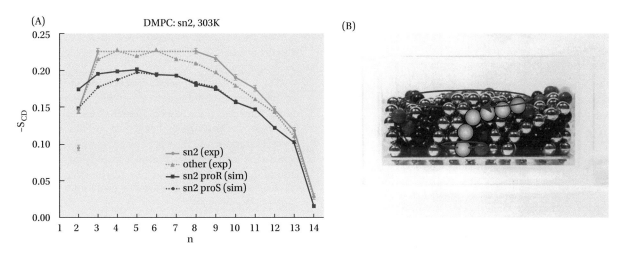

(A) DMPC: sn2, 303K

(B)

- sn2 (exp)
- other (exp)
- sn2 proR (sim)
- sn2 proS (sim)

Figure 2.21 Interpretation of NMR order parameter profiles. (A) Measured order parameter profiles for the sn-2 chain of DMPC measured across one monolayer of the bilayer compared with profiles computed from an MD simulation. The profiles have the same shape, but the simulated profile has systematically lowered order parameters due to the limitations of force-field parameterization (Figure 2.20). (B) A simple physical simulation of the order parameter profile obtained by confining chains of linked beads within a clear plastic box. The beads are attached to a metal plate representing the bilayer interface. Beads close to the metal plate have limited motions, but beads at the ends of the chain (colored blue) can explore large volumes of the box. (A, From Vermeer LS, de Groot BL, Réat V et al. [2007] *Eur Biophys J* 36: 919–931. With permission from Springer Nature. B, Based on White SH [1977] *Ann NY Acad Sci* 303:243-265.)

in the center results from looser packing. This is not correct. **Figure 2.21B** shows a simple model that explains the greater motion: during a fixed time period, the terminal methyl groups can explore more space than methylene groups anchored at the membrane interface by the phospholipid headgroups.

2.2.9 Cholesterol Affects the Packing of Lipid Chains, Keeping Them in a Fluid State

We learned in Chapter 1 that the lipid composition of the cell membranes of mycoplasmas depends entirely on the lipids available in the growth medium. This feature was crucial to proving that the structural fabric of membranes is the lipid bilayer (Figure 1.20). If mycoplasmas are provided with fatty acids with fully saturated chains, then they are viable only when the growth temperature exceeds the thermal phase-transition temperature of the lipids. That is, to survive, the organism's plasma membrane must be in a fluid state. This is true of all organisms. One approach that cells can use to overcome this limitation is to control the lipid acyl chain composition. The key strategy of eukaryotes is to include cholesterol or cholesterol-like molecules in their membranes. How does this help? In essence, the presence of cholesterol inhibits the formation of gel-like states and keeps the membrane in a liquid-like state (**Figure 2.22**).

2.3 MACROSCOPIC DESCRIPTIONS OF BILAYERS DEFINE FUNDAMENTAL PROPERTIES OF BIOLOGICAL MEMBRANES

2.3.1 Lipid Bilayers Can Be Described Grossly Using a Simple Macroscopic Model

What are the macroscopic physical properties of lipid bilayers, and what do those properties tell us about biological membranes? We can begin to answer these questions by considering a simple macroscopic model for the bilayer. We

Some Atomic Nuclei Act Like Small Magnets As a Result of Spin

Spin angular momentum (I) is a fundamental property of subatomic particles, like mass or electrical charge, and individual unpaired electrons and protons possess a spin of 1/2, which is quantized and can be positive or negative. Two or more particles with spins having opposite signs can pair up to eliminate observable spin. Because of the organization of the nucleus, some isotopes have net spins and others do not—those with net spin act as tiny magnets and interact with an external magnetic field if one is present. The magnetism of the nucleus, μ, is proportional to the spin angular momentum: $\mu = \gamma I$, where γ is the gyromagnetic ratio. Some of the useful nuclei for biological studies are ^{1}H, ^{2}H, ^{13}C, and ^{15}N. These and others are the "Nuclear Magnetic" part of NMR. The rules for determining the net spin of a nucleus are:

- If the number of neutrons and the number of protons are both even, then the nucleus has no spin.

- If the number of neutrons plus the number of protons is odd, then the nucleus has a half-integer spin (i.e., $I = 1/2, 3/2, 5/2, ...$)

- If the number of neutrons and the number of protons are both odd, then the nucleus has an integer spin (i.e., $I = 1, 2, 3, ...$)

Interaction with Radio Waves at the Right Frequency Gives Resonance

Consider a group of identical nuclei placed in a static magnetic field of strength B_0. Interaction with the field will cause the nuclei to orient, more in the low-energy direction ground state (parallel to the field in a macroscopic analogy) but with a smaller number having higher-energy positions as a result of thermal energy. Quantum effects cause the higher-energy orientation to be quantized into a discrete number of energy levels equal to $2I + 1$. For a spin-½ nucleus, there will be 2 states identified by their spin quantum numbers m (**Figure 1**). These numbers identify the states and always differ by integers (that's what quantum mechanics is about). For our spin-½ proton, the quantum numbers will be m = +½ and –½ (note that the difference is 1). The sign indicates the preferred alignment

with the magnetic field B, with +½ being more favorable (lower energy) than –½.

We can think of the nuclear spin in the same way that we think about a child's spinning top. The top has angular momentum directed along the spinning axis. This momentum counters Earth's gravitational field and causes the top to precess around the vertical direction of the gravitational field. In the same manner, the nuclear spin precesses around B_0 with a certain frequency, referred to as the Larmor frequency. The frequency is given by the Larmor equation: $\nu = \gamma B_0/2\pi$.

It is possible to apply an electromagnetic field of frequency ν_1 in such a way that the magnetic component of the field, B_1, is perpendicular to the static field B_0. If we change the frequency of the field to match the Larmor frequency ($\nu_1 = \nu$) of the spinning nucleus, then the spin precesses about B_1 as well as B_0, causing the spin to "flip" around B_1. When the flip occurs, energy is absorbed from the electromagnetic field. The change in energy ΔE is determined by Planck's universal constant h: $\Delta E = h\nu$. That is, the flipped spin absorbed a quantum of energy $h\nu$ from the electromagnetic field. This absorption is the "Resonance" in NMR, because the spinning nucleus resonates with the electromagnetic field. When the electromagnetic field is switched off, or the frequency of the field is no longer equal to the Larmor frequency, the spin relaxes back to its starting state, emitting the absorbed energy at frequency ν. Thus, we have **Nuclear** spin interacting with a **Magnetic** field to split energy levels that exhibit **Resonance** with a photon at exactly the right energy = **NMR**.

In the simplest kind of NMR experiment, called continuous-wave NMR, the sample is placed in a high magnetic field using an arrangement like that of **Figure 2**. In this experiment, the electromagnetic field frequency supplied by the transmitter is steadily increased through the resonance frequency and beyond. The receiver picks up the electromagnetic waves emitted by the precessing nuclei. The output of this kind of experiment is a plot of the intensity of the emitted radiation as a function of frequency (**Figure 3**). Another way of doing this experiment is to fix ν_1 and change (sweep) B_0 continuously, causing a resonance when $\gamma B_0/2\pi = \nu_1$. The resonant frequency of the experimental sample ν_s is generally measured relative to the known resonant frequency ν_c of a calibration sample, because NMR machines can differ in the local magnetic fields ν_0 at the sample. In essence, the parameter ν_s is an instrumental constant. We thus report resonant frequencies using the chemical-shift parameter $\delta = (\nu_c - \nu_s)/\nu_0$, which is reported in dimensionless units of parts per million.

Continuous-wave NMR is rarely used with modern NMR equipment. Instead, pulse Fourier transform NMR is used. This method allows a wide range of chemical shifts to be measured simultaneously by taking advantage of the inverse Fourier relation between time t and frequency ν. It is exactly the same idea as Fourier transformation between

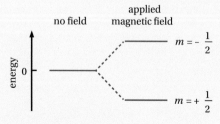

Figure 1 Splitting of energy levels.

(Continued)

SPREAD 2.1 WHAT IS NUCLEAR MAGNETIC RESONANCE AND WHY IS IT USEFUL? (CONTINUED)

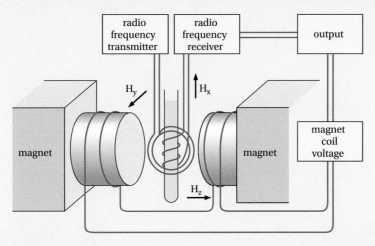

Figure 2 Continuous-scan NMR experiment.

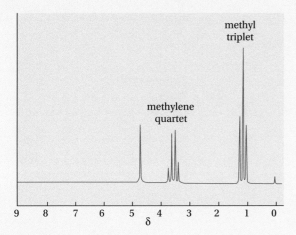

Figure 3 Proton NMR of ethanol.

real space and reciprocal space; the math is the same (see Box 3.4):

$$f(v) = \sum_{-t}^{t} f(t) \exp(ivt) \Delta t$$

In this equation, $f(v)$ and $f(t)$ are the frequency-domain and time-domain data and Δt the pulse width. The term $\exp(ivt)$ is just another way of writing sines and cosines, as explained in books on Fourier transformation. In an actual experiment, the resonance is measured by delivering a radio frequency pulse to the sample that is composed of a wide range of frequencies. Fourier transformation provides the equivalent of a continuous-wave time-domain experiment (**Figure 4**).

The Chemical Shift Identifies Specific Atoms in a Molecule

Moving charges have magnetic fields associated with them, so it is not a surprise to find that the motion of electrons in an atom or molecule will influence the magnetic field seen by a

nucleus. The electronic fields will, generally, oppose the external field and reduce the effective field seen by a given nucleus. Thus, each nucleus is responsive in its resonance frequency to its electronic environment, which will in turn vary with the nearby chemical bonding if the atom is part of a molecule. Chemical shifts, then, report on the local chemistry seen by each nucleus. Since the measured frequencies depend on the field, they are reported as a ratio to the resonance of a standard, in parts per million. This scaling allows observations at different field strengths to be accurately compared.

J-coupling Arises from Nucleus–Nucleus Interactions

In addition to the field effects from electron motions, there are smaller influences from nearby nuclei that are of the same type. The field of one nucleus influences the field seen by a nearby nucleus, resulting in a fine splitting of the resonances, as seen in the spectrum for ethanol in Figure 3. Because the spin states are quantized, there are discrete states from the interactions, and a set of peaks is observed. These field effects are weaker than the chemical shifts, and so the variation of resonance frequency is smaller.

Example: ¹H NMR of Ethanol

The spectrum in Figure 3 is for the ¹H nuclei (protons, in this case) of ethanol, CH_3CH_2OH. Three clusters of peaks are seen, one for the three methyl protons, one for the two methylene protons, and one for the hydroxyl proton. The total signal in each of the three is integrated and found to be in the ratio of 3:2:1, following the number of protons. The chemical shifts give the separation of the clusters, and the J-coupling splitting from nuclear spin interactions gives the fine structure.

Deuterium (²H) NMR Provides Information on Order Parameters

We discussed in the main text that information on the order parameters of phospholipid alkyl chains can be obtained

(Continued)

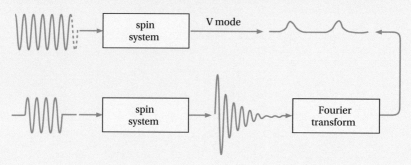

Figure 4 Two methods for obtaining NMR spectra: Continuous-wave (top) and pulse excitation followed by Fourier transformation.

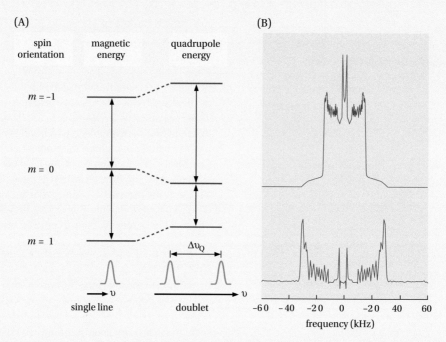

Figure 5 Deuterium energy levels (A) and deuterium NMR signals from perdeuterated phospholipid alkyl chains (B). (A, Seelig J [1978] *Prog Colloid and Polymer Science* 65:172–179. B, From Thurmond RL, Dodd SW, Brown MF [1991] *Biophys J* 59:108–113. Used with permission of Elsevier.)

from perdeuterated chains in which the hydrogens are replaced by deuteriums and that the orientational mobility of each C–D bond provided a measure of the order parameter at each position. How does NMR provide the data, and what do the data look like? According to our rules, the deuteron (1 proton + 1 neutron) has a spin $I = 1$, meaning that there are three energy levels with quantum numbers m = +1, m = 0, m = –1. The theoretical energy levels in the absence of any other effects are shown on the left-hand side of the left panel. In this ideal case, one would observe a single resonance peak, just as for a proton, because the difference in energy between m = +1 and m = 0 and between m = 0 and m = –1 are equal; this is called degeneracy. It turns out, though, that the deuteron has a distorted shape that produces what is called an electrical quadrupole moment, which is sensitive to the electrical field gradient of the surrounding electrons. This perturbs the magnetic

energy levels, which are no longer equal. We say that the degeneracy has been removed. Because there are now two distinct energies, there will be two resonances, called a doublet (**Figure 5A**, right panel).

George Pake discovered this so-called quadripolar splitting. Now, because of the anisotropic motion of alkyl chains, there will be a doublet for each of the deuteriums along the perdeuterated chain. These sum together to give the spectrum shown in Figure 5B. The peaks are very close together because the spectra were taken using multilamellar lipid dispersions, which cause a loss of orientation information. If the oriented multilamellar samples had been used, then the peaks would have been sharper and better separated, as shown in the lower panel of the right-hand figure. A method has been developed for converting the unoriented spectra of the upper panel into spectra like those of the lower panel. The method is called "de-Paking" in honor of George Pake. How

(Continued)

do we interpret these spectra? Those C–D bonds that have the greatest motion have the smallest doublet splittings. So, those in the signal from the terminal methyl region are closest to the center of the spectrum. It has been shown by specific deuteration (i.e., single deuteriums at each position of the chain) that doublet splittings become wider as one moves closer to the headgroup region, where there is least motion (Figure 2.21).

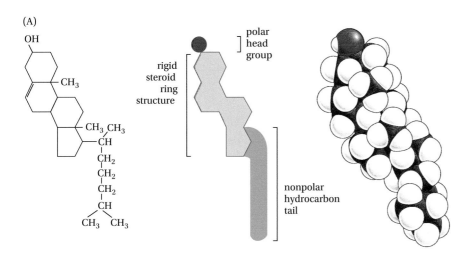

(A)

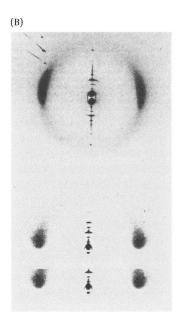

(B)

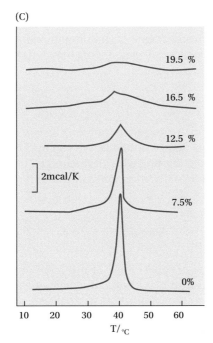

(C)

Figure 2.22 The structure of cholesterol and its effect on phospholipid bilayers. (A) Cholesterol's polar headgroup is a hydroxyl group, which is part of a rigid steroid ring system. Connected to the steroid ring is a short hydrocarbon tail. The amphipathic nature of cholesterol allows it to reside comfortably in phospholipid bilayers. (B) X-ray diffraction patterns of an oriented cholesterol-free egg lecithin (phosphatidylcholine with mixed chains) bilayer (upper panel) and a lecithin:cholesterol (1:1) bilayer. We see in the upper panel diffuse chain scattering peaks along the equator that have an average spacing of 4.6 Å. In the lower panel, the average spacing is higher (4.75 Å) because intervening cholesterol molecules push the lipid alkyl chains apart. Importantly, the diffraction pattern shows greater orientation of the acyl chains normal to the substrate due to the orienting effect of the rigid cholesterol steroid group. (C) Cholesterol added to pure lipid bilayers causes the gradual disappearance of the gel-to-liquid crystal phase transition until it disappears entirely at about 20 mol%. Bilayers in this state are referred to as being in a liquid-ordered (L_o) state, because the alkyl chains remain thermally disordered. (A, From Alberts B, Gray D, Hopkin K et al. [2013] *Essential Cell Biology*, 4th Edition. Garland Science, New York. With permission from W. W. Norton. B, From Levine Y & Wilkins M (1971) *Nature New Biology* 230:69–72. With permission from Springer Nature. C, From Genz A, Holzwarth JF, Tsong TY [1986] *Biophys J* 50: 1043. Used with permission from Elsevier.)

can arrive at a useful model by considering the bilayer immersed in a salt solution as a thin insulating slab (the hydrocarbon core) separating two electrically conducting layers (headgroups and salt solution). This macroscopic model is equivalent to a capacitor with capacitance C_m, familiar from basic physics. If ions can penetrate the bilayer in some way, then we can add in parallel to the capacitor a resistor R_m, representing the conductance of ions that might penetrate the bilayer. This simple model (**Figure 2.23**) is very useful for understanding the key properties of bilayers that are relevant for living systems, such as the conduction of nervous impulses. Electrical measurements on black lipid membranes (Figure 1.19) allow these parameters to be determined directly.

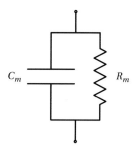

Figure 2.23 A representation of the lipid bilayer as a capacitor and resistor, called an equivalent electrical circuit.

2.3.2 The Bilayer Is a Capacitor with Capacitance C_m

Because a bilayer is a good insulator, it can sustain a separation of charges across it for significant periods of time determined by its time constant $\tau = R_m C_m$. Ion channels or other transporters can affect how long the membrane can hold a charge, because they determine R_m. The ability to store charge is a key to energy storage and conversion by biological membranes.

Capacitance is a measure of the amount of electric charge stored (or separated) for a given electric potential. If the charges on either side of the bilayer are +Q and –Q, and V is the voltage across it, then the capacitance is given by

$$C_m = \frac{Q}{V} \tag{2.4}$$

The SI unit of capacitance is the farad (1 farad = 1 coulomb per volt). The capacitance C_m can be calculated if the geometry of the conductors and the dielectric properties of the insulator between the conductors are known. For example, the capacitance of a bilayer of area A and thickness d is given approximately by

$$C_m = \varepsilon \varepsilon_0 \frac{A}{d} \tag{2.5}$$

The capacitance is measured in farads (F) and the area of the bilayer in cm². ε is the dielectric constant of the hydrocarbon core and has a value of about 2. ε_0 is the permittivity of free space (8.854×10^{-10} Fd cm⁻¹), which is a constant. Note that the capacitance increases as the thickness decreases, so a bilayer, which is very thin, can have a large capacitance. The hydrocarbon core thickness of a bilayer is about 30 Å, which translates to a capacitance of about 0.6 µF cm⁻².

The energy stored across a bilayer is equal to the work done to charge it. Consider a capacitance C_m holding a charge +Q on one plate and –Q on the other. Moving a small element of charge dQ from one plate to the other against the potential difference $V = Q/C$ requires work dW:

$$dW = \frac{Q}{C_m} dQ \tag{2.6}$$

W is the work measured in joules, Q is the charge measured in coulombs, and C is the capacitance in farads. We can find the energy stored by integration of this equation. Starting with an uncharged bilayer (Q = 0) and moving charges from one side to the other until the plates have charges +Q and –Q requires the work W:

$$W_{\text{charging}} = \int_0^Q \frac{Q}{C} dQ = \frac{1}{2} \frac{Q^2}{C} = \frac{1}{2} CV^2 = W_{\text{stored}} \tag{2.7}$$

Thus, the large capacitance of a bilayer equips it to store large amounts of energy. This feature is a key to the energetics of cellular life and is used in chloroplasts and mitochondria, as we shall see.

2.3.3 Lipid Bilayers Have a Very High Electrical Resistance R_m

The influence of the molecular structure of an uncharged solute on its permeation across biological membranes has been evaluated by various methods. Figure 1.9, for example, shows how properties of molecules influence the uncatalyzed transit movement of molecules across membranes. Data for the permeation of these molecules across bilayers are very similar and have historically been taken as support for the presence of lipid barriers in membranes. As expected, larger or more polar molecules permeate more slowly than smaller or less polar molecules. The relative sizes of some atoms are shown in **Figure 2.24**. The size of an atom or molecule influences its permeability and is a key aspect of selectivity in ion channel behavior. Charged molecules crossing

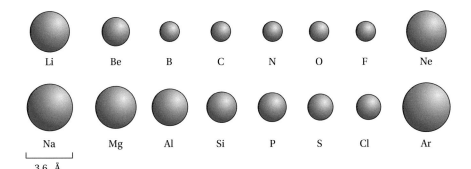

Li Be B C N O F Ne

Na Mg Al Si P S Cl Ar

3.6 Å

Figure 2.24 The relative sizes of atoms in their uncharged forms frames the discussion of their permeability, since the Born radius varies.

the insulating bilayer face an energy barrier, because it is energetically costly to dehydrate them during passage. However, the height of the energy barrier depends upon the size of the molecule. The effect of size can be estimated from the Born energy, which is the energy required to charge an object in a medium with dielectric constant ε:

$$\Delta G_{Bom} = \frac{Q^2}{2a\varepsilon} \qquad (2.8)$$

In this equation, a is the Born radius and Q is the charge. The extra electron(s) in an anion make it larger, while the electron deficiency of a cation makes it smaller (**Figure 2.25**). As a reference point, the Born energy of a monovalent ion of radius 2 Å in water ($\varepsilon \approx 80$) is about 1 kcal mol^{-1}. But if that ion is transferred to an alkane ($\varepsilon \approx 2$), the Born energy is about 25 kcal mol^{-1}. This is the expected value for permeation through a lipid bilayer, because the bilayer's hydrocarbon core is equivalent to a liquid alkane. Indeed, MD simulations yield a similar value. This huge Born energy for a low-dielectric medium tells us that lipid bilayers are excellent insulators and that R_m must be exceedingly large. Indeed, the resistance of a 1 cm^2 bilayer is about 10^{12} ohms. This resistance can be dramatically lowered, however, if the ion radius is increased due to Equation 2.8. This effect can be intuitively understood as spreading the charge over a larger surface. In our example, changing the radius of the ion to 10 Å reduces the Born energy in a low-dielectric membrane to 5 kcal mol^{-1}. Although the Born energy and the radius are conceptually useful, they are not easily calculated for complex molecules.

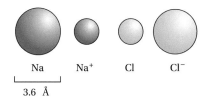

Na Na$^+$ Cl Cl$^-$

3.6 Å

Figure 2.25 The size of an ion can be larger or smaller than the uncharged atom, depending on whether an electron is gained or lost, because the size is determined by the electronic structure.

2.3.4 Bilayers Are More Permeable to Anions than Cations Because of the Membrane Dipole Potential

Negatively charged molecules diffuse across phospholipid membranes more easily than positively charged ones of similar size. For example, the tetraphenylborate anion (TPB$^-$) has a permeability coefficient about 3×10^7 larger than the identically-sized tetraphenylarsonium cation (TPA$^+$) in phosphatidylcholine vesicles. How can this be? A remarkably large positive electrostatic potential in the interior of the membrane accounts for much of this permeability difference, and a model of the potential profiles for cations and anions is shown in **Figure 2.26A**, based on the permeability studies. The size of the potential is little influenced by the charge or nature of the polar headgroups; it is thought mainly to arise from orientation of the dipole moments of the ester linkages of the hydrocarbon chains (**Figure 2.26B**). The dipole potential can affect the structure and function of membrane-incorporated proteins and may account for the evolutionary choice of natural membranes to store energy using positive charges.

MD calculations have been used to predict the size and spatial dependence of the dipole potential. While MD simulations have been quite successful in reproducing the structure of lipid bilayers, as measured with x-ray and neutron scattering, predictions of electrostatic details are less reliable, partly because current models use fixed point charges and do not allow for the polarization

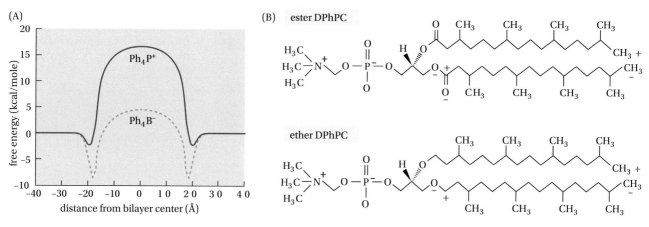

Figure 2.26 Transbilayer energy profiles. (A) The energy profiles seen by positive (solid line) and negative (dashed line) ions crossing a lipid bilayer, from measurements using "hydrophobic ions," that is, ions with large Born radii. Note that positive ions encounter a much larger barrier than do negative ions. The ions respond to the electrostatic potentials created by the systematic orientations of dipoles at the bilayer backbones as well as other small effects. (B) Key features of lipids involved in determining the dipole potential. The upper lipid has the chains esterified, as in eukaryotes and bacteria, while the lower lipid has ether links, as found in archaea. (A, From Franklin JC, Cafiso DS [1993] *Biophys J* 65: 289–299. Used with permission from Elsevier.)

of atoms and bonds in response to electric fields. MD simulations yield peak dipole potentials of about 500–1000 mV. Significantly, both experiments and MD simulations yield much smaller dipole potentials for phospholipids with an ether linkage (found in archaeal lipids) than for those with a conventional ester linkage, suggesting experimental use of these lipids for comparison with each other. This difference can be taken advantage of to measure dipole potentials using electron microscopy (**Box 2.5**), where the potential for a phosphatidyl-choline (PC) with ester links has been estimated as ~500 mV!

2.4 LIPID INTERACTIONS SHAPE MEMBRANE PROPERTIES

2.4.1 Lipids Diffuse Rapidly in the Plane of a Bilayer but Slowly across the Bilayer: Bilayer Lipid Asymmetry

Lateral diffusion. Given that proteins diffuse readily in membranes (Figure 1.25), it is not surprising to find that bilayer lipids can also move rapidly in the plane (**Figure 2.27**). Lateral diffusion measurements have been developed using a variety of tools, one of the most informative being fluorescence recovery after photobleaching (FRAP). FRAP is an optical technique capable of quantifying the 2-D lateral diffusion of a molecularly thin film containing fluorescently labeled probes. This technique is used in studies of not only lipid diffusion but also the diffusion of labeled proteins. Similar techniques have been developed to investigate the 3-D diffusion of molecules inside a cell; they are also referred to as FRAP.

A typical measurement visualizes a lipid surface using an optical microscope, a source of light, and a fluorescent probe attached to a subpopulation of molecules of interest, for example, by labeling a small percentage of the lipids in a bilayer (**Figure 2.28**). A measurement begins by recording a background image of the sample before photobleaching. Next, the light source is focused onto a small region of the viewable area either by switching to a higher-magnification microscope objective or with laser light of the appropriate wavelength, exposing the probe molecules to high-intensity illumination that causes their fluorescence lifetime to quickly elapse, bleaching them so that they no longer fluoresce. The image in the microscope at the original magnification now is that of a fluorescent field with a dark spot produced by the bleaching. With time, the

BOX 2.5   MEASURING THE MEMBRANE DIPOLE POTENTIAL USING ELECTRON MICROSCOPY

While a number of methods have been used to measure the dipole potential, a recent use of electron microscopy of frozen vesicles in ice is particularly compelling. It has few assumptions and has been applied to find the dipole potentials of bilayer membranes with either ester or ether-linked chains. The main idea is that the electrons scatter from atoms, but the paths of the electrons are also influenced by electrostatic fields, since they are charged. The lipids were chosen to have branched alkyl chain structures, as found in archaea, because these will not exhibit thermal phase changes that might complicate observations in ice. The magnitudes and the profiles of the atomic scattering and the dipole potential scattering were obtained using MD models and fitting them, by adjusting their magnitudes, to the observed data. The peak dipole potential was estimated to be 510 and 260 mV for diphytanoyl phosphatidylcholine (ester-DPhPC) and diphytanoyl phosphatidylcholine (ether-DPhPC), respectively (**Table 1**).

Figure 1 summarizes modeling the image of a phospholipid vesicle embedded in ice and viewed with the electron microscope.

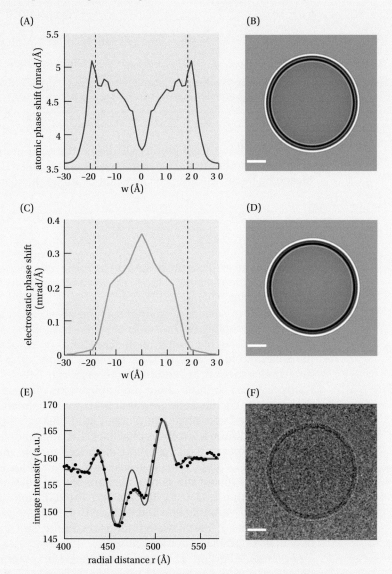

Figure 1 Modeling the electron microscopy image of a phospholipid vesicle embedded in ice. (A) The scattering from the individual atoms in a bilayer. Dashed lines indicate the approximate boundary of the headgroup region. (B) A simulated image of a spherical vesicle of radius 470 Å having the profile of A as would be seen if there were only atomic scattering. Scale bars: 200 Å. (C) Profile of the scattering from the electrostatic potential within an ester-DPhPC membrane. (D) Computed image resulting from the sum of the atomic scattering of A and the electrostatic scattering of C, showing a less prominent ring in the center of the membrane image. (E) The experimental image intensity (black dots) is compared with a fit by the atomic potential alone (red curve) and a fit by the sum of the atomic potential and dipole potential (green curve), which matches the experimental points very well. (F) The experimental image. (From Wang L, Pulkit S Bose, Sigworth FJ [2006] 103: 18528-18533. Copyright (2006) National Academy of Sciences, U.S.A.)

(Continued)

Table 1 Dimyristoyl, Dipalmitoyl, and Diphytanoyl Phosphatidylcholine Membrane Dipole Potentials

Method	Dipole potential (mV)	Lipid
Bilayer	227	DPPC
	228	DPhPC
Monolayer	449	DMPC
MD simulation (CHARMM)	950	DMPC (512 lipids)
MD simulation (CHARMM)	1002	DPhPC (72 lipids)
MD simulation (AMBER)	600	DPPC (64 lipids)
Cryo-EM	510	DPhPC

(Data from Wang L, Pulkit S Bose, Sigworth FJ [2006] *Proc Natl Acad Sci* **103:** 18528–18533. Copyright (2006) National Academy of Sciences, U.S.A.)

unbleached fluorescent probes will diffuse throughout the sample, diluting the inactivated probes in the bleached region. Assuming a Gaussian profile for the bleaching beam, the diffusion constant D can be calculated from

$$D = \frac{w^2}{4t_{1/2}} \tag{2.9}$$

where w is the width of the beam and $t_{1/2}$ is the time required for the bleach spot to recover half of its initial intensity. Related equations have been derived for differently shaped spots.

The result of such measurements is the observation that lipids in the fluid state diffuse rapidly in the plane of pure lipid bilayers. For example, using a photobleach spot of about 6 µm in diameter, $t_{1/2} \approx 2$s for a fluorescently labeled phospholipid in lipid multilayers. From Equation 2.9, $D \approx 4 \times 10^{-8}$ cm^2 s^{-1}. The total time required for complete recovery can be estimated by multiplying $t_{1/2}$ by 5. A way to think about the implications of these numbers is to realize that many eukaryote cells are about 5–10 µm in diameter. If the plasma membrane behaved as a single-component lipid bilayer, a lipid molecule could be expected to diffuse from one end of the cell to the other in about 10 s. Although one might expect lateral lipid diffusion in cell membranes to be slower due to more complex composition and structure, lateral diffusion of lipids in plasma membranes is about the same as in simple lipid bilayers. For mouse erythrocytes, for example, $D \approx 3 \times 10^{-8}$ cm^2 s^{-1}. The diffusion of proteins, however, is much slower. Mu-ming Poo and Richard Cone, who made the first quantitative measurements of lateral protein diffusion, found $D \approx 4 \times 10^{-9}$ cm^2 s^{-1} for rhodopsin in amphibian photoreceptors.

Flip-Flop: "Flip-flop" is the generally agreed-upon term for moving a bilayer lipid from one bilayer leaflet to the other (Figure 2.27). Because the headgroup reorients to face the opposite aqueous region, the term is appropriate. For biological membrane bilayer lipids, the flip requires an intermediate that places the polar head inside the non-polar region of the bilayer as it moves from one monolayer to the other. This process is expected to be unfavorable but strongly dependent on lipid structure. The rate for phosphatidylcholines is very slow and hard to measure because many of the approaches used perturb the rate, for example by appending a probe to the lipid head or acyl chains. The most reliable measurements of flip-flop of unmodified DMPC across LUV bilayers indicate that $t_{1/2}$ is around 9 hours—four orders of magnitude slower than lateral

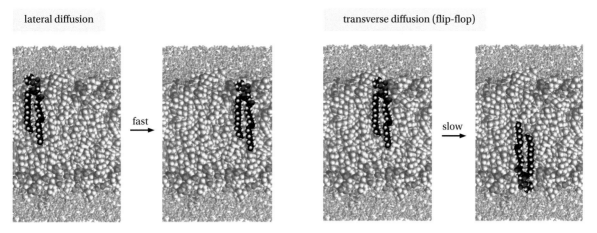

lateral diffusion   transverse diffusion (flip-flop)

fast   slow

Figure 2.27 Lipids diffuse rapidly in the planes of bilayers but flip-flop across the bilayer very infrequently due to the high polarity of the lipid headgroup.

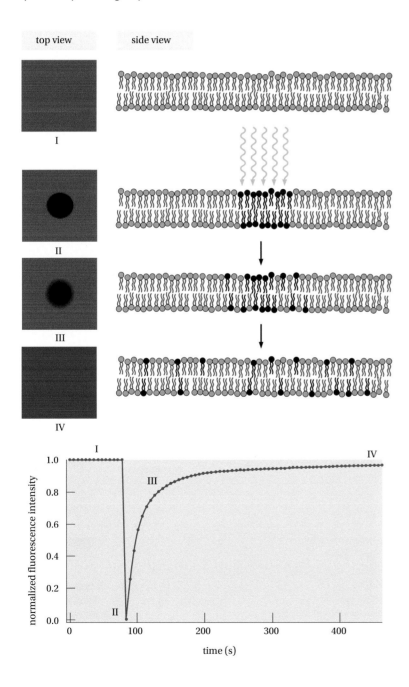

top view   side view

I

II

III

IV

Figure 2.28 Fluorescence Recovery After Photobleaching, or FRAP, experiment for measuring lipid diffusion rates. The fluorescent probes in a small area of a bilayer are bleached by a laser to eliminate their fluorescence; then, the diffusion of unbleached fluorescent molecules from the adjacent bilayer into the area is followed as a recovery of the fluorescence. The time course of fluorescence recovery gives a measure of the diffusion of the lipids.

diffusion. The $t_{1/2}$ flip-flop of cholesterol across phosphatidylcholine bilayers is much faster because of its diminutive headgroup. But its flip-flop is still slow relative to its lateral diffusion. Other lipids with small headgroups, such as ceramides and diacylglycerol, flip-flop with half-times of less than a minute. Proteins and peptides incorporated into lipid bilayers can speed up transbilayer lipid movements, presumably by creating defects that lower the free energy cost of flip-flop. What about biological membranes?

Lipid Asymmetry: Slow flip-flop rates would help maintain any lipid compositional asymmetry that might have arisen during membrane synthesis in the endoplasmic reticulum. But even a flip-flop half-time of hours is fast relative to the many-weeks lifetime of erythrocytes, which have highly asymmetrical transbilayer lipid compositions (**Figure 2.29**). Flip-flop must be slower in the erythrocyte than in the model systems, probably as a result of the influences of proteins, cholesterol, and the cytoskeleton. Note that in the figure, the inside-facing lipids are mainly negatively charged, while the outward-facing lipids are neutral. Most eukaryotic membranes, including plasma membranes and organelles, have asymmetric lipid distributions that are similar to that of erythrocytes. Interestingly, even though the endoplasmic reticulum is the ultimate source of cell membranes, it does not appear to have an asymmetric transbilayer lipid distribution. The asymmetries generally mirror those of the erythrocyte: the cytosolic leaflet is enriched in phosphatidylserines (PS) and phosphatidylethanolamines (PE), and the opposing leaflet enriched in sphingomyelin (SM) and phosphatidylcholine (PC). These asymmetries must be physiologically important, because cells expend ATP to maintain the gradients by means of lipid transporters called flippases (see Chapter 12). The physiological roles of the asymmetry are not fully understood yet. One role may be to facilitate changes in membrane curvature, as discussed in the next section. Another may be to facilitate transmembrane signaling; the appearance of PS on the external surface of a cell appears to be a harbinger of programmed cell death (apoptosis).

2.4.2 A Bilayer Can Be Curved If Its Monolayers Differ in Area

A remarkable property of lipids is the extent to which the area per lipid molecule is conserved for a given headgroup. As mentioned earlier, if the hydrocarbon chains in a bilayer with a given headgroup are varied in length, the thickness of the bilayer varies linearly with the chain length. Bilayer vesicles that are subjected to osmotic stress rupture and reseal to relieve the stress rather than thinning to increase area by >5%. Small increases (<1%) in the area of one monolayer, which can be achieved by the addition of amphiphiles (**Figure 2.30**), cause gross shape changes in bilayer vesicles and even cells (**Figure 2.31**), because the bilayer leaflet with the partitioned amphiphile has a larger area than the other leaflet, and the lipids cannot compensate by changing their areas per molecule.

Figure 2.29 Asymmetric distributions of phospholipids in erythrocyte membranes. Lipid distributions have mainly been measured by comparing intact erythrocytes and open erythrocyte membranes ("ghosts") exposed to impermeant chemical reagents, phospholipases, and exchange proteins that can swap specific lipids. Differences in reactivity, access to degradation, and access to exchange give variable quantitative results but a consistent overall picture as shown: PE, phosphatidylinositol (PI), and PS predominate in the inner leaflet, resulting in a net negative charge on the surface of the inner leaflet; PC and SM predominate in the outer leaflet, resulting in approximate charge neutrality. Studies of other cells give similar results; some cytoplasmic membranes are also asymmetric. (Data from Daleke DL (2008) *Cur Opinion in Hematology* 15:191–195.)

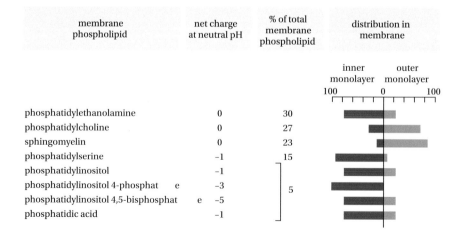

membrane phospholipid	net charge at neutral pH	% of total membrane phospholipid	distribution in membrane
phosphatidylethanolamine	0	30	
phosphatidylcholine	0	27	
sphingomyelin	0	23	
phosphatidylserine	−1	15	
phosphatidylinositol	−1		
phosphatidylinositol 4-phosphate	−3	5	
phosphatidylinositol 4,5-bisphosphate	−5		
phosphatidic acid	−1		

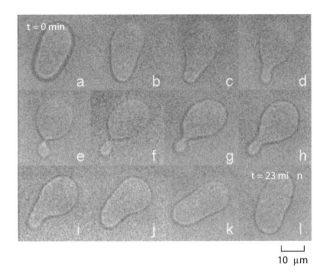

Figure 2.30 Lipid flip-flop causes changes in shape of a giant unilamellar vesicle (GUV) after injection of a 16-carbon ceramide (C16-Cer) into the GUV suspension at t = 0. C16-Cer initially partitions into the outer bilayer leaflet, causing shape distortions due to the area inequality of the two bilayer leaflets. As C16-Cer flip-flops to equilibrate between the two leaflets, the area inequality is removed. This causes the starting shape of the GUV to be restored, providing a measure of the flip-flop rate. (From López-Montero I, Rogriquez N, Cribier S et al. [2005] *J Biol Chem* 280: 25811–25819. With permission from Elsevier.)

This effect can be used to measure flip-flop half-times for amphiphiles injected into a suspension of GUVs made from PC extracted from egg yolks (egg yolk PC or EPC). Figure 2.30 shows such an experiment performed by Philippe Devaux. A 16-carbon ceramide (C16-Cer), injected into the suspension at $t = 0$ min, partitions into the outer monolayer of a GUV. The C16-Cer causes the shape of the GUV to change because of the area imbalance between the outer and inner leaflets of the GUV bilayer. As the C16-Cer equilibrates across the membrane, the areas of the two leaflets become equal, causing the starting shape to be restored after 23 min. A quantitative analysis of data such as these allows the flip-flop half-time to be computed. In this case, $t_{1/2} = 2.4$ min. Similar experiments can be done using erythrocytes (Figure 2.31).

The conserved area arises from the balance of the three regions of interaction we discussed earlier: dynamic forces between headgroups (acting to increase area per lipid), exposure of hydrophobic areas of chains to water (acting to decrease area per lipid), and the internal lateral pressure of the chains against each other (acting to increase area per lipid). The energy of the hydrophobic exposure dominates. Consequences include the general orientation of the chains across a bilayer and the localization of terminal methyl groups near its center. There is evidence that lateral pressure, and its variations as a function of the depth in the bilayer, may affect both membrane protein folding and function, but this is an emerging area of study and not yet well understood except in general terms.

Area conservation reinforces the importance of the energy of exposing lipid hydrophobic area to the aqueous environment—interpreting an area increase of

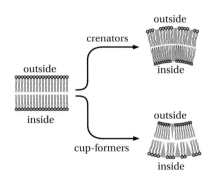
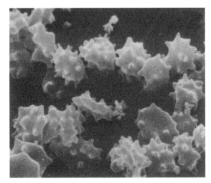
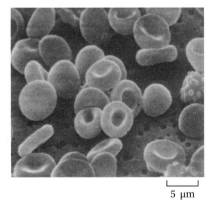

Figure 2.31 Curving of erythrocyte membranes by selective partitioning of surfactants into the inner and outer monolayer of membranes. (From Sheetz MP, Singer SJ [1974] *PNAS* 71: 4457–4461. With permission from Michael Sheetz.)

10 Å^2 per molecule as purely hydrophobic surface, we would expect an increase of energy of a little less than half a kcal mol^{-1} for each lipid. In a small lipid vesicle with ~10^5 lipids, the energy becomes quite large, so it becomes immediately understandable that the area per lipid would be minimized by the hydrophobic effect. This understanding guides several ideas regarding membranes.

2.4.3 Lipid Interactions in the Plane of a Bilayer Can Generate Regions of Different Composition

A bilayer in the fluid phase that predominates in biological membranes has often been considered as a random 2-D liquid. This can be argued as unlikely. Consider that the distribution of the many lipid components in a membrane can be random only if all pair-wise interaction energies of the species are within thermal energies (~0.6 kcal mol^{-1} at room temperature) of each other. Given that there are many lipid components, and that the number of pairs goes as the square of the number, such a narrow range of interaction energies is an extremely unlikely condition. Thus, it should be expected that regions of different lipid composition will exist in membranes and cause the environments of proteins to vary, because protein–lipid interaction energies will also fail to match each other across all protein and lipid species in a membrane. One thus expects heterogeneous distributions, or regionalization, of lipids and proteins in membranes. Additionally, however, the distributions of the proteins in membranes are also determined by interactions with cytoplasmic components, as we discuss later.

The number of possible regions of different composition is limited by the number of different components, as noted by J.W. Gibbs, who defined a "phase rule" that is widely used by physical chemists: for a system in equilibrium, the phase rule relates the number of components (substances), variables (temperature, pressure), and phases to something called the degrees of freedom, more logically referred to as the number of free variables (see Box 2.2):

$$f = C - P + n \qquad (2.10)$$

In this equation, f = number of free variables (also called the degrees of freedom), which is the number of independent variables that can be arbitrarily fixed to establish the intensive state of a system, C = the number of components, P = the number of phases, and n = the number of non-compositional variables such as temperature and pressure. In simple model systems, phases in bilayers have been studied extensively, but it is challenging to apply the lessons to biological membranes where there are many kinds of lipids and other molecules, except to note that there can be many regions of different composition separated in the plane of a membrane. Furthermore, the phase rule only applies to systems in thermodynamic equilibrium—which is unlikely to be true for living cells consuming metabolic energy.

For example, experimental observations show that differential interactions can lead to significant planar separations in bilayers formed from binary or ternary mixtures of pure lipids. An extensively studied example is the case of PC–SM–cholesterol interactions (Box 2.2), where separated regions of SM-cholesterol may arise in part from the presence of an NH group in the SM backbone that can form a hydrogen bond to the OH on cholesterol—an example of the preferential interactions discussed earlier. Such observations give rise to the idea that there are "rafts" of lipid of different composition that may also correlate with functional specialization of membrane regions. In biological membranes, with many kinds of lipids and where much of the area is occupied by a diversity of proteins, the interactions are vastly more complex, and the very large number of different kinds and energies of interactions is expected to lead to considerable heterogeneity in the plane of the structure. Much current work is focused on the idea that regional specialization of areas

(A)

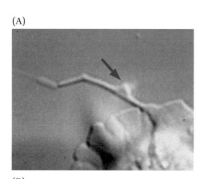

(B)

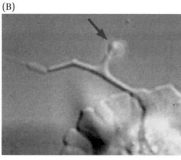

(C)

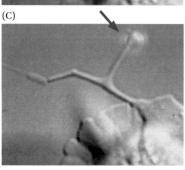

Figure 2.32 Pulling on the surface of a neuronal cell with optical tweezers (red arrows) easily deforms the cell membrane, forming a tubular extension. This reveals the fluidity and deformability of plasma membranes. (From Alberts B, Johnson A, Lewis JH et al. [2007] *Molecular Biology of the Cell*, 5th Edition. Garland Science, New York. With permission from W. W. Norton.)

in a membrane can facilitate biological functions, for example by clustering of receptors to improve signaling.

2.4.4 Bilayer Fluidity Allows Deformation When Strained

From our discussion so far, one might be tempted to think of bilayers as defined, somewhat static structures, but this is far from the case—composition and shape are constantly changing. The composition of any membrane in a living cell changes as membranes fuse and separate, as ligands from the environment or cytoplasm are bound and released, and as different components are synthesized and degraded. The fluidity mentioned at several points allows bilayers, and membranes, to deform easily when challenged by external forces. This flexibility allows cells to shape membranes by exerting forces on them, using a variety of protein structures—special protein domains that pinch off small vesicles, cytoskeleton anchors that push or pull, and proteins that anchor to extracellular environmental features. Pulling on a cell membrane by grabbing it with optical tweezers (a sophisticated version of Janet Plowe's glass needles discussed in Chapter 1) results in the immediate formation of a threadlike extension that is a small tube of bilayer membrane (**Figure 2.32**).

KEY CONCEPTS

- The lipid bilayer is the organizing principle of cell membranes.

- Amphiphiles are molecules that have two spatially separated and distinct chemical moieties.

- Three physical principles make the lipid bilayer possible: the hydrophobic effect, the amphiphilicity of lipids, and lipid shape.

- Because of the hydrophobic effect, the non-polar moieties aggregate to minimize exposure to water and are shielded by the polar moieties, which interact favorably with water.

- Aggregation is characterized by the critical micelle concentration (CMC), which indicates the aqueous concentration at which aggregates appear.

- Simple amphiphiles, such as sodium dodecyl sulfate (SDS), form small spheroidal particles made up of 50 or so molecules.

- Most structurally important membrane lipids have extraordinarily low CMCs due to their two acyl chains, which typically contain 16–18 methyl groups each. Their aggregates tend to prefer planar sheet-like bilayer structures because their cylindrical geometry prevents them from forming tightly curved surfaces like those of SDS.

- The bilayer is extraordinarily stable and versatile and can be formed from mixtures of many kinds of lipids.

- Cholesterol plays an exceptionally important role in bilayer mixtures by maintaining the bilayer in a fluid-like state.

- At equilibrium, bilayer mixtures adhere to the Gibbs phase rule.

- The bilayer's inherent stability and versatility are apparent from the fact that it can form oriented multilamellar stacks on solid surfaces at low hydration or single-walled vesicles with diameters as large as several μm in excess water.

- The principal methods for describing bilayers at the molecular level are x-ray and neutron diffraction, but much information can also be obtained by NMR and molecular dynamics simulations.

- The bilayer is a formidable barrier for ions except those that are organic or very large in diameter.

- Negatively charged organic molecules will penetrate bilayers more readily than positively charged ones because the interior of the bilayer has a positive electrostatic potential relative to the water, due in part to dipoles in the polar regions.

- Lipids can diffuse rapidly along the membrane plane, but transverse movement across membranes (flip-flop) is slow without the participation of membrane proteins.

- Eukaryotic membranes generally have different lipid compositions in their inner and outer leaflets. Energy-consuming membrane protein transporters are required to maintain the composition difference.

- Regions of different composition are expected to arise spontaneously in the membrane plane as a consequence of diverse lipid–lipid, lipid–protein, and protein–protein interaction energies.

FURTHER READING

Als-Nielsen, J., and McMorrow, D. (2001) *Elements of Modern X-ray Physics*, John Wiley & Sons, New York. 318 pp.

Blaurock, A.E. (1982) Evidence of bilayer structure and of membrane interactions from X-ray diffraction analysis. *Biochim. Biophys. Acta* 650:167–207.

Devaux, P.F. (1991) Static and dynamic lipid asymmetry in cell membranes. *Biochemistry* 30:1163–1173.

Edidin, M. (1974) Rotational and translational diffusion in membranes. *Annu. Rev. Biophys. Bioeng.* 3:179–201.

Franks, N.P., and Levine, Y.K. (1981) Low-angle X-ray diffraction, in *Membrane Spectroscopy* (Grell, E., Ed.), pp. 437–487, Springer-Verlag, Berlin.

Kondepudi, D., and Prigogine, I. (1998) *Modern Thermodynamics*, John Wiley & Sons, Chichester.

Lee, A.G. (1977) Lipid phase transitions and phase diagrams. I. Lipid phase transitions. *Biochim. Biophys. Acta* 472:237–281.

McLaughlin, S. (1989) The electrostatic properties of membranes. *Ann. Rev. Biophys. Biophys. Chem.* 18:113–136.

Munro, S. (2003) Lipid rafts: Elusive or illusive? *Cell* 115:377–388.

Richards, F.M. (1977) Areas, volumes, packing, and protein structure. *Annu. Rev. Biophys. Bioeng.* 6:151–176.

Sanders, C.E., and Prosser, R.S. (1998) Bicelles: A model membrane system for all seasons? *Structure* 6:1227–1234.

Serdyuk, I.N., Zaccai, N.R., and Zaccai, J. (2007) *Methods in Modern Molecular Biophysics*, Cambridge University Press, New York. 1120 pp.

Simons, K., and Vaz, W.L.C. (2004) Model systems, lipid rafts, and cell membranes. *Annu. Rev. Biophys. Biomol. Struc.* 33:269–295.

Levental, I., and Veatch, S.L. (2016) The continuing mystery of lipid rafts. *J Mol Biol* 428:4749–4764.

Tanford, C. (1980) *The Hydrophobic Effect: Formation of Micelles and Biological Membranes*, 2nd ed., John Wiley & Sons, New York.

Warren, B.E. (1969) *X-ray Diffraction*, Addison-Wesley, Reading. 381 pp.

KEY LITERATURE

Axelrod, D., Koppel, D.E., Schlessinger, J., Elson, E., and Webb, W.W. (1976) Mobility measurement by analysis of fluorescence photobleaching recovery kinetics. *Biophys. J.* 16:1055–1069.

Bretscher, M.S. (1972) Asymmetrical lipid bilayer structure for biological membranes. *Nature New Biol.* 236:11–12.

Blasie, J.K., Schoenborn, B.P., and Zaccai, G. (1975) Direct methods for the analysis of lamellar neutron diffraction from oriented multilayers: A difference Patterson deconvolution approach. *Brookhaven Symp. Biol.* 27:III-58–III-67.

Boggs, J.M. (1987) Lipid intermolecular hydrogen bonding: Influence on structural organization and membrane function. *Biochim Biophys Acta* 906:353–404

Büldt, G., Gally, H.U., Seelig, A., Seelig, J., and Zaccai, G. (1978) Neutron diffraction studies on selectively deuterated phospholipid bilayers. *Nature* 271:182–184.

Everitt, C.T., and Haydon, D.A. (1968) Electrical capacitance of a lipid membrane separating two aqueous phases. *J. Theor. Biol.* 18:371–379.

Flewelling, R.F., and Hubbell, W.L. (1986) Hydrophobic ion interactions with membranes. Thermodynamic analysis of tetraphenylphosphonium binding to vesicles. *Biophys. J.* 49:531–540.

Flewelling, R.F., and Hubbell, W.L. (1986) The membrane dipole potential in a total membrane potential model. Applications to hydrophobic ion interactions with membranes. *Biophys. J.* 49:541–552.

Kornberg, R.D., and McConnell, H.M. (1971) Inside-outside transitions of phospholipids in vesicle membranes. *Biochemistry* 10:1111–1120.

Levine, Y.K., and Wilkins, M.H.F. (1971) Structure of oriented lipid bilayers. *Nature New Biol.* 230:69–76.

Lewis, B.A., and Engelman, D.M. (1983) Lipid bilayer thickness varies linearly with acyl chain length in fluid phosphatidylcholine vesicles. *J. Mol. Biol.* 166:211–217.

Parsegian, A. (1969) Energy of an ion crossing a low dielectric membrane: Solutions to four relevant electrostatic problems. *Nature* 221:844–846.

Seelig, J. (1977) Deuterium magnetic resonance: Theory and application to lipid membranes. *Q. Rev. Biophys.* 10:353–418.

Sheetz, M.P., and Singer, S.J. (1974) Biological membranes as bilayer couples. A molecular mechanism of drug-erythrocyte interactions. *Proc. Natl. Acad. Sci. U.S.A.* 71:4457–4461.

Vaz, W.L.C., Clegg, R.M., and Hallmann, D. (1985) Translational diffusion of lipids in liquid crystalline phase phosphatidylcholine multibilayers. A comparison of experiment with theory. *Biochemistry* 24:781–786.

Verkleij, A.J., Zwaal, R.F., Roelofsen, B., Comfurius, P., Kastelijn, D., and van Deenen, L.L. (1973) The asymmetric distribution of phospholipids in the human red cell membrane. A combined study using phospholipases and freeze-etch electron microscopy. *Biochim. Biophys. Acta* 323:178–193.

Wang, L., Bose, P.S., and Sigworth, F.J. (2006) Using cryo-EM to measure the dipole potential of a lipid membrane. *Proc. Natl. Acad. Sci. U.S.A.* 103:18528–18533.

Wiener, M.C., and White, S.H. (1991) Fluid bilayer structure determination by the combined use of X-ray and neutron diffraction. I. Fluid bilayer models and the limits of resolution. *Biophys. J.* 59:162–173.

Wiener, M.C., and White, S.H. (1991) Fluid bilayer structure determination by the combined use of X-ray and neutron diffraction. II. "Composition-space" refinement method. *Biophys. J.* 59:174–185.

Wiener, M.C., and White, S.H. (1992) Structure of a fluid dioleoylphosphatidylcholine bilayer determined by joint refinement of x-ray and neutron diffraction data. III. Complete structure. *Biophys. J.* 61:434–447.

Wilkins, M.H.F., Blaurock, A.E., and Engelman, D.M. (1971) Bilayer structure in membranes. *Nature New Biol.* 230:72–76.

Wimley, W.C., and Thompson, T.E. (1990) Exchange and flip-flop of dimyristoylphosphatidylcholine in liquid-crystalline, gel, and two-component, two-phase large unilamellar vesicles. *Biochemistry* 29:1296–1303.

Wu, H., Su, K., Guan, X., Sublette, M.E., and Stark, R.E. (2010) Assessing the size, stability, and utility of isotropically tumbling bicelle systems for structural biology. *Biochim Biophys Acta* 1798:482–488.

EXERCISES

1. I If a cell is a sphere 20 μm in diameter, how long will it take for a lipid to diffuse in the bilayer from one side to the other, assuming no barriers? (Hint: $x \sim (4Dt)^{1/2}$.)

2. a. Draw the structure of any bilayer-forming phospholipid from memory, showing all atoms and charges.

    b. What are the three principal features that lead it to form a lipid bilayer?

    c. The lipid in (a) is exposed to a phospholipase that cuts off an esterified hydrocarbon chain, and the larger product is purified. When the product is suspended in buffer, it is found to form assemblies. Schematically diagram the kind of assembly you would expect it to form and explain why in a short phrase or sentence.

    d. A molecule of bacteriorhodopsin is found to be soluble in a solution containing significant amounts (>2%) of the molecule in (c) and to retain its native form as a protein with seven transmembrane helices. Applying basic principles, diagram your hypothesis of what has occurred to make the protein soluble. What would you expect to happen to the protein if the concentration of the phospholipid product is reduced to very low values? (One sentence.)

3. To calculate the capacitance of a bilayer, we need a value for the dielectric constant of the hydrocarbon layer.

    a. Assume that the bilayer interior is like hydrocarbon, and therefore, take a dielectric of n-decane (~2.0) and an area of 150 square microns (about the area of an erythrocyte) and calculate the capacitance.

    b. If there is a membrane potential of 100 mV, how much energy is stored across the membrane? If the energy required to cleave one phosphodiester in a molecule of ATP is ~30 kJ mol^{-1}, how many ATP equivalents does this represent?

    c. It is found that the octanol:water partition coefficient represents the movement of peptides into a bilayer very well. If the octanol dielectric (~10) is used, how does the energy of the above example change?

4. A lipid vesicle is made from palmitoyloleoylphosphatidylcholine and is 1000 Å in diameter measured from bilayer center to bilayer center.

    a. How many lipid molecules are in the vesicle?

    b. If the diameter between bilayer centers is only 300 Å, how many lipids are in the inner and outer monolayers, respectively?

    c. What is the diameter of a synaptic vesicle measured from bilayer centers?

    d. Would you expect problems in fusing two vesicles as in (b)? Which problems and why?

5. Reading the difference in barrier height from Figure 2.26:

    a. How much difference would there be in the average crossing time for a positive vs. a negative ion if each is monovalent and has a 1 Å radius? (Assume that the Arrhenius equation applies.)

    b. How much larger in radius would the positive ion have to be for the rates to be the same?

    c. Phosphorylation is used by cells to trap molecules in their cytoplasm. Estimate a radius for a phosphate group and calculate its rate of permeation across a lipid bilayer to gain insight on, for example, the role of hexokinase in retaining glucose.

6. Lipid bilayers are increasingly disordered toward the midplane, while micelles are increasingly ordered toward their centers.

   a. Why would a micelle be most ordered at its center? (Hint: think of what sets the maximum diameter of a micelle.)

   b. How might chain dynamics tend to keep the monolayers of a bilayer from interdigitating their chains?

   c. Propose a role for the hydrophobic effect. (Hint: consider lipid area at the bilayer surface.)

Interactions of Peptides with Lipid Bilayers

3

We have learned that membrane proteins have two principal structural motifs for their transmembrane (TM) domains: β-barrels and α-helical bundles (Chapter 1). The β-barrel proteins are less frequent than the helical proteins and are located exclusively in the outer membranes of bacteria, mitochondria, and chloroplasts. Bacterial outer membrane proteins sometimes have an α-helix buried within the barrel, but otherwise, TM domains are not mixtures of barrels and bundles. We have also learned that biology must operate within the limits permitted by thermodynamics and physical chemistry. We will learn in the course of this book that these limits often permit interesting structural variations in order to carry out biological function. For now, we consider the physical principles underlying the two principal structural motifs. What are those principles? This chapter and Chapter 4 are devoted to exploring this question. The fundamental concepts of membrane protein stability are revealed by first considering what is known about the interactions of simple peptides with lipid bilayers, which is the focus of this chapter.

The questions we need to answer are revealed by considering bacteriorhodopsin as it sits fully folded and functional in a lipid bilayer (**Figure 3.1A**). What role does the bilayer interface play in lipid–protein interactions? What are the interactions of α-helices with the membrane bilayer? The bilayer is stable in a free energy minimum, as we learned in Chapter 2. Does the presence of protein perturb the stability and properties of the lipid bilayer? These questions, which are difficult to answer using multi-span membrane proteins, are more easily studied by examining the interactions of membrane-active water-soluble peptides. To guide our study, we can use the thermodynamic scheme shown in **Figure 3.1B**.

The biophysical events cataloged in Figure 3.1B have emerged mostly from studies of the membrane interactions of small water-soluble proteins and peptides. These include antimicrobial peptides, membrane-active toxins, proteins involved in cell signaling, and importantly, synthetic peptides designed to address specific biophysical questions. Broadly, lipid–protein interactions are controlled by the hydrophobic effect and electrostatics (including van der Waals and related interactions as well as Coulombic interactions). We therefore focus at the beginning of this chapter on those subjects in the context of lipid bilayers. Although we tend to think of hydrophobic interactions in terms of partitioning of molecules from water into bulk organic phases, e.g., Box 1.2, a different mindset is required for membranes; we show that bilayers are not necessarily equivalent to bulk phases. The electrical properties of charged interfaces of all sorts have been studied extensively by physical chemists for more than two centuries. Those principles are useful for understanding electrostatic interactions of peptides with membranes, but of course, fluid lipid bilayers present special challenges and new insights.

DOI: 10.1201/9780429341328-4

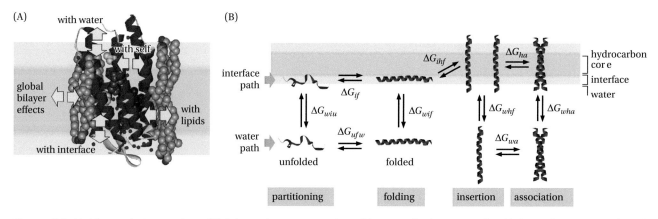

Figure 3.1 Lipid–protein interactions. (A) Schematic representation of bacteriorhodopsin in a lipid bilayer illustrating in broad terms the interactions underlying the stability of membrane proteins in lipid bilayers. (B) Summary of the possible folding pathways of membrane-active peptides. The various pathways, not all of which are experimentally accessible, provide an organizing framework for studying lipid-protein interactions. (A, From White SH et al. [2001] *J Biol Chem* 276: 32395. B, From White SH, Wimley WC [1999] *Annu Rev Biophys Biomol Struct* 28: 319.)

3.1 GIBBS PARTITIONING ENERGIES PROVIDE A FOUNDATION FOR DESCRIBING LIPID– PROTEIN INTERACTIONS

3.1.1 Partitioning Free Energies of Amino Acids Provide Hydrophobicity Scales

The foundation for the biophysical principles of lipid–protein interactions derives, for the most part, from the idea of partitioning of molecules between two immiscible phases, as summarized in Box 1.2. But there are many variations on that theme. The first three-dimensional (3-D) structures of soluble proteins revealed that non-polar amino acids tend to be buried in the protein interior with little aqueous contact, while polar residues tend to be on the surface with full aqueous exposure. This led the pioneers of the field of protein folding and stability to think of a protein's amino acid residues as partitioning between polar surface and non-polar interior. To quantify the idea, Joël Janin computed the mole fraction of each of the 20 amino acids located in the interior (X_i) and on the surface (X_s) of 22 soluble proteins. "Surface" or "interior" was defined according to the water-accessible surface area of each amino acid determined using the method developed by Richards (**Box 3.1**); interior residues were defined as those having less than 20 Å² exposed at the surface. Thinking of protein structure in terms of partitioning, Janin computed partition coefficients $f \equiv X_i/X_s$, which he converted to apparent transfer free energies $\Delta G_t = RT \ln f$, representing transfers from the less polar interior to the polar surface (**Figure 3.2**). He rank-ordered the exposure free energies with the most unfavorable ones (e.g., Leu) at the top and the most favorable at the bottom (e.g., Lys). Such a rank-ordered list is called a hydrophobicity scale. Because in this case, the list is based upon a statistical analysis, it is a statistical hydrophobicity scale. There are now more than 50 statistical scales derived using many different approaches to define buried and exposed residues and many different levels of statistical sophistication.

Considering the protein interior as a bulk phase of low polarity, another approach to hydrophobicity scales has been to measure the partitioning of analogs of amino acid side chains between low-polarity and high-polarity phases. This approach yields physical hydrophobicity scales. Some commonly used and historically important scales are summarized in **Figure 3.3**. The first of the physical scales was determined in the laboratory of Charles Tanford using mixtures of

BOX 3.1   MEASURING ACCESSIBLE SURFACE AREAS OF MOLECULES

Methods to describe and measure surface features and areas of proteins are fundamental tools because the functions of most proteins involve their surface features and because the hydrophobic effect is related to surfaces exposed to water (Box 1.2). The most important of these methods, devised by Byunkook Lee and Frederic M. Richards (1925–2009), is widely used to evaluate surface areas of molecules of known structure.

In the Lee and Richards method, each atom of a molecule is represented as a hard sphere with a choice for the van der Waals radius of the atom, although hydrogen atoms are sometimes combined with the atom to which they are bonded (with a corresponding increase of radius). The surface for a given conformation of the molecule is then probed computationally by rolling a test sphere over the van der Waals surface and recording the path of the sphere as it contacts the molecule. The test sphere is generally given the equivalent radius of a water molecule, which is about 1.4 Å. But different choices can be made. Useful probes should have dimensions comparable to the features of the surface being examined.

What is measured? A cross-section of a part of a hypothetical molecular surface is shown in **Figure 1**. A spherical probe of radius *R* is rolled over the molecule while maintaining contact with the van der Waals surface. Note that the probe does not touch atoms 3, 9, or 11, which are interior atoms.

How is the surface defined and quantitated? There are several choices. One might use the continuous surface defined by the movement of the probe center, which Richards defined as the "accessible surface." An alternative is to consider the "contact surface," defined as those parts of the molecular surface that can actually be in contact with the surface of the probe, giving a series of disconnected

patches, interspersed with the "reentrant surface," a series of patches defined by the interior-facing part of the probe when it is simultaneously in contact with more than one atom. Considered together, the contact and reentrant surfaces represent a continuous surface, which Richards defined as the "molecular surface."

Use of the accessible surface leaves out some information, because the ratio of contact to reentrant surface may be a useful measure of molecular surface roughness. This can be seen qualitatively by inspecting Figure 1. The molecular surface also has the advantage that the area approaches a finite limiting value as the size of the probe increases. In spite of these possible advantages, only the accessible surface is generally calculated.

Choice of probe. With any of the surface definitions, the value for surface measurement will depend on the radius chosen for the probe. In going from *R1* to *R2* in Figure 1, the number of non-contact or interior atoms increases from three to eight, and the accessible surface becomes much smoother. One important use of the surface measurement is to quantify the hydrophobic effect when molecular surfaces are removed from or into contact with water, so the most common probe radius is based on a spherical approximation for a water molecule, taken from the volume (30 Å^3) it occupies in liquid water—about 1.4 to 1.5 Å.

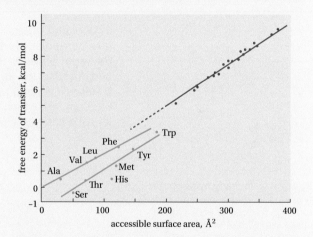

Figure 2 The free energy of transfer from 100% organic solvent to water is plotted as a function of accessible molecular surface area. Data for a series of hydrocarbons are shown as the red dots and line in the upper right, and extrapolate to the origin with a slope of 25 cal Å^{-2}. The data shown in the lower left refer to the side chains of the indicated amino acids: accessible surface areas versus contributions of the side chains to amino acid solubility. The line passing through Ala, Val, Leu, and Phe has a slope of 22 cal Å^{-2}, and the difference for the other series is ascribed to the polar interactions of these side chains with water molecules. (Figure adapted with permission from Richards FM [1977] *Annu Rev Biophys Bioeng* 6: 151–176. Copyright 1977 Annual Reviews.)

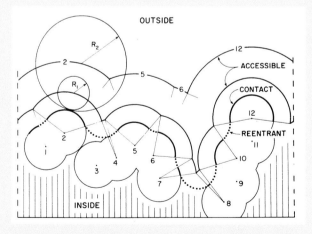

Figure 1 Measuring accessible surface area. (Figure reproduced with permission from Richards FM [1977] *Annu Rev Biophys Bioeng* 6: 151–176. Copyright 1977 Annual Reviews.)

(Continued)

BOX 3.1   MEASURING ACCESSIBLE SURFACE AREAS OF MOLECULES (CONTINUED)

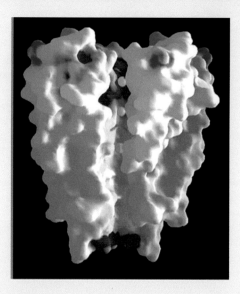

Figure 3 Structure of the KcsA potassium channel surface with a subunit removed to reveal the interior. K⁺ ions are shown in green, electrostatic features in red (negative) and blue (positive), and water-exposed channel regions in yellow. (Doyle DA et al. [1998] *Science* 280: 69–77. With permission from AAAS.)

The hydrophobic effect (again!). By measuring the accessible surface areas of a series of hydrocarbons and graphing the areas versus their solubility in water, a linear relationship is found (**Figure 2** and Box 1.2). The small hydrocarbons (solid dots) fall on a straight line with a slope of 25 cal mol⁻¹ Å⁻², while the non-polar amino acid side chains Ala, Val, Leu, and Phe yield a slope of 22 cal Å⁻². The best modern value for amino acid side chains determined from the partitioning of small peptides is 23 cal Å⁻². Note that the energies are in calories, not kilocalories, but the areas are also in very small units.

Molecular surfaces. In the modern context, the Richards method is used to produce descriptions of molecular surfaces, often with the addition of electrostatic features. As an example, the structure of the potassium channel is shown in Box **Figure 3**.

alcohol and water. Another approach is to think of partitioning of amino acids into non-polar phases as a process of dehydration and consequently, to use the vapor phase as the non-polar phase. Cyclohexane and *n*-octanol have also been used as the non-polar phase. The use of *n*-octanol is important because it solves a general problem for physical scales, which is to find a low-polarity medium into which all amino acids and their analogs partition readily regardless of polarity or charge.

Figure 3.2 An early hydrophobicity scale for soluble proteins based upon the fraction of each type of amino acid buried in the protein interior or accessible from the aqueous environment. (From Janin J [1979] *Nature* 277: 491–492. With permission from Springer Nature.)

amino acid composition of the inside and surface

residue	molar fraction		free energy (kcal mol⁻¹)	
	buried	accessible	f	ΔG_t
Leu	11.7	4.8	2.4	0.5
Val	12.9	4.5	2.9	0.6
Ile	8.6	2.8	3.1	0.7
Phe	5.1	2.4	2.2	0.5
Cys	4.1	0.9	4.6	0.9
Met	1.9	1.0	1.9	0.4
Ala	11.2	6.6	1.7	0.3
Gly	11.8	6.7	1.8	0.3
Trp	2.2	1.4	1.6	0.3
Ser	8.0	9.4	0.8	−0.1
Thr	4.9	7.0	0.7	−0.2
His	2.0	2.5	0.8	−0.1
Tyr	2.6	5.1	0.5	−0.4
Pro	2.7	4.8	0.6	−0.3
Asn	2.9	6.7	0.4	−0.5
Asp	2.9	7.7	0.4	−0.6
Gln	1.6	5.2	0.3	−0.7
Glu	1.8	5.7	0.3	−0.7
Arg	0.5	4.5	0.1	−1.4
Lys	0.5	10.3	0.05	−1.8

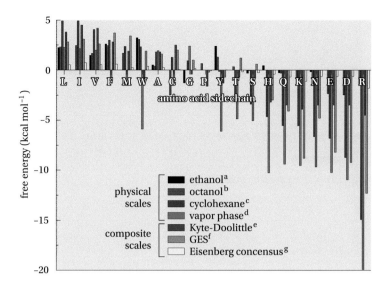

Figure 3.3 Comparison of popular hydrophobicity scales. The free energies shown are for transfer of the amino acid side chains from bulk non-polar phases to water. The ethanol, octanol, and cyclohexane, and the vapor-phase numbers, are based upon experimental measurements (see Box 1.2); these are physical hydrophobicity scales. The Kyte–Doolittle, GES, and Eisenberg-Consensus scales are composite scales derived from physical scales, statistics, and shrewd guesses. The scales are highly correlated. ([a]Nozaki & Tanford (1971) *J Biol Chem* **246**:221. [b]Fauchère & Pliška (1983) *Eur J Med Chem – Chim Ther* **18**:369. [c,d]Radzicka & Woolfenden (1988) *Biochemistry* **27**:1664. [e]Kyte & Doolittle (1982) *J Mol Biol* **157**:105. [f]Engelman, Steitz, & Goldman (1986) *Ann Rev Biophys Chem* **15**:321. [g]Eisenberg et al. (1982) *Faraday Symp Chem Soc* **17**:109.)

The water/octanol system is widely used by drug designers to measure the relative hydrophobicity of candidate drugs. X-ray scattering experiments show that the water-saturated *n*-octanol phase is composed of small, densely packed inverted micelles in which the hydroxyl groups surround pockets of water. The result is that a wide range of solutes can be accommodated within the micelles: polar compounds within the water region, non-polar regions in the alkyl chain region, and amphiphilic compounds at the hydroxyl/water interface. An early comprehensive *n*-octanol/water partitioning study of amino acid hydrophobicity used single amino acids with blocked N- and C- termini (N-acetyl-amino-acid amides). With a complete set of transfer free energies, relative side chain hydrophobicities were determined by subtracting the transfer free energy of the glycine peptide from the others to produce a relative side chain scale (red bars, Figure 3.3). This approach is very common in hydrophobicity scales, emphasizing that most physical hydrophobicity scales are side chain–only scales because the partitioning free energy of the peptide bond is not included. As we will see, accounting for the peptide-bond partitioning free energy and the reference non-polar environment is extremely important.

A comparison of various statistical and physical scales (Figure 3.3) shows, not surprisingly, that they are highly correlated with one another. As interest grew in using hydrophobicity scales to detect amino acids most likely to be buried in protein interiors or to form TM helices, a competition developed to create scales tailored to the identification of these regions of amino acid sequences. Such tailored scales are called composite hydrophobicity scales; they use more or less made-up hydrophobicity values derived from physical scales, statistics, and shrewd guesswork about the disposition of side chains in proteins and membranes. Perhaps the most widely used such scales are those created by Jack Kyte and Russel Doolittle (the Kyte–Doolittle or K-D scale) and by Donald Engelman, Adrian Goldman, and Thomas Steitz (the GES scale, which is a play on words suggesting, appropriately, "guess"). Through their 1982 paper, Kyte and Doolittle popularized so-called sliding-window hydropathy plots (see also Chapter 8), which was an idea used earlier by George Rose for detecting reverse turns in soluble proteins based upon hydrophobicity values. An example of a hydropathy plot published by Kyte and Doolittle for the soluble protein chymotrypsinogen is shown in **Figure 3.4** along with an introduction to the sliding-window hydropathy plot method.

Notice the signs of the free energies in the Kyte–Doolittle and other scales. All these scales assume the transfer of amino acid side chains *from* the low-polarity to the high-polarity solvent, which means that the free energy values for the transfer of non-polar side chains have positive signs, and those for polar side chains have negative signs. We adhere to this tradition here (see **Box 3.2**).

(A)

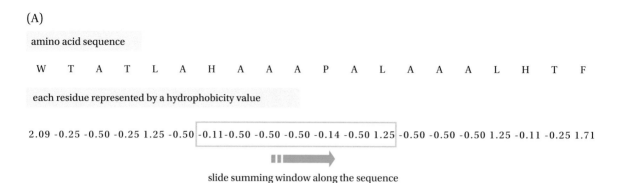

(B)

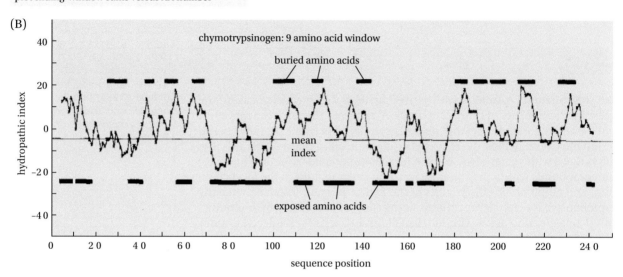

Figure 3.4 Hydropathy plot analysis of protein sequences. The general method, which is widely used, is summarized in (A). (B) shows a hydropathy plot of the soluble protein chymotrypsinogen made using the Kyte–Doolittle (K-D) hydropathy scale. The plot reveals the approximate sequence positions of buried and exposed amino acids. The K-D scale is a frequently used composite scale. It is useful for identifying "greasy" amino acid segments, but like all such scales, it is only loosely connected to direct thermodynamic measurements. (A, From White SH [1982] *J Mol Biol* 157: 105. B, From Kyte J, Doolittle RF [1982] *J Mol Biol* 157: 105–132. With permission from Elsevier.)

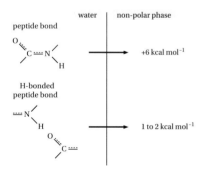

Figure 3.5 The cost of partitioning free peptide bonds from water to a non-polar phase is very high (upper panel) because the H-bonds to water are lost. If the peptide bonds participate in hydrogen bonds in both phases, then the partitioning cost is greatly reduced (lower panel). The free energies are difficult to measure. The values shown summarize an extensive literature on the subject.

3.1.2 The Cost of Partitioning Peptide Bonds into Bulk Non-Polar Phases Is Very High

Missing from all the hydrophobicity scales discussed above is the cost of partitioning the peptide bond (-C=O, -NH). How big is this energetic cost, and does it matter (**Figure 3.5**)? There have been numerous efforts over the years to measure or estimate the cost. Estimates from the partitioning of model compounds into carbon tetrachloride from water yield a value of about +6 kcal mol^{-1}. Another experimental estimate based upon vapor-phase partitioning yielded +10 kcal mol^{-1}. Some theoretical estimates yield +8 kcal mol^{-1}. The most hydrophobic side chain based on partitioning into cyclohexane has a value of −4.9 kcal mol^{-1}, emphasizing that the unfavorable cost of partitioning the peptide bond cannot be overcome by the favorable cost of partitioning the most hydrophobic amino side chain; the cost of partitioning the peptide bond is dominant. But suppose that the peptide bond is hydrogen bonded (H-bonded) to another peptide bond, as in α-helices. Theoretical estimates vary widely, but it appears that H-bonding reduces the peptide-bond partitioning cost considerably, perhaps by as much as 5 kcal mol^{-1}. This hints at the extreme importance of secondary structure H-bonding in the energetics of partitioning. These conclusions are summarized in Figure 3.5.

BOX 3.2   GIBBS FREE ENERGY AND HYDROPATHY PLOT SIGN CONVENTIONS

Traditionally, i.e., throughout the published literature for more than 25 years, the regions in hydropathy plots (Figure 3.3) identifying either TM segments of membrane proteins or the interior of soluble proteins have been identified as positive peaks, as shown in **Figure 1** for bacteriorhodopsin, which has seven TM helices (Figure 1.23). Plotted in this way, the signs of the underlying free energies represent the transfer of an amino acid *from bilayer* or soluble protein interior *to water*. This means that the signs of the free energies favorable for non-polar regions are *positive*. This convention began with Joel Janin's studies of the distribution of amino acids between the interiors and surfaces of soluble proteins (Figure 3.3). Subsequent hydrophobicity scales have generally followed this convention (Figures 3.4 and 3.5).

The zero levels of hydropathy plots constructed using the hydrophobic-index scales of Figure 3-3 are arbitrary because, unlike the Wimley–White (WW) hydrophobicity scales (Figure 3-6) used in Figure 1, the partitioning cost of the peptide bond is not included. An important goal, and one that is essential for thermodynamic accuracy, is to place free energy values on an absolute scale so that the zero level means $\Delta G = 0$, as in Figure 1.

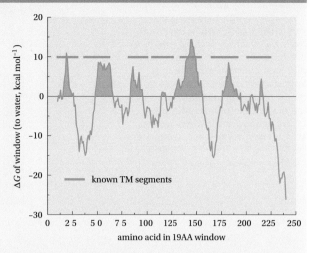

Figure 1 Hydropathy plot of bacteriorhodopsin, which has seven transmembrane helices, prepared using Membrane Protein Explorer (MPEx), available at blanco.biomol.uci.edu/mpex. Notice that identified peaks have positive values, indicating an unfavorable free energy of transfer *from* membrane *to* water. The hydrophobicity scale used is the Wimley–White whole-residue scale, which is an experimentally determined scale that includes the energetic contribution of the peptide bond. It is an absolute scale, meaning that $\Delta G = 0$ corresponds to a true thermodynamic zero rather than a relative value characteristic of side chain-only scales.

3.1.3 Partitioning of Pentapeptides Provides Whole-Residue Amino Acid Hydrophobicities

Physical hydrophobicity scales determined using side chain–only amino acid analogs not only ignore the cost of partitioning the peptide bond; they also ignore the obvious fact that amino acids are part of longer sequences and so may have contact with each other in the polar or non-polar phases or both, thus influencing the combined transfer free energy. Biological studies require information about amino acid hydrophobicities in the context of realistic sequence environments. This information can be obtained from so-called host–guest experiments in which position X of an otherwise fixed sequence is systematically varied in a way that permits extraction of the free energy of X from a series of measurements. The length of the sequence must be carefully chosen, however. Very long sequences are difficult to synthesize, and their secondary structure may depend strongly on X. Very short sequences may not provide sufficient sequence context. With an eye to sequences similar to those found in the TM segments of membrane protein, William Wimley and Stephen White settled on pentapeptides of the form Ac-WL-X-LL and chose to partition them between octanol and water. They showed that none of the sequences had regular secondary structure. From partitioning the free energies of 20 such peptides with each of the common amino acids in the -X- position, side chain hydrophobicities were determined relative to X = Gly. But, what about the peptide bond? To obtain values for it, they measured the partitioning of sequences of the form Ac-WL$_n$, where *n* was varied from 1 to 6. This permitted the incremental change for each added Leu to be determined. Because the incremental free energy change included the peptide bond, its cost could be extracted using the Leu side chain hydrophobicity. The free energy of partitioning whole residues (i.e., peptide bond plus side chains) was thus obtained by adding the peptide-bond

Figure 3.6 The experiment-based Wimley–White (WW) whole-residue hydrophobicity scale that describes the partitioning of amino acid side chains and peptide backbone from water into *n*-octanol. See Box 3.2 for discussion of the ΔG and plot sign conventions. The most important feature of the scale is that it reveals the peptide bond as the major determinant of transfer free energies for whole amino acids. (From Wimley WC, White SH [1996] *Biochemistry* 35: 5109–5124 <author's own figure>.)

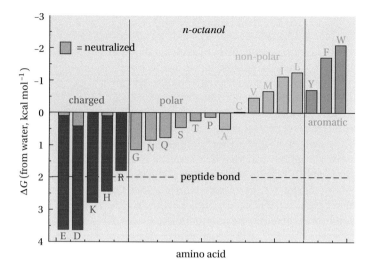

free energies to the side chain free energies. The free energy cost of transferring the peptide bond is shown in **Figure 3.6** by the dashed line. The unfavorable peptide-bond cost dominates the whole-residue transfer free energies; the side chain contributions merely modulate the fundamental backbone contribution.

We noted earlier that hydrated octanol is an interfacial phase rather than a pure bulk phase like cyclohexane because of its micelle-like organization. This could mean that the hydrophobic free energy of partitioning non-polar side chains, judged by the solvation parameter (Box 1.2), would be different compared with a pure hydrocarbon phase. But that is not case. The solvation parameter for partitioning non-polar surfaces of the pentapeptides into octanol is the same as for hydrocarbon phases: 23 cal mol^{-1} Å^{-2}. This shows the ability of octanol to solvate the non-polar side chains selectively via the octane-like interior of the *n*-octanol micelles without interference from the octanol hydroxyl groups.

3.2 LIPID BILAYERS HAVE DISTINCT PROPERTIES COMPARED WITH BULK ORGANIC PHASES

3.2.1 Lipid Bilayers Are Not Equivalent to Bulk Non-Polar Phases

An important question is whether any of the bulk non-polar phases accurately quantify the partitioning of amino acids or other molecules into lipid bilayers (Figure 3.3). Suppose we transfer a non-polar solute molecule (e.g., hexane) from water to either a bulk non-polar solvent (e.g., hexane) or a lipid bilayer. How can we decide if the lipid bilayer interior is equivalent to a bulk non-polar solvent? One might think that if the transfer free energies from water are equal for the bulk and bilayer phases, then the two phases must be equivalent. But this is not necessarily correct, because the Gibbs free energy (ΔG) is composed (Box 1.3) of both an enthalpic (ΔH) component related to interaction energies and an entropic component (ΔS):

$$\Delta G = \Delta H - T\Delta S \qquad (3.1)$$

Obviously, many different combinations of enthalpy and entropy can give the same free energy. To compare the solvent properties of the two phases, we therefore need to know something about the contributions of ΔH and ΔS. Because of the work of many distinguished physical chemists, we know exactly

BOX 3.3 SOLVATION OF HYDROPHOBIC MOLECULES: REVISITING THE HYDROPHOBIC EFFECT

We showed in Box 1.2 that the partitioning free energy of a hydrocarbon solute between bulk water and hydrocarbon phases is proportional to the solute's accessible surface area because of the formation of a shell of waters around the solute. These hydration-shell waters, which are responsible for the hydrophobic effect, have a heat capacity due to stronger water H-bonds in the hydration shell that are expressed as an increased enthalpy. Consequently, the transfer of the solute from water to the hydrocarbon phase is accompanied by a decrease in heat capacity, $-\Delta C_p$, which is proportional to the accessible surface area (**Figure 1A**). How does this heat capacity arise?

The shell of waters cannot H-bond to the hydrocarbon molecule, so they compensate by reorganizing around the molecule to maximize the strength of their H-bonds. The reorganization lowers the entropy; the increased H-bond strength increases enthalpy and consequently, the heat capacity. The strength of H-bonds between waters depends upon the bond distance d and the bond angle Θ (panel B); the smaller d and Θ, the greater the bond strength. Water in bulk tends to form tetragonal clusters (panel C) with linear H-bonds ($\Theta \approx 0$). However, this ideal is not achieved because thermal motion disrupts H-bonds and the perfect tetragonal angle of 104° is frustrated by the 109° bond angle between the two oxygen–hydrogen covalent bonds. Kim Sharp at the University of Pennsylvania has shown by means of simulations that pure water has two broad populations of bond angles (panel D): a large one centered at about 12° (close to linear) with strong H-bonds and a smaller one centered at about 52° (one-half the tetragonal

angle) with weaker H-bonds. The presence of non-polar solutes such as methane or methyl groups shifts waters from the 52° population to the 12° population, which have stronger H-bonds.

What are the consequences of the heat capacity of the hydration shell? Broadly, the heat capacity causes the free energy of partitioning to be temperature dependent. Experiments reveal that there is a temperature T_h at which $\Delta H = 0$ and a temperature T_s at which the entropy change $\Delta S = 0$. Using the equations presented in Chapter 0 and the observation that ΔC_p is independent of temperature, one can show that

$$\Delta H(T) = \Delta H(T_1) + \int_{T_1}^{T} \Delta C_p dT = \Delta H(T_1) + \Delta C_P(T - T_1) \quad (1)$$

$$\Delta S(T) = \Delta H(T_2) + \int_{T_2}^{T} \frac{\Delta C_p dT}{T} dT = \Delta S(T_2) + \Delta C_p \ln\frac{T}{T_2} \quad (2)$$

T_1 and T_2 are arbitrary temperatures; we may therefore choose $T_1 = T_h$ and $T_2 = T_s$. Because $\Delta H(T_h) = 0$ and $\Delta S(T_s) = 0$, the definition of Gibbs energy yields

$$\Delta G(T) = \Delta C_p \left[(T - T_h) - T \ln\left(\frac{T}{T_s}\right) \right] \quad (3)$$

Clearly, the Gibbs energy is independent of temperature if the heat capacity change of partitioning is 0. One can readily calculate $\Delta G(T)$ given T_h (≈ 25 °C), T_s (≈ 112 °C). The temperature dependence of the Gibbs energy for partitioning n-hexane from water into bulk n-hexane ($\Delta C_p = -105$ cal mol^{-1} K^{-1}) is shown in Figure 3.7.

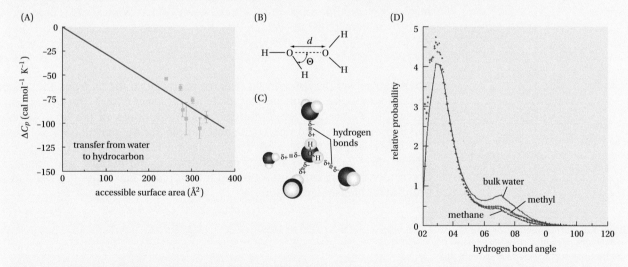

Figure 1 Understanding the hydrophobic effect at the atomic level. (A, Spolar RS et al. [1989] *PNAS* 86: 8382–8385. With permission. B and D, From Sharp KA, Madan B [1997] *J Phys Chem B* 101: 4343–4348. With permission from American Chemical Society. C, From Wikimedia Commons.)

what to expect for the partitioning of non-polar molecules from water to a bulk non-polar phase. We noted earlier (Box 1.2) that one indicator of simple hydrophobic partitioning for a series of non-polar solutes is the linear dependence of ΔG on the solutes' accessible surface areas (Box 3.1). But more important is the linear dependence of the heat capacity change (ΔC_p) on accessible surface

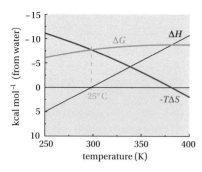

Figure 3.7 Temperature dependence of the free energy (ΔG), enthalpy (ΔH), and entropy (ΔS) of the partitioning of hexane from water to bulk hexane. The temperature dependence is due to the heat capacity of the of the hydration shell associated with hexane in water (Box 3.3).

associated with partitioning (**Box 3.3**), which shows that $\Delta C_p \neq 0$ causes ΔG, ΔH, and ΔS to be temperature dependent. This is shown in **Figure 3.7** for the partitioning of hexane from water to bulk hexane.

As for most other non-polar solutes, Figure 3.7 reveals that near room temperature (25 °C), ΔH is close to zero, causing the favorable free energy to be entirely due to an increase in entropy. This explains why hydrophobic partitioning is widely referred to as being "entropy driven." But, the figure makes clear that hydrophobic partitioning is entropy driven only near room temperature. Notice in the figure that ΔG is roughly constant because $\Delta H(T)$ and $-T\Delta S(T)$ have opposite slopes. Such behavior, referred to as enthalpy–entropy compensation, is often observed in biochemical reactions.

Is partitioning into bilayers equivalent to partitioning into a bulk phase? The answer is no, except for some fatty acid–like molecules, as we discuss later. For transfer of hexane to bulk hexane from water at room temperature, $\Delta G = -7.7$ kcal mol^{-1} with $\Delta H = 0$ (Figure 3.7), while for transfer into DOPC bilayers, $\Delta G = -6.2$ kcal mol^{-1} with $\Delta H = -1.7$ kcal mol^{-1}. We can thus conclude that a bulk hydrocarbon liquid is not an accurate model for lipid bilayers. One reason is that the bilayer is an interfacial phase with the lipids anchored at an interface, which causes the acyl chains to be aligned relative to the membrane normal (Figure 2.21). One might therefore expect that the thermodynamics of partitioning would depend upon the interfacial area per lipid. That is exactly what is observed; the partition coefficient of hexane depends strongly on the average area in the bilayer interface occupied by the phospholipids (**Figure 3.8**). The area per lipid is represented in the figure by the surface density parameter $\sigma = A/A_0$, which can be determined from nuclear magnetic resonance (NMR) measurements of acyl chain order parameters (Spread 2.1). A_0 is the area/lipid in crystals determined by x-ray diffraction (~40 Å^2 per lipid).

Despite the differences between bulk-phase and bilayer partitioning, the primary driving force for partitioning is nevertheless the hydrophobic effect, which can be thought of as excluding hydrocarbons from water. The difference arises from the fact that changes in the bilayer contribute to the free energy of partitioning, because the system free energy minimizes itself in every possible way in the partitioning process, including re-minimizing the free energy of the bilayer. We can thus represent the free energy of partitioning ΔG_{par} of a solute into an uncharged bilayer as

$$\Delta G_{par} = \Delta G_{H\Phi} + \Delta G_{bil} \qquad (3.2)$$

where the subscripts HΦ and *bil* refer, respectively, to the hydrophobic effect and changes in the organization of the bilayer (bilayer effect). The idea is summarized in **Figure 3.9**, where the relative contributions of the hydrophobic and bilayer effect are compared, as an example, for the partitioning of indole (the side chain of tryptophan). For indole and many other compounds, the bilayer effect makes a dominant contribution to partitioning free energies. Consequently, care must be exercised if bulk-phase partitioning is used as a stand-in for bilayer partitioning.

Figure 3.8 Partitioning of hexane into bilayers depends strongly on the phospholipid surface density, which is defined as A/A_0, where A is the area per lipid in the bilayer and A_0 is area/lipid in crystals determined by x-ray diffraction. The data show that A/A_0 rather than acyl chain length is the primary determinant of partitioning. Blue, DLPC; red, DMPC; yellow, DPPC. (From De Young LR, Dill KA [1990] *J Phys Chem* 94: 801–809. With permission from the American Chemical Society.)

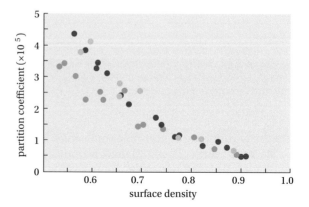

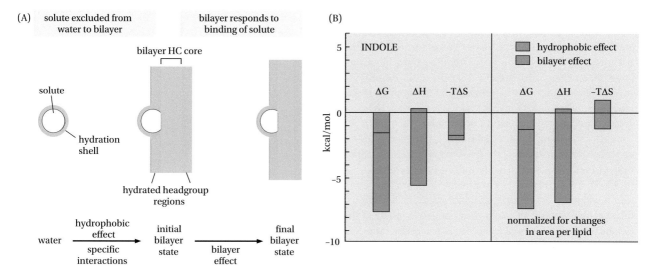

Figure 3.9 Changes in bilayer organization make a significant contribution to the free energy of partitioning of small molecules, as in the example of indole partitioning. (A) The concept of hydrophobic and bilayer effects contributing to partitioning free energy. The bilayer on the right is shown as being thinner as way of indicating a change in bilayer free energy upon partitioning of a solute whose partitioning is initially driven by the hydrophobic effect. (B) In the case of indole, and many other solutes, the favorable response of the bilayer can be more important than the hydrophobic effect. The data on the right-hand side have been corrected for changes in lipid area per lipid (see Figure 3.8). (From Wimley WC, White SH [1996] *Biochemistry* 35: 5109–5124.)

3.2.2 Partitioning of Pentapeptides Provides Whole-Residue Amino Acid Hydrophobicities for Neutral Bilayer Interfaces

Bilayer effects on free energies can be accounted for by measuring the partitioning free energies of the WW host–guest pentapeptides into POPC bilayers (**Box 3.4**). Neutron diffraction and fluorescence measurements show that these pentapeptides partition into the membrane interface. This makes sense because the cost of moving the peptide backbone into the non-polar bilayer interior is very high (Figure 3.6). The WW whole-residue interfacial hydrophobicity scale is shown in **Figure 3.10A**. The scale can only be used for peptides that lack regular secondary structure (i.e., α-helix or β-sheets) in the bilayer, because the pentapeptides have no regular secondary structure in solution or when partitioned into the bilayer. Nor can the scale be used for describing partitioning of charged peptides into bilayers containing charged lipids, because POPC is electrically neutral.

Just as for the WW octanol scale (Figure 3.6), partitioning free energies are dominated by the peptide bond; even though the interface is well hydrated, the energetic cost for transferring peptide bonds into the bilayer interface is still significant. An interesting result from comparing the WW octanol and interface scales (**Figure 3.10B**) is that amino acid partitioning from water into the bilayer is roughly half as favorable as into octanol, although not uniformly; partitioning free energies for the aromatic residues are about the same for bilayers and octanol. For the non-polar side chains, the solvation parameter (Box 1.2) is one-half the value for octanol partitioning (12 cal mol^{-1}Å^{-2}). And notice that the peptide bond falls precisely on the slope = 0.5 line of Figure 3.10B. The explanation for this effect is not clear, but the smaller solvation parameter likely reflects a general property of the complex interface (Figure 2.19), such as altered hydrogen bonding of water arising from the limited concentration of water in the headgroup region.

3.2.3 Aromatic Amino Acids Have a Special Affinity for the Membrane Interface

The aromatic amino acids, Phe, Tyr, and Trp, have the three highest affinities for the POPC interface (Figure 3.10A). This is significant because the first

BOX 3.4   MEASURING PEPTIDE PARTITIONING INTO LARGE UNILAMELLAR VESICLES

Several methods can be used to determine the mole-fraction partition coefficients K_x for the partitioning of peptides into large unilamellar vesicles (LUV). Two direct, commonly used methods are shown in panels A and B. Although some labs use small unilamellar vesicles (SUV), these are to be avoided, because they are non-equilibrium structures. Complete thermodynamic equilibrium is paramount for accurate determinations of the free energy of transfer from buffer to vesicle bilayers. Furthermore, peptide concentration should be kept as low as possible to approximate infinite dilution, which is the standard state. The infinite dilution ideal cannot, of course, be obtained in practice. The practical solution is to make measurements over a range of concentrations to be sure that K_x is independent of peptide concentration in both the aqueous and bilayer phases. The chemical potentials of the peptide in buffer and bilayer are equal at equilibrium. As described in Box 1.2, the standard mole fraction (x) free energy of transfer from buffer (buf) to bilayer (bil) will be given by (**Figure 1**)

$$\Delta G_X^0 = \mu_0^{bil} - \mu_0^{buf} = -RT \ln K_X \qquad (1)$$

How do we extract K_x from the measurements summarized in the figure? The mole-fraction partition coefficient is defined as the ratio of the mole fraction of peptide in bilayer (χ_{bil}) to the mole fraction of peptide in the buffer (χ_{buf}):

$$K_X = \frac{\chi_{bil}}{\chi_{buf}} = \frac{[P]_{bil}/([L]+[P]_{bil})}{[P]_{buf}/([W]+[P]_{buf})} \qquad (2)$$

where $[P]_{bil}$ and $[P]_{buf}$ are the bulk molar concentrations of peptide attributable to the bilayer and buffer phases, respectively, and $[L]$ and $[W]$ are the molar concentrations of lipid and water, respectively. It is always true in partitioning experiments that $[W] = 55.3\ M \gg [P]_{buf}$, so that $\chi_{buf} = [P]_{buf}/[W]$. Because the concentration of the peptide in the bilayer is kept low to avoid concentration-dependent effects, it is also

true that $[L] \gg [P]_{bil}$, so that $\chi_{bil} = [P]_{bil}/[L]$. Consequently, Equation 2 can be written to good accuracy as

$$K_X = \frac{[P]_{bil}/[L]}{[P]_{buf}/[W]} \qquad (3)$$

Notice that no assumptions are made about the disposition of the peptide in the bilayer; one only assumes that there is some peptide associated preferentially with bilayer. The experiment in panel A begins with vesicles suspended in buffer on one side of the dialysis membrane, which is impermeable to the vesicles. At the beginning, the peptide concentration on the other side of the dialysis membrane is known and represents the total peptide concentration $[P]_{total}$. At the end of the experiment, the concentration of the peptide in the vesicle-free compartment is measured. This concentration is $[P]_{buf}$, so that $[P]_{bil} = [P]_{total} - [P]_{buf}$. It is often useful to know the fraction of peptide associated with the bilayer, f_P. It is easy to show that

$$f_P = \frac{K_X[L]}{[W]+K_X[L]} \qquad (4)$$

Another definition of the partition coefficient encountered in the literature is the molar partition coefficient K_M, defined by the equation

$$X_{bil} = K_M[P]_{buf} \qquad (5)$$

K_M has units of (moles/liter)$^{-1}$. Because $[P]_{buf} = \chi_{buf}[W]$, the relation between the mole-fraction partition coefficient and the molar partition coefficient is given by

$$K_X = [W]K_M = 55.3K_M \qquad (6)$$

The fraction of peptide bound is given in terms of K_M by

$$f_P = \frac{K_M[L]}{1+K_M[L]} \qquad (7)$$

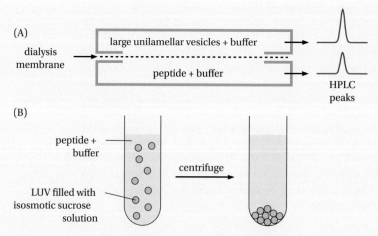

Figure 1 Methods of measuring the partitioning of peptides into large unilamellar vesicles. In (A), a dialysis membrane that is impermeable to vesicles is used to separate the vesicles from the bulk aqueous solution. In (B), the vesicles are separated by centrifugation of the dense sucrose-filled vesicles. Concentrations of peptide can be determined in many ways, but the simplest is by high-performance liquid chromatography. The advantage of this method is that the ratio of the areas under the peaks is related simply to the partition coefficient.

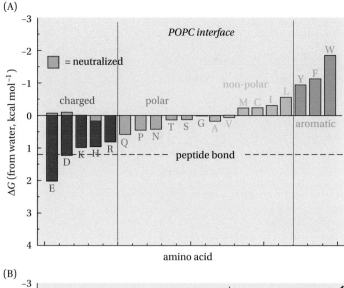

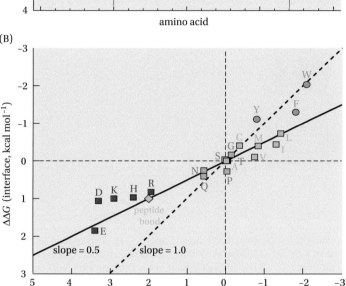

Figure 3.10 Thermodynamics of the partitioning of amino acids into the POPC bilayer interface. (A) The Wimley–White whole-residue interfacial hydrophobicity scale. See Box 3.1 for discussion of the ΔG sign conventions. (B) Comparison of the side chain-only values of the interface and octanol scales using the Ac-WLALL peptide as the reference. (From Wimley WC, White SH [1996] *Biochemistry* 35: 5109–5124.)

high-resolution crystallographic structure of a membrane protein revealed belts of Tyr and Trp residues located at positions in the protein corresponding to the membrane interface (**Figure 3.11A**). Since that first high-resolution structure was determined in 1985 (Chapter 1), every subsequent TM protein structure has been found to have similar belts. It is generally believed that Tyr and Trp stabilize proteins in membranes, and we will see later that they are important in the synthesis of membrane proteins. Interestingly, Phe does not play a distinctive role in interfacial stabilization, possibly because it lacks an H-bond-forming group. Why do Tyr and Trp have a special affinity for the membrane interface? Initially, it was thought that the affinity was due solely to their amphiphilic character, defined by the covalent linkage of a hydrophilic, H-bonding -OH or -NH group to the aromatic and presumably hydrophobic ring. The idea was that the hydrocarbon aromatic ring would bury itself in the bilayer hydrocarbon core with the -OH or -NH group exposed to the membrane interface. Although the amphiphilic character is important, experimental studies suggest a much more interesting picture.

Neutron diffraction measurements of lipid bilayers containing tripeptides (A-W-A-*O-tert*-butyl) with ring-deuterated Trp revealed that Trp is not located in the hydrocarbon core. Rather, it is largely confined to the headgroup region (**Figure 3.11B**). This could be because the strongly polar peptide bonds prevent Trp from penetrating into the hydrocarbon core. But, that conclusion is nulled by NMR observations of Trp side chain analogs lacking peptide bonds

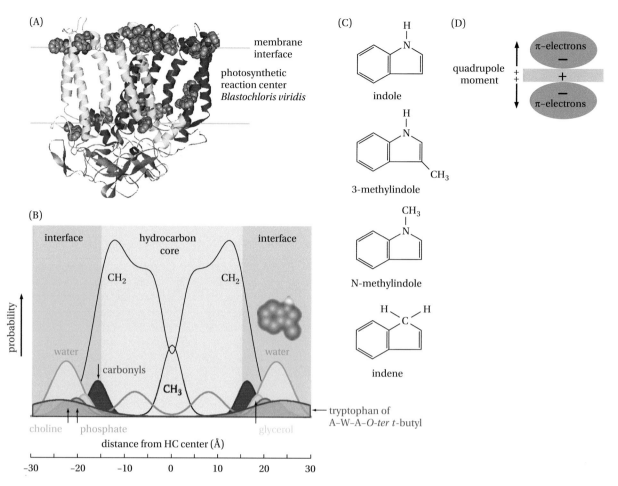

Figure 3.11 Tryptophan (shown in grey) has a special affinity for membrane interfaces. (A) Trp distribution in the photosynthetic reaction center (PDB: 1PRC). (B) Location of Trp in a DOPC interface as determined by neutron diffraction. (C) Trp side chain analogs. Each of these is located in the POPC interface. (D) The complex electrostatic structure of Trp causes it to prefer the equally complicated membrane interface (Figure 2.19). (B, From Yau WM, Wimley WC, Gawrisch K [1998] *Biochemistry* 37: 14713–14718.)

(**Figure 3.11C**). All the analogs, including indene, which lacks an NH group, are located solely in the headgroup region of POPC without penetrating into the hydrocarbon core beyond the first three carbons of the acyl chains, which generally have high water exposure (Figure 2.9). Although atomic details are missing, the emergent picture suggests that interfacial affinity is dominated by Trp's flat rigid shape and its "aromaticity." Insertion of the rigid ring into the fluid acyl chain region is entropically unfavorable because the ring would tend to order the acyl chains. The aromaticity is defined by Trp's π-electronic structure, which gives rise to both dipolar and quadrupolar moments (**Figure 3.11D**). Whereas electrical dipoles respond to electric fields, quadrupoles interact with electric field gradients, which must be severe in the membrane interface (Figure 2.19). In short, electrically complex Trp and, presumably, Tyr prefer the electrically complex membrane interface rather than the bilayer interior.

3.2.4 The Interfacial Partitioning Free Energies of Unstructured Peptides Can Be Computed Accurately Using Interfacial Hydrophobicity Scales

An important goal of research on peptide–bilayer interactions is to be able to predict from a peptide's amino acid sequence its partitioning free energy into membrane interfaces, because this allows us to test our understanding of the principles. Predictions are complicated by the presence of charged lipids and the induction of secondary structure that often accompanies partitioning. We will tackle the complications later in this chapter, but as a starting point, it is useful to consider the prediction of partitioning free energies of unstructured peptides into the zwitterionic interfaces of phosphatidylcholine (PC) bilayers. The whole-residue interfacial scale (Figure 3.10), which is derived from partitioning measurements of small unstructured peptides, offers this possibility. To do the job properly, one must account for the partitioning properties of the N- and C-termini of the peptides, which, depending upon pH, can be charged ($-NH_3^+$, $-COO^-$) or neutral ($-NH_2$, $-COOH$). Furthermore, it is common for the termini of synthetic peptides to have blocking groups to eliminate pH-dependent charge effects. The N-terminus is typically blocked by covalently linking an acetyl group; the C-terminus is typically blocked by an amine group. The free energy contributions of the various end groups and charge states for the WW pentapeptides are summarized in **Figure 3.12A**, which presents an algorithm for predicting partitioning. A comparison of the calculated and measured free energies for the WW pentapeptides shows excellent agreement; the least-squares fit of a straight line through the data reveals, within small experimental uncertainty, a slope of 1.0 and an intercept of 0 (**Figure 3.12B**). Also included, as an independent test, are data for the partitioning of a family of unstructured peptides based upon the sequence of the 13-residue tryptophan-rich antimicrobial peptide indolicidin (see later).

The contribution of the C-terminal group is quite strong, whereas the N-terminus makes little difference. This seems odd at first thought. One would think, by symmetry, that the partitioning of $-NH_3^+$ into the PC interface would be just as unfavorable as the partitioning of $-COO^-$. The likely explanation is

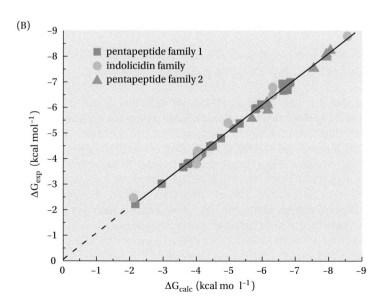

(A) partitioning free energy from water to PC bilayer interface

+

sum of whole-residue
IF scale free energies

+

(B)

Figure 3.12 An experiment-based algorithm for predicting the partitioning of unfolded peptides into the POPC bilayer interface. (A) The algorithm accounts for the differing contributions of the peptide-terminating groups. Changes in the charge state of the C-terminus have a large effect on partitioning compared with the N-terminus. (B) A comparison of predicted (ΔG_{calc}) and measured (ΔG_{exp}) partitioning free energies. Within experimental error, the slope is 1 and the intercept is 0, meaning that the algorithm accurately predicts partitioning free energies for peptides that do not adopt regular secondary structure in water or bilayer. (From Hristova K, White SH [2005] *Biochemistry* 44:12614–12619, provided by author.)

that positive charge of the interface is provided by the large choline group ($-N(CH_3)_3^+$) on the lipid, while negative charge is provided by the lipid phosphate ($-PO_4^-$) group. A major structural difference between the two compounds is that the methyl groups of the choline prevent close approach of the peptides to the charged nitrogen, thus reducing its hydrogen bonding potential; a low Born energy (Equation 2.8) due to the larger diameter of the choline is probably important as well.

3.3 CHARGED PEPTIDES INTERACT STRONGLY, AND PREDICTABLY, WITH CHARGED INTERFACES

Hydrophobic "bonding" was a common thread in the early literature on the hydrophobic effect. Bonding implies specific, distance-dependent interactions between non-polar side chains. We now know that this thinking is misleading. A hexane molecule in water doesn't know about other hexanes except through chance encounters. Partitioning or aggregation of non-polar molecules in water results from stochastic sampling of the two phases and the free energy cost of being in one phase or another. Coulombic interactions, on the other hand, involve long-range interactions between charged molecules. The electric field of charged particles or surfaces attracts or repels charged molecules over large distances; attraction or repulsion is determined by the direction of the electric field, the sign of the charge on the molecule, and the dielectric environment of the charges. The force exerted between two charged particles is described by the Coulomb inverse-square law described in basic physics textbooks (Appendix 1).

Much of the theory of Coulombic interactions is devoted to calculating forces between charged objects of various geometries. For biological membranes, the main geometry of interest is point charges interacting with extended, planar surfaces (**Figure 3.13**). From a charge's point of view, say a sodium ion, the surface of a cell is essentially flat, because the radius of a cell (~1 μm) is very large compared with the charge's radius (~1 Å). The best non-biological equivalent is the parallel-plate capacitor, which provides a basis for describing electrostatic principles necessary for understanding Coulombic interactions of charged molecules in solution with the electric field of charged membranes. Most problems in biological electrostatics involve Coulombic potentials that can be measured with voltmeters of varying degrees of sophistication. We will learn later that a typical transmembrane potential of muscle cells is about –90 mV. Given a bilayer thickness $d = 30$ Å, the electric field across the membrane will be a hefty 30,000 volts cm^{-1}.

3.3.1 Electrostatic Properties of Charged Lipid Bilayers Are Described by Gouy–Chapman Theory

Immerse an object, any object, in an electrolyte solution. If the object carries fixed charges on its surface, there are predictable consequences described by the theory of interfacial double layers. Because of its surface charges, the object will generate an electric field at its surface that extends into the electrolyte (Figure 3.13). If the surface is negatively charged, which is the only case for biological membranes, the electric field will cause cations to be attracted to the surface and anions to be repelled. This causes a separation of charges at the surface, referred to as an electrical double layer. Because the ions are undergoing incessant movement due to their thermal energy, they are constantly working to escape the influence of the field. The charges thus form a diffuse layer near the surface. Hermann Helmholtz (1821–1894) described the diffuse double layer as being equivalent to an electrical capacitor (Figure 3.13). While it's true that the double layer is similar to a capacitor in the sense that there is charge separation,

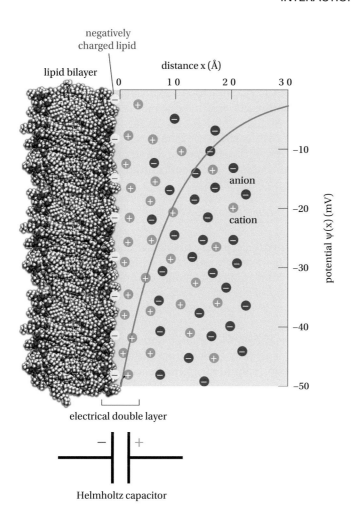

negatively charged lipid

lipid bilayer

distance x (Å)

anion

cation

potential ψ(x) (mV)

electrical double layer

Helmholtz capacitor

Figure 3.13 The electrical diffuse layer close to a charged surface. The positive ions in solution are attracted to the negatively charged lipid surface, whereas negative ions are repelled. The ions have thermal energy that causes a disorderly distribution of the ions in the surface's vicinity. This cloud-like distribution is referred to as the diffuse double layer, because within the layer, there are more positive charges on one side than another. Because there is a separation of positive and negative charges, similar to a capacitor, the layer is sometimes referred to as a Helmholtz capacitor.

the model is inappropriately static. Our modern view of diffuse double layers comes largely from the defining work of Louis Georges Gouy (1854–1926) and David Leonard Chapman (1869–1958). Consequently, when studying electrical double layers, we generally use Gouy–Chapman theory, which accounts for the membrane's attraction and repulsion of anions and cations, respectively, and the thermal motions of the ions in solution. The theory has been criticized in various ways, e.g., it treats ions as point charges, but nearly exhaustive experimental measurements have shown that Gouy–Chapman theory provides a remarkably accurate and robust description of charged lipid bilayers and their interactions with ions and peptides.

The equations for describing the electrostatics of charged membranes are rather cumbersome (Appendix 1), but as is often true in mathematical physics, conditions can be found that yield simpler and more manageable equations. Happily, for physiological conditions (~0.1 M salt solutions and surface potentials $\psi_0 \leq 100$ mV), the potential as a function of distance away from the membrane $\psi(x)$ has a simple exponential dependence on x:

$$\psi(x) = \psi_0 \exp(-\kappa x) \tag{3.3}$$

For most situations encountered in biology, this equation describes $\psi(x)$ with good accuracy. The constant κ, which must have units of inverse length to make the exponent dimensionless, determines how rapidly $\psi(x)$ decays with distance. Its numerical value can be computed from

$$\kappa = \left(\frac{2z^2 F^2 C}{\varepsilon \varepsilon_0 RT} \right)^{\frac{1}{2}} \tag{3.4}$$

in which z is the valence, C the molar salt concentration, ε the dielectric constant of water ($\approx$80), ε_0 the permittivity of free space, F the Faraday (Box 1.1), R the gas constant, and T the absolute temperature in Kelvins (see Appendix 1). The inverse of κ, $1/\kappa$, is called the Debye length; for a 100 mM salt solution, the Debye length is about 11 Å, while for a 1 mM salt solution, it is about 340 Å. The Debye length can be thought of as the distance at which mobile charge carriers screen out the electric field of the fixed charges. The length defines the radius of a Debye sphere, outside which the charge is screened. Because $1/\kappa$ varies as $1/\sqrt{C}$, the higher the salt concentration, the smaller the Debye length. For T = 25 °C and monovalent salts of molar concentration C, the Debye length λ_d in Å can be calculated from (Appendix 1).

$$\frac{1}{\kappa} \equiv \lambda_d = \frac{3.432}{\sqrt{C}} \tag{3.5}$$

The exponential form of $\psi(x)$ is easy to deal with and is good for most practical problems, but what is the magnitude of ψ_0, which is the potential at the surface of a membrane? Its value must be calculated from

$$\sigma = \left(8C\varepsilon\varepsilon_0 RT\right)^{\frac{1}{2}} \sinh\left(\frac{zF\Psi_0}{2RT}\right) \tag{3.6}$$

In this equation, σ is the surface charge density of the membrane (electronic charges/Å²), which is usually known from the lipid composition of the membrane. An iterative computational approach is usually used to find ψ_0; for given values of C and σ, vary ψ_0 until the two sides of Equation 3.6 agree. This is easily done with common computer programs such as MathCad®, Mathematica®, MatLab®, or Origin®. Gathering together the various constants, Equation 3.6 can be written:

$$\sigma = \frac{\sqrt{C}}{A} \sinh\left(\frac{\Psi_0}{2RT/zF}\right) \tag{3.7}$$

With C expressed in molar concentration, σ in electronic charges per Å², and $T = 25$ °C, then $A = 136.6$ and $RT/F = 25.7$ mV. If the potential is not too large, Equation 3.7 can be simplified to

$$\sigma = \varepsilon\varepsilon_0 \kappa\Psi_0 \approx 7.1\times10^{-10}\cdot\kappa\Psi_0 \tag{3.8}$$

We show in **Figure 3.14** the application of these equations to explore the effects of salt concentration and surface charge density on the shape and extent of the electric double layer. While we tend to think of salt concentrations in terms of mammals (~100 mM), keep in mind that life exists in a wide range of salt conditions, ranging from freshwater ponds to salt water marshes.

3.3.2 Gouy–Chapman Theory Describes the Behavior of Simple Charged Peptides in the Vicinity of Charged Membranes

Basic Gouy–Chapman theory predicts, not surprisingly, that positively charged peptides should bind, i.e., partition on to, membranes that contain negatively charged lipids. The simplest case is the partitioning of positively charged polylysine peptides, which lack hydrophobic amino acids. For exactly the same reason that ions near the membrane surface form a diffuse double layer, there should be a diffuse layer of charged peptides near the membrane surface. The concentration of peptides as a function of distance from the charged membrane should thus be given by

$$\left[P(x)\right] = \left[P(\infty)\right]\exp\left(\frac{-zF\Psi(x)}{RT}\right) \tag{3.9}$$

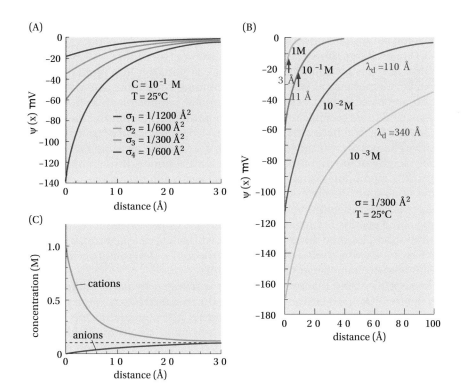

Figure 3.14 Summary of diffuse double layers according to Gouy–Chapman theory. (A) Electrical potential as a function of distance away from a charged membrane for different concentrations of negatively charged lipids in the membrane. (B) Electrical potential as a function of distance away from a charged membrane for different salt concentrations. (C) Concentrations of cations and anions as a function of distance away from a negatively charged membrane. Debye lengths λ_d are indicated in red. (From McLaughlin S [1977] *Current Topics in Membranes and Transport* 9: 71–144. With permission from Academic Press and Elsevier.)

where $\psi(x)$ is given for small potentials by Equation 3.3 and $[P(\infty)]$ is the peptide concentration far from the membrane. How can we judge the amount of peptide associated with the membrane if it is diffusely associated with the membrane? Gibbs solved this problem long ago by inventing the so-called surface excess, which is computed by integrating $[P(x)] - [P(\infty)]$ over distance. These ideas are summarized in **Figure 3.15A,B** for the association of trilysine ($z = 3$) with bilayers formed from PC:PG membranes.

Our analysis so far has treated peptide–bilayer interactions in a simplistic manner; the membrane surface is treated as a flat plane rather than a complex interface, and the charged peptides are treated as point charges. Does this matter? This question has been studied exhaustively for many years by many laboratories. The answer is that detailed atomic treatments of the problem give numbers that are not much different from those expected from the simple Gouy–Chapman approach (**Figure 3.15C**).

3.3.3 Partitioning of Charged Hydrophobic Peptides into Membranes Reveals the Non-Additivity of Electrostatic and Hydrophobic Interactions

So far, we have considered hydrophobicity-driven partitioning separately from electrostatic-driven partitioning. What about peptides that are both charged and hydrophobic? Are the partitioning free energies of the two processes simply additive? To examine that question, we take our first look at partitioning of natural peptides, many of which have antimicrobial activity. An especially useful peptide is melittin, which is the principal toxin of the European honey bee *Apis mellifera*. This 26-residue peptide, which can disrupt membranes at high concentrations, is perhaps the most widely studied membrane-active peptide. One of its most important properties, as we will see, is that it binds strongly to membranes as an α-helix despite being a random coil in solution. It is of interest here, however, because it has both charged (blue) and hydrophobic (green) amino acids:

GIGAVLKVLTTGLPALISWIKRKRQQ-CONH$_2$

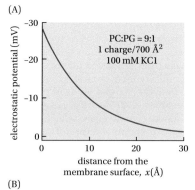

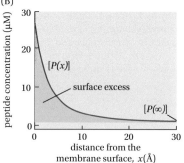

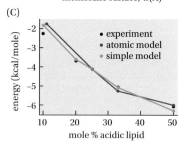

Figure 3.15 Adsorption of a positively charged peptide, in this case Lys$_3$, to a negatively charged membrane with a diffuse double layer (panel A) described by Gouy–Chapman theory. Just as for ions in solution, the peptides must also be distributed according to the Boltzmann equation. Consequently, there is a diffuse cloud of peptides in the vicinity of the membrane (B). The amount of peptide bound is given by the surface excess, which is the integrated difference in charged peptide concentration [P(x)] near the surface relative to a distant location where the concentration is [P(∞)]. (C) compares atomic and simple Gouy–Chapman results with experimental results. (From Ben-Tai N, Honig B, Peitzsch RM et al. [1996] *Biophys J* 71: 561–575. With permission from the Biophysical Society and Elsevier.)

Including the N-terminal amino group, melittin has six positive charges ($z = +6$). Some authors, for uncertain reasons, do not include the N-terminal amino group in the charge count (i.e., $z = +5$), while others vacillate ($z = +5$–6). Because melittin also has hydrophobic residues, it can partition into neutral (zwitterionic) PC membranes as well as into membranes containing acidic lipids. The membrane partitioning of charged hydrophobic peptides such as melittin can be expected to be complicated. If melittin partitions into neutral membranes, the membrane will become charged. Gouy–Chapman theory predicts that the positive charges of partitioned melittin will decrease the distribution of melittin in the vicinity of the membrane and hence, partitioning. If melittin partitions into acidic membranes, some of the lipid's negative charge will be neutralized; at the same time, the electrical double layer will be affected. For both neutral and acidic lipid membranes, melittin partitioning is further complicated by burial of the peptide in the membrane interface, where the reduced water concentration and lowered dielectric constant will shift the pK$_a$ of titratable groups. It is easy to see why there is no adequate theory for describing accurately the partitioning of charged hydrophobic peptides.

The severity of the theoretical problem is revealed by partitioning melittin into neutral POPC vesicles and measuring the so-called ζ (zeta) potential using microelectrophoresis. The ζ potential—the surface potential at the hydrodynamic plane of shear for bilayer vesicles moving in an electric field—can be used to estimate the surface potential $\psi(0)$. If one knows how much melittin is bound and consequently, its surface concentration c_s, one can estimate the surface charge density of a vesicle of surface area A. The expected surface charge density is zc_s/A, where $z = 6$ is the formal charge of melittin. When the experiment is carried out and the calculations done (**Figure 3.16**), the data are not consistent with $z = 6$ but rather, with $z = 2.2$, now called the effective charge z_{eff}. The effective charge of melittin and many other peptides has been measured in various ways; invariably, $z_{eff} < z$. This means that the valence z must be replaced by z_{eff} in Gouy–Chapman calculations; for example, the electrostatic work is written:

$$\Delta W(x) = z_{eff} F \Psi(x) \tag{3.10}$$

This phenomenon has never been satisfactorily explained, but it is deeply connected in some way with the balance between the electrostatic and hydrophobic interaction properties of membrane-active peptides. One of the complicating effects, which is probably important, is that melittin forms an α-helix when it binds to the membrane interface, perturbing the bilayer interface in the process.

A qualitative connection between hydrophobic and electrostatic interactions of charged hydrophobic peptides has been established using another antimicrobial peptide, indolicidin (**Figure 3.17**), which is a cationic peptide isolated from bovine neutrophils. It is unusual in several respects: it is small (13 residues) and is rich in proline (3 copies), tryptophan (5 copies), and basic residues (3 copies). An important characteristic is that it does not form regular secondary structure when it binds to membranes. Hydrophobic-electrostatic additivity can be examined by synthesizing numerous variants with different amounts of charge and hydrophobicity. For example, in one set of variants, the Trp residues are replaced with Phe, Tyr, or Leu. To assess the connection between electrostatic and hydrophobic interactions, experimentally observed partitioning free energies (ΔG_{obs}) were compared with expected values using

$$\Delta G_{obs} = \Delta G_{H\Phi} + \Delta G_{ES} \tag{3.11}$$

where

$$\Delta G_{ES} = zF\Psi(0) \tag{3.12}$$

and

$$\Delta G_{H\Phi} = -\sigma_{np} A_{np} \qquad (3.13)$$

For reasons outlined earlier, although Equation 3.12 is commonly used for calculating electrostatic free energies, it is only an approximation. Equation 3.13, on the other hand, provides an accurate measure of the hydrophobic contribution to partitioning (Box 1.2 and Box 2.1); σ_{np} is the non-polar solvation parameter (~12 cal mol^{-1} Å^{-2} for the membrane interface), and A_{np} is the accessible non-polar surface area. Measurements of partitioning as a function of surface potential for membranes formed from mixtures of POPC and POPG yield values of z_{eff} that are always smaller than the formal valence of the peptides, just as for melittin. To estimate the effect of hydrophobicity on electrostatic interactions, z_{eff}/z can be plotted against $\Delta G_{H\Phi}$. **Figure 3.18** shows that there is a strong inverse correlation between z_{eff}/z and $\Delta G_{H\Phi}$; the greater $\Delta G_{H\Phi}$, the smaller z_{eff}/z. Consistent with the measurements of the ideal partitioning of polylysine peptides (Figure 3.15), peptides without a hydrophobic component are expected to have $z_{eff}/z = 1$.

3.3.4 Some Physiologically Important Proteins Bind to Membranes with the Help of Covalently Linked Lipids

We concluded earlier that for the most part, partitioning of non-polar solutes into bilayers is not equivalent thermodynamically to partitioning into bulk phases. The major exception is the partitioning of free fatty acids, which can aid the tethering of proteins to membranes to facilitate, for example, signaling pathways. As shown in **Box 3.5**, bilayers and bulk phases seem equivalent for low concentrations of fatty acids. The partitioning of fatty acids and related compounds into lipid bilayers is important because many cellular proteins involved in signaling and regulatory processes are modified covalently, either co- or post-translationally, to include lipids such as fatty acids or isoprenoids (**Figure 3.19**). These are peripheral membrane proteins because they partition into the interface without forming TM structures.

A particularly important class of eukaryotic peripheral proteins is the GPI-anchored proteins that have covalently attached glycosylphosphatidylinositol (GPI) (see Chapter 6). Another is the Src ("sarc") tyrosine protein kinase family, which is part of a large web of biochemical processes that help cells respond to biochemical environmental changes. Since many responses are initiated at the plasma membrane, it makes sense to locate the proteins involved close to their sites of action. Kinases catalyze the phosphorylation of hydroxyl groups on serine, threonine, or tyrosine by transfer of phosphate from ATP.

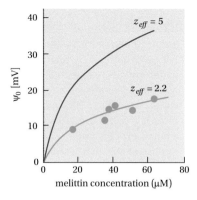

Figure 3.16 Measurements (blue filled data points) of the surface potential ψ_0 of POPC vesicles with bound melittin show that the effective charge of melittin is 2.2 rather than the expected value of 5. (From Beschiaschvilli G, Baeurele H-D [1991] *Biochim Biophys Acta* 1068: 195–200. With permission from Elsevier.)

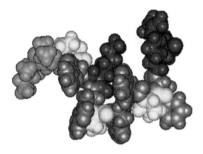

ILPWKWPWWPWRR – CONH$_2$

Figure 3.17 Structure and amino acid sequence of the antimicrobial peptide indolicidin. (From Ladokhin AS, White SH. [2001] *J Mol Biol* 309: 543–552.)

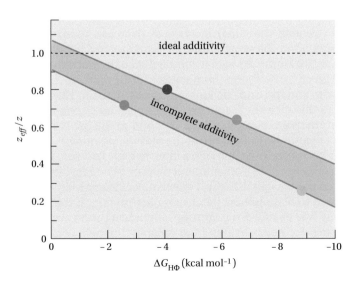

Figure 3.18 The effective charge z_{eff} of charged hydrophobic peptides depends upon hydrophobicity. The different symbol colors refer to different families of peptides. The brown zone is meant to highlight the general trend of the data from the several families examined. As hydrophobicity decreases toward 0, z_{eff} approaches the formal valence z for a set of peptides modeled on 13-residue indolicidin, which has no regular secondary structure when partitioned into bilayers. (From Ladokhin AS, White SH [2001] *J Mol Biol* 309: 543–552.)

BOX 3.5 FATTY ACID PARTITIONING INTO BULK AND BILAYER PHASES

Some proteins have covalently linked fatty acids or related compounds to aid partitioning into cell membranes. We earlier considered hydrophobic partitioning between water and non-polar phases in terms of the solute's accessible surface area (Box 1.2). For fatty acids and alkyl hydrocarbons, it is useful to consider the partitioning free energy as a function of the number of methylene (-CH_2-) groups. An important question is whether the partitioning of fatty acids between water and lipid bilayers is energetically different from bulk-phase partitioning, e.g., water-to-heptane. The preceding data allow us to answer that question. Because all curves are linear in the number of methylenes, one can easily extract the free energy (cal mol^{-1}) for transferring a single methylene from water, as summarized in the following. In the formulae, the superscripts *np* and *w* mean the non-polar phase and water, respectively.

Water-to-bulk hydrocarbon

n-alkanes (**Figure 1**, panel A):

$$\mu_0^{np} - \mu_0^{w} = -2102 n_{CH_3} - 884 n_{CH_2} \qquad (1)$$

Water-to-heptane

fatty acids (panel B):

$$\mu_0^{np} - \mu_0^{w} = -4260 n_{CH_3} - 825 n_{CH_2} \qquad (2)$$

Water-to-lipid bilayers

fatty acids (panel C):

$$\mu_0^{np} - \mu_0^{w} = -4200 n_{CH_3} - 825 n_{CH_2} \qquad (3)$$

The important conclusion from these data is that the free energy of partitioning fatty acids between bulk and bilayer phases is about the same on a per-methylene basis.

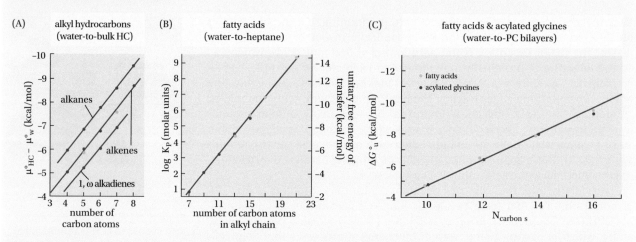

Figure 1 Measurements of the partitioning of fatty acids between water and various non-polar phases. (A, From Tanford C [1974] *The Hydrophobic Effect*. With permission from John Wiley and Sons, New York. B, Based on Smith R, Tanford C [1973] *PNAS* [1973] 289. With permission. C, From Peitzsch RM, McLaughlin S [1973] *Biochemistry* 32: 10436–10443. With permission from American Chemical Society.)

The phosphorylation can either increase or decrease a target protein's activity, depending upon the protein and its structure. Src self-phosphorylates to become an active kinase through a complex set of interactions involving three domains (**Figure 3.19A**): a Src homology 3 (SH3) domain, a Src homology 2 (SH2) domain, and the catalytic kinase catalytic domain. How does lipidation of Src fit into this picture? A somewhat complicated enzymatic process adds a 14-carbon myristate through an amide bond to the Src N-terminal glycine. The myristate group helps bind Src at the plasma membrane interior, which is required for kinase activity. Although a lone myristic fatty acid binds strongly to lipid bilayers (**Figure 3.20**), other interactions must be considered in binding of myristoylated proteins, such as Src. For example, the myristate chain might fold back and bind to hydrophobic patches on the protein. If the myristate group were the only way of anchoring Src, only about half the Src would be bound to the membrane; Src needs help to bind reliably. The help comes from electrostatic interactions between the negatively charged membranes and six lysine residues near the N-terminus of Src, as summarized in **Figure 3.19B**,C. Src

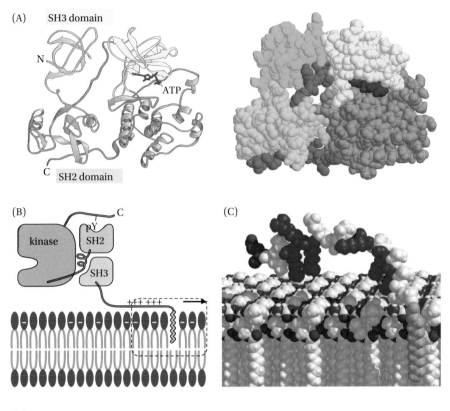

(D)

Src(2–16)	myristate-GS SKSKPKDPSQRRR
MARCKS(151–175)	KKKKKRFSFKKSFKLSGFSFKKNKK
HIV-1 Gag (2–31)	myristate-GA RASVLSGG ELDRWEKIRLRPGGKKKYKL
K-Ras 4B (174–185)	GKKKKKKSKTSC–farnesyl

Figure 3.19 The structure and membrane interactions of Src proteins. (A) Molecular structure of a Src protein shown in the ribbon and space-filling formats. The kinase domain (orange/yellow structure), shown with an ATP molecule, phosphorylates the SH2 domain, causing the SH2 domain to change structure to admit a substrate molecule (binding partner). (B) A cartoon representation of the interaction of the myristoylated N-terminus interacting with a lipid bilayer. Binding to the bilayer is strengthened by electrostatic interactions with the negatively charged inner monolayer of the asymmetric membrane bilayer. (C) Atomic model showing the interaction of the six N-terminal lysines with the anionic bilayer. (D) Examples of proteins that require both lipidation and charges to bind to membranes. (A, From Alberts B, Johnson A, Lewis JH et al. [2007] *Molecular Biology of the Cell*, 5th Edition. Garland Science, New York. With permission from W. W. Norton. B–D From Murray D, Ben-Tai N, Honig B et al. [1997] *Structure* 5: 985–989. With permission from Elsevier.)

reveals a general theme for the binding of many important peripheral membranes (**Figure 3.19D**): Partitioning of covalently linked lipids acts in concert with basic amino acids to bind to the anionic membrane lipids on the inside surface of a plasma membrane.

3.4 MEMBRANE-ACTIVE PEPTIDES PROVIDE CLUES TO SECONDARY STRUCTURE FORMATION AT MEMBRANE INTERFACES

3.4.1 The Bilayer Interface Induces Secondary Structure Formation

When LUVs are added to a solution of melittin, the peptide partitions into the membrane interface and adopts an α-helical conformation in a process referred to as partitioning–folding coupling. The clearest evidence for this process comes from circular dichroism (CD) measurements, illustrated in **Figure 3.21A**. The CD spectrum of melittin in the absence of vesicles reveals an ensemble of conformations that have little secondary structure. As lipid vesicles are added to the solution, a progressive increase in secondary structure typical of helices appears; the greater the amount of vesicles, the more helical the CD spectrum becomes. Importantly, there is a so-called isodichroic point at 205 nm, meaning

modification	chemistry	substrates	membrane affinity	K_x	ΔG_0 (kcal mol^{-1})
N-myristoylation 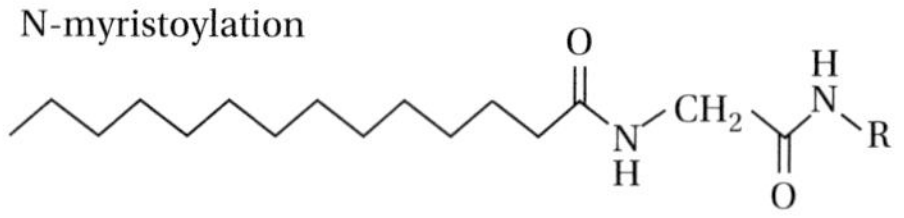	amide; stable	myristoyl-CoA nascent polypeptide (co-transitional)	$K_d^{\text{eff}} = 80$ μM	6.9×10^5	−7.9
prenylation: farnesylation:	thioester; stable	cytosolic protein (posttransitional)			
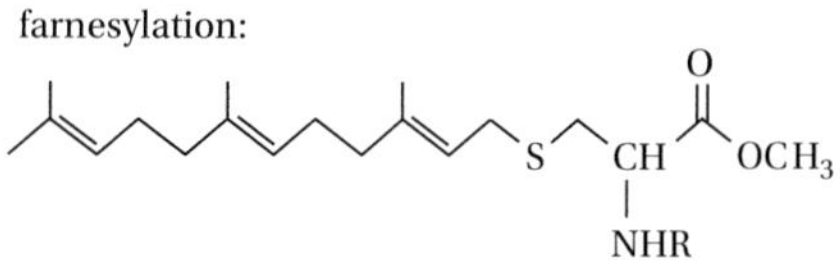		farnesyldiphosphate	$K_d^{\text{eff}} = 100$ μM (unmethylated)	5.5×10^5	−7.8
			$K_d^{\text{eff}} = 8$ μM (methylated)	6.9×10^6	−9.3
geranylgeranylation: 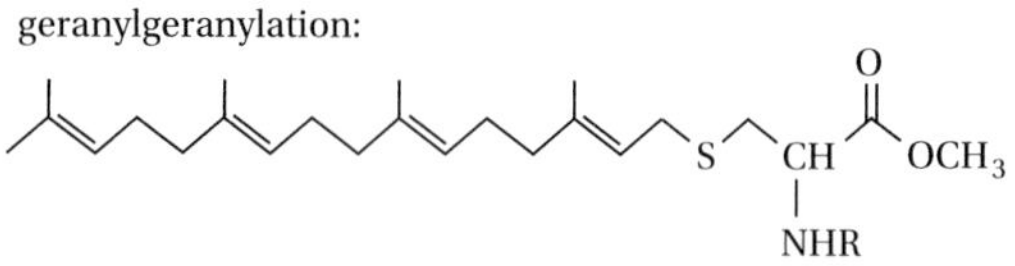		geranylgeranyldiphosphate	$K_d^{\text{eff}} = 2$ μM (unmethylated)	2.8×10^7	−10.1
			$K_d^{\text{eff}} = 0.2$ μM (methylated)	2.8×10^8	−11.5
palmitoylation: 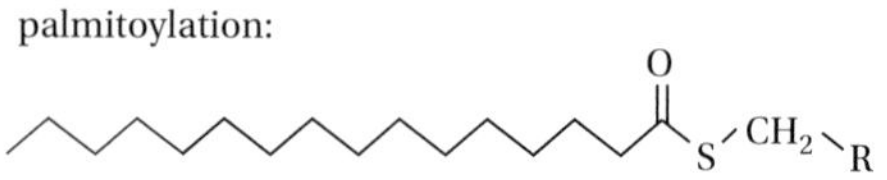	thioester; labile	palmitoyl-CoA membrane associated protein (posttranslational)	$K_d^{\text{eff}} = 5$ μM	1.1×10^7	−9.6

$K_d^{\text{eff}} = 55.3/K_x$

Figure 3.20 Summary of common lipid modifications of proteins. Membrane affinities, partition coefficients, and partitioning free energies for lipid bilayers are shown. See Box 3.3 for the relationship between membrane affinity and the partition coefficient.

that all spectra have the same ellipticity at that wavelength. This sort of behavior is typical of two-state processes in which the observed spectra are the sum of two states, here a spectrum of the unfolded peptides in solution and a spectrum of folded peptide on the membrane. Said another way, we see two conformational states simultaneously, one coming from largely unfolded melittin in solution and the other arising from highly structured melittin on the membrane (**Figure 3.21B**). But it is important to remember that these are ensembles of peptide conformations; we see only the average ensemble conformation in the experiment. The general CD signature of α-helices is the strong minimum at 222 nm. The free energy of binding of melittin to a membrane can be determined by plotting the relative helicity at 222 nm against the lipid concentration (**Figure 3.21C**). Using the nomenclature of Box 3.4, the data points suggest that at a sufficiently high lipid concentration [L], all the peptides will have partitioned into the bilayer. Let S, the signal strength, be the relative ellipticity at 222 nm. Because of the two-state equilibrium, S is given by the signal s_{bound} from the partitioned peptides and the signal s_{free} due to unpartitioned peptides, weighted, respectively, by the fractions of partitioned peptides (f_p) and the unpartitioned peptides ($1 - f_p$). S can then be computed from

$$S = \left\{ f_p s_{bound} + \left(1 - f_p\right) s_{free} \right\} \left[P\right]_{total} \tag{3.14}$$

Lurking in this equation is the partition coefficient K_X, because the fraction of peptide partitioned defined in Box 3.4 is

$$f_p = \frac{K_X[\text{L}]}{[\text{W}] + K_X[\text{L}]} \tag{3.15}$$

The result is that K_x, and consequently, the mole-fraction partitioning free energy, can be obtained by finding the best-fit curve using Equations 3.14 and 3.15, and non-linear least-squares computer fitting routines. Such a curve is the red curve shown in **Figure 3.21C**, which corresponds to $\Delta G_X^0 \approx -7$ kcal mol^{-1}.

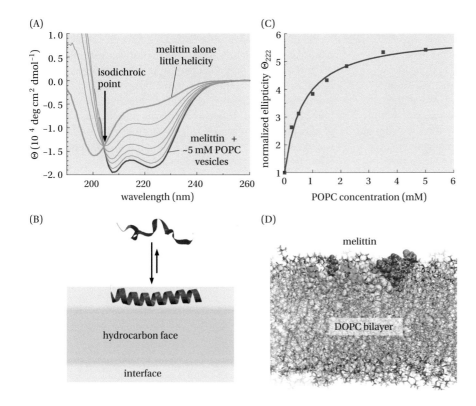

Figure 3.21 Partitioning of melittin into lipid bilayers. (A) The addition of lipid vesicles to a solution of melittin causes the appearance of CD spectra characteristic of α-helices. This spectrum is due to melittin adopting an α-helical structure upon binding to the vesicles. (B) The isodichroic point in the CD spectra shown in panel A indicates the presence of two average conformers of melittin: unfolded melittin in solution and folded melittin on the membrane. (C) The progressive increase in absorbance at 222 nm allows the partition coefficient of melittin to be determined using Equations 3.14 and 3.15. Extrapolation of the binding curve to infinite lipid concentration indicates that fully bound melittin is about 70% helical. (D) Image from a molecular dynamics simulation of melittin in a DOPC bilayer. (A, From White SH, Wimley WC, Ladohkhin AS et al. [1998] *Meth Enzymol* 295: 62–87. B, From Ladokhin AS, Fernández VM, White SH [2010] *J Membrane Biology* 236: 247–253. C, SH White, unpublished. D, From Fernández VM, Jayasinghe S, Ladokhin AS et al. [2007] *J Mol Biol* 370: 459–470. With permission from Elsevier.)

Extrapolation of the red curve to very high lipid concentrations allows one to estimate the average helicity of bound melittin as about 70%. This helicity corresponds to 18 out of 26 residues being fully α-helical. Researchers commonly assume that every melittin peptide bound on the surface is 70% helical, but in reality, there is probably a conformational ensemble whose average helicity is 70%. Molecular dynamics simulations indicate that the distribution of conformations in the ensemble is relatively narrow, centered on 70% helicity.

We said that melittin partitions into the bilayer interface. How do we know that? Melittin contains a tryptophan residue whose fluorescence signal depends strongly on the polarity of the local environment (**Box 3.6**). The fluorescence can be inhibited (quenched) by certain molecules that are soluble in the aqueous phase, e.g., acrylamide, or in the lipid bilayer phase, e.g., bromine attached to lipid acyl chains. If Trp is fully buried and has no exposure to water, then aqueous quenchers will have little effect, but if Trp has some exposure to water, then it can be quenched. If the Trp of melittin is at the interface, then bromines at chain positions close to the glycerol backbone will have a large quenching effect, whereas bromines located near the terminal groups will have a lesser effect. We conclude from such measurements that melittin's Trp, and therefore, the melittin α-helix, is located in the interface. X-ray diffraction measurements in conjunction with molecular dynamics simulations identify precisely where melittin resides in bilayers at very low melittin concentrations (**Figure 3.21D**). At very high concentrations, such as those delivered by angry bees, aggregates of melittin can destabilize the bilayer, which leads to cell lysis.

3.4.2 The High Cost of Partitioning Peptide Bonds into the Interface Explains Partitioning–Folding Coupling

Why do melittin and many other peptides fold into a helix when they partition? Figure 3.11A provides the primary clue: the cost of partitioning a peptide bond into the interface that does not participate in hydrogen-bond interactions is about 1 kcal mol^{-1}. This seems modest until one realizes that the costs are additive; the cost of partitioning unfolded melittin into the POPC interface would be

BOX 3.6 TRYPTOPHAN FLUORESCENCE: A VERSATILE TOOL FOR STUDYING PROTEIN-LIPID INTERACTIONS

The indole ring of tryptophan is one of many aromatic compounds that can be excited optically to fluoresce. Light that has a wavelength corresponding to the difference between the ground-state electronic orbital and a higher-state electronic orbital excites the molecule to the higher-energy state, which contains numerous vibrational substates. Through a process called internal conversion, non-radiative energy loss transfers the excitation to a lower-energy vibrational state, which then decays to the ground state by emission of a photon. The time constant for decay is typically 3 ns. There are two ways to record steady-state fluorescence: record emission at a fixed wavelength while varying the excitation wavelength (an excitation spectrum) or, more commonly, excite at a fixed wavelength and scan emission wavelength (emission spectrum). For tryptophan, emission spectra such as shown in **Figure 1**

panel A are obtained by exciting at a fixed wavelength of 280 nm. An important feature of tryptophan fluorescence is that the emission spectrum depends upon its local environment. Generally, as the environment becomes less polar, fluorescence intensity increases, and the wavelength of maximum fluorescence (λ_{max}) shifts toward shorter wavelengths (blue shift). Because water is an efficient quencher of fluorescence, the transfer of Trp into a lipid bilayer often results in increased fluorescence intensity. These features are observed for melittin, which tells us that melittin's tryptophan is in the bilayer. By measuring the intensity of the fluorescence at a fixed wavelength (334 nm in panel A) as a function of lipid concentration, the partition coefficient of melittin can be determined (panel B) using the same formalism as for CD measurements (Figure 3.21C; Equations 3.14 and 3.15).

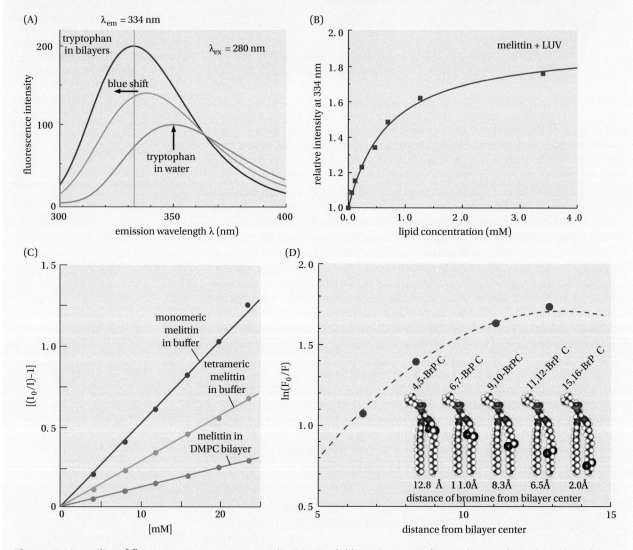

Figure 1 Versatility of fluorescence measurements. (B, From Ladokhin AS, Jayasinghe S, White SH [2000] *Anal Biochem* 285: 235–245. C, From Vogel H [1981] *FEBS Lett* 134: 37. With permission from Elsevier. D, Based on Ladokhin AS [1997] *Meth Enzymol* 278: 462–473.)

(Continued)

BOX 3.6   TRYPTOPHAN FLUORESCENCE: A VERSATILE TOOL FOR STUDYING PROTEIN–LIPID INTERACTIONS (CONTINUED)

Tryptophan's fluorescence blue shift suggests that tryptophan is located in the bilayer. But, where in the bilayer is it? Another fluorescence method, fluorescence quenching, provides answers. Certain chemicals, such as bromine, acrylamide, and nitrate, can quench tryptophan's fluorescence through collisions that cause non-radiative transitions to the ground state and thus reduce the fluorescence. The extent of quenching depends in part upon how close the quencher can get to the tryptophan. The formalism used is attributable to Otto Stern (1880–1969) and Max Volmer (1885–1965). Stern is perhaps most famous for his co-discovery of spin quantization, for which he was awarded the 1943 Nobel prize. The idea of the Stern–Volmer relation is simple: measure the decrease in fluorescence as a function of quencher concentration [Q], and quantify the extent of quenching by means of the Stern–Volmer constant K using the relation

$$\frac{F_0}{F} = 1 + K[Q] \text{ or } \left(\frac{F_0}{F} - 1\right) = K[Q] \qquad (1)$$

An example for the quenching of melittin from the aqueous phase as a function of concentration of the quencher NO_3^- is shown in **Figure 1C**. Notice that aqueous-phase quenching of melittin is much higher for melittin free in solution or as tetramers in solution than for membrane-associated melittin. (Melittin forms tetramers at high salt concentrations due to shielding of repulsive charge interactions, as expected from Equation 3.4.) We can conclude that tryptophan is in the membrane but accessible to water, implying that it is in the membrane interface. To determine the location more precisely, one can quench from within the bilayer using bromine-labeled lipids (inset, panel D). The analytical procedure is somewhat different, however. In broad sweep, fluorescence $F(h)$ of melittin in bilayers that contain lipids brominated at position h is determined

and compared with the fluorescence F_0 of tryptophan in the membrane in the absence of quencher; the closer bromine is to tryptophan, the greater the quenching. There are two methods for estimating the depth of tryptophan from measurements of F_0/F as a function of quencher position; the parallax method, invented by Erwin London, and distribution analysis, invented by Alexey Ladokhin. The latter method takes into account that the transbilayer distribution of bromines is broad and can be described by a Gaussian distribution, similar to the distribution of DOPC double bonds (see Figure 2.19). Because the transbilayer distributions of bromines have been determined by x-ray diffraction measurements in the laboratory by Thomas McIntosh, a quantitative analysis of quenching can be carried out. If h is the known position of a particular bromine quencher, then $F_0/F(h)$ can be described by the Gaussian function

$$\ln\left[\frac{F_0}{F(h)}\right] = \frac{S}{\sigma\sqrt{2\pi}} \exp\left[-\frac{(h-h_m)^2}{2\sigma^2}\right] \qquad (2)$$

that captures the idea of transbilayer Gaussian-distributed quenchers. This equation is fitted to data points using three parameters: h_m, S, and σ. The parameter h_m describes the most likely position of tryptophan. The dispersion, σ, accounts for the finite sizes of the quencher and tryptophan and the thermally disordered transbilayer distributions of quencher and probe. S is a scale factor that can be used to examine the degree of Trp exposure to lipid phase. The data in panel D show that the Trp of melittin is ~13 Å from the bilayer center. This is quite consistent with the transbilayer position of melittin determined by x-ray diffraction (**Figure 3.23**). Because Equation 2 has three parameters, one obviously must measure F(h) using at least three probes, each with a different h.

+26 kcal mol^{-1} rather than the –7 kcal mol^{-1} observed. The logical conclusion is that the formation of secondary structure dramatically reduces the cost of partitioning. That is, the cost of partitioning a hydrogen-bonded peptide bond is much, much lower than the cost of partitioning an un-bonded one. This raises the question of just how much H-bonding contributes to the free energy of folding in the interface.

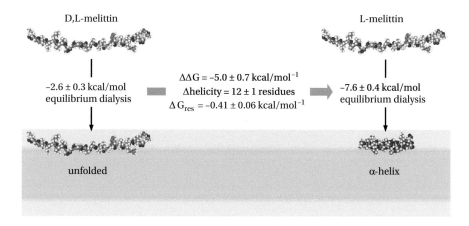

D,L-melittin

L-melittin

$\Delta\Delta G = -5.0 \pm 0.7$ kcal/mol^{-1}
Δhelicity = 12 ± 1 residues
$\Delta G_{res} = -0.41 \pm 0.06$ kcal/mol^{-1}

–2.6 ± 0.3 kcal/mol
equilibrium dialysis

–7.6 ± 0.4 kcal/mol
equilibrium dialysis

unfolded

α-helix

Figure 3.22 Energetics of folding peptides into bilayer interfaces. An approximate method of determining the folding free energy of melittin using diastereomeric peptides that contain both D- and L-amino acids, which prevents folding. As illustrated, the free energy reduction per residue in a helical conformation is about 0.4 kcal mol^{-1}. (From Ladokhin AS, White SH [1999] *J Mol Biol* 285: 1363–1369.)

How does one determine the effect of H-bond formation on partitioning free energy? The problem is to estimate the cost of partitioning the unfolded peptide for comparison with the cost of partitioning the folded peptide. Although some unfolded peptide might be present on the membrane, its signal will be swamped by the signal from the folded peptide. A simple, approximate approach for determining the folding free energy is shown in **Figure 3.22** based upon the observation that diastereomeric peptides containing both D- and L-amino acids cannot readily form regular secondary structure. As illustrated, the CD spectra and partitioning free energies of native melittin and a diastereomeric melittin containing four D-amino acids allow one to estimate the free energy reduction for secondary structure formation. This experiment yields about -0.4 kcal mol^{-1} per residue. The small amount of data presently available from other peptides suggests that ΔG_{res} will always lie between -0.3 and -0.5 kcal mol^{-1}, depending upon amino acid sequence and lipid. These seem like small numbers, but they add up to become a strong driving force for folding, as the example shows. But why, then, is melittin only 70% helical? That is because other interactions are also at work. For example, the lysine side chains in the peptide may interact more strongly with the lipids when not in an α-helical conformation. And of course, thermal motion acts to increase entropy.

3.4.3 The Hydrophobic Moment of Structured Peptides Is of Little Energetic Importance in Interfacial Partitioning

A significant feature of melittin, found in many membrane-active peptides, is that it is amphiphilic when in an α-helical conformation. The most common representation of an amphiphilic helix shows polar residues along the length

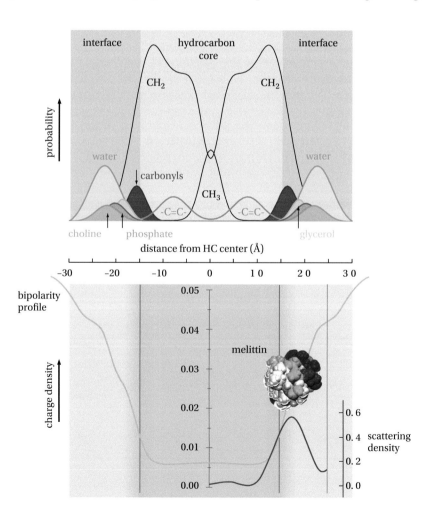

Figure 3.23 Position of melittin in the interface of a DOPC bilayer determined by absolute-scale x-ray diffraction methods. The long axis of melittin is located at the steepest point of the bilayer polarity gradient.

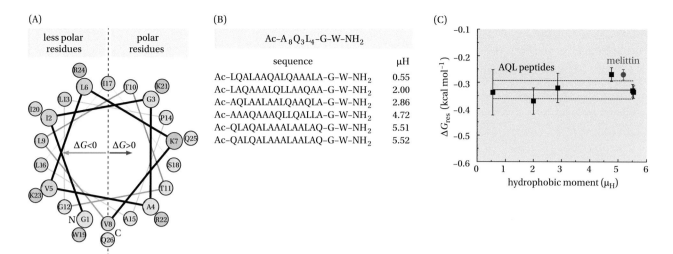

Figure 3.24 The hydrophobic moment of amphiphilic helices has little effect on peptide partitioning into the membrane interface. (A) Helical wheel representation of an α-helix. This view down the helix axis of melittin shows polar residues (yellow, blue) on one surface and non-polar residues (green) on the other. The hydrophobic moment μH is calculated as described in Box 3.6. (B) A family of uncharged peptides that have the identical hydrophobicities but different hydrophobic moments. (C) Partitioning free energies of the peptides in panel B plotted against μH. These data show that the hydrophobic moment has little effect on partitioning free energy. (C, From Almeida PF, Ladokhin AS, White SH [2012] *Biochim Biophys Acta Biomembranes* 1818: 178–182. With permission from Elsevier.)

of one-half of the helix surface and non-polar residues along the opposite surface. This is especially apparent in helical wheel representations such as the one illustrated for melittin in **Figure 3.24A**. This representation is equivalent to looking down the axis of the helix. Helical peptide amphiphilicity can be quantitated by calculating the hydrophobic moment (μH) (**Box 3.6**). The relation between bilayer partitioning free energy (determined as in Figure 3.21) and μH can be examined using a series of peptides that have the same amino acid composition but different sequences and consequently, different values of μH (**Figure 3.24B**). Although one instinctively thinks that μH should have a significant effect on partitioning free energy, it does not (**Figure 3.24C**). Why, then, is the hydrophobic moment such a prominent feature of many membrane-active peptides? One explanation may be that the hydrophobic moment encourages association/aggregation of the peptides in the membrane to form physiologically significant structures such as pores.

3.4.4 Amphiphilic Helices Can Be Used as Scaffolds for Forming Nanodisc Membranes

Amphiphilic helices do more than just partition into bilayer interfaces. If of sufficient length, they can form helical scaffolds around a disc of bilayer to form high-density lipoproteins (HDLs) involved in cholesterol transport from cells through the blood stream to the liver (**Figure 3.25A**). The protein component, called apolipoprotein A-I (apoA-I), consists of ten or more amphiphilic helices connected in tandem and arranged as amphiphilic belts that shield phospholipid acyl chains and cholesterol compounds from water. These simple disc-like structures are an early structural step, which leads to increasingly complex spherical structures with three apoA-I belts that create different topological arrangements as cholesterol esters are taken up from the blood stream for delivery to the liver.

Starting with the basic principles encompassed in HDLs, nanodiscs (**Figure 3.25B**) have been engineered to serve as hosts for membrane proteins. They are formed by solubilizing engineered membrane scaffold proteins (MSPs) and phospholipids in detergents and then removing the detergent by dialysis. If detergent-solubilized membrane proteins are present, they too will

BOX 3.7   THE HYDROPHOBIC MOMENT IDEA

The hydrophobic moment μ, invented by David Eisenberg, is a useful way of characterizing the amphipathic property of α-helices (panel A). In the most general terms, any protein with N residues whose 3-D structure is known can be assigned a total hydrophobic moment defined as

$$\vec{\mu} = \sum_{n=1}^{N} H_n \vec{s}_n \qquad (1)$$

The unit vector is drawn from the nucleus of the nth alpha-carbon to the center of mass of the carbon's side chain, whose hydrophobicity has a value H_n (panel B). Although this vector representation for whole proteins is rarely used, it forms the foundation for quantitating the hydrophobic moment of amphipathic α-helices, illustrated in the helical wheel plot, in which the helix is projected onto a plane that is normal to the axis of the helix and in this case, viewed from the N-terminal looking toward the C-terminal (panel C). This representation shows the helix as a periodic structure. Why periodic? An important structural feature of α-helices is that for each full turn of the helix, there are 3.6 residues. This means that a side chain emerges from the helix every $360°/3.6 = 100°$, each with its unit vector approximately perpendicular to the helix axis. A perfect amphipathic helix is created by distributing the non-polar and polar residues in the sequence so that the non-polar (e.g., Leu) residues are on one surface and the polar residues (e.g., Lys) are on the opposing surface. Each of the unit vectors can be decomposed into two vectors (red vectors,

panel C), which gives rise to an x- and a y-component of the hydrophobic moment, given by $H_n\sin\omega$ and $H_n\cos\omega$. As one progresses along the axis of a helix of arbitrary length N, one can think just as well of traveling along the helical backbone in angular space, so that a side chain will be encountered every $100°$. Simple application of the Pythagorean theorem allows us to determine the magnitude of the total hydrophobic moment μ as

$$\mu = \left\{ \left[\sum_{n=1}^{N} H_n \sin(100n) \right]^2 + \left[\sum_{n=1}^{N} H_n \cos(100n) \right]^2 \right\}^{1/2} \qquad (2)$$

Ah ha! Because of the $100°$ periodicity, this looks something like a Fourier series. Indeed, this is special case for $100°$ periodicity of the spectral density (or modulus) of a Fourier series. The equation can be generalized to any periodicity. Let the periodicity be represented by $\omega = 2\pi/m$; for the α-helix, $m = 3.6$. This allows us to express the hydrophobic moment in terms of ω as

$$\mu(\omega) = \left\{ \left[\sum_{n=1}^{N} H_n \sin(\omega n) \right]^2 + \left[\sum_{n=1}^{N} H_n \cos(\omega n) \right]^2 \right\}^{1/2} \qquad (3)$$

To illustrate the periodicity of an amphipathic helix, replace L and K in panel C with the values 1 and 0, respectively, and compute $\mu(\omega)$. The data are plotted in panel D. The major peak occurs at $100°$, indicating that the sequence has the properties of an amphipathic α-helix (**Figure 1**).

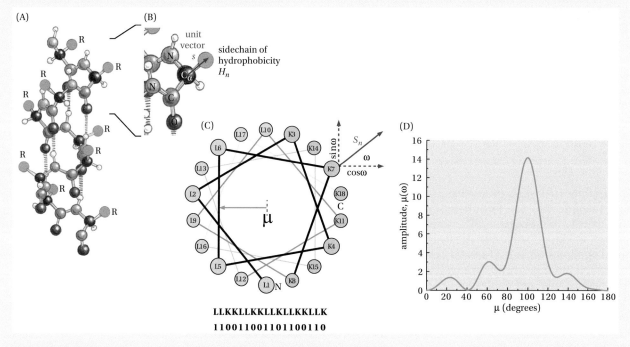

LLKKLLKKLLKLLKKLLK
110011001101100110

Figure 1 Defining and computing the hydrophobic moment of an alpha-helix. (A,B, Stryer L [1988] *Biochemistry*. W. H. Freeman, New York. With permission.)

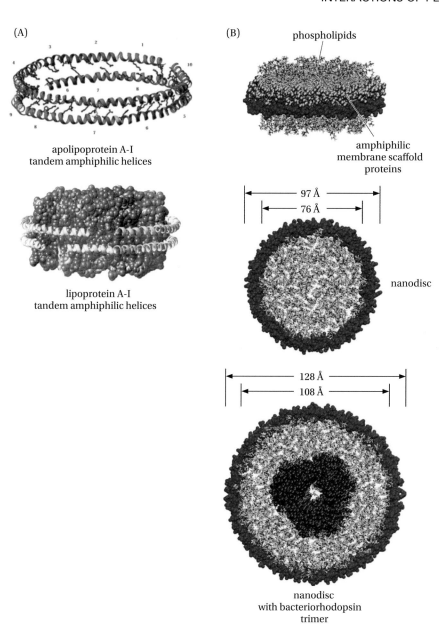

(A)

apolipoprotein A-I
tandem amphiphilic helices

lipoprotein A-I
tandem amphiphilic helices

(B)

phospholipids

amphiphilic
membrane scaffold
proteins

97 Å

76 Å

nanodisc

128 Å

108 Å

nanodisc
with bacteriorhodopsin
trimer

Figure 3.25 Natural high-density lipoproteins (HDLs) that transport cholesterol in the blood inspired the invention of nanodiscs for solubilizing membrane proteins. (A) At an early stage of formation, HDLs consist of two belts of tandem amphiphilic helices that capture phospholipids and cholesterol in their hydrophobic interiors. (B) Nanodiscs formed from phospholipids and engineered amphiphilic membrane scaffold proteins (MSPs). (A, From Klon AE, Jones MK, Segrest JP et al. [2000] 79: 1679–1685. With permission from The Biophysical Society and Elsevier. From Segrest JP, Harvey SC, Zannis V et al. [2000] *Trends Cardiovasc Med* 10: 246–252. With permission from Elsevier. B, From Bayburt RH, Sligar SG [2010] *FEBS Lett* 584:1721–1727. With permission from Elsevier.)

be incorporated into the nanodiscs at the single-molecule level for structural and biophysical studies (Chapter 9).

3.4.5 The Stability of Transmembrane Helices Results from the High Cost of Partitioning Peptide Bonds into Lipid Bilayers

One of the earliest membrane proteins to be studied intensively by biochemists, particularly Jere Segrest, was single-span glycophorin A (GpA), found in red blood cells. A simple analysis of GpA using the *n*-octanol hydrophobicity scale (Figure 3.6) reveals why TM helices are extremely stable and consequently resist calorimetric unfolding (Chapter 4). As shown in **Figure 3.26**, there are two opposing contributions to TM stability. The side chains of GpA are quite hydrophobic and provide a very favorable free energy of insertion of –32 kcal mol^{-1}. Insertion of the helical backbone (think of a polyglycine helix), in contrast, is very unfavorable, contributing about +23 kcal mol^{-1} to stability. The net stability, resulting from the favorable hydrophobic effect of the side chains and the

Figure 3.26 Thermodynamic summary of TM helix stability. The free energy of insertion of a glycophorin transmembrane helix can be separated into two components: side chain and helical-backbone insertion free energies. The free energies of insertion are computed using the WW octanol scale (Figure 3.7). Although bulk octanol is not a perfect stand-in for the bilayer, the free energies for side chains and backbone are so large that errors in the difference free energies are unlikely to affect the general features of TM helix stability. As in many biochemical reactions, net free energies result from small differences between large opposing free energies. If the backbone of the helix is unfolded, the energetic cost of exposing non-H-bonded peptide bonds is so immense that the only TM structure possible is a helix. This observation explains why helices cannot be unfolded in calorimetric measurements (Figure 4.8).

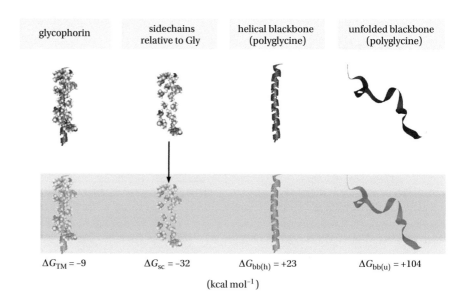

glycophorin

sidechains relative to Gly

helical blackbone (polyglycine)

unfolded blackbone (polyglycine)

$\Delta G_{TM} = -9$ $\Delta G_{sc} = -32$ $\Delta G_{bb(h)} = +23$ $\Delta G_{bb(u)} = +104$

(kcal mol^{-1})

unfavorable cost of dehydrating the helix backbone, is −9 kcal mol^{-1}. As is true of most proteins, soluble or membrane, the relatively small net-favorable free energy results from the difference between two large numbers. Although the partitioning of the intrahelical hydrogen-bonded peptide groups is unfavorable (about 1.2 kcal mol^{-1} per H-bond), the partitioning cost of free peptide bonds is considerably higher. If a backbone H-bond is broken, an additional unfavorable cost of at least 4 kcal mol^{-1} per peptide bond is incurred (Figure 3.5). The result is that the cost of partitioning an unfolded 21-residue peptide into the bilayer interior exceeds 100 kcal mol^{-1}! Said another way, the cost of unfolding an α-helix completely in a lipid environment is about 80 kcal mol^{-1}. To appreciate a free energy of this size, consider that it means that there is only one chance in 10^{70} that a helix would be unfolded in the lipid bilayer, or essentially, zero. To put that number in perspective, the age of the Universe is about 10^{17} seconds!

3.4.6 pHLIP Peptides Can Partition Spontaneously as Transmembrane Helices

The reversible folding of α-helical membrane proteins into lipid bilayers is problematical, if not impossible. But perhaps, one could reversibly refold a single TM helix with the idea of measuring directly the free energy of insertion along the lines of Figure 3.26. This has also proven difficult, because helices that are hydrophobic enough to partition into the bilayer aggregate in the water phase, which makes it difficult to determine partitioning free energies directly. One major exception is the peptide pHLIP (**Figure 3.27**), derived from the amino acid sequence of bacteriorhodopsin's helix C, which inserts spontaneously into lipid bilayers at acid pH values. At high pH values, several Asp residues are charged, causing the mostly disordered surface helix precursor to be reasonably soluble, equilibrating between a bilayer surface and the aqueous phase. Lowering the pH to protonate the Asp residues causes the precursor to aggregate, unless lipid bilayer vesicles are present, in which case, the peptide spontaneously forms a surface helix and inserts into and across the bilayer. This makes sense because protonation of Asp and Glu dramatically lowers their cost of partitioning into non-polar phases (Figure 3.6). Several families of pHLIP, based upon helix C, are being intensively studied. The thermodynamics of insertion can be determined, but establishing a thermodynamic connection to the scheme of Figure 3.26 has proven difficult because insertion begins with an unfolded peptide on the interface. Kinetic studies show that the peptide forms a helix at the surface when the pH is dropped and that the helix then inserts. So,

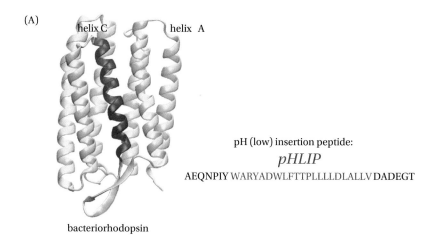

(A)

helix C   helix A

pH (low) insertion peptide:

pHLIP

AEQNPIY WARYADWLFTTPLLLLDLALLV DADEGT

bacteriorhodopsin

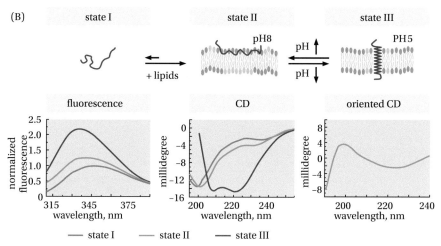

(B)

state I   state II   state III

pH8   PH5

+ lipids   pH ↑ / pH ↓

fluorescence   CD   oriented CD

— state I   — state II   — state III

Figure 3.27 Reversible insertion of the "pH-Low Insertion Peptide" (pHLIP) is controlled by pH. The pHLIP peptide is based upon helix C of bacteriorhodopsin (A). At high pH values, the aspartates are deprotonated (charged) and cannot enter the membrane, but they can bind to the membrane interface. This conclusion is supported by the fluorescence and CD spectra shown in (B). When the pH is lowered, the aspartates are protonated (neutral charge), which allows pHLIP to fold and insert into the membrane as an α-helix (indicated by the CD spectrum of State III). The trans-membrane configuration is verified by the oriented CD spectrum on the lower right. Oriented CD involves measuring the CD spectrum of bilayers oriented perpendicular to the spectrometer's optical axis. The spectra of helices oriented parallel and perpendicular to the bilayer are characteristically different. (Adapted from Andreev OA, Karabadzhak AG, Weerakkody D et al. [2010] *PNAS* 107: 4081.)

the connection to Figure 3.26 is not so remote after all; it is the kinetics of the process that stand in the way of thermodynamic analysis. An emergent property of pHLIP peptides is that cargos, such as fluorescent dyes, attached to the C-terminus can be transported across the membrane, because pHLIP peptides insert vectorially. This property makes pHLIP peptides useful for targeting cargos to acidic diseased tissues, such as cancers.

3.4.7 Hydrophobic Mismatch and the Single Helix

Our primary focus so far has been on peptides that tend to prefer the bilayer interface. Another approach that has proved very useful is to study peptides designed to capture particular features of TM proteins, such as changes in bilayer thickness in the vicinity of membrane proteins. Two families of peptides that have been particularly useful are the WALP and KALP peptides (**Figure 3.28**), which capture important features of membrane proteins, in particular the belts of aromatic and positively charged residues located at the membrane interface. The WALP peptides have a membrane-spanning hydrophobic domain comprised of Leu and Ala that is bounded by pairs of tryptophans. The KALP peptides are similar except that lysines replace the tryptophans.

Because the WALP and KALP peptides are insoluble in water, bilayers containing them are prepared by mixing the lipids and peptides in a polar organic solvent, evaporating the solvent, and then dispersing the mixtures in water. NMR is the principal, and a particularly useful, tool for studying the behavior of both peptides and lipids. Phosphorus NMR, for example, allows one to establish the phase behavior of the lipid–peptide mixtures using ^{31}P-NMR (**Figure 3.29A**).

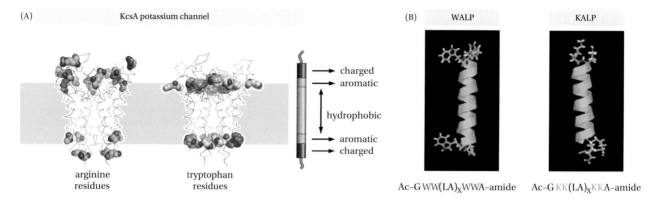

Figure 3.28 Peptides that mimic features of transmembrane helices. (A) Illustration of the design principle behind the WALP and KALP peptide families. (B) The WALP and KALP peptides. (A, Modified from de Planque MRR, Killian JA [2003] *Mol Membrane Biol* 20: 271–284. B, From Killian JA [2003] *FEBS Lett* 555: 134–138. With permission from Elsevier.)

This is important because phosphorus NMR studies of mixtures of WALP peptides with phospholipids with different chain lengths provided some of the earliest evidence of the importance of hydrophobic mismatch in the energetics of lipid–protein interactions. For positive mismatch, i.e., if the helix is longer than the bilayer thickness, matching can be achieved at little energetic cost simply by tilting relative to the bilayer normal. Consequently, long helices and

Figure 3.29 WALP peptides of different lengths affect the phase behavior of lipids with different acyl chain lengths. (A) The shape of ^{31}P NMR signals distinguish bilayer, isotropic, and hexagonal phases in lipid–peptide mixtures. Isotropic phases, believed to be due to cubic lipid phases (see Pn3m, Figure 2.12), have a single symmetric peak. (B) A proposed structure of the hexagonal phase. Rather than spanning the bilayer, the helices span between the hexagonal rods. (C) Short helices in thick bilayers are not favored because the bilayer must distort near the helix. Rather than pay that price, a hexagonal phase forms. There is little energetic cost to match long helices to thin bilayers; the helices can simply tilt. (A, From Killian JA, Salemink I, de Planque MRR et al. [1996] *Biochemistry* 35: 1037–1045. With permission from American Chemical Society. (inset) From Ramadurai S, Holt A, Schafter LV et al. [2010] 99: 1447–1454. With permission from The Biophysical Society and Elsevier.)

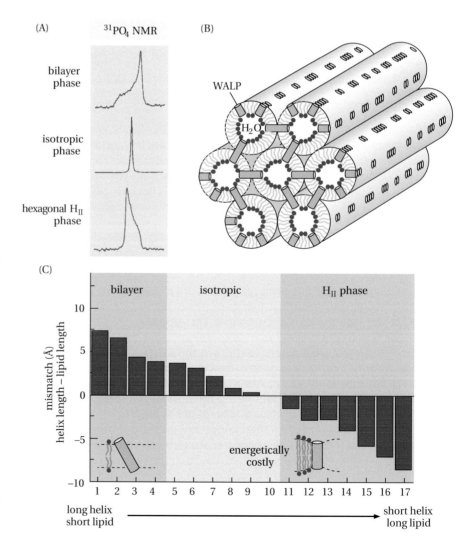

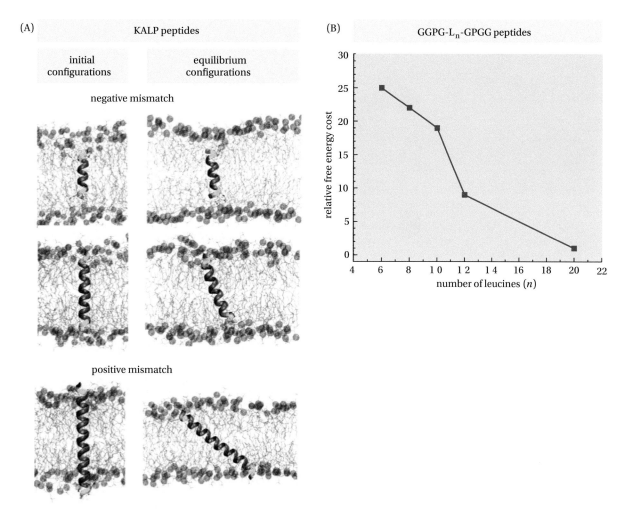

Figure 3.30 Simulations of hydrophobic mismatch for single TM helices. (A) Negative mismatch for short KALP helices causes thinning distortion near the helix, whereas for positive mismatch, long helices simply tilt. (B) The relative free energy cost of thinning distortions for negative mismatch rises dramatically as the helix becomes shorter. This may explain why negative mismatch can lead to hexagonal phases (Figure 3.29). (A, From Kandasamy Sk, Larson RG [2006] *Biophys J* 90: 2326–2343. With permission from Elsevier. B, From Jaud S, Fernández FM, Nilsson IM et al. [2009] 106: 11588–11593.)

lipids with short acyl chains prefer bilayer phases (**Figure 3.29B**). For negative mismatch, i.e., the helix is shorter than the bilayer thickness, the bilayer must distort and become thinner near the helix. If this thinning distortion were not energetically costly, then one would expect to observe only bilayer phases. However, hexagonal phases are preferred for negative mismatch, which implies that it is energetically costly to distort the bilayer by thinning it close to the helix (**Figure 3.29C**).

Molecular dynamics simulations of single KALP helices in bilayers are consistent with the idea that positive mismatch is relieved by simple tilting, whereas bilayer thinning in the helix neighborhood is required for negative mismatch (**Figure 3.30A**). The question raised by phase behavior measurements of WALP-lipid mixtures is the cost of negative mismatch–thinning distortions in the vicinity of helices. The relative cost of negative mismatch has been computed from molecular dynamics simulations of another model TM helix peptide, GGPG-L_n-GPGG, in POPC bilayers (**Figure 3.30B**). Relative to neutral mismatch, the energetic cost of thinning increases dramatically as the number of leucines decreases. For the bilayer to adapt to a helix with 12 leucines, the energetic cost of distortion increases tenfold relative to the neutral mismatch with 20 leucines. It is no wonder, then, that WALP–lipid mixtures form hexagonal phases rather than bilayer phases with extreme negative mismatch.

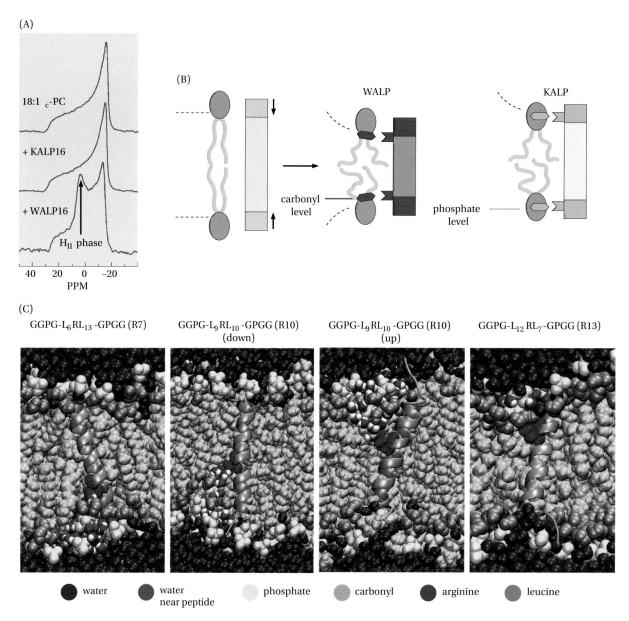

Figure 3.31 Aromatic and basic residues snorkel into the bilayer interface. (A) NMR spectra reveal the appearance of hexagonal H_{II} phase for the WALP16 peptide in DOPC but not for the KALP16 peptide. This implies different effective lengths of the two peptides even though their hydrophobic domains are identical. The logical explanation is that the long, flexible lysine side chain can "reach" easily for the lipid phosphate group, making the peptide effectively longer. (B) A schematic representation of the differences in snorkeling by aromatic and basic residues. (C) Molecular dynamics simulations reveal how bilayers might distort in the vicinity of arginine residues in order to optimize arginine–phosphate interactions. The simulations also reveal that the phosphate–arginine interaction can be mediated by carbonyl groups and water. (A, From Maurits RR, de Planque MRR, Kruijtzer JA et al. [1999] *J Biol Chem* 274: 20839–20846. With permission from Elsevier. B, From Killian JA, von Heijne G [2000] *Trends Biochem Sci* 255: 492–433. With permission from Elsevier. C, From Schow EV, Freites JA, et al. [2011] *J Membrane Biol* 239:35–48.)

3.4.8 Side Chain Snorkeling Is Important in Transmembrane Helix–Bilayer Interactions

Membrane proteins are enriched in the membrane interfacial region with tryptophan, tyrosine, arginine, and lysine. This is not unexpected, because Trp and Tyr have a particular affinity for the interface (Figure 3.11), and Arg and Lys have strong interactions with phosphate groups. If the affinities are strong enough, it would be reasonable to think that the peptide and the lipids would arrange themselves to optimize these interactions. This is exactly what is observed by ^{31}P NMR (**Figure 3.31A**). The two peptides WALP16 and KALP16 are identical except for the substitution of lysine for tryptophan. However, lipid ^{31}P NMR spectra for mixtures of WALP16 and KALP16 with DOPC are not the same: H$_{II}$ phase is present in the WALP16 system but not in the KALP16 system. Extensive measurements of this type have led to the conclusion that the WALP peptides are effectively shorter than KALP peptides even though their hydrophobic segments are identical. This is a manifestation of "snorkeling," in which Trp or Lys can "reach" for their preferred locations in the bilayer interface, shown schematically in **Figure 3.31B**; Trp likes to reside roughly at the level of the lipid carbonyl groups, whereas Lys likes to reside close to the phosphate groups. Because Lys is rather long and flexible due to its methylene groups, it can reach further than Trp. The difference in reach causes the KALP16 peptides to have a longer effective length than the WALP16 peptides.

The cartoon of Figure 3.31B illustrates the idea of snorkeling, but it does not give a sense of the dynamics of snorkeling or the deformation of the bilayer. These features of snorkeling are best appreciated through molecular dynamics simulations. Frames from simulations of a leucine TM peptide with a single arginine in different positions along the helix are shown in **Figure 3.31C**. The snorkeling is quite apparent. The arginine interacts strongly with the phosphates, but frequently, the interaction is mediated through water and the carbonyl groups. The effect of snorkeling on the bilayer structure becomes apparent when the arginine is moved closer to the bilayer center. The arginine–phosphate interaction is so strong that the bilayer distorts locally to optimize the interaction, as expected from the NMR measurements. Furthermore, this reveals the stability of the helix relative to the bilayer.

KEY CONCEPTS

- Gibbs partitioning energies provide a foundation for describing lipid–protein interactions.

- Partitioning free energies of amino acids provide hydrophobicity scales.

- The energetic cost of partitioning peptide bonds into bulk non-polar phases is very high if they do not participate in H-bonding.

- Lipid bilayers are not equivalent to bulk non-polar phases.

- Partitioning of pentapeptides between water and octanol and between water and zwitterionic membrane interfaces provide whole-residue amino acid scales.

- Tryptophan and tyrosine amino acids have a special affinity for membrane interfaces.

- Charged peptides interact strongly and predictably with charged membrane interfaces.

- Electrostatic properties of charged lipid bilayers are described by Gouy–Chapman theory.

- Partitioning of charged hydrophobic peptides into membranes reveals the non-additivity of electrostatic and hydrophobic interactions.

- Some proteins bind to membranes with the help of covalently linked lipids.

- Membrane-active peptides provide important insights into secondary structure formation at membrane interfaces.

- The bilayer interface induces secondary structure formation through partitioning-folding coupling.

- The high energetic cost of partitioning peptide bonds into membrane interfaces explains partitioning-folding coupling.

- Peptide-bond energetics cause α-helices and β-barrels to be very stable transmembrane structures.

- The stability of transmembrane helices results from the high cost of partitioning peptide bonds into lipid bilayers.

- Some special peptides can partition spontaneously as transmembrane helices.

- Snorkeling of aromatic and charged amino acids is an important feature of helix–bilayer interactions.

FURTHER READING

Bayburt, T.H., and Sligar, S.G. (2010) Membrane protein assembly into nanodiscs. *FEBS Lett* 584:1721–1727.

Jensen, M.Ø., and Mouritsen, O.G. (2004) Lipids do influence protein function: The hydrophobic matching hypothesis revisited. *Biochim. Biophys. Acta* 1666:205–226.

Killian, J.A., and von Heijne, G. (2000) How proteins adapt to a membrane-water interface. *Trends Biochem.Sci.* 25:429–434.

Killian, J.A., and Nyholm, T.K.M. (2006) Peptides in lipid bilayers: The power of simple models. *Curr. Opin. Struct. Biol.* 16:473–479.

London, E., and Ladokhin, A.S. (2002) Measuring the depth of amino acid residues in membrane-inserted peptides by fluorescence quenching. *Curr. Top. Membr.* 52:89–115.

McLaughlin, S. (1989) The electrostatic properties of membranes. *Ann. Rev. Biophys. Biophys. Chem.* 18:113–136.

Murray, D., Ben-Tal, N., Honig, B., and McLaughlin, S. (1997) Electrostatic interaction of myristoylated proteins with membranes: Simple physics, complicated biology. *Structure* 5:985–989.

Nyholm, T.K.M., Ozdirekcan, S., and Killian, J.A. (2007) How protein transmembrane segments sense the lipid environment. *Biochemistry* 46:1457–1465.

Segrest, J.P., Harvey, S.C., and Zannis, V. (2000) Detailed molecular model of apolipoprotein A-I on the surface of high-density lipoproteins and its functional implications. *Trends Cardiovascular Med.* 10:246–252.

Tamm, L.K. (2005) *Protein-Lipid Interactions. From Membrane Domains to Cellular Networks*, pp. 444, WILEY-VCH Verlag GmbH & Co KGaA, Weinheim.

White, S.H., and Wimley, W.C. (1998) Hydrophobic interactions of peptides with membrane interfaces. *Biochim. Biophys. Acta* 1376:339–352.

White, S.H., Ladokhin, A.S., Jayasinghe, S., and Hristova, K. (2001) How membranes shape protein structure. *J. Biol. Chem.* 276:32395–32398.

White, S.H., Wimley, W.C., Ladokhin, A.S., and Hristova, K. (1998) Protein folding in membranes: Determining energetics of peptide-bilayer interactions. *Methods Enzymol.* 295:62–87.

White, S.H., and Wimley, W.C. (1999) Membrane protein folding and stability: Physical principles. *Annu. Rev. Biophys. Biomol. Struc.* 28:319–365.

KEY LITERATURE

Andreev, O.A., Karabadzhak, A.G., Weerakkody, D., Andreev, G.O., Engelman, D.M., and Reshetnyak, Y.K. (2010) pH (low) insertion peptide (pHLIP) inserts across a lipid bilayer as a helix and exits by a different path. *Proc. Natl. Acad. Sci. U.S.A.* 107:4081–4086.

Ben-Tal, N., Honig, B., Peitzsch, R.M., Denisov, G., and McLaughlin, S. (1996) Binding of small basic peptides to membranes containing acidic lipids: Theoretical models and experimental results. *Biophys. J.* 71:561–575.

Beschiaschvili, G., and Baeuerle, H.-D. (1991) Effective charge of melittin upon interaction with POPC vesicles. *Biochim. Biophys. Acta* 1068:195–200.

de Planque, M.R.R., Kruijtze, J.A.W., Liskamp, R.M.J., Marsh, D., Greathouse, D.V., Koeppe, R.E., II, de Kruijff, B., and Killian, J.A. (1999) Different membrane anchoring positions of tryptophan and lysine in synthetic transmembrane α-helical peptides. *Biochemistry* 274:20839–20846.

De Young, L.R., and Dill, K.A. (1990) Partitioning of nonpolar solutes into bilayers and amorphous n-alkanes. *J. Phys. Chem.* 94:801–809.

Eisenberg, D., Weiss, R.M., and Terwilliger, T.C. (1982) The helical hydrophobic moment: A measure of the amphiphilicity of a helix. *Nature* 299:371–374.

Engelman, D.M., Steitz, T.A., and Goldman, A. (1986) Identifying nonpolar transbilayer helices in amino acid sequences of membrane proteins. *Annu. Rev. Biophys. Biophys. Chem.* 15:321–353.

Hristova, K., Dempsey, C.E., and White, S.H. (2001) Structure, location, and lipid perturbations of melittin at the membrane interface. *Biophys. J.* 80:801–811.

Hristova, K., and White, S.H. (2005) An experiment-based algorithm for predicting the partitioning of unfolded peptides into phosphatidylcholine bilayer interfaces. *Biochemistry* 44:12614–12619.

Hunt, J.F., Rath, P., Rothschild, K.J., and Engelman, D.M. (1997) Spontaneous, pH-dependent membrane insertion of a transbilayer α-helix. *Biochemistry* 36:15177–15192.

Jaud, S., Fernández-Vidal, M., Nillson, I., Meindl-Beinker, N.M., Hübner, N.C., Tobias, D.J., von Heijne, G., and White, S.H. (2009) Insertion of short transmembrane helices by the Sec61 translocon. *Proc. Natl. Acad. Sci. USA* 106:11588–11593.

Janin, J. (1979) Surface and inside volumes in globular proteins. *Nature* 277:491–492.

Killian, J.A., Salemink, I., de Planque, M.R.R., Lindblom, G., Koeppe, R.E., II, and Greathouse, D.V. (1996) Induction of nonbilayer structures in diacylphosphatidylcholine model membranes by transmembrane α-helical peptides: Importance of hydrophobic mismatch and proposed role of tryptophans. *Biochemistry* 35:1037–1045.

Kyte, J., and Doolittle, R.F. (1982) A simple method for displaying the hydropathic character of a protein. *J. Mol. Biol.* 157:105–132.

Ladokhin, A.S. (1997) Distribution analysis of depth-dependent fluorescence quenching in membranes: A practical guide. *Meth. Enzymol.* 278:462–473.

Ladokhin, A.S., and White, S.H. (1999) Folding of amphipathic α-helices on membranes: Energetics of helix formation by melittin. *J. Mol. Biol.* 285:1363–1369.

Ladokhin, A.S., Jayasinghe, S., and White, S.H. (2000) How to measure and analyze tryptophan fluorescence in membranes properly, and why bother? *Anal. Biochem.* 285:235–245.

Ladokhin, A.S., and White, S.H. (2001) Protein chemistry at membrane interfaces: Non-additivity of electrostatic and hydrophobic interactions. *J. Mol. Biol.* 309:543–552.

Mouritsen, O.G., and Bloom, M. (1984) Mattress model of lipid-protein interactions in membranes. *Biophys. J.* 46:141–153.

Nozaki, Y., and Tanford, C. (1971) The solubility of amino acids and two glycine peptides in aqueous ethanol and dioxane solutions. Establishment of a hydrophobicity scale. *J. Biol. Chem.* 246:2211–2217.

Radzicka, A., Pedersen, L., and Wolfenden, R. (1988) Influences of solvent water on protein folding: Free energies of solvation of cis and trans peptides are nearly identical. *Biochemistry* 27:4538–4541.

Reshetnyak, Y.K., Andreev, O.A., Segala, M., Markin, V.S., and Engelman, D.M. (2008) Energetics of peptide (pHLIP) binding to and folding across a lipid bilayer membrane. *Proc. Natl. Acad. Sci. U.S.A.* 105:15340–15345.

Schow, E.V., Freites, J.A., Cheng, P., Bernsel, A., von Heijne, G., White, S.H., and Tobias, D.J. (2011) Arginine in membranes: The connection between molecular dynamics simulations and translocon-mediated insertion experiments. *Journal of Membrane Biology* 239:35–48.

Segrest, J.P. Jackson, R.L., Morrisett, J.D. et al. (1974) A molecular theory of lipid-protein interactions in the plasma lipoproteins. *FEBS Lett* 38:247–258.

Segrest, J.P., Jackson, R.L., Marchesi, V.T., Guyer, R.B., and Terry, W. (1972) Red cell membrane glycoprotein: Amino acid sequence of an intramembranous region. *Biochem Biophys Res Commun.* 49:964–969

Yau, W.-M., Wimley, W.C., Gawrisch, K., and White, S.H. (1998) The preference of tryptophan for membrane interfaces. *Biochemistry* 37:14713–14718.

Wimley, W.C., and White, S.H. (1993) Membrane partitioning: Distinguishing bilayer effects from the hydrophobic effect. *Biochemistry* 32:6307–6312.

Wimley, W.C., Creamer, T.P., and White, S.H. (1996) Solvation energies of amino acid sidechains and backbone in a family of host-guest pentapeptides. *Biochemistry* 35:5109–5124.

Wimley, W.C., and White, S.H. (1996) Experimentally determined hydrophobicity scale for proteins at membrane interfaces. *Nat. Struct. Biol.* 3:842–848.

EXERCISES

1. Treat the membrane as a slab of octanol, and a helix as a cylinder with successive segments representing the amino acids. Using the Wimley–White scale, consider the interactions of the following sequences with the membrane:

   A. $(NQ)_5(LV)_5(NQ)_5$

   B. $(NQ)_5(LV)_{10}(NQ)_5$

   C. $(NQ)_5(LV)_{15}(NQ)_5$

      a. If the slab of octanol is 30 Å thick and the peptides are α-helices perpendicular to the plane of the slab, and assuming that the slab does not distort, what are the association energies of each peptide with the slab? (Hint: remember to include the water–peptide interactions as well as the lipid–peptide interactions.)

      b. If we allow multiple copies of the peptide in each case, what would you expect their planar distribution to look like?

      c. If we allow the slab to distort, and the helix remains perpendicular to the plane, sketch the distortion you would expect. Give a one-sentence explanation for each case. (Hint: recall the black film torus.)

      d. If we allow the helix to tilt, sketch the change you would expect for each case and give a one-sentence explanation.

2. It has been suggested that the inside surface of the plasma membrane might help soluble proteins to fold, acting as a chaperone.

   a. Why might this idea work? (short answer)

   b. Name two problems that might limit (or even exclude) the widespread use of such a role.

   c. If such a function were observed *in vitro*, would large or small vesicles be likely to work best?

3. Negatively charged lipids can facilitate the binding of positively charged peptides, as shown for the binding of melittin to bilayers. In cell membranes, most of the negatively charged lipids are on the cytoplasmic surface of the bilayer. Would you expect that these lipids would affect the binding of positively charged peptides to the outer cell surface? How and why? (Hint: remember that the charges across a capacitor "see" each other.)

4. The use of partitioning scales to predict the interactions of helices with lipid bilayers has generally assumed that the helices are continuous α-helices. What if the helices are not continuous or not α-helices?

    a. What might the effect of a proline be? Would you expect that the absence of a backbone hydrogen bond would be influential? (Note that prolines are relatively abundant in TM helices.)

    b. What if the helix is a 3-10 helix? How many amino acids would be included in the calculation for a 30 Å hydrophobic region? Would you expect the backbone polarity to be more or less than for an α-helix? (Hint: consider the exposure and linearity of the H-bonds.)

Membrane Protein Folding and Stability

4

The physicochemical principles underlying simple peptide–bilayer interactions discussed in Chapter 3 provide an important foundation for describing the thermodynamic stability of membrane proteins (MPs), but they are not sufficient for understanding the formation of three-dimensional (3-D) structure or the stability of multi-span MPs. For example, they do not account for interactions between helical elements within the bilayer environment. When we begin to think about the physical principles that determine MP structure, an important question immediately arises: what do we mean by "stability"? As we will see, the answer is not simple. But certainly, the answer must include the energetics of the interactions between transmembrane helices and between β-strands in the environment of the lipid bilayer.

Our view of the requirements for describing the stability of membrane proteins has grown as new structural features have emerged from the ever-increasing number of crystallographic, nuclear magnetic resonance (NMR), and electron cryomicroscopic structures. Until the early 1990s, the only examples of helical bundle membrane proteins were bacteriorhodopsin (bR) and the photosynthetic reaction center (Figures 1.27, 1.28). These structures suggested that the transmembrane helices of MPs were packed very regularly and were close to being parallel to the membrane normal, as observed for bacteriorhodopsin (**Figure 4.1A**). It seemed in those early days that the main problem was to understand the energetics and geometry of helix packing. This view was reinforced by insightful experiments demonstrating that functional bR could be reassembled in lipid vesicles by fusing together vesicles containing only bR helices AB with vesicles containing only helices CDEFG (**Figure 4.1B**). This led to the Two-Stage model of membrane protein folding: after the insertion of TM helices by biological processes (**Figure 4.1C**), the helices spontaneously assemble to form the native protein and bind the prosthetic group retinal. Of course, in the 1990s, we were at the very early stages of understanding of membrane protein biogenesis (Chapter 6) and did not appreciate the complexities of the process. Although subsequent work showed that helix-bundle proteins can be assembled in living cells by co-expression of MP fragments, the *in vivo* assembly of MPs is generally more complex and interesting than implied by the Two-Stage model, which must be used carefully in light of our modern understanding of biogenesis (Chapter 6). Nevertheless, the early experiments reinforced the view that MP assembly and stability must adhere to physicochemical rules.

As more protein structures have been determined, the application of the rules has become more complicated. For example, many helix-bundle membrane proteins (including bacteriorhodopsin, Figure 4.1A) have charges buried deep in their hearts, while others have large water-filled compartments (**Figure 4.2A**). Even the translocon central to MP biogenesis (Chapter 6) is water filled, containing more than 400 water molecules (**Figure 4.2B**). Furthermore, some MPs, such as bacteriorhodopsin and the photosynthetic reaction center, bind

DOI: 10.1201/9780429341328-5

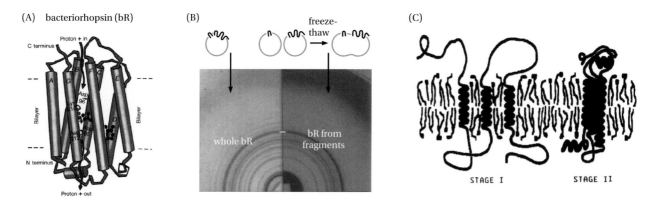

Figure 4.1 Early biophysical approaches to membrane protein stability. Until the 1990s, most α-helical membrane proteins were thought to be similar to bacteriorhodopsin (bR; panel A). An important step toward understanding membrane protein stability was taken by Popot and Engelman, who used x-ray diffraction to show that functional bR can be assembled in membranes from fragments (panel B). This discovery led to the Two-Stage model of MP assembly (panel C) and also suggests a mechanism for the formation of MP oligomers. See Chapter 1 and Chapter 2 for a discussion of x-ray and neutron diffraction studies of lipid bilayers. (A, From Caspar DLD [1990] *Nature* 345: 666–667. With permission from Springer Nature. B, From Popot J-L, Gerchman, S-E, Engelman DM [1987] *J Mol Biol* 198: 655–676. C, From Popot J-L, Engelman DM [1990] *Biochemistry* 29:4031–4037.)

prosthetic groups that contribute their binding energies to stability. Whereas the Two-Stage model focused solely on helix–helix interactions in the bilayer, subsequent schemes necessarily must include a broader array of lipid–protein interactions (Figure 3.1A).

We begin our discussion of MP stability by considering the stability of α-helical membrane proteins (α-MPs), starting with helix–helix interactions within folded proteins and insights provided by understanding the dimerization of the single-span MP glycophorin A (GpA). These are essential for understanding the stability of folded α-MPs. We then turn to the question of multi-span α-MPs by asking what information might be obtained using traditional soluble protein approaches to MP folding and stability, such as temperature and denaturants. The answer to that question reveals the fundamental problem of defining the thermodynamic reference state for describing α-MP stability. For soluble proteins, we can readily measure stability using the free energy difference between unfolded and folded states in aqueous solution. How are we to think about the relative free energies of a fully folded α-MP in the lipid bilayer and an unfolded or partially unfolded protein in water or aqueous detergent solution? New methods have in recent years allowed us to examine stability within the bilayer without using detergents, reflecting the idea that stability should be measured using the free energy difference between the folded and unfolded (or at least partially unfolded) states within the bilayer.

The choice of reference states for beta-barrel membrane proteins (β-MPs) is conceptually much simpler, because β-MPs generally enter the membrane from an aqueous compartment, whereas α-MPs generally enter the membrane with the aid of ribosomes and membrane insertases. As we will see, however, measuring the stability of β-MPs presents formidable challenges.

4.1 INTERACTIONS BETWEEN α-HELICES ARE CENTRAL TO 3D STRUCTURE FORMATION

The insertion of transmembrane helices into the membrane is only one step on the folding pathway. What kinds of interactions drive helix–helix assembly and ultimately stabilize the folded structure? How can these interactions be studied? What are the energies involved? The simplest system that can be used to

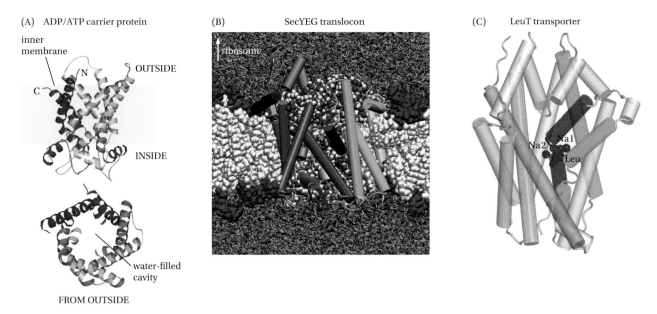

Figure 4.2 TM helical proteins are often not simple, close-packed helix bundles. Some MPs, such as the ADP/ADP carrier protein (A) and the SecYEG translocon (B), contain large numbers of water molecules, creating aqueous interfaces within the membrane. One must thus be cautious when assuming that transmembrane helices are exposed only to the bilayer interior. Further, charges can be stably positioned in the interiors of MPs such as the Leu transporter (C), so the energy penalty for placing a charge within the bilayer is not applicable inside the protein. This point should be obvious from the existence of ion channels. (A, From Pebay-Peyroula E et al. [2003] *Nature* 426: 39–44. With permission from Springer Nature. B, From Cymer F, von Heijne G, White SH [2015] *J Mol Biol* 437: 999–1022. With permission from Elsevier. C, From Yamashita A et al. [2005] *Nature* 437: 215–223. With permission from Springer Nature.)

address these questions is a transmembrane helix dimer, but we can also examine interactions within pairs of helices in multi-spanning α-MPs.

4.1.1 van der Waals Interactions Underlie Favorable Helix–Helix Interactions in α-Helical MPs

The hydrophobic effect explains why greasy α-helices prefer to reside in the bilayer rather than water. Once in the bilayer, however, the hydrophobic effect is no longer available. Why, then, should helices tend to pack together to form a structure? The answer is that helix packing is driven strongly by van der Waals dispersion forces, which also help stabilize soluble proteins. It is a short-range force (referred to technically as an induced-dipole:induced-dipole interaction) that arises in the interactions between all atoms and molecules. The strength of van der Waals dispersion energies can be estimated by creating packing voids by site-directed mutagenesis in both soluble and membrane proteins. For example, suppose a bulky isoleucine in a protein is replaced by a glycine, thus creating a void (poor packing). The cost of creating voids can be quantitated from the surface area of the voids and the change in folding free energy. For soluble proteins, the cost of creating packing voids has been determined experimentally to be 20 ± 5 cal $mol^{-1}Å^{-2}$. Interestingly, this is approximately equal to the atomic solvation parameter that underlies the hydrophobic effect. The value for membrane proteins is very similar: 18 ± 10 cal $mol^{-1}Å^{-2}$. Of course, the structure, stability, and function of a particular protein or a class of proteins depend upon many subtle variations in packing efficiencies and other interactions. Because van der Waals forces work to minimize voids in proteins, the packing efficiency of neighboring helices in proteins must play a significant role in structure formation. Francis Crick and subsequently Cyrus Chothia showed that neighboring helices in soluble proteins are packed tightly using a "ridges-into-grooves" arrangement, also referred to as "knobs-into-holes" packing.

4.1.2 Transmembrane Helix Dimers Are Stabilized by Close Packing

Many of our current ideas about interactions between transmembrane helices derive from studies of one particular protein, glycophorin A (GpA). GpA is a single-span membrane protein found on human red cells that carries blood type epitopes. A peculiar but very useful feature of the GpA transmembrane helix is that it self-dimerizes even in relatively harsh detergents such as sodium dodecyl sulfate (SDS). By random mutagenesis, seven residues in the GpA transmembrane helix have been identified as mediating dimerization (**Figure 4.3A**). In particular, Gly79 and Gly83 are highly susceptible to mutation. These results make perfect sense in the context of the NMR structure of the GpA dimer, also determined in detergent solution (**Figure 4.3B**).

The GpA data led to the identification of the first widespread interaction motif for transmembrane helices, the GXXXG motif. This motif is overrepresented in natural transmembrane helices and is often found in high-resolution structures. But, further studies have shown that not all GXXXG motifs mediate strong helix–helix interactions and that residues outside the motif can be important modulators of the interaction. The GpA dimer is held together by van der Waals interactions and tight shape complementarity due to knobs-into-holes packing. Possibly, weak hydrogen bonds between the Gly α hydrogens and carbonyl oxygens on the opposing monomer also contribute. Single mutations of Gly to large hydrophobic residues in the dimer interface can de-stabilize the dimer by up to nearly 4 kcal mol^{-1}, showing how important the close packing between the two helices is.

The association free energies of GpA dimerization have been determined by different methods in a wide variety of detergents and lipid bilayers. In SDS micelles, the dimerization free energy is about –9 kcal mol^{-1}, which is comparable to estimates for the free energy of transfer of a GpA monomer from water to bilayer (Figure 3.26). Within a natural cell membrane, however, the interaction energy is only –3 kcal mol^{-1}, while for simple lipid bilayers, the interaction energies range from –12 to –3 kcal mol^{-1}, depending on the lipid. This difference shows the importance of the lipid environment in helix–helix interactions.

4.1.3 A Small Number of Structural Motifs Describe the Majority of Helix–Helix Interactions in Membrane Proteins

Packing motifs with small residues (Gly, Ala, Ser) spaced either four or seven residues apart are frequent not only in simple dimers but also in multi-span

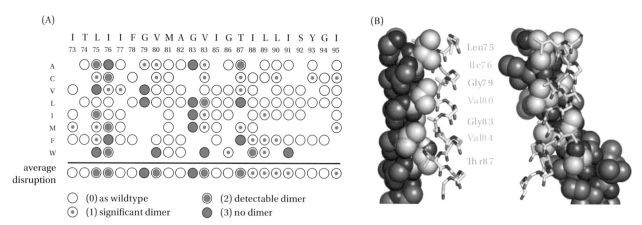

Figure 4.3 Dimerization motif of glycophorin A (GpA). (A) Effects of conservative mutations on the formation of GpA dimers in SDS. The data suggests a dimer interface motif—LIxxGVxxGVxxT—and a right-handed dimer packing. (B) NMR structure of the GpA dimer (PDB: 1AFO). One monomer is shown in spacefill, the other as a stick model. The two intermolecular Gly-Gly pairs allow the two helices to come very close together. (A, From Lemmon MA, Flanagan JM, Treutlein HR et al. [1992] *Biochemistry* 31: 12719–12725. With permission from American Chemical Society. B, From MacKenzie KR [2006] *Chem Rev* 106: 1931–1977. With permission from American Chemical Society.)

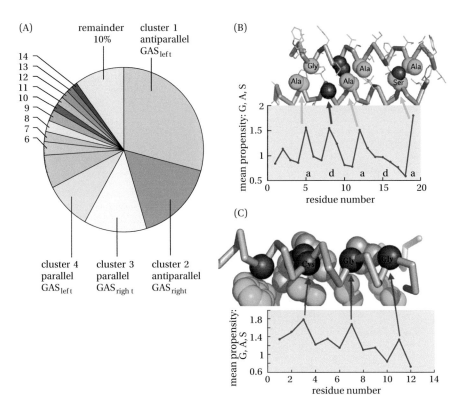

Figure 4.4 Helix–helix interactions in membrane proteins can be clustered into a small number of structural motifs (A). (B) Clusters 1 and 4 contain antiparallel and parallel helix pairs with a left-handed crossing angle and a preponderance of small Gly, Ala, and Ser (G,A,S) residues spaced seven residues apart. (C) Clusters 2 and 3 contain antiparallel and parallel helix pairs with a right-handed crossing angle and small residues spaced four residues apart. (From Walters RFS, DeGrado WF [2006] PNAS 103: 13658-13662. With permission of the author.)

α-MPs. Approximately two-thirds of all pairwise transmembrane helix–helix interactions that one can extract from the known high-resolution membrane protein structures can be clustered into four major groups (**Figure 4.4**). The first and largest cluster contains antiparallel helices with a small left-handed crossing angle and small residues spaced seven residues apart (a so-called Ala-coil interaction). Cluster 4 is a parallel version of cluster 1. Clusters 2 and 3 contain helices that have a right-handed crossing angle and interact via antiparallel and parallel GXXXG motifs; the GpA motif belongs to cluster 3.

4.1.4 Hydrogen Bonds between Polar Residues Can Stabilize Transmembrane Helix–Helix Interactions

In model systems, strongly polar residues such as Asn, Gln, Asp, Glu, and His can drive the formation of dimers or trimers of transmembrane helices, presumably through hydrogen bonding across the dimer interface. Because only a single residue is involved on each helix, this type of interaction is of low specificity and may therefore not be very useful in a biological context. Instead, multiple, more weakly interacting polar residues (Ser, Thr) may be more appropriate as mediators of helix–helix packing in natural proteins.

How much do inter-helical hydrogen bonds contribute to the stability of membrane proteins? Measurements on bacteriorhodopsin using double-mutant cycles (see later) yield values between –1.7 and –0.1 kcal mol^{-1} for eight different hydrogen-bond interactions in the protein (**Figure 4.5**). The average value is –0.7 kcal mol^{-1}, i.e., individual inter-helical hydrogen bonds tend to be weakly stabilizing. These values are quite similar to those found for hydrogen bonds in water-soluble proteins, and the hydrogen-bond distances in the interior of membrane proteins and soluble proteins are also similar. Of course, because a given structure may contain multiple hydrogen-bond interactions, the summed contribution of these bonds to the overall stability can be substantial. Recall the energetics of helix formation of melittin partitioned into membrane interfaces (Chapter 3). For each residue that participates in the formation of a backbone hydrogen bond, the partitioning free energy is reduced by about 0.4 kcal mol^{-1}, which by itself is not large. But 70% of the melittin amino acids

Figure 4.5 Hydrogen bonds in bacteriorhodopsin. The contribution to the protein's total stability of each hydrogen bond shown was determined using double-mutant cycle thermodynamic measurements (see Figure 4.10). While individual energies are not large, the sum can be considerable. (From Joh NH et al. [2008] *Nature* 453: 1266–1270. With permission from Nature Publishing Group.)

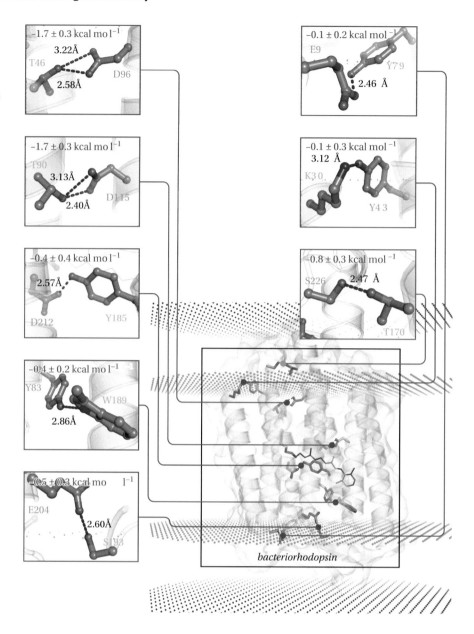

are helical, meaning that the total free energy reduction driving partitioning is a substantial –7 kcal mol^{-1}.

The H-bond interactions between the CONH groups of the helix backbone in transmembrane helices are much more difficult to determine. Theoretical calculations suggest values 10 times more favorable than H-bonds between side chains in hydrophobic environments (Figure 3.26). These estimates and experimental measurements suggest values of –4 to –5 kcal mol^{-1} per backbone H-bond.

4.1.5 Helix Connecting Links Are Not Necessary for the Formation of Native Structures

Interactions between the helices of α-helical membrane proteins strongly stabilize the proteins. Of course, the helices are connected by inter-helix sequences located in the extracellular spaces. And some proteins, such as bacteriorhodopsin, also have prosthetic groups within the protein that can stabilize the protein and that are crucial for function. How do these additional features of helix bundles contribute to stability? The simple answer is that the connecting links and prosthetic groups have a relatively small effect on stability compared with the helix–helix van der Waals interactions. A number of experiments on different proteins support this conclusion. The early Popot–Engelman experiments (Figure 4.1B) showed that the connecting link between helices B and C (Figure 4.1A) and the link between A and B need not be intact for stable bR reassembly. But what about the other connecting links? Thomas Marti at the Bernhard Nocht Institute in Hamburg reassembled bR from pairs of complementary bR fragments' helices. For example, helices AB and helices CDEFG (indicated by the shorthand AB•CG) could be recombined in dimyristoylphosphatidylcholine (DMPC)/CHAPS/SDS micelles along with the retinal chromophore. Other complementary fragments included AC•DG, AD•EG, and AE•FG. The optical spectrum of the retinal was essentially the same in all cases, meaning that bR had been correctly folded for all fragment combinations. Intact connecting links are clearly not required for proper assembly. But what about protein stability compared with wild-type protein under similar conditions? The stability of reassembled protein could be examined by measuring the chromophore spectrum as a function of increasing SDS concentration (**Figure 4.6**). The figure shows that the reassembled fragments are less stable than native bR, but in this experiment, it is difficult to determine the stability differences quantitatively. Nevertheless, it is clear that direct helix packing interactions are the primary determinants of 3-D structure formation.

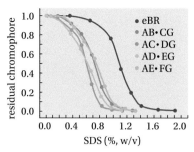

Figure 4.6 Stability of bacteriorhodopsin reassembled from fragments. Each fragment is indicated by the first and last helix it contains, e.g. AD=ABCD. At low SDS concentrations, the optical absorption of the retinal chromophore (residual chromophore = 1.0) indicates fully folded and functional bR. As the SDS concentration is increased, the residual chromophore decreases. The native bR (eBR) is clearly more stable than bR assembled from fragments. Nevertheless, at low SDS concentrations, eBR and bR assembled from fragments have the same structure and are stable. (From Marti T [1998] *J Biol Chem* 273: 9312–9322. With permission of Elsevier.)

4.1.6 Proline Residues Permit Kinks in Transmembrane Helices

From studies of water-soluble proteins, proline is generally considered a strong helix-breaking residue. Nevertheless, proline is found at an appreciable frequency in transmembrane helices, where it often serves important structural roles or imparts conformational flexibility to a helix. Because of its cyclic side chain structure, proline has a locked phi rotation angle that is compatible with being in a helix but has lost a stabilizing helix hydrogen bond. Thus, it can fit in an α-helix while providing a weak spot for kinks or bends in the helix backbone at a reduced energy penalty.

Still, many of the kinks and bends seen in high-resolution structures do not involve proline residues, and mutation of Pro residues to Ala in bacteriorhodopsin has a surprisingly small effect on the structure (**Figure 4.7**). A likely explanation is that the parts of the structure that surround the proline kink have evolved to provide a good fit to the kinked helix that will stabilize the kinked conformation even if the proline is mutated. It is possible that a similar explanation holds for the evolution of non-proline kinks in membrane proteins, and one

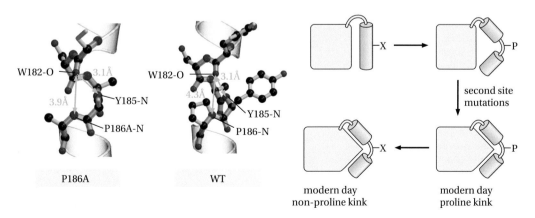

Figure 4.7 Left: $N_i+4...O_i$ distances around Pro186 in transmembrane helix F from bacteriorhodopsin (right) and in the Pro186-Ala mutant. The backbone changes are very small, despite the greater conformational flexibility and hydrogen-bonding capacity of Ala. Right: Evolutionary scenario for the evolution of non-proline kinks. (From Yohannan S, Faham S, Yang D et al. [2004] *PNAS* 101: 959-963. With permission from J.U. Bowie.)

can imagine an evolutionary scenario where a transmembrane helix initially becomes kinked as a result of an X-to-proline mutation, second-site mutations in surrounding parts of the structure then lead to improved packing interactions with the kinked helix, and finally, the proline mutates to another residue but the kink is left intact (Figure 4.7, right).

4.2 α-HELICAL MEMBRANE PROTEINS CAN BE DESTABILIZED USING HEAT, DETERGENTS, AND DENATURANTS

4.2.1 Calorimetric Measurements Reveal the High Stability of Transmembrane α-Helices

Soluble proteins can be fully and reversibly unfolded by means of differential scanning calorimetry (DSC) (Box 1.6). DSC measurements yield a different unfolding picture for bacteriorhodopsin (**Figure 4.8A**). Two transitions are seen. The lower-temperature transition is reversible and is due to the melting of the hexagonal lattice of the purple membranes formed by bR trimers (**Figure 4.8B**). The higher-temperature transition observed at about 100 °C is irreversible, because bR and the lipids form some kind of complex from which the structure of the protein and the lipid bilayer cannot be recovered by cooling. Circular dichroism measurements show that bR, unlike soluble proteins, retains at least 50% of its secondary structure in the denatured state. As shown

Figure 4.8 Thermal denaturation of membrane proteins. (A) DSC scan of bacteriorhodopsin (bR) in native membranes shows two thermal transitions. The lower-temperature transition is due to melting of the hexagonal lattice of bR trimers. The higher-temperature transition is due to irreversible denaturation. CD measurements and the enthalpy of unfolding indicate that even in the denatured state, bR may retain 50% of its helical secondary structure, emphasizing TM stability as a principle. (B) Trimeric arrangement of bR in native membranes. One monomer of the trimer is outlined. In this high-resolution structure, one can also see the arrangement of lipids within and surrounding the trimer; the lipids are a part of the lattice organization and the protein. (C) Comparison of the unfolding enthalpy of soluble and membrane proteins. The specific enthalpy of the melting of membrane proteins is generally lower than for soluble proteins, because their hydrophobic side chains are not fully exposed to water; they remain buried in a lipid environment and retain considerable secondary structure. (A, C, From Jackson M, Sturtevant J [1978] *Biochemistry* 17: 911–915. B, From Beirhali H [1999] *Structure* 15: 909–917. With permission from Elsevier.)

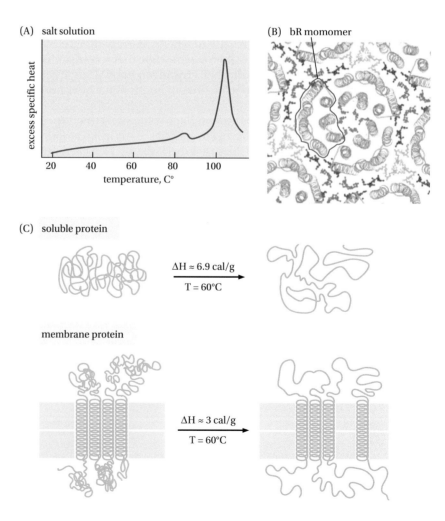

in **Figure 4.8C**, the enthalpy of membrane protein unfolding is generally much smaller (3 cal g^{-1}) compared with soluble proteins (6.9 cal g^{-1}). This is explained by the great stability of α-helices in lipid environments; the major share of the unfolding enthalpy is due to the unfolding of polypeptide segments that are external to the lipid bilayer (Figure 4.8C).

α-helical proteins have been the primary focus of calorimetric measurements; there have been very few measurements made on β-barrels. The limited amount of data available indicates that β-barrels also undergo two phase transitions, consisting of a lower transition that is reversible (probably due to external loops) and a higher irreversible transition due to unfolding of the membrane-buried domain.

4.2.2 α-Helical Proteins Can Be Denatured and Refolded Using Detergents and Lipids

Given the high stability of α-helices in lipid bilayers, it is not surprising that they present special challenges for non-thermal denaturing/refolding studies of the sort that are common in soluble proteins (**Box 4.1**). The earliest studies (1980s) showed that bacteriorhodopsin could be denatured with SDS and refolded to a native state in the presence of deoxycholate/phospholipid mixtures. Denaturation in the context of membrane proteins requires clarification, however. As with thermal denaturation, denaturation in SDS and other detergents is not complete; typically, around 50% of the native helicity remains. Bacteriorhodopsin can be completely unfolded by means of unpleasant solvents such as formic acid, but these are not generally useful for quantitative reversible unfolding.

Quantitative studies of the denaturation and refolding of α-helical membrane proteins have proven to be challenging but not impossible. Because detergents mix rather well with one another, a common approach is first to extract the protein from membranes into a non-denaturing detergent (e.g., *n*-decyl *β*-D-maltoside, abbreviated DM) and then denature by the addition of SDS. Examples of denaturation curves for diacyl glycerol kinase (DGK) and bacteriorhodopsin are shown in **Figure 4.9A** and **Figure 4.9B**, respectively. What thermodynamic information do we obtain from studies such as these? Using analyses like those of Box 4.1, the values obtained for bR unfolding from bicelles (see Figure 2.14) are ΔG_U^0 = 16 kcal mol^{-1} and m = –22 kcal mol^{-1}. Similar numbers are obtained for DGK, meaning that the proteins are extremely stable, although mystery remains as to the structural interpretation of unfolding. One idea that is supported by small-angle neutron scattering of SDS-denatured proteins is a beads-on-a-chain model in which structural components, e.g., α-helices, are solubilized individually by SDS, resulting in a chain of SDS-encrusted secondary structure elements. Further complicating the problem is that denaturing rearrangements of the protein must be accompanied by SDS rearrangements as well. Exactly what ΔG_U^0 means in terms of protein stability in a lipid bilayer is thus not known with certainty.

Given the fuzziness of our knowledge of the unfolded state, how can we extract useful information about biological membrane protein synthesis and stability from detergent-induced unfolding measurements? One way is not to worry about the meaning of ΔG_U^0 but rather, focus on changes in ΔG_U^0 caused by mutations. In the simplest case, one assumes that the energetic difference between the wild-type and mutant proteins in the denatured states is small, thus attributing changes in free energy, ΔG_U^0, to energetic differences between the folded proteins. This is risky, however. A better approach is to use so-called double-mutant cycles that allow one to gauge interactions of particular interest. An example of this procedure for estimating the interaction energy between two amino acids forming a hydrogen bond is shown in **Figure 4.10**. This approach assumes, however, that the effects of the mutations are independent and additive. Whether they are or not is uncertain.

BOX 4.1 MEASURING THE STABILITY OF SOLUBLE PROTEINS

All soluble proteins that have stable secondary and tertiary structures under normal physiological conditions reside in free energy minima of –5 to –10 kcal mol⁻¹ pretty much independently of molecular weight. How do we measure the stability (free energy) of a soluble protein, and what is its significance? More germane to this book: what is the stabilizing free energy of a membrane protein? The answer to the last question is simple: we do not really know, because unlike soluble proteins, membrane proteins cannot generally be completely unfolded reversibly in the bilayer environment. Nevertheless, some progress is being made by using approaches modified from those used to study the stability of soluble proteins. Thus, a basic knowledge of soluble protein folding is useful. Two commonly used methods are chemical and thermal denaturation.

Chemical denaturation requires using a suitable measure of protein conformation and a chemical agent that can cause unfolding without covalent damage. Several measures of folding are in common use, including tryptophan fluorescence and circular dichroism (CD). Tryptophans fully exposed to water have a fluorescence maximum at 350 nm. As tryptophans lose water exposure during folding into a compact structure with a less polar interior, tryptophan fluorescence shifts toward shorter wavelengths (blue shift). At the same time, the quantum yield—the ratio of emitted to absorbed photons—increases. Similarly, an unfolded protein has a random-coil CD spectrum, whereas a folded protein has strong absorbance in some region characteristic of the folded protein, e.g., absorbance at 220 nm, which is characteristic of α-helices. The idea, then, is to follow one these spectroscopic measures of conformation as a function of folded-state perturbation. The most commonly used destabilizers are temperature, urea, and guanidinium

chloride (GdmCl). A hypothetical example of unfolding induced by urea is shown in Box **Figure 1A**. The mechanism of urea denaturation is still not resolved fully. There are two hypotheses: urea binds strongly to the unfolded protein, or it modifies the solvent environment. The "data" are examples of the two-state unfolding of a protein in which there is an equilibrium between the native (N) and unfolded (U) states:

$$N \rightleftarrows U \tag{1}$$

These experimental data must be converted to a plot of the fraction of unfolded protein (f_u) as a function of denaturant concentration (Box **Figure 1B**). To do this, one must know the value of the observed parameter at each urea concentration for the native and unfolded forms. The general approach is to extrapolate the baseline regions to where the native and unfolded states are predominant. However, if the baseline regions are short, because the protein is exceptionally stable or unstable, the extrapolations can be a source of error.

There will be an equilibrium between the folded and unfolded forms of the protein at each denaturant concentration, which allows the free energy of unfolding ΔG_u to be calculated at each concentration:

$$\Delta G_U = -RT \ln \frac{[U]}{[N]} = -RT \ln \frac{f_U}{1-f_U} \tag{2}$$

In this equation, the ratio [U]/[N] defines the equilibrium constant K_u for unfolding for a particular denaturant concentration. Often, site-specific mutations (*mut*) cause changes in stability. The change in unfolding free energy is computed from $\Delta\Delta G_U = \Delta G_U^{wt} - \Delta G_U^{mut}$.

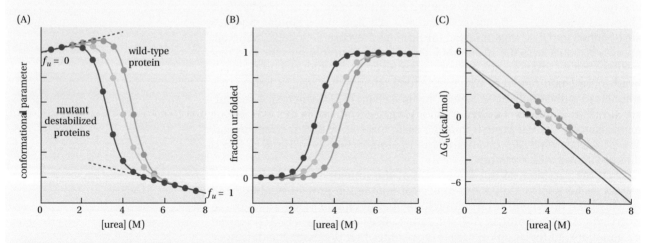

Figure 1 Measuring the stability of a soluble protein using urea denaturation. (A) A conformational parameter, such as Trp fluorescence, is measured as a function of urea concentration. (B) The data are normalized relative to the red dashed lines in panel A. (C) The free energy of unfolding ΔG_u is computed from Equation 2, assuming a linear dependence on denaturant concentration (Equation 3). Because only the steepest part of the curve in panel B is the most accurately determined region, only those data are used in making the plot. Extrapolation of the linear curves to urea concentration = 0 provides an estimate of ΔG_U^0. But notice the long extrapolation. (From Goldenberg D [1992], in Creighton TE (eds), *Protein Folding*. WH Freeman, Oxford. With permission from John Wiley and Sons.)

(Continued)

BOX 4.1   MEASURING THE STABILITY OF SOLUBLE PROTEINS (CONTINUED)

The assumption is generally made that the unfolding free energy is linear in denaturant concentration C (which may not always be true), so that the dependence of ΔG_u can be written

$$\Delta G_U(C) = \Delta G_U^0 - mC \qquad (3)$$

The unfolding free energy at zero denaturant (ΔG_U^0) is estimated by fitting values of ΔG_U in the transition region to Equation 3. An example is shown in **Figure 1C**. The m in Equation 3 is strongly correlated with the unfolded protein's accessible surface area.

The method of Equation 3 permits differences in the stability of mutant proteins with respect to the wild-type protein to be estimated by comparing the intercepts of the extrapolations in Figure 1C, providing that the wild-type and mutant proteins have the same value of m. But if the m-values differ, the interpretation becomes problematic.

Just as temperature can be used to melt lipid bilayers (Box 1.6), it can also be used to unfold soluble proteins. An example of a DSC measurement of the heat capacity of hen egg-white lysozyme is shown in **Figure 2**. Such measurements were brought to perfection by Peter Privalov. Much thermodynamic information about stability can be obtained from such measurements. For example, the enthalpy of unfolding can be obtained by integration of the heat capacity curve. Particularly interesting is the

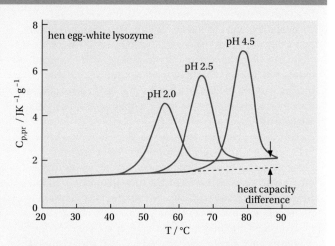

Figure 2 Thermal unfolding of hen egg-white lysozyme. As the protein unfolds, the heat capacity changes due to exposure of buried non-polar amino acid chains to water; the net change is indicated by the arrows. Notice that melting curves depend upon pH, indicating that stability also depends on pH. Indeed, pH changes can be used to denature proteins. The widths of the peaks are a measure of the cooperativity of the unfolding transition; the narrower the transition, the higher the cooperativity. (From Privalov PL in Creighton TE (eds), *Protein Folding*. WH Freeman, Oxford. With permission from John Wiley and Sons.)

observation that the heat capacity of the protein increases upon denaturation; this is a result of the hydrophobic effect due to exposure of buried non-polar groups (arrows).

An important lesson emerges from the difficulties encountered in denaturing and refolding α-helical membrane proteins into lipid bilayers. After synthesis by ribosomes, soluble proteins can fold spontaneously into their native active forms. If the same process were followed for membrane proteins that are insoluble in water, the likelihood of reliable incorporation of membrane proteins into lipid bilayers would be very small. This low probability must mean that specialized auxiliary structures and processes have evolved to manage the folding and insertion of helical membrane proteins. We will learn about these in Chapter 6.

4.2.3 α-Helical Membrane Proteins Can Be Unfolded Using Atomic Force Microscopy

Atomic force microscopy (AFM) makes it possible to pull transmembrane helices out of the membrane one by one. In the classical atomic force microscope, a long, flexible cantilever with an atomically sharp probe attached to one end is used to scan a surface. The deflection of the cantilever as it moves across the surface is recorded, producing a contour of the surface (**Figure 4.11A–D**). The vertical displacements of the cantilever can be measured within a few Å, making it possible to image the surfaces of membrane proteins protruding from a lipid bilayer.

The AFM can also be used to pull on molecules while recording the force exerted on the cantilever (Figure 4.11E). By tethering the end of a membrane protein to the AFM probe, the protein can be pulled out of the membrane, starting from the tethered end, usually the C-terminus. Because both the force F on the cantilever and the distance d between the tip and the membrane can

BOX 4.2 BILAYER MISMATCH AND GETTING RID OF THE TRASH

In all membranes, the turnover of proteins is essential. Particularly in the plasma membrane, space is limited, and many functions must be constantly adjusted to cope with changes in functional demands and to get rid of damaged or inactive proteins. We have seen that transmembrane helices are very stable and form the main structures of most transmembrane regions, so how can they be removed when they are no longer needed?

Part of the answer is to create enzymes that can cleave helices inside the bilayer hydrophobic region, creating unstable partial helices that move out of the bilayer to the membrane surfaces (based on the principles in this chapter, why would a solitary half-helix be unstable?). Rhomboid proteases are a widespread example of one way to accomplish this task and may make use of mismatch not only to enable proteolysis but also to facilitate rapid surface mobility to find substrate helices. Having single helices as substrates may make sense—as noted in this chapter, unfolding will likely result in transient or permanent single helices as a protein is destabilized by damage or modification of the parts exposed to the cytoplasm.

How can a transmembrane helix be hydrolyzed inside the bilayer? Serine proteases require water molecules as part of their catalytic cycles, so the Rhomboid molecule may distort the membrane thickness by mismatch to place its catalytic serine (yellow in **Figure 1**, left) in a favorable environment. A substrate helix (Figure 1, right) does not distort the bilayer, so it is thought that the binding to the protease must move its cleavage sites to the environment of the serine in the catalytic complex, although the complex has not yet been structurally documented.

Another consequence of the mismatch thinning is the fast diffusion of the protease in a membrane, which could improve function by allowing the protease to rapidly explore the membrane to remove unwanted helices. Using

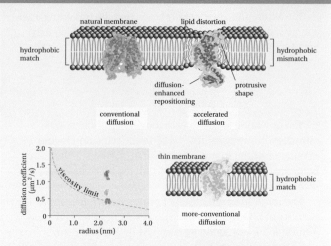

Figure 2 Mismatch allows rapid diffusion of rhomboid in membranes. Using experimental measurements in lipid bilayers and in natural membranes, it has been observed that the rhomboid protein diffuses more rapidly in a natural membrane than it should, given the viscosity of the membrane (the "viscosity limit"). When placed in a matching bilayer, diffusion is slower. Experiments suggest that the key to the fast diffusion is a mismatch with the lipids. (Kreutzberger AJB, Ji M, Aaron J et al. [2019] *Science* 363: eaao0076. With permission from AAAS.)

single molecule tracking of fluorescently labeled Rhomboid in bilayers and in cell membranes, Sinisa Urban and colleagues found that the mismatch speeds surface diffusion (**Figure 2**), presumably helping the protease to find its targets in a crowded membrane environment.

The Rhomboid serine protease superfamily is widely distributed in biology and is likely to be a highly refined evolutionary design. Finding that there is a synergy between catalytic mechanism and diffusion to find substrates is a fascinating example of efficiency.

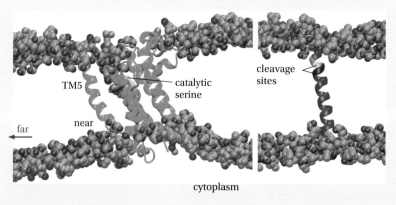

Figure 1 Simulation of a rhomboid protease and a substrate helix in a bilayer. The headgroup positions are distorted to thin the bilayer next to the protein (left), particularly near the catalytic serine. The thinning may allow access to the cleavage sites on the substrate helix (right), which is otherwise well matched to the bilayer. A relatively water-accessible environment facilitates catalysis in serine proteases. (From Bondar A-N, del Val C, White SH [2009] *Structure* 17: 395–405.)

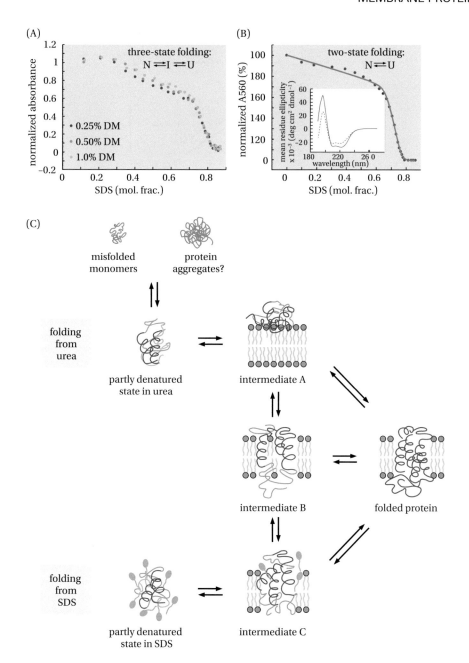

Figure 4.9 Using detergents to unfold membrane proteins. (A) Denaturation curves for diacylglycerol kinase, which has three detectable states. In this case, the refolding is in the presence of the non-denaturing detergent decyl β-D-maltoside (DM). (B) Denaturation of bacteriorhodopsin, which has two states. CD spectra of the two states (inset) show that the denatured state retains most of the helical structure of the native state. (C) A suggested general scheme for folding α-helical membrane proteins out of detergents and urea. (A, From Lau FW, Bowie JU [1997] *Biochemistry* 36: 5884–5892. With permission from the American Chemical Society. B, From Curnow P, Booth PJ [2007] *PNAS* 104: 18970–18975. With permission from PJ Booth. C, From Booth PJ, Curnow P [2006] *Curr Opin Struct Biol* 16: 480–488. With permission from Elsevier.)

be measured simultaneously, the unfolding of the protein can be visualized as a force–distance (*F-d*) curve, panel F in Figure 4.11. The *F-d* curve shown in Figure 4.11F is for bacteriorhodopsin. The peaks correspond to the pulling of successive pairs of helices: as the tether is pulled taut, the force *F* increases until a helix pair snaps out of the membrane, relaxing the tension on the tether. The same process is repeated again for the next pair of helices, and finally, the last helix is pulled out. The fact that helices are removed in pairs at moderate pulling speeds shows the importance of helix–helix interactions for stability. It also shows that backbone H-bonds are not easily broken in the interior of the bilayers. With sufficient force, however, helices can be pulled out individually by "unwinding" the helices by breaking backbone H-bonds.

This kind of unfolding process is obviously very different from unfolding with heat or detergents. A delightful feature of AFM is that single molecules are studied, one at a time. Unfortunately, the unfolding is a non-equilibrium process and is frequently not reversible. Furthermore, the unfolded state is an

Figure 4.10 Double-mutant-cycle analysis of hydrogen-bond (HB) energetics between two amino acid side chains, X and Y. If Y is mutated, the free energy change relative to the wild-type residue will be $\Delta\Delta G_u(Y)$, which includes the HB contribution ΔG_{HB} and any other contributions ΔG_{Yother} of Y to free energy in the folded or unfolded state. The same formalism applies to free energy changes related to mutations of X, $\Delta\Delta G_u(X)$. If both residues are mutated, then $\Delta\Delta G_u(XY)$ includes ΔG_{HB} and any other changes in free energy of X and Y in the folded or unfolded state: ΔG_{Yother} and ΔG_{Xother}. The nub of the double-mutant cycle, shown in the figure, is to add the single-mutant effects and to subtract the double-mutant effects, which on paper leads to cancellation of the "other" contributions, so that only ΔG_{HB} remains. This approach thus assumes that the effects of the mutants are additive and independent. Is the assumption correct? Maybe, maybe not; no one is sure. (From Bowie JU [2011] *Curr Opin Struct Biol* 21: 42–49. With permission from Elsevier.)

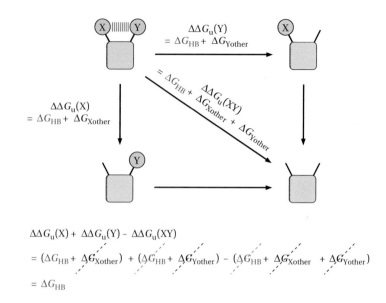

$$\Delta\Delta G_u(X) + \Delta\Delta G_u(Y) - \Delta\Delta G_u(XY)$$

$$= (\Delta G_{HB} + \Delta G_{Xother}) + (\Delta G_{HB} + \Delta G_{Yother}) - (\Delta G_{HB} + \Delta G_{Xother} + \Delta G_{Yother})$$

$$= \Delta G_{HB}$$

extended polypeptide chain in aqueous buffer, meaning that the underlying thermodynamic states are the folded protein in the bilayer and the unfolded protein in water. Nevertheless, the data are extremely useful, because the non-equilibrium problem can be solved using Jarzynski's equality. The equality, established from statistical mechanics, connects free energy differences between two equilibrium states to measurements made via a non-equilibrium process, such as pulling helices out of the bilayer by AFM. The careful application of the Jarzynski equality to *F–d* curves for bR collected over a wide range of temperatures and pulling rates yields a free energy difference between membrane-folded and water-unfolded states of about 300 ± 50 kcal mol^{-1}, or an average free energy change of about 1.2 kcal mol^{-1} per residue. Does the observed total free energy make sense based upon what we have learned so far in this Chapter, especially considering the much lower free energy values determined by detergent denaturation?

The huge bR extraction free energy of 300 kcal mol^{-1} is at first difficult to comprehend. The net bilayer-to-water free energy of transfer of the seven individual TM helices of bR estimated from the Wimley–White whole-residue *n*-octanol scale (Figure 3-6) is about –14 kcal mol^{-1}. How can these dramatically different free energies be reconciled? The likely explanation is that the energetic cost of breaking backbone H-bonds in a non-polar solvent is about +5 kcal mol^{-1}, whereas that of breaking the H-bonds in water is only about +1 kcal mol^{-1}. In the AFM pulling experiments, the helices are exposed to water as they unwind from the bilayer, suggesting that the per-residue free energies measured in the AFM experiments represent the breaking of backbone H-bonds in an aqueous environment. Ignoring the hydrocarbon/water transfer free energies of the 248 bR amino acid residues, the approximate total free energy cost of breaking the backbone H-bonds in water would be 248 kcal mol^{-1}. Inclusion of the cost of partitioning the sidechains of the seven TM helices into the bilayer would raise the total extraction free energy change to 262 kcal mol^{-1}, which agrees within computed error with the Jarzynski-based calculation of 290 ± 48 kcal mol^{-1}. This number serves as a reminder of the immense stability of α-MPs in lipid membranes.

We learned earlier that denaturation–refolding studies of bR yield free energies of unfolding of about 16 kcal mol^{-1} but that about 50% of the α-helicity is retained. The difference between the detergent-induced unfolding free energy and the AFM-determined free energy is consistent with the idea that individual intact helices are solubilized by the detergent during the unfolding of bR without breaking backbone H-bonds.

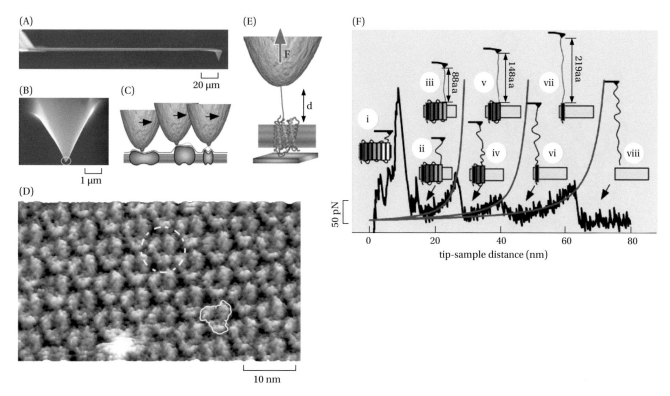

Figure 4.11 Atomic force microscopy (AFM) can be used to study the unfolding of single membrane proteins in a lipid bilayer. (A) A probe is attached to a cantilever (the probe is at the right-hand end). The probe has an atomically sharp tip (B) that can be scanned across a surface, mapping out a contour line (C). Repeated parallel offset scans of the surface result in an image of the surface (D), in this case bacteriorhodopsin in the purple membrane. The trimeric structure of bR in purple membranes is clearly resolved. By tethering the C-terminus of the membrane protein to the probe tip (E) and pulling on the cantilever with a force F while measuring the tip–membrane distance d, one can visualize the force profile of the extraction of the protein from the membrane as a force-distance (F–d) curve (F). (From Müller DJ, Kedrov A, Tanuj K et al. [2008] *ChemPhysChem* 9: 954–955. With permission from Wiley VCH Verlag.)

4.3 β-BARREL MEMBRANE PROTEINS CAN BE REVERSIBLY UNFOLDED USING DENATURANTS

4.3.1 β-Barrels Can Fold Reversibly into Membranes with the Help of Urea or Guanidinium Hydrochloride

Virtually all outer membrane proteins (OMPs) of Gram-negative bacteria are β-barrels. The protein OmpA is a classic example (**Figure 4.12A**). It is a small protein with eight β-strands that participates in bacterial conjugation. As we will learn in Chapter 6, OMPs are secreted across the inner (plasma) membrane into the periplasmic space, where they fold into the outer membrane with the help of chaperones and a membrane-bound insertase. OMPs might be expected, then, to be better suited to reversible denaturation and refolding into membranes than are α-helical membrane proteins. That is exactly what is observed. As summarized in Figure 4.12, OmpA can be denatured with urea in the presence of lipid vesicles. When the urea is diluted, OmpA partitions into lipid vesicles and refolds into its native structure. An important feature of OmpA is that its apparent molecular weight on polyacrylamide-SDS gels depends upon whether it is folded or unfolded; the apparent molecular weight of unfolded form is 5 kD heavier than the folded form. This property allows the reversibility of folding to be proven (Figure 4.12C). The happy consequence is that the folding of OmpA (Figure 4.12D) can be analyzed in the manner of soluble proteins (Box 4.1). And

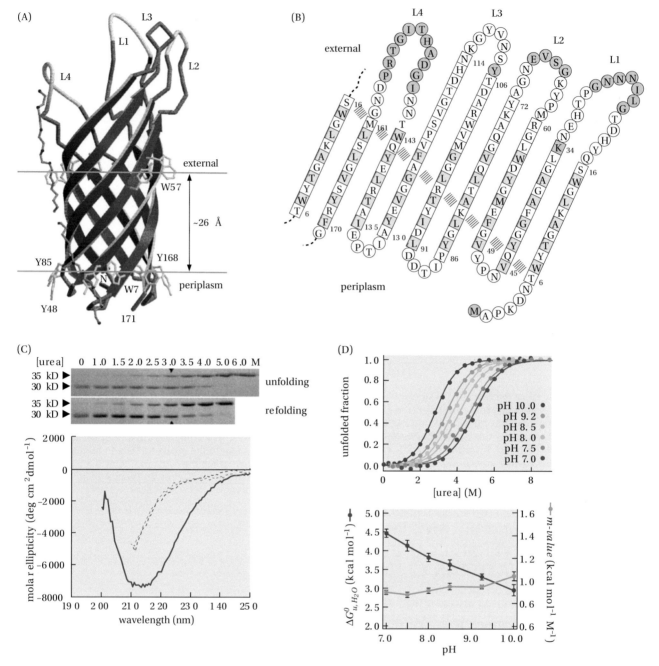

Figure 4.12 Reversible folding of the β-barrel protein OmpA into bilayers using urea denaturation. The structure of OmpA and its sequence are shown in panels A and B (the red connections show a set of mutually hydrogen-bonded positions). (C) Unlike helical membrane proteins, OmpA can be solubilized by urea in a denatured state with little secondary structure. Unfolded and folded OmpA migrate to distinct locations on SDS gels, allowing the two states to be distinguished (upper panel). The reversibility of folding is also seen in CD, where the minimum at ~213 nm (lower panel, red curve) indicates β-structure in the absence of urea, and its absence (dashed curves) shows a lack of regular structure in the presence of urea. Panel D reveals the pH dependence of folding. (A and B, From Pautsch A, Schultz GE [1998] *Nat Struct Biol* 5: 1913-1017. With permission from Nature America Inc. C and D, From Tamm LK, Hong H, Liang B [2004] *Biochim Biophys Acta* 1666: 250-263. With permission from Elsevier.)

just as for many soluble proteins, the urea concentration required for unfolding depends upon pH. At pH 7, the free energy of unfolding for OmpA in POPC/POPG bilayers is about 5 kcal mol⁻¹. Interestingly, this value for a whole β-barrel is not too different from the free energy of insertion estimated for a single α-helix (Figure 3.29). Measurements of the kinetics of insertion suggest that refolding and insertion of OmpA happen simultaneously when denatured OmpA refolds into lipid bilayers.

In addition to OmpA, Karen Fleming and her colleagues have found that other β-barrel proteins can also be reversibly refolded, particularly PagP and OmpLA, but using guanidinium HCl (GdnHCl), which is another denaturant commonly used in soluble protein folding. These two proteins are examples of β-barrel enzymes. PagP, with 8 β-strands, is a palmitoyl transferase; OmpLA, with 12 β-strands, is a phospholipase. The free energies of unfolding of PagP (14 kcal mol⁻¹) and OmpLA (32 kcal mol⁻¹) are much higher than for OmpA. The reason for the higher values may be related to the facts that OmpA tends to self-associate and has a soluble periplasmic domain that has about the same molecular weight as the membrane domain. In any case, these favorable folding free energies are huge.

Why are β-barrels so much easier to unfold and refold reversibly? Two features of OmpA and other OMPs are probably responsible. First, the spacing of the amino acid along a β-strand is about 3.3 Å per residue, meaning that only seven or eight residues are required to span the bilayer's hydrocarbon core. Second, the polarity of the residues along the strands of the barrel alternates between polar and non-polar (Figure 4.12B), so that the average polarity of a strand is much higher than an equal number of residues in a TM α-helix. These two features make OMPs much more amenable to urea or GdnHCl denaturation. Biologically, these features also allow proteins involved in membrane-protein assembly to distinguish between α-helical and β-barrel membrane proteins. In Gram-negative bacteria, SecYEG translocons located in the plasma membrane (Chapter 6) secrete nascent β-barrel chains into the periplasm, where they must fold and insert into the outer membrane. Our ability to denature and refold β-barrel membrane proteins shows that refolding into membranes is a natural tendency exploited by biology with the help of chaperones and insertases.

4.3.2 Reversible Refolding of OmpLA Can Be Used to Derive an Amino Acid Hydrophobicity Scale

OmpLA can be reversibly refolded into DLPC LUV out of guanidinium HCl, obeying a strict three-state equilibrium process (**Figure 4.13A**, black curve). This provides an opportunity to determine a side chain hydrophobicity scale in the context of a membrane protein in a lipid bilayer. Scale values can be obtained using the host-guest method utilized for determining an n-octanol hydrophobicity scale for pentapeptides (Figure 3.6), except that here, the host is OmpLA. An alanine in sequence position 210 lies in the center of the membrane, based upon the crystallographic structure (inset, **Figure 4.13B**). Changes in the reversible folding free energy that accompany substitutions of the 20 common amino acids at position 210 leads to a hydrophobicity scale for amino acids at the center of the bilayer in the context of the environment of OmpLA and the surrounding lipids. If the folding free energy of OmpLA with Ala at 210 is ΔG_{Ala}, then the free energy contribution relative to Ala of any amino acid X is given by $\Delta\Delta G_{AX} = \Delta G_{Ala} - \Delta G_X$. The set of $\Delta\Delta G_{AX}$ values is shown in Figure 4.13B, rank-ordered by magnitude.

How does one know if the OmpLA side chain hydrophobicity scale makes sense? First, one should expect the non-polar residues to be at the favorable end of the scale ($\Delta\Delta G_{AX} < 0$) and very polar residues at the opposite end ($\Delta\Delta G_{AX} > 0$). Figure 4.13B shows this to be the case. Second, one might reasonably expect for the non-polar side chains a linear dependence of $\Delta\Delta G_{AX}$ on the accessible sidechain surface area (Box 1.2). **Figure 4.13C** shows this to be the case as well. Furthermore, the resulting solvation parameter is –23 cal mol⁻¹ Å⁻², which is exactly the same as the value obtained for pentapeptide partitioning in n-octanol. A third test is whether the hydrophobicity scale is linearly related to other scales. **Figure 4.13D** shows that the OmpLA scale is linearly correlated with the n-octanol scale (Figure 3.7) with a highly significant correlation coefficient R of 0.89. The slope of the fitted line, however, is 0.5 rather than 1. What does this mean? The answer is, of course, that the contexts for the two scales are different.

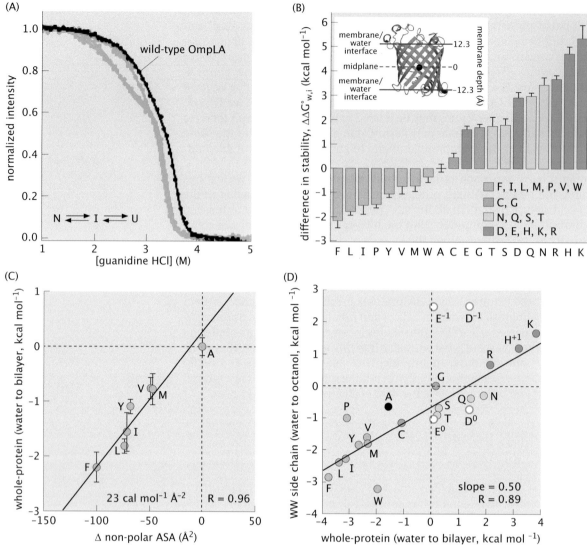

Figure 4.13 Determination of a side chain hydrophobicity scale based on the reversible three-state folding of the β-barrel protein OmpLA. (A) Folding curves for OmpLA measured using Trp fluorescence. The black curve is for wild-type OmpLA that has an Ala residue at sequence position 210 located near the center of the bilayer and on the lipid-exposed surface of the protein. The teal-colored curves are for substitutions of Asn, Gln, Ser, and Thr at position 210. (B) The inset shows a schematic of OmpLA in a bilayer, showing that Ala210 is located at the bilayer center. The bars show rank-ordered values of $\Delta\Delta G_{AX}$. The colors refer to the different classes of amino acids shown. (C) A plot of $\Delta\Delta G_{AX}$ for non-polar amino acids against accessible surface area. The slope of the curve yields a solvation parameter, which is identical to the value obtained for pentapeptide partitioning into *n*-octanol. (D) A plot of the OmpLA hydrophobicity values against the side chain values obtained from *n*-octanol partitioning (derived from Figure 3.6). Although the correlation coefficient is high, the slope is only 0.5, reflecting the different contexts of the partitioning measurements. (From Moon CP, Fleming KG [2011] *PNAS* 108: 10174-10177. With permission from KG Fleming.)

4.4 LIPID BILAYERS AND MEMBRANE PROTEINS ADAPT TO EACH OTHER

4.4.1 The Refolding of β-Barrels Depends upon Lipid Bilayer Properties

An important observation from the studies of OmpA folding by Lukas Tamm is that the energetics of OmpA folding depend strongly on the lipid composition of the bilayers into which OmpA is refolded (**Figure 4.14**). ΔG_U^0 and *m* depend upon both the length of the lipid acyl chains and the number of double bonds in the chain; these determine the thickness of the bilayer hydrocarbon core,

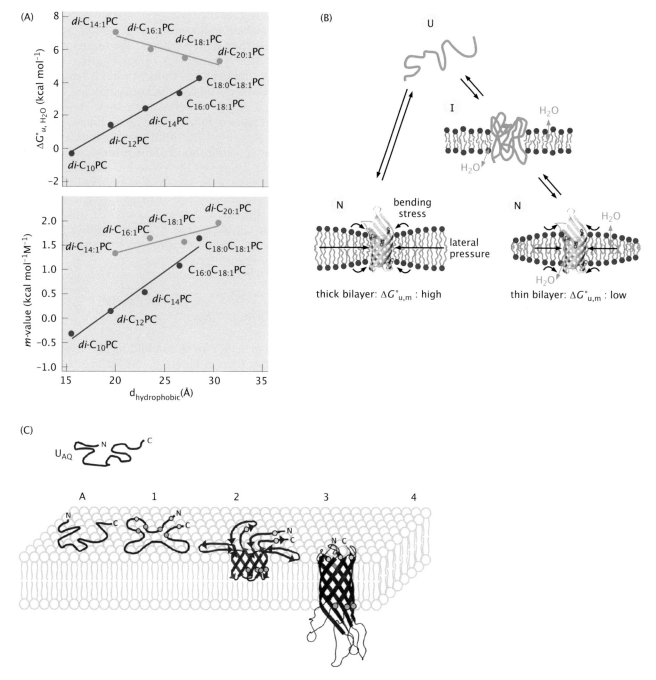

Figure 4.14 Folding free energies of the OmpA β-barrel depend upon lipid bilayer thickness. (A) The dependence of ΔG and m on lipid chain length and pattern of unsaturation. (B) A schematic interpretation of the data based upon the idea that bilayer distortion in the vicinity of OmpA is a major contributor to folding free energies. (C) A scheme for the folding pathway of OmpA based upon extensive thermodynamic measurements in the laboratory of Karen Fleming. (A and B, From Hong H, Tamm LK [2004] *PNAS* 101: 4065–4070. With permission from LK Tamm. C, From Danoff EJ, Fleming KG [2017] *Biochemistry* 56: 47–60. With permission from the American Chemical Society.)

which is shown as the x-axis of **Figure 4.14A**. This makes it clear that the adaptation of OmpA and the lipid bilayer to each other has energetic consequences for protein stability. Although the free energies of refolding OmpA into bilayers are relatively small in the experiments summarized in Figure 4.14 (0 to –8 kcal mol^{-1}), they nevertheless are strongly affected by the lipid type. ΔG_U^0 increases linearly with thickness if the chains lack double bonds or contain only a single double bond, but if both chains have double bonds, then ΔG_U^0 increases as bilayer thickness decreases. At a thickness of about 30 Å, the two types of lipids are equivalent. Why does the number of double bonds matter? As discussed in Chapter 2, short-chain lipids with two *cis* double bonds tend to prefer hexagonal phases rather than bilayer phases. The tendency to form hexagonal phases is quantitated by a parameter called negative curvature stress, discussed in Chapter 2. A simple way to think of what is happening to ΔG_U^0 as lipid type changes is how well the bilayer thickness matches the length of the transmembrane domain of OmpA, which is about 26 Å (Figure 4.13A). **Figure 4.14A** shows that the two types of lipids converge for bilayers of about this thickness. As suggested in **Figure 4.14B**, thick bilayers must deform to become thinner in the immediate neighborhood of the protein, whereas thin bilayers must deform to become thicker in the protein's neighborhood. As noted earlier, it is the total free energy of the system that matters in lipid–protein interactions; the system acts to minimize free energy through changes in the lipid bilayer as well as in the protein or peptide. As a result, deformations in the lipid bilayer have energetic consequences for OmpA folding. An experiment-based scheme for how OmpA might fold into a membrane is shown in **Figure 4.14C**.

4.4.2 Hydrophobic Mismatch Is a General Feature of Protein–Bilayer Interactions

Not only does the bilayer thickness matter for OmpA folding; it can also matter for membrane protein function, as seen in **Figure 4.15A** for melibiose transport by the melibiose permease, which is an α-helical membrane protein. The dependence of folding energetics and melibiose transport on bilayer thickness validates the concept of hydrophobic mismatch introduced on theoretical grounds by Ole Mouritsen and Myer Bloom long before we knew much about the atomic structure of lipid bilayers or membrane proteins. Mouritsen and Bloom, thinking of membrane proteins as rigid hydrophobic cylinders (**Figure 4.15B**), realized that deformations of bilayers close to proteins would have energetic consequences for lipid–protein interactions. They derived an equation for the change in the free energy of a bilayer of thickness d_L that contained a protein of length d_P:

$$\Delta G = \kappa \left(\frac{\rho_P}{\pi \xi_L} + 1 \right) (d_P - d_L)^2$$

In this equation, ρ_P is the circumference of the "cylindrical" protein. Bilayers in excess water have thickness fluctuations arising from thermal motion that depend upon the properties of the bilayer. These fluctuations are characterized by the parameter ξ_L, which describes the persistence length of the fluctuations, which is related to interactions between the lipids in the bilayer. The phenomenological parameter κ characterizes how easily the area per lipid can be changed in the lipid bilayer. The major significance of the equation is that it connects protein-induced bilayer free energy changes to the physical properties of the bilayer. This is another aspect of the bilayer effect discussed earlier (see Figure 3.9).

How can we know whether hydrophobic mismatch is a real feature of protein–lipid interactions, given that crystallographic structures of membrane proteins do not generally include images of the protein's native bilayer? Numerous molecular dynamics simulations of membrane proteins in lipid bilayers reveal clearly the existence of mismatch around membrane proteins. An example of mismatch between POPC and POPE bilayers and the glycerol facilitator channel GlpF is shown in **Figure 4.15C and D**. The POPC and POPE have different physical properties, particularly regarding their hydrogen-bonding capabilities, which are revealed in the mismatch profiles. Hydrophobic mismatch can be important for how proteins function. A good example is increased in-plane membrane diffusion of the GlpG protease of *E. coli* (**Box 4.2**).

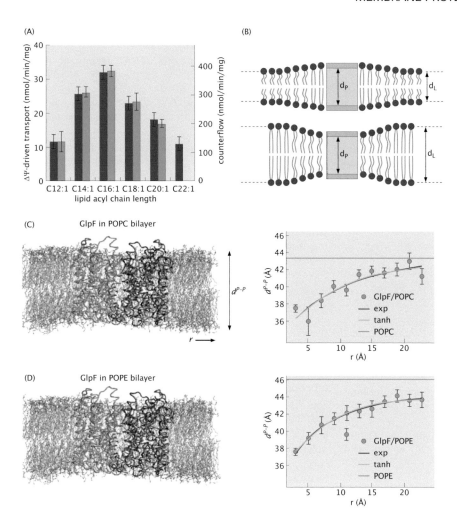

Figure 4.15 Hydrophobic mismatch associated with α-helical membrane proteins. (A) The functional importance of mismatch is revealed by the dependence of melibiose transport on lipid acyl-chain saturation. (B) A schematic representation of the responses of lipid bilayers of different thicknesses to the presence of a membrane protein. (C) A molecular dynamics simulation of the GlpF glycerol facilitator channel in a POPC bilayer. The graph shows the thickness variation of the bilayer in the vicinity of GlpF. (D) The same as in panel C except that the lipid is POPE rather than POPC. (From Jensen MØ, Jensen O, Mouritsen OG [2004] *Biochim Biophys Acta* 1666: 205–226. With permission from Elsevier.)

KEY CONCEPTS

- Calorimetric measurements show that transmembrane helices are very stable.

- The stability of transmembrane helices results from the high cost of breaking backbone hydrogen bonds in non-polar environments.

- Association of transmembrane helices in bilayers is favored by van der Waals interactions and can also be stabilized by interhelical H-bonds and polar interactions.

- While inter-helix connecting links contribute to α-helix membrane protein stability, they are not necessary for the formation of native 3-D structure.

- α-helical peptides can be denatured and refolded using detergents and lipids, but not necessarily completely or reversibly.

- α-helical membrane proteins can be unfolded using atomic force microscopy, resulting in estimates of transmembrane helix stability.

- β-barrel membrane proteins can be folded reversibly into membranes with the help of denaturants.

- Reversible folding of β-barrels into bilayers depends upon lipid bilayer thickness because of hydrophobic mismatch.

- Hydrophobic mismatch is a general feature of protein–bilayer interactions.

FURTHER READING

Booth, P.J., and Curnow, P. (2009) Folding scene investigation: Membrane proteins. *Curr. Opin. Struct. Biol.* 19:8–13.

Bayburt, T.H., and Sligar, S.G. (2010) Membrane protein assembly into nanodiscs. *FEBS Lett* 584:1721–1727.

Bowie, J.U. (2001) Stabilizing membrane proteins. *Curr. Opin. Struct. Biol.* 11:397–402

Bowie, J.U. (2005) Solving the membrane protein folding problem. *Nature* 438:581–589.

Fleming, K.G. (2014) Energetics of membrane protein folding. *Annu Rev Biophys.* 43:233–255.

Haltia, T., and Freire, E. (1995) Forces and factors that contribute to the structural stability of membrane proteins. *Biochim. Biophys. Acta* 1241:295–322.

Hong, H., and Tamm, L.K. (2004) Elastic coupling of integral membrane protein stability to lipid bilayer forces. *Proc. Natl. Acad. Sci. U.S.A.* 101:4065–4070.

Jensen, M.Ø., and Mouritsen, O.G. (2004) Lipids do influence protein function: The hydrophobic matching hypothesis revisited. *Biochim. Biophys. Acta* 1666:205–226.

Muller, D.J. (2008) AFM: A nanotool in membrane biology. *Biochemistry* 47:7986–7998.

Popot, J.-L., and Engelman, D.M. (1990) Membrane protein folding and oligomerization: The 2-stage model. *Biochemistry* 29:4031–4037.

Popot, J.-L., and Engelman, D.M. (2000) Helical membrane protein folding, stability, and evolution. *Annu. Rev. Biochem.* 69:881–922.

Tamm, L.K., Hong, H., and Liang, B. (2004) Folding and assembly of β-barrel membrane proteins. *Biochim. Biophys. Acta* 1666:250–263.

Tamm, L.K. (2005) *Protein-Lipid Interactions. From Membrane Domains to Cellular Networks*, pp. 444, WILEY-VCH Verlag GmbH & Co KGaA, Weinheim.

White, S.H., and Wimley, W.C. (1999) Membrane protein folding and stability: Physical principles. *Annu. Rev. Biophys. Biomol. Struc.* 28:319–365.

KEY LITERATURE

Curnow, P., and Booth, P.J. (2007) Combined kinetic and thermodynamic analysis of α-helical membrane protein unfolding. *Proc. Natl. Acad. Sci. U.S.A.* 104:18970–18975.

Curnow, P., and Booth, P.J. (2009) The transition state for integral membrane protein folding. *Proc. Natl. Acad. Sci. U.S.A.* 106:773–778.

Engelman, D.M., Steitz, T.A., and Goldman, A. (1986) Identifying nonpolar transbilayer helices in amino acid sequences of membrane proteins. *Annu. Rev. Biophys. Biophys. Chem.* 15:321–353.

Kahn, T.W., Sturtevant, J.M., and Engelman, D.M. (1992) Thermodynamic measurements of the contributions of helix-connecting loops and of retinal to the the stability of bacteriorhodopsin. *Biochemistry* 31:8829–8839.

Lau, F.W., and Bowie, J.U. (1997) A method for assessing the stability of a membrane protein. *Biochemistry* 36:5884–5892.

Marti, T. (1998) Refolding of bacteriorhodopsin from expressed polypeptide fragments. *J Biol Chem* 273:9312–9322.

Moon, C.P., and Fleming, K.G. (2011) Side-chain hydrophobicity scale derived from transmembrane protein folding into lipid bilayers. *Proc. Natl. Acad. Sci. USA* 108:10174–10177.

Mouritsen, O.G., and Bloom, M. (1984) Mattress model of lipid-protein interactions in membranes. *Biophys. J.* 46:141–153.

EXERCISES

1. Helix–helix interactions are a major theme in the higher-order structures of proteins in general and of membrane proteins in particular. The constraints of the bilayer limit the interactions such that there are mainly four kinds, right and left handed, and parallel and antiparallel.

    (a) Satisfy yourself that you understand the cases of right- and left-handed interactions, reviewing literature as needed. (Hint: as you move along the axis of a helix dimer, think "righty-tighty, lefty-loosey" as an analogy from driving screws.)

    (b) Why might the electrostatics of a helix favor antiparallel versus parallel interactions?

    (c) A "heptad repeat" describes the contacts between helices in a dimer. Would this be a left-handed or right-handed case? Why would it not work for both cases?

    (d) Why might you expect antiparallel dimers to be more abundant than parallel dimers in polytopic proteins?

2. Why would the binding of a prosthetic group tend to stabilize a membrane protein structure? (Hint: think of the binding energy.)

3. Oligomer formation is a common theme in membrane proteins—in fact, it is rare to find a protein that is a monomer. For the following origins of helical membrane protein stability, discuss the ways in which it might be more or less important for the formation and stability of oligomers. Give reasons for each of your choices:

    (a) Prosthetic group binding

    (b) Helix–helix interactions

    (c) Loop regions

    (d) Water interactions

    (e) Helix–lipid interactions

4. Repeat the analysis of Exercise 3 for β-barrel membrane proteins:

    (a) Prosthetic group binding

    (b) Strand–strand interactions

    (c) Loop regions

    (d) Water interactions

    (e) Helix–lipid interactions

Protein Trafficking in Cells

5

The biophysical principles underlying the behavior of lipids and proteins discussed in the previous chapters are ultimately played out in the context of cellular life. Biological membranes are formed by an intimate interplay between lipids and proteins, and the compartmentalization that is made possible by the existence of such membranes is a fundamental aspect of living systems at every level of organization.

At the cellular level, as we learned in Chapter 1, the plasma membrane provides the cell's interface to the outside world. It helps to concentrate and protect the cell's vital macromolecules, small metabolites, and ions, and makes it possible for the cell to control the continuous inward/outward flux of molecules and ions and to mediate cell-to-cell signaling events. Importantly, it also serves as an electrochemical energy storage device. In eukaryotic cells, many important biochemical reactions and pathways are sequestered within intracellular membrane-enclosed organelles, such as mitochondria, chloroplasts, the endoplasmic reticulum (ER), and the Golgi apparatus. The specialized activities of these compartments themselves depend upon sub-compartmentalization as well as upon their organelle-specific proteins.

While these compartments are advantageous, they create the problem of sorting and delivering the specialized membrane and soluble proteins that reside within them. There are two sorting and delivery problems. The first is the sorting and delivery of membranes. In eukaryotic cells, proteins and lipids can be sent back and forth between different cellular membrane systems by vesicular transport. This applies not only to proteins that follow the secretory pathway from their site of synthesis at the ER via the Golgi apparatus to the plasma membrane, but also to cell-surface proteins that can be recycled back into intracellular compartments or even carried between the apical and the basolateral plasma membranes in polarized cells. These interesting and important problems are the subject of Chapter 7.

The second sorting and delivery problem, which is the subject of this chapter, is the targeting and delivery of proteins to the various membrane compartments. To reach their intended locations in the cell, water-soluble proteins must first be targeted to the correct organelle and then translocated across one or more organelle membranes (**Figure 5.1**). Other soluble proteins are secreted from the cell into the surrounding medium or tissue. With few exceptions, the targeting step depends on targeting sequences (also called signal sequences) in the protein. Targeting sequences are stretches of polypeptide that bind to targeting receptors present in the cytoplasm or on the membrane of the target organelle. Depending on the target organelle, the receptor may hand over the protein to a protein-conducting channel in the target membrane—a translocase—or may accompany the cargo protein into the organelle, whereupon it dissociates from the cargo protein and recycles back across the membrane. Integral membrane proteins in most cases use the same receptors as do soluble

Figure 5.1 The main protein trafficking pathways in eukaryotic cells. The various organelles shown were discussed in Chapter 1. This figure emphasizes the trafficking of proteins from the cytosol to organelles. Targeting to organelles depends upon short sequence motifs called signal sequences.

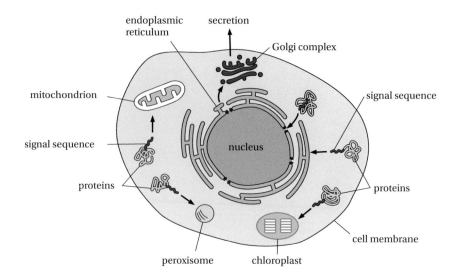

proteins for targeting to the organelle but are then integrated into, rather than translocated across, the target membrane.

Many questions must be answered to understand protein trafficking. What are the general features of targeting sequences? What features distinguish one type of targeting sequence from another? What is the fate of the targeting sequence after proteins have reached their targets? How are proteins translocated across or into target membranes? Are the translocation processes the same for all organisms and organelles? We answer these questions in this chapter by discussing targeting to and translocation across the inner and outer membranes of bacteria, the ER membrane, the mitochondrial inner and outer membranes, the chloroplast envelope and thylakoid membranes, the peroxisomal membrane, and the nuclear envelope. The answers form a necessary background for our discussion of the biosynthesis of membrane proteins in Chapter 6.

5.1 TRANSLOCASES MEDIATE THE TRANSPORT OF PROTEINS ACROSS THE ENDOPLASMIC RETICULUM MEMBRANE

5.1.1 Proteins Selected for Transport Are Identified by N-Terminal Signal Peptides: The Signal Hypothesis

Because nearly all proteins are made by cytoplasmic ribosomes, a fundamental problem faced by all cells is how to route different proteins to their destination compartments. The problem is particularly acute in the highly organized eukaryotic cell, but even bacteria must be able to distinguish between proteins destined for the cytoplasm, the periplasm, and the extracellular environment.

The basic principle that guides protein targeting within the cell was first enunciated as the "signal hypothesis," formulated by Günter Blobel and collaborators at the Rockefeller University in New York in the early 1970s (see Box 6.1). Electron microscopy images and subcellular fractionation experiments showed that ribosomes are found either free in the cytoplasm or bound to the ER membrane. Furthermore, certain secreted proteins, such as antibody light chains, were found to be synthesized by ER-bound ribosomes (see **Box 5.1** for a primer on the ribosome). To rationalize these results, Günter Blobel (1936–2018) and David Sabatini proposed in 1971 that secreted proteins contain an N-terminal extension—a signal peptide—that mediates binding of the ribosome–nascent

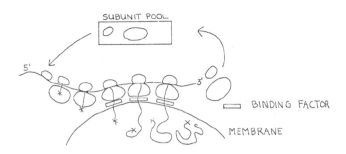

Figure 5.2 The first published sketch presenting the signal hypothesis. X indicates the (at the time hypothetical) N-terminal "signal peptide." The idea that the signal peptide is eventually cleaved off from the mature protein was not part of the model at this time. (From Blobel G, Sabatini DD [1971] *Biomembranes*, ed. LA Mason, Springer, New York. With permission from Plenum Press, New York.)

chain complex (RNC) to a receptor on the ER membrane, eventually leading to the opening of a protein-conducting channel through which the nascent chain can be translocated into the lumen of the ER (**Figure 5.2**).

Serendipitously, César Milstein (1927–2002) in Cambridge (UK) was at the same time studying the biosynthesis of antibodies using cell-free translation reactions that either included or excluded ER-derived membranes (so-called rough microsomes); see **Box 5.2**. He made the crucial observation that an antibody light chain was made as a slightly larger precursor protein when no ER membrane was present during translation (**Figure 5.3**). But when membranes were present, light chains of the same molecular weight as the mature form were obtained. He correctly surmised that the additional peptide segment present in the precursor form was somehow involved in targeting the protein for secretion, i.e., that it constituted a signal peptide. Further studies eventually clarified the conserved sequence features of signal peptides (**Box 5.3**).

5.1.2 Proteins Are Transported across the ER Membrane through the SRP/Sec61 Pathway

Today, the mechanism responsible for translocation of nascent polypeptides across the ER membrane has been worked out in considerable detail. The main experimental approaches that have been used are genetics, biochemical studies in cell-free systems, and structural studies of the components involved. Initially, translocation across the ER membrane and the bacterial inner membrane were thought to be based on quite different principles, but it later turned out that the two systems are very similar (and also similar to the translocation system in archaea) and that one can very often transfer knowledge obtained in one system to the other. We will therefore discuss them in parallel.

In eukaryotic cells, translocation across the ER membrane is nearly always co-translational, i.e., the nascent polypeptide chain is translocated as it comes out of the ribosome. Membrane proteins use the same process except that transmembrane segments are inserted into the membrane during the translocation reaction (see Chapter 6). Although co-translational translocation also occurs in prokaryotes, most secreted bacterial proteins are translocated across the inner membrane post-translationally, using an ATPase called SecA rather than the ribosome to drive translocation (see below).

The breakthrough discoveries were made in the early 1980s using fractionation of a cell-free *in vitro* translocation system similar to the one used by Milstein. This led to the purification of a complex, initially called the signal recognition protein (SRP), that was shown to be required for co-translational translocation of model proteins into rough microsomes. Subsequent studies revealed that SRP also contained an RNA molecule, causing the SRP to be renamed the signal recognition particle, maintaining the acronym SRP. Further work led to the discovery of an ER-bound SRP receptor (SR), and finally, a protein-conducting protein in the ER membrane—the Sec61 complex—was discovered in yeast. The name came from the observation that the largest protein in the complex has a molecular weight of 61 kDa.

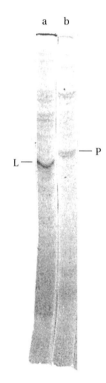

Figure 5.3 The first observation of a precursor form of a secreted protein. Lane a shows the mature antibody light chain product (L) obtained by *in vitro* translation in the presence of ER-derived membranes, and lane b shows the precursor form (P) obtained in the absence of membranes. (From Milstein C, Brownlee G, Harrison T et al. [1072] *Nature* 239: 117-120.)

BOX 5.1 THE RIBOSOME—A PROTEIN-SYNTHESIZING MACHINE

The ribosome is the most complex molecular machine yet to have yielded to high-resolution structural studies, and its various functions are understood in great detail. Bacterial ribosomes have a molecular mass of around 2.5 MDa (the yeast ribosome is 3.3 MDa) and are composed of two subunits—the small and the large—assembled from three RNA molecules totaling some 4500 nucleotides, and about 50 different proteins. The job of the ribosome is to translate the information encoded in messenger RNAs (mRNAs) into polypeptide sequences, thereby producing all the thousands of different proteins present in a cell.

The overall architecture of a bacterial ribosome is shown in **Figure 1**. The two subunits carry out different functions. The small subunit guides the mRNA through the ribosome, and its decoding center ensures the proper alignment of

the tRNA anti-codons with the codons in the mRNA such that correct and incorrect codon–anti-codon pairings can be discriminated. The large subunit contains the active site where amino acids are transferred from the tRNAs onto the growing polypeptide chain, called the peptidyl-transferase center or PTC. The nascent polypeptide exits the ribosome through a ~100 Å-long tunnel that traverses the large subunit.

Polypeptide synthesis requires three different kinds of events on the ribosome: initiation, elongation, and termination (**Figure 2**). During the initiation step, the interaction between a segment in the ribosomal 16S RNA and a short stretch in the mRNA—the Shine-Dalgarno sequence—helps position the start codon in the mRNA in the small subunit. With the help of three initiation factors (IF1–3: IF2

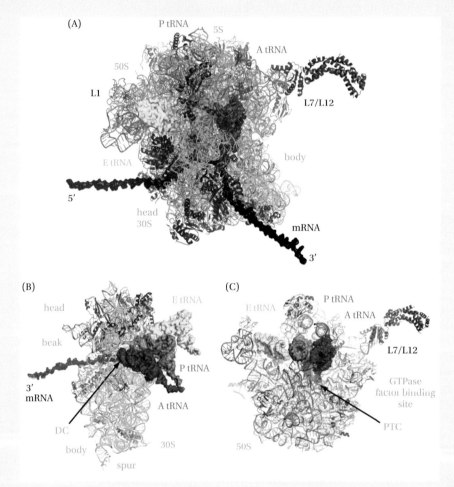

Figure 1 The bacterial ribosome. (A) Intact ribosome with a bound mRNA and three tRNAs. The small subunit is shown in blue and the large subunit in gold. (B, C) The small and the large subunits splayed apart and viewed from their common interface. There are three tRNA-binding sites: the A-site that holds an incoming amino-acylated tRNA, the P-site that holds a peptidyl-tRNA, which is still attached to its amino acid and via this to the growing polypeptide chain, and the E-site that holds an empty, deacylated tRNA. The three tRNAs all bind to the mRNA near the decoding center (DC) in the small subunit. The addition of the incoming amino acid to the growing polypeptide takes place in the peptidyl-transferase center (PTC) in the large subunit. The tunnel in the large subunit that holds the nascent polypeptide chain starts at the PTC and ends at the backside of the subunit. (From Schmeing TM et al. [2009] *Nature* 461: 1234. Used with permission from Springer Nature.)

(Continued)

BOX 5.1   THE RIBOSOME—A PROTEIN-SYNTHESIZING MACHINE (CONTINUED)

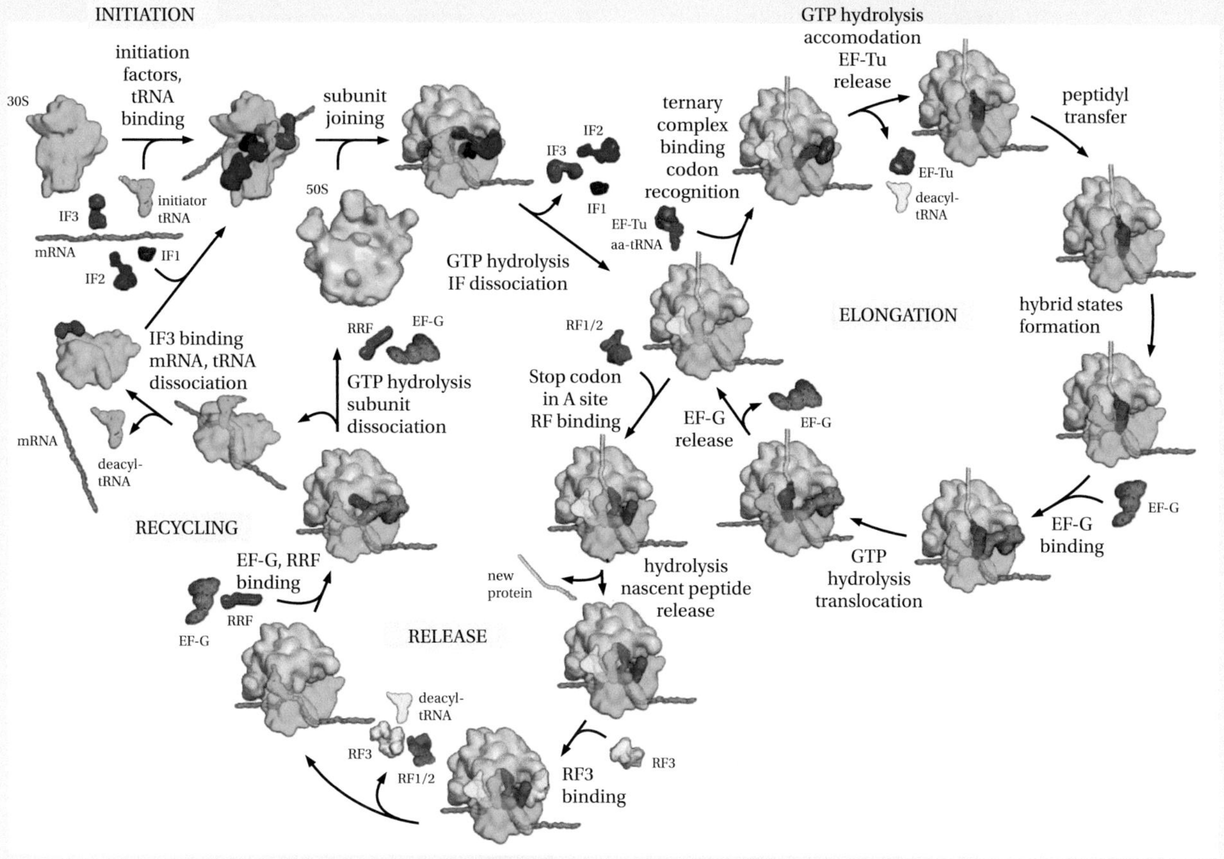

Figure 2 The complete initiation-elongation-termination/release cycle of the bacterial ribosome. (From Schmeing TM et al. [2009] *Nature* 461: 1234. Used with permission from Springer Nature.)

is a GTPase), the acylated initiator tRNA (f-Met-tRNA$^{f\text{-Met}}$) binds to the initiator codon on the small subunit, the large subunit joins, and the initiator tRNA moves into the P-site with the f-Met moiety placed in the PTC, ready to react with the next acylated tRNA that enters the A-site.

In each elongation step, an acylated tRNA with the correct anti-codon is delivered to the ribosome in complex with the GTP-bound form of elongation factor Tu (EF-Tu). After GTP hydrolysis on EF-Tu, the tRNA moves into the A-site, and its amino-acylated end is positioned in the PTC such that it can react with the nascent chain-carrying tRNA (the peptidyl-tRNA) present in the P-site. After the new peptide bond has formed, elongation factor G (EF-G) uses GTP hydrolysis to drive both the mRNA and the two tRNAs to translocate one step (three nucleotides in the mRNA, corresponding to one codon) relative to the ribosome,

thereby placing the former P-site tRNA into the E site and the former A-site tRNA into the P-site. A new codon is now placed in the A-site, and the elongation cycle repeats itself until the ribosome reaches a stop codon in the mRNA. Bacterial ribosomes can sustain an elongation rate of up to 20 amino acids per second. Eukaryotic ribosomes are about 10-fold slower.

There is no tRNA that can bind to stop codons (except for mutated forms called suppressor tRNAs: see Box 6.5). Instead, when a stop codon is present in the A-site, a protein that mimics a tRNA, called release factor (RF), binds in the A-site and catalyzes the release of the newly synthesized polypeptide from the peptidyl-tRNA in the P-site. Finally, recycling factors induce the disassembly of the two ribosomal subunits and the mRNA, priming the system for a new round of protein synthesis.

Genetic studies in yeast and *E. coli* led to the discovery of genes coding for proteins that are necessary for protein translocation. To everyone's delight, these genetically defined proteins turned out to be homologous to the SRP, SR, and Sec61 proteins that the biochemists had purified. Finally, as DNA sequencing started to produce sequences of genes encoding secreted proteins, amino acid sequences of signal peptides started appearing in the literature, and statistical

BOX 5.2 CELL-FREE TRANSLATION

Cell-free translation techniques have played and still play an important role in molecular cell biology, because they make it possible to make proteins in a test tube to which one can add different purified components (proteins, RNA, small molecules) to study their interactions with the newly made protein either during its synthesis on the ribosome (co-translational interactions) or after it has been released from the ribosome (post-translational interactions).

In the early days, mRNA was obtained by purification of polysomes, i.e., mRNAs with multiple bound ribosomes, from cell lysates. The mRNA was then added to lysate prepared from red blood cells (which contains all the components required for protein synthesis, such as ribosomes, tRNAs, aminoacyl-tRNA synthetases, initiation-, elongation-, and termination factors, ATP, GTP, etc.) together with a radioactive amino acid (most often [^{35}S]-Met), leading to translation of the mRNA and synthesis of radioactively labeled protein that could then be visualized by gel electrophoresis and autoradiography (i.e., placing the gel with the radioactive protein bands on top of an x-ray film).

For studies of protein secretion and membrane protein biogenesis, a particularly important component to add to the cell-free system is so-called rough microsomes, i.e.,

membrane vesicles derived from the ER that contain Sec61 translocon complexes. Rough microsomes are usually prepared by homogenization of pancreatic cells followed by a series of ultra-centrifugation steps to obtain a fraction enriched in ER membrane. When rough microsomes are present during the translation reaction, ribosomes synthesizing secreted or membrane proteins are targeted to the microsomal membrane, and the protein is translocated through the Sec61 translocon into the lumen of the microsomal vesicle, where it is protected from, e.g., proteolysis by proteases added to the reaction mix.

Today, one does not need to isolate the mRNA and add it to the translation reaction. Systems have been developed where transcription and translation are done in the same test tube, so that all one needs to do is to add a plasmid carrying the gene of interest, or a polymerase chain reaction (PCR)-amplified DNA fragment with the gene, to the cell-free synthesis system. RNA polymerase will then make the mRNA from the DNA, and ribosomes will translate the mRNA, making the protein (**Figure 1**). But the basic methodology is very much the same as the one used by the early pioneers.

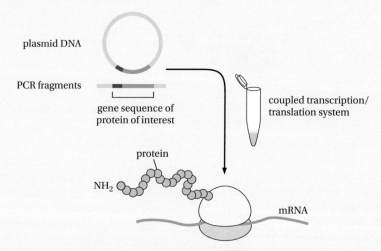

Figure 1 A coupled transcription–translation system for cell-free synthesis of proteins. In studies of protein secretion, rough microsomes are added at the start of the reaction.

MRVPAQLLGLLLLWLPGARCAIRMTQSPSS...

Figure 5.4 A typical ER-targeting signal peptide. It is composed of a positively charged N-terminal part (n-region; red), a central hydrophobic part (h-region; green). and a C-terminal polar part (c-region; blue). The arrow indicates the site where the signal peptide is cleaved off from the mature protein (black) by the signal peptidase enzyme located in the lumen of the ER.

analyses of these uncovered a conserved tripartite architecture with a short, positively charged N-terminal segment (n-region), a central hydrophobic segment (h-region), and a more polar C-terminal segment (c-region) that defines the cleavage site between the signal peptide and the mature protein (**Figure 5.4** and Box 5.3).

After these pioneering studies, the availability of purified components and a range of biochemical and biophysical techniques, including x-ray crystallography and cryo-EM, has made it possible to arrive at a rather detailed understanding of the targeting and translocation processes. The evolved three-part strategy is (a) recognition during protein synthesis that a particular protein is destined for the ER, (b) arrest of protein elongation until the ribosome reaches the Sec61 translocase (also referred to as a translocon), and (c) resumption of chain elongation followed by translocation across the membrane.

BOX 5.3   THE DISCOVERY OF SIGNAL PEPTIDES

ER-targeting signal peptides were the first kind of targeting signal to be discovered. Their existence was postulated by Günter Blobel and David Sabatini in 1971 (Figure 5.2), and the first experimental indication that secreted proteins are synthesized with an extra N-terminal extension was obtained by César Milstein and his collaborators in 1972 (Figure 5.3). Blobel was awarded the 1999 Nobel Prize in Physiology or Medicine for his discovery.

But what do signal peptides look like? Obviously, the amino acid sequence of mature secreted proteins that could be purified from blood or other extracellular sources was no help, since the signal peptide is cleaved off already in the ER. Cell-free protein synthesis of the kind used by Milstein came to the rescue: when translation was carried out in the absence of rough microsomes, the full-length protein with

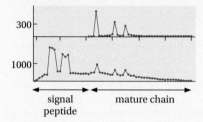

Figure 1 Radioactivity in each sequencing cycle of [³H]-Leu labeled mature light chain prepared by growing intact cells in a medium containing [³H]-Leu (top panel) and in uncleaved, full-length light chain precursor prepared by cell-free translation in the presence of [³H]-Leu (bottom panel). The signal peptide and mature chain are indicated.

its signal peptide was obtained, but only in minute amounts not suitable for standard protein sequencing techniques. However, by adding one radioactive amino acid, such as [³H]-Leu, to the translation mix and then analyzing the resulting product in an amino acid sequencer, the positions of the Leu residues could be determined. This was first done in 1975 by Schechter et al., using mRNA coding for a mouse antibody light chain. As seen in Box **Figure 1**, [³H]-Leu was found in positions 6–8 and 11–13 in the signal peptide and in positions 4, 11, and 15 in the mature light chain. This showed that the signal peptide had a hydrophobic character, with many Leu residues.

Over the next decade, many more signal peptides were sequenced, mostly by cDNA sequencing, and it became clear that they were all markedly hydrophobic and often had a positively charged lysine or arginine near the N-terminus.

The first thorough statistical analysis of a reasonably large collection of 78 eukaryotic signal peptides was published by von Heijne in 1983. He noted a number of conserved features, such as the exclusive presence of amino acids with small side chains in positions –1 and –3 relative to the cleavage site and a high incidence of helix-breaking residues such as proline in position –5. Based on these patterns, he proposed a simple scheme for predicting signal-peptide cleavage sites and a model for how the signal peptide binds to the signal peptidase enzyme during the cleavage reaction (**Figure 2**). This model was later seen to be essentially correct when the x-ray structure of the signal peptidase from *E. coli* was determined (Figure 5.11).

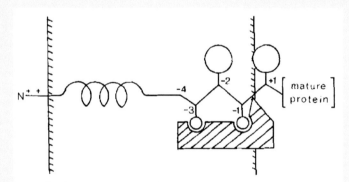

Figure 2 An early model for how the C-terminal end of the signal peptide fits into the active site of the signal peptidase enzyme. (From von Heijne G [1983] *Eur J Biochem* 133: 17–21.)

The basic co-translational pathway is outlined in **Figure 5.5**. Step 1: The N-terminal signal peptide emerges from the tunnel in the large ribosomal subunit and binds to a GTP-loaded SRP that docks to the ribosome and arrests translation. Step 2: The ribosome-SRP–nascent chain complex diffuses to and interacts with the GTP-loaded SR on the ER membrane. Step 3: The SRP–SR complex undergoes a conformational change from an "early" to a "closed" form. Step 4: the signal peptide is transferred from SRP into the channel of the Sec61 translocon, the channel opens across the membrane, translocation of the nascent chain is initiated, and the SRP–SR complex undergoes a second conformational change converting it into an "activated" form. Step 5: The two GTP

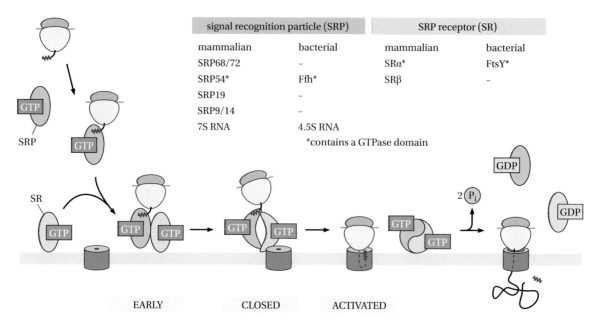

signal recognition particle (SRP)		SRP receptor (SR)	
mammalian	bacterial	mammalian	bacterial
SRP68/72	–	SRα*	FtsY*
SRP54*	Ffh*	SRβ	–
SRP19	–		
SRP9/14	–		
7S RNA	4.5S RNA		

*contains a GTPase domain

EARLY   CLOSED   ACTIVATED

Figure 5.5 A schematic representation of protein translocation across the ER membrane through the SRP/SRP-receptor/Sec61 pathway. Sketches are not to scale. Substrate proteins can be rejected from the pathway at each step, i.e., there are multiple checkpoints to ensure that only the right substrates are translocated. The same pathway is used for insertion of helix-bundle integral membrane proteins into the ER (see Chapter 6). The inset table compares the subunit compositions of mammalian and bacterial SRP and SR. (From Saraogi I [2011] Protein Sci 29; 1790-1795. With permission from Wiley-Blackwell. Table adapted from Walter P, Johnson AE. 1994. *Annu Rev Cell Biol* 10: 87-119 and reproduced with permission from Keenan RJ et al. [2001] *Annu Rev Biochem* 70: 755-775. Copyright 2001 Annual Reviews.)

molecules in the SRP–SR complex are hydrolyzed, and the complex disassociates. Translocation of the nascent polypeptide chain continues until the ribosome reaches the stop codon and the process is terminated.

Figure 5.6 shows a structure of a eukaryotic ribosome–nascent chain–SRP complex obtained by single-particle electron cryo-microscopy (cryo-EM). A long 7S RNA molecule forms the backbone of the SRP and binds the signal peptide–binding protein SRP54 at one end and additional proteins at the other end. Upon binding to the signal peptide, SRP induces a temporary elongation arrest in the ribosome as a result of the interaction of the SRP's Alu domain, which blocks the binding of necessary elongation factors (Box 5.1). At this early stage, the signal peptide is buried between the ribosome and the SRP54 protein. The elongation arrest is lifted once the ribosome has docked to the Sec61 translocon and the SRP has detached from the complex.

The rate of polypeptide translocation through the Sec61 translocase is set by the rate of polypeptide elongation on the ribosome, which is 3–5 amino acids per second in eukaryotic cells and 10–20 amino acids per second in bacteria. Protein synthesis is thus not a speedy process; about a minute will be required to synthesize a 300–amino acid protein in eukaryotes! Of course, cells have many thousands of ribosomes at work to produce the required amounts of protein steadily, even if slowly. Because bacterial cells grow and divide rather rapidly, the bacterial synthetic machinery appears to have evolved to be an order of magnitude faster than its eukaryotic counterpart.

The bacterial SRP and SR are simplified versions of the eukaryotic complexes (inset, Figure 5.5), perhaps as a means of speeding up protein synthesis. In bacteria, SRP has only two components: the protein Ffh and a 4.5S RNA (S is the Svedberg unit, a measure of the sedimentation rate of a molecule in an ultracentrifuge). Lacking an Alu domain, the bacterial SRP does not arrest elongation. The eukaryotic SR has two subunits, the GTPase SRα and the membrane-bound SRβ that anchors SRα to the membrane. The bacterial SR, called FtsY, is a peripheral membrane protein that partitions between the cytoplasm

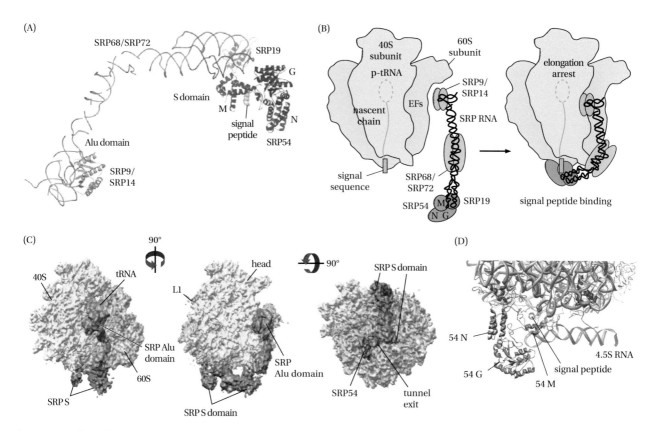

Figure 5.6 The eukaryotic ribosome–SRP complex. (A) The components of the signal recognition particle as determined by X-ray crystallography. (B) A cartoon that summarizes the observations that the SRP recognizes the emerging N-terminal signal peptide, binds to the ribosome, and causes arrest of protein elongation by interfering with tRNA entry. SRP reaches from the emerging signal peptide to the vicinity of the peptidyl-transferase center, where elongation factors (EFs) would normally bind. (C) Cryo-EM reconstructions of an SRP–ribosome complex. The small and large ribosomal subunits are shown in yellow and blue, SRP in red, and P-site tRNA in green. (D) An atomic-level view of the SRP (red) at the exit tunnel of the ribosome (blue) with a bound signal peptide (green). (A and B, From Egea PF, Stroug RM, Walter P [2005] *Curr Opinion Struct Biol* 15: 213–220. With permission from Elsevier. C and D, From Halic M et al. [2006] *Nature* 444: 507–511. With permission from Springer Nature.)

and the inner membrane due to an amphiphilic surface helix. Consequently, FtsY is not always membrane-bound.

5.1.3 The SRP and the SR Ensure Proper Targeting of the Ribosome to the Sec61/SecYEG Translocons

The signal peptide–binding proteins SRP54 (in eukaryotes) and Ffh (in prokaryotes) both carry three equivalent domains, called the M-, N-, and G-domains (**Figure 5.6A**). The M-domain binds the signal peptide, the N-domain interacts with the SR, and the G-domain can bind and hydrolyze GTP. The SR is composed of an N- and a G-domain, both similar in structure to their counterparts in the SRP54 and Ffh proteins, but lacks the M-domain.

Figure 5.7A shows the structure of a bacterial SRP in complex with its cognate SR. Cryo-EM images reveal the structure of the ribosome–SRP complex in the absence of SR (**Figure 5.7B**) and the dramatic conformational change that takes place in SRP when it docks with SR (**Figure 5.7C**). The signal peptide binds in a hydrophobic crevice in the Ffh M domain (closed by a loop structure when a signal peptide is absent), as summarized in **Figure 5.8**. The crevice is lined by aliphatic Leu, Ile, Val, and Met residues and two aromatic Trp residues. A stretch of about nine hydrophobic residues in the h-region of the signal peptide can bind in the crevice. The aliphatic side chains in the crevice can adapt

Figure 5.7 Bacterial SRP–SR interactions. (A) Structure of the bacterial SRP (4.5S RNA plus Ffh) bound to the SR (FtsY) (PDB: 5NCO). Two GDP molecules are bound at the Ffh–FtsY interface. (B) A structural view of bacterial SRP with Ffh bound near the ribosome exit tunnel. The signal peptide is largely obscured by Ffh in this view. (C) Structural view of the ribosome-bound FtsY–SRP complex. Note that the binding of FtsY to Ffh has caused a very large conformation change in Ffh, exposing the signal peptide. (B and C, Courtesy of Nenad Ban, ETH, Zurich.)

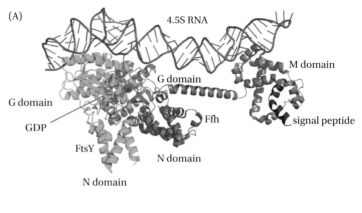

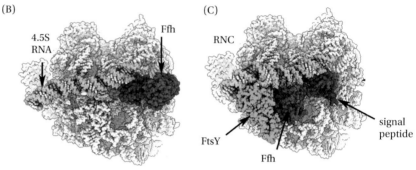

to many different h-region sequences, explaining the wide sequence variation found in natural signal peptides.

Structural and biochemical studies have further clarified how the G-domains of the SRP Ffh protein and the SR FtsY mutually activate each other's GTPase activity (**Figure 5.9**). The two GTP binding sites are brought

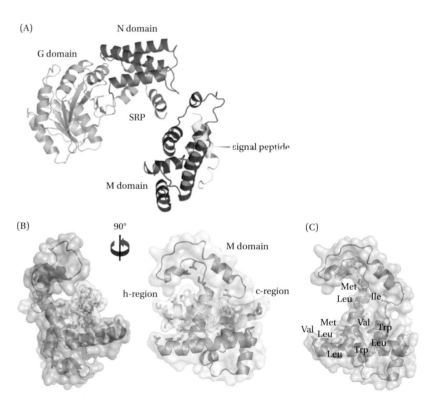

Figure 5.8 Interaction of a signal peptide with the signal peptide-binding protein Ffh in the bacterial SRP (PDB 3KL4). (A) Overall structure of the Ffh–signal peptide complex. (B) Space-filling representations of the signal peptide (yellow) in the M-domain binding pocket. (C) View of the binding pocket with the signal peptide removed. The binding pocket is rich in hydrophobic and aromatic residues that interact with the hydrophobic portion of the signal peptide. (From Janda CY et al. [2010] *Nature* 465: 507–510. With permission from Springer Nature.)

in contact in the "closed" state (Figure 5.5), and the activated GTPase conformation is achieved after the SRP–SR complex is released from the ribosome. Residues from both Ffh and FtsY are parts of both active sites, and consequently, the GTPase activities of the two proteins on their own are very low. This is an elegant way to ensure that GTP hydrolysis happens only when a ribosome-bound SRP with bound signal peptide is correctly docked to the SR–translocon complex.

The SRP–SR–Sec61/SecY targeting and translocation pathway is both accurate and efficient due to multiple checkpoints before the final ribosome–Sec61/SecY docking event, as indicated in Figures 5.5 and **5.9**. The binding between the signal peptide and the SRP represents one such checkpoint. Suboptimal, signal peptide–like N-terminal segments may survive this first step but are discarded in later steps based on their effects on SRP–SR complex formation and on the kinetic competition between disassociation of the suboptimal segment from the SRP–SR complex and GTP hydrolysis.

In the translocon-docked state, SRP dissociates from the ribosome and a ~150 Å-long, continuous nascent chain conduit is formed, leading all the way from the PTC at the entry to the ribosomal tunnel in the large ribosomal subunit to the luminal exit site from the Sec61 translocase. A structure of the docked state derived from single-particle cryo-EM is shown in **Figure 5.10**. Two long loops from Sec61 penetrate into the outermost part of the ribosomal tunnel, apparently to help guide the nascent chain into the translocon.

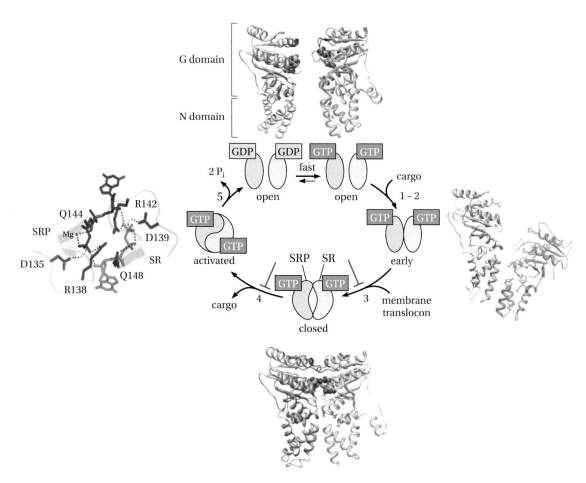

Figure 5.9 Proper docking between Ffh and FtsY is required to activate their respective GTPase sites. An arginine from SRP (R138) is part of the active site of SR, and the corresponding R142 of SR is part of the active site of SRP. See Figure 5.5. (From Saraogi I [2011] *Protein Sci* 29; 1790–1795. With permission from Wiley-Blackwell.)

Figure 5.10 Structure of the mammalian ribosome–Sec61 complex obtained by cryo-EM. (A) The location of the peptidyl-transferase center (PTC) is indicated. (B) Section through the complex, showing the tRNA-attached nascent chain in the ribosome exit tunnel. (From Voorhees RM, Fernández IS, Scheres SHW et al. [2014] *Cell* 157: 1632–1643. With permission from Elsevier.)

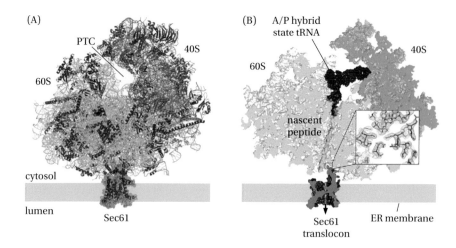

5.1.4 Signal Peptidase (SPase) Cleaves Signal Peptides

The signal peptide is cleaved from the mature protein by the signal peptidase enzyme (called leader peptidase in bacteria). The recognition motif for signal peptidase is a five to six amino acids-long stretch (the c-region) at the C-terminus of the signal peptide (Figure 5.4) characterized by amino acids with small side chains (Gly, Ala, Ser) in positions –1 and –3 counting from the cleavage site. In bacteria, the c-region fits as an extended chain in a cleft on the surface of the bacterial signal peptidase (**Figure 5.11**) and makes backbone hydrogen bonds to a β-strand in the enzyme. Signal peptidase is anchored to the periplasmic surface of the inner membrane by an N-terminal transmembrane domain (not shown in the figure), and the rather hydrophobic surface that surrounds the active site cleft is thought to be buried in the membrane–water interface region such that the signal peptide does not have to be completely pulled out of the membrane for cleavage. In eukaryotes, signal peptide cleavage is carried out by a multi-subunit complex, but only Sec11p within the complex is essential for peptide cleavage. Although Sec11p has low sequence identity with the bacterial signal peptidase, it can cleave bacterial signal peptides. Similarly, bacterial SPase can recognize and cleave eukaryotic signal peptides.

After they have been cleaved by signal peptidase, signal peptides are generally removed from the membrane. If a peptide is short and not strongly hydrophobic, it is likely to be unstable in the membrane. In that case, it diffuses away from the membrane and can be digested by proteases in the cytoplasm or periplasm. Signal peptides that are stable in the membrane must be degraded by intramembrane peptidases to prevent their accumulation in the membrane. In

Figure 5.11 The catalytic domain of signal peptidase from *E. coli* (PDB: 1B12). (A) A surface representation of the protein viewed parallel to the membrane. Hydrophobic residues facing the membrane are shaded in green. Substrate binding pockets are labeled S1 and S3. The catalytic site is indicated by Ser90. Two N-terminal transmembrane helices that anchor the enzyme in the inner membrane were removed before the enzyme was crystallized and are therefore not shown. (B) The active site of the protein containing a tetrapeptide mimic (cyan). The catalytic site is formed by Ser90 and Lys145. The small P1 and P3 side chains are buried in the shallow S1 and S3 binding pockets, while the P2 side chain points into solution. (From Paetzel M et al. [1998] *Nature* 396: 186–190. With permission from Springer Nature.)

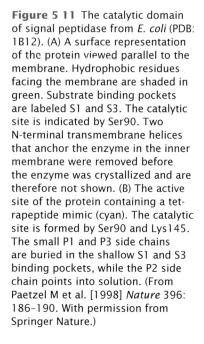

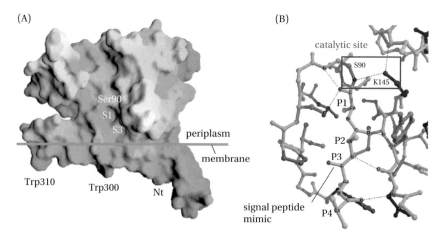

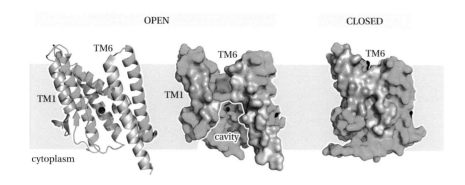

OPEN   CLOSED

Figure 5.12 Open and closed forms of the RseP signal peptide peptidase from *M. jannaschii* (PDB: 3B4R). Note the deep, water-filled cavity that allows both peptide substrates and water molecules to reach near the Zn²⁺ ion (red) in the active site when the protein is in its open state. Water is necessary for the hydrolytic cleavage of the peptide substrates.

eukaryotes, membrane-bound signal peptides are degraded by the signal peptide peptidase, which is a multi-spanning integral membrane protein that has its active site located well below the membrane surface. It belongs to the so-called presenilin family of proteases.

The main signal peptide peptidase in *E. coli* seems to be the RseP intramembrane protease (**Figure 5.12**). The protein has six TM helices, and the active site is near the bilayer midplane. In its open state, the protein has a water-filled pocket containing the active site, which permits hydrolysis of peptide bonds and consequently, cleavage of signal peptides, causing them to be expelled from the membrane for further degradation by soluble proteases.

5.1.5 Most Secretory Proteins in Bacteria Are Translocated Post-Translationally across the Inner Membrane

The main translocase in the inner membrane of bacteria is homologous to the eukaryotic Sec61 translocon and is composed of the SecY, SecE, and SecG subunits (corresponding to the α, γ, and β subunits in the eukaryotic Sec61 complex; see Box 6.1). Bacteria also have SRP and SR homologs that target membrane proteins to the SecYEG translocon for co-translational insertion into the inner membrane (inset, Figure 5.5). Unlike in eukaryotes, translocation of bacterial secreted proteins is in most cases post-translational, i.e., the polypeptide is fully synthesized and released from the ribosome before it is targeted to the Sec translocon. The generally accepted model for post-translational secretion is summarized in **Figure 5.13**. Because translocation occurs post-translationally, the ribosome is out of the picture; it is replaced by an ATP-driven motor protein known as SecA. Like the ribosome, SecA binds to the SecYEG translocon in order to push its passenger peptide through the membrane.

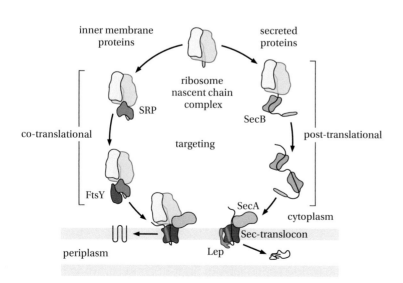

Figure 5.13 Schemes for co- and post-translational membrane targeting and translocation in *E. coli*. In the co-translational pathway, an emerging hydrophobic segment in the nascent chain triggers tight SRP binding to the ribosome–nascent chain complex, followed by FtsY-mediated targeting of the complex to the SecYEG translocon. This pathway is used mainly by inner membrane proteins. In the post-translational pathway, the SecB chaperone binds to the nascent chain as it emerges from the ribosome and keeps it in an unfolded state. The nascent chain is then handed over to the SecA/YEG translocon. This pathway is used mainly by soluble periplasmic proteins and outer membrane proteins. (From Lurink J, Yu Z, Wagner S et al. [2012] *Biochimica et Biophysica Acta (BBA) Bioenergetics* 1817: 965–976. With permission from Elsevier.)

Figure 5.14 A model based on NMR studies of an unfolded secretory protein (PhoA) bound to the SecB chaperone (in grey). The bound segments of PhoA are color-coded as indicated at the top. (From Huang C et al. [2016] *Nature* 537: 202–206. With permission from Springer Nature.)

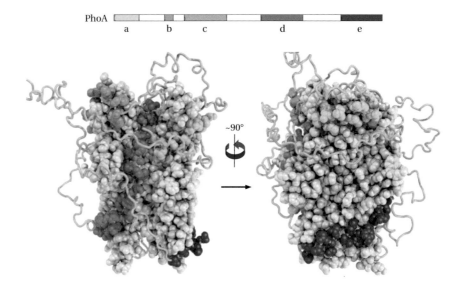

SecA can also bind to the ribosome and interact with nascent chains as they emerge from the exit tunnel. In addition, many nascent chains bind to the cytoplasmic chaperone SecB, which prevents premature folding of the protein in the cytoplasm. Each monomer in the homo-tetrameric SecB chaperone has two long hydrophobic grooves on its surface that can bind a wide variety of short peptide sequences that are enriched in hydrophobic and aromatic residues (**Figure 5.14**. However, SecB does not appear to bind directly to the signal peptide. A C-terminal region in SecA can bind to SecB with low affinity, but the binding affinity is greatly enhanced if SecA is simultaneously bound to SecYEG and SecB is bound to an unfolded protein with a signal peptide (a pre-protein). This ensures that SecB delivers pre-proteins to SecA–SecYEG complexes in the inner membrane. ATP binding to SecA weakens the affinity and releases SecB, leaving the pre-protein bound to the SecA–SecYEG translocon complex.

SecA uses the energy released by ATP hydrolysis to push the polypeptide chain through the SecYEG channel (**Figure 5.15**). The unfolded polypeptide is thought to pass through a clamp in SecA that prevents backsliding, but how SecA converts the energy released by ATP hydrolysis to polypeptide movement is not understood in detail.

There is not always a clear separation of co- and post-translational transport. An important observation is that only a few bacterial signal peptides have a strong preference for the SRP pathway, whereas the majority have a strong preference for the SecB/SecA pathway, and some proteins seem to be secreted with the help of both pathways. It remains an interesting challenge to understand the reasons for having these two mechanisms.

Genes encoding components of the Sec61/SecYEG translocon have been found in all fully sequenced prokaryotic and eukaryotic genomes, and the Sec-type translocon is thus one of the very few membrane protein complexes that seem to be invariably present throughout the living world. Sec61-like translocases are not found in the inner membrane of some mitochondria, however. In these membranes, the translocase used is called Oxa1 (a homolog of the bacterial FtsY; see Chapter 6). The appearance of translocase proteins with the ability to facilitate polypeptide translocation across, and insertion into, membranes must have been a critical step in the evolution of the very first cells.

5.2 BACTERIA HAVE MANY SYSTEMS FOR EXPORTING PROTEINS

As described in Chapter 1, Gram-negative bacteria have both an inner and an outer membrane with a peptidoglycan layer in between, while Gram-positive

(A)

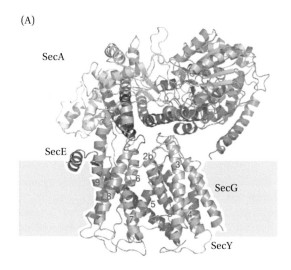

SecA

SecE

2b

9
6

8

5

7

4

SecG

SecY

Figure 5.15 SecA-driven translocation of proteins through the bacterial SecYEG translocon. (A) The SecA–SecYEG complex (PDB: 5EUL). Note the "two-helix finger" (brown) that penetrates into the cytoplasmic funnel of the translocon channel and is thought to drive translocation by pushing on the nascent chain. (B) Cartoon model of SecA-driven translocation with the clamp in green and the two-helix finger in brown. (From Zimmer J et al. [2008] *Nature* 455: 936–943. With permission from Springer Nature.)

(B)

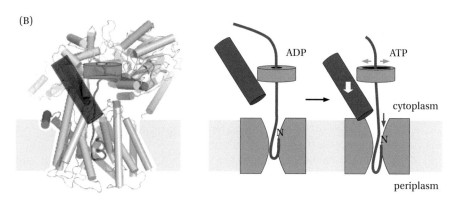

ADP

ATP

cytoplasm

N

N

periplasm

bacteria and archaea lack an outer membrane (but have a tough cell wall composed of peptidoglycan). Among other functions, the Gram-negative outer membrane and peptidoglycan protect against osmotic changes seen by a bacterium, but they also create a separate compartment: the periplasm. As with the cytoplasmic compartments in eukaryotes, additional transport and control mechanisms are thus needed. Therefore, despite the fact that the SecYEG complex is a ubiquitous translocon found in all cells, bacteria have evolved additional systems for translocating proteins across both the inner and outer membranes. This may be a result of evolutionary pressure to secrete rapidly certain proteins necessary for cell survival.

5.2.1 Bacterial Lipoproteins Are Sorted between the Inner and Outer Membranes

We saw in Chapter 3 that many otherwise soluble proteins can be anchored to membranes by covalently attached fatty acids (Figure 3.19). Eukaryotic lipoproteins are particularly important in signaling and regulatory processes. Bacteria also have lipoproteins that are important for maintenance of cell-surface structures, transport, and drug efflux. Some lipoproteins localized at the cell surface can cause inflammatory responses in host cells. Bacterial lipoproteins are associated with both the inner and the outer membranes of Gram-negative bacteria. They generally face the periplasm and are anchored in the inner or outer membrane by a diacylglyceryl moiety attached to an N-terminal Cys residue. Like other proteins found in the periplasm, most bacterial lipoproteins are secreted across the inner membrane by SecYEG, although some are secreted via the Tat pathway (see later).

MKATKLVLGAVILGSTLLAGCSSNAKIDQL

Figure 5.16 Typical lipoprotein signal peptide. The positively charged n-region is in red, the hydrophobic c-region is in green, and the cleavage site for lipoprotein signal peptidase is indicated by the arrow. The conserved "lipobox" that defines the cleavage site is boxed.

But, how are lipoproteins identified by the cell, and how do they receive their lipid moieties? The answer is special signal peptides, which look much like normal Sec-type signal peptides but have a different C-terminal cleavage site that includes a defining and critical Cys residue (**Figure 5.16**). After translocation of the protein through the SecYEG translocon, the diacylglyceryl group is attached to the Cys residue by the Lgt enzyme, whereupon the signal peptide is cleaved just before the Cys residue by a dedicated lipoprotein signal peptidase, LspA (**Figure 5.17A**). The cleavage site is defined by the conserved lipobox motif L(A,G)G↓C.

Lipoproteins that are destined for the outer membrane are extracted from the inner membrane by the ABC transporter LolCDE (Chapter 12) and transported across the periplasm by the "localization of lipoproteins" (Lol) system (**Figure 5.17B–D**). Lipoproteins that remain anchored in the inner membrane have an Asp residue—a "Lol avoidance signal"—next to the N-terminal Cys that prevents them from being recognized by the Lol system. Nature has evolved clever ways to use single amino acids to encode protein destinations!

5.2.2 Fully Folded Proteins Can Be Exported by the Bacterial Twin-Arginine Translocation System

The twin-arginine translocation (Tat) system is an alternative pathway for translocation across the inner membrane that enables some proteins to be translocated in a folded form. Typically, periplasmic proteins requiring cofactors that are present only in the cytoplasm use the Tat pathway, as do metalloproteins that contain metals that are found at a high concentration in the cytoplasm but not in the periplasm and therefore must be incorporated into the protein before it is translocated across the inner membrane.

Proteins targeted to the Tat system are made with signal peptides that look much like the standard Sec signal peptides except that they have a diagnostic Arg-Arg motif near the N-terminus, a somewhat lower overall hydrophobicity, and often a "Sec-avoidance" motif in the form of one or a couple of positively charged residues at the C-terminal end or immediately downstream of the signal peptide (**Figure 5.18A**).

The Tat translocase is composed of three subunits, TatA, TatB, and TatC. TatB and TatC are thought to form a receptor complex that recognizes Tat signal peptides, while TatA forms a large, multimeric translocation pore of variable stoichiometry (**Figure 5.18B**). Transport through the Tat complex is driven by the membrane proton-motive force, but the mechanistic details are not well understood.

5.2.3 Some Bacterial Proteins Require Specialized Secretion Systems

In addition to the Sec and Tat systems, bacteria contain a number of specialized secretion systems that act on a restricted set of substrates. Gram-negative bacteria such as *E. coli* contain at least six distinct groups of secretion systems, called, aptly, types I–VI (**Figure 5.19**). While SecYEG provides a function essential for life, the additional systems are presumed to have added selective advantages during evolution.

Type I systems can transport substrate proteins directly from the cytoplasm to the extracellular medium. They are generally composed of three components: an ATP-driven transporter that moves the substrate protein across the inner membrane, a periplasmic connector protein, and an outer membrane channel that extends into the periplasm. The outer membrane channel is called TolC and is common to many different type I systems, while the other two components are specific to different substrates. An example is shown in **Figure 5.20**; in this particular case, substrate transport across the inner membrane is driven by proton translocation in the AcrB subunit rather than ATP hydrolysis.

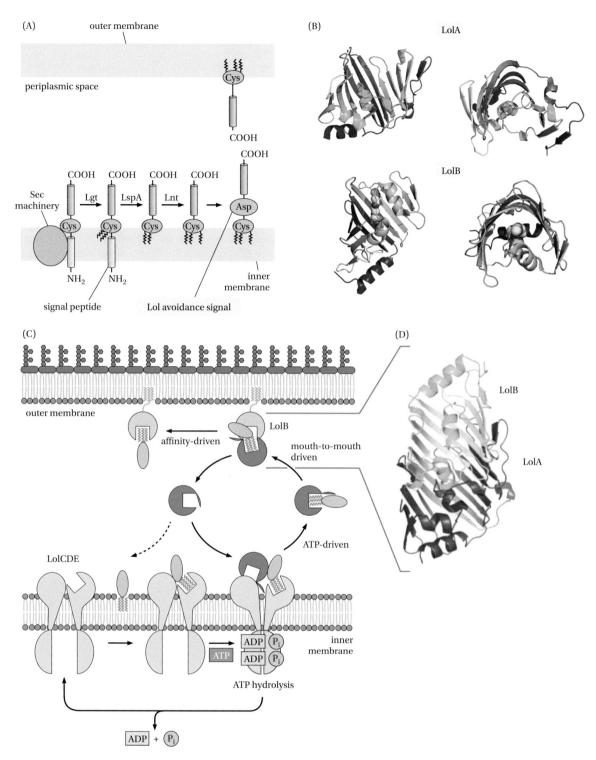

Figure 5.17 Targeting and transport of lipoproteins. (A) Pre-lipoproteins are translocated through the SecYEG translocon and are then modified by a series of enzymes (Lgt, LspA, Lnt). Lipoproteins that have an Asp immediately after the lipid-modified Cys residue cannot enter the outer membrane transport pathway and thus, remain anchored to the inner membrane. (B) Structures of the transport proteins LolA (PDB: 1IWL) and LolB (PDB: 1IWN). The two proteins have similar structures, even though their sequence identity is only about 10%. They both have hydrophobic pockets believed to bind lipid, as seen from the PEG molecule (yellow/red) bound in LolB. (C) Scheme for the transport of lipoproteins to the outer membrane. LolCDE (yellow) is an ABC transporter (see Chapter 12) that requires ATP. Apparently, energy is required to extract the lipoprotein (blue) from the membrane for handoff to LolA (purple). LolA and LolB form a dimer in the process of handing the lipoprotein to LolB (green), which is itself a lipoprotein anchored to the outer membrane. (D) A model of the LolA-LolB dimer. (B, From Takeda, K et al. [2003] *EMBO J* 22: 3199–3209. With permission from John Wiley and Sons. C and D, From Okuda S et al. [2011] *Annu Rev Microbiol* 65: 239–259. With permission.)

Figure 5.18 The twin-arginine translocation system. (A) Typical Tat signal peptide (*E. coli* CueO protein). The N-terminal twin-arginine motif is in red, the h-region is in green, the C-terminal Sec-avoidance motif is in red, and the signal peptidase cleavage site is indicated by the arrow. (B) Cartoon representation of the Tat translocase system. Precursor-bound TatBC assembles with TatA subunits into an active translocon. (From Robinson C, Matos CFRO, Beck D et al. [2011] *Biochimica et Biophysica Acta (BBA) Biomembranes* 1808: 876–884. With permission from Elsevier.)

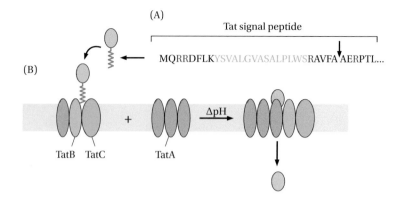

Type II systems move proteins from the periplasm across the outer membrane. Substrate proteins are first translocated across the inner membrane by the SecAYEG translocon and are then driven across a large ring-shaped protein complex in the outer membrane. The system is energized by a cytoplasmic ATPase that interacts with a complex of inner membrane proteins, which in turn is coupled to the outer membrane channel.

Type III systems are used by pathogenic bacteria to inject toxic effector proteins into target eukaryotic cells. These systems are related to the flagellar systems that bacteria use for propulsion. A ring-like basal body in the inner membrane is connected to a cytoplasmic, energy-supplying ATPase and to an inner rod that protrudes into the periplasm (**Figure 5.21A**). A second ring-like structure in the outer membrane guides a central needle into the extracellular

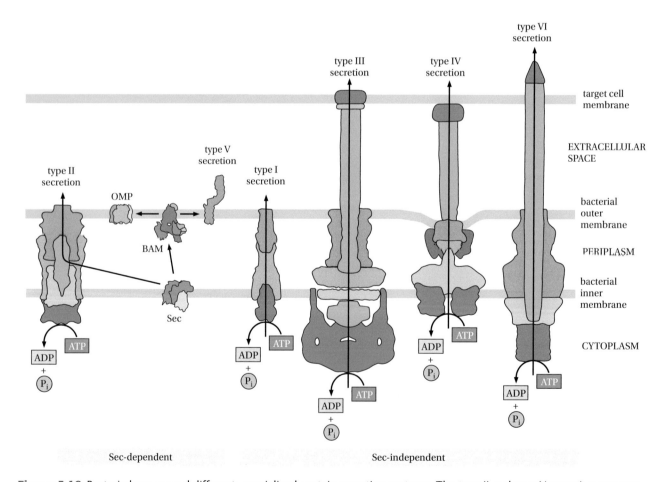

Figure 5.19 Bacteria have several different, specialized protein secretion systems. The type II and type V secretion systems work in concert with the SecA/SecYEG system. Types I, III, IV, and VI are self-contained.

space. The length of the needle is tightly controlled, and at its tip is a needle extension that makes contact with a translocase complex, composed of bacterial proteins, that is integrated into the host cell membrane. The effector protein is translocated through the entire system as an unfolded polypeptide and folds once inside the target cell.

Type IV systems are superficially similar to type III systems but are composed of a distinct set of proteins. They generally export unfolded proteins into target cells and are thought to have evolved from bacterial conjugation machines that transport DNA between cells. A cryo-EM structure of a type IV secretion system is shown in **Figure 5.21B**.

Type V systems are relatively simple, consisting of a passenger module and a translocation module with specificity for the passenger. There are two type V families. In the two-partner family, the passenger and translocation modules are separate proteins. In the autotransporter family, on the other hand, the passenger is the N-terminal part of a single two-domain protein. Both types of proteins are translocated across the inner membrane by the SecAYEG translocon system.

The type V autotransporters are particularly interesting. The layout of the sequence of a typical autotransporter, EspP, is shown in **Figure 5.22A**. The C-terminal translocator domain forms a β-barrel in the outer membrane and helps the N-terminal passenger domain to translocate into the extracellular medium. There are two types of autotransporters. One type, such as EspP, has peptidase activity and cleaves the passenger, allowing it to be secreted into the extracellular space (**Figure 5.22B**). In the case of EspP, the secreted passenger is a protease that has a prominent β-helix structural motif; a motif frequently found in cleaved passenger domains. The other type of autotransporter, typified by EstA, has a non-cleavable passenger domain (**Figure 5.22C**). The transported passenger of EstA is a lipase. Autotransporters are one-shot translocators, with only a single passenger domain being translocated by each translocator domain.

How is the passenger domain transported? For EspP, it has been found that the passenger domain is transported through a hybrid β-barrel channel formed between EspP and BamA, an outer membrane protein "insertase" that catalyzes the membrane insertion of other outer membrane proteins (such as EspP). The Bam complex is discussed in Chapter 6, and a BamA hybrid β-barrel is shown in Figure 6.11B. One hypothesis is that the free energy of folding of the passenger is responsible for dragging the passenger through the transiently formed BamA-EspP hybrid β-barrel. This hypothesis may explain why the β-helix is a common passenger structural motif: one can imagine that the zippering of the β-helix provides the motive force. Once the passenger has been fully translocated, EspP detaches from BamA and folds into its final structure (Figure 5.22B).

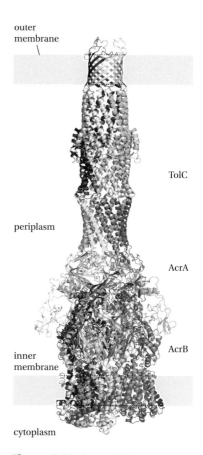

Figure 5.20 A cryo-EM structure of the type I AcrB-AcrA-TolC efflux pump (PDB: 5O66). The system provides a controlled excretion path across two membranes, from the cytoplasm to the external environment.

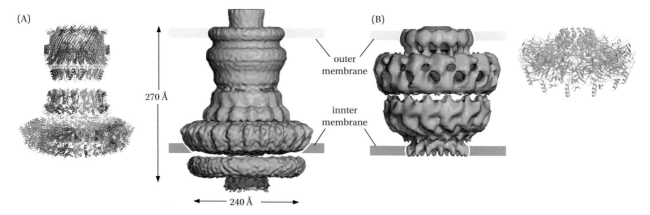

Figure 5.21 Structures of type III and type IV secretion systems. (A) Atomic model and electron density cryo-EM structure of the *Salmonella typhimurium* type III secretion injectisome. (B) Cryo-EM electron density of a type IV secretion complex from *Agrobacterium tumefaciens* and high-resolution atomic model of the outer membrane channel part of the type IV complex. (A, From Worrall LJ et al. [2016] *Nature* 540: 5597-5601. Used with permission from Springer Nature. B, From Chandran V et al. [2009] *Nature* 462: 1011-1015. With permission from Springer Nature.)

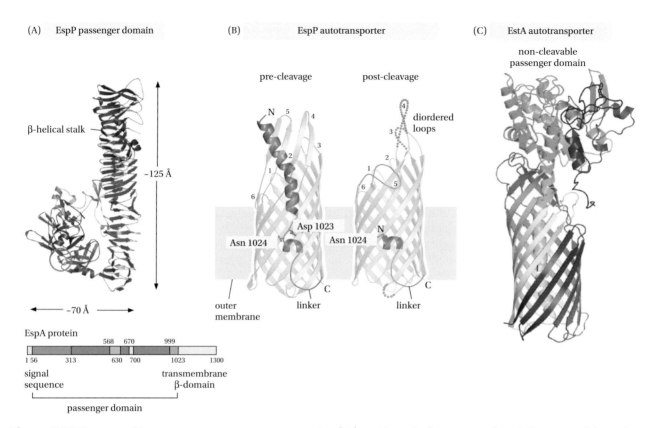

(A)   EspP passenger domain

(B)   EspP autotransporter

(C)   EstA autotransporter

Figure 5.22 Some type V transporters are autotransporters, in which an N-terminal passenger domain is exported through a β-barrel formed by a C-terminal domain. The passenger can be cleaved from the barrel or not, depending upon whether the β-barrel contains an autocatalytic peptidase activity. (A) The passenger domain of EspP from *E. coli* (PDB: 3SZE), a secreted serine protease. Like many cleaved passenger proteins, a β-helix is a prominent structural motif in EspP. (B) The EspP β-barrel transporter domain includes a catalytic site that cleaves the passenger for release into the extracellular space (PDB: 3SLT, 2QOM). (C) The EstA autotransporter from Pseudomonas aeruginosa (PDB: 3KVN) does not cleave its passenger but simply exposes the passenger on the outer surface of the bacterium. In this case, the passenger is a lipase. (A, From Khan S, Mian HS, Sandercock LE et al. [2011] *J Mol Biol* 413: 989–1000. With permission from Elsevier. B, From Barnard JG, Peterson JH, Noinaj N et al. [2012] *J Mol Biol* 415: 128–142. With permission from Elsevier. C, From van den Berg B [2010] *J Mol Biol* 396: 627–633. With permission from Elsevier.)

Type VI systems, finally, are evolutionarily related to certain DNA injection systems found in bacteriophages (bacteriophages are viruses that infect bacteria). They have a central hollow tube or needle held within an outer tube that can contract against the bacterium's inner membrane. This contraction pushes the inner tube through the membrane of the target cell (**Figure 5.23**), creating a conduit for DNA in the case of a bacteriophage or leading to injection of an effector protein in the case of a bacterial type VI secretion system. There are type VI secretion systems that can inject effector proteins both into eukaryotic cells and into the periplasm of a target bacterium.

5.2.4 Usher Proteins Secrete Bacterial Pili

Gram-negative bacteria are coated by fibers called variously pili, fimbriae, or fibrillae (**Figure 5.24**). These coats are essential for pathogenesis, because they are responsible for attaching bacteria to their host tissues. They are secreted by a special transport system called the usher pathway. The pathway is more complicated than the type I–VI systems, because the system must not only regulate the assembly of fimbriae; it must also regulate their secretion onto the extracellular surface of the outer membrane.

The system consists of chaperones that deliver the pili subunits to the usher proteins (**Figure 5.25**). The pili subunits are essentially modified

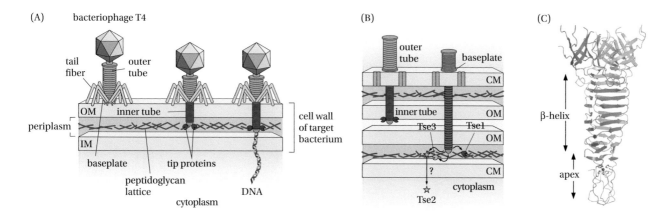

Figure 5.23 Type VI secretion systems (panel B) are related to bacteriophage DNA injection systems (panel A). Contraction of the outer tube pushes the inner tube into the target cell, allowing DNA or effector proteins (Tse1–3) to enter. Shown in panel C is the trimeric structure of the membrane-piercing tail spike of the bacteriophage φ92 (PDB: 3PQI). (A, B, From Cotter P [2011] *Nature* 475: 301–303. With permission from Springer Nature. C, From Browning C, Schneider MM, Bowman VD et al. [2012] *Structure* 20: 326–339. With permission from Elsevier.)

immunoglobulin (Ig) domains comprised solely of β-strands. The Ig fold has been re-purposed many times in the course of evolution. For example, it is used as a structural element in many different biological systems, such as titin—composed of more than 200 Ig-like domains—which is responsible for the passive elasticity of muscle fibers.

There are different types of pili, which are characterized by the type of subunits used in their construction. P pili have a PapA base structure, while Type I pili have a FimA base structure (shown in blue in Figure 5.25). Pathogenic *E. coli* that invade the urinary bladder and form pain-inducing biofilms use Type 1 pili to adhere to bladder cells. The base structures of pili are constructed from more than 1000 PapA or FimA subunits linked together by strand exchange (**Figure 5.25B**). That is, a β-strand of one subunit pairs with the strand of an adjacent subunit to form chains of subunits.

High-resolution x-ray structures have been determined for most of the components of the usher pathway (**Figure 5.26**). As we have noted before, there is no ATP or GTP in the periplasmic space, which means that this remarkable assembly process must occur spontaneously, driven only by the decrease in free energy arising from the preferred associations of the components.

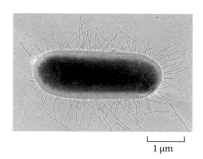

$1\ \mu m$

Figure 5.24 *E. coli* cell surrounded by a halo of fibrous pili that are secreted by usher proteins. The pili allow bacteria to adhere to host cells. (From Capitani G, Eldam O, Glockshuber R et al. [2006] *Microbes and Infection* 8: 2284–2290. With permission from Elsevier.)

5.3 PROTEINS ARE IMPORTED INTO MITOCHONDRIA, CHLOROPLASTS, AND PEROXISOMES

5.3.1 Proteins Are Imported into Mitochondria through the TOM and TIM Translocases

Current estimates based on proteomics studies put the number of mitochondrial proteins in the yeast *Saccharomyces cerevisiae* at around 1000, of which the overwhelming majority are encoded in the nuclear DNA, translated in the cytoplasm, and imported into the organelle; human mitochondria contain at least this many proteins. Only a handful of proteins (13 in human mitochondria) are encoded in the mitochondrial DNA, are synthesized by mitochondrial ribosomes and hence, do not need to be imported.

The majority of the proteins imported into the matrix of the mitochondrion contain positively charged, amphiphilic N-terminal targeting sequences called

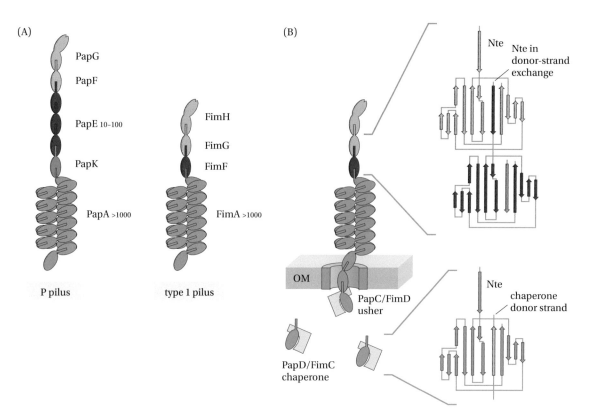

Figure 5.25 Pilus assembly though the usher pathway. (A) Structural scheme for P and type 1 pili. PapA and FimA are base structures that oligomerize by donor-strand exchange of their N-terminal strand (Nte, panel B). Other types of Pap and Fim protomers are assembled at the pili termini to imbue the pili with characteristic properties. (B) Schematic of the assembly of P and Type 1 pili. Each type of pilus subunit is brought to the usher by specific chaperones. Pili are assembled by the usher proteins and pushed through the outer membrane. (From Renaut H, Tang C, Henderson NS et al. [2008] *Cell* 133: 640–652. With permission from Elsevier.)

pre-sequences (**Figure 5.27**), while most of the mitochondrial inner and outer membrane proteins have internal targeting sequences that are not cleaved off after targeting.

The mitochondrial protein import system is significantly more complex than the ER system. Around 35 different yeast proteins have been implicated in import to date, and they are organized into at least 10 distinct but possibly interacting complexes. Some components in these complexes clearly trace their ancestry to bacterial proteins, while others have an unclear evolutionary history. Unfortunately, there are only a few high-resolution structures available for any of these proteins, so one must rely mainly on cartoon representations of the various components.

A simplified overview of the mitochondrial protein import pathways is shown in **Figure 5.28**. Here, we will focus on the import of soluble protein into the matrix and the intermembrane space; the import of membrane proteins destined for the outer and inner membranes is discussed in Chapter 6.

The TOM complex (TOM stands for Translocase Outer Membrane) in the mitochondrial outer membrane contains import receptors as well as a transmembrane channel and is the common gateway into the mitochondrion (**Figure 5.29A**). The TOM20 receptor has a hydrophobic groove for binding to N-terminal mitochondrial pre-sequences; the bound pre-sequence forms an amphiphilic α-helix **Figure 5.29C**. Proteins are handed from the TOM20 receptor to the TOM channel. The TOM channel is a double β-barrel integral membrane protein held together by small helical transmembrane subunits (**Figure 5.29B**).

Proteins destined for the matrix are transferred from the TOM complex to the TIM23 complex in the inner membrane, which also houses both receptors and a channel protein (Figure 5.29A). The electrical potential across the inner

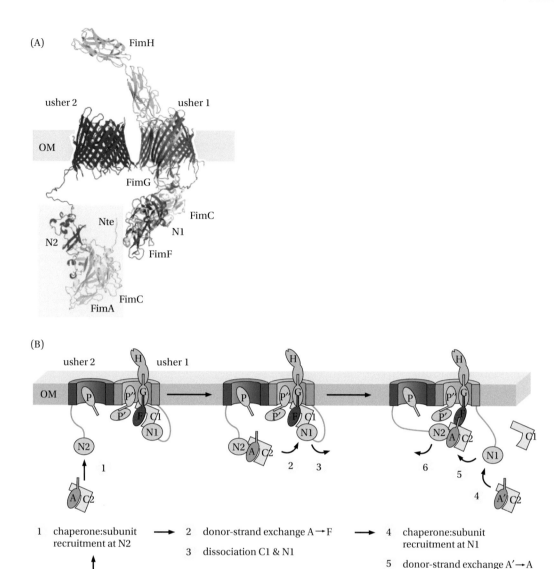

Figure 5.26 Structures of usher proteins and a schematic view of the assembly of pili. (A) Structure associated with Type 1 pili. The ushers (PDB: 2VQI) are pairs of β-barrel outer membrane proteins. Usher 1 assists in the assembly of the fiber that usher 2 pushes across the membrane. (B) Schematic describing the process of pilus assembly by the usher system. (From Renaut H, Tang C, Henderson NS et al. [2008] *Cell* 133: 640–652. With permission from Elsevier.)

membrane (Δψ) drives the positively charged N-terminal targeting peptide through the TIM23 channel. A matrix-localized motor protein, the PAM (presequence translocase-associated motor) complex, then binds the incoming chain and pulls it into the matrix, using ATP hydrolysis as the energy source. Finally, the pre-sequence is removed by a matrix-localized protease (MPP), and the protein folds, often with the help of matrix chaperones.

The intermembrane space is a more oxidizing environment than the cytoplasm and the mitochondrial matrix, and many soluble intermembrane space proteins contain disulfide bonds that need to be properly formed during folding. The redox-active MIA40 protein forms transient mixed disulfides with such proteins as they exit from the TOM channel and cooperates with another sulfhydryl oxidase, Erv1, and with complex IV in the respiratory chain (Chapter 14) to facilitate their folding and assembly (**Figure 5.30**).

Figure 5.27 A typical mitochondrial pre-sequence (from human malate dehydrogenase). It lacks negatively charged amino acids and is enriched for positively charged residues (red). The arrow indicates the site where the pre-sequence is cleaved off from the mature protein (black) by the mitochondrial processing peptidase enzyme located in the mitochondrial matrix.

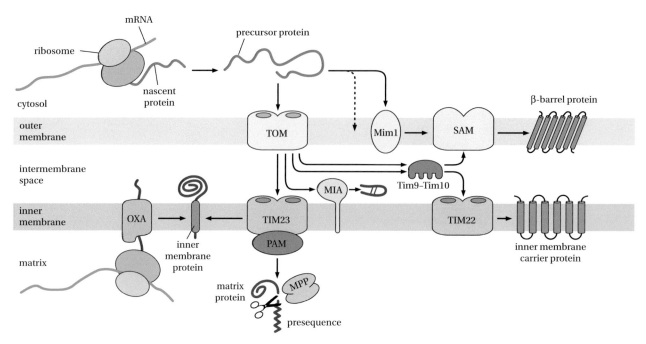

Figure 5.28 Overview of the mitochondrial protein import pathways. Soluble matrix proteins are imported through the TOM and TIM23 channels with the help of the PAM motor complex. Some inner membrane proteins are also imported through the TOM-TIM23 pathway, while others (e.g., the carrier proteins) use the TIM22 complex. Inner membrane proteins can also be inserted from the matrix, using the OXA insertase. Outer membrane β-barrel proteins use the MIM-SAM pathway. Disulfide bond formation in intermembrane space proteins is catalyzed by the MIA complex. (From Schmidt O et al. [2010] *Nature Rev Mol Cell Biol* 11: 655–557. Used with permission from Springer Nature.)

5.3.2 Proteins Are Imported into Chloroplasts through the TOC and TIC Translocons

As for mitochondria, nearly all chloroplast proteins are encoded in nuclear DNA, translated in the cytoplasm, and imported into the chloroplast. Chloroplast targeting sequences are called transit peptides; they are rich in hydroxylated amino acids, generally lack acidic residues, and contain no obvious secondary structure motifs (**Figure 5.31**).

Superficially, the chloroplast import system is similar to its mitochondrial counterpart, with receptor–channel complexes in both the outer and inner envelope membranes (**Figure 5.32**). The Toc75 protein in the outer membrane TOC complex belongs to the same protein family as the bacterial outer membrane protein BamA and the mitochondrial outer membrane protein Sam50, which are both involved in the assembly of β-barrel integral membrane proteins (Chapter 6). The TIC complex in the inner envelope membrane is composed of a number of subunits, with Tic110, a protein with two transmembrane α-helices, being the most likely to form the channel itself.

Protein import across the outer and inner envelope membranes into the stroma requires GTP hydrolysis by TOC components and ATP hydrolysis by TIC components as well as by molecular chaperones in the stroma. There is apparently no need for a membrane potential across the inner envelope membrane. Transit peptides are removed from the imported proteins by the stromal processing peptidase (SPP).

In addition to the outer and inner envelope membranes, chloroplasts have a third membrane system, the thylakoid. Some proteins, mainly those involved in photosynthesis, have to be inserted into or translocated across the thylakoid membrane. Thylakoids contain both Sec-type, Tat-type, and SRP-dependent protein import systems and hence, closely resemble bacteria in the way they translocate proteins across the membrane (**Figure 5.32B**). Nuclear-encoded proteins destined for the lumen of the thylakoids have bi-partite N-terminal

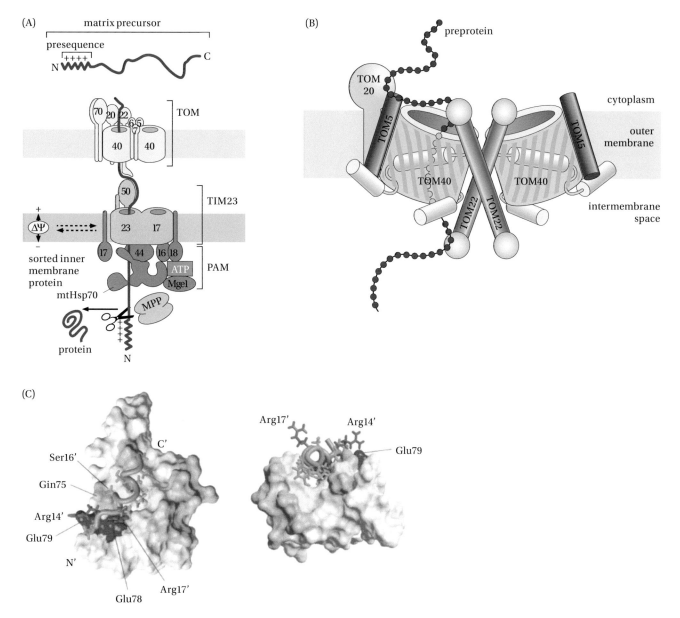

Figure 5.29 Import of matrix proteins into mitochondria. (A) General scheme for protein import. Import proceeds through the TOM complex in the outer membrane and the TIM23 complex in the inner membrane. Pre-sequences are removed by the MPP protease. (B) Model of the TOM complex based on crosslinking and cryo-EM studies. (C) The peptide-biding part of the TOM20 receptor (PDB: 1OM2). The bound pre-sequence (blue, amino acids are primed) is folded into an amphiphilic α-helix with its hydrophobic side buried in the binding groove. Two positively charged Arg residues (14′ and 17′) in the pre-sequence interact with negatively charged Glu residues on the receptor. (A, From Schmidt O et al. [2010] *Nature Rev Mol Cell Biol* 11: 655–667. Used with permission from Springer Nature. B, From Bausewein T, Mills DJ, Langer JD et al. [2017] *Cell* 170: 693–700. Used with permission from Elsevier. C, From Abe Y, Shodai T, Muto T et al. [2000] *Cell* 100: 551–560. Used with permission from Cell Press and Elsevier.)

targeting peptides composed of a stroma-targeting transit peptide followed by a typical Sec or Tat signal peptide.

5.3.3 Proteins Are Imported into Peroxisomes by a "Piggy-Back" Mechanism

Peroxisomes do not contain any DNA and hence, must import all of their proteins. They can import fully folded proteins and even protein complexes, which means that a protein lacking a peroxisomal targeting sequence (PTS) can "piggy-back" on a PTS-containing protein.

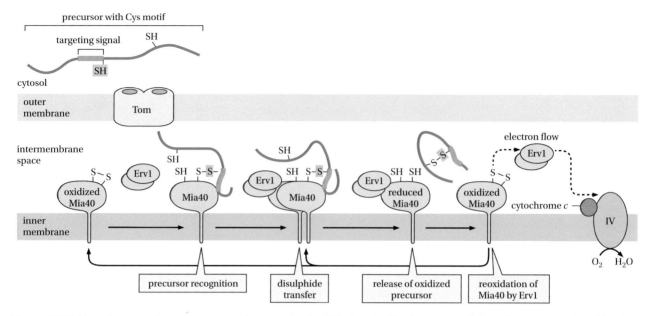

Figure 5.30 Many intermembrane space proteins contain disulfide bonds. The formation of these bonds is catalyzed by the Mia40 and Erv1 disulfide oxidoreductases in the intermembrane space. The final electron acceptor is complex IV in the mitochondrial respiratory chain. (From Schmidt O et al. [2010] *Nature Rev Mol Cell Biol* 11: 655–557. Used with permission from Springer Nature.)

There are two kinds of PTSs: the highly conserved C-terminal tri-peptide Ser-Lys-Leu (PTS1) and a much less conserved N-terminal targeting sequence (PTS2); either of these is sufficient for import. Both the PTS1 and PTS2 receptors are cytoplasmic proteins that form complexes with the folded cargo protein.

The cargo-carrying PTS-receptors associate with a docking complex in the peroxisomal membrane (**Figure 5.33**). The cargo–receptor complex is translocated towards the lumen of the peroxisome by an unknown mechanism and then dissociates. The receptor is then ubiquitinated and recycled into the cytoplasm in a step that requires ATP hydrolysis. Peroxisomal membrane proteins are handled differently, and it is presently thought that many, if not all, such proteins are first sorted to the ER and then delivered by vesicular transport (see Chapter 7) to the peroxisome.

5.4 PROTEINS ARE TRANSPORTED IN AND OUT OF THE NUCLEUS THROUGH THE NUCLEAR PORE COMPLEX

The nucleus contains the cell's DNA, and all RNA molecules, including the mRNAs, are synthesized in the nucleus. On the other hand, translation of mRNAs by ribosomes takes place in the cytoplasm, and mRNAs as well as other RNAs such as micro-RNAs and tRNAs must be exported out of the nucleus. At the same time, many proteins, including DNA-binding histones, transcription factors, and RNA polymerases, must be imported from the cytoplasm into the nucleus. Ribosomes go through a particularly elaborate assembly process, where mRNAs encoding ribosomal proteins are first exported from the nucleus and translated in the cytoplasm. The ribosomal proteins are then imported into

Figure 5.31 A typical chloroplast transit peptide (from ferredoxin). It lacks negatively charged amino acids and is enriched for hydroxylated amino acids (yellow). The arrow indicates the site where the transit peptide is cleaved off from the mature protein (black) by the stromal processing peptidase located in the chloroplast stroma.

MASTLSTLSVSASLLPKQQPMVASSLPTNMGQALFGLKAGSRGRVTAM**ATYKV**

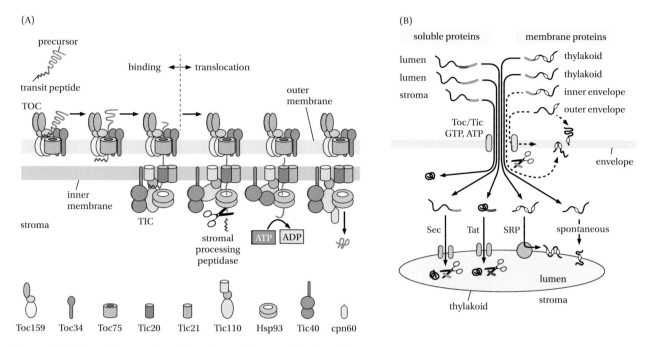

Figure 5.32 Protein import into chloroplasts. (A) General scheme for import. Both the TOC and TIC translocons are composed of multiple proteins. (B) The thylakoid contains four different import pathways: Sec, Tat, SRP-dependent, and a "spontaneous" membrane insertion pathway with no known protein components. (A, From Li H et al. [2010] *Annu Rev Plant Biol* 61: 157–180. With permission. B, From Jarvis P, Robinson C [2004] *Curr Biol* 14: R1064–R1077. Used with permission from Elsevier.)

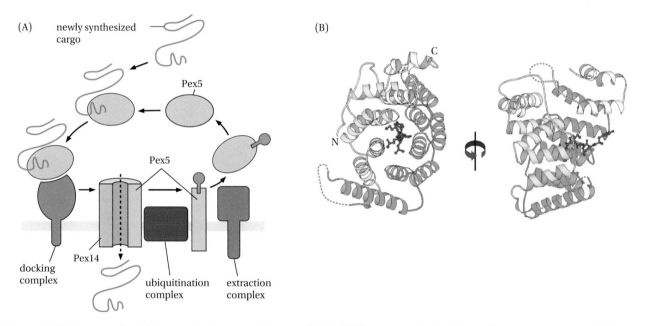

Figure 5.33 Import of soluble proteins into peroxisomes. (A) The PTS1 receptor Pex5 delivers the cargo protein to a docking complex that includes the Pex14 protein. Pex5 and Pex14 together form the import channel. After cargo import, Pex5 is mono-ubiquitinated (blue lollipop), extracted from the membrane, and de-ubiquitinated in preparation for another round of import. (B) The human Pex5 receptor (PDB: 1FCH) with a bound PTS1 peptide (in purple) of sequence LQSKL. The C-terminus of the peptide penetrates deep into the receptor. (A, From Nuttall JM, Motley A, Hettema E [2011] *Curr Opinion Cell Biol* 23: 421–426. Used with permission from Elsevier. B, From Gatto GJ Jr et al. [2000] *Nature Struct Biol* 7:1091–1095. Used with permission from Springer Nature.)

the nucleus, where they assemble with ribosomal RNA into the large and small ribosomal subunits, which in turn are exported back to the cytoplasm.

The nucleus is surrounded by a double membrane (**Figure 5.34**). The outer nuclear membrane is continuous with the ER membrane. Macromolecules can enter and exit the nucleus through nuclear pore complexes (NPCs). The NPC is

Figure 5.34 The nuclear pore complex. (A) The nuclear pore complex (NPC) connects the inner and outer nuclear membranes. (B) A high-resolution cryo-EM structure of a fungal NPC. Although it looks empty in this structure, the central channel is filled with long flexible protein chains containing so-called FG (phenylalanine-glycine) repeats. (A) From Alberts B, Johnson A, Lewis JH et al. [2007] *Molecular Biology of the Cell*, 5th Edition. Garland Science, New York. With permission from W. W. Norton. (B) From Lin DH, Stuwe T, Schillbach S et al. [2016] *Science* 352:308. With permission from AAAS.

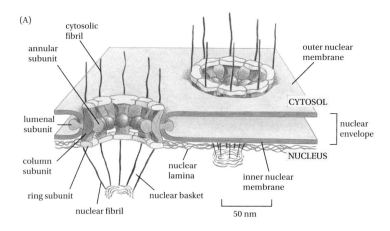

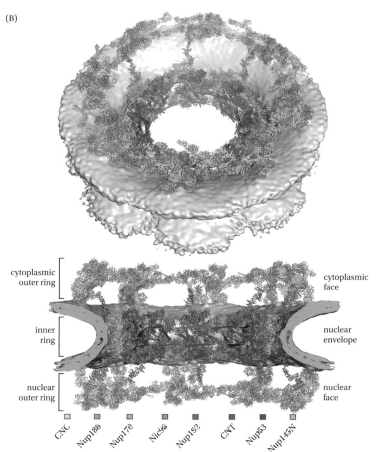

...SPPKAVKRPAATKKAGQAKKKKLDK...

Figure 5.35 A bi-partite nuclear localization sequence (from *Xenopus laevis* nucleoplasmin). Critical positively charged residues are marked in red. The sequence is found inside the protein and is not cleaved off after import into the nucleus.

one of the largest protein complexes in a eukaryotic cell. The molecular weight of the mammalian NPC is around 110 MDa, and it is composed of multiple copies of some 30 different proteins.

Proteins with a molecular weight lower than 60 kDa can move through the NPC by passive diffusion, but bigger proteins, protein complexes, and RNA molecules can only move in and out of the nucleus by active transport through the NPC. Protein-RNA complexes as big as the large subunit of the ribosome translocate across the NPC, and hundreds of macromolecules can be transported through a single NPC per second, with mean passage times on the order of 10 ms.

Active transport requires that a nuclear localization signal (NLS) be present in the cargo protein. NLSs are composed of one, or sometimes two, closely spaced, short stretches of positively charged amino acids (**Figure 5.35**). The

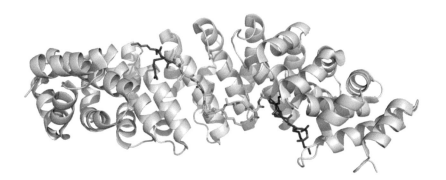

Figure 5.36 A bi-partite nuclear localization signal (NLS; in green) bound to an importin (PDB: 1EJY). The positively charged residues in the NLS that bind to the importin are in red.

NLS binds to nuclear import receptors (also called importins) in the cytoplasm (**Figure 5.36**) that mediate transport through the NPC. Proteins and RNA can also be exported from the nucleus if they contain a nuclear export signal (NES) that binds to nuclear export receptors (also called exportins).

An overview of the mechanism of nuclear import and export is shown in **Figure 5.37**. The central player is the GTPase Ran, which is found both in the nucleus and in the cytoplasm. The GTPase activity of Ran is stimulated by the

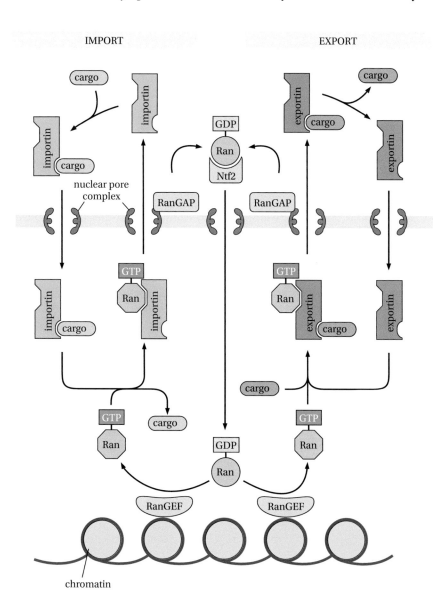

Figure 5.37 Nuclear protein import and export processes are driven by GTP hydrolysis by the RanGAP-activated Ran GTPase. (From Mosammaparast N, Pemberton LF [2004] *Trends Cell Biol* 14: 547–556. Used with permission from Elsevier.)

Figure 5.38 Importin β ferries cargo through FG-repeat hydrogels. (A) A fluorescent cargo protein is normally excluded from the FG-repeat hydrogel (left) but readily enters and diffuses through the FG-repeat hydrogel when bound to the import receptor importin β (right). In this experiment, a small piece of hydrogel (grey part in the "hydrogel" images) is placed in buffer (black part in the "hydrogel" images), and the buffer–hydrogel boundary region is imaged in a confocal microscope. The cargo protein is labeled with a red-fluorescent molecule (RedStar), and laser-excited red fluorescence is imaged as a function of time after cargo addition to the buffer ("RedStar channel" images). (B) An FG-repeat-covered nanopore manufactured in a thin polycarbonate membrane. The chart shows the flux of the Ntf2 importin through the nanopore (red) compared with the control protein bovine serum albumin (BSA) that lacks an NLS (blue). (A, From Frey S, Görlich D [2007] *Cell* 13: 512–523. Used with permission from Elsevier. B, From Jovanovic-Talisman T et al. [2009] *Nature* 457: 1023–1027. Used with permission from Springer Nature.)

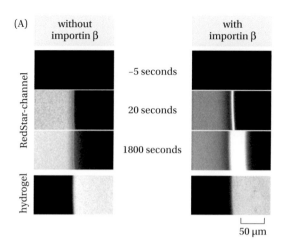

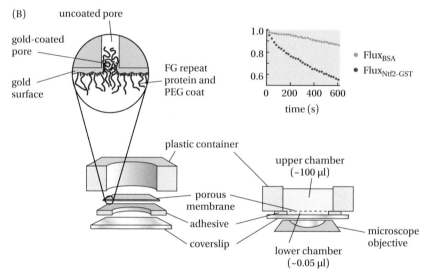

cytoplasmic Ran GTPase-activating protein (RanGAP), so that the cytoplasm contains mainly RanGDP. There is no RanGAP in the nucleus; instead, a nuclear protein, the Ran guanine nucleotide exchange factor (RanGEF), stimulates the exchange of GDP for more abundant GTP, so that the nucleus contains mostly RanGTP.

Import of cargo proteins proceeds as follows. First, an importin binds to an NLS in the cargo protein, and the cargo–importin complex moves through the NPC. Once in the nucleus, the GTP-loaded form of the Ran protein, which has a high affinity for binding the importin, displaces the cargo from the importin. The RanGTP–importin complex then exits the nucleus and meets up with the Ran GTPase-activating protein (RanGAP) in the cytoplasm. This activates the GTPase activity of Ran, producing RanGDP, which has no affinity for the importin. The importin is now free to load another cargo, while RanGDP is translocated back into the nucleus bound to its own, specific importin Ntf2. Finally, RanGDP encounters the Ran guanine nucleotide exchange factor protein (RanGEF) in the nucleus, and GDP is exchanged for GTP, priming Ran for interaction with a new cargo–importin complex.

Export of macromolecules from the nucleus follows similar principles, except that in this case, RanGTP promotes the formation of a RAN–exportin–cargo complex, which disassembles in the cytoplasm when RanGTP is converted to RanGDP. The empty exportin is then translocated back into the nucleus.

How does the NPC selectively recognize importin/exportin–cargo complexes, and how do these complexes move across the nuclear envelope? The prevailing model is based on the observation that many of the proteins lining the NPC pore contain extensive regions of so-called phenylalanine-glycine (FG) repeats. The flexible FG-repeat segments are thought to form a tangled web in the center of the NPC that serves as a diffusion barrier to macromolecules. Importins and exportins, however, have multiple binding sites for FG-repeats on their surface and become "dissolved" in the center of the NPC, essentially diffusing through the FG-repeat network. Such diffusion has been observed in artificially produced FG-repeat hydrogels and in artificial nanopores covered with FG-repeat proteins (**Figure 5.38**).

KEY CONCEPTS

- Translocases are fundamental assemblies, used in all living organisms, that move proteins into and across membranes.

- Protein translocation through the membrane of the endoplasmic reticulum (ER) is mediated by the Sec61 translocon, which is structurally homologous to the bacterial SecYEG translocon.

- Signal peptides that target ribosome–nascent chain complexes for translocation across the ER membrane are recognized by the signal recognition particle (SRP) as the nascent chain emerges from the ribosome. SRP causes elongation arrest in eukaryotes—but not in bacteria—that persists until the ribosome docks to the Sec translocon with the help of the SRP receptor (SR).

- Signal peptides are cleaved from the parent sequence by signal peptidase (SPase) and then degraded by intramembrane proteases, such as RseP in bacteria.

- In bacteria, protein translocation across the inner membrane is carried out by two general translocation systems: the SecYEG translocon system and the twin-arginine translocation (Tat) system.

- The SecYEG system in bacteria can mediate both co- and post-translational translocation of unfolded polypeptides. Post-translational translocation requires the participation of the ATP-driven SecA motor protein.

- The Tat system mediates post-translational translocation of folded proteins and protein complexes and is driven by the transmembrane electrochemical potential.

- Bacteria contain a number of specialized secretion systems, each of which acts on only a small number of substrates.

- Proteins are imported into mitochondria via translocases in the outer and inner mitochondrial membranes (TOM and TIM complexes).

- Proteins are imported into chloroplasts via translocases in the outer and inner envelope membranes (TOC and TIC complexes).

- Proteins and RNA can enter and exit the nucleus via nuclear pore complexes.

- Nuclear import and export of proteins and RNA is driven by the Ran GTPase.

- Cargo–receptor complexes destined for nuclear import or export diffuse through the FG-repeat meshwork in the center of the nuclear pore complex.

FURTHER READING

Araiso, Y., Imai, K., and Endo, T. (2021) Role of the TOM complex in protein import into mitochondria: Structural views. *Annu. Rev. Biochem.*, in press

Beck, M., and Hurt, E. (2017) The nuclear pore complex: Understanding its function through structural insight. *Nat Rev Mol Cell Biol.* 18:73–89.

Ben-Shem, A., Jenner, L., Yusupova, G., and Yusupov, M. (2010) Crystal structure of the eukaryotic ribosome. *Science* 330:1203–1209.

Berks, B.C. (2015) The twin-arginine protein translocation pathway. *Annu Rev Biochem.* 84:843–864.

Bos, M.P., Viviane Robert, V., and Jan Tommassen, J. (2007) Biogenesis of the gram-negative bacterial outer membrane. *Annu. Rev. Microbiol.* 61:191–214.

Chotewutmontri, P., Holbrook, K., and Bruce, B.D. (2017) Plastid protein targeting: Preprotein recognition and translocation. *Int Rev Cell Mol Biol.* 330:227–294.

Costa, T.R., Felisberto-Rodrigues, C., Meir, A., Prevost, M.S., Redzej, A., Trokter, M., and Waksman, G. (2015) Secretion systems in Gram-negative bacteria: Structural and mechanistic insights. *Nat Rev Microbiol.* 13:343–359.

Park, E., and Rapoport, T.A. (2012) Mechanisms of Sec61/SecY-mediated protein translocation across membranes. *Annu Rev Biophys.* 41:21–40.

Schleiff, E., and Becker, T. (2011) Common ground for protein translocation: Access control for mitochondria and chloroplasts. *Nature Rev. Mol. Cell Biol.* 12:48–59.

Schmeing, T.M., and Ramakrishnan, V. (2009) What recent ribosome structures have revealed about the mechanism of translation. *Nature* 461:1234-1242.

Schwerter, D.P., Grimm, I., Platta, H.W., and Erdmann, R. (2017) ATP-driven processes of peroxisomal matrix protein import. *Biol. Chem.* 398:607-624.

Sloan, K.E., Gleizes, P.E., and Bohnsack, M.T. (2016) Nucleocytoplasmic transport of RNAs and RNA-Protein complexes. *J Mol Biol.* 428:2040-2059.

Wiedemann, N., and Pfanner, N. (2017) Mitochondrial machineries for protein import and assembly. *Annu Rev Biochem.* 86:685-714.

Zhang, X., and Shan, S. (2014) Fidelity of cotranslational protein targeting by the signal recognition particle. *Annu. Rev. Biophys.* 43:381-408.

KEY LITERATURE

Blobel, G., and Sabatini, D.D. (1971) *Ribosome-membrane Interaction in Eukaryotic Cells in Biomembranes* (Manson, L.A., Ed.), Vol. 2, pp. 193-195, Plenum, New York.

Blobel, G., and Dobberstein, B. (1975) Transfer of proteins across membranes. I. Presence of proteolytically processed and unprocessed nascent immunoglobulin light chains on membrane-bound ribosomes of murine myeloma. *J. Cell Biol.* 67:835-851.

Blobel, G., and Dobberstein, B. (1975) Transfer of proteins across membranes. II. Reconstitution of functional rough microsomes from heterologous components. *J. Cell Biol.* 67:852-862.

Deshaies, R.J., and Schekman, R. (1987) A yeast mutant defective at an early stage in import of secretory protein precursors into the endoplasmic reticulum. *J. Cell Biol.* 105:633-645.

Dingwall, C., Sharnick, S.V., and Laskey, R.A. (1982) A polypeptide domain that specifies migration of nucleoplasmin into the nucleus. *Cell* 30:449-458.

Emr, S.D., Hanley-Way, S., and Silhavy, T.J. (1981) Suppressor mutations that restore export of a protein with a defective signal sequence. *Cell* 23:79-88.

Maccecchini, M.L., Rudin, Y., Blobel, G., and Schatz, G. (1979) Import of proteins into mitochondria: Precursor forms of the extramitochondrially made F1-ATPase subunits in yeast. *Proc. Natl. Acad. Sci. USA* 76:343-347.

Harmey, M.A., Gerhard Hallermeyer, G., Harald Korb, G.H., and Walter Neupert, W. (1977) Transport of cytoplasmically synthesized proteins into the mitochondria in a cell free system from *Neurospora crassa. Eur. J. Biochem* 81:533-544.

Milstein, C., Brownlee, G.G., Harrison, T.M., and Mathews, M.B. (1972) A possible precursor of immunoglobulin light chains. *Nature New Biol.* 239:117-120.

Oliver, D.B., and Beckwith, J. (1981) *E. coli* mutant pleiotropically defective in the export of secreted proteins. *Cell* 25:765-772.

von Heijne, G. (1983) Patterns of amino acids near signal-sequence cleavage sites. *Eur.J.Biochem.* 133:17-21.

van den Berg, B., Clemons, W.M., Collinson, I., Modis, Y., Hartmann, E., Stephen C. Harrison, S.C., and Rapoport, T.A. (2004) X-ray structure of a protein-conducting channel. *Nature* 427:36-44.

Walter, P., Ibrahimi, I., and Blobel, G. (1981) Translocation of proteins across the endoplasmic reticulum. I. Signal recognition protein (SRP) binds to in-vitro-assembled polysomes synthesizing secretory protein. *J. Cell Biol.* 91:545-550.

EXERCISES

1. Why do most bacterial inner membrane proteins follow a co-translational translocation/ insertion pathway? Why is this not a practical solution for bacterial outer membrane proteins?

2. Describe how the Ran protein can drive unidirectional import or export of proteins and RNA molecules across the nuclear membrane. Can you think of a good reason why nuclear proteins have targeting sequences that are not cleaved after import into the nucleus?

3. Why might it be advantageous to have additional protein secretion mechanisms in bacteria that are relatively fast compared with the rate of translation on a ribosome by SecYEG? (Hint: think about surface to volume ratios.)

4. In Figure 5.1, identify the steps in the delivery of membrane proteins where chaperones must be involved and the steps where they are not. List the protein pathways discussed in the chapter that are missing from this simplified diagram.

5. Two strains of *E. coli*, A and B, are placed at equal initial concentrations in a rich medium. Strain A grows at 30 minutes per division, Strain B at 40 minutes per division. What will the ratio of A to B be after 24 hours? This result

emphasizes the need for speed in the competition among strains and the selection pressure for efficiency. Now, review your answer to Question 3.

6. Biosynthetic pathways generally balance speed and control. What might two of the advantages be for eukaryotic cells to have so many cytoplasmic compartments and to employ the sorting mechanisms that are used (there are more than two—the aim is to get you to think about the observed facts).

Biosynthesis and Assembly of Membrane Proteins

6

In all cells, regardless of type or species, ribosomes translate mRNAs into soluble cytoplasmic proteins, soluble secreted proteins, and water-insoluble membrane proteins. Ribosomes are universal machines, blind to the fate of the proteins they produce. Soluble and membrane proteins carry the sequence information to fold but must find their destinations and avoid misfolding. How do membrane proteins manage this remarkable feat? The answer is that they have the help of protein and nucleoprotein partners that determine the fates of the different classes of proteins using signals encoded in the proteins' sequences. We discussed this marvelous system in Chapter 5; here, we focus on how translocons—the same ones that secrete proteins—manage the insertion of membrane proteins into target membranes. We will learn that translocons are members of a broader class of proteins referred to as translocases or insertases. We will see that translocons, besides acting as protein-conducting channels for secreted proteins and for large loops in membrane proteins, also guide the insertion of transmembrane helices into the lipid bilayer, helping them to avoid misfolding pathways.

How do translocons distinguish proteins destined for secretion from proteins destined for insertion? The brief answer is that they take advantage of the physical chemistry of lipid–protein interactions (Chapter 3) to identify nascent transmembrane helices as the elongating chain interacts with the translocon. Important questions are how and when the identified helices are passed into the lipid bilayer and how to understand the energy landscape the chain encounters during biogenesis. The synthesis of amino acid chains by the ribosome is remarkably slow: in bacteria, the synthesis of a typical transmembrane helix of 20 amino acids requires about 1 second, and the synthesis rate is even slower in eukaryotes. These rates provide ample time for a nascent chain emerging from the ribosome to explore the complex and dynamic energy landscape formed by the bilayer, the ribosome, and the translocon. Little is known about this energy landscape or how to describe it in a useful way.

What about the β-barrel membrane proteins found in the outer membranes of Gram-negative bacteria? Are outer membrane translocons involved? The answer is yes. We will answer the question of just how it is done. For eukaryotes, the situation is more complicated, because most membrane proteins destined for mitochondria and chloroplasts are synthesized in the cytoplasm and trafficked to the organelles where other types of translocases are located that are distinct from those of the endoplasmic reticulum (ER). How do those translocases work, and how can greasy proteins find their way to them without aggregating?

With these many questions in mind, we focus in this chapter on how translocases guide the insertion of membrane proteins into the different cellular membranes, and the sequence characteristics of the nascent chain that determine the final structure of the protein. Along the way, we will compare the cellular insertion processes with what we know about protein–lipid interactions from

DOI: 10.1201/9780429341328-7

biophysical studies of simple model systems, discussed in Chapter 3. We begin by asking if there are any membrane proteins that insert and fold without the aid of an insertase.

6.1 CELLULAR MECHANISMS OF MEMBRANE PROTEIN ASSEMBLY

6.1.1 How Do Cells Make Integral Membrane Proteins?

In the test tube, membrane proteins need to be pampered to prevent denaturation, aggregation, and precipitation. We do this by keeping them in a water-soluble state using detergents or polymers that shield the proteins' extensive hydrophobic surfaces from contact with water. Reconstitution of detergent-solubilized proteins into lipid bilayers requires careful removal of the detergent in the presence of bilayer-forming lipids to form so-called proteoliposomes. In keeping with their finicky nature, the formation of proteoliposomes must be optimized for each membrane protein. The cell, of course, is much more adept at handling membrane proteins than are scientists.

Most of a cell's proteins are water-soluble and are extruded from the ribosome into the aqueous cytosol, where they fold and can oligomerize, often with the help of chaperones. Insoluble integral membrane proteins, in contrast, fold correctly to their native states only in the non-aqueous environment of lipid bilayers, yet they are made by the same ribosomes that produce the water-soluble proteins. The key to understanding how ribosomes handle membrane proteins was the seminal observation that they are made by membrane-bound ribosomes, so that they can be inserted co-translationally into the membrane. Certain classes of membrane proteins, however, are not co-translationally inserted but instead, exist transiently as water-soluble species. For example, β-barrel proteins destined for the outer membrane of Gram-negative bacteria are secreted into the periplasm and transit to the outer membrane with the help of chaperones; likewise, mitochondrial outer and inner membrane proteins encoded in the nuclear genome and made in the cytosol can be kept in a water-soluble state by cytosolic chaperones. Many bacterial toxins are secreted by bacteria in a water-soluble form that is converted to a membrane-inserted state only upon oligomerization on the target membrane. Toxins and other proteins that can exist in both a stable water-soluble form and a stable membrane-inserted form are referred to as adventitious membrane proteins, because they originate from a source other than the target organism.

6.1.2 Bacterial Toxins Can Insert Spontaneously into Membranes

It is difficult to prove that a protein inserts completely spontaneously into a membrane *in vivo*; if a protein can be shown not to require translocon A or chaperone B, there is always the lingering possibility that some other, as yet unknown, protein may be needed. Bacterial membrane-targeting toxins, however, can often be purified in a water-soluble form, allowing the requirements for insertion into liposomes or biological membranes to be studied in detail.

An extensively studied example is α-hemolysin. This toxin, secreted by the bacterium *Staphylococcus aureus*, attacks the plasma membrane of target mammalian cells, where it forms ion-conducting pores, which kill the cell. The mushroom-shaped pore is made from seven copies of the α-hemolysin monomer that oligomerize on the target membrane (**Figure 6.1**). The membrane-embedded stem is a 14-strand transmembrane β-barrel formed from 7 long β-hairpins. Every second residue of the strands is hydrophobic and faces the outside of the barrel. In the assembled barrel, all the backbone hydrogen bonds

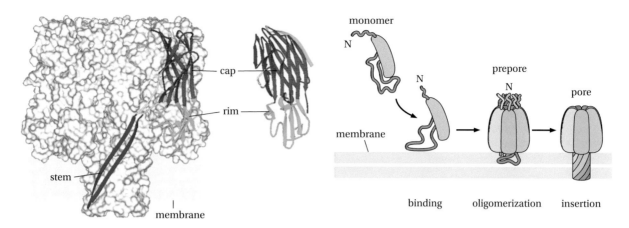

Figure 6.1 α-hemolysin. (A) The assembled pore with one monomer highlighted. The long β-hairpin is shown in blue. On the right, the β-hairpin is tucked away in the water-soluble monomeric form. (B) Model for the conversion of monomeric α-hemolysin to the heptameric membrane pore. (A From Prévost G, Monteil H, Colin DA et al. [2003] *FEBS Lett* 552: 54–60. With permission from John Wiley and Sons. B, From Engleman D [1996] *Science* 274: 1850–1851.)

are satisfied internally, which reduces the high cost of partitioning peptide bonds (see Section 3.4). This alternation between hydrophobic residues on the outer surface and polar residues in the barrel interior creates a large hydrophobic surface that interacts favorably with the surrounding lipid, and an internal water-filled pore that makes the membrane leaky to ions and small molecules.

In the water-soluble monomeric state, the toxin's β-hairpin is folded back against the body of the protein, thus hiding its hydrophobic residues. The available evidence indicates that α-hemolysin can insert and assemble spontaneously into liposomes of the right composition without the help of translocases or chaperones—a strong argument, but we still cannot exclude help from other factors *in vivo*. Spontaneous assembly must mean that the free energy of the strands is lower when in the membrane-embedded barrel configuration than when folded against the protein body. Apparently, once a monomer is on the target membrane, the hydrophobic residues in the β-hairpin interact with the lipids, pulling the β-hairpin out of the folded monomer. Multiple membrane-bound monomers then oligomerize, driving the conformational transition from the monomeric to the membrane-inserted heptameric state (Figure 6.1). The precise reaction pathway is not known, but presumably, the β-hairpins pair up as they penetrate the membrane, driven by the free energy reduction that accompanies hydrogen bonding of peptide bonds.

The formation of β-barrels is not the only way that toxins have for folding into membranes. The toxin ClyA from *E. coli*, for example, forms pores using the helix-bundle motif (**Figure 6.2**). Like many toxins, it also exists in two forms: a soluble monomeric form and a membrane-inserted form in which 12 monomers assemble into a large, elegant structure sporting a helix-lined pore. Nevertheless, β-hairpins play a role in assembly. As illustrated in the figure, a large conformational change occurs as the monomer converts from the soluble to the membrane-inserted state, in which a β-hairpin converts into a short α-helix. At the same time, in an astoundingly large conformational change, the long A1 α-helix flips from one end of the molecule to the other. Again, because this is a spontaneous process, free energy reduction must drive the formation of the beautiful structure seen in the figure.

6.1.3 Most Membrane Proteins Are Co-translationally Inserted into the Membrane

The elaborate structural transitions seen in membrane-targeting toxins, which are necessary to accommodate assembly of the pore structure, are dramatically

Figure 6.2 The ClyA toxin (PDB: 2WCD). (A) The soluble monomeric form undergoes a major conformational rearrangement during assembly of the pore on the target membrane. (B) The assembled, membrane-inserted pore. Note that only the tip of the pore penetrates the membrane. (From Mueller M et al. [2009] *Nature* 459: 726–720. With permission from Springer Nature.)

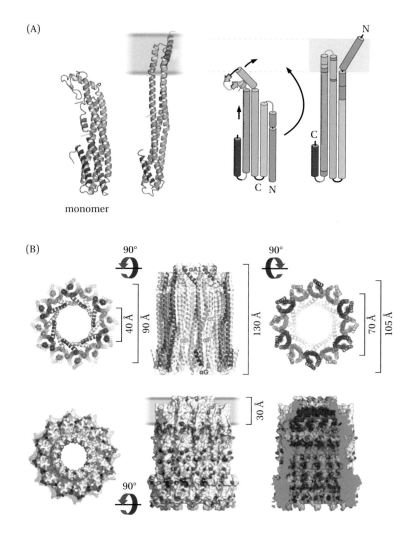

different from what goes on during membrane insertion of typical helix-bundle membrane proteins, such as the photosynthetic reaction center (Figure 1.28). The overwhelming majority of cells' helix-bundle membrane proteins must insert into their target membranes co-translationally as they come off the ribosome to avoid misfolding and aggregation. This is true for bacterial inner membrane proteins as well as for most of the organellar and plasma membrane proteins in eukaryotic cells.

The insertion process is best understood for proteins that insert into the ER membrane (see **Box 6.1** for a short history lesson and a summary of the structure of the Sec61/SecY translocon). A useful technique for studying co-translational insertion is site-specific UV-inducible cross-linking between a nascent polypeptide on the ribosome and components of the Sec61 translocon in the ER (see **Box 6.2**). This and other biochemical techniques have led to the basic cartoon model for membrane protein insertion into the ER membrane shown in **Figure 6.3**.

According to this model, when the N-terminal signal peptide (or the most N-terminal transmembrane helix) on the protein emerges from the ribosome, it binds to the SRP (see Chapter 5). Upon binding, SRP both reduces the translation rate of the ribosome and targets the RNC to the Sec61 translocon due to its affinity for the SRP receptor in the translocon complex. Correct docking of SRP to the SRP receptor leads to mutual activation of GTPase activities in the two molecules and release of SRP from the translocon complex. The signal peptide apparently inserts into the bilayer near the "lateral gate" in the translocon, the translocon channel opens up across the membrane, and translocation of the nascent chain commences.

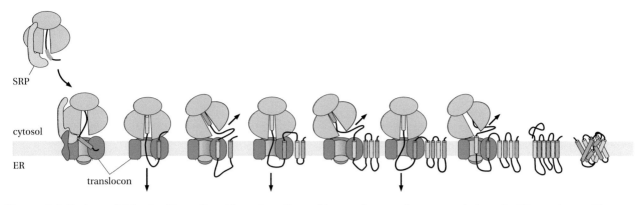

Figure 6.3 Basic model for Sec61-mediated insertion of a multi-spanning membrane protein into the ER membrane. The arrows show how successive hydrophilic segments of the chain are extruded alternately on the lumenal and cytosolic sides of the membrane. (From Skach WR [2009] *Nature Struct Mol Biol* 16: 606–612. With permission from Springer Nature.)

As the first transmembrane helix in the nascent chain enters the translocon, it engages the lateral gate in the channel wall and slides into the surrounding lipid bilayer (this critical step is discussed in detail later). At the same time, the transverse conduit leading across the membrane opens to allow the first helix-connecting loop to emerge into the lumen of the ER. Closure of the translocon after the insertion of the next transmembrane helix confines the second helix-connecting loop to the cytosolic side of the membrane. Successive transmembrane helices in the nascent chain all exit through the lateral gate and alternately trigger the transverse conduit to open (to allow the following connecting loop to become translocated across the membrane) or induce its closure (to confine the following connecting loop to the cytosolic side).

Clearly, a key player in this mechanism is the lateral gate in the Sec61 translocon (Box 6.1). The existence of the lateral gate was initially inferred from biochemical experiments on Sec61 and the bacterial SecYEG translocons, but the best proof has been obtained from structural studies of homologous translocons from archaea (**Figure 6.4**). A particularly striking structure of an archaeal

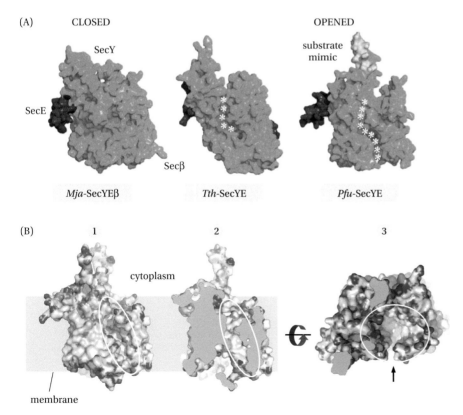

Figure 6.4 Structures of SecY translocons reveal different degrees of opening toward the lipid bilayer. (A) Three translocon structures with different degrees of opening of the lateral gate. The view is from the plane of the membrane. The lateral gate is indicated by stars. Mja, *Methanococcus jannaschii* (PDB: 2YXQ); Tth, *Thermus thermophilus* (PDB: 2ZQP); Pfu, *Pyrococcus furiosus* (PDB: 3MP7). (B) Three views of the Pfu-SecYE structure. (1) View along the membrane plane, showing the partially open lateral gate (white oval). (2) Same view as in (1), but with the front of the molecule removed to show the rear wall of the channel. (3) View of the upper parts of the channel (white oval) seen from the cytoplasmic end. The lateral gate is indicated by the arrow. (A, From Egea PF, Stroud RM [2010] *PNAS* 107: 17182–17187. With permission from RM Stroud. B, From Öjemalm K, Higuchi T, Jiang Y et al. [2011] *PNAS* 108: E359–E364.)

BOX 6.1   DISCOVERY AND STRUCTURE OF THE SEC-TYPE TRANSLOCONS

Discovery

Protein secretion and membrane protein biogenesis are managed in all cells by secretion (Sec) proteins that form translocons. The existence of translocons in the ER was proposed by Günter Blobel and David Sabatini in 1971 to explain how ribosomes could bind to the ER membrane and secrete proteins into the ER lumen. But, they asked, why aren't all proteins secreted; what's special about secreted proteins? They postulated that proteins destined for secretion must have N-terminal signal peptides (see Chapter 5, and especially Figure 5.2).

Like everything else having to do with membranes and membrane proteins in the 1970s (see Chapter 1), the idea of membrane-embedded translocons was controversial. The first experimental evidence for such proteins finally came in 1981 when a genetic selection in *E. coli* was used to identify a mutation in a gene (then named *prlA*, for "protein localization") that allowed export to the periplasm of the outer membrane protein maltoporin (LamB) with a defective signal peptide (**Figure 1**). The gene was found in an operon (a cluster of genes whose expression is controlled by a single promoter) that encodes many ribosomal proteins but *prlA* was shown in 1983 not to correspond to a ribosomal protein. At about the same time, other researchers found a temperature-sensitive *E. coli* mutant with a defect in protein export at high temperatures, which they named *secY*. Because the mutation mapped to the same gene locus as *prlA*, the two mutations were in the same gene.

The *secY* gene was cloned in 1983, but even though it was noted in passing that the SecY protein had a high percentage of hydrophobic amino acids, it was not until 1987 that the protein was shown to have 10 transmembrane helices (**Figure 2**), and the suggestion was made (based on analogy with bacteriorhodopsin and lactose permease) that SecY

LamB signal peptide

MMITLRKPLAVAVAAGVMSAQAMA VDFHGYARSGIG

deletion

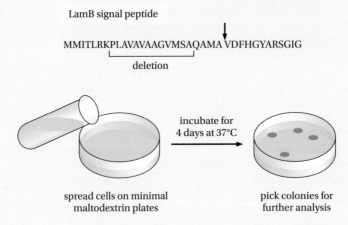

spread cells on minimal
maltodextrin plates

incubate for
4 days at 37°C

pick colonies for
further analysis

Figure 1 The original selection strategy used to isolate the *prlA* (= *secY*) gene. *E. coli* cells from a previously isolated strain in which the *lamB* gene has a deletion near its 5′ end that removes a big chunk of the signal peptide (underlined) are spread on growth plates that have maltodextrin as the only carbon source. The *lamB* gene encodes an outer membrane porin specific for uptake of maltodextrin sugars, and the cells will grow on the maltodextrin plates only if the LamB protein is correctly exported to the outer membrane. Since most of the LamB signal peptide has been deleted, the protein stays in the cytoplasm. The rationale behind the selection strategy is that if there are proteins involved in the recognition of signal peptides, mutations in their corresponding genes may make them less discriminative and hence, support export of some amount of the LamB protein to the outer membrane, allowing colonies to grow on the plates. Out of 52 such mutants isolated in the selection, 50 were mapped to the *prlA* gene.

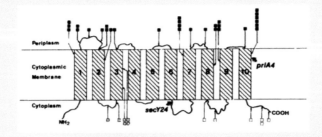

Figure 2 The first topology model of SecY. The topology was deduced from SecY-PhoA fusions (see Box 6.5). Filled squares indicate the fusion joints for highly active PhoA fusions, while dotted and open squares indicate fusions with intermediate and low PhoA activities, respectively. (From Ito K, Akiyama Y [1987] *EMBO J* 6: 3465–3470. With permission from John Wiley and Sons.)

(Continued)

might be a channel for polypeptide translocation, i.e., a translocon.

translocon name conventions

mammals	Sec61α	Sec61β	Sec61γ
yeast	Sec61p	Sbh	Sss1p
bacteria	SecY	SecG	SecE
archaea	SecY	Secβ	SecE

Figure 3 Name conventions for Sec proteins in different organisms.

In 1987, a genetic selection in yeast identified a temperature-sensitive mutation in a gene called *sec61* (the corresponding protein was later renamed Sec61α; see **Figure 3**) that had a severe defect in protein translocation from the cytosol to the lumen of the ER when grown at high temperatures. The sequences of the yeast and human sec61 genes (determined in 1991 and 1992) showed similarity to bacterial *secY* genes, establishing that the Sec-type translocon was conserved from bacteria to man. Other components of the Sec-type translocon complexes, such as SecE and SecG in bacteria and Sec61β, Sec61γ, Sec62, and Sec63 in eukaryotes, were likewise discovered through genetic screens.

Further work showed that ribosomes engaged in the synthesis of secreted and membrane proteins in mammalian cells were bound to the Sec61 complex and that nascent polypeptide chains emerging from the bound ribosomes could be chemically cross-linked to translocon

components (see Box 6.2). This strengthened the idea that the Sec complexes are indeed protein-conducting channels, serving to provide a conduit for unfolded polypeptides through membranes.

Sec-Translocon Structure

Although the discovery of the Sec translocons spawned a large body of work in many laboratories across the world, it took a very long time before the first high-resolution x-ray structures were determined. Increasingly better EM and cryo-EM images of purified Sec61 oligomers and ribosome-Sec61 complexes were obtained starting in 1996 (**Figure 4**).

It was only in 2004, with the publication of the structure of the SecYEβ complex from the hyperthermophilic bacterium *Methanococcus jannaschii,* that we could first see a translocon in atomic detail (**Figure 5**). It is an unremarkable protein in terms of size; it has ten transmembrane helices (just as predicted by PhoA mapping, Figure 2), two fewer than the lactose transporter LacY. Like many other transporters, SecY has two five-helix domains related symmetrically by rotation around an axis parallel to the membrane plane (see Figure 6.29).

A key feature of SecYEβ is a latent passageway (gate) formed between helices TM2 and TM7, which are in the first (TM1–TM5) and second (TM6–TM10) five-helix bundles, respectively. This is the only possible exit route for proteins to enter the lipid bilayer, because the two bundles are joined between TM5 and TM6. Furthermore, the single transmembrane helix of SecE blocks any conceivable passage between TM5 and TM6. Exactly how the opening of the gate is regulated and how nascent helices pass through it is

(A)

(B)

(C)

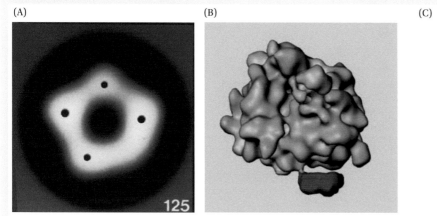
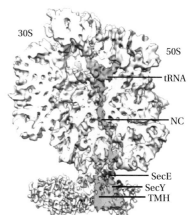

Figure 4 Images of Sec61/SecYEG translocons obtained by negative stain EM (A) and cryo-EM (B, C). The ring-like structure in panel (A) is an oligomeric form of detergent-solubilized mammalian Sec61. Panel (B) shows monomeric yeast Sec61 (red) bound to a ribosome (blue), and panel (C) shows a cutaway view of monomeric *E. coli* SecYEG bound to a ribosome. A tRNA-bound nascent polypeptide chain (NC) with an N-terminal transmembrane helix (TMH) is visible (green). SecYEG is inserted into a lipid nanodisc (grey). (A, From Hanein D, Matlack KES, Jungnickel B et al. [1996] *Cell* 87: 721-732. Used with permission from Elsevier. B, From Beckmann R, Bubeck D, Grassucci R et al. [1997] *Science* 278: 2123-2126. Used with permission from the American Association for the Advancement of Science. C, From Frauenfeld J et al. [2011] *Nature* 18: 614-621. Used with permission from Springer Nature.)

(Continued)

BOX 6.1   DISCOVERY AND STRUCTURE OF THE SEC-TYPE TRANSLOCONS (CONTINUED)

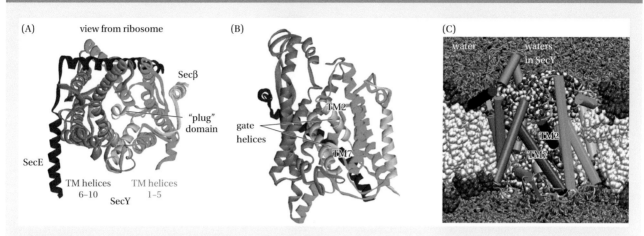

Figure 5 Structural features of the SecYEβ structure from *Methanococcus jannaschii* (PDB: 1RH5). Views of the hetero-trimeric protein from the perspectives of the ribosome (A) and the membrane bilayer (B). In panel (A), notice that SecE blocks any possibility of transmembrane segments exiting from the upper and left sides of the protein. Instead, the lateral gate is between the gate helices TM2 and TM7. SecY has two anti-parallel internal repeats (brown and green) that are rotated 180° with respect to an axis lying in the membrane plane. (C) A snapshot from a molecular dynamics simulation of SecYEβ embedded in a lipid bilayer. The translocon brings into contact the nascent chain, the SecY protein, water, and the lipid bilayer, thus enabling thermodynamic selection of potential transmembrane helices in the nascent chain sequence. (From Cymer F, von Heijne G, White SH [2015] *J Mol Biol* 427: 999–1022. With permission from Elsevier.)

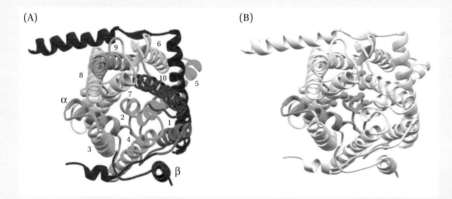

Figure 6 The SecYEβ complex from *M. jannaschii*. (A) View from the cytoplasm, looking down onto the membrane as in Figure 5A, B, except for a 90° clockwise rotation. The short "plug domain" (light blue) that closes the channel from the periplasmic side is especially apparent. During protein translocation, the plug domain moves out of the way as the nascent polypeptide is pushed down the center of the channel. (B) The location of the known *prlA* mutations. Notice that they are located in the vicinity of the plug domain and on the face of helix TM7, where they can somehow stabilize the open form of the channel. (From van den Berg B et al. [2004] *Nature* 427: 36–44. Used with permission from Springer Nature.)

obscure, although additional structures from other species in different states are starting to provide clues (Figure 6.4).

A wonderful thing about science is that eventually, many pieces of a scientific puzzle come together to give a complete view. The first SecY translocon structure revealed that the *prlA* mutants, which allowed export of secreted proteins with defective signal peptides, are located in the heart of the translocon, where nascent chains are expected

to pass (**Figure 6**). The known *prlA* mutations in the *secY* gene map mostly to the center of the channel, particularly in the region of the plug domain and the face of the TM7 gate helix that lines the channel. The mutations are perfectly situated to stabilize, in some as yet unknown way, the open state of the channel, thereby making it possible for proteins with defective signal peptides to gain access.

BOX 6.2   CHEMICAL CROSS-LINKING CAN BE USED TO STUDY PROTEIN TRANSLOCATION

Which components of the translocation pathway do nascent chains interact with during passage? To answer this question, we must probe the interactions between the nascent polypeptide chain and its environment during targeting and translocation. One particularly useful technique is based on site-specific chemical cross-linking between the nascent chain and surrounding proteins. *In vitro* translation of model secreted or membrane proteins offers a possibility to include tRNAs charged with non-natural amino acids in the translation mix. If the side chain of the non-natural amino acid includes a reactive chemical moiety that can be activated by, for example, UV light, one can induce covalent cross-links between nascent chains carrying the reactive amino acid and any protein in its immediate surroundings. The cross-linked proteins can then be identified by using a panel of antibodies directed against proteins known to be involved in the process, or by mass spectrometry.

With a few extra tricks, we can even position the reactive amino acid relative to the ribosome and other components, such as signal recognition particle (SRP) or the Sec61 translocon, before turning on the UV lamp (**Figure 1**). If the mRNA encoding the nascent chain is engineered such that it lacks a stop codon, the ribosome will stall on the last codon and will not release the nascent chain. This generates a stalled ribosome–nascent chain complex (RNC) or, if rough microsomes are present during the translation reaction, a stalled RNC–Sec61 complex.

Another trick is to introduce the reactive amino acid in a specific position in the nascent chain. There are two ways to do this. One is to use a nascent chain that contains a single lysine residue and replace tRNA-Lys by tRNA-Lys* (where Lys* is the reactive amino acid) in the translation mix. The other way is to place a stop codon in the mRNA (by mutating the gene) in the position where one wants the reactive amino acid and include a so-called suppressor tRNA (i.e.,

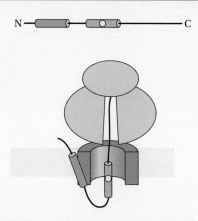

Figure 1 Chemical cross-linking of an RNC-Sec61 complex. The mRNA lacks a 3′ stop codon, causing the ribosome to stall on the last codon at the end of the mRNA. The nascent chain has two transmembrane helices, and the position where the reactive amino acid is incorporated is indicated by the yellow dot. The reactive amino acid is located in the translocon channel in the stalled RNC-Sec61 complex and will form covalent cross-links to translocon components upon UV irradiation.

a tRNA that can read the stop codon) carrying the reactive amino acid in the translation mix (see Box 6.4). Either of these ways gives full control over the location of the reactive amino acid in the RNC–Sec61 complex when the cross-linking reaction is induced by UV illumination, although the precision of the location is influenced by the size and flexibility of the derivatized amino acid.

Using this technique, researchers have identified all the components present in the RNC–translocon complex and have been able to map the stage during the targeting and translocation process when each component is near different parts of the nascent chain.

translocon is that of SecYE from *Pyrococcus furiosus* (Pfu-SecYE), in which the channel has been caught in a conformation with the lateral gate open along its entire length. Not much imagination is required to picture the lateral gate as even more open, allowing an incoming transmembrane helix direct access to the lipid bilayer.

The detailed mechanics of the opening and closing of the transverse conduit, called the protein-conducting channel by Blobel (Box 6.1), and the lateral gate in the translocon are unclear. How does the RNC affect the conformational dynamics of the translocon? Does the lateral gate remain open during chain translocation, does it flicker between an open and a closed state, or is opening somehow triggered by the transmembrane segments themselves? Or, is it simply a highly adaptable pathway across the membrane? Does the closed structure form immediately upon insertion of the last transmembrane segment, or is there an annealing phase? Can interactions between the nascent chain and the ribosome inside the ribosomal tunnel be transmitted to the translocon? These are still open questions, because we are so far unable to follow the passage of nascent chains from ribosomes across or into the membrane bilayer at

Figure 6.5 Two models for translocon-mediated insertion of a transmembrane helix. (A) The "In-out" model. The transmembrane helix (in black) first moves all the way into the central translocon channel and then exits through the lateral gate. (B) The "sliding" model. The transmembrane helix slides along the outer part of the lateral gate into the membrane. The leading polar segment penetrates through the lateral gate and is shielded from lipid contact. (From Cymer F, von Heijne G, White SH [2015] *J Mol Biol* 427: 999–1022. With permission from Elsevier.)

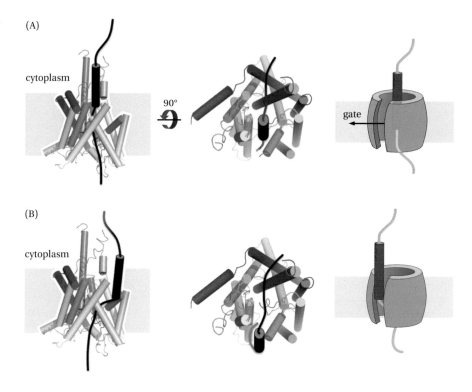

the single-residue level. Keep in mind that nascent chains elongate at the rate of about 10–20 residues per second in *E. coli* and 3–5 residues per second in mammalian cells, which means there is ample time for a nascent chain to explore even very complex energy landscapes during the membrane-insertion process. An example of two different models for how the translocon may guide a transmembrane helix into the membrane is shown in **Figure 6.5**.

The model in Figure 6.3 is strictly sequential, i.e., the transmembrane helices leave the translocon one by one and independently of each other. This is at best a first approximation to the real mechanism. It has been shown for a number of multi-spanning membrane proteins that the insertion of a transmembrane helix can be influenced by other transmembrane helices in the protein, implying that transmembrane helices may exit the translocon in pairs or even as higher oligomers (**Figure 6.6**). This possibility raises a fundamental question regarding the size of the translocon channel: is there really room for multiple transmembrane helices? The central channel in the structures shown in Figure 6.4 is certainly too narrow to fit more than one helix at a time, and there may be space in the lateral gate region for an additional helix. But to fit more

Figure 6.6 Transmembrane helices may exit the translocon in pairs during the co-translational insertion of complex multi-spanning membrane proteins. (From Skach WR [2009] *Nature Struct Mol Biol* 16: 606–612. With permission from Springer Nature.)

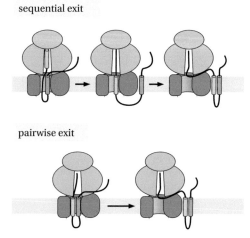

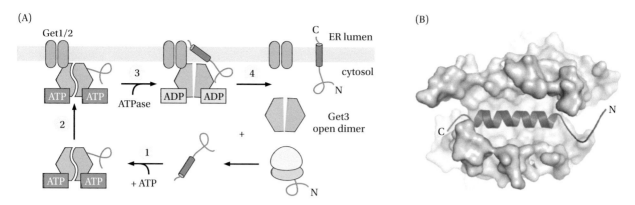

Figure 6.7 Insertion of tail-anchored proteins into the ER. (A) The hydrophobic C-terminal anchor segment binds tightly to the ATP-bound state of the Get3 receptor (PDB: 2WOJ). Upon making contact with the Get1/2 complex in the ER membrane, Get3 is induced to hydrolyze its ATP, which reduces the affinity for the anchor segment. The anchor segment then inserts across the membrane. (B) Model of Get3 with a bound anchor segment (purple). (From Mateja A et al. [2009] *Nature* 461: 361–366. Used with permission from Springer Nature.)

than two helices, either the translocon must expand dramatically in size, multiple copies of the translocon must assemble to form a "superchannel," or helices must be able to remain in a "holding position" just outside the lateral gate from which they can interact with incoming helices. There is no indication of a large expansion in the channel or the formation of multimeric super-channels in the available electron microscopy structures of ribosome–nascent chain–translocon complexes, making such models less likely to be correct.

6.1.4 Tail-Anchored Membrane Proteins Use a Distinct Insertion Machinery in the ER

The inherent logic of the co-translational insertion model is that the ribosome must be targeted to the translocon at an early stage of translation to ensure that the hydrophobic transmembrane helices are never exposed to the cytosol. However, "tail-anchored" membrane proteins do not have an N-terminal signal peptide, and their only hydrophobic segment is at the extreme C-terminal end of the protein. This means that the transmembrane segment is still inside the ribosome when translation terminates and therefore cannot be recognized by SRP for co-translational targeting to the Sec61 translocon.

Instead, tail-anchored proteins are extruded from the ribosome into the cytosol, where the C-terminal hydrophobic transmembrane anchor is recognized by dedicated chaperones, and the protein is delivered to a soluble receptor protein, Get3 (**Figure 6.7**). Get3 most likely binds the anchor in a long hydrophobic crevice and thereby shields it from the cytosol. Get3 is structurally related to SRP but is an ATPase rather than a GTPase. Get3 then binds to the Get1/2 receptor complex in the ER membrane, triggering ATP hydrolysis and release of the bound anchor segment. Finally, the anchor helix inserts across the membrane with a short C-terminal tail exposed to the ER lumen. Whether this last step is spontaneous or requires some kind of translocon function by Get1/2 is not known.

6.1.5 GPI-Anchored Proteins Are Synthesized with a Hydrophobic C-Terminal Tail

Certain proteins exposed on the surfaces of eukaryotic cells are anchored in the membrane by a C-terminal glycosylphosphatidylinositol (GPI) moiety (see Chapter 3). GPI is a glycolipid composed of a phospholipid, a glycan core, and a phosphoethanolamine linker that attaches it to the protein (**Figure 6.8**). GP-anchored proteins are important players in signal transduction, the immune

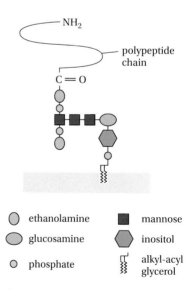

Figure 6.8 The GPI anchor. The core GPI structure is ethanolamine-phosphate-(mannose)3-glucosamine-phosphatidylinositol. It is covalently linked to the C-terminus of the polypeptide chain and partitions into the outer leaflet of the lipid bilayer.

Figure 6.9 Structural features of GPI-anchored proteins. (A) GPI-anchored proteins have an N-terminal signal peptide and a C-terminal GPI-attachment signal. (B) The GPI-attachment signal. The transamidase cleavage site is preceded by a polar, possibly unstructured, segment and is connected to the C-terminal hydrophobic segment by a spacer region approximately seven residues long. (From Poisson G, Chauve C, Chen X et al. [2007] *Genomics, Proteomics and Bioinformatics* 5: 121–130. With permission from Elsevier.)

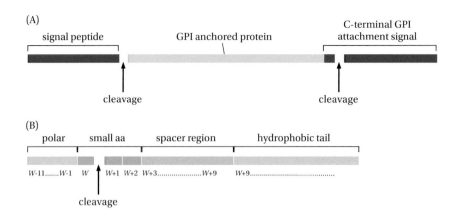

response, and other cellular processes. GPI-anchored proteins on the cell surface can be released to the extracellular space through the action of phosphatidylinositol-phospholipases that cleave the GPI anchor from the protein.

GPI-anchored proteins are synthesized with an N-terminal signal peptide, a water-soluble extracellular domain, and a C-terminal "GPI-attachment signal" (**Figure 6.9**). The GPI signal is cleaved off, and the GPI anchor is added in the ER by a GPI transamidase enzyme that transfers the GPI moiety *en bloc* to the cleaved C-terminus of the protein.

The GPI-attachment signal is rather complex and includes an unstructured region upstream of the cleavage site, a cleavage-site region rich in amino acids with small side chains (Ala, Ser, Gly), a moderately polar and probably flexible spacer region, and a C-terminal hydrophobic tail. The role of the hydrophobic tail is unclear, and it is uncertain whether it becomes integrated into the ER membrane or is translocated into the lumen of the ER before the GPI anchor is attached.

6.1.6 β-Barrel Membrane Proteins Are Chaperoned to the Outer Membrane in Gram-Negative Bacteria

Bacterial β-barrel proteins by necessity can only insert into the outer bacterial membrane post-translationally because they must first pass through the inner membrane and the aqueous periplasmic space, a 160–170 Å -wide region between the inner and outer membranes. Similarly to periplasmic proteins, they are made with an N-terminal signal peptide that targets them to the SecYEG translocon in the inner membrane. Because the transmembrane β-strands are quite short (five to eight residues), and only every second residue needs to be hydrophobic in order to make the lipid-exposed surface of the barrel match the bilayer, there are no long hydrophobic segments in the polypeptide chain that can form transmembrane α-helices during the chain's passage through the SecYEG channel. The lack of long hydrophobic segments also helps to explain why β-barrel membrane proteins can be unfolded in water and then refolded into lipid vesicles *in vitro* using denaturants such as urea and guanidinium hydrochloride (see Chapter 3).

Because the lipid compositions of the inner leaflet of the outer membrane and the outer leaflet of the inner membrane are similar, outer membrane proteins must be prevented from inserting spontaneously into the inner membrane upon exit from the SecYEG channel. This prevention appears to be achieved by a combination of a large activation barrier for insertion into bilayers of this particular lipid composition and a stoichiometric excess of periplasmic chaperones that sequester the outer membrane proteins, maintaining them in an unfolded or partially folded state (**Figure 6.10**). The major chaperone for outer membrane proteins is SurA, but proteins appear to be able to shuttle back and forth between SurA and at least three other chaperones: Skp, DegP, and FkpA.

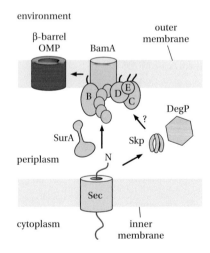

Figure 6.10 Outer membrane proteins are chaperoned through the periplasm and integrated into the outer membrane with the aid of the barrel assembly machine (Bam) complex.

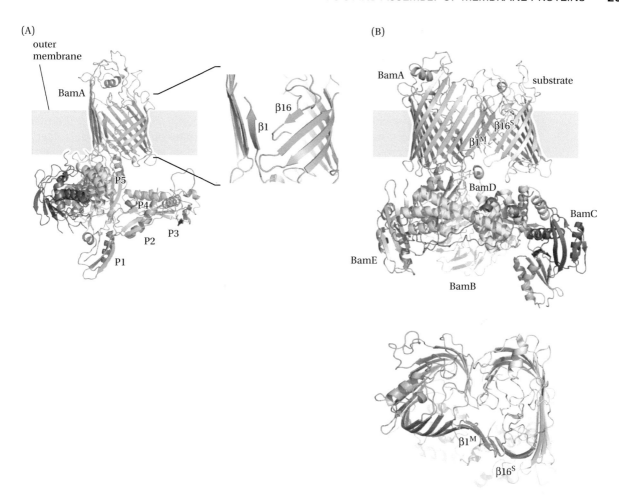

Figure 6.11 The Bam complex. (A) BamACDE (PDB: 5EKQ). The POTRA domains in BamA are labeled P1–P5. The weak connection between the first and last β-strands in BamA is shown in close-up on the right. (B) Side and top views of BamABCDE bound to an almost fully folded substrate (BamA itself, in fact, in grey; PDB: 6V05). The first β-strand in BamA (β1^M) is hydrogen-bonded to the last β-strand in the substrate (β16^S). Note that the β-strands at the other ends of the two β-barrels do not make hydrogen bonds to each other but rather, curl into the central space. This may facilitate the pairing-up of the first and last β-strands in the substrate in the last step of folding, and the ultimate release of the substrate from the Bam complex.

Integration into the outer membrane is mediated by the barrel assembly machine (Bam) complex, which receives outer membrane proteins from the periplasmic chaperones and lowers the kinetic barrier for insertion and folding in the outer membrane. The central component of the complex is the BamA protein, itself an outer membrane β-barrel protein with a periplasmic appendage containing a number of so-called POTRA domains. BamA associates with four outer membrane lipoproteins to form the full Bam complex (**Figure 6.11A**).

BamA forms a large, hollow β-barrel, closed at the top. The five POTRA domains protrude into the periplasm and are sandwiched between the BamCDEF proteins. An unfolded substrate β-barrel protein is thought to interact with the POTRA domains by β-strand augmentation, i.e., by adding one or more of its β-strands to one or more of the POTRA domain β-sheets. An unusual feature in the BamA β-barrel is that the first and last strands in the barrel are held together by no more than a couple of hydrogen bonds, possibly allowing the opening of a lateral gate between the center of the barrel and the bilayer. Keep in mind, however, that breaking even a single hydrogen bond in a hydrophobic environment is energetically costly (Chapter 3). The lipid bilayer may be distorted by the less-than-perfect packing between the two β-strands, possibly facilitating the membrane insertion of the substrate outer membrane protein. An attractive model is that pairs of β-strands in the substrate protein are successively transferred from the POTRA domains and insert between the first and last strands of BamA, allowing the new β-barrel to grow step-wise by "bulging out" from the BamA barrel until it is finally released into the outer membrane and the BamA barrel snaps back into its original state. A structure of the Bam complex, "caught in the act" bound to a nearly fully folded substrate protein, supports this model (**Figure 6.11B**).

The absence of energy sources in the periplasm means that the membrane insertion of β-barrel proteins must be spontaneous, and the final folded form of the protein must be thermodynamically more stable than the chaperone- and Bam-bound states. This is borne out by measurements of the thermodynamic stability of β-barrel proteins in lipid bilayers (see Chapter 4).

6.1.7 Mitochondrial Membrane Proteins Are Either Imported from the Cytosol or Made *In Situ*

Large-scale proteomics analyses have identified nearly 1000 different proteins in yeast mitochondria. Around 30% of these are predicted to be integral membrane proteins located in either the outer or the inner mitochondrial membrane. Of these 300-odd membrane proteins, only 7 are encoded in the mitochondrial genome; the great majority are encoded in the nuclear genome and are imported post-translationally into the organelle.

Except for a small number of outer membrane proteins anchored to the membrane by an N-terminal transmembrane α-helix, all the imported mitochondrial membrane proteins transit through the TOM translocon (TOM stands for Translocase Outer Membrane; see Chapter 5). β-barrel outer membrane proteins are then chaperoned through the intermembrane space to the SAM (Sorting And Assembly) complex, which helps them fold and integrate into the membrane (**Figure 6.12**). The SAM complex is homologous to the bacterial Bam complex discussed earlier.

All known mitochondrial inner membrane proteins have transmembrane α-helices. Inner membrane proteins that are imported from the cytosol use the TOM complex to get across the outer membrane, but then, the import pathways diverge. Some inner membrane proteins, mainly metabolite carriers that ensure the influx and efflux of small metabolites across the inner membrane, are integrated into the membrane with the help of the TIM22 complex (TIM stands for Translocase Inner Membrane) (**Figure 6.13**). The metabolite carrier proteins have six transmembrane helices and internal targeting sequences in the loops between the helices. These sequences are recognized by the TOM70

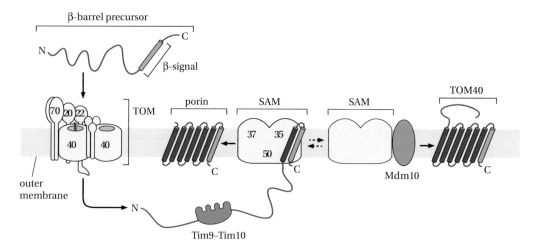

Figure 6.12 Outer membrane β-barrel proteins are synthesized in the cytosol and are imported across the outer membrane through the TOM complex. The Tim9–Tim10 chaperone in the intermembrane space hands them over to the SAM complex for membrane integration. (From Schmidt O et al. [2010] *Nature Rev Mol Cell Biol* 11: 655–657. With permission from Springer Nature.)

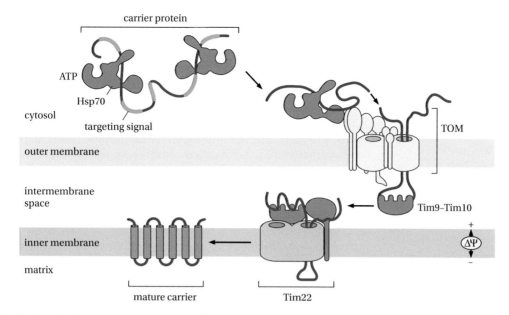

Figure 6.13 Mitochondrial carrier proteins transit through the TOM complex and are then targeted to the TIM22 complex, which mediates their insertion into the inner membrane. (From Schmidt O et al. [2010] *Nature Rev Mol Cell Biol* 11: 655–657. With permission from Springer Nature.)

receptor, which is a part of the TOM complex in the outer membrane. The carrier proteins are chaperoned by Tim9–Tim10 across the intermembrane space and are delivered to the TIM22 complex for membrane insertion. This poorly understood insertion process is driven by the electrical potential ($\Delta\psi$) across the inner membrane.

The great majority of inner membrane proteins are synthesized with N-terminal targeting sequences that direct them to the same import pathway used by soluble matrix proteins; i.e., they go through the TOM and TIM23 complexes (**Figure 6.14**) (see Chapter 5). For proteins with only a single transmembrane helix placed not too far from the N-terminal targeting sequence, $\Delta\psi$ is

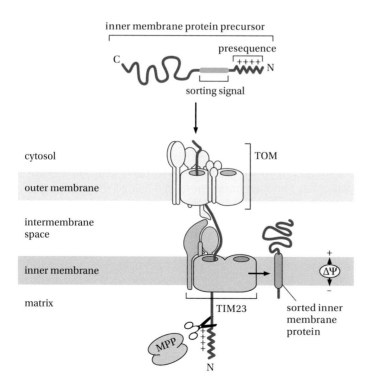

Figure 6.14 Single-spanning inner membrane proteins with the transmembrane helix close to the N-terminal targeting sequence go directly into the membrane from the TIM23 complex. (From Schmidt O et al. [2010] *Nature Rev Mol Cell Biol* 11: 655–657. With permission from Springer Nature.)

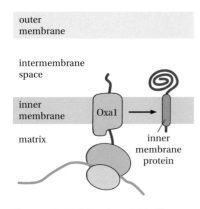

Figure 6.15 Mitochondrial ribosomes synthesizing inner membrane proteins bind to the Oxa1 translocon, allowing co-translational insertion into the inner membrane. (From Schmidt O et al. [2010] *Nature Rev Mol Cell Biol* 11: 655–657. With permission from Springer Nature.)

the sole energy source required to drive import and membrane insertion. When the transmembrane helix is further away from the N-terminus, or when there are multiple transmembrane helices in the protein, the Presequence protein Associated Motor (PAM motor) is required to generate the driving force for translocation of matrix-localized parts of the protein through the TIM23 channel (see Chapter 5).

Multi-spanning inner membrane proteins may insert into the membrane by exiting laterally from the TIM23 complex or may first be fully imported into the matrix and then inserted into the inner membrane from the matrix side. The latter mechanism has been called conservative sorting because it recapitulates the way inner membrane proteins are inserted in the bacterial progenitors of the mitochondrion, i.e., from the cytoplasmic side of the membrane (equivalent to the matrix side in mitochondria).

This is also the way the small number of inner membrane proteins that are encoded in the mitochondrial genome are inserted. Mitochondria in organisms such as baker's yeast, plants, and mammals do not contain a homolog of the bacterial SecYEG translocon, but they do contain an inner membrane protein called Oxa1 that is also found in bacteria and thylakoids (**Box 6.3**). The bacterial Oxa1 homologue, YidC, can associate with the SecYEG complex, but can also mediate membrane insertion of certain small inner membrane proteins by itself. Mitochondrial Oxa1 can bind to ribosomes and therefore ensure co-translational membrane insertion of mitochondria-encoded inner membrane proteins (**Figure 6.15**).

6.2 ENERGETICS OF MEMBRANE PROTEIN ASSEMBLY

6.2.1 Membrane-Insertion Propensities Can Be Measured *In Vivo*

From everything we know about integral membrane proteins—the hydrophobic exterior surfaces facing the lipid, co-translational or chaperone-dependent membrane-insertion mechanisms to avoid exposing hydrophobic surfaces to an aqueous environment, the spontaneous insertion of model hydrophobic peptides into liposomes—we conclude that "higher than average hydrophobicity" of exposed surfaces is the main characteristic that sets membrane proteins apart from water-soluble proteins. This is certainly fine as far as it goes, and many quite successful methods to predict transmembrane helices from amino acid sequence are based on this simple concept (see Chapter 8). But, how can we measure and quantify hydrophobicity? One way, discussed in Chapter 3, is to carry out physical chemistry experiments with well-defined chemical compounds (peptides, liposomes or membrane-mimetics, buffers) and measure things like partitioning equilibria and binding/unbinding kinetics.

Another approach is to try to find ways to carry out quantitative measurements of protein–membrane interactions in a living cell, or in a cell-free system that approximates the complexity of the living cell. This has been done for single-spanning transmembrane helices, both in the ER of mammalian cells and in the inner membranes of bacteria and mitochondria. Ideally, direct comparison between data generated in such biological systems and the detailed thermodynamic analyses possible using physical chemistry techniques should tell us something about the physical basis for membrane protein assembly *in vivo*.

One way to carry out measurements of translocon-mediated helix-insertion propensities is based on designed protein constructs where engineered glycosylation sites serve as markers for membrane translocation (**Box 6.4**). The membrane-insertion propensities of the 20 natural amino acids can be calculated from such data (**Figure 6.16A**). These values show that only the most

BOX 6.3   MEMBRANE INSERTASES THAT REPLACE OR SUPPLEMENT SEC TRANSLOCONS

The inner membranes of many mitochondria, despite an abundance of transmembrane α-helical membrane proteins, do not have Sec-type translocons. Instead, a simpler translocase containing five or six transmembrane helices handles co-translational insertion. This insertase, Oxa1, is highly homologous to the YidC insertase of bacteria, the Get1,2 and EMC insertases in the ER, and the Alb3 insertase of chloroplasts (**Figure 1**). An even simpler insertase of

this type containing only three transmembrane segments is found in the plasma membrane of *Methanococcus jannaschii*. YidC has no structural features suggestive of a pore as seen in Sec61/SecY translocons. Instead, based upon nascent chain cross-linking studies, the protein apparently provides a pathway consisting primarily of a hydrophilic, water-filled groove that extends from the cytoplasm halfway across the membrane, bordered by two transmembrane

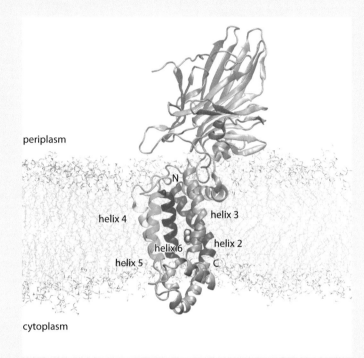

Figure 1 Structure of *E. coli* YidC (PDB: 6AL2) in a lipid bilayer as determined from a molecular dynamics simulation of the crystallographic structure in a POPE:POPG lipid bilayer. YidC distorts the bilayer, possibly facilitating the insertion of substrate TMs. From Chen Y, Capponi S, Zhu L et al. [2017] *Structure* 25: 1403–1414. Used with permission from Elsevier

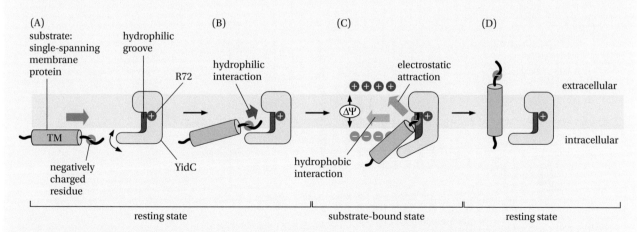

Figure 2 YidC does not form a protein-conducting channel. Rather, it provides a transmembrane "sliding" pathway for transmembrane helices carrying negative charges at one terminus. (A, B) binding of single pass protein to Yid C, (C) electrostatic and hydrophobic interactions facilitate insertion, (D) inserted helix is released from YidC, which returns to the resting state. (From Kumazaki K et al. [2014] *Nature* 509: 516–520. With permission from Springer Nature.)

(Continued)

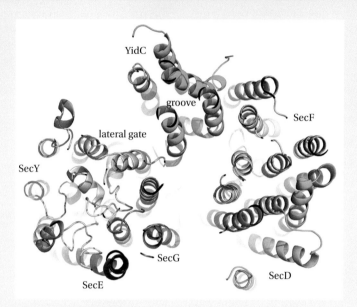

Figure 3 A slice through the membrane domain of the *E. coli* SecYEGDF-YidC holo-translocon (PDB: 5MG3) based on a low-resolution (14 Å) cryo-EM reconstruction. The lateral gate in SecY faces the entrance to the hydrophilic groove in YidC.

helices (TM3 and TM5) forming a "greasy slide" that may help transmembrane helices in the substrate protein cross the membrane. Buried within the protein is a single water-accessible arginine. This arginine encourages the passage across the membrane of transmembrane helices carrying negative charges at the N-terminus (**Figure 2**).

There are several single-span membrane proteins in *E. coli* that utilize the YidC pathway exclusively. More often, however, YidC acts in concert with SecYEG to aid the insertion of many different membrane proteins. Although there are many more copies of YidC than SecY in *E. coli*, cross-linking studies reveal that some of the YidC molecules associate closely with SecYEG as part of so-called holo-translocons. The association is such that the YidC greasy slide faces the SecY lateral gate (**Figure 3**). The SecD and SecF proteins, thought to enhance protein secretion through SecYEG by pulling on the nascent chain from the periplasmic side, also associate with the holo-translocon.

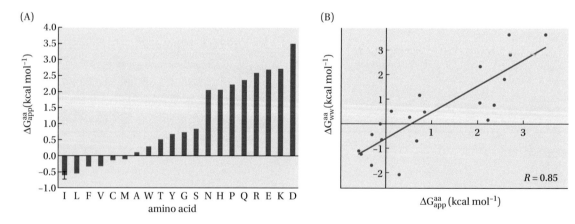

Figure 6.16 The "biological" hydrophobicity scale. (A) Propensities for insertion into the ER membrane for the 20 natural amino acids (given as the free energy difference between a membrane-inserted state and a non-inserted state inside the Sec61 translocon). (B) Correlation between the values in panel (A) and partitioning free energies between water and octanol. (A, From Hessa T et al. [2005] *Nature* 433: 377–381. With permission from Springer Nature. B, From Moon CP, Fleming KJ [2011] *PNAS* 108: 10174–10177. With permission from KJ Fleming.)

BOX 6.4   ASN-LINKED GLYCOSYLATION CAN BE USED AS A TOPOLOGICAL MARKER FOR MAMMALIAN PROTEINS

Almost all secreted and membrane proteins that pass through the SRP-Sec61 pathway become glycosylated in the ER and hence, are glycoproteins. The ER contains a luminally disposed integral membrane enzyme complex, the oligosaccharide transferase (OST), that attaches a high-mannose oligosaccharide (five to nine mannose residues) to Asn residues in nascent proteins as they enter the lumen of the ER. To be a good substrate for the OST, the Asn residue must be in the context of an Asn-X-Thr/Ser-Y tetrapeptide, where X and Y cannot be Pro. To be accessible to the OST, the Asn residue must be at least 10–15 residues away from the nearest transmembrane helix.

A single oligosaccharide adds about 2.5 kDa to the molecular mass of the protein, an increase that for medium-size proteins can be detected readily as a migration shift on an SDS-PAGE gel (**Figure 1**). Because Asn-linked glycosylation can only happen in the lumen of the ER, the addition of an oligosaccharide to a specific Asn residue in a protein points to a luminal location of that part of the protein.

Naturally occurring or engineered Asn glycosylation sites can therefore be used to infer the topology of membrane proteins. Typically, one first removes any natural sites by mutating Asn residues to Gln and then introduces new sites in strategic positions. The mutant protein variants are then expressed either *in vivo* or more often, *in vitro* in the presence of rough microsomes (i.e., ER-derived membrane vesicles) and are analyzed as migration shifts in SDS-PAGE. Glycosylated sites are assumed to be located in the lumen and non-glycosylated sites in the cytosol, providing a good starting point for topology modeling using appropriate topology-prediction software (see Chapter 8).

In another application, the membrane-insertion efficiency of a transmembrane helix can be measured using two engineered glycosylation sites, one near each end of the helix (**Figure 2**). This is how the data shown in Figure 6.15 and Figure 6.16 were obtained.

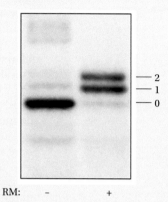

RM: –   +

Figure 1 *In vitro* transcription/translation of a protein containing two acceptor sites for Asn-linked glycosylation in its lumenal domain. When translated in the absence of rough microsomes (–RM, left lane), the non-modified protein (no attached oligosaccharides) is produced. When rough microsomes are present during translation, the protein is co-translationally inserted into the membrane, and oligosaccharide chains are added to the two acceptor sites (+RM, right lane). In this particular case, one of the two acceptor sites is only partially glycosylated, giving rise to a mixture of molecules carrying one or two oligosaccharides. (From Hessa T et al. [2005] *Nature* 433: 377–381. With permission from Springer Nature.)

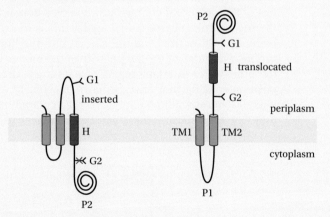

Figure 2 The membrane-insertion efficiency of a transmembrane helix (H) can be measured by placing acceptor sites for Asn-linked glycosylation on both sides of the helix. The construct is translated in vitro in the presence of rough microsomes and analyzed by SDS-PAGE. The membrane-insertion efficiency is $I = f_1/(f_1 + f_2)$, where f_1 is the intensity of the band representing molecules carrying one oligosaccharide chain (i.e., molecules where H is inserted, left) and f_2 the intensity of the band representing molecules carrying two oligosaccharide chains (i.e., molecules where H is not inserted, right). The red x signifies that the G2 glycosylation acceptor site is not modified when the H segment is inserted (left). (From Hessa T et al. [2005] *Nature* 433: 377–381. With permission from Springer Nature.)

hydrophobic amino acids (Leu, Val, Ile, Phe, and Met) promote membrane insertion, while charged and strongly polar amino acids (Asp, Glu, Arg, Lys, His, Asn, Gln, and Pro) reduce insertion. Given the values shown in Figure 6.16A, to compensate for just one or two charged residues in a transmembrane helix, one needs quite a few hydrophobic residues.

The values in Figure 6.16A correlate rather well with hydrophobicities calculated from the partitioning of short peptides between water and a membrane-mimetic solvent such as octanol (**Figure 6.16B**). An important distinction to keep in mind, though, is that the partitioning is between translocon and bilayer rather than between bilayer and water as in the case of the octanol hydrophobicity scale.

A more stringent test of the role of hydrophobicity in membrane insertion of transmembrane helices is to incorporate non-natural amino acids carrying different-sized hydrophobic side chains into the transmembrane helix. Non-natural amino acids can be inserted into proteins using so-called suppressor tRNAs that can read the stop codon AUG (**Box 6.5**). By charging the suppressor tRNA with amino acids carrying linear alkyl side chains of increasing length and incorporating these amino acids into a transmembrane helix, one can very precisely measure the change in the free energy of insertion into the ER membrane as a function of the size of the side chain. As seen in **Figure 6.17**, the measured free energy is directly proportional to the accessible surface area of the side chain. This is true also for non-polar aromatic side chains of varying size, but the proportionality constants are slightly different: -10 cal mol^{-1} Å^{-2} for the linear alkynes and -7 cal mol^{-1} Å^{-2} for the aromatic side chains.

As noted in Chapter 3, the solvation parameter for partitioning non-polar solutes from water into hexane or octanol is -23 cal mol^{-1} Å^{-2}, i.e., a factor ~2.5 larger. Why is this? First, the environment in the translocon channel is likely to be considerably less polar than pure water, since the space cannot contain pools of bulk water and the internal protein surfaces are not very polar; second, the ER membrane has a high protein content, meaning that it is more polar than a pure hydrocarbon phase. We should therefore expect that the solvation parameter for translocon-to-membrane partitioning should be reduced compared with water-to-hexane partitioning.

The values in Figure 6.16 are for residues placed near the middle of the membrane. For hydrophobic residues, these values do not change much if the residue is moved closer to the membrane surface. In contrast, the insertion propensities of the polar and charged amino acids depend strongly on position within the membrane (**Figure 6.18**). Again, this is to be expected if transmembrane helices partition from the translocon into the lipid bilayer according to basic thermodynamics, since the polarity of the bilayer increases as one approaches the lipid–water interface region (Chapter 2). The experimentally determined position-specific free energies correlate nicely with so-called statistical free energies. Statistical free energies are calculated from position-specific residue frequencies in natural transmembrane helices using the formula $\Delta G_{stat} = -RT$

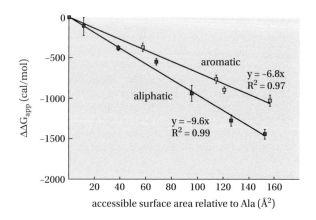

Figure 6.17 The apparent free energy of membrane insertion (ΔG_{app}) decreases in proportion to the increase in accessible surface area of hydrophobic non-natural amino acids. Red data points are for linear alkyl side chains with up to eight carbons; green data points are for non-polar aromatic side chains ranging in size from phenylalanine to biphenyl. (From Öjemalm K, Higuchi T, Jiang Y et al. [2011] *PNAS* 108: E359–E364. With permission from von Heijne.)

BOX 6.5 HOW TO INCORPORATE NON-NATURAL AMINO ACIDS INTO PROTEINS USING SUPPRESSOR tRNA

During protein synthesis, aminoacylated tRNAs are selected by the ribosome in response to the codons in the mRNA. Normally, this allows the incorporation of the 20 naturally occurring amino acids into the growing peptide chain. However, techniques have been developed that make it possible to expand the genetic code to include non-natural amino acids.

These techniques take advantage of the fact that there are three different stop codons in the standard genetic code: UAA, UGA, and UAG. Of these, UAG is not used very frequently in nature. Special tRNAs, so-called suppressor tRNAs, have been found that can read the UAG stop codon, leading to the insertion of an amino acid rather than chain termination (**Figure 1**). If we could somehow charge such a suppressor tRNA with a non-natural amino acid, this amino acid would be incorporated into the growing peptide chain in response to a UAG codon.

Suppressor tRNAs can be charged *in vitro* using an artificial enzyme called Flexizyme, which is a ribozyme composed of RNA rather than protein. The suppressor tRNA can then be added to an *in vitro* translation system together with an mRNA in which a UAG stop codon has been engineered into the position where one wants the non-natural amino acid to be incorporated. This is how the data in Figure 6.16 were obtained.

A more sophisticated version of the method has been developed for incorporating non-natural amino acids into proteins *in vivo*. Again, one uses a UAG stop codon in the mRNA to specify the site of incorporation, but charging of the suppressor tRNA is now done using an engineered aminoacyl tRNA synthetase enzyme designed to recognize the non-natural amino acid and to add it to the suppressor tRNA. The genes encoding the suppressor tRNA and the new aminoacyl tRNA synthetase are inserted into the cell, while the non-natural amino acid is added to the growth medium.

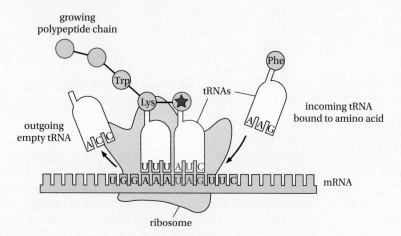

Figure 1 Incorporation of a non-natural amino acid (*) in response to the stop codon UAG by use of a suppressor tRNA with an anticodon (AUC) that can base-pair with the UAG codon.

$\ln(f_i/f_{i,tot})$ where f_i is the frequency of residue type i in the given position and $f_{i,tot}$ is the average frequency of residue type i in the set of proteins used in the analysis.

The data in Figure 6.18 and from measurements of how the free energy of insertion into the ER membrane varies with the overall length of the hydrophobic segment can be approximated by a simple phenomenological expression for the insertion free energy in kcal mol^{-1}:

$$\Delta G_{app}^{pred} = \sum_{t=1}^{L} \Delta G_{app}^{aa(i)} + 0.3\mu + 9 - 0.7L + 0.008L^2$$

where L is the length of the segment, $\Delta G_{app}^{aa(i)}$ is the contribution from the residue in position i in the segment (read off from Figure 6.18), and μ is the hydrophobic moment of the segment when folded into an α-helix (see Chapter 3). We will discuss computational methods to identify transmembrane helices in protein sequences in Chapter 8.

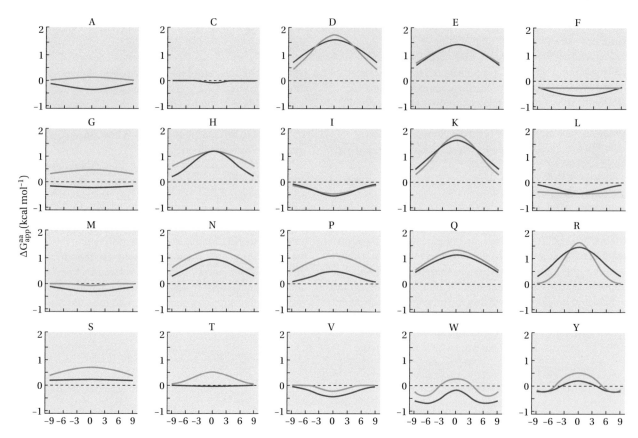

Figure 6.18 The blue curves show residue free energies (ΔG_{app}) for insertion into the ER membrane as a function of position within the membrane (position zero is the middle of the membrane). The red curves show "statistical" free energies calculated from position-specific residue frequencies in known transmembrane helices. (From Hessa T et al. [2007] *Nature* 450: 1026–1030. With permission from Springer Nature.)

Residues outside the hydrophobic core of a transmembrane helix can affect its ability to insert into the membrane. In particular, positively charged residues increase the propensity of a helix to insert into the ER membrane if they are located near the cytoplasmic end of the hydrophobic stretch. This effect is so strong that it has received its own name: the positive-inside rule, discussed further later. This rule applies not only to the ER but also to the bacterial inner membrane, the thylakoid membrane of chloroplasts, and the inner mitochondrial membrane (but only to proteins that are inserted into the inner mitochondrial membrane from the matrix side).

6.2.2 Some Transmembrane Helices Cannot Insert by Themselves

One interesting conclusion from studies such as those presented in Figure 6.18 is that not all transmembrane helices in multi-spanning membrane proteins have a sufficiently high content of hydrophobic residues (or a sufficient number of flanking positively charged residues) to be able to insert stably in the lipid bilayer by themselves (**Figure 6.19**). Instead, they somehow depend on other parts of the protein for stable insertion.

In a few cases, specific interactions between different transmembrane helices in a protein have been found to be required for one or both helices to insert. This suggests that transmembrane helices can interact with each other during the membrane-insertion stage, i.e., while they are still in or near the translocon. Another way that nature has discovered to build proteins with one or more not very hydrophobic transmembrane helices is to place a more hydrophobic segment next to the segment that will ultimately span the membrane in the final

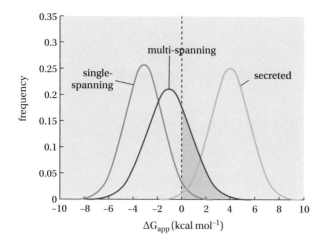

Figure 6.19 Distributions of predicted membrane-insertion free energies (ΔG_{app}) for transmembrane helices in natural proteins. The purple curve shows data for transmembrane helices in single-spanning membrane proteins, the red curve for transmembrane helices in multi-spanning membrane proteins, and the green curve for the most hydrophobic segments in a set of soluble secreted proteins. Only transmembrane helices for which $\Delta G_{app} < 0$ kcal mol^{-1} are predicted to insert stably across the membrane by themselves. The shaded region indicates transmembrane helices from multi-spanning proteins with $\Delta G_{app} > 0$ kcal mol^{-1}. (From Hessa T et al. [2007] Nature 450: 1026-1030. With permission from Springer Nature.)

structure. The more hydrophobic segment forms the initial transmembrane helix but is then displaced from the membrane by the more polar segment as the protein folds (**Figure 6.20**). A third way is to place a weakly hydrophobic segment between two strongly hydrophobic transmembrane helices that both have a strong preference for the same membrane orientation, such that the weakly hydrophobic segment is forced to insert across the membrane (**Figure 6.21**). Finally, studies on model proteins have shown that sequence elements that cannot fold into stable globular structures located both upstream and downstream of a transmembrane helix can affect the efficiency of membrane insertion, suggesting that marginally hydrophobic transmembrane helices remain in the translocon for a certain time, during which flanking sequence elements can affect the insertion process.

6.2.3 Many Membrane Proteins Form Homo- or Hetero-Oligomeric Complexes, but Multi-domain Architectures Are Rare

Once inserted into the membrane, many, if not most, integral membrane proteins form either homo- or hetero-oligomeric complexes; two examples are shown in **Figure 6.22**. An often-used method to identify complexes is blue-native polyacrylamide gel electrophoresis (BN-PAGE). Separation of protein complexes by BN-PAGE can be combined with a second-dimension separation using denaturing SDS-PAGE, as described in **Box 6.6**. Since only a small fraction of all membrane proteins have been analyzed biochemically, we do not know exactly how many are monomeric and how many form oligomers.

In eukaryotes, globular proteins are often composed of structurally and functionally distinct domains that have become fused to each other during evolution. Membrane proteins are different: domain recombination involving different types of membrane domains does not seem to be common in either

Figure 6.20 One monomer of the trimeric transporter protein Glt$_P$ (PDB: 1XFH). (A) The transmembrane helices are colored from red to blue according to increasing hydrophobicity (blue is more hydrophobic). (B) The most hydrophobic segments that overlap each of the transmembrane helices are similarly colored. Note that TM4 has low hydrophobicity (A) but that it overlaps a much more hydrophobic segment that is partly displaced out of the membrane in the folded structure (B). During assembly, the blue segment forms the initial transmembrane helix. In the assembled trimer, TM4 is largely buried in the trimer interface and is not lipid-exposed. (From Kauko A, Hedin LE, Thebaud E et al. [2010] *J Mol Biol* 397: 190–201. Used with permission from Elsevier.)

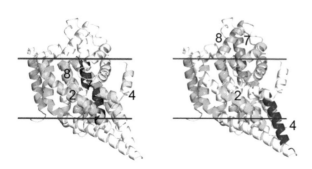

Figure 6.21 A transmembrane helix with a strong orientational preference (blue) can change the hydrophobicity threshold for membrane insertion of an upstream transmembrane helix (red). The topology of the protein can be deduced from the number of glycans (Y) that are added to the protein by the lumenal oligosaccharyl transferase enzyme (see Box 6.4; non-glycosylated glycan acceptor sites are indicated by red crosses in the topology diagrams). The red helix is composed of n Leu and (19-n) Ala residues, and the threshold for membrane insertion is defined as the number of Leu residues required to get 50% insertion (n_{50}). The fraction of molecules with either of the three topologies shown on top (which carry, respectively, one, three, and two glycans) is indicated by the corresponding colored curves in the panels below as a function of n. When the blue helix has no orientational preference, $n_{50} = 2.8$ for the red helix (B). Adding five positively charged Lys residues to the C-terminus of the blue helix makes it prefer the N_{lum}-C_{cyt} orientation and reduces n_{50} for the red helix to 0.7 (A). Adding positively charged residues to the N-terminus of the blue helix has the opposite effect and increases n_{50} for the red helix to 3.5 (C). (From Öjemalm K, Halling KK, Nilson IM et al. [2012] *Molecular Cell* 45: 529–540. With permission from Elsevier.)

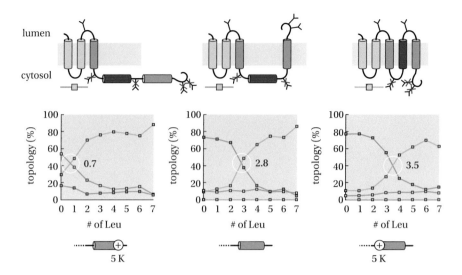

prokaryotes or eukaryotes (**Figure 6.23A, B**). Rather, complex membrane protein structures are mostly assembled through oligomerization of different subunits. Perhaps, the diffusion-collision constraint of the bilayer relaxes the need for covalent domain mixing by orienting subunits and confining diffusion to two dimensions, reducing the entropy loss needed for oligomer formation.

Internal duplication of a single domain is common in membrane proteins, however. Exact numbers are difficult to obtain, as many such duplications are apparent only in the three-dimensional (3-D) structure and cannot be discerned by sequence analysis, but estimates suggest that nearly half of all bacterial and nearly one-quarter of all human membrane proteins with six or more transmembrane helices contain internal duplications. An example is shown in **Figure 6.23C**), where a homo-tetrameric bacterial Na⁺ channel is compared with a eukaryotic Na⁺ channel with an internally fourfold repeated domain structure. We will return to such duplications later in the chapter.

6.3 MEMBRANE PROTEIN TOPOLOGY

6.3.1 Most Membrane Proteins Have a Unique Topology

By the "topology" of a membrane protein, we mean a specification of both the segments in the polypeptide chain that span the membrane and the overall

(A)   (B)

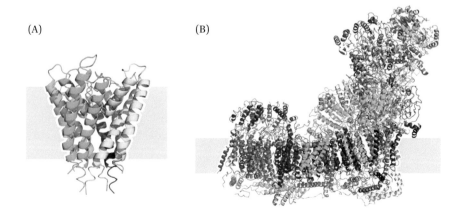

Figure 6.22 Multimeric membrane proteins. (A) The homo-tetrameric bacterial potassium channel KcsA. (B) The hetero-oligomeric mitochondrial Complex I. Individual subunits are indicated by coloring.

BOX 6.6   BLUE-NATIVE POLYACRYLAMIDE GEL ELECTROPHORESIS ALLOWS IDENTIFICATION OF MEMBRANE PROTEIN COMPLEXES

Many membrane proteins are found in complexes with other membrane proteins. Notable examples of membrane protein complexes are photosystems I and II, which are responsible for photosynthesis. All the proteins of the respiratory chain that produce ATP also exist as complexes. How can we identify membrane protein complexes in a cell? First, the cell membrane must be dissolved in such a way that complexes do not fall apart. Second, the complexes must be separated from each other. And third, the component proteins in each complex must be identified and their stoichiometry determined.

A technique that meets these requirements is BN-PAGE. When it is applied to the inner membrane of a bacterium such as *E. coli*, inner membrane vesicles are first purified by ultra-centrifugation in a sucrose gradient and collected from the appropriate part of the gradient. Next, a mild detergent such as dodecyl maltoside (DDM) is used to solubilize the membrane. Most membrane protein complexes are sufficiently stable to survive in DDM. Then comes the crucial step: the dye Coomassie blue is added to the sample, and the sample is electrophoresed down a polyacrylamide gel (the "first dimension" gel). Coomassie blue binds nonspecifically to proteins and gives them a negative net charge but does not disrupt protein complexes. In the gel, the complexes are separated according to size and shape (**Figure 1**).

In the next step, a strip of the one-dimensional blue-native gel containing the sample is cut from the gel and soaked in a buffer containing the harsh detergent SDS, which causes the dissociation of all protein complexes into individual protein chains. The gel strip is then placed across the top of a second gel that also contains SDS, and the sample is again electrophoresed down the gel (the "second dimension" gel) and stained with either Coomassie blue or silver to visualize the proteins. The individual protein chains are now separated according their molecular weight, and the components of each complex run in vertical "channels" according to the position of the corresponding complex in the blue-native gel strip (**Figure 2**). Finally, after cutting the protein spots out of the gel, the identity of each protein is determined by mass spectrometry. The intensity of each spot can be used to estimate the amount of each protein and hence, the stoichiometry of the complex.

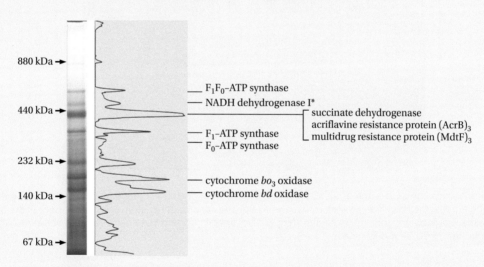

Figure 1 First-dimension separation of *E. coli* inner membrane protein complexes in a Coomassie blue gel. The positions of the more abundant complexes are indicated. (From Stenberg F, Chovanec P, Maslen SL et al. [2005] *J Biol Chem* 280: 34409-34419. Published under CC BY 4.0 license.)

(Continued)

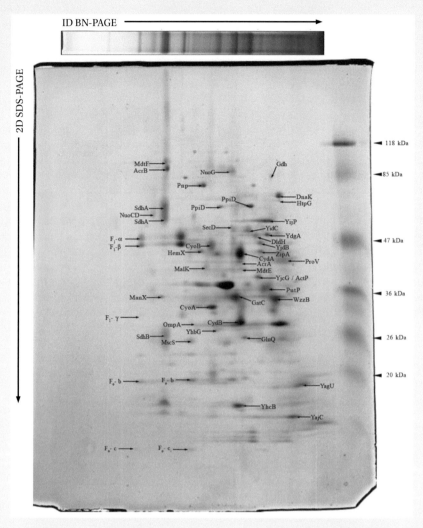

Figure 2 Second-dimension separation in an SDS-gel of the individual proteins in the complexes from the gel slice in Figure 1. The blue-native gel slice was placed across the top of the second-dimension gel as indicated. Protein spots identified by mass spectrometry are shown. (From Stenberg F, Chovanec P, Maslen SL et al. [2005] *J Biol Chem* 280: 34409-34419. Published under CC BY 4.0 license.)

orientation of the protein relative to the membrane. Almost (but not quite) all membrane proteins have a unique topology both in terms of the number of transmembrane segments and in terms of orientation. A number of different experimental methods have been developed to map membrane protein topology (**Box 6.7**).

Different topologies are not populated to the same extent in nature (**Figure 6.24**). On a global level, proteins with an even number of transmembrane helices and both the N- and C-terminus in the cytoplasm are heavily over-represented. Proteins with 10 or 12 transmembrane helices are particularly over-represented; as shown by the color-code, these proteins are mostly small-molecule transporters. In higher eukaryotes, there is a marked expansion of the superfamily of G-protein coupled receptors with seven transmembrane helices (note that data for higher eukaryotes are not included in Figure 6.24).

β-barrel proteins have a more restricted distribution of topologies (**Figure 6.25**). Among the bacterial outer membrane proteins, there is a very

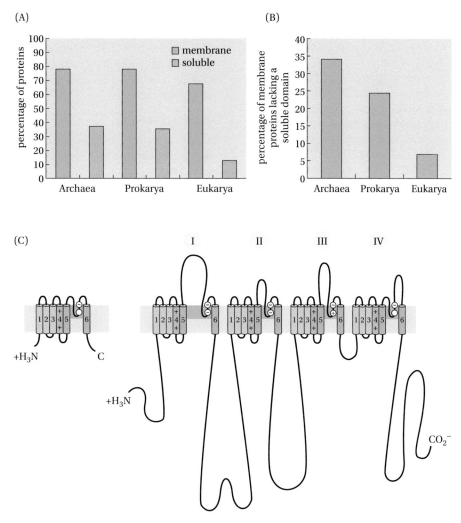

Figure 6.23 Complex membrane protein structures are mostly assembled through oligomerization of different subunits. (A) Percentage of membrane and soluble proteins that contain only one single type of protein domain (this includes proteins with an internal duplication of a domain) in different types of organisms. (B) Percentage of membrane proteins that lack an attached soluble domain in different types of organisms. (C) One subunit of a bacterial homo-tetrameric Na^+ channel (left) compared with a eukaryotic multi-domain Na^+ channel (right). A potential evolutionary advantage of the multi-domain architecture is that the four domains can evolve independently. (A,B, From Liu Y, Gerstein M, Engelman DM [2004] Proc Natl Acad Sci USA 101: 3495–3497. With permission from DM Engelman. C, From Yu FH, Catterall WA [2003] *Genome Biol* 4: 207. https://doi.org/10.1186/gb-2003-4-3-207. With permission from Springer Nature.)

strong preference for even numbers of β-strands and both the N- and the C-terminus in the periplasm. In fact, to date, there is only one protein known that has an odd number of β-strands: the VDAC voltage-dependent anion channel, found in the outer membrane of mitochondria.

6.3.2 The Distribution of Positively Charged Residues Correlates with Membrane Orientation

The main driving force for membrane integration is hydrophobicity, but how is the orientation of the protein in the membrane decided? Statistical studies of helix-bundle proteins show that there is a very strong tendency for positively charged residues (Arg and Lys) to be concentrated in loops and N- and C-terminal tails that face the cytoplasm (or more generally, the side of the membrane from which the protein is inserted) (**Figure 6.26**). This "positive-inside rule" holds for helix-bundle proteins from nearly all biological membranes, including the plasma membrane, the membranes along the exocytic and

BOX 6.7 THE TOPOLOGY OF A MEMBRANE PROTEIN OF UNKNOWN STRUCTURE PROVIDES INSIGHTS INTO FUNCTION

In the absence of a high-resolution 3-D structure of one's favorite membrane protein, a topology model provides a good basis for planning mutagenesis experiments and inferring functionally important regions. Depending on whether the protein is of prokaryotic or eukaryotic origin, different methods are available to produce experimental data that help construct a topology model.

Reporter Fusions

Topology mapping of prokaryotic proteins is usually carried out in *E. coli*. The most widely used technique is based on reporter fusions. Topology reporters are proteins that are active or inactive depending on whether they are localized in the cytoplasm or the periplasm. Cytoplasmic reporters that are active in the cytoplasm but inactive in the periplasm include green fluorescent protein (GFP; encoded by the *gfp* gene) and the enzyme β-galactosidase (LacZ). In contrast, the enzymes alkaline phosphatase (PhoA) and β-lactamase (Bla) are active in the periplasm but not in the cytoplasm. There are simple assays available for scoring the activity of these reporter proteins in live *E. coli* cultures.

In a typical experiment, the *phoA* gene is fused in frame at the 3′ end of increasingly long fragments of the target gene, creating fusion proteins in which the fusion junction is ideally placed near the C-terminal end of loops between two predicted transmembrane helices. The alkaline phosphatase activity in cells expressing the fusion proteins can be measured spectrophotometrically by adding a substrate that alkaline phosphatase can convert into a colored product in a growing culture. This same reaction can also be monitored by a simple plate assay using the same substrate (**Figure 1**).

In a similar manner, a corresponding set of target gene–*gfp* in-frame fusions is constructed, and GFP fluorescence is measured on cell cultures. In the best case, one observes a complementary pattern of active and inactive PhoA and GFP fusion proteins, which makes it easy to infer the topology of the target protein. Topology reporters have also been developed for yeast.

Cysteine Labeling

Topology reporters usually work surprisingly well given the rather drastic changes imposed on the protein when making the fusions. Chemical labeling of protein constructs that contain only a single, unique cysteine provides an alternative way of assessing the topology of a membrane protein and can be used in both prokaryotic and eukaryotic cells.

In this method, one first mutates away any naturally occurring cysteines in the protein and then introduces, one by one, a single cysteine in each of the predicted loops between the transmembrane helices. Each variant is then expressed in *E. coli*, and the cells are treated with a chemical reagent such as 3-(N-maleimido-propionyl biocytin) (MPB). MPB can pass through pores in the outer membrane but cannot readily pass through the inner membrane. If the unique cysteine is located in a periplasmic loop it will react with MPB, but not if it is located in a cytoplasmic loop. The MPB molecule can be visualized using a chemiluminescent reagent.

As shown in the example in **Figure 2**, cysteines in cytoplasmic loops react with MPB only when the inner membrane is first disrupted by sonication of the cells, while those in periplasmic loops react with MPB even without sonication. Cysteine labeling can also be used on plasma membrane proteins in eukaryotic cells.

Glycosylation Mapping

Topology determination by glycosylation mapping is similar in spirit to the cysteine labeling method but works only in eukaryotic cells. The general method, described in Box 6.3, takes advantage of the fact that N-linked glycans are covalently attached to proteins only in the lumen of the

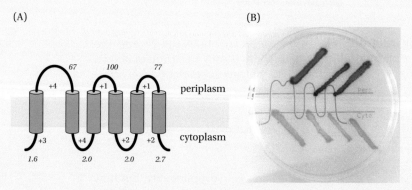

(A)   (B)

Figure 1 PhoA-based topology mapping of the *E. coli* inner membrane protein MalG. (A) The numbers in italics show the PhoA activities of the corresponding fusion proteins. The number of positively charged residues in each loop is indicated. (B) An indicator plate with cells expressing MalG–PhoA fusions streaked out in locations corresponding to the fusion joints in the MalG protein. Dark blue cells have high PhoA activity. (With permission from Jon Beckwith.)

(Continued)

BOX 6.7   THE TOPOLOGY OF A MEMBRANE PROTEIN OF UNKNOWN STRUCTURE PROVIDES INSIGHTS INTO FUNCTION (CONTINUED)

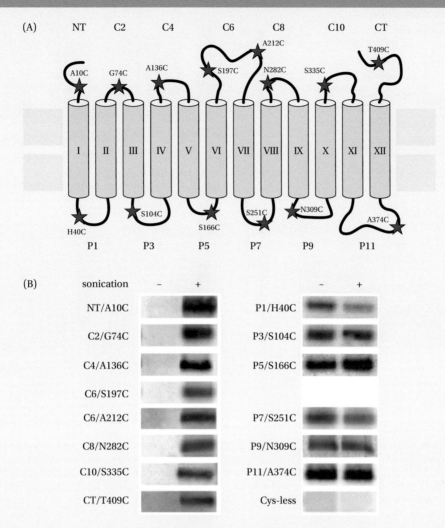

Figure 2 Topology mapping of the *E. coli* CscB transporter by cysteine labeling. Single cysteine residues placed in cytoplasmic (C) loops react with the cysteine-specific reagent MPB only if the inner membrane is first disrupted by sonication, while single cysteines introduced into periplasmic (P) loops are labeled by MPB even without sonication. A variant lacking all cysteines (Cys-less) is not labeled at all. (From Vitrac H et al. [2011] *J Biol Chem* 285: 15182-15194. Published under CC BY 4.0 license.)

ER. To determine topology, single glycan acceptor sites are introduced, one at time, into predicted loops. After expressing the protein in cells (or an *in vitro* translation system supplemented with ER-derived rough microsomes as in Box 6.4), the diagnostic shift in molecular weight reveals whether a loop is located in the cytoplasm or the ER lumen.

endocytic pathways, the nuclear membrane, the mitochondrial inner membrane, and the thylakoid membrane, as well as for the inner membrane of prokaryotes. It does not hold for β-barrel membrane proteins, however. Because there is no comparable bias in the distribution of the negatively charged amino acids, it is the accumulated positive charge rather than net charge that correlates best with membrane orientation.

6.3.3 Membrane Orientation Can Be Manipulated by Sequence Alterations

The positive-inside rule suggests that at the most basic level, membrane protein topology may be controlled by a combination of hydrophobic segments that

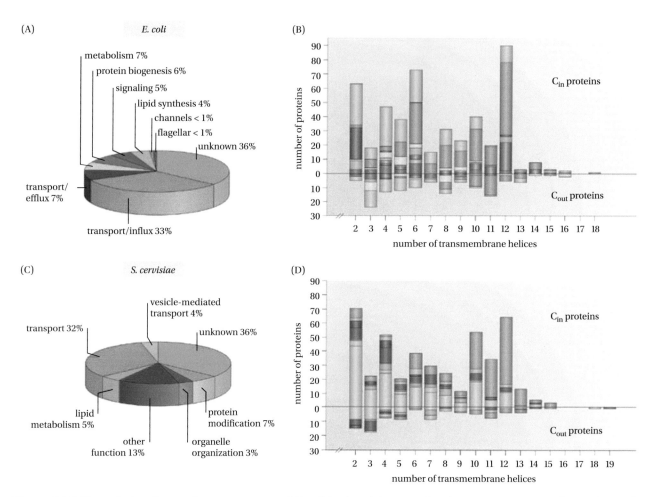

Figure 6.24 The helix-bundle membrane proteomes of *E. coli* (top) and the yeast *Saccharomyces cerevisiae* (bottom). Panels (A) and (C) show the distributions among different functional groups. Panels (B) and (D) show the distributions of different topologies; bars above the abscissa represent proteins with their C-terminus in the cytoplasm, while bars below the abscissa are for proteins with the C-terminus on the extra-cytoplasmic side. Proteins with a single transmembrane helix are not included in the statistics. (From von Heijne G [2006] *Nature* 7: 909–918.)

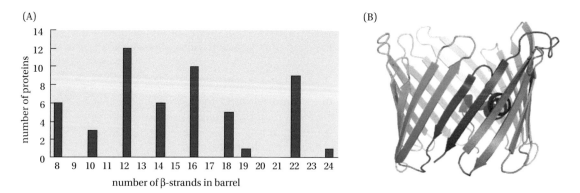

Figure 6.25 Nearly all β-barrel membrane proteins have an even number of β-strands. (A) The diagram shows the number of proteins of known 3-D structure with a given number of strands in their β-barrel. (B) The only known example of a protein with an odd number of β-strands is the mitochondrial VDAC channel, which has 19 β-strands (PDB: 3EMN). Note that the most N-terminal β-strand (dark blue) is parallel to the most C-terminal β-strand (red).

drive membrane integration and loops with a high or low content of Arg and Lys residues that control the in/out orientation of the protein in the membrane. Indeed, topology can be manipulated both by inserting or deleting hydrophobic segments and by changing the distribution of positively charged residues in the loops and the N- and C-terminal tails. An early example of how the orientation of an *E. coli* inner membrane protein can be re-engineered is shown in **Figure 6.27**.

Because most helix-bundle membrane proteins are inserted into the membrane co-translationally, we might think that the first transmembrane helix will determine the orientation of all the following helices regardless of how positively charged residues are distributed. This is not so, however. Instead, each transmembrane helix tends to orient with the end flanked by the highest number of positively charged residues in the cytoplasm. Therefore, loops with high numbers of Arg and Lys alternate with loops containing few such residues throughout the protein. If this alternating pattern is disturbed, e.g., if both ends of a hydrophobic segment are flanked by high numbers of Arg and Lys, membrane insertion of the hydrophobic segment may be compromised ("topological frustration").

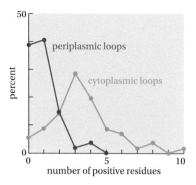

Figure 6.26 The positive-inside rule was first proposed in 1986, based on a statistical analysis of 20 membrane proteins. The graph shows frequency distributions of periplasmic loops (red) and cytoplasmic loops (blue) containing different numbers of positively charged residues. (From von Heijne G [1986] *EMBO J* 5: 3021–3027. With permission from John Wiley and Sons.)

6.3.4 The Translocon, the Membrane Potential, and Lipid Composition Can All Affect Protein Topology

How can positively charged residues have such a strong effect on membrane protein topology? Although the positive-inside rule was discovered in the mid-1980s, it still lacks a simple mechanistic explanation. Different studies have

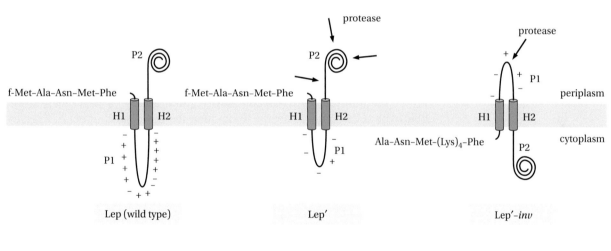

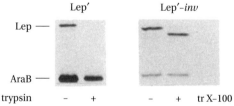

Figure 6.27 Re-engineering the membrane orientation of the *E. coli* inner membrane protein Lep from N_{out}-C_{out} to N_{in}-C_{in}. The wild-type Lep protein has a highly positively charged cytoplasmic loop (P1). In a first step, most of these charges were removed by a deletion (Lep′). In a second step, four extra Lys residues were added to the N-terminal tail (Lep′-inv), thereby inverting the bias in the distribution of positively charged residues across the first transmembrane helix (H1). The topology of the different constructs was probed by adding a protease (trypsin) to the periplasm. As seen in the gel pictures, trypsin digests the large C-terminal P2 domain of Lep′ but cleaves Lep′-inv in the short loop between the two transmembrane helices, showing that Lep′ and Lep′-inv have opposite orientations in the inner membrane. AraB is a cytoplasmic protein that serves as a control to show that the inner membrane is intact and does not let the protease penetrate into the cytoplasm. Addition of the detergent Triton X-100 dissolves the inner membrane and gives the protease access to the cytoplasm; hence, all the proteins are completely digested. (From von Heijne G [1989] *Nature* 341: 456–458. With permission from Springer Nature.)

demonstrated that the topology of membrane proteins can be affected by mutations in the Sec61 translocon, by a lowering of the electrochemical gradient across the inner membrane of *E. coli*, and by changes in the lipid composition of the bacterial inner membrane. Certain mutations in the Sec61α subunit of the yeast ER translocon can promote either the N_{in} or the N_{out} orientation of single-span membrane proteins, but the mechanistic basis for these effects is unclear.

A reduction in the electrochemical gradient across the inner membrane of *E. coli* reduces the topological effect of Arg and Lys residues, suggesting that electrostatics may play a role: the periplasmic side of the inner membrane is more acidic and has a positive electrical potential relative to the cytoplasmic side, and positively charged residues in a translocating nascent chain thus move "uphill" against the potential. However, the positive-inside rule appears to hold also in acidophilic bacteria, i.e., bacteria that live at such low ambient pH that the electrical component $\Delta\Psi$ of the electrochemical gradient is inverted compared with *E. coli*, with the periplasmic side being more negative than the cytoplasmic side. Finally, the ER in eukaryotic cells is thought not to have a sizable $\Delta\Psi$, yet the positive-inside rule holds. It is not clear how these observations can be reconciled.

Studies on lactose permease (LacY) from *E. coli* have uncovered an important role of lipid composition in determining membrane protein topology. Using suitable mutant strains, it is possible to drastically change the lipid composition of the *E. coli* inner membrane from the natural one, which is 70% PE (phosphatidylethanolamine, a zwitterionic lipid), 20% PG (phosphatidylglycerol, a negatively charged lipid), and 5% CL (cardiolipin, a lipid with four acyl chains and two negative charges in its headgroup), to one lacking PE altogether. In the absence of PE, the inner membrane has a very large negative surface charge density, and cells can only grow in media containing high levels of Mg^{2+}. Under these extreme conditions, the N-terminal six transmembrane helices of LacY assemble in the membrane with an inverted orientation, the seventh helix does not insert at all across the membrane, and the five C-terminal helices attain the same orientation as in the normal protein (**Figure 6.28**). Mutations of negatively charged residues present in the N-terminal domain can counteract the effects of PE-depletion; i.e., both a negative-outside and a positive-inside rule are in operation under these conditions. The presence of high levels of PE in wild-type cells therefore seems to dampen the topological effect of negatively charged residues, leaving positively charged Arg and Lys residues as the main topological determinants.

6.3.5 Membrane Proteins Can Evolve by Gene Duplication and Addition/Deletion of Terminal Helices

Large multi spanning helix-bundle membrane proteins seem to evolve mainly by internal gene duplication and not by domain recombination, as noted earlier in this chapter. Surveys based on detecting sequence similarity between the N- and C-terminal parts of membrane proteins show that the most common type of internal gene duplication is one where the entire protein is duplicated, leading to a doubling of the number of transmembrane helices. Depending on whether the original protein has an even or odd number of transmembrane helices, the outcome of the duplication is radically different: in the former case, the two duplicated domains in the resulting protein have the same topology, i.e., they are orientated in parallel in the membrane, whereas in the latter case, the two domains will have an anti-parallel orientation relative to each other (**Figure 6.29**).

Obviously, the anti-parallel domain orientation resulting from an internal duplication of a protein with an odd number of transmembrane helices is problematic from the point of view of the positive-inside rule, as the two domains will have opposing charge distributions. Therefore, we may be inclined to think that such duplications are rare. This is not the case, however, as a rather large fraction of all multi-spanning membrane proteins of known structure are built

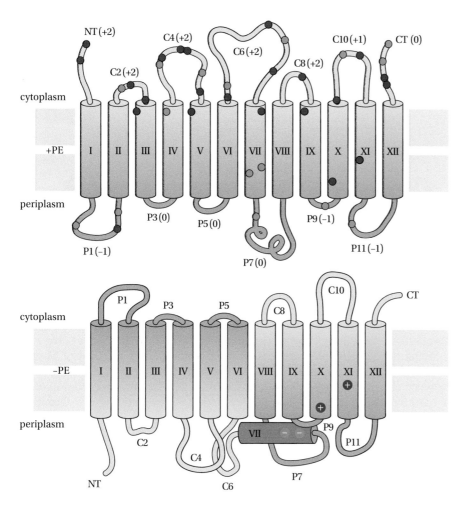

Figure 6.28 The topology of lactose permease (LacY) depends on the lipid composition of the inner membrane. (A) Topology In wild-type *E. coli* cells where zwitterionic PE accounts for ~70% of the inner membrane lipids. (B) Topology in cells lacking PE and with strongly increased levels of negatively charged lipids. The orientation of the TM I–VI domain is inverted, and TM VII is not inserted across the membrane. Positively (red) and negatively (blue) charged residues in the different loops are indicated in panel (A). (Figure reproduced with permission from Dowhan W, Bogdanov M. 2009. *Annu Rev Biochem* 78: 515–540. Copyright 2009 Annual Reviews.)

from two anti-parallel domains. Because both domains follow the positive-inside rule in these proteins, there must have been an evolutionary pressure to redistribute positively charged residues toward the cytoplasmic loops in one of the two domains after the internal duplication occurred. For transport proteins in particular, single polypeptide chains with internal domain duplications seem

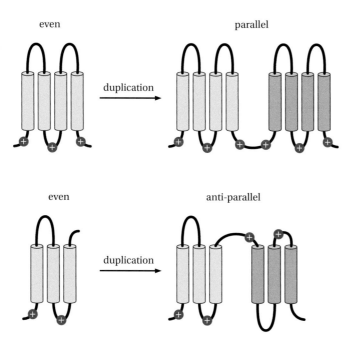

Figure 6.29 Evolution of membrane proteins via an internal gene duplication. When the original protein has an even number of transmembrane helices, the two domains in the resulting duplicated protein will be orientated in parallel in the membrane. With an odd number of transmembrane helices in the original protein, the domain orientation in the duplicated protein will be anti-parallel, creating a problem in terms of the positive-inside rule.

Figure 6.30 Lactose permease (panel A) is composed of two parallel, homologous domains (red and green), each with six transmembrane helices. Each of the six-helix domains is in turn composed of two anti-parallel three-helix domains (A and B, C and D). Aquaporin 0 (panel B) is composed of two anti-parallel domains, each with three transmembrane helices and one reentrant loop. The two reentrant loops meet in the middle of the membrane (circle). (From Radestock S, Forrest LR [2011] *J Mol Biol* 407: 698–715. With permission from Elsevier.)

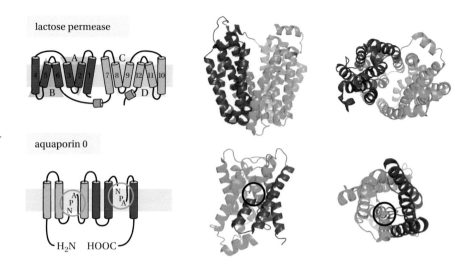

to have some evolutionary advantage over dimers of single domains, as very few protein families have been found that contain both internally duplicated family members as well as single-domain family members. Examples of proteins with parallel and anti-parallel domain organizations are shown in **Figure 6.30**.

Membrane proteins can also evolve by the addition or removal of N- and C-terminal transmembrane helices but rarely by the insertion of extra helices between already existing helices. A variation on this theme is the addition of an N-terminal signal peptide to a membrane protein with a cytoplasmic N-terminus. While this is most likely not enough to cause the reorientation of a protein with many transmembrane helices, it can cause inversion of topology for single-spanning proteins. An example where an N-terminal signal peptide has this effect on an ion channel protein with two transmembrane helices is shown in **Figure 6.31**.

6.3.6 Some Proteins Have a Dual Topology

What happens when a protein does not follow the positive-inside rule, i.e., when there are similar numbers of Arg and Lys residues in the extracellular and intracellular loops? From protein engineering experiments of the kind shown

Figure 6.31 K+ channels are formed from four identical subunits, each with two transmembrane (TM) helices, a reentrant loop (P-loop), and N- and C-terminal cytoplasmic domains. Glutamate receptors are also K+ channels, but their orientation in the plasma membrane is inverted due to the presence of an N-terminal signal peptide (SP) that induces translocation of the N-terminal globular domain. The signal peptide is cleaved by the signal peptidase enzyme in the lumen of the ER.

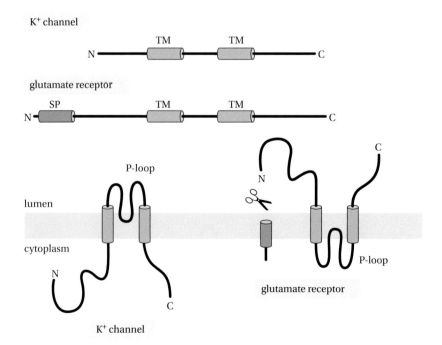

in Figure 6.27, we have learned that model proteins with a balanced charge distribution can insert into the membrane with a mixed, or "dual," topology. Such proteins are "undecided" about their orientation, and each molecule presumably makes a random choice on which way to insert.

In a few cases, nature has taken advantage of this possibility to evolve "dual topology" proteins. The known dual topology proteins tend to be rather small, with fewer than six transmembrane helices, and most of them seem to be small-molecule or ion transporters. The best-studied examples are found in a family of prokaryotic proteins called the small multidrug-resistance proteins (SMRs). SMR proteins have four transmembrane helices and form anti-parallel dimers in the membrane. In some members of the SMR family, the two oppositely orientated monomers are encoded by different genes and the active protein is an anti-parallel heterodimer, but other family members are encoded by a single gene and the active protein is a dual topology homodimer. As expected from the positive-inside rule, the single-gene dual topology SMR proteins have a balanced distribution of Arg and Lys residues, while the heterodimeric pairs have oppositely skewed distributions (**Figure 6.32**).

6.3.7 Membrane Proteins Can Have Multiple Topologies

The great majority of all membrane proteins have a single topology. Indeed, given the intricate functions carried out by membrane proteins, it would make little sense for a protein to have multiple topological forms with different numbers of transmembrane helices and different orientations in the membrane. Still, in addition to the dual topology proteins, a few cases are known where a single polypeptide generates multiple topologies. An intriguing example is the Scrapie prion protein (PrP) (**Figure 6.33**). PrP has an inefficiently recognized N-terminal signal peptide, an inefficiently recognized transmembrane helix in the middle of the protein, and a C-terminal GPI-attachment signal. This results in three distinct forms of the protein: a GPI-anchored form, where the signal peptide but not the transmembrane segment is recognized ([Sec]PrP), and two single-spanning forms with opposite orientations, where either both the signal peptide and the transmembrane segment are recognized ([Ntm]PrP) or only the latter is recognized ([Ctm]PrP).

6.3.8 Membrane Proteins Can Undergo Dynamic Changes in Topology

Is topology always determined during the initial membrane-integration step, or can proteins undergo dynamic changes in topology during their lifetime? There are a handful of cases known where the initial topology changes long after the translocon-mediated insertion step. As an example, some bacteriophages (i.e., viruses that infect bacteria) produce proteins called holins that, as the name implies, can make holes in the bacterial inner membrane that allow cytoplasmic viral enzymes to reach and digest the cell wall and cause cell lysis.

Figure 6.32 Charge biases in homo- and heterodimeric members of the small multidrug-resistance proteins (SMR) family. (A) The distribution of charge bias (i.e., the net difference in the number of Arg+Lys residues) between the periplasmic and cytoplasmic loops clusters around zero for single-gene dual topology proteins (green) and around ±5 for paired-gene oppositely orientated proteins (red). (B) The charge biases in the two members of each SMR heterodimer are opposite one another. If the first member of the pair has a negative bias, then the second member has a positive bias, and vice versa. The numbers indicate the number of protein pairs found with given charge biases (e.g., one pair with charge biases of +15 and −5 was found). (From Rapp M, Granseth E, Seppälä S et al. [2006] *Nature Struct Molec Biol* 13: 112–116. With permission from Springer Nature.)

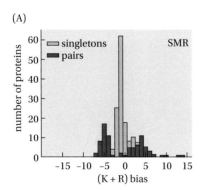

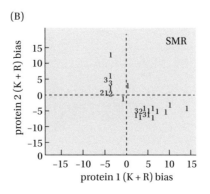

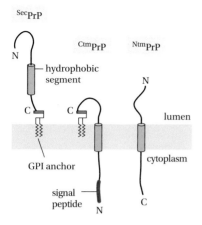

Figure 6.33 The Scrapie prion protein PrP. The predominating form is ^{Sec}PrP, but the two minor topological variants ^{Ctm}PrP and ^{Ntm}PrP are also made in the cell. In the figure, the N-terminal signal peptide has been clipped off by signal peptidase in ^{Sec}PrP and ^{Ntm}PrP. (From Harris DA [2003] *British Med Bull* 66: 71–85. With permission from Oxford University Press.)

Holins initially insert two C-terminal transmembrane helices into the inner membrane, leaving a third hydrophobic N-terminal segment in the cytoplasm (**Figure 6.34**). The insertion of this segment is blocked by N-terminally placed positively charged residues that prevent the N-terminus from translocating across the membrane. During infection, other phage proteins cause a slow dissipation of the electrochemical gradient across the inner membrane, eventually lowering the electrostatic barrier to translocation of the charged N-terminus sufficiently for the N-terminal hydrophobic segment to insert into the membrane and activate the lytic function of the protein.

Aquaporin-1 (AQP1), a human water-channel protein, provides an example where the topology change is more intricate and happens soon after the initial insertion event. In its final folded form, AQP1 has six transmembrane helices and two "reentrant loops"—short hairpin-like structural elements that penetrate only halfway across the membrane (see Chapter 10). The two reentrant loops meet in the middle of the membrane and are critical for the formation of the water-conducting channel. During the co-translational membrane-insertion process, the second and fourth transmembrane helices do not insert into the membrane. The full six-helix topology is formed later when the third helix flips its orientation, thereby bringing the second and fourth helices into their proper positions (**Figure 6.35**).

6.3.9 Misfolded Membrane Proteins Are Degraded

One consequence of the rather complex membrane-insertion mechanisms that the cell uses is that membrane proteins often do not insert and fold with 100% efficiency. A particularly problematic example is the cystic fibrosis transmembrane conductance regulator protein (CFTR), for which only 25%–50% of all molecules fold properly and get transported to the plasma membrane; the remaining fraction is rapidly degraded. This makes CFTR especially sensitive to mutations that compromise its stability and hence, prevent its appearance on the plasma membrane, resulting in cystic fibrosis. CFTR is discussed in Chapter 12.

Degradation of misfolded membrane proteins takes place in the ER through a process called ER-associated degradation (ERAD). Membrane proteins can either be directly extracted from the membrane by the combined action of a membrane-integral "retro-translocase," minimally formed by the Hrd1 and Hrd3 proteins and an ATP-driven hexameric AAA-family protein called p97 that pulls on the protein chain, or clipped in one of their transmembrane helices by intramembrane proteases, causing release of the fragments from the membrane (**Figure 6.36**). In either case, ubiquitination of the released protein (Hrd1 is a ubiquitin ligase) then targets it for degradation by the proteasome.

What features are used by the cell to distinguish between misfolded and properly folded membrane proteins? This is not completely clear. For luminal protein domains that carry one or more N-linked glycans (see Box 6.3),

Figure 6.34 The topology of holins changes depending on the membrane electrochemical potential ($\Delta\psi$). The N-terminus of TM1 carries a positive charge and is prevented from flipping across the membrane by the opposing $\Delta\psi$ (which is positive on the periplasmic side; left). When $\Delta\psi$ is reduced, TM1 can insert into the membrane, and the holin becomes active (right). (Figure reproduced with permission from Wang I-N, Smith DL, Young R [2000] *Annu Rev Microbiol* 54: 799–825. Copyright 2000 Annual Reviews.)

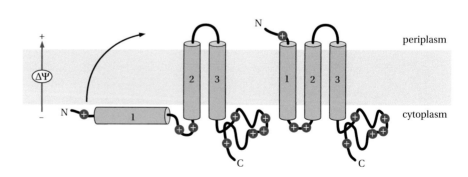

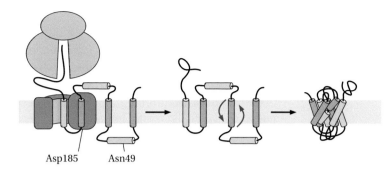

Figure 6.35 Reorientation of the AQP1 topology. The process depends on a critical interaction between residues in TM2 (Asn49) and TM5 (Asp185). (From Skach WR [2009] *Nature Struct Mol Biol* 16: 606–612. With permission from Springer Nature.)

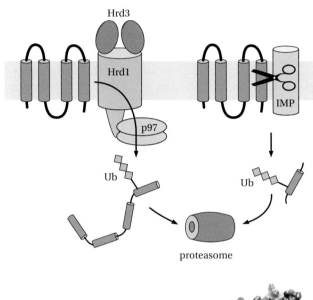

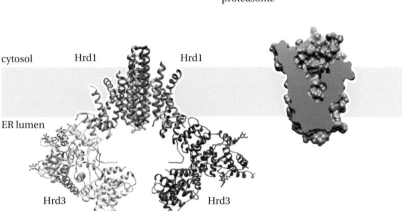

Figure 6.36 Degradation of misfolded proteins. (A) Misfolded membrane proteins in the ER are degraded either via a pathway that involves recognition by Hrd3 and retro-translocation through Hrd1 driven by the AAA ATPase p97, or by cleavage of transmembrane helices by an intramembrane protease (IMP) causing release from the membrane. In both cases, ubiquitinated (Ub) chains are degraded by the proteasome. (B) structure of the membrane domain of yeast Hrd1 in complex with Hrd3 (lacking its C-terminal transmembrane helix, left; PDB 6VJY). A possible substrate-conducting channel runs down the middle of Hrd1 (right). (From Schoebel S et al. [2017] *Nature* 548: 352–355. With permission from Springer Nature.)

de- and re-glucosylation of the glycan and recognition of misfolded glyco-protein by the ER-resident chaperones calnexin and calreticulin help mark proteins for ERAD. Cleavage by intramembrane proteases usually takes place at "weak spots" in the transmembrane helices where helix-destabilizing residues such as Pro or Gly allow local disturbances in the helical structure. Incorporation of such a "weak" helix into the properly folded structure protects it from proteolysis.

KEY CONCEPTS

- Nearly all integral membrane proteins depend on specialized protein complexes for membrane insertion.

- Membrane insertion can be either co- or post-translational.

- Certain toxins, such as α-hemolysin, can insert spontaneously into their target membrane without the aid of insertion machineries.

- ER insertion of membrane proteins is mediated by the SRP-SR-Sec61 pathway. In prokaryotes, insertion into the inner membrane is mediated by the homologous Ffh-FtsY-SecYEG pathway.

- Tail-anchored membrane proteins are inserted post-translationally into the ER membrane with the help of the ATP-dependent Get machinery.

- Some plasma membrane proteins are anchored in the membrane by a C-terminal glycosylphosphatidylinositol (GPI) anchor.

- β-barrel proteins in Gram-negative bacteria are chaperoned across the periplasmic space and then inserted into the outer membrane by the Bam machinery.

- Mitochondrial inner membrane proteins are imported through the TOM complex in the outer membrane and are then inserted into the inner membrane by either the TIM22 or the TIM23 complex.

- Mitochondrial outer membrane proteins are imported through the TOM complex into the intermembrane space and are then inserted into the outer membrane by the SAM complex (which is related to the Bam machinery in Gram-negative bacteria).

- The few mitochondrial inner membrane proteins that are synthesized in the mitochondrial matrix are inserted co-translationally into the inner membrane by the Oxa1 translocon (which is related to the bacterial YidC and chloroplast Alb3 translocons).

- Only the most hydrophobic amino acids—Leu, Val, Ile, Phe, and Met—promote translocon-mediated membrane insertion of transmembrane helices, while charged and strongly polar amino acids—Asp, Glu, Arg, Lys, His, Asn, Gln, and Pro—reduce the efficiency of membrane insertion.

- Complex membrane protein structures tend to be built as assemblies of smaller subunits rather than constructed as multi-domain polypeptide chains.

- For nearly all kinds of helix-bundle membrane proteins, the positively charged residues Arg and Lys tend to be concentrated in loops and N- and C-terminal tails that face the cytoplasm. This is known as the "positive-inside rule."

- Helix-bundle membrane proteins evolve by gene duplication, and by the addition or deletion of N- and C-terminal transmembrane helices, rather than by domain recombination.

- Misfolded membrane proteins can be degraded through the ER-associated degradation pathway.

FURTHER READING

Bohnert, M., Pfanner, N., and van der Laan, M. (2015) Mitochondrial machineries for insertion of membrane proteins. *Curr Opin Struct Biol.* 33:92–102.

Hegde, R S., and Keenan, R.J. (2011) Tail-anchored membrane protein insertion into the endoplasmic reticulum. *Nat Rev Mol Cell Biol.* 12:787–798.

Hegde, R.S. (2022) The function, structure, and origins of the ER membrane protein complex (EMC). *Annu Rev Biochem*, in press.

Cymer, F., von Heijne, G., and White, S.H. (2015) Mechanisms of integral membrane protein insertion and folding. *J Mol Biol* 427:999–1022.

Kuhn, A., and Kiefer, D. (2017) Membrane protein insertase YidC in bacteria and archaea. *Mol Microbiol.* 103:590–594.

Bogdanov, M., Dowhan, W., and Vitrac, H. (2014) Lipids and topological rules governing membrane protein assembly. *Biochim Biophys Acta.* 1843:1475–1488.

Noinaj, N., Gumbart, J.C., and Buchanan, S.K. (2017) The β-barrel assembly machinery in motion. *Nat Rev Microbiol.* 15:197–204.

Hennerdal, A., Falk, J., Lindahl, E., and Elofsson, A. (2010) Internal duplications in α-helical membrane protein topologies are common but the non-duplicated forms are rare. *Prot. Sci.* 19:2305–2318.

Iacovache, I , Bischofberger, M., and van der Goot, F.G. (2010) Structure and assembly of pore-forming proteins. *Curr. Opin. Struct. Biol.* 20:241–246.

Lam, V.H., Lee, J.H., Silverio, A., Chan, H., Gomolplitinant, K.M., Povolotsky, T.L., Orlova, E., Sun, E.I., Welliver, C.H., and Saier, M.H. Jr. (2011) Pathways of transport protein evolution: recent advances. *Biol. Chem.* 392:5–12.

Paulick, M.G., and Bertozzi, C.R. (2008) The glycosylphosphatidylinositol anchor: A complex membrane-anchoring structure for proteins. *Biochemistry* 47:6991–7000.

Shao, S., and Hegde, R.S. (2011) Membrane protein insertion at the endoplasmic reticulum. *Annu Rev Cell Dev Biol.* 27:25–56.

von Heijne, G. (2006) Membrane-protein topology. *Nature Rev. Mol. Cell. Biol.* 7:909–918.

KEY LITERATURE

Bogdanov, M., Heacock, P.N., and Dowhan, W. (2002) A polytopic membrane protein displays a reversible topology dependent on membrane lipid composition. *EMBO J.* 21:2107-2116.

Buck, T.M., Wagner, J., Grund, S., and Skach, W.R. (2007) A novel tripartite motif involved in aquaporin topogenesis, monomer folding and tetramerization. *Nat Struct Mol Biol.* 14:762-769.

Carvalho, P., Goder, V., and Rapoport, T.A. (2006) Distinct ubiquitin-ligase complexes define convergent pathways for the degradation of ER proteins. *Cell* 126:361-373.

Hessa, T., Meindl-Beinker, N.M., Bernsel, A., Kim, H., Sato, Y., Lerch-Bader, M., Nilsson, I.M., White, S.H., and von Heijne, G. (2007) Molecular code for transmembrane-helix recognition by the Sec61 translocon. *Nature* 450:1026-1030.

Liu, Y., Gerstein, M., and Engelman, D.M. (2004) Transmembrane protein domains rarely use covalent domain recombination as an evolutionary mechanism. *Proc. Natl. Acad. Sci. USA* 101:3495-3497.

McDowell, M.A., Heimes, M., Fiorentino, F., Mehmood, S., Farkas, A., Coy-Vergara, J., Wu, D., Bolla, J.R., Schmid, V., Heinze, R., Wild, K., Fleming, D., Pfeffer, S., Schwappach, B., Robinson, C.V., and Sinning, I. (2020) Structural basis of tail-anchored membrane protein biogenesis by the GET Insertase complex. *Mol. Cell* 80:72-86.

Müsch, A., Wiedmann, M., and Rapoport, T.A. (1992) Yeast Sec proteins interact with polypeptides traversing the endoplasmic reticulum membrane. *Cell* 69:343-352.

Rapp, M., Seppälä, S., Granseth, E., and von Heijne, G. (2006) Identification and evolution of dual topology membrane proteins. *Nature Struct.Mol.Biol.* 13:112-116.

Schoebel, S., Mi, W., Stein, A., Ovchinnikov, S., Pavlovicz, R., DiMaio, F., Baker, D., Chambers, M.G., Su, H., Li, D., Rapoport, T.A., and Liao, M. (2017) Cryo-EM structure of the protein-conducting ERAD channel Hrd1 in complex with Hrd3. *Nature* 548:352-355.

Tomasek, D., Rawson, S., Lee, J., Wzorek, J.S., Harrison, S.C., Li, Z., and Kahne, D. (2020) Structure of a nascent membrane protein as it folds on the BAM complex. *Nature* 583:473-478.

Stefanovic, S., and Hegde, R.S. (2007) Identification of a targeting factor for post-translational membrane protein insertion into the ER. *Cell* 128:1147-1159.

van den Berg, B., Clemons, W.M., Collinson, I., Modis, Y., Hartmann, E., Harrison, S.C., and Rapoport, T.A. (2004) X-ray structure of a protein-conducting channel. *Nature* 427:36-44.

von Heijne, G. (1986) The distribution of positively charged residues in bacterial inner membrane proteins correlates with the trans-membrane topology. *EMBO J.* 5:3021-3027.

EXERCISES

1. Most β-barrel proteins in bacteria have an even number of strands.

   a. What might this fact suggest about the insertion pathway? (Hint: consider α-hemolysin.)

   b. Given your suggestion, what would you predict for the distribution of the N- and C- termini at the inner or outer surfaces of the outer membrane?

2. As noted, cytoplasmic proteins are frequently evolved to gain new functions using covalently connected domain recombination, but transmembrane regions much less so. Rather, membrane proteins tend to use oligomerization to create new and complex structures.

   a. Propose an explanation for this distinction, invoking geometric and diffusional considerations. (Hint: consider diffusion and collision of molecules in the two cases.)

   b. Active sites in soluble enzymes are frequently formed from clefts at domain boundaries, so new functions may evolve by mixing domains. How might this consideration bear on the formation of new membrane protein functions? (Hint: consider where the functional sites on membrane proteins are located.)

   c. Soluble domains are frequently attached covalently to transmembrane domains. Would you expect these soluble domains to use the soluble domain rules (frequent recombination) or the transmembrane domain rules (infrequent recombination)? Why?

3. *E. coli* can replicate itself in ~0.5 h under favorable conditions. The bacterium is typically a rod about 0.5 micrometers in diameter and about 2 micrometers long. Assuming that the average membrane protein is 500 kDa in size and that the membrane is 50% protein by area, calculate:

   a. The total area of the membrane in nm^2.

   b. The number of proteins that can fit in it. Make an assumption, for example that a protein is a cylinder of a certain height, and calculate volume as $v = v_0 M/N$ = (partial specific volume) (molecular weight)/ (Avogadro's Number). A good average for v_0 is $0.73 \ cm^3 \ g^{-1}$.

   c. The number of active translocons needed to insert these proteins into the membrane in half an hour (assume a translation rate of ten amino acids per second).

   d. Compare your answer with the number of ribosomes in a rapidly growing *E. coli* cell.

4. The translocon is thought to open via a lateral gate to allow transmembrane helices to move from its channel to enter the lipid bilayer. Propose a plausible idea for why the lipid molecules do not appear to move through the gate into the channel and possibly block it. Note that there is not a clear answer at present, and several mechanisms might work.

5. The hydrophobic moment of a helix is a way to represent the sidedness of its hydrophilic/hydrophobic amino acids, that is, whether polar amino acids are evenly positioned around the helix axis or mostly on one side. Why might this parameter appear in the equation for the free energy of insertion of helices from the translocon into the endoplasmic reticulum membrane?

6. It is observed that the membrane-insertion propensity of a polar or charged amino acid is much less unfavorable if it is located near a helix end than in the middle section of the helix. Review the water profile across a lipid bilayer in Chapter 2 to understand a factor that may help to explain this observation.

7. For years, it was assumed that a protein could not reorient across a membrane because of the energy barrier to moving its polar regions across a bilayer. The argument is similar to the argument about lipid flip-flop (see Chapter 2). Yet, flip-flop happens. What is the key to thinking about protein flip-flop? (Hint: consider kinetics vs. equilibrium thermodynamics.)

7

How Proteins Shape Membranes

The phospholipids of cell membranes spontaneously form liposomes and vesicles when dispersed in water (Chapter 2). A suspension of vesicles compartmentalizes the aqueous phase into inside and outside phases largely isolated from one another by the bilayer barrier. This bilayer-based compartmentalization is a basic aspect of living systems (Figure 1.4). Just as for vesicles, there is an inside and an outside. Of course, unlike vesicles, cells must grow and divide, which means that organelle and plasma membranes must grow and divide as well, under metabolic control. How can the cell implement this seemingly simple division into two equal daughter cells? As we will learn in this chapter, the division involves the coordinated actions of numerous proteins that reshape a cell's membranes. Cell division is only one aspect of membrane shaping. Membranes are reshaped throughout a cell in the course of its biological activities. Membrane vesicles are formed by budding events, trafficked to specific locations within the cell, and there reabsorbed by membrane fusion events. These events are managed by specific membrane-shaping proteins that we discuss in this chapter.

Electron micrographs (EM) of eukaryotic cells reveal a dizzying array of membranes (**Figure 7.1A**). The plasma membrane, besides forming a barrier between the cytoplasm and the cell exterior, encloses other membranous compartments that define organelles such as mitochondria (Figure 1.4). The specialized activities of these compartments themselves depend upon sub-compartmentalization as well as upon their organelle-specific membrane proteins. As we mentioned in Chapter 1, most if not all of the compartments are connected biosynthetically (**Figure 7.1B**). How are the structures and compositions of the vesicle-like compartments created, and how are they related? Some kind of communication between the compartments is necessary to ensure orderly maintenance of cellular function and metabolism. What are the mechanisms of communication? These questions provide the themes for this chapter, which describes how lipid–protein and protein–protein interactions shape membranes and consequently, cells.

Because, as we learned in Chapter 5, protein trafficking and signaling in cells are devoted to delivering proteins to the various cell compartments, we focus on the proteins and processes that maintain the organelles of eukaryotic cells. Protein and lipid trafficking are inherent in organelle maintenance. Although lipids can be trafficked by specialized transfer proteins, trafficking of proteins and lipids occurs mainly through the movement of membrane vesicles and their contents. Cartoons are often used to show the relationships and trafficking of the vesicles (**Figure 7.1B**), but cellular life is far more complicated—and interesting—than can possibly be captured by a cartoon. The directional movements of organelle compartments are shown in the figure by the arrows, but the cause of the movements is not. To understand the trafficking of vesicles, we must answer a number of questions. What distinguishes the membrane

DOI: 10.1201/9780429341328-8

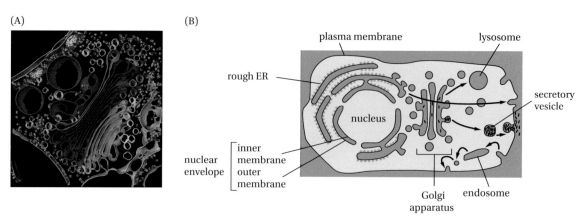

Figure 7.1 Summary of membrane systems in eukaryotic cells. (A) Slice from a cryo-tomogram of a *Chlamydomonas* cell. Different membrane systems are highlighted: ER (yellow), *cis*-Golgi (green), *medial*-Golgi (magenta), *trans*-Golgi (blue), and *trans*-Golgi network (dark blue). Transport vesicles budding from the respective membranes are also indicated. (B) Biosynthetically related eukaryotic organelle compartments. An important feature of cells is the topological relationships of the organelles. If one considers a secretory vesicle, for example, if it fuses with the plasma membrane, its interior becomes one with the exterior of the cell. The interior of the secretory vesicle is thus topologically equivalent to the cell exterior. In this figure, the spaces shown in red are all topologically equivalent. Green dots on the rough ER signify ribosomes. (A, From Bykov YS [2017] *eLife* 6: E32493. B, From Alberts B, Johnson A, Lewis JH et al. [2007] *Molecular Biology of the Cell*, 5th Edition. Garland Science, New York. With permission from W. W. Norton.)

vesicles familiar to the physical chemist from the vesicles involved in organelle trafficking? Are all vesicles equivalent, or do their compositions vary as they move through cells? How can we isolate and study these vesicles? How are the vesicles created? How are they moved, and what determines their direction of movement? We will address these questions and learn that lipid–protein interactions (Chapter 3) underlie virtually all of the intracellular vesicle-trafficking processes.

7.1 THE CYTOSKELETON PROVIDES A FRAMEWORK FOR CELL SHAPE AND VESICLE TRANSPORT

Most cells, across all kingdoms of life, have some kind of supporting framework for controlling cell shape and integrity. In *E. coli* and other Gram-negative bacteria, for example, the peptidoglycan layer serves this purpose by protecting the cells from lysis (Chapter 1). In plants, a tough cellulose cell wall serves a similar role. Animals and other "soft" multicellular organisms, which lack these tough outer surfaces, have evolved a different approach to controlling shape and integrity: a cytoplasmic skeleton or "cytoskeleton." The need for a cytoskeleton is perhaps most obvious in red blood cells (RBCs) from their biconcave shape (**Figure 7.2A**) and their ability to deform reversibly as they pass through capillaries that are smaller than the RBC diameter (**Figure 7.2B**). Cytoskeletons similar to those of the RBC play important roles as scaffolds and in the assembly of membrane domains in all eukaryotic cells. But here, we are most interested in how this elaborate network of spectrin polymers, actin filaments, microtubules, and intermediate filaments controls animal cell motility, shape, and particularly, intracellular vesicular transport.

7.1.1 How the Red Blood Cell Gets Its Shape

RBCs are about 10 μm in diameter, yet they can readily pass through capillaries that are only 4 μm in diameter (**Figure 7.2B**). RBCs can undergo the remarkable

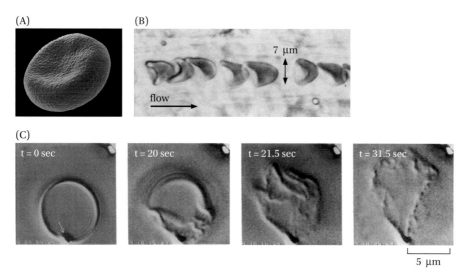

Figure 7.2 Red blood cells (RBCs) are extremely flexible and deformable. (A) Scanning EM image of a human RBC of diameter ≈10 μm. Red color added during image processing. Permission from Shutterstock / royaltystockphoto.com. (B) *In vivo* micrograph of RBCs passing through a 7 μm diameter capillary. Arrow indicates direction of blood flow. Note the parachute-like shape. From Skalak R, Branemark PI [1969] *Science* 164: 717–719. With permission from The American Association for the Advancement of Science. (C) The progressive change in appearance of an RBC as the membrane lipids are extracted with Triton X-100 detergent, applied at *t* = 0. The last panel reveals that RBCs have a cytoskeleton. Without this skeleton, RBCs would be spherical and rapidly destroyed during passage through capillaries. (A, Permission from Shutterstock/ royaltystockphoto.com. B, From Skalak R, Branemark PI [1969] Science 164: 717–719. With permission from The American Association for the Advancement of Science. C, From Svoboda K, Schmidt CF, Branton D et al. [1992] *Biophys J* 63: 784–793. With permission from Elsevier.)

distortions seen during passage through narrow capillaries because of their cytoskeletons. A balance of interactions between the cytoskeleton and the fluid lipid membrane results in the characteristic biconcave shape of RBCs. The existence of the RBC cytoskeleton is apparent from the discoidal remnants of RBCs whose lipids have been extracted with detergents (**Figure 7.2C**). Cryo-electron tomography reveals that the skeleton is a net-like structure on the cytoplasmic surfaces of RBCs (**Figure 7.3**). The net consists of chain-like spectrin polymers connected to junctional complexes attached to transmembrane proteins. Through extensive biochemical studies, in which gel electrophoresis played a critical role, we now know much about the proteins that form the cytoskeleton. The complexity of the protein composition of the skeleton was apparent from the very first sodium dodecyl sulfate (SDS)-gel electrophoretograms (Figure 1.16).

Many years of biochemical and biophysical studies of the protein network have yielded a reasonably complete understanding of the composition and organization of the cytoskeleton and its connections to RBC membrane proteins. These connections maintain a uniform distribution of transporter proteins over the RBC surface. The importance of the cytoskeleton net is dramatically apparent in RBCs with mutant spectrins or ankyrins that prevent the formation of the cytoskeleton net, causing a disease called hereditary spherocytosis because of the resulting spherical shape of the RBC. The lack of the net causes the mutant RBCs to be so fragile that they cannot survive capillary passage, leading to hemolytic anemia.

7.1.2 Cytoskeletal Filaments Organize Cells Spatially and Determine Mechanical Properties

The cytoskeleton of animal cells has three kinds of filaments. The intracellular distribution of two of these, actin filaments and microtubules, is revealed in **Figure 7.4** by means of fluorescently labeled antibodies. Notice that the actin filaments (red) are arranged concentrically around the cell nucleus, whereas

(A)

(B)

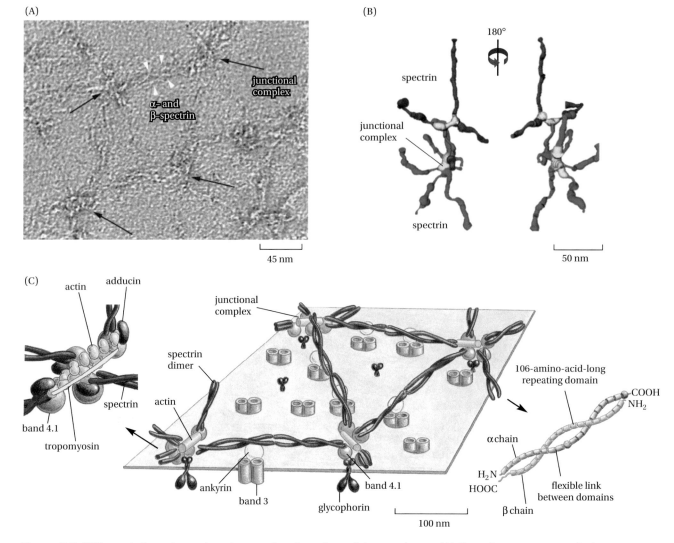

(C)

Figure 7.3 RBC cytoskeleton located on the cytoplasmic surface of the membrane. (A) Cryo-electron tomography image of lipid-extracted RBCs reveals the net-like organization of the cytoskeleton. (B) 3-D reconstructions of the junctional complex "knots" that link together double strands of α- and β-spectrins that form the "threads" of the net. (A, B) From Nans A, Mohandas N, Stokes DL [2011] *Biophys J* 101: 2341–3250. With permission from Elsevier. (C) Schematic of the cytoskeleton net showing that the junctional complexes bind to numerous membrane proteins via ankyrin proteins. (From Alberts B, Johnson A, Lewis JH et al. [2007] *Molecular Biology of the Cell*, 5th Edition. Garland Science, New York. With permission from W. W. Norton.)

the microtubules (green) are arranged radially. The actin filaments determine cell shape and underlie whole-cell locomotion. The microtubules radiate outward from the nucleus, serving as tracks for the orderly movement of vesicles carried by molecular motors from endoplasmic reticulum (ER) to Golgi to plasma membrane. The radial microtubules thus determine the locations of organelles within cells (Figure 7.1). Not shown in the figure are the intermediate filaments that provide mechanical strength. As in RBCs, the plasma membrane in eukaryotic cells is attached to the underlying cytoskeleton and consequently, responds to dynamic changes of the cytoskeleton. Unlike in RBCs, the cytoskeleton not only serves to control the shape of the cell but can be rapidly disassembled and reassembled during the cell cycle and in response to various internal and external stimuli; this is further discussed in Chapter 15. The general properties of actin, microtubules, and intermediate filaments are summarized in **Box 7.1**.

BOX 7.1 CELLULAR FILAMENTS CONTROL CELL SHAPE AND MOTILITY

Actin filaments (also known as microfilaments) are helical polymers of the protein actin. They are flexible structures with a diameter of 8 nm that organize into a variety of linear bundles, two-dimensional networks, and three-dimensional gels. Although actin filaments are dispersed throughout the cell, they are most highly concentrated in the cortex, just beneath the plasma membrane.

Microtubules are long, hollow cylinders made of the protein tubulin. With an outer diameter of 25 nm, they are much more rigid than actin filaments. Microtubules are long and straight and frequently have one end attached to a microtubule-organizing center (MTOC) called a centrosome.

Intermediate filaments are ropelike fibers with a diameter of about 10 nm; they are made of intermediate filament proteins, which constitute a large and heterogeneous family. One type of intermediate filament forms a meshwork called the nuclear lamina just beneath the inner nuclear membrane. Some of the proteins are keratins, related to the structure of hair. Structural studies are emerging concerning these filaments. Other types extend across the cytoplasm, giving cells mechanical strength. In an epithelial tissue, they span the cytoplasm from one cell–cell junction to another, thereby strengthening the entire epithelium (**Figure 1**).

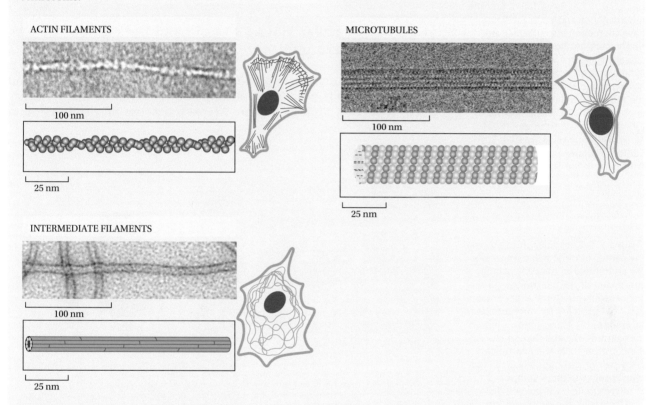

Figure 1 Filaments and microtubules within the cell. (From Alberts B, Johnson A, Lewis JH et al. [2007] *Molecular Biology of the Cell*, 5th Edition. Garland Science, New York. With permission from W. W. Norton.)

7.1.3 Microtubules Formed from α- and β-tubulin Are Polarized

Near the center of most eukaryotic cells—except for higher plants and fungi—are two orthogonally arranged cylindrical structures called, appropriately, centrioles (Figure 1.1, **Figure 7.5A,B**). They are surrounded by the pericentriolar material (PCM) that contains proteins, such as γ-tubulin, necessary for microtubule nucleation. The PCM and the centrioles together constitute the centrosome, which serves as the MTOC. The centrosome plays three key roles in the lives of cells: nucleation of microtubules for guiding motor proteins, management of the microtubule network used in cell division (**Figure 7.5C**), and the formation of cilia used for cell locomotion (**Figure 7.5D**).

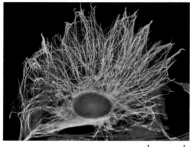

L————L
10 μm

Figure 7.4 The cytoskeleton of a eukaryotic cell. Microtubules are in green and actin filaments in red. The nucleus is in blue. The cell in this picture has been fixed, and the microtubules and actin filaments have then been visualized using fluorescent antibodies that bind specifically to each kind of structure. Notice that microtubules are generally arranged radially with respect to the nucleus, while actin is arranged circumferentially. (From Alberts B, Johnson A, Lewis JH et al. [2007] *Molecular Biology of the Cell*, 5th Edition. Garland Science, New York. With permission from W. W. Norton.)

Figure 7.5 Centrioles, located near the center of most eukaryotic cells, are embedded in the centrosome, composed of the centrioles and the pericentriolar material that contains essential proteins for microtubule formation. (A) Thin-section electron micrograph of a cell that shows the centrioles. (B) The centrioles are a pair of orthogonally arranged cylindrical structures composed of microtubules. During cell division (mitosis), the centrioles are reproduced to produce a pair of centrosomes that are subsequently responsible for separating mother and daughter chromosomes during division. (C) Metaphase of the cell-division cycle showing how the centrosomes and associated microtubules guide separation of the mother and daughter chromosomes. (D) Centrioles become basal bodies in the formation of cilia formed from microtubules. (A and C, From Alberts B, Johnson A, Lewis JH et al. [2007] *Molecular Biology of the Cell*, 5th Edition. Garland Science, New York. With permission from W. W. Norton. B, Nigg EA, Raff JW [2009] *Cell* 139: 663–678. With permission from Elsevier. D, From https://en.wikipedia.org/wiki/Cilium.)

Microtubules are formed by spontaneous, but regulated, assembly of heterodimers of α- and β-tubulins (**Figure 7.6**) into protofilaments that in turn assemble into microtubules containing 13 protofilaments. The heterodimers (panel A) always assemble into protofilaments in a head-to-tail fashion (panels B–D). Although α–β contacts are strongly favored in the protofilaments, in the microtubules, α–α and β–β contacts are preferred between protofilaments. The structural interactions amongst the tubulins cause the microtubules to have a spiral arrangement with α-tubulins at one end (defined as minus) and β-tubulins at the other (defined as plus). The microtubules are thus polarized, i.e., all the α–β dimers within a microtubule are aligned in the same direction. The nucleation sites for microtubule formation are so-called ring complexes—formed from γ-tubulin—located at the ill-defined surface of the centrosome (panel E). Because the ring complexes have a strong affinity for α-tubulin, the microtubules grow outward, forming a starburst cluster of polarized microtubules (panel F).

7.1.4 The Molecular Motors Kinesin and Dynein "Walk" along Microtubules to Transport Membrane Vesicles

Among the many marvels of cell life are the molecular motors kinesin and dynein that literally walk along microtubules powered by the hydrolysis of ATP (**Box 7.2**). Both motors are ATPases, because they split ATP into

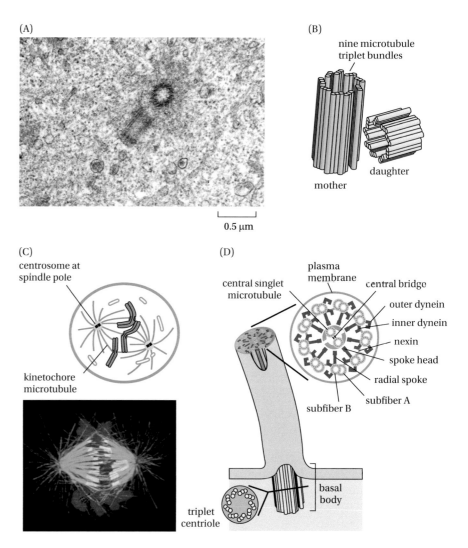

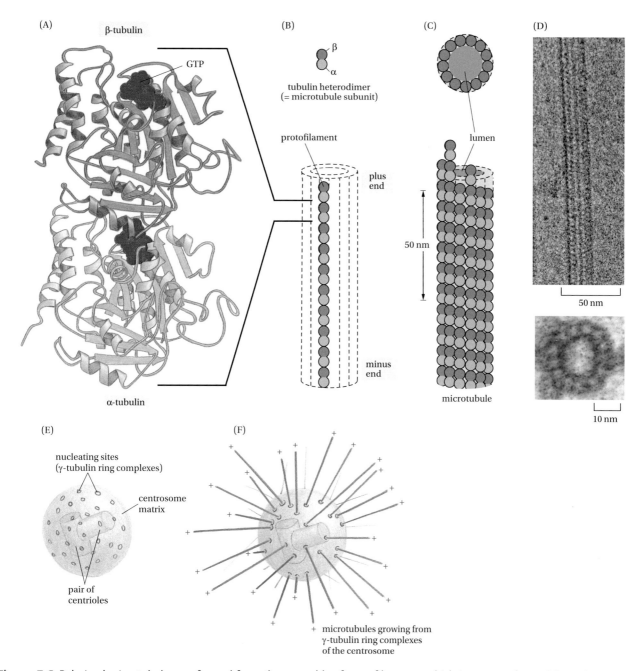

Figure 7.6 Polarized microtubules are formed from the assembly of protofilaments, which in turn are formed from the assembly of α-β tubulin heterodimers. (A) The structure of a heterodimer of α-β tubulin. (B) The heterodimers assemble into a protofilament. Because the dimers always form head-to-tail, the filaments are polarized with α-tubulin at the "minus" end and β-tubulin at the "plus" end. (C) 13 protofilaments assemble to from a microtubule. (D) Electron micrograph of the hollow microtubule. (E) Centrosome complex showing microtubule nucleation sites consisting of γ-tubulin ring complexes. (F) Sketch showing microtubules emanating from the centrosome matrix. Because microtubules are polarized, the positive poles are located at the microtubule tips distal to the matrix. (From Alberts B, Johnson A, Lewis JH et al. [2007] *Molecular Biology of the Cell*, 5th Edition. Garland Science, New York. With permission from W. W. Norton.)

ADP and phosphate to produce propelling forces, but they have different evolutionary origins. Kinesins are members of the myosin ATPase family (familiar from skeletal-muscle contraction), whereas dyneins belong to the AAA ATPase family (AAA means ATPases Associated with diverse cellular Activities) that typically assemble into ring-like hexameric structures to carry out a diverse number of motor-like activities. Both types of ATPases are

BOX 7.2   DYNEIN AND KINESIN MOTORS

Eukaryote cells must move vesicular cargos to and from the plasma membrane. The vesicular trafficking is enhanced by movement of cargo-bearing motor proteins along microtubules, as discussed in the main text. There are two evolutionarily unrelated motor proteins, kinesins and dyneins (**Figure 1A**), that transport cargo toward the positive (+) or negative (−) end of microtubules, respectively. Dyneins are related to hexameric AAA ATPases that serve many functions throughout the cell, whereas kinesins are closely related to muscle myosin (B). Modern microscopic methods and fluorescent molecules attached to dyneins and kinesins allow us to observe movements along microtubules (C; courtesy of Jennifer L. Ross and Leslie Conway). Utilizing the energy stored in ATP, both types of motors quite literally walk along the tubules, as illustrated in (D) for kinesins. Both motors have very high molecular weights and are comprised of many different components that have evolved to transport specific cargos. (**Figure 1**)

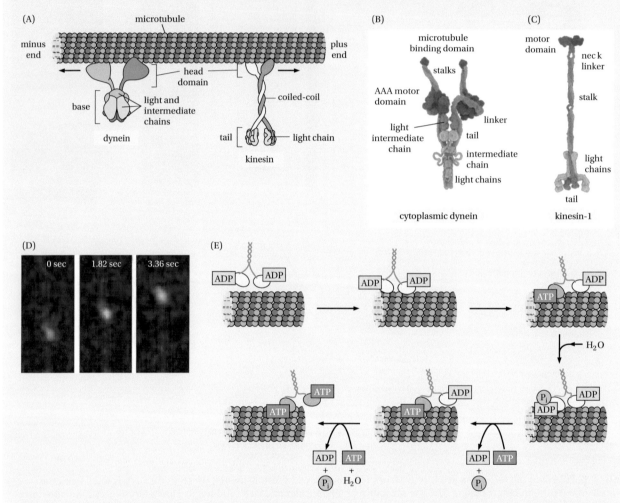

Figure 1 Tracking the movement of a fluorescently labeled motor protein. (A, From Cooper GM [2000] *The Cell: A Molecular Approach*. Sinauer Associates, New York. With permission from Oxford University Press. B and C, From Carter AP [2013] *J Cell Sci* 126: 705–713. With permission from Taylor & Francis. (D) Courtesy of Leslie Conway and Jennifer L. Ross.)

able to serve motor functions because of the large conformational changes driven by ATP. These conformational changes power the motor movements and can produce forces on the order of 5 pN (measured by optical tweezers; see later).

The primary difference between the movements of these two structurally different motors is that kinesins generally move their cargos along the protofilaments toward the plus end of the microtubules, whereas the dyneins always move cargos toward the minus end. These opposite movements underlie vesicle trafficking in cells. Broadly, there are two types of dyneins: cytoplasmic dyneins that carry a range of cargos toward the cell nucleus and specialized dyneins that transport materials along nerve axons between the cell body and synapses. The cell bodies are located in the central nervous system, while the axons project to, e.g., peripheral muscle cells, meaning that vesicles must be transported by dynein and kinesin motors many centimeters between their peripheral and central destinations.

Dyneins are large protein complexes containing heavy chains that include the motor domain and various accessory chains that apparently adapt the motor to different cargos and play various regulatory roles, such as suppression of motility and strength of binding to microtubules. The kinesins, on the other hand, are more specialized and comprise an extremely large superfamily containing 15 families. Mice, for example, have 45 kinesin genes. The families are broadly grouped into three types depending on the location of the motor domain, which can be at the N-terminus, the C-terminus, or in the middle of the chain.

The roles of kinesin and dyneins can be seen, for example, in the vesicular transport of calumenins, which are low-affinity calcium binding proteins secreted into the extracellular space by cells and may play a role in signaling. Like many other secreted proteins, calumenin is encapsulated in vesicles and transported along the secretory pathway to the plasma membrane, where, through membrane fusion, the vesicles empty their contents into the extracellular space. The secretion process can be followed using a variety of fluorescent probes and time-lapse photography (**Figure 7.7**). Two motors participate in the process. Cytoplasmic dynein transports the vesicles from the ER to the Golgi apparatus, while a member of the kinesin family transports the vesicles from the Golgi to the plasma membrane. In the case of calumenin, smaller vesicles merge into larger ones prior to fusing with the plasma membrane to discharge their contents into the extracellular space.

7.2 MANY DIFFERENT PROTEINS ACT TOGETHER TO CREATE VESICULAR COMPARTMENTS IN CELLS

Subcellular compartments grow and divide, and transport material from one compartment to another by pinching off and fusing with small membrane vesicles. In addition to *de novo* synthesis of lipids, proteins, and DNA, these events require molecular processes that can drive membrane fission and fusion processes in a controlled manner. In many instances, membranes must be locally bent or curved into tubular or vesicular structures. This is done by proteins that bind at surfaces of lipid bilayers to encapsulate vesicles or to reshape vesicles by forcing the bilayers to distort into non-spherical shapes, such as tubes. The principal encapsulating/reshaping proteins that will concern us here are summarized in **Figure 7.8**.

The main currency of cell shaping is vesicles, which transport lipids and membrane proteins in their lipid bilayer and soluble proteins and other materials within their aqueous compartment. Most vesicles are encapsulated in protein coats that prevent uncontrolled release of their contents. Delivery of vesicles by motor proteins ensures orderly vesicle traffic flow, but that is not sufficient to assure precise targeting. The different types of protein-coated vesicles also include proteins that provide for orderly engagement of vesicles with target membranes and control of fusion and fission events at the target site. Hundreds of proteins are devoted to targeting. Here, we are concerned primarily with the proteins that shape vesicles directly.

COPI and COPII are **co**at **p**rotein (COP) complexes that coat vesicles responsible for transporting proteins between the ER and the Golgi apparatus. COPII

Figure 7.7 The protein calumenin, a low-affinity calcium binding protein, is moved along the cellular secretory pathway after synthesis and encapsulation in the endoplasmic reticulum (ER). Two isoforms, 1 and 2, are synthesized and are symbolized in the figure as Calu-1/2. (A) Micrographs of HeLa cell transport vesicles containing Calu-1/2. The protein has been labeled with green fluorescent protein to reveal the vesicles within the ER and at the cell boundary. An α-tubulin antibody carrying a red fluorophore reveals the microtubules. (B) The progress of a single vesicle on its way to the plasma membrane is revealed by time-lapse photography. (C) Schematic representation of the secretion of Calu-1/2. Dynein carries the vesicles along microtubules from the ER to the Golgi. From the Golgi to the plasma membrane, the vesicles are transported by Kif5b kinesins. (D) Smaller vesicles merge into larger ones prior to fusion with plasma membrane, leading to release of Calu-1/2 into the extracellular space. (From Wang Q, Feng H, Zheng P et al. [2012] *PlosOne* 7: e34344.)

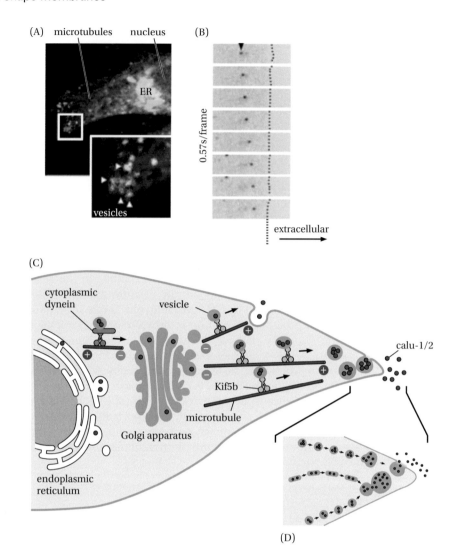

vesicles move proteins from the ER to the Golgi complex (anterograde transport), whereas COPI vesicles transport proteins from the Golgi back to the ER (retrograde transport). Clathrin is the major protein responsible for the formation of coated vesicles for endocytosis and intracellular transport. Another protein implicated in endocytosis is caveolin, which forms small invaginations (caveolae; Latin for "little caves") in the plasma membrane of many cells. Caveolae can progress to full-fledged vesicles for endocytosis, but they are the least understood of the coat systems.

The actions of COPI, COPII, clathrin, and caveolin all produce vesicular structures. The general scheme for the first three systems involves coat formation, pinching off the coated vesicles from the parent membrane, intracellular transport, uncoating, and fusion with the target membrane. Pinching off of vesicles, often referred to as fission or vesicle scission (like scissors; from the Latin word *scindere*, meaning to cut or cleave) is invariably the responsibility of dynamins, which are large GTPases. Although caveolae also must undergo some kind of bud/pinch-off/un-coat processes, the exact mechanisms are presently unclear. Regardless of the coat type, all of the uncoated vesicles fuse with the target membrane by means of SNARE proteins.

In the course of their lives, cells also require curved non-vesicular, tubule-like structures. A very large family of relatively rigid banana-shaped scaffold proteins known as BAR domains do this job. In the ER, this job falls primarily to reticulons with the aid of an additional protein family called DP1 (mammals) or Yop1p (yeast). The tubular structures formed by reticulons and DP1/Yop1p are

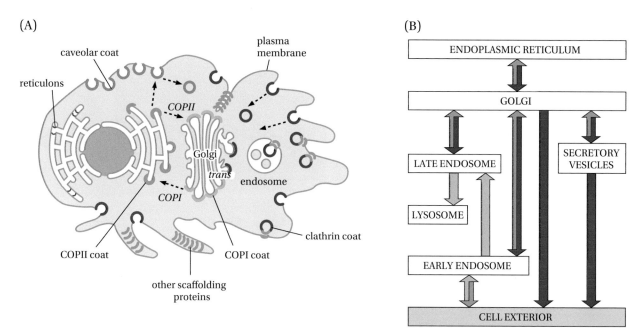

Figure 7.8 Cells contain an extensive vesicular network. Major proteins that shape membranes of the network and their sites of action are summarized in panel (A). The dashed arrows indicate the general directions of trafficking with cells. The general traffic pattern for the movement of vesicular compartments is indicated in panel (B). Red arrows indicate outward movements from the ER and the Golgi network, while blue arrows show retrograde movements back toward the ER. Green arrows indicate the endocytic pathway in which molecules are ingested into plasma membrane-derived vesicles. Secretory vesicles are confined largely to specialized cells that secrete hormones, neurotransmitters, or digestive enzymes. (From Alberts B, Johnson A, Lewis JH et al. [2007] *Molecular Biology of the Cell*, 5th Edition. Garland Science, New York. With permission from W. W. Norton.)

caused to fuse into a network by the atlastins (ATLs), which are GTPases that belong to the dynamin family.

7.2.1 Clathrins Form Membrane Vesicles from the Plasma Membrane for Endocytosis

The complexity of the formation of cellular vesicles was apparent in EM of yolk-protein uptake by mosquito oocytes made in the early 1960s by Keith R. Porter (1912–1997) and Thomas F. Roth (**Figure 7.9A**). The vesicles were clearly not simple lipid vesicles of the sort we can make from phospholipids. Rather, they bristled with electron-dense material covering the vesicles inside and out. Barbara M.F. Pearse, working with isolated vesicles (**Figure 7.9B**), discovered in 1975 that the endocytic vesicles were covered with a spherical coat of hexagonally arranged proteins that she named clathrin, derived from the Greek word for lattice (**Figure 7.9C**). Because of the cage-like clathrin coat, the vesicles have been referred to as a "vesicle in a basket." The clathrin coats are built from triskelions (from Greek, meaning "three-legged") composed of three heavy chains and three light chains (**Figure 7.10A**). The triskelions are assembled into cages of different sizes by forming pentagonal and hexagonal patterns (**Figure 7.10B and C**). Clathrin-coated vesicles can vary widely in diameter, typically between 60 and 200 nm, because the triskelions are physically flexible enough to form cages with vertex angles of 60° (hexagons) or 72° (pentagons). How clever early humans—and much later, R. Buckminster Fuller (1895–1983)—must have felt designing triskelions (**Figure 7.10D**) and geodesic domes, but nature was there first!

The assembly and disassembly of clathrin coats do not occur by chance; rather, many different proteins participate in a multi-step process that begins with nucleation at a plasma membrane invagination (**Figure 7.11**). Within these so-called "pits" are nucleation domains comprised of several proteins that have a preference for the lipid phosphatidylinositol(4,5)P_2. Among them

Figure 7.9 The discovery of clathrin-coated vesicles. (A) Early electron micrographs showing the exocytosis of vesicles from a cell. Notice the complex of structured protein surrounding the vesicles. (B) Electron micrographs of vesicles isolated by Barbara M.F. Pearce. (C) The coat structure was surmised to be a hexagonal lattice formed from a protein that Barbara Pearce named "clathrin." (A, From Alberts B, Johnson A, Lewis JH et al. [2014] *Molecular Biology of the Cell*, 6th Edition. Garland Science, New York. With permission from W. W. Norton. B, C Pearce B [1976] *PNAS* 73: 1255–1259. With permission from Barbara Pearce.)

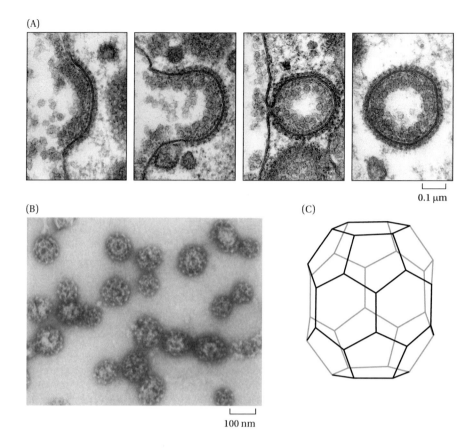

are F-BAR domains (see later) that underlie the gentle curvature of the pit. In the next step, the nucleation domain recruits adapter protein 2 (AP2) and a cargo-specific adapter protein. AP2 is a hetero-tetrameric complex, which is the most abundant component of clathrin-coated vesicles after clathrin itself. Naturally, cells have numerous cargo-specific adapters. The triskelia are then recruited from the cytoplasm to the adapter complex (step 3), where they polymerize to form the invaginating cage aided by BAR domains (see later). The crucial step 4 is GTP-driven vesicle scission by dynamin (see later) that releases the vesicle into the cytoplasm for motor transport to its destination. There, the vesicle is uncoated with the help of the ATPase heat-shock-related protein Hsc70 and its cofactor auxilin. Hsc70 begins its work by binding to the terminal domains of the triskelia (Figure 7.10A).

7.2.2 Caveolins also Carry Out Endocytosis but May Be More Important as Membrane-Status Sensors

Caveolae are apparent in EM of cells, especially in endothelial cells and adipocytes (**Figure 7.12A**). They account for as much as 50% of the cell surface in some cells. Unlike clathrin-coated vesicles, however, their shapes are variable—ranging from open to crater-like—and significantly, no heavy electron-dense coat is apparent. The only thing we know with certainty is that at least two classes of proteins are required for caveolae formation: caveolins (CAVs) and cavins. Caveolae apparently originate in the trans-Golgi network and are subsequently trafficked between the plasma membrane and endosomes (**Figure 7.12B**). Vesicles formed in the presence of cavins form complexes with the caveolins, especially when cholesterol is present. Vesicles are not formed in the absence of cavins. Caveolins are integral membrane proteins lacking typical transmembrane segments. Rather, they are tethered to the lipid vesicle by a wedge-shaped

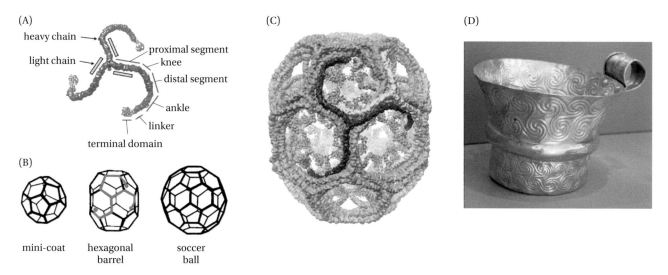

Figure 7.10 Triskelion protein complexes of clathrin assemble to form the clathrin coat. (A) Cryo-EM structure of a triskelion composed of three heavy chains and three light chains. (B) Triskelia can assemble into cages of different sizes by forming hexagonal and pentagonal patterns. (C) Structure of a clathrin cage determined using cryo-EM. (D) The triskelion pattern is a recurring artistic motif dating back to at least 4000 BCE. Here, the design is found on an ancient gold cup from Mycenae. (A, B, C, From Fotin A et al. [2004] *Nature* 432: 573–579. Used with permission from Springer Nature. D, From the National Archaeological Museum of Athens. Picture by Giovanni Dall'Orto, November 10 2009.)

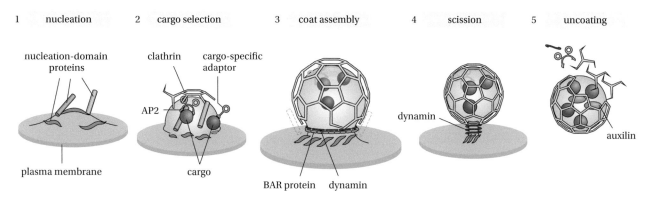

Figure 7.11 Five steps in the assembly and disassembly of clathrin coats. 1. Nucleation at a plasma membrane invagination where nucleation domains comprised of several proteins that have a preference for the lipid phosphatidylinositol(4,5)P$_2$ are found. Among the domains are BAR domains that create and define the gentle curvature of the invagination. 2. The adapter protein AP2 and a cargo-specific adapter protein are recruited to the nucleation site. 3. Triskelia are then recruited from the cytoplasm to the adapter complex, where they polymerize to form the invaginating cage, aided by BAR domain proteins. 4. GTP-driven scission by dynamin releases the coated vesicle into the cytoplasm for motor transport to its destination. 5. The vesicle is uncoated at its destination with the help of the ATPase heat-shock-related protein Hsc70 and its cofactor auxilin. (From McMahon HT et al. [2011] *Nature* 12: 517–533. With permission from Springer Nature.)

helical hairpin (**Figure 7.12C**). Nuclear magnetic resonance (NMR) and molecular dynamics simulations of caveolin are consistent with adhesion to bilayers via a helical hairpin. Crystallographic measurements have revealed that cavins form trimers via a coiled-coil arrangement and may wrap around vesicles in a belt-like fashion (**Figure 7.13**).

The binding of cavins to vesicles requires the presence of charged phosphatidylserine, phosphatidylinositol, and cholesterol located on the cytoplasmic surfaces of asymmetric bilayers (Chapter 2). Vesicle formation appears to be exquisitely sensitive to membrane tension and cholesterol, and it is believed that the primary roles of caveolae are to sense the status of the plasma membrane and to regulate the plasma membrane area and cholesterol content.

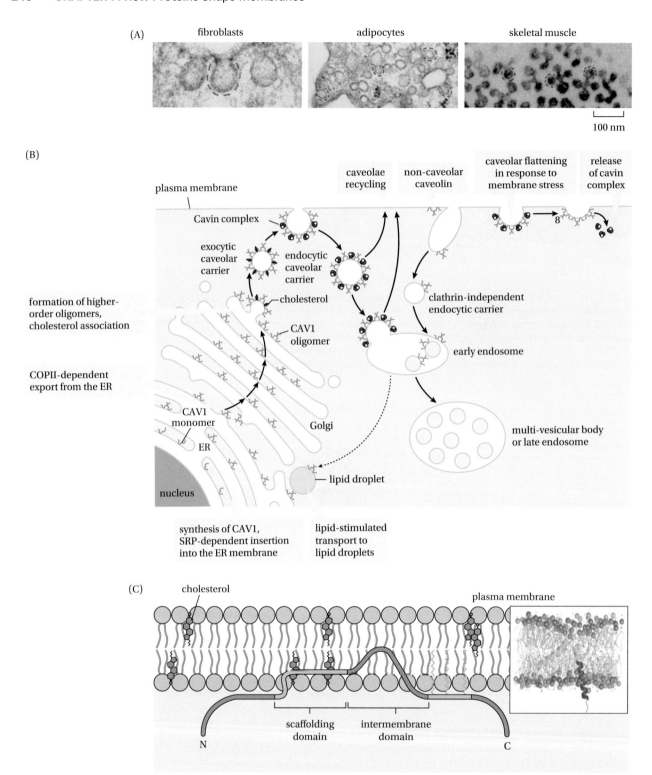

Figure 7.12 Overview of caveolae. (A) Electron micrographs of caveolae observed in several different cell types. (B) The ultimate origin of caveolae is the trans-Golgi network. Lipid vesicles are formed in the presence of cavins and cholesterol, which can cycle between the plasma membrane and early endosomes. They are apparently very sensitive to membrane stress and the concentration of cholesterol, suggesting that they play an important sensory role in plasma membrane homeostasis. (C) Although caveolin is an integral membrane protein, it does not have normal transmembrane segments. Rather, it attaches to membranes by a helical hairpin whose structure has been determined by NMR and molecular dynamics simulations (inset). (From Parton RG et al. [2013] *Nature* 14: 98–112. With permission from Springer Nature.)

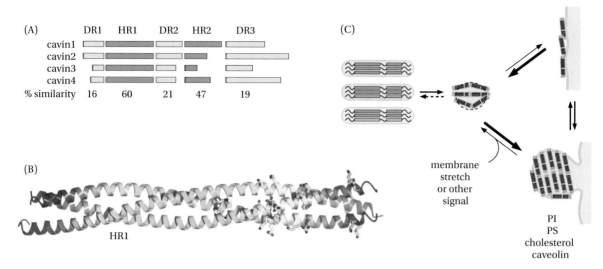

Figure 7.13 Sequence patterns and partial structures of cavins. (A) Sequence patterns of four cavins from various species reveal rigid elements separated by flexible links. The blue regions are helical (HR), and the yellow regions are disordered (DR) and hence flexible. (B) Crystallographic studies reveal that the HR domains form helical trimers (PDB: 4QKV). (C) Little is known about how these proteins form coats on vesicles. Some electron microscope studies reveal striations on caveolae, suggesting the arrangement shown in the lower right of the panel. Phosphatidylserine and phosphatidylinositol are required for vesicle binding (upper right), but cholesterol and caveolin are necessary for vesicle formation. (From Kovtun et al. [2014] *Dev Cell* 31:405–419.)

7.2.3 COPI and COPII Vesicles Maintain the Endoplasmic Reticulum and the Golgi Apparatus

The formation of COPI and COPII vesicles follows the general principles observed in the formation of clathrin-coated vesicles, except for the proteins involved. The formation and transport of these vesicles are crucial for moving proteins and lipids from their sites of synthesis in the ER and for passage through the Golgi apparatus, during which secreted proteins are covalently modified on their way to their final destinations. Another important role for COPI and COPII vesicular transport is to maintain the structure and integrity of the Golgi apparatus.

7.2.4 COPII Coats Are Adaptable to Cargos of Odd Shapes

COPII vesicles principally move cargos out of the ER to the Golgi network. They have inner and outer coats and typically have diameters of 60–80 nm, but the assembly mechanism is sufficiently adaptable that coats of many different sizes and shapes can be formed as required to accommodate the cargo. Five cytosolic proteins (Sar1, Sec23, Sec24, Sec13, and Sec31) are required for construction of a COPII vesicle. The inner coat is built from Sec23 and Sec24, which occur in tight heterodimers. Sec24 is believed to serve as the primary subunit responsible for binding membrane cargo proteins; other proteins are transported in the vesicle lumen. The outer coat is formed from Sec13 and Sec31, which form stable hetero-tetramers of two subunits each.

The structure of COPII vesicles has been determined from crystal structures of the individual proteins and cryo-EM reconstructions of assembled complexes (**Figure 7.14**). Assembly begins with the binding of Sar1 to the ER membrane via an amphipathic helix. Sar1 is a small GTPase that is in its GDP-bound form when in solution. In the first assembly step, the amphipathic helix becomes exposed when Sar1 exchanges GDP for GTP. GDP–GTP exchange on Sar1 is catalyzed by the Sec12 guanine nucleotide exchange factor (GEF). Because Sec12 is present only in the ER, the initiation of COPII formation can only occur there. Sar1 becomes the ER anchor for the recruitment of the Sec23/Sec24 heterodimer.

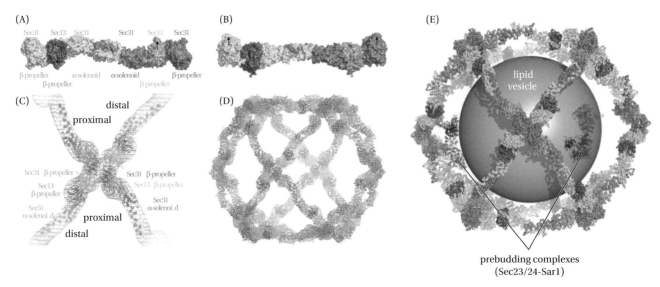

Figure 7.14 Structure of COPII vesicles determined by crystallography and cryo-EM. In contrast to the triskelia of clathrin, the basic assembly unit is tetrameric rather than trimeric. (A) The Sec13/Sec31 complex consists of a linear array of alternating Sec13 and Sec31 subunits. (B) Same as panel A but rotated about its long axis by 90°. (C) Molecular model of the hetero-tetrameric assembly unit that forms the lattice-work vertex of the outer coat. (D) The hetero-tetrameric assembly units assembled into an outer coat. (E) A model of the fully assembled COPII coat. Here, the inner coat components, consisting of complexes of Sec23, Sec 24, and Sar1, are included. A hypothetical lipid vesicle (grey) is shown. The model is built from 24 of the hetero-tetrameric assembly units. Cells, however, use larger or smaller numbers of assembly units, depending on the cargo. Even oblong loads can be accommodated by the formation of tube-like structures (see Figure 7.15). (From Fath S, Mancias JD, Goldberg XB [2007] *Cell* 129: 1325–1336. Used with permission of Elsevier.)

Coat assembly proceeds by recruitment of the inner coat Sec23/Sec24 complexes via a direct interaction between Sar1 and Sec23. The reaction proceeds rapidly, because Sec23 is a GTPase-activating protein (GAP) that increases Sar1 GTPase activity. Sar1 and the Sec23/Sec24 complex together form a pre-budding complex that initiates Sec13/Sec31 activity to build the outer coat. The "arms" of the outer coat have a tetrameric symmetry rather than the trimeric symmetry of the triskelia of clathrin (Figure 7.10). This symmetry of the COPII outer coat allows the accommodation of "odd sized" loads (**Figure 7.15**).

7.2.5 COPI-Coated Vesicles Are Constructed from Preassembled Coatamers

Directional vesicular transport requires proteins that reside in the donor compartment but are not needed in the recipient compartment after delivery. Thus, there is a need to recycle many membrane proteins between compartments. The function of COPI vesicles is the retrieval of escaped ER-resident proteins and proteins that have participated in COPII fusion events. This is possible because ER-resident membrane proteins contain signals that enable direct binding to COPI coats. The packaging of many soluble proteins, on the other hand, requires that they bind to specialized membrane-bound receptors. A good example is the KDEL receptor, which is a multi-span membrane protein that binds the C-terminal amino acid sequence KDEL on escaped ER-resident soluble proteins in the Golgi and brings them into COPI-coated vesicles for return to the ER.

The COPI coatamers form spontaneously from seven COP proteins, generally described as a 3-subunit B-subcomplex and 4-subunit F-complex. The complexes and their individual protein components have been studied extensively by cryo-EM tomography and crystallography, respectively. The structural arrangement of the complexes in the coatamer is triangular (**Figure 7.16A**), which allows them to be assembled in many different ways to coat vesicles of varying size (**Figure 7.16B**). COPI vesicles are typically 50 nm in diameter, ranging from 30 to 70 nm (**Figure 7.16C**).

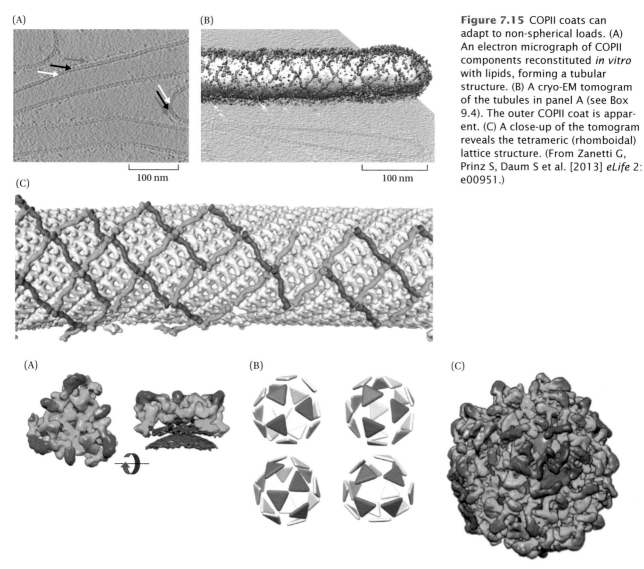

Figure 7.15 COPII coats can adapt to non-spherical loads. (A) An electron micrograph of COPII components reconstituted *in vitro* with lipids, forming a tubular structure. (B) A cryo-EM tomogram of the tubules in panel A (see Box 9.4). The outer COPII coat is apparent. (C) A close-up of the tomogram reveals the tetrameric (rhomboidal) lattice structure. (From Zanetti G, Prinz S, Daum S et al. [2013] *eLife* 2: e00951.)

Figure 7.16 COPI vesicle triad construction scheme. (A) Electron tomogram of the COPI triad, a complex of seven proteins. The complex is water-soluble and present in high concentrations in the cytoplasm. In this image, the bilayer is colored red and the protein components in shades of green and blue. (B) Schematic showing the versatility of arranging triangles (triads) to cover spherical surfaces. (C) Electron tomogram of a complete COPI vesicle. Here, non-triad accessory proteins are indicated by purple, red, and orange coloring. (A and C, From Dodonova SO, Diestelkoetter-Bachert P, von Appen A et al. [2015] *Science* 349: 195–198. With permission from The American Association for the Advancement of Science. B, From Faini M, Prinz S, Beck R et al. [2012] *Science* 336: 1451–1545. With permission from The American Association for the Advancement of Science.)

What causes the coatomers to assemble around Golgi membranes to form vesicles? The answer was discovered by James Rothman in 1991. Extensive biochemical studies of COPI vesicles revealed that three copies of ADP-ribosylation factor (ARF) protein were present in each coatomer. Finding that ARF binding was GTP-dependent, he proposed a simple scheme for the control of COPI vesicle formation (**Figure 7.17**). Subsequent work by many laboratories showed that GTPases are widely used by cells in the control of vesicle coating and uncoating.

7.3 SURFACE-BINDING PROTEINS MODULATE MEMBRANE CURVATURE AND CAUSE VESICLE SCISSION

As illustrated by the formation of COPI and COPII vesicles, surface-binding proteins bend membranes through the energetics of lipid–protein (Chapter 3) and

Figure 7.17 GTPases control vesicle coating and uncoating. The figure summarizes James Rothman's early hypothesis about how vesicle coating and uncoating is regulated in cells. (A) ARF proteins (green) in solution with bound GDP are not membrane-active. Displacement of GDP by GTP, catalyzed by a GDP–GTP exchange factor on the membrane (blue), converts ARF to a membrane-active state by, for example, causing a conformational change such as the exposure of a membrane-binding amphipathic helix. (B) Activated ARF, now bound to the membrane, nucleates coat formation. During uncoating, ARF hydrolyzes GTP to GDP + P_i, unbinds, and is released back into the cytoplasm, causing disassembly of the coat. (From Serafini T, Orci L, Amherdt M et al. [1991] Cell 67: 239-253. With permission from Elsevier.)

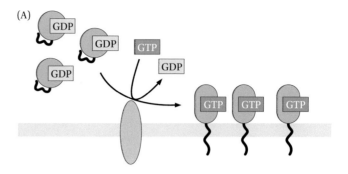

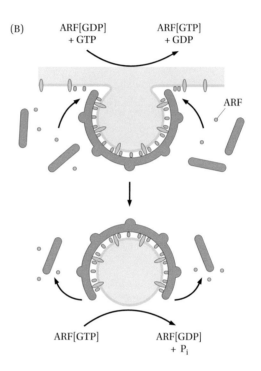

protein–protein interactions. Cells have a large armamentarium of proteins for shaping membranes into different curved structures necessary for life.

7.3.1 BAR Domains Shape Vesicle Curvature

BAR domains were discovered by Pietro De Camilli through studies of the interaction of amphiphysin with dynamin in nerve terminals. This was done using dual-immunofluorescence techniques, which revealed that amphiphysin and dynamin co-localized in the nerve terminals (**Figure 7.18A**). Furthermore, the two proteins co-migrated on non-denaturing SDS gels (**Figure 7.18B**), indicating direct association of the two proteins *in situ*.

Subsequent work led to the discovery of many different proteins containing BAR domains that are associated with the formation and fusion of coated vesicles. (BAR is a mercifully short abbreviation for the protein names of the founding members of the class: **B**in1, **A**mphiphysin, and **R**vs167.) Most BAR domains have a concave, positively charged surface (**Figure 7.18C**) that can stabilize curved membrane surfaces through electrostatic interactions with lipids (**Figure 7.19A**). The superfamily of BAR domains is exceedingly large, and crucially, BAR domain proteins provide a wide range of curvatures that allow them to form sharply or gently curved surfaces (**Box 7.3**).

In vitro reconstitution and electron tomography show the strong ability of BAR domains to reshape lipid membranes into tube-like structures (**Figure 7.19B**). The biophysics and physical chemistry of tube formation can be studied by

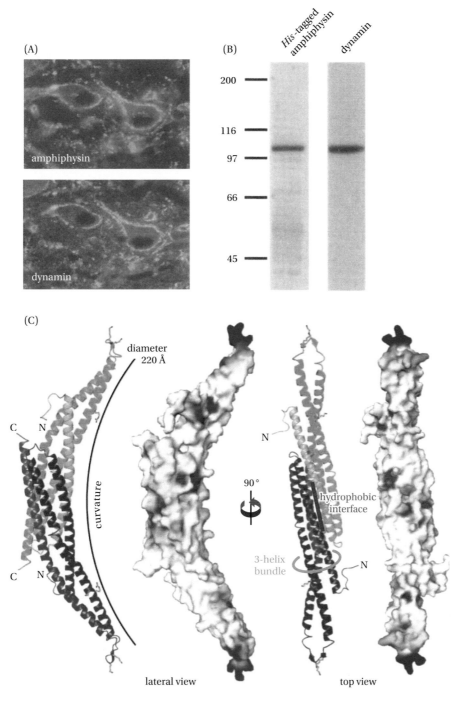

Figure 7.18 Discovery and structure of BAR domain proteins. (A) Frozen sections of rat cerebellum double-stained with fluo-rescently labeled antibodies to amphiphysin and to dynamin. The stains co-localized in the sections. (B) Non-denaturing SDS-gel electrophoresis revealed that amphiphysin and dynamin co-migrated on the gels, indicating that they formed a complex. (C) The crystallographic structure of the amphiphysin BAR domain from *Drosophila* (PDB: 1URU) shows that the three-helix bundle BAR domains dimerize at a hydrophobic interface to form a banana-shaped six-helix bundle with a concave membrane-binding surface. (A and B, From David C, McPherson PS, Mundigl O et al. [1996] *PNAS* 93: 331–335. With permission from DeCamilli.)

pulling BAR domain-stabilized tubules from lipid vesicles sucked partially into glass pipettes (**Figure 7.20A**). The tubules are pulled from the vesicle by a small bead attached to a micromanipulator. Protein- and lipid-specific fluorescent dyes reveal delightful images of the process (**Figure 7.20B**). The procedure allows many different BAR domain proteins to be studied (**Figure 7.20C**).

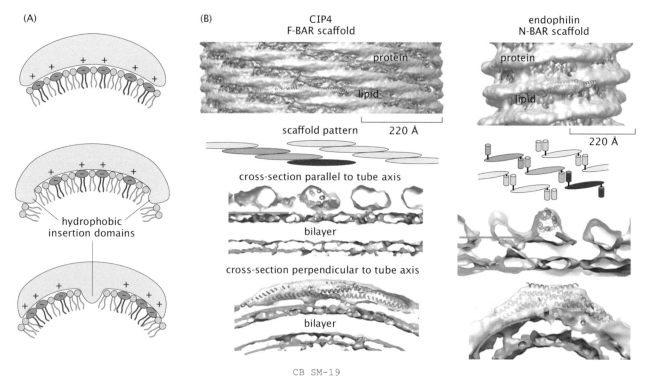

CB SM-19

Figure 7.19 BAR domains are versatile shapers of vesicles. (A) Banana-shaped BAR domains interact with lipid bilayers by means of electrostatic and hydrophobic interactions. Negatively charged lipids principally face the cytoplasmic surface of a plasma membrane bilayer. (B) Their relatively rigid structures can bend lipid vesicles into tubular structures, as illustrated by these electron tomograms of tubules formed *in vitro*. (From Mim C, Unger VM [2012] *Trends Biochem Sci* 37: 526–533. With permission from Elsevier.)

7.3.2 Dynamins Pinch Off Vesicles Shaped by BAR Domains

The favorable free energy of lipid–protein interactions allows BAR domains to reshape bilayers into flask shapes and tubules (Box 7.3), but such interactions cannot cause the orderly pinch-off phenomenon required for vesicle scission. Pinch-off requires work. From what we have learned so far about vesicle trafficking, the most likely source of energy is GTP hydrolysis catalyzed by a GTPase. The first insight into the scission problem came from the discovery of temperature-sensitive fruit fly mutants with a paralytic phenotype arising from the loss of synaptic transmission. It turned out that the mutant gene encoded dynamin, which was already known to be a GTPase that could associate with microtubules. EMs revealed tubules in synapses encircled by rings (**Figure 7.21A**). *In vitro* EM studies of dynamin interacting with lipid vesicles showed that dynamins induced tubular structures similarly encircled by rings that appeared to have a helical pattern (**Figure 7.21B**). This work, carried out by Sandra Schmid in 1995, gave a basic understanding of the vesicle-scission process (**Figure 7.21C**). Since then, a huge amount of work in many different laboratories has led to a good understanding of vesicle scission (**Figure 7.22**).

There is more to pinching off than just dynamin. Even though BAR domains cannot scissor the vesicles free, they are nevertheless an important part of the overall process. Using real-time internal reflection microscopy measurements in combination with specific fluorescence labeling of multiple BAR domain proteins thought to be involved in vesicle transport, fusion, and scission, one can follow the appearance and disappearance of clathrin-coated vesicles before, during, and after cleavage (**Figure 7.23**). There is a clear sequential binding and unbinding of at least four BAR domain proteins during the scission process.

BOX 7.3   BAR DOMAIN MENAGERIE

BAR domains can sculpt vesicles into novel shapes during, for example, endocytosis (**Figure 1**). This is possible due to the wide range of curvatures of the banana-shaped molecules (**Figure 2**).

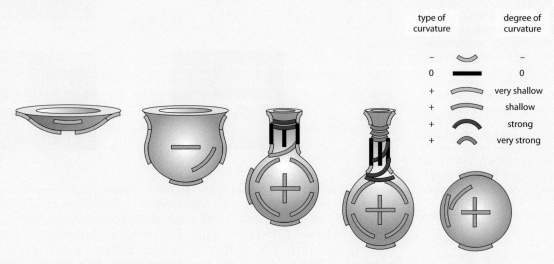

Figure 1 Example of how BAR domains sculpt vesicles. Because BAR domains have a wide range of curvatures, they are well adapted for subtly curving vesicle surfaces. (From Kessels MM, Koch D, Qualmann B [2011] *EMBO J* 30: 2501–2515. With permission from John Wiley and Sons.)

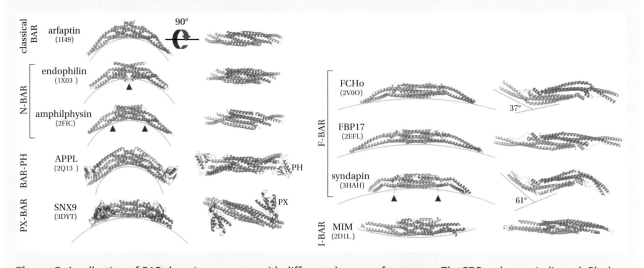

Figure 2 A collection of BAR domain structures with different degrees of curvature. The PDB codes are indicated. Black arrowheads indicate hydrophobic segments within modules that drive insertion into the membrane bilayer. (From Kessels MM, Koch D, Qualmann B [2011] *EMBO J* 30: 2501–2515. With permission from John Wiley and Sons.)

7.3.3 SNARE Complexes Control Fusion of Vesicles with Membranes

After the vesicle has formed, the coat proteins are stripped away from the surface. Another complex set of reactions then drives vesicle targeting to and fusion with the target membrane, so that the vesicle will fuse only with the correct target membrane. The donor and acceptor membranes contain two kinds of proteins: Rab proteins and SNARE proteins (**Figure 7.24**). The name SNARE is a shorthand for **S**oluble **N**SF **A**ttachment Protein **RE**ceptor, in which NSF is short for <u>N</u>-ethylmaleimide-<u>s</u>ensitive <u>f</u>actor.

Figure 7.20 Pulling BAR domain–stabilized tubules from lipid vesicles. (A) The experimental set-up involves stabilizing a lipid vesicle on a micropipette in a protein-containing solution and pulling the tubule out using a small adherent bead attached to a micromanipulator. The various physical parameters that can be varied to study the biophysics of tubule formation are indicated. (B) Lipid- and protein-specific fluorescent labels allow the process to be visualized. (C) Examples of pulling tubules in the presence of different BAR domains and dynamin. The giant unilamellar vesicles (GUV) are revealed by including Texas Red fluorophore covalently linked to the GUV lipids. The BAR domains were labeled with the green fluorophore AlexFluor 488. The labeled BAR domains partition only onto bilayers with high curvature (the tubules in panel A). Consequently, only the tubules fluoresce green, whereas the vesicles of low curvature show only red fluorescence. (A, B, From Tian A, Baumgart T [2009] *Biophys J* 96: 2676–2688. With permission from Elsevier. C, From Baumgart et al. [2011] *Annu Rev Phys Chem* 62:483–506 and Roux et al. [2010] PNAS 107:4141–4146. With permission from Annual Reviews and Patricia Bassereau.)

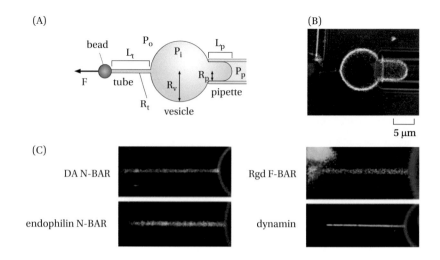

Rab proteins, represented by more than 60 genes in the human genome, serve as "address labels" and give each organelle an identity. They are GTPases that can be switched between an inactive, soluble GDP-bound state and an active, membrane-bound GTP-bound state. The switch is controlled by GDP–GTP exchange factors (Rab-GEFs) that are found on both the vesicle and the target membranes. In their active, membrane-bound form, Rab proteins can recruit other proteins, called Rab effectors, to the membrane. Rab effectors can have a variety of functions: they can help tether vesicles to the target membrane or help in the membrane fusion reaction.

The key proteins for the actual fusion event are the α-helical SNARE proteins. Early *in vitro* experiments with lipid vesicles showed that pure lipid vesicles rarely fuse when vesicles tethered to pipettes are pushed together. One of the main reasons is that bringing the lipid headgroups close enough together to allow inter-bilayer mixing prior to fusion requires dehydration of the opposing bilayer surfaces. This is energetically costly. And if the bilayers are charged, the cost is even greater. The high energy barrier is advantageous to cells, because haphazard, uncontrolled fusion events cannot be permitted *in vivo*. An important biological principle is that cells exert control by selectively lowering high energy barriers. In the case of fusion, the selective lowering of the fusion barrier is the job of the SNARE proteins that perform work to fuse the opposing membranes. A crucial issue is the source of energy for fusion.

As indicated schematically in **Figure 7.24A**, complementary pairs of SNAREs are present on the vesicle and target membranes: a v-SNARE (synaptobrevin) on the vesicle and t SNARE complex on the target (syntaxin 1A + SNAP-25A). Synaptobrevin and syntaxin 1A are each anchored by a C-terminal transmembrane helix and project a long N-terminal helical segment into the cytosol. The structure of a v-SNARE/t-SNARE complex, as it looks after the fusion reaction has occurred, suggests the underlying power source for fusion (**Figure 7.24B**). The complex forms a tightly folded four-helix bundle stabilized by a hydrophobic core. The current hypothesis is that it is the formation of the v-SNARE/t-SNARE complex, called "zippering," that provides the energy required to fuse the membrane vesicle with the target membrane.

To find out how much energy can be provided for fusion, scientists have examined the zipping and unzipping of the four-helix bundle using magnetic and optical tweezer technology to measure directly the force required to unzip the v-SNARE/t-SNARE complex (**Box 7.4**). In essence, the SNARE complex is grabbed at each end and ripped apart while measuring the force required. The result is a force–extension trace from which the energetics of unfolding/folding

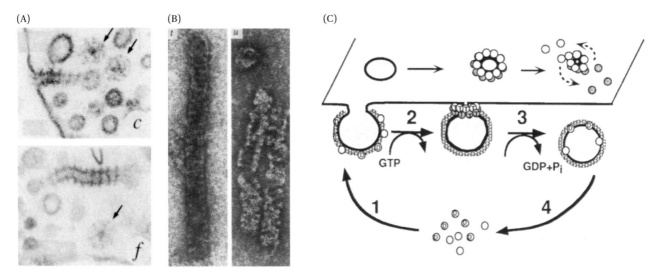

Figure 7.21 Early evidence for the involvement of dynamin in vesicle cleavage. (A) Electron micrographs of synaptic terminals with tubular vesicles and ring structures, now known to be dynamin. (B) Electron micrographs of tubular structures reconstituted from lipid vesicles and dynamin. The right-hand image is the same except that the preparation was partially digested with subtilisin. Partial unwinding of the dynamin coat is seen. Close inspection suggests a helical structure. (C) The model for vesicle scission proposed by Sandra Schmid based upon the data shown in panels A and B. (A, From Takei K et al. [1996] *Nature* 374: 186-190. With permission from Springer Nature. B and C, From Hinshaw JE et al. [1995] *Nature* 374: 190-192. With permission from Springer Nature.)

can be determined (**Figure 7.25**). The estimated free energy released during formation of the v-SNARE/t-SNARE complex is about 40 k_BT, roughly matching estimates of the energetic cost of dehydrating the interface between two opposing lipid bilayers.

Finally, if the v-SNARE/t-SNARE complex is so stable, how is it disassembled? This is accomplished by an AAA ATPase that unwinds the complex into its components. The complete disassembly of a SNARE complex requires the hydrolysis of about 50 ATP molecules, corresponding to ~1000 k_BT of free energy.

Figure 7.22 Summary of scission by dynamin. (A) Domain structure of dynamin. (B) Crystallographic structure of a dynamin dimer (PDB: 6DLU) and cryo-EM structure of a membrane-bound dimer in which the membrane-binding pleckstrin-homology domain (PHD, yellow) is visible. (C) Cartoon representation of the structure and summary of dynamin organization in preparation for cleavage. (D) Cartoon summary of scission. (From Ferguson SM et al. [2012] *Nature* 13: 75–88. With permission from Springer Nature.)

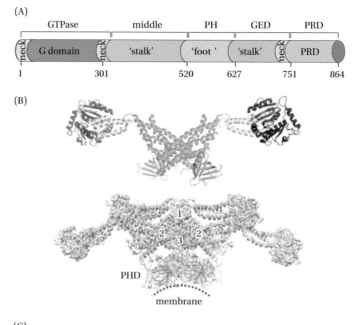

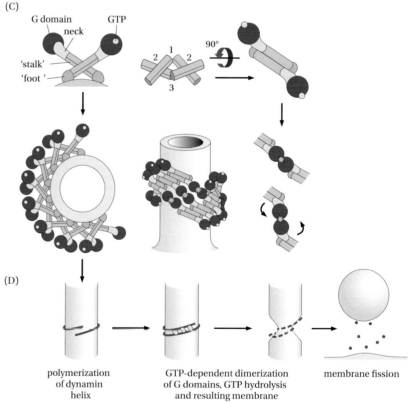

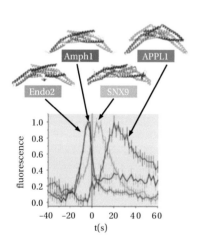

Figure 7.23 Timing of events underlying clathrin-coated vesicle fission in live cells determined by real-time internal reflection microscopy and protein-specific labeling. Different BAR domains bind and unbind in a specific order relative to the time of scission (*t* = 0). (From Taylor MJ, Perrais D, Merrifield CJ [2011] *Plos Biol* 9: e1000604.)

polymerization of dynamin helix

GTP-dependent dimerization of G domains, GTP hydrolysis and resulting membrane constriction

membrane fission

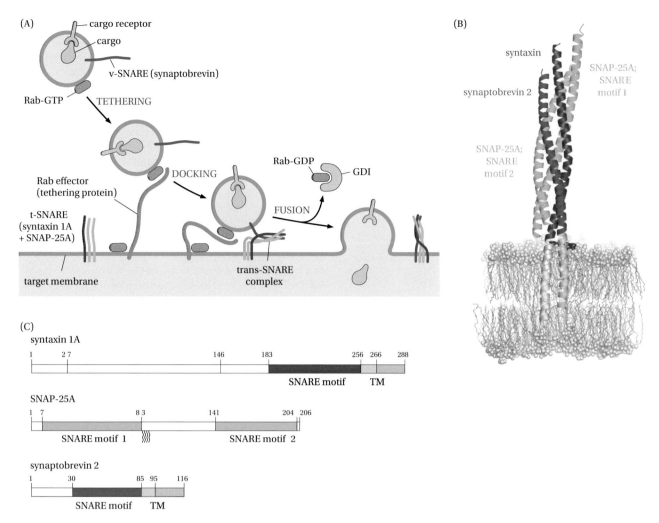

Figure 7.24 Control of membrane fusion events. (A) Tethering, docking, and fusion reactions are controlled by Rab and SNARE proteins. (B) The structure of the post-fusion SNARE complex embedded in a lipid bilayer (PDB: 3HD7). (C) The domain structures of syntaxin, SNAP-25A, and synaptobrevin. There are four so-called SNARE motifs that stabilize the four-helix bundle. (A, From Alberts B, Johnson A, Lewis JH et al. [2014] *Molecular Biology of the Cell*, 6th Edition. Garland Science, New York. With permission from W. W. Norton. B and C, From Stein A et al. [2009] *Nature* 460: 525–528. With permission from Springer Nature.)

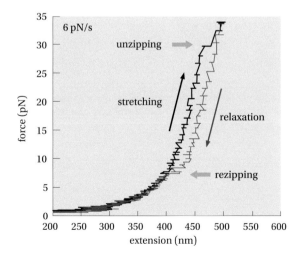

Figure 7.25 Single-molecule force–extension curve for unzipping and zipping a SNARE complex, obtained using magnetic tweezers. The force was ramped up or down by 6 pN s⁻¹. (From Min D et al. [2013] *Nature Communications* 4: 1705. With permission from Springer Nature.)

BOX 7.4   OPTICAL AND MAGNETIC TWEEZERS

Optical and magnetic tweezers make it possible to pull on single molecules or molecular complexes with forces in the pN range. As shown schematically in **Figure 1**, a fully assembled v-SNARE/t-SNARE complex is tethered via two DNA handles between a glass slide and a magnetic

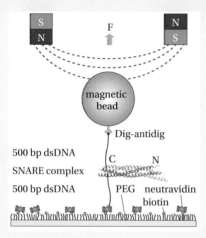

Figure 1 A magnetic tweezer set-up. The magnetic bead is pulled by the magnetic field toward the point where the strength of the field is maximal. (From Min D et al. [2013] *Nature Communications* 4: 1705. With permission from Springer Nature.)

bead. Two magnets placed near the magnetic bead create a magnetic field, which pulls on the bead with a force F that depends on the distance between the bead and the magnets. In a typical experiment, F is gradually increased by moving the magnets closer to the bead, and the position of the bead at each instance in time is recorded by optical monitoring. Unzipping events give rise to a sudden stepwise increase in the extension and can be identified from force–extension curves such as the one shown in Figure 7.25. Thus, the force required to unzip the v-SNAR/t-SNARE bundle can be determined.

In an optical trap, a single molecule is also tethered between two surfaces, either two beads or a bead and a glass surface, using DNA handles (**Figure 2**). One of the beads is placed near a focused laser beam. The intense laser light will pull the bead toward the beam focus (the "optical trap"); the force F exerted on the bead is proportional to the distance between the bead center and the focal point of the laser beam. Similarly to the magnetic tweezers, force–extension curves can be measured by increasing the distance between the stationary bead and the optical trap while recording the displacement of the trapped bead from the center of the trap. Optical tweezers can also be used to pull thin membrane tubes from a lipid vesicle, as shown in Figure 7.20.

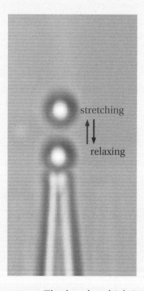

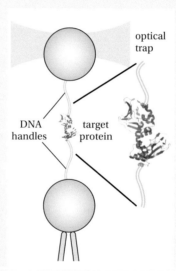

Figure 2 An optical tweezer set-up. The bead—which is made of a dielectric material—is pulled toward the point where the strength of the electromagnetic light field is maximal. The second bead is held at the end of a micropipette by suction. (From Mallardiab.org.)

KEY CONCEPTS

- Eukaryotic cells, surrounded by the plasma membrane, have a number of membrane-delimited organelle compartments, including the endoplasmic reticulum and the Golgi apparatus.

- These compartments are maintained and communicate by the trafficking of lipid vesicles transported by ATP-powered molecular motors that travel along polarized microtubules.

- Dynein motors carry vesicle cargos toward the cell nucleus, while kinesins carry cargos outward from the nucleus.

- The transported vesicles have coat proteins that carry, in essence, protein address labels indicating the target membrane. Orderly transport and loading/unloading of vesicles is ensured by distinctive protein coats.

- COPII vesicles move proteins from the ER to the Golgi complex, whereas COPI vesicles transport proteins from the Golgi back to the ER.

- Clathrin-coated (CC) vesicles are used for endocytosis and intracellular transport. Caveolin proteins form small invaginations (caveolae) in the plasma membrane of many cells that can become full-fledged vesicles for transport. Their main role seems to be to regulate the chemical composition of the plasma membrane.

- The general scheme for the principal vesicle systems COPI, COPII, and CC involves coat formation, pinching off the coated vesicles from the parent membrane, intracellular transport, and uncoating followed by

fusion at the target membrane. Pinching off vesicles is the responsibility of dynamins, which are large GTPases.

- The vesicles are shaped for the scission process by stiff banana-shaped BAR domains, which are also important for forming tube-shaped compartments. Whereas CC and COPI vesicles are spherical, COPII vesicles can be tubular. This is possible because the coat is based upon a tetrameric arrangement of oblong structural units rather than the triangular units of COPI and CC vesicles.

- CC and COPI vesicle coats are based upon triangular subunits. Unlike COPII and CC vesicles, which are assembled from oblong proteins that form a net around the vesicles, COPI vesicles are assembled from preformed protein complexes called coatamers.

- Regardless of how membranous vesicles are formed, they fuse with the target membrane after the protein coat has been disassembled.

- Fusion is controlled and mediated by α-helical SNARE complexes. During fusion, the complex forms a tightly folded four-helix bundle stabilized by a hydrophobic core.

- It is generally accepted that the folding of the SNARE complex, called "zippering," provides the energy for fusing the membrane vesicle with the plasma membrane. A process such as zippering that produces work is necessary for fusion, because pure lipid vesicles rarely fuse when vesicles are pushed together.

FURTHER READING

General Background

Alberts et al. (2015) *Molecular Biology of the Cell*, 6th ed., Garland Science, New York.

Shibata et al. (2009) Mechanisms shaping the membranes of cellular organelles. *Ann Rev Cell Dev Biol* 25:329–354.

Guo et al. (2014) Protein sorting at the trans-Golgi Network. *Ann Rev Cell Dev Biol* 30:169–206.

BAR domains

Qualmann et al. (2011) Let's go bananas: Revisiting the endocytic BAR code. *EMBO J* 30:3501–3515.

Simunovic et al. (2015) When physics takes over: BAR proteins and membrane curvature. *Trends Cell Biol* 25:780–792

Mim, and Unger (2012) Membrane curvature and its generation by BAR proteins. *Trends Biochem Sci* 37:526–533.

Baumgart et al. (2011) Thermodynamics and mechanics of membrane curvature generation and sensing by proteins and lipids. *Annu Rev Phys Chem* 62:483–506.

Erythrocyte Membranes

Mohandas, and Evans (1994) Mechanical properties of the red cell membrane in relation to molecular structure and genetic defects. *Annu Rev Biophys Biomol Struct* 23:787-818.

Baines (2010) The spectrin-ankyrin-4.1-adducin membrane skeleton: Adapting eukaryotic cells to the demands of animal life. *Protoplasma* 244:99-131.

Elgsaeter et al. (1986) The molecular basis of erythrocyte shape. *Science* 234:1217-1223.

Cytoskeleton

Brinkley (1985) Microtubule organizing centers. *Ann Rev Cell Biol* 1:145-172.

Kelly (1990) Microtubules, membrane traffic, and cell organization. *Cell* 61:5-7.

Bennett, and Healy (2007) Organizing the fluid membrane bilayer: Diseases linked to spectrin and ankyrin. *Trends Mol Med* 14:28-36.

Molecular Motors & Trafficking

Caviston, and Holzbaur (2006) Microtubule motors at the intersection of trafficking and transport. *Trends Cell Biol* 16:530-537.

Gennerich, and Vale (2009) Walking the walk: How kinesin and dynein coordinate their steps. *Cur Opin Cell Biol* 21:59-67.

Schliwa, and Woehlke (2003) Molecular motors: Insight review. *Nature* 422:759-765.

Karcher et al. (2002) Motor-cargo interactions: The key to transport specificity. *Trends Cell Biol* 12:21-27.

McLaughlin et al. (2016) Collective dynamics of processive cytoskeletal motors. *Soft Matter* 12:14-21.

Caveolae & Caveolin

Kovtun et al. (2015) Cavin family proteins and the assembly of caveolae. *J Cell Sci* 128:1269-1278.

Stan (2005) Structure of caveolae. *Biochim Biophys Acta* 1746:334-348.

Parton, and del Pozo (2013) Caveolae as plasma membrane sensors, protectors and organizers. *Nature Reviews Mol Cell Biol* 14:98-112.

Clathrin-Coated Vesicles

Kirkhausen et al. (2014) Molecular structure, function, and dynamics of Clathrin-mediated membrane traffic. *Cold Spring Harbor Perspectives Biol* 6:a016725.

McMahon, and Boucrot (2011) Molecular mechanism and physiological functions of clathrin-mediated endocytosis. *Nature Reviews Mol Cell Biol* 12:517-533.

COPI & COPII Vesicles

Szul, and Sztul (2011) COPII and COPI traffic at the ER-Golgi interface. *Physiology* 26:348-364.

Popoff et al. (2011) COPI Budding within the Golgi stack. *Cold Spring Harb Perspect Biol* 3:a005231.

Jackson (2014) Structure and mechanism of COPI vesicle biogenesis. *Cur Opin Cell Biol* 29:67-73.

Lee, and Miller (2007) Molecular mechanisms of COPII vesicle formation. *Seminar Cell Devlop Biol* 18:242-434.

Gürkan et al. (2006) The COPII cage: unifying principles of vesicle coat assembly. *Nature Reviews Mol Cell Biol* 7:727-738.

Key literature

BAR domains

David et al. (1996) A role of amphiphysin in synaptic vesicle endocytosis suggested by its binding to dynamin in nerve terminals. *Proc Natl Acad Sci USA* 93:331-335.

Taylor et al. (2011) A high precision survey of the molecular dynamics of mammalian Clathrin-Mediated endocytosis. *PLoS Biology* 9:e1000604.

Peter et al. (2004) BAR domains as sensors of membrane curvature: The Amphiphysin BAR structure. *Science* 303:495-499.

Takei et al. (1999) Functional partnership between amphiphysin and dynamin in clathrin- mediated endocytosis. *Nature Cell Biol* 1:33-39.

Chen et al. (2015) Regulation of membrane-shape transitions induced by I-BAR domains. *Biophys J* 109:298-307.

Erythrocyte Membranes

Evans, and La Celle (1975) Intrinsic material properties of the erythrocyte membrane indicated by mechanical analysis of deformation. *Blood* 45:29-43.

Fairbanks et al. (1971) Electrophoretic analysis of the major Polypeptides of the human erythrocyte. *Membrane Biochemistry* 10:2606-2617.

Nans et al. (2011) Native ultrastructure of the red cell cytoskeleton by cryo-electron tomography. *Biophys J* 101:2341-2350

Jay (1973) Viscoelastic properties of the human Red Blood Cell membrane.1. Deformation, volume loss, and rupture of Red Cells in Micropipettes. *Biophys J* 13:1166-1182.

Skalak, and Branemark (1969) Deformation of Red Blood Cells in capillaries. *Science* 164:717-719.

Deuticke (1968) Transformation and restoration of biconcave shape of human erythrocytes induced by amphiphilic agents and changes of ionic environment. *Biochim Biophys Acta* 163:494-500.

Cytoskeleton

Wang et al. (2012) Structure of the ZU5-ZU5-UPA-DD tandem of ankyrin-B reveals interaction surfaces necessary for ankyrin function. *Proc Natl Acad Sci USA* 109:4822-4827.

Lange et al. (1982) Role of the Reticulum in the stability and shape of the isolated Human Erythrocyte Membrane. *J Cell Biol* 92:714-721.

Kerssemakers et al. (2006) Assembly dynamics of microtubules at molecular resolution. *Nature* 442:709-712.

Hotani, and Horio (1988) Dynamics of microtubules visualized by darkfield microscopy: Treadmilling and dynamic Instability. *Cell Motility and the Cytoskeleton* 10:229-236.

Galjart (2010) Plus-end-tracking proteins and their interactions at microtubule ends. *Cur Biol* 20:R528-R537.

Molecular Motors & Trafficking

Jeppesen, and Hoerber (2012) The mechanical properties of kinesin-1: A holistic approach. *Biochem Soc Transactions* 40:438-443.

Soldati, and Schliwa (2006) Powering membrane traffic in endocytosis and recycling. *Nature Rev Mol Cell Biol* 7:897-908.

Rice et al. (1999) A structural change in the kinesin motor protein that drives motility. *Nature* 402:778-784.

Svoboda et al. (1993) Direct observation of kinesin stepping by optical trapping interferometry. *Nature* 365:721-727.

Caveolae & Caveolin

Hayashi et al. (2009) Human PTRF mutations cause secondary deficiency of caveolins resulting in muscular dystrophy with generalized lipodystrophy. *J Clin Invest* 119:2623-2633.

Walser et al. (2012) Constitutive formation of caveolae in a bacterium. *Cell* 150:752-763.

Lu et al. (2008) Deletion of cavin/PTRF causes global loss of caveolae, dyslipidemia, and glucose intolerance. *Cell Metab* 8:310-317.

Sinha et al. (2011) Cells respond to mechanical stress by rapid disassembly of caveolae. *Cell* 144:402-413.

Echarri et al. (2012) Caveolar domain organization and trafficking is regulated by Abl kinases and mDia1. *J Cell Sci* 125:3097-3133.

Clathrin-Coated vesicles

Pearse (1976) Clathrin: A unique protein associated with intracellular transfer of membrane by coated vesicles. *Proc Natl Acad Sci USA* 73:1255-1259.

Fotin et al. (2004) Molecular model for a complete clathrin lattice from electron cryomicroscopy. *Nature* 432:573-579.

Woods et al. Common features of coated vesicles from dissimilar tissues: Composition and structure. *J Cell Sci* 30:87-97.

Henne et al. (2010) FCHo Proteins are nucleators of clathrin-mediated endocytosis. *Science* 328:1281-1284.

Matsui, and Kirchhausen (1990) Stabilization of clathrin coats by the core of the clathrin-associated protein complex AP-2. *Biochemistry* 29:10791-10798.

COPI & COPII vesicles

Bigay et al. (2003) Lipid packing sensed by ArfGAP1 couples COPI coat disassembly to membrane bilayer curvature. *Nature* 426:563-566.

Faini et al. (2012) The structures of COPI-coated vesicles reveal alternate coatomer conformations and interactions. *Science* 336:1451-1454.

Dodonova et al. (2015) A structure of the COPI coat and the role of coat protein in membrane vesicle assembly. *Science* 349:195-198.

Whittle, and Schwartz (2010) Structure of the Sec13-Sec16 edge element, a template for assembly of the COPII vesicle coat. *J Cell Biol* 190:347-361.

Jin et al. (2012) Ubiquitin-dependent regulation of COPII coat size and function. *Nature* 482:495-500.

Stagg et al. (2006) Structure of the Sec13/31 COPII coat cage. *Nature* 439:234-238.

EXERCISES

1. Why are microtubules better suited structurally than actin filaments for defining the radial dimensions of cells?

2. Hereditary spherocytosis is a disease in which the red cells form spheres rather than the biconcave disc shapes that they normally have. The spherical shape is the result of disruptions in the formation of the red blood cell cytoskeleton. Based on Figure 7.3, name three proteins that might cause spherocytosis if mutated.

3. Why do COPI complexes invariably form spherical structures, whereas COPII complexes can form either tubular structures or spherical structures?

4. Consider the tubule pulling experiment of Figure 7.20. Would tubules pulled in the presence of F-BAR domains have a larger or smaller diameter than tubules pulled in the presence of N-BAR domains (see Box 7.3)? Why?

Membrane Protein Bioinformatics

8

Evolution of organisms involves DNA mutations that lead to changes in the amino acid sequences of proteins or in the sequences of functional RNAs, causing homologous proteins in different organisms to have different amino acid sequences even though the structure and function of the proteins may be similar. For example, hemoglobins from, say, whales and humans are quite similar in structure, but their amino acid sequences differ. Bioinformatics takes advantage of the similarities and differences in sequences to construct evolutionary histories of proteins and to gain insights into the relationships among sequence, three-dimensional (3-D) structure, and protein function.

Today, there are millions of protein sequences and tens of thousands of protein structures available in the large international database repositories, which are growing at ever increasing rates. The only way to come to grips with such volumes of data is by using computers. Indeed, the definition of bioinformatics is the application of computer and information theory science to fundamental problems in molecular biology and medicine. The rapidly growing field of bioinformatics has evolved in part in an attempt to meet the challenge of extracting knowledge about biological processes from the millions of protein sequences available in databases.

Bioinformatics has turned out to be an especially powerful tool in the area of membrane proteins. Issues that have been successfully addressed by bioinformaticians include prediction of the subcellular localization of proteins, identification of membrane protein encoding genes in genome sequences, topology prediction for both helix-bundle and β-barrel membrane proteins, identification of sequence patterns that suggest structure–function relationships, and 3-D structure prediction.

Bioinformaticians typically ask questions such as: How do we align protein sequences to best detect similarities stemming from common ancestry? How do we best use amino acid sequence to predict structure and function? How do we use extant protein sequences and structures to study molecular evolution? In this chapter, we present the main methods used in membrane protein bioinformatics to answer these kinds of questions. The most important general question that we address is what you might realistically hope to be able to learn about your favorite membrane protein by punching keys on a computer (or tapping on a tablet).

DOI: 10.1201/9780429341328-9

8.1 EVOLUTION OF PROTEIN AMINO ACID SEQUENCES PROVIDES A BASIS FOR BIOINFORMATICS

8.1.1 The Statistical Variations among Related Protein Amino Acid Sequences Provide Insights into Evolutionary Processes

Bioinformatics traces its origins to the time when it first became possible to determine the amino acid sequences of proteins by peptide sequencing (DNA sequencing was invented much later). The first proteins to be sequenced were those that could be purified in large amounts from natural sources. Bovine insulin was sequenced by Fred Sanger in 1951. As more proteins were sequenced, it was discovered that there were slight sequence variations between the same protein isolated from different organisms and that the variations were more numerous when the two organisms were more distantly related in the Tree of Life. But even before sequencing was widely available, analysis of the amino acid composition of proteins by methods such as paper electrophoresis revealed species-specific differences between homologous proteins, most famously for hemoglobin α- and β-chains from primates.

Cytochrome c, which was one of the earliest proteins to arise at the time of the emergence of aerobic life, was one the first proteins whose complete sequence was determined for a wide variety of species. By 1963, the sequences of cytochromes c from human, rabbit, pig, chicken, tuna, and yeast had been determined. Emanuel Margoliash and Walter Fitch realized that for the sequences to have evolved from a common ancestor, the DNA codon composition must have evolved. One of the fundamental questions they addressed was how many nucleotide changes were required to convert, say, a codon for alanine into a codon for asparagine. That is easily done by using the genetic code (Box 1.5) to create a table of mutation values for amino acid pairs. From such a table, one finds, e.g., that two nucleotide changes are required to evolve alanine into asparagine. The next question was how many mutations (called the mutation distance) would be required to convert the sequence of cytochrome c from one species to the sequence of another species. Using this idea and rigorous mathematical principles, Margoliash and Fitch could construct an evolutionary tree for cytochromes c in which the lengths of branches of the tree were given by the mutation distances (**Figure 8.1**).

Margaret Dayhoff built upon the ideas of Fitch and Margoliash, realizing that computers were ideally suited to compare amino acid sequences. She began to collect all published protein sequences into a compendium called the *Atlas of Protein Sequence and Structure* that was published in multiple editions, starting in 1965. She used sequences from the *Atlas* to improve the mutation table of Fitch and Margoliash by recognizing that not all mutations were equally likely. For example, hydrophobic amino acids in the core of proteins are more likely to mutate into other hydrophobic amino acids to preserve the basic fold. Statistical analyses of her database led her to create Point-Accepted Mutation (PAM) matrices in 1978, which describe the evolutionary rates at which different amino acids replace each other during evolution. These contributions formed the basis for modern bioinformatics. Dayhoff has been called "the mother and father of bioinformatics" because she pioneered the construction of sequence databases, the development of algorithms to compare sequences, and the use of sequence data to study evolution on the molecular level.

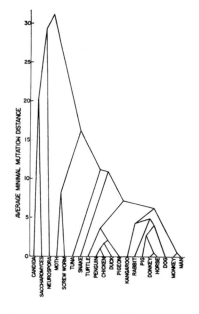

Figure 8.1 Evolution of cytochromes c based upon mutation distances. (From Fitch WM, Margollash E [1967] *Science* 155: 279–284. With permission from The American Association for the Advancement of Science.)

8.1.2 Sequence Alignment Algorithms Are Central to Bioinformatics

The most basic and most easily accessible information about a protein is its amino acid sequence. Therefore, most of the techniques used by bioinformaticians interested in proteins use sequence data as the primary input. The input sequence is usually analyzed using methodology from statistics or algorithms from the machine-learning field. The typical wet-lab user expects some kind of prediction as output, although in many bioinformatics studies, the aim is, rather, to find patterns in large sequence datasets that in turn can be used to produce prediction programs.

The most important and widely used bioinformatics tools are those that align two or more protein sequences. A simple way to represent the problem is the dot-plot. Such a plot is shown in **Figure 8.2**, where two homologs of the SecY translocon protein have been aligned. The prominent diagonal streaks show that the two proteins are very similar and can be aligned for almost their entire lengths. For example, the dot indicated by the arrow represents the boxed local alignment shown below the dot-plot, which is part of a somewhat longer region of high sequence similarity. The gaps between the streaks are regions of lower sequence similarity.

The dot-plot makes it possible to get a visual feeling for whether two proteins can be aligned in a meaningful way, but it clearly does not automatically produce a complete sequence alignment (especially in the white areas between the diagonal streaks), nor is it useful when one wants to scan an entire sequence database such as UniProt to find related sequences. Fortunately, there are fast algorithms that can do this with minimal user intervention.

The BLAST algorithm (**Box 8.1**) is one of the most widely used alignment tools and is available in many implementations on the web. It produces both a set of local sequence alignments between two proteins (roughly corresponding to the diagonal streaks in the dot-plot) and a statistical measure of the significance of the overall sequence similarity between the two proteins.

As a word of caution, transmembrane helices have a more restricted amino acid composition than globular proteins, being composed mostly of hydrophobic amino acids. Many sequence alignment algorithms, including BLAST, have a setting that masks "low-complexity regions" (i.e., regions with a restricted amino acid composition) before making the alignment. Thus, in some analyses, transmembrane helices will be scored as low-complexity regions and hence, will not be included in the alignment if this setting is turned on.

Pairwise sequence alignments give at best a rough indication of which parts of a protein sequence are subjected to strong evolutionary constraints and leave

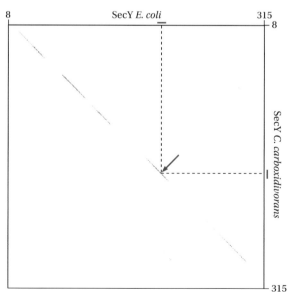

SecY *E. coli* | 185

QSIGIATGLPNMPGMQGLVINPGFAFYFTAVVSLVTGTMFLMWLGEQITERGIGNGISIIIFAGIVAGLPPAIAHTIEQARQGDLHFLVLLLVAVLVFAVTFFVVFVE
AAIQAISTYVIIARANALHDPSNKLNAFLIMLTLTTASTFLIWLGDRITDSGIGNGISLLIFYNIVSRFPTTLNQIVGLQKMETINFIELVALVIVAFAMFVSVVIMT

SecY *C. carboxidivorans* | 175

Figure 8.2 A dot-plot of SecY from *Clostridium carboxidivorans* (on the y-axis) against SecY from *E. coli* (on the x-axis). The window size in this example is 15 and the cutoff is 6, i.e., a dot is placed in the plot whenever the two 15-mers corresponding to the position of the dot have 6 or more matching residues. By varying these parameters, one can change the degree of sequence similarity required and thereby filter the plot at different levels of stringency. The red arrow indicates the dot representing the boxed local alignment shown below the dot-plot (the two segments are indicated on the axes of the dot-plot). The dot-plot is an exploratory tool and does not normally provide any measure of statistical significance.

BOX 8.1   BLAST SEARCHES AND SCORING MATRICES

As the opportunity to mine the data in protein sequence databases has grown, the need for fast algorithms to find homologs to a query sequence in the database has become more and more acute. The most widely used method for fast similarity searches is the Basic Local Alignment Search Tool, or BLAST. BLAST looks for short segments of high sequence similarity between the query sequence and all sequences in a database, based on the empirical observation that statistically significant sequence alignments obtained by exhaustive but much slower alignment algorithms nearly always contain such regions of high local sequence similarity.

Exhaustive alignment algorithms such as the Smith–Waterman and Needleman–Wunsch algorithms are based on a technique called dynamic programming (see any bioinformatics textbook) and guarantee that given a substitution matrix and a gap-score model, the best-scoring alignment will be found. They are too slow for searches across large sequence databases, however. The basic BLAST algorithm was designed to be much faster while still producing good, though not necessarily optimal, alignments. It proceeds in the following steps (see **Figure 1**):

1. Remove low-complexity regions from the query sequence. Regions with long runs of one or a few residues (e.g., ...QQQQQQQQ..., ...PPPPPPPP..., ...GGGWGQPHGGGWGQPH...) that may result from mutational events not related to common ancestry can give rise to spurious hits in the database and are filtered out in this step. This sometimes creates problems for membrane proteins since transmembrane segments, which contain mostly hydrophobic residues, may get filtered out.

2. Partition the query sequence into overlapping 3-letter "words" and expand each word into a set of high-scoring matching words. As an example, the sequence MANMFALILVIATLVTGILWCVD... is partitioned into the overlapping words MAN, ANM, NMF, MFA, FAL, etc. Each word is then expanded into a set of high-scoring matching words using an amino acid substitution matrix such as BLOSUM62 (see later). Each 3-letter word has 20^3 possible matching words, but only some of these are sufficiently similar to the original word to qualify as high-scoring (MAN may be expanded to the related high-scoring words MSN, MGN, and MVN but not to QPP, RSL, etc.).

3. Scan each sequence in the database for words matching each of the expanded word sets from step 2.

4. For each sequence in the database, look for instances of matching words that are the same number of residues D apart (with $D < 40$) in the query sequence

and the database sequence, i.e., matching words that fall close to each other on the same diagonal in a dot-plot between the query sequence and the database sequence.

5. Extend the second of the two matched words in both directions, again using a scoring matrix such as BLOSUM62. Stop extending when the total alignment score decreases by some set amount (typically 20) from the maximum found so far for this match. The resulting region is called a high-scoring segment pair (HSP).

6. Evaluate the statistical significance of each HSP using a so-called Extreme Value Distribution (EVD). The parameters describing the EVD are obtained by an empirical fitting procedure.

7. Calculate a new alignment for each significant HSP match between the query sequence and a database sequence allowing gaps in the alignment, and re-score the match. The final score is called S.

8. Calculate an E-value for each match. The E-value is the number of target sequences with a similarity score $\geq S$ expected to be retrieved from the database by chance alone. An E-value of 0.1 means that there is a 10% chance of finding a database sequence with a similarity score $\geq S$ if one searches the database with a "random" sequence. As the database grows, a higher similarity score is required to meet a given E-value.

All sequence alignment algorithms must have a quantitative measure of amino acid similarity and a quantitative model for scoring gaps. With such a scheme in place, each possible alignment between two sequences can be scored by summing the scores of all pairs of aligned residues and adding the gap scores. Similarity scores, derived from alignments of proteins known to be related by common ancestry, quantitate how often two particular types of residues, such as G and Q, are aligned. Residues with similar chemical properties that can readily substitute for each other in proteins will have positive scores, while dissimilar residues that would tend to disrupt protein structure and impair function will have negative scores.

As an example, the widely used BLOSUM62 matrix is shown in **Figure 2** (BLOSUM stands for BLOcks SUbstitution Matrix.) The BLOSUM matrix is based on the log-odds score, i.e., the natural logarithm of the probability of observing a given pair of residues aligned in an alignment of two homologous sequences. The score s for an observed pair of residues a and b is calculated using Equation 1:

$$s(a,b) = \frac{1}{\lambda} \ln \frac{f_{ab}}{f_a f_b} \tag{1}$$

(Continued)

BOX 8.1   BLAST SEARCHES AND SCORING MATRICES (CONTINUED)

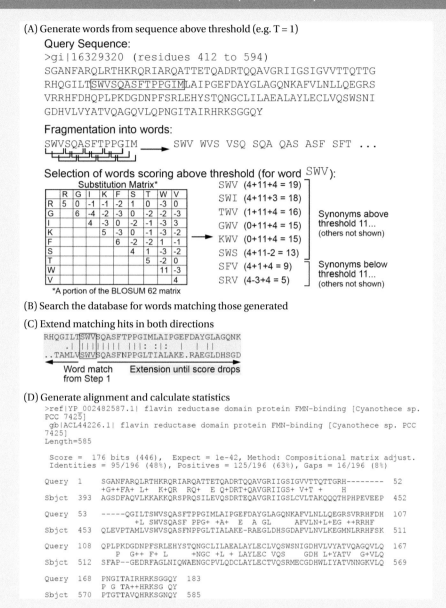

(A) Generate words from sequence above threshold (e.g. T = 1)

Query Sequence:
```
>gi|16329320 (residues 412 to 594)
SGANFARQLRTHKRQRIARQATTETQADRTQQAVGRIIGSIGVVTTQTTG
RHQGILTSWVSQASFTPPGIMLAIPGEFDAYGLAGQNKAFVLNLLQEGRS
VRRHFDHQPLPKDGDNPFSRLEHYSTQNGCLILAEALAYLECLVQSWSNI
GDHVLVYATVQAGQVLQPNGITAIRHRKSGGQY
```

Fragmentation into words:

SWVSQASFTPPGIM ⟶ SWV WVS VSQ SQA QAS ASF SFT ...

Selection of words scoring above threshold (for word SWV):

Substitution Matrix*

	R	G	I	K	F	S	T	W	V
R	5	0	-1	-1	-2	1	0	-3	0
G		6	-4	-2	-3	0	-2	-2	-3
I			4	-3	0	-2	-1	-3	3
K				5	-3	0	-1	-3	-2
F					6	-2	-2	1	-1
S						4	1	-3	-2
T							5	-2	0
W								11	-3
V									4

*A portion of the BLOSUM 62 matrix

SWV (4+11+4 = 19)
SWI (4+11+3 = 18)
TWV (1+11+4 = 16)
GWV (0+11+4 = 15) } Synonyms above threshold 11... (others not shown)
KWV (0+11+4 = 15)
SWS (4+11-2 = 13)
SFV (4+1+4 = 9) } Synonyms below threshold 11... (others not shown)
SRV (4-3+4 = 5)

(B) Search the database for words matching those generated

(C) Extend matching hits in both directions

```
RHQGILTSWVSQASFTPPGIMLAIPGEFDAYGLAGQNK
    .| |||||||||  |||: :|:   |   | ||
..TAMLVSWVSQASFNPPGLTIALAKE.RAEGLDHSGD
```
Word match ⟶   Extension until score drops
from Step 1

(D) Generate alignment and calculate statistics

```
>ref|YP_002482587.1| flavin reductase domain protein FMN-binding [Cyanothece sp.
PCC 7425]
 gb|ACL44226.1| flavin reductase domain protein FMN-binding [Cyanothece sp. PCC
7425]
Length=585

 Score =  176 bits (446),  Expect = 1e-42, Method: Compositional matrix adjust.
 Identities = 95/196 (48%), Positives = 125/196 (63%), Gaps = 16/196 (8%)

Query   1    SGANFARQLRTHKRQRIARQATTETQADRTQQAVGRIIGSIGVVTTQTTGRH--------  52
             +G++FA+ L+  K+QR  RQ+  E Q+DRT+QAVGRIIGS+ V+T +       H
Sbjct   393  AGSDFAQVLKKAKKQRSPRQSILEVQSDRTEQAVGRIIGSLCVLTAKQQQTHPHPEVEEP  452

Query   53   -----QGILTSWVSQASFTPPGIMLAIPGEFDAYGLAGQNKAFVLNLLQEGRSVRRHFDH  107
                  +L SWVSQASF PPG+ +A+  E A GL      AFVLN+L+EG ++RRHF
Sbjct   453  QLEVPTAMLVSWVSQASFNPPGLTIALAKE-RAEGLDHSGDAFVLNVLKEGMNLRRHFSK  511

Query   108  QPLPKDGDNPFSRLEHYSTQNGCLILAEALAYLECLVQSWSNIGDHVLVYATVQAGQVLQ  167
               P   G++ F+ L    +NGC +L + LAYLEC VQS   GDH L+YATV  G+VLQ
Sbjct   512  SFAP--GEDRFAGLNIQWAENGCPVLQDCLAYLECTVQSRMECGDHWLIYATVNNGKVLQ  569

Query   168  PNGITAIRHRKSGGQY  183
              P G TA++HRKSG QY
Sbjct   570  PTGTTAVQHRKSGNQY  585
```

Figure 1 The basic steps in a BLAST search. (From Kerfeld CA, Scott KM [2011] *PloS Biol* 9: e1001014.)

where f_{ab} is the observed frequency of the a,b pair in a large database of trusted sequence alignments (the BLOCKS database) and f_a, f_b are the frequencies of residues a and b in the database. λ is a simple scaling factor chosen to make the rounded-off scores whole integers for ease of computation. As an example, the score for an alignment of two Trp (W) residues is given by plugging in the values for f_{WW} and f_W derived from the BLOCKS database ($f_{WW} = 0.0065$; $f_W =$

0.013) and $\lambda = 0.347$ (chosen by the inventors of BLOSUM) in (1) and rounding off: $s(W,W) = 11$.

There are different versions of the BLOSUM matrix, calculated from alignments of sequences that have diverged to different degrees. The most widely used matrix (BLOSUM62) was obtained using alignments of sequences with a maximal sequence identity of 62%; hence the name.

(Continued)

BOX 8.1   BLAST SEARCHES AND SCORING MATRICES (CONTINUED)

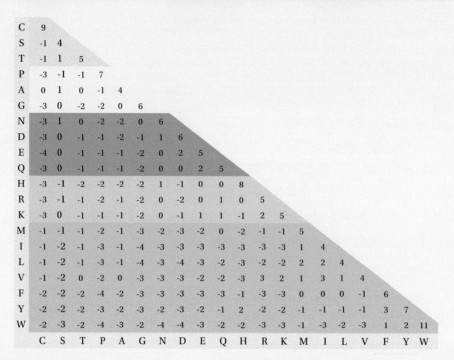

Figure 2 The BLOSUM62 substitution matrix. The score for a given aligned pair of amino acids is read off at the intersection of the two residues, e.g., an E-E match has a score of +5, an E-D match has a score of +2, and an E-C match has a score of −4. Note that amino acids with similar chemical characteristics have been placed near each other, such that the high-scoring pairs tend to be near the diagonal. (From Zvelebil MJ, Baum JO [2007] *Understanding Bioinformatics*. Garland Science, New York. With permission from Taylor & Francis.)

Gaps in the alignment are usually scored using only two parameters, a gap-opening penalty (*I*) and a gap-extension penalty (*E*); since a gap should lower the total score, both *I* and *E* have negative values. The gap score *G* for a gap *n* residues long is defined as

$$G(n) = I + (n-1)E \qquad (2)$$

Obviously, *I* and *E* both must be optimized for any particular substitution matrix in order to produce meaningful alignments.

open the question of whether there are specific residues that are conserved or nearly conserved across a whole protein family—in most protein families, such highly conserved residues are rare and can define functionally important sequence motifs that characterize even distant members. This type of information can be captured in sequence profiles, which show the frequency with which different residues and gaps appear in every position along a multiple sequence alignment (**Box 8.2**). Using databases of pre-computed sequence profiles such as Pfam instead of single sequences makes it possible to detect much weaker sequence similarities than do, e.g., pairwise BLAST alignments.

As a general rule, bioinformatics prediction programs work better if they are applied to a set of aligned sequences all at once rather than to individual sequences. However, many methods used in the membrane protein area have been developed with single-sequence analysis in mind and do not readily lend themselves to multi-sequence analysis.

8.1.3 Structurally Similar Proteins Can Have Dramatically Different Sequences

An old dictum says that protein structure is better conserved than sequence. There are many examples of this among membrane proteins. A particularly striking case is that of a bile-acid transporter homolog (ASBT_{Nm}) from the

BOX 8.2   SEQUENCE PROFILES CAN DETERMINE STRUCTURAL SIMILARITIES

Proteins are rarely unique; they belong to evolutionary families of related proteins. Proteins related by common ancestry can be found both within an organism and in different organisms. During evolution, the individual proteins in a protein family accumulate mutations and tend to drift apart in primary structure, while the tertiary structure and the functional characteristics tend to be better conserved. Different parts or positions in the sequence accumulate mutations at different rates, reflecting the relative structural or functional importance of the different residues in the protein.

A common way to reveal the position-specific variation in mutation rate is to first make a multiple sequence alignment between all the proteins known to belong to a protein family (or to a protein-domain family) and then encode the multiple alignment in the form of a sequence profile. An example from the original publication where the concept of sequence profiles was first introduced is shown in Figure 1. Obviously, the sequence profile contains much more information than any individual sequence, because it reflects the different patterns of sequence conservation along the multiple alignment.

Sequence profiles represent the entire protein family much better than any individual sequence and provide much better templates than single sequences when searching for new instances of the protein family in sequence databases.

Another important use of sequence profiles is to first partition all known protein sequences into families and then construct a sequence profile for each family. Any new protein sequence can then be rapidly searched against these profiles to ascertain whether it belongs to any one of them. A widely used data base of sequence profiles is Pfam. Alternatively, one can conduct a sequence profile search on-the-fly using PSI-BLAST, a program that constructs an initial sequence profile from a set of high-scoring sequences from BLAST and then iteratively improves the profile by using it in additional searches against the database.

In the world of membrane proteins, sequence profiles can be used as input to many topology prediction programs, based on the (often correct) assumption that all members of a protein family have the same topology.

POS	PROBE	CONSENSUS	A	C	D	E	F	G	H	I	K	L	M	N	P	Q	R	S	T	V	W	Y	+/-
1	E G V L	V	3	-2	3	4	0	4	-1	3	-1	4	4	1	1	1	-2	1	2	6	-6	-2	9
2	L L S P	L	2	-2	-2	-1	3	0	-1	3	-1	6	5	-1	3	0	-1	3	1	4	1	-1	9
3	V V V V	V	2	2	-2	-2	2	2	-3	11	-2	8	6	-2	1	-2	-2	0	2	15	-9	-1	9
4	K E A T	A	6	-2	5	6	-5	4	1	0	5	-2	0	3	3	3	1	3	6	0	-6	-4	9
5	A P L P	P	6	-1	0	1	-2	2	0	1	0	2	2	0	8	2	0	2	2	3	-5	-4	9
6	G G G G	G	7	1	7	5	-6	15	-1	-3	0	-4	-3	4	3	2	-3	6	4	2	-11	-7	9
7	S S Q E	D	4	-1	7	7	-6	7	2	-2	2	-3	-2	4	3	6	1	6	2	-1	-6	-5	9
8	S S T P	S	4	4	2	2	-4	4	-1	0	2	-3	-2	2	7	0	1	10	6	0	-2	-4	9
9	V L V A	V	5	0	-1	-1	3	1	-2	7	-2	7	6	-1	1	-1	-3	0	2	10	-5	-1	9
10	K R R S	R	0	-1	1	1	-5	0	2	-2	8	-3	1	3	3	3	10	5	1	-2	7	-5	9
11	M L I I	I	0	-2	-3	-2	7	-3	-3	11	-1	11	10	-2	-1	-2	-2	1	9	-3	1	1	9
12	S S T S	S	4	6	2	2	-3	5	-1	0	2	-3	-2	3	4	-1	1	12	6	0	0	-4	9
13	C C C C	C	3	15	-5	-5	-1	2	-1	3	-5	-8	-6	-3	1	-6	-3	7	3	3	-13	10	9
14	K S Q R	K	1	-2	3	3	-6	1	3	-2	7	-3	0	3	3	5	7	4	1	-2	2	-5	9
15	A A G S	A	10	3	4	4	-5	8	-1	-1	1	-2	-1	3	4	1	-2	7	4	2	-6	-4	9
16	T S D S	S	4	3	5	4	-5	6	0	0	2	-3	-2	4	3	1	1	9	6	0	-3	-4	9
17	G G S Q	G	5	1	6	5	-6	9	1	-2	1	-3	-2	4	3	4	0	6	3	0	-6	-6	9
18	Y F L S	F	-1	2	-4	-3	9	-3	0	4	-3	6	3	-1	-3	-3	-3	1	-1	2	7	7	9
19	T T R L	T	1	-2	0	1	0	0	0	2	2	2	3	1	1	1	3	1	7	2	1	-2	9
20	F F . L	F	-2	-3	-6	-4	10	-4	-1	6	-4	9	6	-3	-4	-4	-3	-2	-1	3	7	8	4
21	S S . D	S	3	2	5	4	-4	5	0	-1	2	-3	-2	4	3	1	1	8	2	-1	-2	-3	4
22	S . . S	S	2	3	1	1	-2	3	-1	0	1	-2	-1	2	2	0	1	8	2	0	1	-2	4
23	. . . G	G	2	0	2	1	-2	4	0	0	0	-1	-1	1	1	1	-1	2	1	1	-3	-2	4
24	. . . D	D	1	-1	4	3	-2	2	1	0	1	-1	-1	2	1	2	0	1	1	0	-3	-1	4
25	. . . G	G	2	0	2	1	-2	4	0	0	0	-1	-1	1	1	1	-1	2	1	1	-3	-2	4
26	. A G N	A	6	0	4	3	-4	6	1	-1	1	-2	-1	5	2	2	-1	3	3	1	-5	-3	4
27	Y N Y T	Y	0	5	0	-1	5	-1	2	1	-1	0	-1	4	-3	-2	-2	0	3	0	3	6	4
28	E D D Y	D	2	-2	9	8	-3	3	4	-1	1	-3	-2	5	-1	4	-1	1	-1	-1	-6	0	9
29	L M A L	L	3	-5	-3	-1	6	-1	-2	6	-1	10	10	-2	0	0	-2	-1	0	6	-1	0	9
30	Y N A W	N	4	1	3	2	0	2	3	-1	1	-1	-1	8	0	1	-1	2	1	-1	-1	2	9
⋮	⋮																						
48	S G N S	S	4	3	5	3	-4	7	0	-2	2	-4	-3	6	3	1	0	10	3	0	-2	-4	9
49	S S N Y	S	2	5	2	1	1	2	1	0	1	-2	-2	5	1	-1	0	8	1	-1	3	1	9

Figure 1 On the left, four 49-residue long sequences from a multiple sequence alignment are shown vertically (positions 31–47 have been omitted). Next, the consensus amino acid (i.e., the most common amino acid) in each position is shown. Finally, the sequence profile is displayed as a 21 × 49 matrix. Each matrix element is a function of the frequency with which the corresponding amino acid is found in that position in the multiple alignment. The 21st column scores the frequency of insertions/deletions in the different positions in the multiple sequence alignment. In the simplest case, each matrix element is put equal to the frequency of the corresponding amino acid in that position in the alignment, but more complicated weighting schemes are often used. In the example shown, frequently occurring amino acids have positive scores, and rare amino acids have negative scores. (From Gribskov M, McLachlan AD, Eisenberg D [1987] *PNAS* 84: 4355–4358. With permission from David Eisenberg.)

Figure 8.3 Structure is better conserved than sequence. (A) Structural superposition of the bile-acid transporter homolog ASBT$_{Nm}$ (blue/red) and the Na$^+$/H$^+$ antiporter NhaA (grey). The N- and C-terminal domains have been superimposed separately. (B) A critical structural element where two broken trans-membrane helices cross over each other, creating binding sites for Na$^+$ ions. (From Hu N-J et al. [2011] *Nature* 478: 408–411. With permission from Springer Nature.)

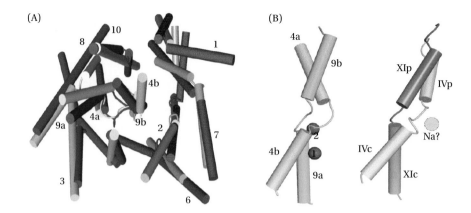

bacterium *Neisseria meningitides* and the Na$^+$/H$^+$ antiporter NhaA from *E. coli* (**Figure 8.3**). The sequence similarity between these two proteins is so low (only 12%) that it is impossible to produce a good sequence alignment, yet the structures of the two proteins overlap almost perfectly. A functionally critical crossover connection involving two transmembrane helices is also perfectly conserved. From examples such as this, it is clear that many cases of structural similarity are missed by sequence alignment algorithms.

8.1.4 Statistical Analysis Can Reveal Motifs in Membrane Helices, Such as GxxxG

A question that has been asked for both soluble and membrane proteins is whether there are amino acids or sequences that recur for structural or functional reasons. Analyses of sequence databases can give insights if the application of statistical methods reveals highly improbable recurrences of patterns. One such pattern is the GxxxG motif in transmembrane helices, which is often found at helix–helix interaction interfaces, and we use it as an example here in order to illustrate how such an analysis can be done.

We begin with the selection of a database that is appropriate for the purpose—here the SwissProt section of the UniProt database is used, since it is adequately large and contains annotated transmembrane helices identified using the methods described later, such as TMHMM. We then eliminate helices that are too similar, for example by requiring that there be less than 50% sequence identity, so that the data are not biased by multiple examples of closely related sequences. Finally, we choose a motif to search for, say AxxL. Some sequences are found that contain this motif, as shown in **Figure 8.4A**.

But there are many As and Ls in the sequences, so how can we be sure that there is any significance to our finding? We must test the observation against the occurrence of random correlations. The most complete way to do this is to computationally examine each sequence, permuting it to give a list of *all* the possible arrangements of the set of amino acids it contains, and test each of them for AxxL. While possible in this case, such a computation is too large to be done for a set of longer sequences, where other statistical methods involving more assumptions would be needed.

We then derive a "parent" distribution function for the probability of finding *n* examples of AxxL, a smooth curve that has a huge number of examples in it (**Figure 8.4B**), and we can compare the mean expected for a random model, 4043 cases in this example, with the observed number of examples in the helix sequences in the database, here 4140. Using the distribution function from the randomized sequences then allows calculation of the probability that the observation could have happened by chance: $p = 7.5 \times 10^{-2}$. This probability is not small enough to be convincing, so we cannot confidently think that the observed cases are not random.

Now, consider the case of the GxxxG motif, originally observed at the interface of the Glycophorin A transmembrane helix homodimer. Here, the finding

(A) **IVMFVAVLLIAAFVAGIL**

LLIVVATIIIVLIVIVWI

FGVGVALVAVLGAALLAL

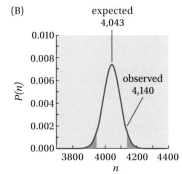

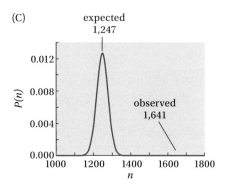

Figure 8.4 Searching for statistically overrepresented sequence motifs in transmembrane helices. (A) Three transmembrane helices, two of which contain AxxL motifs. (B) Expected distribution of the number of AxxL hits (n) obtained by many rounds of permutation of each of the transmembrane sequences in the database. The number of hits found in the original database (4140) is close to the mean number of expected hits (4043). (C) The number of hits to the GxxxG motif in the same sequence database (1641) is significantly higher than the mean number in permuted sequences (1247).

is very different: 1247 cases are expected on average, on a random basis, but 1641 cases are found in the database (**Figure 8.4C**). Calculation gives the probability that this number of observed cases happened by chance as $p = 6.4 \times 10^{-34}$. Thus, we can safely assume that there is a significance to the motif.

In this case, the role of the GxxxG motif in helix association is also supported by two other types of data: the observation of the interface in a nuclear magnetic resonance (NMR) structure of the Glycophorin A helix dimer, and a study of partially randomized sequences using a genetic selection for helix dimerization in *E. coli* plasma membranes. The most common motif found in the assay was GxxxG (and often GxxxGxxxG) (**Figure 8.5**).

8.2 PREDICTION METHODS ALLOW IDENTIFICATION OF FUNCTIONAL AND STRUCTURAL FEATURES OF MEMBRANE PROTEINS

The ultimate source of information about a given protein is experimental data of various kinds. We learn about its sequence with the help of DNA sequencing,

```
LLₗₗGVₗₗGAₗₗTₗ    GPₗₗGGₗₗGGₗₗAₗ    GVₗₗGVₗₗGLₗₗGₗ
LLₗₗGGₗₗGAₗₗTₗ    VGₗₗGVₗₗGIₗₗAₗ    AGₗₗGAₗₗGSₗₗTₗ
LVₗₗGVₗₗGAₗₗTₗ    AVₗₗGVₗₗGSₗₗTₗ    LLₗₗGVₗₗGVₗₗAₗ
VLₗₗGIₗₗGVₗₗSₗ    VLₗₗGGₗₗGAₗₗTₗ    LIₗₗGAₗₗGGₗₗTₗ
TLₗₗGAₗₗGVₗₗTₗ    LLₗₗGAₗₗGAₗₗTₗ    LVₗₗGVₗₗGVₗₗTₗ
LVₗₗGAₗₗGIₗₗTₗ    TIₗₗGVₗₗGSₗₗTₗ    LSₗₗSGₗₗGSₗₗTₗ
LLₗₗGGₗₗGAₗₗTₗ    GVₗₗGVₗₗGSₗₗTₗ    SIₗₗGIₗₗGIₗₗTₗ
LVₗₗGAₗₗGAₗₗTₗ    PLₗₗGVₗₗGIₗₗTₗ    SLₗₗGVₗₗGLₗₗAₗ
LVₗₗGVₗₗGLₗₗGₗ    PGₗₗGLₗₗGAₗₗGₗ    PLₗₗGLₗₗGLₗₗGₗ
SVₗₗGVₗₗGVₗₗTₗ    GIₗₗGIₗₗGIₗₗTₗ    TVₗₗGVₗₗGLₗₗTₗ
SVₗₗGLₗₗGAₗₗTₗ    LVₗₗGAₗₗGSₗₗTₗ    GLₗₗGIₗₗGLₗₗGₗ
VLₗₗGVₗₗGVₗₗAₗ    LLₗₗGVₗₗGLₗₗGₗ    SLₗₗGVₗₗGVₗₗTₗ
LVₗₗGVₗₗGLₗₗAₗ    SLₗₗGVₗₗGLₗₗAₗ    GVₗₗGLₗₗGVₗₗTₗ
ALₗₗGVₗₗGVₗₗAₗ    VLₗₗGIₗₗGVₗₗSₗ    VLₗₗGVₗₗGVₗₗTₗ
LVₗₗGVₗₗGVₗₗSₗ    GVₗₗGVₗₗGSₗₗTₗ    LVₗₗGIₗₗGLₗₗAₗ
SLₗₗGIₗₗGLₗₗGₗ    LVₗₗGVₗₗGLₗₗAₗ    ISₗₗSSₗₗSSₗₗTₗ
GVₗₗGIₗₗGVₗₗTₗ
```

Figure 8.5 GxxxG dimerization motifs found in a genetic screen for transmembrane helices with strong tendency to dimerize. (From Russ WP, Engelman DM [2000] *J Mol Biol* 296: 911–919. With permission from Elsevier.)

about its subcellular localization and physiological role using the wide arsenal of techniques offered by molecular cell biology, and about its structure and molecular function by biochemical, biophysical, and structural studies. Bioinformatics strives to summarize much of the information gleaned from the totality of such studies across large numbers of proteins in the form of algorithms (often available as web servers) that attempt to predict different characteristics of proteins based on their amino acid sequence. For membrane proteins, the currently most important bioinformatics tools predict subcellular localization, topology, and 3-D structure.

8.2.1 The Subcellular Localization of Proteins Can Be Predicted

As we discussed in Chapter 5, cells have intricate processes for sorting proteins between different subcellular compartments, and these machineries are often the same ones that mediate insertion of membrane proteins into organelle membranes. Protein sorting depends on signal peptides in the protein; examples include signal peptides that target proteins to the secretory pathway, mitochondria, peroxisomes, chloroplasts, and the nucleus.

Because signal peptides are short stretches of amino acids found at the termini of, or embedded within, proteins, and since they are recognized by receptors that bind specifically to this short stretch of amino acids, it ought to be possible to develop prediction methods that imitate nature's ability to recognize signal peptides in protein sequences. This is indeed the case, and many such methods are available.

A good example is the SignalP predictor, which has been continually improved since the first version came out in 1997. SignalP uses machine learning—specifically Artificial Neural Networks (ANNs; see **Box 8.3**)—to build predictors that are trained to identify proteins endowed with N-terminal signal peptides that target them to the Sec translocon in the ER (or the homologous SecYEG translocon in the inner membrane of bacteria). As seen in the logo plot in **Figure 8.6**, this kind of signal peptide has a rather well-conserved tripartite structure with a short, positively charged N-terminal part (n-region), a central hydrophobic part (h-region), and a short C-terminal part (c-region) that contains a cleavage site recognized by the signal peptidase enzyme. The cleavage site is characterized by small residues (typically Ala, Gly, Ser, or Val) in positions –1 and –3 and polar residues in positions –4 to –6. Also, Pro is almost never found in position +1.

SignalP and other similar machine-learning predictors are trained in a semiautomatic fashion by presenting them with a collection of true signal peptides together with a similar number of sequences that are not signal peptides. During the training phase, a large set of internal parameters (typically a few hundred

Figure 8.6 Logo plot of a collection of 266 signal peptides from periplasmic and outer membrane proteins in Gram-negative bacteria. The signal peptides have been aligned from the signal peptidase cleavage site (located between positions –1 and +1). Each column in the logo plot consists of stacks of letters representing amino acids. The overall height of a stack indicates the degree of sequence conservation at that position (measured in bits), while the height of each letter within the stack indicates the relative frequency of the corresponding amino acid at that position. The n-region contains positively charged Lys (K) and Arg (R) residues (blue), the h-region is rich in hydrophobic residues such as Leu (L), Ala (A), and Val (V) (black), and the c-region has a conserved motif with small amino acids (mainly Ala (A), Gly (G), and Ser (S)) in positions –1 and –3. Negatively charged Asp (D) and Glu (E) residues (red) are almost completely absent from the signal peptides. (Logo plots are described in *Nucleic Acids Res.* (1990) **18**: 6097. Nielsen H, Engelbrecht J [1997] *Protein Engineer* 10: 1–6. With permission from Oxford University Press.)

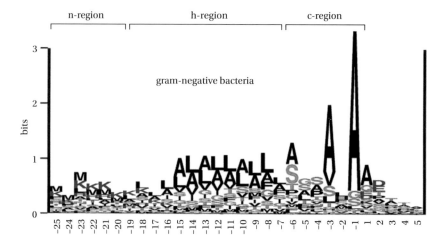

BOX 8.3   ARTIFICIAL NEURAL NETWORKS

Artificial neural networks (ANNs) were originally developed for pattern recognition tasks in, e.g., speech recognition software. The first application in bioinformatics was for prediction of secondary structure in globular proteins. ANNs have also been used for prediction of signal peptides and transmembrane helices.

ANNs are inspired by how memories might be stored as connections between neurons in the brain. The basic idea is to process the information in the input data (an amino acid sequence, say) through an interconnected system of nodes that can send signals to each other. The pattern to be recognized by a predictor based on an ANN is encoded as a set of parameters that define the connection strengths between the nodes.

A schematic diagram of an ANN designed to predict secondary structure (α-helix, β-strand, coil) in globular proteins from amino acid sequence is shown in **Figure 1**. It is composed of nodes organized into an input layer, a hidden layer, and an output layer. The input layer reads a segment of n amino acids from the sequence. The input layer usually consists of 20*n nodes, where the first 20 nodes encode the first amino acid, the next 20 nodes encode the second amino acid, and so on. Among each set of 20 input nodes, each node corresponds to one of the 20 common coded amino acids (omitting selenocysteine and pyrrolysine) and is set to 1 if that amino acid is present in the corresponding position in the amino acid sequence; otherwise, it is set to 0. In this way, a segment of n amino

acids is represented as a string of n 1s and 19*n 0s in the input layer.

As shown in **Figure 1A**, every node in the input layer is connected to every node in the hidden layer, but the strengths of the connections vary. The information in the input layer is propagated to each node in the hidden layer by multiplying the value (1 or 0) of each input node by the connection strength s ($0 < s < 1$) between the two nodes and then summing all the inputs for each hidden layer node. Finally, the information is propagated from the hidden layer nodes to the nodes in the output layer in the same way. In the example shown in **Figure 1B**, the output layer has three nodes corresponding to the three kinds of secondary structure, and the output node with the highest value is chosen as the prediction for the middle amino acid in the n-residue-long segment read by the input nodes. The input layer is then shifted one position over the amino acid sequence, and the whole process is repeated.

There are three critical steps in the construction of an ANN-based predictor: choosing a training set and a test set of protein sequences, choosing an architecture for the ANN, and the actual training of the ANN. The training set should contain a balanced number of validated examples, e.g., sequence segments that are known to form an α-helix, a β-strand, or an irregular structure (coil). The training set should be pruned of segments that are too similar to each other, i.e., it should be non-redundant. The test set should also include validated instances of α-helix, β-strand, and

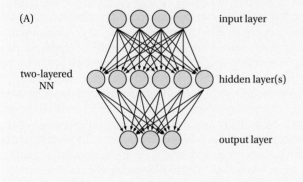

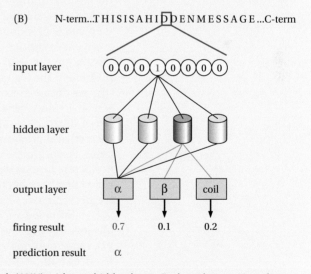

Figure 1 (A) Schematic diagram of an Artificial Neural Network (ANN) with one hidden layer. Each node in a given layer is connected to all nodes in the next layer (arrows). The individual connection strengths between nodes are optimized during the training phase. (B) Part of an ANN designed to predict protein secondary structure. The secondary structure prediction ANN reads in a sequence segment that is n residues long. Each amino acid in the segment is represented by 20 input nodes, one of which is set to 1, so the entire input layer has 20*n nodes (this way of encoding an amino acid sequence is called "sparse encoding" since 95% of the nodes are set to 0). The information fed into the input layer propagates through the network and is finally condensed into the three output nodes that represent the input segment's propensity to form an α-helix, β-strand, or coil structure. The node with the highest value is chosen as the predicted secondary structure of the central amino acid in the segment. (From Zvelebil MJ, Baum JO [2007] *Understanding Bioinformatics*. Garland Science, New York. With permission from Taylor & Francis.)

(Continued)

BOX 8.3   ARTIFICIAL NEURAL NETWORKS (CONTINUED)

coil segments, should not contain sequences with high similarity to the segments in the training set, and should be internally non-redundant. Constructing good training and test sets is absolutely essential and is usually the most time-consuming part of the process.

Next, the ANN architecture must be specified in terms of the number of hidden layers, the number of nodes in each layer, and a function *f* that describes the input–output relation for the nodes. *f* is usually a non-linear function; in the simplest case, it is chosen as a step function such that a given node in the hidden layer or the output layer is set to 1 if the summed inputs are larger than a given threshold (the "node fires") and set to 0 otherwise.

The training of the ANN is normally carried out in a "supervised learning" mode. The training algorithm automatically presents each segment in turn to the ANN, records the values of the output nodes, and makes small adjustments in the internal connection strengths between each node in such a way that the values in the output nodes change slightly in the direction of the correct result. This is repeated until the prediction performance on the training set starts to plateau, whereupon the training session

is terminated. If the training is allowed to proceed too far, the ANN might start encoding particulars of the segments in the training set rather than the general features of the sequence motifs one is looking for, and the ANN may become overtrained.

Since it is difficult to avoid some degree of overtraining, the "true" performance of the ANN must be evaluated on a test set that is independent of the training set. One way to achieve this without sacrificing an important part of the painstakingly assembled, verified set of positive and negative examples is to employ a cross-validation procedure. The idea is to split the original set of non-redundant sequences into five (say) equal-size subsets and then train an ANN using four of the subsets as the training set and the remaining subset as the test set. This is done a total of five times with the five different subsets as the test set, and the overall performance of the global ANN (where the entire set of examples is used for training) is estimated as the mean performance over the five ANNs obtained by cross-validation. ANNs provide a good way to encode simple linear sequence motifs and have found many uses in bioinformatics.

or more) that are defined by the underlying ANN architecture is systematically optimized until there is no further improvement in the predictor's ability to discriminate between the true signal peptides and the other sequences. To estimate how well the predictor will perform on new sequences that were not included in the training set (i.e., its ability to generalize from the training data), it is evaluated on a test set of new sequences that have less than a preset percentage of sequence identity (e.g., 30%) to the sequences in the training set.

A typical output from SignalP is shown in **Figure 8.7**. SignalP correctly identifies the presence of a signal peptide (high S score) and correctly predicts the position of the signal peptidase cleavage site between residues Ala[21] and Arg[22] (the position of maximal Y score). Overall, SignalP predicts signal peptides with sensitivity and specificity values around 0.9 (i.e., on average SignalP finds ~90 out of 100 true signal peptides, and ~90 out of 100 predicted signal peptides are correct) and predicts the correct signal peptidase cleavage site in 70–80% of the cases. Performance can be further improved by, e.g., training organism-specific predictors and by using more advanced machine-learning algorithms. Similar predictors have been developed for many other sorting signals, but none reach the performance levels that SignalP has on ER-targeting signal peptides.

An alternative approach to predicting subcellular localization is based on the observation that the environment differs between organelles in terms of, e.g., pH, redox potential, and the presence of proteases, and assumes that such differences will lead to different selection pressures on the proteins and hence, be reflected in their overall amino acid composition. Prediction methods that analyze the overall amino acid composition of proteins generally perform less well than those that look for specific sorting signals but provide complementary information and can be combined with signal-based methods. They can also be used to predict subcellular localization for compartments for which the sorting signal is unknown.

One can also use sequence similarity to predict subcellular localization based on the assumption that if the location of a close homolog is known, then the query protein is likely to be found in the same location. A variation on this theme is that certain protein domains function only in one specific location;

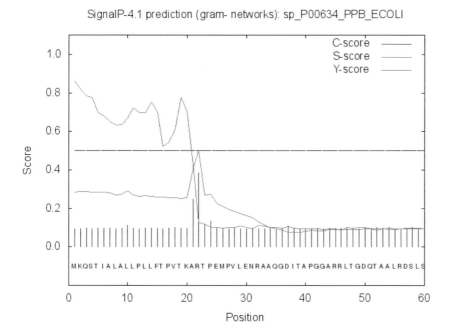

Figure 8.7 SignalP prediction for the N-terminal part of the periplasmic protein PhoA from *E. coli*. The S score (green) distinguishes between parts of the sequence that are predicted to belong to the signal peptide (above the purple cutoff) or not to be part of the signal peptide. The C score (red) is a measure of how similar the local sequence is to a typical cleavage site, and the Y score (blue) is obtained by multiplying the C score with the derivative of the S score, i.e., the Y score is high for positions with a high C score in the region where the S score predicts that the signal peptide ends. The signal peptide is (correctly) predicted to be cleaved between residues Ala[21] and Arg[22].

hence, if such a domain is found in a query protein, it is likely located in that location.

There are also global predictors of subcellular localization that try to distinguish between multiple possible locations at the same time. As an example, SignalP has been combined with two other predictors that predict mitochondrial and chloroplast signal peptides into the global predictor TargetP. The ability to predict the individual locations increases when the three individual predictors have to "compete" to make the final prediction. The PSORT family of predictors predicts a much wider selection of possible subcellular localizations; PSORTb is designed for bacterial proteins, and WoLF PSORT is designed for eukaryotic proteins.

The average lengths of transmembrane segments in single-spanning membrane proteins that are located in different organelles along the secretory pathway differ significantly depending on the organelle (**Figure 8.8**). This observation may reflect differences in the local bilayer thickness of regions in the organelle membranes, although average membrane thicknesses may not follow such a rule. Such differences, together with minor differences in amino acid composition, can be used to predict the target membrane (endoplasmic reticulum (ER), Golgi, or plasma membrane) of single-spanning proteins.

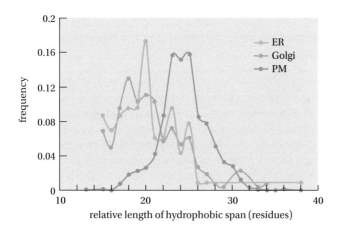

Figure 8.8 Transmembrane helices from vertebrate single-spanning membrane proteins are shorter for proteins retained in the ER (green) or Golgi (orange) than for proteins located in the plasma membrane (PM; blue). (From Sharpe HJ, Stevens TJ, Munro S [2010] *Cell* 142: 158–169.)

8.2.2 Membrane Proteins Can Be Identified in Genomic Data

The best way to identify helix-bundle membrane proteins in sequence databases or proteomes derived from whole-genome sequencing is simply to use topology prediction algorithms (see next section) and classify all sequences with one or more predicted transmembrane helix as integral membrane proteins. With the best topology predictors, you can expect to find more than 90% of all true helix-bundle membrane proteins in this way with less than 10% false-positive predictions. A caveat to keep in mind when analyzing proteomes predicted from full-genome DNA sequences is that short open reading frames (typically <100 codons) are generally not included in the lists because they are highly likely to appear even in random DNA sequences. Nevertheless, many naturally occurring proteins, including membrane proteins, are hidden among these short open reading frames and will only be identified if specifically looked for in the original DNA sequence.

There is one class of proteins that is difficult to distinguish from the true helix-bundle membrane proteins, namely, those that have an N-terminal signal peptide targeting them to the secretory pathway. As is clear from Figure 8.6, such signal peptides look much like transmembrane helices: a stretch of hydrophobic amino acids flanked by positively charged amino acids on one end. This makes it difficult for sequence-based predictors to make a clean distinction between a cleavable signal peptide and an N-terminal transmembrane helix and hence, to separate secreted proteins from N-terminally anchored, single-spanning membrane proteins. On average, the hydrophobic region in signal peptides is shorter than in transmembrane helices, and more signal peptides than transmembrane helices have a strongly predicted signal peptidase cleavage site at their C-terminus, but these features are not always sufficient for a reliable prediction. The SignalP 4.0 predictor has been optimized to discriminate between signal peptides and N-terminal transmembrane helices, and some topology predictors (see next section) include a module to do the same job.

β-barrel membrane proteins are not so easy to identify, since they differ less from water-soluble proteins in terms of amino acid sequence. Only every other amino acid in a transmembrane β-strand faces the lipid and hence, needs to be hydrophobic. Also, the strands are quite short and hence, do not stand out in the sequence in the way that the long hydrophobic transmembrane helices do.

8.2.3 Topology Can Be Predicted for Both Helix-Bundle and β-Barrel Proteins

The topology of a membrane protein, i.e., a specification of which parts of the sequence form transmembrane helices or β-strands and their orientation relative to the membrane (N_{in} or N_{out}), provides a basic description of the protein's secondary structure and its overall membrane orientation. As such, a topology model is a very useful guide to building structural models and to the planning and interpretation of mutagenesis experiments when no high-resolution structure is available.

In the early days of membrane protein bioinformatics, the topology of helix-bundle membrane proteins was predicted simply by trying to identify segments of a high average hydrophobicity that might form transmembrane helices. The classic method dates from 1982 and is due to Jack Kyte and Russel Doolittle. Kyte and Doolittle did two critical things. They constructed a hydrophobicity scale based on the then available data on transfer free energies of amino acid side chains from liquid water to a vapor phase as well as statistics on surface exposure of different residues gleaned from 3-D structures of globular proteins (see Chapters 3 and 6 for more on hydrophobicity scales), and they wrote a simple computer program (37 lines of code in the programming language C) to calculate and plot a running average of the hydrophobicity along the amino acid sequence. Transmembrane helices were predicted by requiring an average hydrophobicity >+1.6 over a 7 to 13 residue averaging window. Later, it was realized that it makes more sense to use a larger window to represent a bilayer hydrophobic thickness, say 19 amino acids. The original Kyte–Doolittle plot for bacteriorhodopsin is shown in **Figure 8.9**.

A major difficulty with topology predictions based only on hydrophobicity turned out to be that the least hydrophobic transmembrane helices in many multi-spanning membrane proteins are less hydrophobic than the most hydrophobic segments found in many secreted proteins or even in some extramembranous regions in membrane proteins. Moreover, hydrophobicity-based methods cannot predict the orientation of proteins relative to the membrane. Both problems could be addressed once the positive-inside rule was discovered (Chapter 6).

The positive-inside rule states that the part of the protein that faces the cytoplasm (or, more generally, the side of the membrane from which the protein is inserted) has a higher content of positively charged residues (Lys, Arg) than the part that faces the other side of the membrane. Obviously, once the transmembrane helices in a protein are known (or predicted), it is trivial to predict the orientation of the protein using the positive-inside rule. But the positive-inside rule can also be used to improve the prediction of the transmembrane helices themselves. This was first done in 1992 in the TopPred algorithm.

In the TopPred algorithm, a Kyte–Doolittle-type hydrophobicity plot is first produced. Then, as shown in **Figure 8.10**, two thresholds are applied to predict a set of "unambiguous" transmembrane helices and a set of "putative" transmembrane helices. The putative helices have a relatively low hydrophobicity and cannot be confidently predicted to be either helices or loops. In the example shown in **Figure 8.10** (the LacY protein from *E. coli*), 11 peaks in the hydrophobicity plot are above the upper threshold and hence classified as unambiguous transmembrane helices, but the 12th peak (indicated by the arrow) is classified only as putative. This leaves two possible topologies, one with 11 and one with 12 transmembrane helices, and the choice between them cannot be made based on hydrophobicity alone. This is where the positive-inside rule is brought into play. The simple idea is to choose the topology that has the highest bias in the distribution of positively charged residues between the two sides of the membrane. For LacY, the model with 12 transmembrane helices has a much higher bias than the one with 11 and is in fact the correct model, as has been shown experimentally. TopPred gave a dramatic improvement in prediction performance, and all modern topology prediction methods basically rely on the idea of simultaneously optimizing hydrophobicity and positive-charge bias.

Today, the best topology prediction methods are built on machine-learning algorithms, most commonly hidden Markov models (HMM, **Box 8.4**). The HMM formalism makes it possible to encode the basic rules of membrane protein topology from the outset and then optimize the internal parameters of the model over a training set of proteins with known topology. As an illustration, the architecture underlying the TMHMM predictor is shown in **Figure 8.11**. It

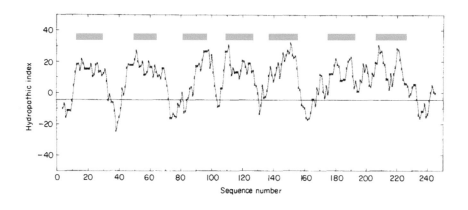

Figure 8.9 Kyte–Doolittle plot for bacteriorhodopsin obtained with a window size of seven residues. The transmembrane helices (as determined by x-ray crystallography) are indicated by the green bars. Five of the seven transmembrane helices are clearly delineated as prominent peaks, while the two C-terminal transmembrane helices are less well defined on the plot. (From Kyte J, Doolittle RF [1982] *J Mol Biol* 157: 105–132. With permission from Elsevier.)

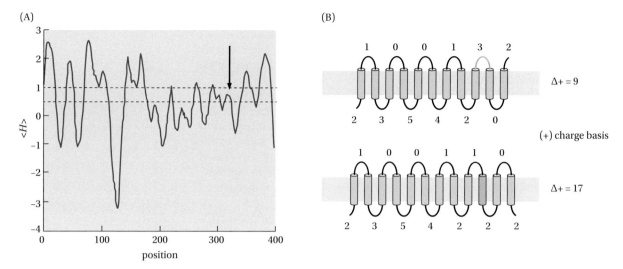

Figure 8.10 The TopPred algorithm applied to the LacY protein. (A) Hydrophobicity plot for LacY. The two hydrophobicity thresholds used to classify hydrophobic segments in the protein as "unambiguous" and "putative" transmembrane helices are shown as dashed lines. (B) Topology models of LacY. The single "putative" transmembrane helix found in the protein sequence is indicated by the arrow in panel A and is shown in green in the two topology models. Also shown on the two topology models is the number of Arg+Lys residues in the loops and N- and C-terminal tails. The total (Arg+Lys) charge bias (Δ+) is indicated. The bottom model with 12 transmembrane helices has the largest charge bias and is the correct one. (From von Heijne G [1982] *J Mol Biol* 225: 487–494.)

Figure 8.11 The TMHMM architecture encodes the membrane protein topology as a series of connected states. Each state describes one type of sequence element: an inside tail or loop, a transmembrane helix (with capping residues on both ends), short and long outside tails or loops, and large globular domains. The internal structure of the transmembrane helix state is shown at the bottom; it has a pattern of allowed transitions designed to fit hydrophobic segments as short as 5 and as long as 25 residues. The cap and loop states have similar internal structures to allow for variation in segment lengths. The optimal threading of the sequence of the protein shown at the top through the TMHMM model is such that the short N-terminal tail fits best into the "short outside loop" state (i.e., it is polar but lacks Arg and Lys residues), the following segment fits the cap-helix-cap states, the next segment fits well into the "inside loop state" (i.e., it is polar and rich in Arg and Lys residues), and so on. TMHMM therefore predicts that the topology is 3TM-Nout. (Top and middle panels: From Zvelebil MJ, Baum JO [2007] *Understanding Bioinformatics.* Garland Science, New York. With permission from Taylor & Francis. Bottom panel: From Krogh A, Larsson B, von Heijne G et al. [2001] *J Mol Biol* 305: 567–580. With permission from Elsevier.)

is based on the premise that there is a simple "grammar" that describes a membrane protein: an N-terminal tail located on the outside of the cell membrane (say) must be followed by a transmembrane helix, which must be followed by an inside loop, which must be followed by a transmembrane segment and outside loop, and so on until the C-terminal tail is reached. This somewhat oversimplified picture (see next section) is recapitulated in the TMHMM model. The transition probabilities between the different states in the model, together with probabilities that describe the typical amino acid composition of the different states, are optimized during the training session.

To produce a prediction, the protein sequence is "threaded" through the model to find the optimal fit between the sequence and the model such that segments of the sequence that best fit the amino acid composition of the "inside

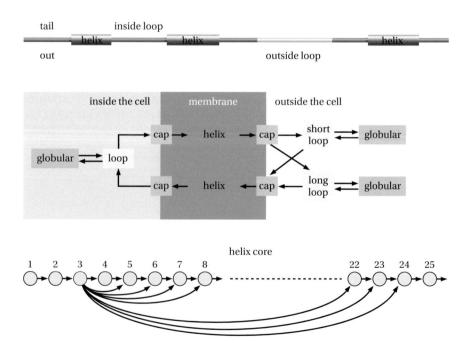

BOX 8.4 HIDDEN MARKOV MODELS

The sequence profile method (Box 8.2) for representing the properties of a multiple sequence alignment is excellent, but an even better representation can be obtained using an HMM. A Markov model (MM) is a statistical model that describes a random variable that fluctuates between different states and where the change in the random variable between one state and the next depends only on its current state, not on previous states (i.e., the system has no memory). A very simple example of an MM that generates a random string of zeroes and ones is shown diagrammatically in **Figure 1**.

For the MM in Figure 1, we can deduce the path that has been taken through the model from the emitted string of

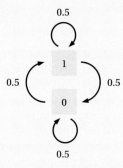

Figure 1 A Markov Model (MM) that generates a random string of ones and zeroes. State "1" outputs ("emits") a one, and state "0" emits a zero. All "transition probabilities" from a state back to itself and between states are 0.5.

zeroes and ones. This is not so for an HMM, where many different paths through the model may result in the same emitted string; the exact path is "hidden" from us. The HMM formalism has found widespread use in science and technology, in areas ranging from speech recognition to bioinformatics.

Within bioinformatics, the basic idea of HMMs can be illustrated by imagining a genie that knows only one thing but knows it well: how to design amino acid sequences that all fold into a functional enzyme (lysozyme, say). The genie

writes out a multitude of such sequences on a computer and e-mails them to a biologist. The genie has made life easy for itself: it has a series of boxes, one for each position in the lysozyme sequence, and in each box there are a large number of balls with the names of the 20 amino acids written on them. In each box, the overall composition of balls reflects the evolutionary conservation of residues in that particular position in lysozyme. To generate a new sequence, the genie pulls a ball from each box in turn and reads off the amino acids one by one. In order to generate insertions and deletions, each time the genie goes from one box to the next, it can choose (with certain probabilities) either to skip over the next box (generating a deletion in the sequence) or to go to an "intermediate" box that contains a random collection of balls (generating an insertion in the sequence). The biologist sees only the final sequence but does not know from which box any given amino acid in the sequence has been pulled (the boxes are "hidden" from the biologist) and thus, cannot know for certain which path through the boxes the genie has taken.

Similar to the genie's method, a typical HMM used to represent multiple sequence alignments is shown in **Figure 2**. There are a number of different states in the model (corresponding to the genie's boxes):

[S] Start state. No amino acid is emitted from this state.

[N] N-terminal unaligned sequence state. Emits an amino acid according to a background probability distribution. Note that one can make a transition from [N] to [B] or one can return to [N] and generate a new amino acid from this state.

[B] Begin state (for entering the main model). No amino acid is emitted from this state.

[Mx] Match state x. Emits an amino acid with probabilities corresponding to the frequencies of the amino acids in position x in the multiple sequence alignment.

[Dx] Delete state x. No amino acid is emitted from this state.

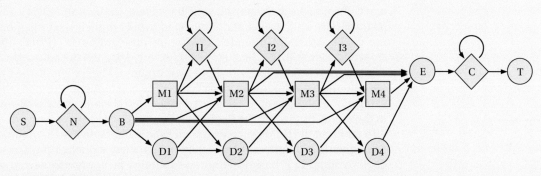

Figure 2 Hidden MM (HMM) architecture used to represent multiple sequence alignments (e.g., protein domain families). Match states are denoted by "M," insert states by "I," and delete states by "D." The "N" and "C" states model variable flanking sequences that are not part of the alignment. "S" and "T" are the start and terminate states.

(Continued)

[Ix] Insert state x. Emits an amino acid according to a background probability distribution.

[E] End state (for exiting the main model). No amino acid is emitted from this state.

[C] C-terminal unaligned sequence state. Emits an amino acid according to a background probability distribution.

[T] Terminate state. No amino acid is emitted from this state.

All the parameters in the HMM (the emission and transition probabilities) are estimated by a systematic training procedure using precompiled training data (such as a multiple sequence alignment).

Similarly to the genie, by moving through an HMM of this kind, a computer can generate sequences that reflect the underlying architecture and probabilities of the HMM, but one can also use the HMM formalism to calculate the probability that a given sequence has been generated by the HMM. This is how HMMs are used in "prediction mode": one first builds an HMM that represents a particular multiple sequence alignment (one including all known lysozyme sequences, say) and then uses this HMM to search for candidate lysozyme sequences in, e.g., a newly sequenced genome.

HMMs can also be used to represent other kinds of sequence characteristics, such as membrane protein topologies. A typical example is the TMHMM topology predictor shown in Figure 8.11.

loop" states (i.e., segments rich in Arg and Lys) tend to end up in that state, segments that best fit the amino acid composition of the helix states (i.e., segments rich in hydrophobic amino acids) tend to end up in those states, etc. The optimal threading is the best prediction for the topology of the protein, given the model. The general principle that information from a multiple sequence alignment improves predictions holds also for topology predictions, and the best current prediction servers, such as TOPCONS, include sequence profiles as input.

With modern methods, one can expect single-sequence topology predictions to be correct for around 70% of all sequences (to count as correct, a prediction should have the right number of transmembrane helices and the right orientation relative to the membrane) and profile-based methods to reach around 80% correct predictions. When using topology predictors to discriminate between integral membrane proteins (i.e., proteins with ≥1 predicted transmembrane helix) and globular proteins (i.e., proteins with no predicted transmembrane helices), one can expect sensitivities and specificities above 90% (but note the complication discussed earlier, that N-terminal signal peptides in secreted globular proteins sometimes score as transmembrane helices). Most topology prediction methods allow constraints based on pre-existing knowledge (such as the inside or outside location of the N- or C-terminus, or a particular loop) to be incorporated from the outset; obviously, this will also increase the prediction accuracy.

Much less work has been done on topology prediction for β-barrel membrane proteins. There are some general structural features that are common to most (but not all) β-barrel proteins: the number of β-strands is even, the N- and C-termini are at the periplasmic barrel end, the β-strand tilt is <45°, and all β-strands are antiparallel and connected locally to their next neighbors along the chain. An HMM intended to capture this "grammar" (PRED-TMBB2) is shown in **Figure 8.12**. The β-strand states are designed to capture the fact that every second residue faces the lipid membrane and hence should be hydrophobic, and the experimentally observed very strong tendency for both the N- and C-termini of the barrel to be located in the periplasmic space is also included in the model by placing the Beginning (B) and End (E) states in the periplasm.

In addition to HMMs such as the one shown in **Figure 8.12**, methods based on sequence similarity and carefully constructed sequence profiles are quite good at spotting β-barrel membrane proteins in newly sequenced genomes or sequence databases. As an example, one such method, HHomp, finds 57 of 59 known outer membrane proteins in the well-studied bacterium *E. coli* and seems to have a low rate of false-positive hits. In contrast to helix-bundle membrane protein predictors such as TMHMM and PRED-TMBB2, this kind of method can only find new proteins that are related by detectable sequence

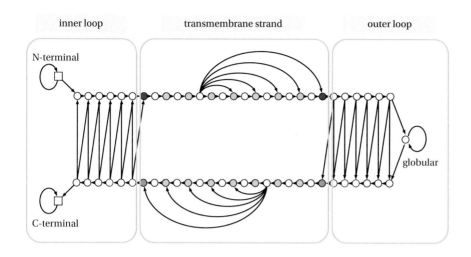

inner loop   transmembrane strand   outer loop

N-terminal

C-terminal

globular

Figure 8.12 The PRED-TMBB2 hidden Markov model for β-barrel membrane proteins. Note the alternating white and pale red states in the transmembrane β-strand part of the model, included to make the model able to encode the typical β-strand pattern where every second residue is hydrophobic. The β-strands are allowed to vary in length between 7 and 17 residues. (Bagos PG, Liakopoulos TD, Spyropoulos IC et al. [2004] *BMC Bioinformatics* 5:29. https://doi.org/10.1186/1471-2105-5-29. With permission from Springer Nature.)

similarity to already known β-barrel membrane proteins, however. The HHomp method does not explicitly predict the topology of the protein, but the output can be used as a guide for homology-based 3-D structure modeling.

8.2.4 Interfacial Helices, Reentrant Loops, and Kinks

The simple description of membrane protein topology given earlier, while very useful, does not completely capture the architectural principles gleaned from the known high-resolution structures. In particular, it ignores important (and possibly "predictable") structural elements such as interfacial helices that lie flat on the surface of the membrane and reentrant loops (short segments that penetrate part of the way across the membrane but then turn back). Proteins in the aquaporin family (see Chapter 10) all have two reentrant loops that meet in the middle of the membrane and together form what might be considered a seventh transmembrane helix (split in two halves) (**Figure 8.13**). Prominent reentrant loops are also found in many ion channels, where they form part of the selectivity filter (see Chapter 11). Reentrant loops tend to be mostly shielded from direct lipid contact in the folded structure and are generally not sufficiently hydrophobic to be scored as potential transmembrane helices by topology predictors. Attempts have been made to train predictors to specifically recognize reentrant loops and interfacial helices, but so far with limited success. Reentrant loops are also found frequently in β-barrel membrane proteins but are not included in the current β-barrel topology prediction programs.

A structural feature that is possible to predict rather well, however, is kinks in transmembrane helices. Kinks are often, but not always, induced by proline residues and are characterized by a sudden break in the direction of the helix axis and loss of backbone hydrogen bonds. Kinks can be predicted from a multiple sequence alignment: if more than 10% of the sequences in the alignment have a Pro residue in a particular position within a predicted transmembrane helix, there is a high probability that all the helices in the alignment have a kink in this position, even if in some of the helices there is no proline. Presumably, the surrounding protein has evolved such that the kink is maintained by packing interactions even if the proline is mutated (Figure 4.7). Neural networks (NNs) have also been applied to this problem with good success. Prolines are more abundant in transmembrane helices than in α-helices in soluble proteins, suggesting that kinks in transmembrane helices often are functionally important.

8.2.5 GPI Anchors Can Be Predicted

As we described in Chapter 6, many cell-surface proteins are anchored to the plasma membrane by a C-terminal glycosylphosphatidylinositol (GPI) moiety.

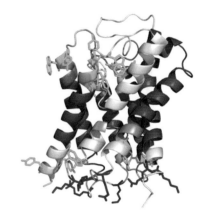

Figure 8.13 Reentrant loops in aquaporin (PDB: 1FQY). The six transmembrane helices are all hydrophobic, as indicated by the red color. The two reentrant loops (in white) are less hydrophobic. The two half-helices (one from each reentrant loop) meet in the middle of the membrane, but neither reentrant loop spans the membrane. Aromatic Trp and Tyr residues near the lipid–water interface region are shown in green, and positively charged residues in blue (the intracellular side is at the bottom; note the positive-inside bias).

Figure 8.14 FragAnchor identification of GPI-anchor signal. (A) The GPI-attachment signal can be decomposed into an anchor site where the GPI moiety is attached by the transamidase while the rest of the signal is cleaved off, a polar spacer region, and a hydrophobic C-terminal tail. (B) The HMM architecture used in FragAnchor to represent the GPI-attachment signal. Note how it is possible to jump from state 3 to any of the states 4 to 13, and from state *i* (where *i* ≥ 16) to states *i* + 1 to 36, making it possible to model spacer and tail regions of different lengths. (From Poisson G, Chauve C, Chen X et al. [2007] *Genomics, Proteomics and Bioinformatics* 5: 121–130. With permission from Elsevier.)

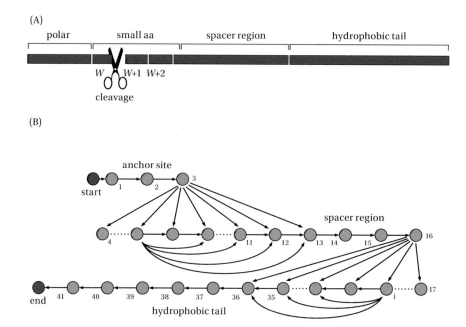

The GPI anchor is added by a transamidase enzyme, located in the lumen of the ER, that recognizes a C-terminal GPI-attachment signal in the substrate protein. The GPI-attachment signal is removed from the protein concomitantly with the addition of the GPI anchor.

Several predictors that try to identify GPI-attachment signals and predict the site of GPI addition are available. One example is FragAnchor (**Figure 8.14**), a predictor that combines an NN and an HMM. The neural network was trained on a positive training set of 79 known, 50-residue-long GPI-attachment signals and a negative training set of 79 50-residue-long C-terminal sequences from proteins known not to be GPI-anchored. Each amino acid in the input sequence is represented by two values: its hydrophobicity (according to the Kyte–Doolittle scale) and its molecular weight. Because the input sequence is 50 residues long, there are thus 100 input nodes in the NN. In the predictor, the NN is used in a first step to find potential GPI-attachment signals, and these potential signals are then fed into the HMM for a refined prediction. The HMM was designed to represent the tri-partite nature of the GPI-attachment signal, as shown in Figure 8.14. The output from the full predictor is a prediction for the GPI-attachment site and a reliability score that indicates how strong the prediction is.

8.2.6 Prediction of Membrane Protein 3-D Structure Is Possible

Because topology can be predicted quite well for helix-bundle membrane proteins, it is reasonable to ask if it is possible to predict the full 3-D structure of membrane proteins from the amino acid sequence. There are two major ways to approach this problem: homology modeling and structure prediction based on evolutionary covariation. Homology modeling is only possible when the protein to be modeled (the target) has a sequence homolog of known 3-D structure (the template). The modeling is done by first aligning the sequence of the target protein to the sequence of the template protein and then exchanging the residues in the target 3-D structure according to the sequence alignment. One then performs an energy minimization calculation to remove steric clashes from the model. Homology modeling appears to work about as well for membrane proteins as for globular proteins. In practice, this means that if the sequence

SPREAD 8.1 BIOINFORMATIC ANALYSIS OF BOVINE AQUAPORIN-1: A WORKED EXAMPLE

Water channels are discussed in Chapter 10. One member of this large protein family is aquaporin 1 (AQP1). Because a high-resolution x-ray structure is available for bovine AQP1 (PDB code 1J4N), topology and structure prediction are obviously not necessary; nevertheless, it is instructive to compare the output from different predictors with the known structure, especially because of the novel topology of AQP1, which has two reentrant helices in addition to six transmembrane helices. We begin by downloading its amino acid sequence from UniProt (**Figure 1**).

We can ask a number of questions regarding this sequence. First, does the protein have an N-terminal signal peptide that will target the protein to any particular organelle (see Chapter 5)? To address this question, we use two predictors: TargetP and SignalP. TargetP is designed to predict whether a protein is targeted to the secretory pathway, to mitochondria, to chloroplasts (in plant cells), or none of the above. If TargetP predicts that the protein will go to the secretory pathway, SignalP will be used to predict whether the signal peptide will be cleaved off or not. Here is the output from TargetP (**Figure 2**).

As we see, the score for the presence of a mitochondrial targeting peptide (mTP) is low, while the score for the presence of a secretory signal peptide (SP) is high. Consequently, the predicted location of AQP is the secretory pathway (LOC = S), and since the score is close to 1, the prediction is in the highest reliability class (RC = 1).

Even though TargetP predicts the presence of an SP, it is not certain whether this peptide will be cleaved off by the signal peptidase in the ER or remain attached to the protein as an N-terminal transmembrane helix. SignalP has been optimized to recognize cleaved N-terminal SPs and to discriminate those from uncleaved transmembrane helices. We use SignalP to find out if AQP1 has a cleavable SP.

SignalP 5.0 predicts cleaved signal peptides (as well as cleaved Tat and lipoprotein signal peptides) but scores, e.g., uncleaved transmembrane helices as "other." For AQP1, SignalP 5.0 predicts no cleaved signal peptide, leading us to

conclude that the first hydrophobic segment in the protein is most likely a transmembrane helix (**Figure 3**).

So, now we know that AQP1 is likely to have an N-terminal transmembrane helix that also serves as a targeting signal for the secretory pathway. But what does the rest of the protein look like? To answer this question, we try some topology predictors. There are many such predictors, and a good way to find out how consistent the predicted topology of the protein is across a number of predictors is to use a consensus method such as TopCons (http://topcons.cbr.su.se), which runs five different topology predictors that search for homologs and base their prediction on a multiple sequence alignment. In the case of AQP1, TOPCONS predicts a topology with six transmembrane helices and the N- and C-termini in the cytosol (i.e., "inside" the cell), but a couple of the individual predictors only predict five helices. TOPCONS also provides an estimate of the free energy of membrane insertion of each 19-residue-long segment along the chain (the ΔG value as calculated from the "biological" hydrophobicity scale described in Chapter 6) (**Figure 4**).

How well does the consensus prediction made by TOPCONS match the structure of AQP1? In the x-ray structure shown, predicted segments of different colors are superimposed on the otherwise all green backbone. The non-green segments correspond to the six predicted transmembrane helices, and the remaining green parts correspond to the predicted connecting loops. Overall, the correspondence is quite good. Interestingly, the two reentrant helices that meet in the middle of the membrane (the two green helices in the foreground) and that contain the two NPA signature motifs (see Chapter 10) are not predicted as transmembrane helices, suggesting that they bury themselves in the membrane only during a late stage of folding and not during the initial membrane insertion step (**Figure 5**).

Many proteins undergo post-translational modifications as they mature, and some of these can be predicted. A

```
>sp|P47865|AQP1_BOVIN Aquaporin-1
MASEFKKKLFWRAVVAEFLAMILFIFISIGSALGFHYPIKSNQTTGAVQDNVKVSLAFGLSIATLAQSV
GHISGAHLNPAVTLGLLLSCQISVLRAIMYIIAQCVGAIVATAILSGITSSLPDNSLGLNALAPGVNSG
QGLGIEIIGTLQLVLCVLATTDRRRRDLGGSGPLAIGFSVALGHLLAIDYTGCGINPARSFGSSVITHN
FQDHWIFWVGPFIGAALAVLIYDFILAPRSSDLTDRVKVWTSGQVEEYDLDADDINSRVEMKPK
```

Figure 1 Amino acid sequence of AQP1.

```
Name                    Len         mTP    SP   other  Loc  RC
-----------------------------------------------------------------
sp_P47865_AQP1_BOVIN    271         0.015  0.947 0.103  S    1
-----------------------------------------------------------------
cutoff                              0.000  0.000 0.000
```

Figure 2 The TargetP output for AQP1. A signal peptide (SP = 0.947) targeting the protein to the ER for transport along the secretory pathway (Loc = S) is predicted with high reliability (RC = 1).

(Continued)

SPREAD 8.1 BIOINFORMATIC ANALYSIS OF BOVINE AQUAPORIN-1: A WORKED EXAMPLE (CONTINUED)

particularly reliable prediction can be made for Asn-linked glycosylation, since the "signal" for this addition is a simple Asn-X-Thr/Ser-Y tetrapeptide, where X and Y can be any amino acid except Pro (see Box 6.3). The NetNGlyc predictor finds only one potential site for Asn-linked glycosylation in AQP1 (**Figure 6**). This is located in a rather long loop predicted to be extracellular by TOPCONS, and the site is in fact glycosylated in the real protein. Note that only extracellular loops can be glycosylated (the enzyme responsible for the modification is located in the lumen of the ER and hence, cannot glycosylate sites located in cytosolic loops); this should be kept in mind when assessing the NetNGlyc predictions, because the predictor has no information about the topology.

Prior knowledge about glycosylated sites in proteins can be used to improve the topology prediction, and many predictors (including TOPCONS) have an option to constrain sites to one or the other side of the membrane *a priori*. In the case of AQP1, we could have run TopCons with the constraint that the glycosylated residue (Asn[42]) should be located in an extracellular loop; since this loop is predicted to be extracellular even without this constraint, it would not have made any difference, but in other cases, additional information of this kind can be critical to obtain a good prediction. Rather reliable predictions can also be made for the attachment of a C-terminal GPI anchor. For AQP1, the FragAnchor predictor is adamant that there is no GPI anchor, as indeed there is not.

Protein type	Signal Peptide (Sec/SPI)	Other
Likelihood	0.0314	0.9686

Figure 3 Output from SignalP 5.0.

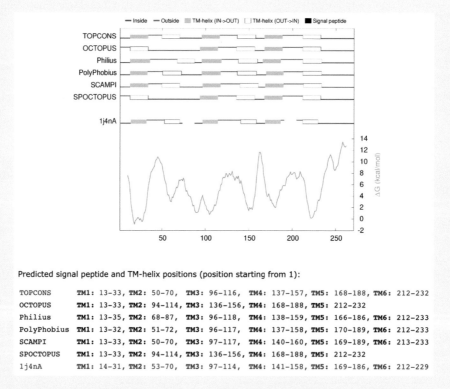

Predicted signal peptide and TM-helix positions (position starting from 1):

```
TOPCONS      TM1: 13-33,  TM2: 50-70,   TM3: 96-116,  TM4: 137-157, TM5: 168-188, TM6: 212-232
OCTOPUS      TM1: 13-33,  TM2: 94-114,  TM3: 136-156, TM4: 168-188, TM5: 212-232
Philius      TM1: 13-35,  TM2: 68-87,   TM3: 96-118,  TM4: 138-159, TM5: 166-186, TM6: 212-233
PolyPhobius  TM1: 13-32,  TM2: 51-72,   TM3: 96-117,  TM4: 137-158, TM5: 170-189, TM6: 212-233
SCAMPI       TM1: 13-33,  TM2: 50-70,   TM3: 97-117,  TM4: 140-160, TM5: 169-189, TM6: 213-233
SPOCTOPUS    TM1: 13-33,  TM2: 94-114,  TM3: 136-156, TM4: 168-188, TM5: 212-232
1j4nA        TM1: 14-31,  TM2: 53-70,   TM3: 97-114,  TM4: 141-158, TM5: 169-186, TM6: 212-229
```

Figure 4 Topology prediction for AQP1 by the consensus method TOPCONS. The experimentally observed topology in the known x-ray structure (PDB 1j4na) is also shown (the two breaks indicate the reentrant helices).

(Continued)

SPREAD 8.1 BIOINFORMATIC ANALYSIS OF BOVINE AQUAPORIN-1: A WORKED EXAMPLE (CONTINUED)

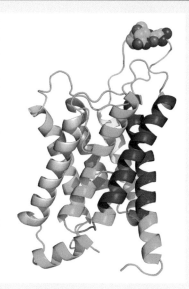

Figure 5 X-ray structure of AQP1. The six non-green transmembrane helices predicted by TOPCONS are overlaid on the green background structure. Note the two green reentrant helices in the foreground. They are not predicted as transmembrane helices by TOPCONS. The Asn-Gln-Thr glycosylation site is indicated in spacefill.

```
Name: sp|P47865|AQP1_BOVIN  Length: 271
MASEFKKKLFWRAVVAEFLAMILFIFISIGSALGFHYPIKSNQTTGAVQDNVKVSLAFGLSIATLAQSV
GHISGAHLNPAVTLGLLLSCQISVLRAIMYIIAQCVGAIVATAILSGITSSLPDNSLGLNALAPGVNSG
QGLGIEIIGTLQLVLCVLATTDRRRRDLGGSGPLAIGFSVALGHLLAIDYTGCGINPARSFGSSVITHN
FQDHWIFWVGPFIGAALAVLIYDFILAPRSSDLTDRVKVWTSGQVEEYDLDADDINSRVEMKPK
```

Figure 6 NetNGlyc predicts a single Asn-linked glycosylation site—NQT—in AQP1.

identity between the template and target proteins is >30%, one can build a rather accurate model for the transmembrane part. Loops are generally more variable and difficult to model.

Structure prediction based only on the amino acid sequence is obviously more difficult than homology modeling. Nevertheless, if a sufficient number of sequence homologs—on the order of 100—are known, it is possible to compute "evolutionary coupling scores" between all pairs of residues in a sequence based on the pairwise co-variation between columns in a multiple sequence alignment. The basic idea behind this method is that residues that are close in space in the 3-D structure tend to co-evolve: if one residue is mutated (to a smaller residue, say), there is an increased probability to observe a compensating mutation (to a larger residue) in a neighboring residue. Such co-varying residues are possible to spot in large sequence alignments. Once the most strongly coupled residue pairs have been identified in this way, one can use distance geometry–based methods (similar to those used to solve protein structures by NMR) to find the 3-D structure of the protein that is best compatible with the identified distance constraints (**Figure 8.15**). The current method of choice for sequence-based 3-D structure prediction is the program AlphaFold2 that uses an advanced neural network architecture trained on very large sets of protein sequences and 3-D structures.

One remaining difficulty in sequence-based structure prediction is that many membrane proteins—transporters in particular—take on a series of

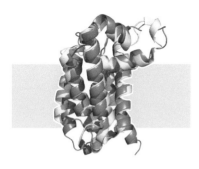

Figure 8.15 A comparison of a structure predicted by the program trRosetta (blue) and the corresponding experimentally solved structure (yellow) of the *E. coli* inner membrane protease GlpG (PDB 3B45).

distinct conformations during the reaction cycle, and it is unclear which, if any, of these a predicted structure represents. If the structure of one conformation of a membrane protein is known at high resolution, it is sometimes possible to model the structure of other conformations, however. Many small-molecule transport proteins work according to an "alternating access" mechanism (see Chapter 13), i.e., they switch between conformations that are open toward one or the other side of the membrane. Moreover, these proteins often contain topologically inverted repeats (**Figure 8.16**). In such cases, it seems reasonable to assume that the inward-facing and outward-facing conformations are related by an approximate two-fold symmetry axis in the plane of the membrane and that if a structure is known for one of the two conformations, the structure of the other can be modeled by homology modeling of the sequence of the first repeat onto the structure of the second, and vice versa. In the example shown in Figure 8.16, the molecule has two inverted-topology repeats (A-B and C-D). A model for the inward-facing conformation built using the known structure of the outward-facing conformation in fact managed to catch the main features of this conformation, as became clear when the structure of the latter was solved.

Figure 8.16 Symmetry-based modeling of the inward-facing conformation of the aspartate transporter Glt$_{Ph}$ using the structure of the outward-facing conformation. (A) The topology-inverted repeats A-B and C-D. (B) The sequence of repeat B is aligned to that of repeat A and then modeled onto the 3-D structure of repeat A, repeat A is modeled onto the 3-D structure of repeat B, and repeats C and D are modeled onto each other in the same way. (C) The known structure of the outward-facing conformation used as a starting point for the modeling is shown on the left. The resulting model for the inward-facing conformation is shown on the right, superimposed on the subsequently determined crystal structure of the inward-facing conformation. The model correctly reproduces the large movements of the HP1 and HP2 segments relative to the rest of the protein. Note that only one Glt$_{Ph}$ monomer is shown; in reality, Glt$_{Ph}$ forms a trimer in the membrane. The HP2 segment is buried inside the trimer and is not exposed to the lipid bilayer in the full structure. (Faraldo-Gómez JD, Forrest LR [2011] *Curr Opinion Struct Biol* 21: 173–179. With permission from Elsevier.)

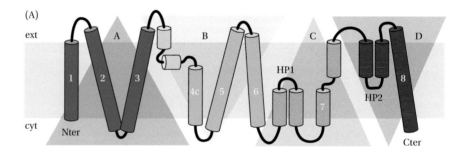

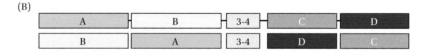

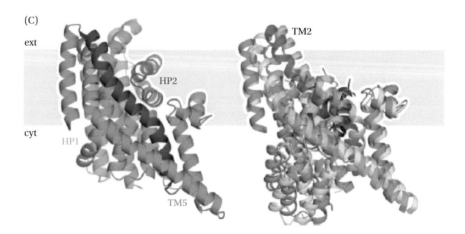

KEY CONCEPTS

- Sequence alignment algorithms such as BLAST are at the heart of bioinformatics.

- Because they are composed mostly of hydrophobic residues, transmembrane helices sometimes score as low-complexity regions by standard sequence alignment programs unless this option is manually turned off.

- The subcellular localization of proteins can be predicted by pattern recognition programs, such as SignalP and TargetP, that try to identify signal peptides. Subcellular localization can also be predicted by sequence similarity.

- Membrane topology of helix-bundle proteins can be predicted by HMM-based programs such as TMHMM and consensus-based servers such as TOPCONS. Programs that use a multiple sequence alignment as input generally perform better than singe-sequence

programs. One way to improve the prediction is to use experimentally derived constraints as additional input to the prediction program.

- Topology prediction programs are good at distinguishing between soluble proteins and helix-bundle membrane proteins.

- Helix interaction structural motifs, such as GxxxG, can be identified.

- Kinks in transmembrane helices can be predicted by neural network–based predictors such as TMKink.

- Homology modeling of 3-D structure works equally well for membrane proteins and soluble proteins. A key aspect of homology modeling is to get a good alignment between the target sequence and the sequence of the template protein.

- 3-D structures of membrane proteins can be predicted by deep-learning methods such as AlphaFold2.

FURTHER READING

Emanuelsson, O., Brunak, S., von Heijne, G., and Nielsen, H. (2007) Locating proteins in the cell using TargetP, SignalP, and related tools. *Nature Protocols* 2:953–971.

Elofsson, A., and von Heijne, G. (2007) Membrane protein structure: Prediction vs. reality. *Annu.Rev.Biochem.* 76:125–140.

Faraldo-Gómez, J.D., and Forrest, L.R. (2011) Modeling and simulation of ion-coupled and ATP-driven membrane proteins. *Curr Opinon Struct Biol* 21:173–179.

Tusnády, G.E., and Simon, I. (2010) Topology prediction of helical transmembrane proteins: how far have we reached? *Curr Protein Pept Sci.* 11:550–561.

KEY LITERATURE

Altschul, S.F., Gish, W., Miller, W., Myers, E.W., and Lipman, D.J. (1990) Basic local alignment search tool. *J Mol Biol* 215:403–410.

Barth, P., Wallner, B., and Baker, D. (2009) Prediction of membrane protein structures with complex topologies using limited constraints. *Proc Natl Acad Sci USA* 106:1409–1414.

Fitch, W.M., and Margoliash, E. (1967) Construction of phylogenetic trees. *Science* 155:279–284.

Forrest, L.R., Tang, C.L., and Honig, B. (2006) On the accuracy of homology modeling and sequence alignment methods applied to membrane proteins. *Biophys J.* 91:508–517.

von Heijne, G. (1986) A new method for predicting signal sequence cleavage sites. *Nucl.Acids Res.* 14:4683–4690.

von Heijne, G. (1992) Membrane protein structure prediction: Hydrophobicity analysis and the 'Positive Inside' rule. *J.Mol. Biol.* 225:487–494.

Hopf, T.A., Colwell, L.J., Sheridan, R., Rost, B., Sander, C., and Marks, D.S. (2012) Three-dimensional structures of membrane proteins from genomic sequencing. *Cell.* 149:1607–1621.

Nakai, K., and Kanehisa, M. (1992) A knowledge base for predicting protein localization sites in eukaryotic cells. *Genomics* 14:897–911.

Nielsen, H., Engelbrecht, J., Brunak, S., and von Heijne, G. (1997) Identification of prokaryotic and eukaryotic signal peptides and prediction of their cleavage sites. *Protein Engineer.* 10:1–6.

Krogh, A., Larsson, B., von Heijne, G., and Sonnhammer, E.L.L. (2001) Predicting transmembrane protein topology with a hidden Markov model: Application to compete genomes. *J.Mol. Biol.* 305:567–580.

Kyte, J., and Doolittle, R.F. (1982) A simple method for displaying the hydropathic character of a protein. *J Mol Biol* 157:105–132.

Meruelo, A., Samish, I., and Bowie, J. (2011) TMKink: A method to predict transmembrane helix kinks. *Prot Sci* 20:1256–1264.

Ovchinnikov, S, Park, H, Varghese, N., Huang, P.S., Pavlopoulos, G.A., Kim, D.E., Kamisetty, H., Kyrpides, N.C., and Baker, D. (2017) Protein structure determination using metagenome sequence data. *Science* 355:294–298.

Senes, A., Gerstein, M., and Engelman, D.M. (2000) Statistical analysis of amino acid patterns in transmembrane helices: The GxxxG motif occurs frequently and in association with beta-branched residues at neighboring positions. *J. Mol. Biol.* 296:921–936.

Remmert, M., Linke, D., Lupas, A., and Soding, J. (2009) HHomp - prediction and classification of outer membrane proteins. *Nucl Acids Res* 37:W446–W451.

Jumper, J., et al. (2021) Highly accurate protein structure prediction with AlphaFold. *Nature* 596:583–589.

EXERCISES

1. Why is it generally better to use a multiple sequence alignment as input to a prediction program than to use a single sequence as input?

2. How can the positive-inside rule help in predicting the transmembrane helices in a protein?

3. Why do prediction programs tend to confuse N-terminal signal peptides and transmembrane helices?

4. Do you think that reentrant loops in membrane proteins form early or late during protein folding?

5. Why is an experimentally verified site of N-linked glycosylation useful in topology prediction? Would an experimentally verified phosphorylation site be equally useful? A verified disulfide bond?

Primer on Biomolecular Structure Determination

9

The relation between structure and biological function has underpinned biological research ever since the Dutch spectacle maker Zacharias Jansen (1580–1638) invented the compound light microscope and Anton van Leeuwenhoek (1632–1723) discovered single-celled organisms. The advent of x-ray crystallography, pioneered by William Henry Bragg (1862–1942) and his son William Lawrence Bragg (1890–1971), and the development of macromolecular crystallography by John Kendrew (1917–1997) and Max Perutz (1914–2002) revolutionized biology by enabling structure–function studies at the atomic level.

Atomic-scale structures are essential for using the concepts of chemistry and physics to understand how membranes and their proteins work. Except for the brief introduction to one-dimensional diffraction measurements for lipid bilayers presented in Chapters 1 and 2, we have said little about how three-dimensional (3-D) structures of biomolecules are determined. Until recently, x-ray crystallography was the sole method for obtaining chemical models of membrane proteins. Now, during the first decades of the 21st century, we are witnessing a revolution in membrane protein structural biology as a result of extraordinarily powerful free-electron laser sources and electron microscopes that can yield structures from minute crystals and even follow changes in structure that accompany biochemical activity. Even more exciting, perhaps, are the revolutionary changes that have taken place in electron cryomicroscopy (abbreviated as "cryo-EM") that are freeing us from the tyranny of "no crystals, no structures."

Although it might not be apparent at first, microscopy and crystallography share a common set of principles related to the physics of the scattering of radiation by matter, which we discuss in this chapter. The mathematics of Fourier series provide a tool for connecting the two. Imaging by light and electron microscopes is easier to understand than imaging using crystallographic methods. We begin this chapter with a brief discussion of imaging and its connection to diffraction-based structural methods. Then, because the imaging of single protein molecules is becoming routinely possible due to revolutionary changes in electron microscopy, we turn to cryo-EM as a starting point for understanding crystallographic methods.

DOI: 10.1201/9780429341328-10

9.1 MICROSCOPY AND CRYSTALLOGRAPHY DEPEND UPON DIFFRACTION AND FOURIER TRANSFORMATION

9.1.1 Understanding How Lenses Magnify Objects Reveals That Fourier Transformation of Images Is an Inherent Feature of Image Formation

A simple convex lens can produce magnified images of objects we encounter in our everyday world (**Figure 9.1A**). This is possible because light rays (described mathematically in **Box 9.1**) from an object passing through the glass lens are bent at the glass surface according to Snell's law of refraction. As described by the thin-lens equation, a magnified image of a well-lit object can be faithfully projected onto a flat surface, called the *image plane*. If one rotates the object (such as a macroscopic model of bacteriorhodopsin; **Figure 9.1**) around its three axes, combing the resulting collection of image-plane projections can provide a 3-D view of the molecule. This simple high school physics description of how lenses magnify fails, however, to show accurately the underlying physics.

The light striking the object in Figure 9.1 is scattered by every point on the surface. A portion of the scattered light from each point is collected by the lens and focused on to the image plane to create the image. The interesting physics occurs at the *back focal plane* (**Figure 9.1B**), however. The light scattered in each direction from all the points intersects at this plane, where the summed constructive and destructive interference from all points produces a diffraction pattern (**Box 9.2**). Each direction of scattered light passes through the back focal plane at a discrete point, and conversely, each point in the back focal plane corresponds to a single direction of light scattered from the object. The relationship between an object and its diffraction pattern is the same as that between a density distribution and its Fourier transform. If one divides the bacteriorhodopsin photograph into a very large number of pieces, one can imagine that

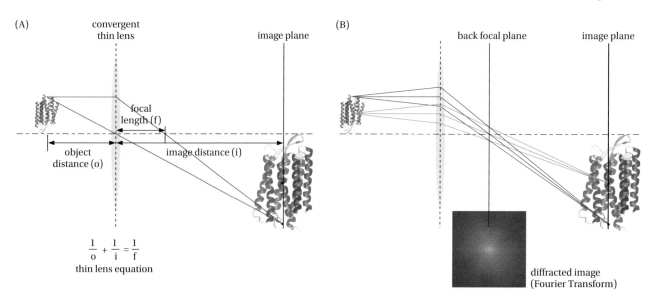

$$\frac{1}{o} + \frac{1}{i} = \frac{1}{f}$$
thin lens equation

Figure 9.1 Principles of microscope imaging. (A) The diagram shows how a convex (convergent) lens projects a magnified image of an object onto a flat plane called the *image plane* according to the thin-lens equation. (B) The image of the object—in this case, a macroscopic 3-D model of bacteriorhodopsin (PDB code 1C3W)—is produced by light scattered from every point on the object that is gathered by the lens and focused onto the image plane. For simplicity, we show here scattered light for only two points. The light rays from all the points on the image intersect at the *back focal plane* and interfere constructively and destructively to produce a *diffracted image* (Box 9.1). This diffracted image is a FT of the image (photo inset at back focal plane).

BOX 9.1 WAVES INTERFERE CONSTRUCTIVELY AND DESTRUCTIVELY

Waves, whether light waves or ripples on a pond, are sinusoidal oscillations that can be described by a cosine function. In time and space, a wave of amplitude U and wavelength λ propagated along the x-axis at a constant velocity c is described by

$$y(x,t) = U \cos\left[2\pi \frac{(x-ct)}{\lambda} + \alpha\right] \quad (1)$$

The parameter α is called the *phase*. If you move along with the wave at constant velocity c, the distance along the x-axis increases linearly with time due to the $(x - ct)$ term. Let's look at the height of the wave at some arbitrary time $t = 0$. In that case,

$$y(x,0) = U \cos\left[2\pi \frac{x}{\lambda} + \alpha\right]$$

Recalling that $\cos(0°) = 1$, setting the phase α equal to zero places the maximum amplitude of the wave, U, at the origin (x = 0, panel A). (Angles are measured in radians: $360° = 2\pi$). Thus,

$$y(x,0) = U \cos\left(\frac{2\pi x}{\lambda}\right) \quad (2)$$

(blue curve, panel A). Using the same coordinate system, another wave of the same amplitude with $\alpha \neq 0$ will be shifted along the x-axis, as shown in panel B (red curve), where, for example, $\alpha = \pi/3$ (60°). The waves interfere constructively to produce a resultant wave of larger amplitude but with the same wavelength (green curve, panel B). If the wave has a phase $\alpha = \pi$, it will be exactly out of phase (red curve, panel C), causing exact cancelation with the "blue" wave in panel A for all values of x. This is destructive interference. In two or three dimensions, the situation is more complex (e.g., Box 9.2), but the principles are the same (**Figure 1**).

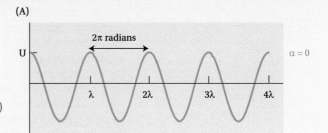

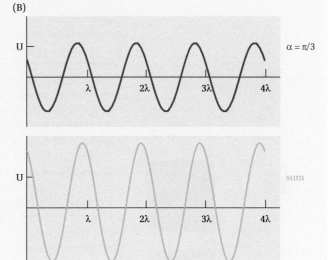

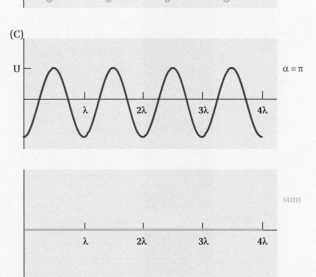

Figure 1 Constructive and destructive interference of electromagnetic waves. A wave is shown in (A). If it is combined with an identical but slightly phase-shifted wave (red), they add as shown in (B). If the peak of one passes a given point at the same time as the trough of the other, they cancel as shown in (C).

the interference of light reflected by the closely spaced pieces—think of them as a huge collection of apertures—will produce a complex diffraction pattern (Figure 9.1B) due to the optical interference of myriad apertures. This pattern can be produced by doing a Fourier transformation of the bacteriorhodopsin image. The objective lens of every microscope, including those of electron microscopes, has a back focal plane and produces a diffracted image that is not apparent to us because our eyes see only the image plane. We can visualize the back focal plane in an optical microscope, however, by replacing the normal eyepiece with a telescopic one that can focus on the back focal plane.

To understand optical interference seen at the back focal plane, consider light passing through a small circular aperture as a stand-in for the light scattered from the atoms of the molecule. As presented in panel B of **Figure 9.2**, light passing through a small circular aperture produces wavelets that interfere to produce a circular diffraction pattern (**Figure 9.2A**). Multiple apertures in various arrangements and patterns behave similarly except that the light from the wavelets from the apertures combines to produce, in each case, a *unique* diffraction pattern (**Figure 9.2B–E**). Returning to the image in Figure 9.1, imagine the image of bacteriorhodopsin as composed of zillions of closely spaced elements of different sizes. The diffraction pattern in this case will be very complicated indeed!

The unique diffraction patterns observed for each object in Figure 9.2 can be computed by Fourier transformation (FT) of the images of the apertures! Image analysis software (ImageJ) for doing FT of any image is available from the U.S. National Institutes of Health (http://imagej.nih.gov/ij/). Using the software, you will find that inverse Fourier transformations (FT^{-1}) of the Fourier-transformed images produce the original images of the apertures. How cool is that?!

An important feature of electron microscopes is that they can be configured to observe directly either the object at the image plane or its FT at the back focal plane. This choice helps to make the electron microscope extremely versatile for determinations of molecular structure.

9.1.2 Radiation Wavelength Limits What We Can See

All forms of radiation follow the same imaging rules, but the imaging method is determined by the size of the imaged particle, radiation wavelength λ, and how the radiation interacts with matter. The usefulness of a particular type of radiation is determined by the experimenter's ability to resolve two distinguishable radiating points on the object. Think of observing at the image plane the radiation emanating from two points on the object. How close can the points be and still be distinguishable? Rayleigh's criterion gives the answer (**Box 9.2**). The maximum value of sinθ is 1 ($\theta = \pi/2$ radians). Hence, the minimum separation d between two distinguishable points is

$$d = 1.22\lambda$$

For visible light, $\lambda \approx 5000$ Å, which is far too large to allow individual atoms (~1 Å in diameter) to be discerned. However, x-ray generators of one sort or another produce x-rays with short wavelengths, typically ≈1 Å. Detection becomes quite practical, because then $\lambda/d \sim 1$. It would be great if convex lenses were able to refract x-rays to focus them as in Figure 9.1. Unfortunately, they cannot, because the index of refraction for x-rays is less than 1 for most materials. This means that x-rays are bent away from the normal and cannot be focused. In principle, one could overcome this problem using a concave lens, except that most lens materials are opaque to x-rays. Instead, one determines 3-D structures from molecules organized as crystalline arrays. The great discovery of the Braggs and others was that these crystalline arrays produce diffraction patterns that represent unique FTs of the crystalline structures. Hence, given adequate diffraction patterns, the inverse FT of the patterns leads to the determination of models of molecular structures that include chemical details.

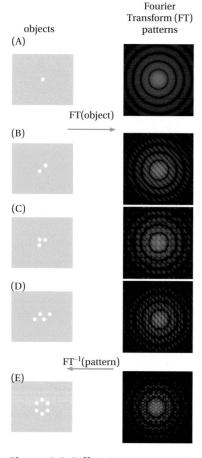

objects
(A)

Fourier
Transform (FT)
patterns

FT(object)

(B)

(C)

(D)

FT^{-1}(pattern)

(E)

Figure 9.2 Diffraction patterns and their Fourier transforms produced by small circular apertures arranged in various two-dimensional patterns. Each object on the left has a unique Fourier transform (FT), shown on the right. Inverse Fourier transformation (FT^{-1}) of the diffraction patterns reproduces the object patterns. The relation between object and transform is unique.

BOX 9.2   BASICS OF DIFFRACTION

When light passes through an aperture whose dimensions are comparable to the wavelength of light, a diffraction pattern is produced. **Figure 1A** compares the diffraction of waves from two apertures, one very narrow and another wider one. In each case, a plane wave strikes the aperture from the left and produces the wave pattern shown on the right. Notice that the larger aperture on the right produces a more complex pattern. Light passing through a circular aperture and projected onto a surface produces an Airy pattern (**Figure 1B**) that has the intensity distribution shown in **Figure 1C**. As shown in **Figure 1D**, the detailed structure of the Airy diffraction pattern changes with the diameter of the aperture, as expected from panel A. If there are two apertures that are close enough together, then the wave patterns from the two apertures produce interference patterns that depend upon the separation of the apertures (**Figure 1E**). Notice that as the distance between the apertures becomes smaller, it becomes more difficult to discern the existence of two apertures. This phenomenon is the basis for the Rayleigh criterion for **lateral resolution** of two *points*. If the pattern of the two-spot diffraction is projected onto a screen, the minimum observable separation d will be determined by the angle θ between the rays from the two spots: $\sin\theta = 1.22\lambda/d$. For small angles, $\sin\theta \approx \theta$. Lord Rayleigh derived this formula using Bessel functions by considering the overlap of the Airy patterns (panel B) produced by each spot. (For the mathematical cognoscenti, the factor of 1.22 is the first zero of the order-one Bessel function of the first kind.)

This definition of resolution differs from the definition of resolution R used in optical microscopes, which is $R = 1.221\lambda/(\mathrm{NA}_{condenser} + \mathrm{NA}_{objective})$, where the NAs are the numerical apertures of the condenser and objective lenses, respectively. If the NAs of the condenser and objective are equal, then $R = 0.61\lambda/\mathrm{NA}$.

We can think of Figure 1 as a model for scattering x-rays by two atoms. The only difference is that the atoms are much smaller than the wavelength of visible light, and x-rays have correspondingly shorter wavelengths; the x-ray diffraction patterns of a pair of atoms should behave in the same way as in Figure 1E. The situation changes dramatically for large numbers of atoms arranged on a regular lattice. Suppose the atoms are distributed uniformly along an axis, as shown in **Figure 2**.

Just as in Figure 1A, plane waves impinge on the atoms, which re-radiate the energy as spherical waves that can interfere with one another constructively or destructively (Box 9.1). The waves largely interfere destructively except in certain directions where the waves interfere constructively. Those "certain directions" are defined by the Bragg condition, discussed in Box 1.4 and illustrated in Figure 2 for $h = 1$ and $h = 2$. We showed in Box 1.4 that the Bragg condition can be expressed in reciprocal space by the expression $1/\lambda = h/d$. Consequently, when considering Bragg diffraction, the wave equation (2) in Box 9.1 can be rewritten:

$$y(x,0) = U\cos\left(\frac{2\pi x}{\lambda}\right) = U\cos\left(2\pi\frac{h}{d}x\right) \qquad (1)$$

In general, of course, we must include the phase α_h. Hence,

$$y(x,0) = U\cos\left(2\pi\frac{h}{d} + \alpha_h\right) \qquad (2)$$

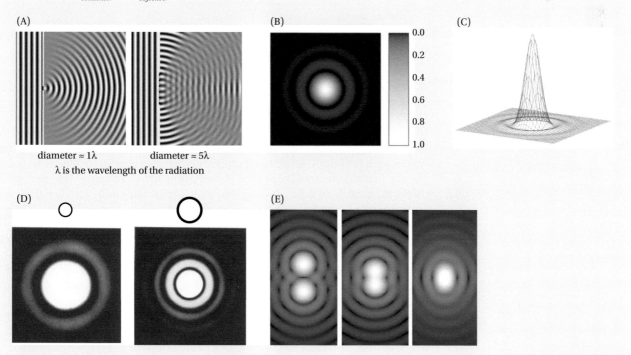

(A)

diameter ≈ 1λ   diameter ≈ 5λ
λ is the wavelength of the radiation

(B)

(C)

(D)

(E)

Figure 1 Diffraction of waves passing through two apertures.

(Continued)

BOX 9.2   BASICS OF DIFFRACTION (CONTINUED)

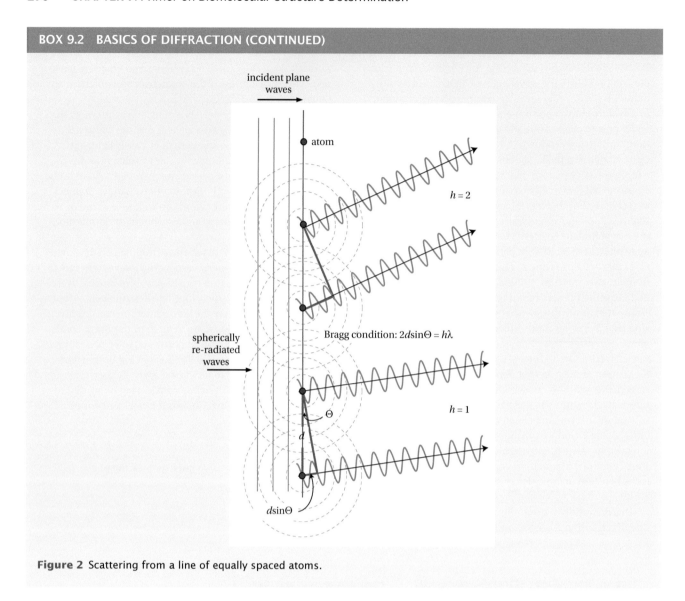

Figure 2 Scattering from a line of equally spaced atoms.

What about other forms of radiation, such as electrons and neutrons? Louis de Broglie postulated a wave–particle duality, in which all matter has wave-like behavior. The de Broglie equation describes the duality with the equation $\lambda = h/p$, where λ is the wavelength, h is Planck's constant, and p = momentum [mass (m) × velocity (v)]. Electrons can be accelerated fast enough by the electric field inside an electron microscope to gain sufficient momentum to behave as short-wavelength radiation. Modern electron microscopes accelerate electrons using electric fields generated by electric potentials of 200 kV or more. This translates into wavelengths of 0.025 Å or lower, which is considerably smaller than the diameter of a hydrogen atom. If one could hold a biological molecule perfectly still and not destroy it by collisions with the high-energy electrons, direct images could, in principle, be obtained by electron microscopy in the manner of optical imaging (Figure 9.1). Electron and optical imagining are exactly equivalent except that the electron rays are focused using magnetic lenses that bend the electron beams. This seemingly fantastical idea is now being realized by electron cryomicroscopic methods, discussed later.

We mentioned in Chapter 2 the use of neutron diffraction for studying the structure of lipid bilayers. Depending on the details of production, neutrons

BOX 9.3   CHARACTERISTICS OF RADIATION USED IN STRUCTURE DETERMINATIONS

The study of biomolecular structure, as noted in the main text, involves the use of probing radiation at the appropriate wavelengths. Several choices with useful wavelengths have been developed for this purpose: x-rays are produced for use in the laboratory or at synchrotrons; neutrons are produced at reactors or spallation sources; electrons are produced and used in electron microscopes. Each of these has its own mechanism of scattering, so the view gained of a structure will vary, and each produces damage from inelastic and other scattering events, summarized in the following table.

	Microscope electrons	1.5 Å x-rays	1.8Å neutrons
Ratio of inelastic/ elastic scattering	3	10	0.08
Damage mechanism	secondary electron emission	photoelectric electron emission	nuclear reactions and recoil energy
Energy deposited per inelastic event (eV)	20	8,000	2,000
Energy deposited per elastic event (eV)	60	80,000	160
Energy deposited relative to electrons: inelastic	1	400	100
Energy deposited relative to electrons: elastic	1	1,000	2.5

(Data derived from Henderson [1995] *Quarterly Revs Biophys* **28**:171–183.)

If the scattering by an atom is elastic, the scattered radiation will have the same wavelength as the incoming radiation and will be useful for structure determination, because the scattered waves will interfere coherently. If the scattering is inelastic, energy is lost from the scattered wave, and its wavelength is changed. Hence, the first row has significance: Neutrons give little inelastic scattering but have much more incoherent scattering, as background. Incoherent scattering arises when the scattered waves have lost their phase information and cannot give structural information. Incoherent scattering arises mainly from hydrogen nuclei and can be improved by fully deuterating the sample, but some of the contrast advantages of using neutrons would then be lost.

We see from the last row that relative to damage, each scattering event by an electron is much more informative than the scattering of an x-ray and somewhat more informative than a scattering event by a neutron. But, there remain reasons to use all of these, since each has particular advantages: for example, x-rays measure an average over many molecules in a crystal, spreading the damage and reducing its effect on the quality of a structural map, and neutrons principally scatter from atomic nuclei, so atomic isotopes vary in scattering, allowing isotopic labeling. In particular, the difference of scattering by hydrogen versus deuterium can give views of proton locations in a biological sample, including water, and so can provide a view that may inform ideas of mechanism.

This analysis suggests that the electron microscope will continue to evolve in importance as the technical barriers are overcome.

have wavelengths of a few ångströms; they have huge mass compared with electrons but travel very slowly (remember, $p = mv$ in the de Broglie equation). An advantage of neutrons is that they generally cause less damage than x-rays (but more than electrons) during passage through matter (**Box 9.3**); the disadvantage is that the intensity of neutron beams is very low compared with x-ray or electron sources. Happily, modern spallation neutron sources are becoming powerful enough to carry out diffraction experiments on protein crystals. Perhaps the most important feature of neutrons is that substitution of deuterons for hydrogens produces very high image contrast (Section 9.2). Box 9.3 gives a comparison of the characteristics of different kinds of radiation used for structural studies, and it emerges that electrons are best!

9.2 ELECTRON MICROSCOPY IS A VERSATILE TOOL FOR MOLECULAR STRUCTURE DETERMINATION

The electron microscope has been used as a tool in cell biology since the 1940s but has only recently emerged as a means to obtain structural information from single particles at chemical resolution. Three kinds of uses of the microscope have been developed: crystallography, tomography, and single-particle analysis. Tomography is discussed in **Box 9.4** and crystallography later in the chapter. While there is still considerable space for more progress, single-particle imaging has now emerged as the method of choice for many kinds of structures, especially membrane protein assemblies of high molecular weight. For this reason, we focus here on the study of unstained particles embedded in vitreous ice at low temperatures.

9.2.1 Electron Microscopes Can Produce Direct Images of Biological Molecules

EM imaging occurs just as described earlier for visible light (Figure 9.1). We showed that atomic-scale imaging is theoretically possible with modern electron microscopes because wavelengths of ~0.02 Å can be produced. We noted, however, that the molecule would have to be held stationary in space. A fundamental problem to be overcome is that high-energy electrons transfer momentum to the molecule, causing it to move. Furthermore, to achieve 3-D structures, we would have to take many images of the molecule from many different directions. Finally, radiation damage caused by the high-energy electrons would have to be minimized. Fortunately, these issues are being addressed with considerable success. With the new advances, the resolution of a structural map has moved from ~7 Å in favorable cases to ≤1.5 Å. Favorable cases are those where the individual structures are nearly identical, so many cases will not allow high resolution, but the improvements will also allow more detailed classification of particles in different classes of conformation, improving the visualization of multiple structural conformations. One way to think about the new advances is in terms of the information they contain. Information content varies with the inverse cube of the resolution, so a 3 Å structure has more than 10 times the information content of a 7 Å structure, or 125 times that of a 15 Å structure obtained by negative staining.

9.2.2 Images Can Be Formed from Single Molecules

While the electron microscope is capable of imaging molecules at chemical resolution, there have been practical problems that have limited its use. The main difficulties were radiation damage, movement of the sample under irradiation by the beam, and the identification of multiple views of structures seen as fields of spread-out particles so that they can be averaged. The damage issue is (partly) addressed by freezing the structures in vitreous ice to the temperature of liquid nitrogen (77 K). Vitreous ice has no long-range crystal structure and is obtained using very rapid freezing rates ($>10^6$ degrees Kelvin per second!) that immobilize water molecules before they can form large crystals. While damage caused by the electron beam to the structure is still present, the frozen state keeps it from falling apart too easily. Thus, the "cryo" in electron cryomicroscopy refers to the sample, not the electrons (frequently, it is called cryo-electron microscopy—inaccurate, but it may be too late to change. We will settle for "cryo-EM"). Similar cryo methods are used in x-ray crystallography, for the same reason—the effects of damage are slow to appear at low temperatures, lengthening the period during which observations can be made.

BOX 9.4   ELECTRON TOMOGRAPHY

Tomography is a way to study the distribution of matter in a large, irregular 3-D object, such as an organelle, cell, or organism, by combining multiple images of its parts viewed at different angles. Very high-resolution views of structure provide the arrangements of atoms and give a view of chemistry; lower-resolution views give a larger view of how the parts fit together. Unification of these views will give the best understanding of life in different states and levels of organization, basically using stereo views and tying the datasets together. An important bridge from the chemistry level (x-ray crystallography, cryo-EM) to the light microscopy level is the use of electron tomography. Tomography can be done with the same equipment as cryo-EM but using thin sections of the biological material (fixed, embedded,

or frozen). Each section is imaged in a series of tilted views to get multiple two-dimensional (2-D) images, which are then combined to generate a 3-D image of the sectioned object. The resolutions attainable are more modest than for single-particle imaging but are being improved (now ~3.5 Å in the best cases). The general procedure is summarized in **Figure 1A**. Examples of oriented slices of the ribosome/Sec61 translocon holocomplex from HeLa cells' endoplasmic reticulum are shown in **Figure 1B**. The image reconstruction from such slices is shown in **Figure 1C**. The Sec61 translocon is colored blue, the translocon associated protein (TRAP) is colored green, and oligosaccharyltransferase (OST) is colored red.

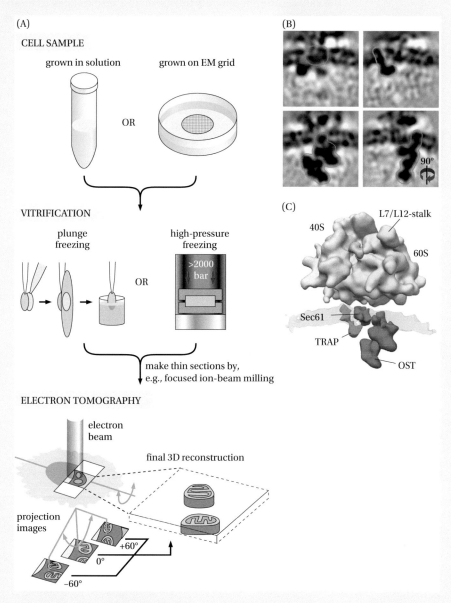

Figure 1 (A, From Asano S et al. [2016] *J Mol Biol* 428: 332–343. B,C, From Pfeffer S et al. [2014] *Nature Communications* 2014 5: 3072. With permission from Springer Nature.)

Until the recent invention of high-speed direct electron detectors, the drift of the specimen was a major problem that blurred the images and limited the resolution that could be achieved. But, data are now collected through a series of snapshots—like a movie—and the motion can be taken out of the analysis by shifting the single frames. This process is illustrated in **Figure 9.3**.

Methods for classifying and averaging structural views have evolved for years and are now very useful. Each snapshot will generally contain images of a field of molecules in different orientations relative to the beam, and each molecule is seen as a projected image (**Figure 9.4A**). The computational task is to classify the views into groups that are the same view, align the views in a group to improve signal strength relative to noise, average the signals (images) from many examples of each view (**Figure 9.4B**), and finally, to combine the averaged views to obtain a three-dimensional image, using a process known as 3-D image reconstruction. The resulting image is a map of the distribution of the electron scattering strength; note that the scattering density map for electrons is not the same as for x-rays (see later), although they are often assumed to be the same. Enough detail is now seen in these images to build chemical models of molecules using knowledge of the covalent structures present (**Figure 9.4C, D** and **Figure 9.5**). The strengths and weakness of structure determination by cryo-EM are summarized in **Box 9.5**.

If we wish to study the structure of a membrane protein or complex, an additional difficulty is present: making the structure soluble or visualizing it in a lipid environment. In the study of the TRPv1 channel shown in Figure 9.4 and Figure 9.5, the channels were stabilized using amphipols, but detergents and nanodiscs have also been used (**Box 9.6**). For crystallographic structure determination, detergent solubilization is an essential first step.

9.3 X-RAY CRYSTALLOGRAPHY IS THE MAINSTAY OF STRUCTURAL BIOLOGY

We introduced the basic ideas of diffraction and Fourier reconstruction when we discussed the structure of lipid bilayers in Chapters 1 and 2. As we learned, diffraction from stacks of fluid lipid bilayers is a one-dimensional problem because of the lack of order in the bilayer plane. Further, because of the great thermal disorder of fluid bilayers, only a limited number of diffraction peaks

Figure 9.3 A key to single particle imaging: a fast camera. One of the great leaps allowing the current EM revolution is the realization that specimen motion imposes a major limit on resolution. A new camera with improved efficiency in detecting and recording every electron allows short exposures that can "freeze" the motion. In (A), the consequences of motion are illustrated—the image of each object in a field is blurred by the motion induced by the electron beam heating, charging, and damaging the support film on which the objects are arrayed. In (B), a series of snapshots taken at short time intervals is used to track and correct for the motion of each object, and the resulting images are then averaged and used to reconstruct a high-resolution image. In favorable cases, one can separate and image classes of particles with different conformations. (From Bai X-c, McMullan G, Scheres SHW [2015] *Trends Biochem Sci* 40: 49–57. With permission from Elsevier.)

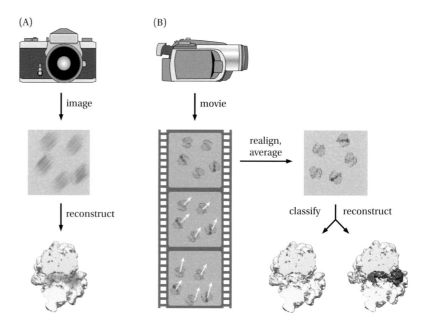

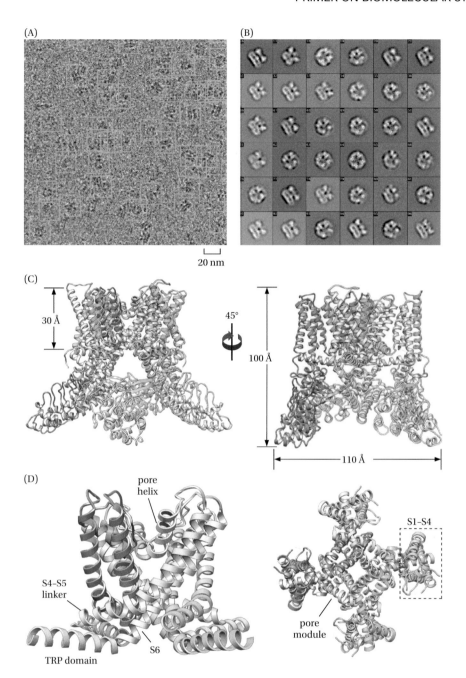

Figure 9.4 Cryo-EM views are classified and averaged to obtain the structure of the TRPv1 channel (PDB: 3J5P). (A) A cryo-EM image field. Individual particle images in the field are indicated by the green boxes. The images are faint because low doses of the electron beam are used to minimize damage and jiggling. The particle images are 2-D projections of the structure. (B) Thousands of particle images are gathered and classified into groups with the same view, averaged to suppress noise, and combined to arrive at the TRPv1 3-D structure. The data used to reconstruct the images shown here have a resolution that is significant to about 3 Å, allowing the chain to be traced and sidechains to be positioned. Look closely at the particle image in the red square. Which structure in panels C and D does the image resemble most closely? (C) The overall structure of the channel. (D) Views of the pore module of the channel. (From Liao M et al. [2013] *Nature* 504: 107–112. With permission from Springer Nature.)

are produced—usually fewer than 10. The situation for proteins is quite different. Foremost, we want to see the molecule as a 3-D object so that we can view it from any perspective. This is achieved in cryo-EM by using thousands of images, each of which shows the protein frozen in time from a particular direction. In crystallography, the molecules are arranged on a 3-D lattice. To see the molecule in three dimensions, we must first determine diffraction patterns by rotating the crystal in the x-ray beam in order to have enough data to reconstruct the structure in three dimensions.

9.3.1 Why Crystals of Proteins?

Why not do an x-ray structure on a single molecule? Although this goal might be reached using Free-Electron Lasers (FEL), single molecules are presently difficult to isolate, hold, and orient, and the structure might be obliterated by

BOX 9.5 CRYO-EM: STRENGTHS AND WEAKNESSES

Strengths

- Only a tiny amount of (pure) sample is needed.

- A crystal is not required.

- The phase problem does not arise, since lenses produce images, and phasing is much better than for x-ray structures (see later).

- The resolution can be as good as ~1.5 Å, enough to build a chemical model in most cases, and improvements continue.

- The environment is thought to be relatively benign (frozen water).

- Methods for evaluating the quality of a structure, such as Fourier Shell Correlation, are being developed.

Weaknesses

- The particles must be rigid enough so that particle averaging can be done (if they are not rigid, they will have many conformations, and high resolution is not possible; but if they have only a few, the states can be separated in the image analysis step, which may yield important functional information).

- The ~3 Å resolution typical of current membrane protein structures, while very good, is not enough to model side chain conformations or water molecules with confidence, and they can be important in mechanisms. However, in recent cases, it has proved possible to get structures of soluble proteins at ~1.5 Å, where these features are seen.

- Structures are at 77 K or lower, with unknown consequences.

- It has been difficult to get structures of small objects (MW < ~100 kDa) because the image classification relies on seeing distinct structural features for sorting into classes of different orientation.

damage long before enough information was obtained. The advantage of crystals is that they hold trillions of molecules in fixed, repeated spatial relationships to each other, and they are macroscopic objects that can be mounted and viewed from different angles. In a perfect crystal, all the molecules are nearly identical (not counting thermal motion) and are in identical environments. In this way, the problem of holding and viewing a molecule is solved by the molecules being in a stable crystalline array. The scattering from each of the N molecules interferes constructively to give a measurable diffraction pattern (enhanced ~N^2-fold). Further, the radiation damage is distributed approximately randomly over all the molecules, and the effects of damage can be suppressed somewhat by freezing the crystal, as noted earlier. Even as damage occurs in the course of the experiment that progressively destroys the crystal, the undamaged parts continue to diffract, albeit at reduced intensity.

The repeating entity in a crystal is the "unit cell," which can generate the whole crystal by positioning identical versions related by translations to a single cell used as the starting point and filling the space occupied by the crystal (**Figure 9.6**). The repeated arrangement of unit cells in the crystal is called the lattice and can be represented as a set of points at corresponding

Figure 9.5 Fitting the electron scattering density map using chemical information. (A) The experimental result of the cryo-EM study (Figure 9.4) is the density map (grey mesh), which must then be interpreted by building a chemical model (orange). (B) The ion channel of the TRPv1 structure at 3.4 Å is shown after interpreting the map of panel (A). Note that sidechains are suggested by the map but not specified; if it were not for the expectations that these are polypeptide helices, that the conformations are restricted, and that the sequence limits choices, it would be hard to interpret the map. The overall structure of the channel is seen to be very similar to many other ion channels (Chapter 11). (From Liao M et al. [2013] *Nature* 504: 107–112. With permission from Springer Nature.)

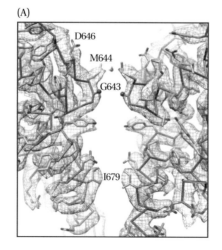

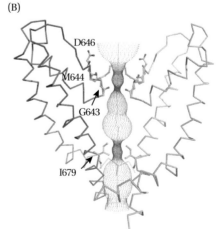

BOX 9.6   SOLUBILIZING MEMBRANE PROTEINS FOR STRUCTURAL STUDIES

The crucial issues for structural studies of α-helical membrane proteins (MPs) is to isolate, purify, and solubilize them without destroying their tertiary structure. MPs are generally over-expressed in a suitable expression system that includes a His tag at the N- or C-terminus, solubilized in detergent, and purified using immobilized metal affinity chromatography (IMAC) in which the His tag interacts strongly with the immobilized metal (typically Ni^{2+}). Except for one emerging method described later, detergent solubilization is the crucial step in all methods of sample preparation.

We learned in Chapter 3 that even the harshest detergents, such as sodium dodecyl sulfate (SDS), are unable to disrupt α-helices. But, harsh detergents can easily disrupt the native structure of inter-helical loops and extracellular domains. Consequently, the choice of detergent becomes crucial. The most common ones, which account for the majority of MP crystallographic structures, are shown in **Figure 1**. Although quantitative details are lacking, the free energy of the protein in the detergent micelle must be more favorable than in the native membrane in order to extract it. Thus, the interaction of these detergents with the membrane bilayer as well as the protein is likely to be a crucial issue (see Box 2.1).

Crystallographic structures do not generally provide an image of the protein in a membrane environment unless the protein crystals are type I (Box 9.11). While the lipid is rarely visualized in type I crystals, at least the proteins of the unit cell have a packing arrangement closer to that expected of a membrane-embedded protein. The use of the lipid cubic phase for crystallization ensures this kind of arrangement (**Figure 2**). Structures of both α-helical and β-barrel MPs have been solved using lipid cubic phase crystallization.

High-resolution cryo-EM for structure determination has revolutionized MP structure determination. The method can be used with detergent-solubilized proteins directly without the need for crystals. Given that the main

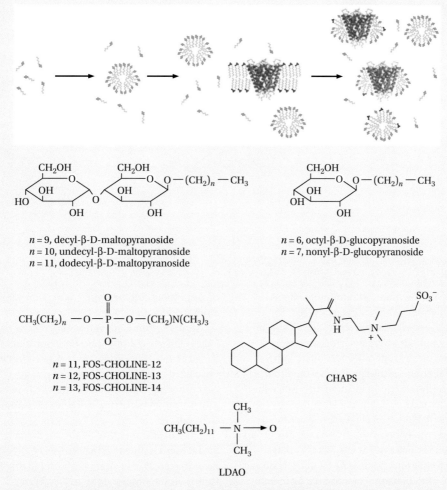

n = 9, decyl-β-D-maltopyranoside
n = 10, undecyl-β-D-maltopyranoside
n = 11, dodecyl-β-D-maltopyranoside

n = 6, octyl-β-D-glucopyranoside
n = 7, nonyl-β-D-glucopyranoside

n = 11, FOS-CHOLINE-12
n = 12, FOS-CHOLINE-13
n = 13, FOS-CHOLINE-14

CHAPS

LDAO

Figure 1 The most commonly used detergents for solubilizing membrane proteins for structural studies. The cartoon in the upper panel shows the extraction and solubilization of a membrane protein. (From Newby ZER et al. [2009] *Nature* 4: 619–637. With permission from Springer Nature.)

(Continued)

BOX 9.6   SOLUBILIZING MEMBRANE PROTEINS FOR STRUCTURAL STUDIES (CONTINUED)

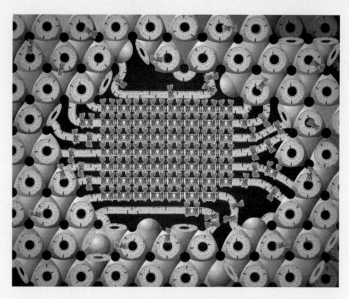

Figure 2 Schematic of the formation of type I crystals of membrane proteins using the lipidic cubic phase. (From Caffrey M, Cherezov V [2009] *Nature* 4: 706–731. With permission from Springer Nature.)

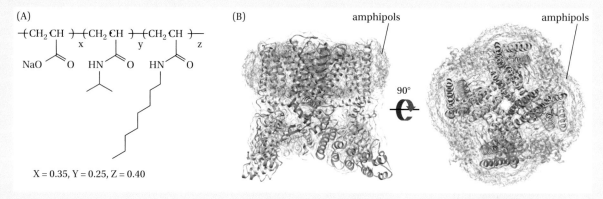

Figure 3 (A) Amphipol structure. (B) Visualization by cryo-EM of amphipols on the surface of the TRPv1 channel. (A, Based on Le Bon C et al. [2014] *Journal of Membrane Biology* 237: 797–814. B, From Liao M, Cao E, Julius D et al. [2014] *Curr Opin Struct Biol* 27: 1–7. With permission from Elsevier.)

goal is to keep the MP in solution with minimal effects on structure, considerable work has been devoted to developing alternative solubilization tools such as amphipols, which are amphiphilic polymers with extended, unbranched chains that are randomly decorated with polar and non-polar groups. Amphipols can wrap around the non-polar hydrophobic region of an MP to replace the lipid bilayer contacts and to present a polar surface to an aqueous environment. **Figure 3A** shows the structure of the A8-35 amphipol, a polymer with MW ~8 kDa. The group composition of A8-35 is 35% carboxylate (X), 25% octyl chains (Y), and 40% isopropyl groups (Z). Figure 9.4 shows the structure of the TRPv1 channel determined by means

of amphipol solubilization. Figure 3B reveals the amphipol shell covering the channel.

Another promising approach for cryo-EM studies of MPs is the use of lipid nanodiscs introduced in Chapter 3 (see Figure 3.25). Their advantage is that the protein is visualized in a lipid bilayer environment, which can be manipulated. **Figure 4** summarizes the use of nanodiscs and a cryo-EM structure of TRPv1 determined using the method.

The disadvantage of all the methods so far described is that there is always an intermediate detergent solubilization step. A new method of solubilizing MPs directly from cells has emerged (**Figure 5**). Water-soluble styrene maleic acid polymers (SMAs) added to suspensions of

(Continued)

BOX 9.6   SOLUBILIZING MEMBRANE PROTEINS FOR STRUCTURAL STUDIES (CONTINUED)

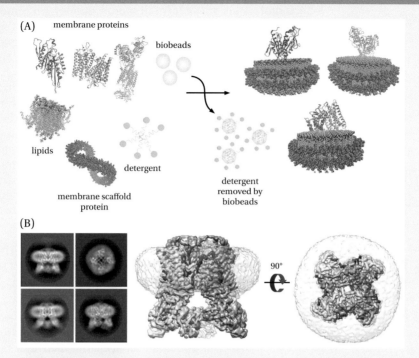

Figure 4 Solubilization of membrane proteins using nanodiscs. (A) General scheme for sample preparation. (B) Left image shows representative cryo-EM images of TRPv1 embedded in nanodiscs. Right image shows the TRPv1 structure embedded in the nanodisc. (A, From Rouck JE, Kraph JE, Roy J et al. [2017] *FEBS Lett* 591: 2057–2088. With permission from John Wiley and Sons. B, From Gao Y et al. [2009] *Nature* 534: 347–351. With permission from Springer Nature.)

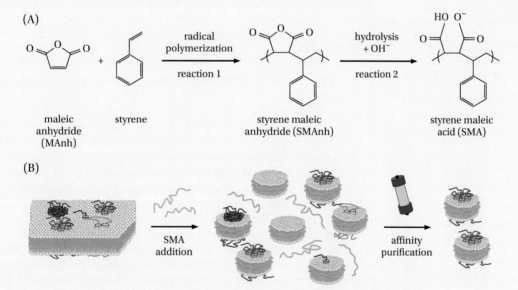

Figure 5 Membrane proteins can be removed directly from isolated membranes using styrene maleic acid polymers (A). The molecular details of the process are largely unknown. But what is certain is that not only is the protein solubilized; the lipids surrounding the protein in the membrane are also solubilized (B). (From Dorr JM et al. [2016] *Eur Biophys J* 45: 3–21.)

isolated membranes solubilize the MPs along with their neighboring lipids. In essence, the SMAs solubilize all the MPs directly without an intervening detergent step. The protein–lipid complexes take the form of nanodiscs. Nanodiscs containing a single protein species can be isolated from the heterogeneous mixture using affinity chromatography. These SMALP (SMA lipid particles) have so far been found particularly useful for spectroscopic, electron paramagnetic resonance (EPR), and NMR studies of MPs and their lipid surrounds.

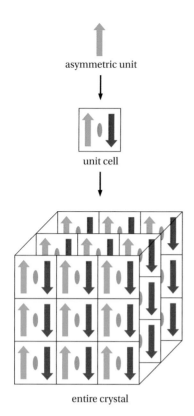

asymmetric unit

unit cell

entire crystal

Figure 9.6 The recurring pattern of proteins in a crystal. One copy of a molecule or complex is indicated as the asymmetric unit, which may be repeated locally if the structure is a dimer or other oligomer in which the structure can be generated by repeating the structure after rotation and/or translation. A unit cell may have more than one asymmetric unit—in this example, the unit cell has two, related by a two-fold rotation to each other. Repeating the unit cell by regular translations in space generates the crystal.

loci in the unit cells, for example at the center of each cell. X-rays can be produced with wavelengths ~1 Å and so are suitable for visualizing biological molecules at the level of atoms, enabling an understanding of their chemistry.

Crystals produce radiation in specific, concentrated spots, called reflections, that arise from the crystal lattice (**Figure 9.7**). This concentration allows good measurements, because the energy of scattered radiation is concentrated in the spots, elevating the signal above background and allowing the background to be subtracted using the spaces between the spots. However, these virtues come at a price: information about the relative phases of waves in different reflections is lost in measuring the intensities of the reflections, and a major challenge is to recover the phase information, which provides information on the location of atoms within the unit cell. The spots in the crystal diffraction arise from the repeating diffraction by the lattice itself and can be conceptualized by knowing that the FT of a lattice of points is also a lattice of points, as illustrated in **Figure 9.8**.

9.3.2 Diffraction from a Crystal Combines the Influence of the Contents of the Unit Cell with the Influence of the Crystal Lattice

Diffraction from a crystal combines the influence of the scattering from a unit cell with the influence of the lattice. Formally, we can think of the crystal as having these two components that are combined in a mathematical operation called *convolution*. Basically, the convolution of two functions, $f(x)$ and $g(x)$, involves the integration of the point-wise multiplication of the two functions as a function of the amount by which one of the functions is translated. Stripping away the mathematical formalism, convolving the unit cell with the lattice essentially places a copy of the unit cell at each lattice point. We write this as *crystal = UnitCell∗lattice*, where ∗ is the symbol for convolution. This concept leads us to an important theorem, which is one of the key ideas in crystallography:

$$FT(\text{crystal}) = FT(\text{UnitCell} * \text{lattice}) = FT(\text{UnitCell}) \times FT(\text{lattice})$$

In other words, the FT of the crystal is the product of the FTs of the unit cell and the FT of the lattice. The idea is illustrated schematically in **Figure 9.9**. An important part of obtaining the structure of a macromolecule is to determine the structure of the lattice, which allows *FT(UnitCell)* to be obtained. Finally, the structure of the unit cell is obtained from the inverse FT (Figure 9.2). It should now be intuitive that the pattern of intensities of the spots reflects the structure in the unit cell. This is the main idea of crystallography, which is to generate a map of the unit cell from the intensities of the spots.

Fourier analysis of diffraction data from crystals is computationally demanding and demands a strong mathematical framework. The first step in the development of the framework is to understand how to describe electromagnetic waves using complex numbers (**Box 9.7**), which in turn, allows the formalism of Fourier analysis to be described compactly (**Box 9.8**). But as has been mentioned, we do not have all the information by measuring the intensities of the spots—the phase information is lost in the measurement but is needed for the inverse transformation (Box 9.8). A large part of the job in macromolecular crystallography is to obtain phase information, and generally, getting the phases is considered to be "solving" the structure, as we discuss later.

(A)

(B)

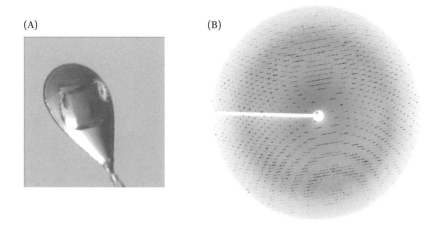

Figure 9.7 Diffraction from a protein crystal. (A) A small protein crystal and its crystallization solution (the "mother liquor") are trapped in a monofilament loop, usually frozen, and placed in the x-ray beam. (B) A diffraction pattern from a crystal that is rotated systematically in the x-ray beam so that diffraction spots are obtained from all of the diffracting planes in the crystal. The white spot in the center is the shadow of the beam stop placed between the crystal and the detector to prevent damage to the detector. The exact pattern observed depends upon the structural arrangement of the proteins within the crystal as well as the structure of the protein itself. (Courtesy of LibreTexts.)

9.4 WHAT IS THE "PHASE PROBLEM," AND HOW DO WE CONQUER IT?

Waves influence each other via two features: their amplitudes and their phases. The reason why phases are important is that they provide information on the positions of the atoms in the unit cell (Box 9.8). Wave interference interactions are described by the principle that if two or more propagating waves of the same type (x-rays, electrons, light, etc.) are incident on the same point, the total amplitude at that point is equal to the sum of the amplitudes of the individual waves. If the crest of a wave encounters the crest of another wave of the same wavelength at that point, then the resulting amplitude is the sum of the individual amplitudes and is called constructive interference, and the waves are said to be "in phase" (Box 9.2). If the crest of one wave meets the trough of another wave, addition gives an amplitude that is the sum of a positive and a negative amplitude, and the result is zero if they have equal amplitudes—this situation is called destructive interference, and the waves are said to be "out of phase." Intermediate phase relations lead to intermediate resulting waves, given by the sum (Box 9.1 and Box 9.2).

If there are two sources, such as atoms in a structure, scattering from the atoms will define an interference pattern like the one in **Figure 9.10**. If we look at the points indicated across the bottom of the figure, at A, the waves are out of phase and cancel, at B and C, the waves are in phase and add, but the resulting

lattice   lattice diffraction

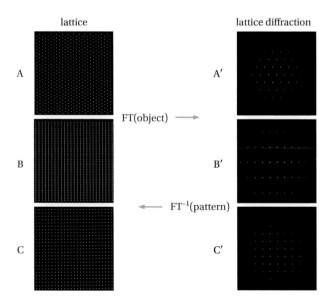

A   A′

FT(object) ⟶

B   B′

⟵ FT⁻¹(pattern)

C   C′

Figure 9.8 Three different two-dimensional lattices are shown as arrays of points with their corresponding Fourier Transforms. Note that the transforms are also arrays of points. In A,A′ the lattice has hexagonal symmetry, and its transform preserves the symmetry. In B,B′ the lattice has rectangular cells, and as we now expect, the long dimension of the rectangles is the short dimension of the transform. A simple square lattice in C,C′ results in a simple square lattice in the transform. (From Harburn G et al. [1975] *Atlas of Optical Transforms*, Cornell University Press, Ithaca, NY. With permission from Taylor & Francis.)

Figure 9.9 A convolution in object space is a product in wave/Fourier space. In (A), a molecule that will fill the unit cell is convoluted with a rectangular lattice. This operation places a unit cell at each point in the lattice, giving the crystal. (B) The diffraction pattern of the crystal is given by the product of the Fourier transform (FT) of the molecule with the FT of the lattice. Note that the result is that the pattern from the unit cell is seen as a modulation of the intensities of the spots. (From Harburn G et al. [1975] *Atlas of Optical Transforms*, Cornell University Press, Ithaca, NY. With permission from Taylor & Francis.)

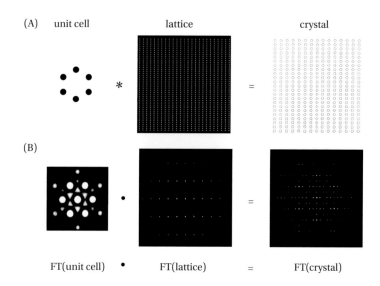

Figure 9.10 Interference of waves emitted by two point sources, such as atoms. The dashed line represents a nodal line where there is always destructive interference. Yellow and blue spots represent wave maxima and minima that are propagating outward from the two sources. Waves arriving at B and C are out of phase with one another.

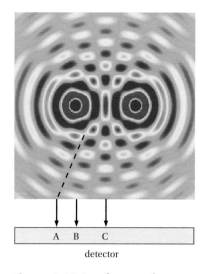

wave crests arrive at the detector at different times—the crest arriving at B (yellow) is nearly out of phase with the crest that will soon arrive at C, and a trough (blue) is at C. The detector only measures the number of photons arriving at a particular point and not the relative arrival times. Consequently, the detector, represented by the box, will measure intensities and lose the phase information. This loss of information is the principal challenge of deriving a structural map from the measurement of x-ray intensities, from which we get amplitudes as the square root of the intensities (Box 9.8). If we can determine the phases of these amplitudes, we can get a map of the structure.

The diffraction data, measured as the intensities, together represent the squared amplitudes of the FT of the unit cell structure. If the phases are also known, the inverse FT can be used to obtain an electron density map of the structure (Box 9.8). The electron density is the aspect of the structure that is mainly responsible for the scattering of the x-rays. There are several methods that have been developed for recovering phase information.

One approach is to place a known scattering center, such as an atom with many electrons (a "heavy" atom), in the structure. Scattered radiation from the heavy atom will interact with the scattering from other atoms. This causes changes in scattering from the unit cell. By knowing the position(s) of the heavy atom(s), which often can be found from the amplitude data, the phase shifts can be used to obtain experimental values for the unshifted phases. An assumption is that the binding of the heavy atoms does not perturb the structure, and the method is generally called "multiple isomorphous replacement" when several different heavy atom derivatives are used.

A related method, Multi-wavelength Anomalous Diffraction (MAD), also relies upon perturbing the phases but has a different basis. MAD phasing uses the fact that the inner electrons of an atom can absorb x-rays differently at particular wavelengths above and below an electronic energy transition and reemit the x-rays after a delay, inducing a phase shift in all of the reflections, known as the anomalous dispersion effect. By measuring diffraction patterns using wavelengths on either side of the transition, the phase shifts can be used to obtain the phases for the crystal. With the availability of precise wavelength variation using synchrotron x-radiation, it became possible to measure data using several wavelengths on a single crystal, often exploiting analogs of amino acids, such as seleno-methionine (Met with Se instead of S), incorporated into a protein. MAD phasing, using a single heavy atom constituent, is now in widespread use.

If the structure of a close homolog of a protein is known, a process known as "molecular replacement" can be used. In this method, the phases of the homolog are used with the intensities from the unknown structure to obtain a preliminary map, which is then refined (see later). Generally, this approach is

BOX 9.7 WAVES CAN BE DESCRIBED USING COMPLEX NUMBERS

The electromagnetic waves diffracted by a sample, such as simple apertures (Box 9.2) or protein crystals (Figure 9.7), can be described as cosine waves with amplitude U and phase α (Box 9.1). Calculation of Fourier diffraction patterns arising from the interference of waves requires that both U and α be accounted for meticulously. The same holds for computing structures from diffraction patterns by inverse FT. These computations quickly become very complicated, even for simple geometries; a more compact and computationally tractable form of Equation 1 shown in Box 9.1 is needed. This form is provided by the formalism of complex numbers, which allows both the amplitude and the phase of a wave to be stated compactly. Amplitude and phase are represented as a vector **U** on the complex plane (boldface font means a vector), as shown in **Figure 1**.

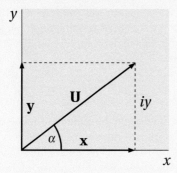

Figure 1 Composition of complex numbers.

Using this nomenclature, as described in most calculus textbooks, the vector $\vec{U}$ is described by its magnitude U and its coordinates in the complex plane:

$$\vec{U} = \vec{x} + i\vec{y}$$

where i is the imaginary unit that satisfies the equation $i^2 = -1$. This equation may be expressed as

$$\vec{U} = U(\cos\alpha + i\sin\alpha) = U\exp[i\alpha]$$

where U represents the magnitude of the vector $\vec{U}$ and $\exp[i\alpha]$ is defined by

$$\exp[i\alpha] = \cos\alpha + i\sin\alpha$$

Figure 2 shows a wave with phase $\alpha = 60°$, represented as a vector of magnitude U that rotates through 2π during one cycle. The y coordinate of the wave for some value of x is given by the projection of the vector onto the y-axis, as shown. Using the exponential form of the complex variable, we can describe the wave by

$$y(x,0) = \mathrm{Re}\left\{U\exp\left[i\left(\frac{2\pi x}{\lambda}\right) + \alpha\right]\right\}$$

where Re means "the real part of." As an exercise, show that this representation of the wave is equivalent to Equation 1 in Box 9.1.

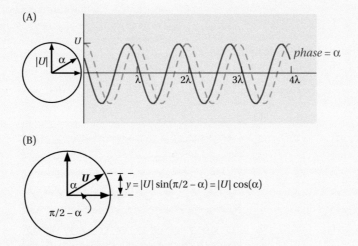

Figure 2 Representation of complex numbers as an Argand diagram.

less preferred, since the phases from the known structure will strongly bias the result, and errors have been made. But, there are many examples where it has proven useful.

Usually, when an initial set of phases has been determined, a process of "refinement" is carried out to improve the structural map. Using the initial map, portions of the structure are fitted into it using the known sequence and properties of polypeptides. Common themes, such as α-helices and β-sheets, can be very helpful as guides. The partial model is used to calculate a new set of

BOX 9.8   FOURIER TRANSFORMS AND THE IMPORTANCE OF PHASE INFORMATION

We discussed FTs for one-dimensional diffraction in Chapter 2 (Box 2.4). Proteins are three-dimensional, of course, but the one-dimensional problem suffices for indicating the general principles. Reviewing, any continuous or piece-wise continuous real function $w(z)$ can be written as

$$w(z) = \sum_h \left\{ A_h \cos 2\pi h \frac{z}{d} + B_h \sin 2\pi h \frac{z}{d} \right\} \quad (1)$$

where d is the lattice repeat distance, A_h and B_h are the Fourier coefficients, and h is a sequence of integers ($h = 0$, 1, …, ∞). This Fourier series must be modified for practical diffraction applications. The first step is to include phase information explicitly (Box 9.1):

$$w(z) = \sum_h \left\{ A_h \cos \left(2\pi h \frac{z}{d} + \alpha_h \right) + B_h \sin \left(2\pi h \frac{z}{d} + \alpha_h \right) \right\} \quad (2)$$

Notice that each of the h terms of the series has an associated phase α_h. The second step is to represent the series using the complex-number representation (Box 9.7):

$$w(z) = \sum_h A_h \exp i \left(2\pi \frac{h}{d} z + \alpha_h \right) \quad (3)$$

This equation can further be written

$$\mathbf{w}(z) = \sum_h \mathbf{A}_h \exp i \left(2\pi \frac{h}{d} z \right) \quad (4)$$

in which $\mathbf{A}_h$ is $|\mathbf{A_h}|\exp[i\alpha_h]$, or $A_h \exp[i\alpha_h]$. This format represents the Fourier coefficient as a complex vector to associate A_h explicitly with α_h.

The purpose of the FT is, given measured diffracted intensities, to determine the distribution of the atoms in the unit cell as distributions of electron density or more accurately, electron scattering-length density. The scattering length of an atom determines the strength of its interaction with the incident x-rays and is proportional to the number of electrons in the atom. Oxygen ($Z = 8$) and carbon ($Z = 6$) atoms scatter much more strongly than hydrogen atoms ($Z = 1$), which is why hydrogens are not apparent in diffraction-determined structures except at extraordinarily high resolutions. What is sought in crystallography are structure factors F_h, which are simply the Fourier coefficients suitably scaled to include the atomic scattering lengths. We can thus rewrite Equation 4 in terms of electron scattering-length density as

$$\rho(z) = \sum_h F_h \exp[i\alpha_h] \exp i \left(2\pi \frac{h}{d} z \right) \quad (5)$$

The intensity of x-rays scattered by electrons is proportional to the square of the scattering length, which means that the intensity will be proportional to the square of the structure factor. You may recall from the algebra of complex numbers that the square of a complex number is given by **FF***, i.e., the product of the complex number with its complex conjugate:

$$F_h \exp[i\alpha_h] \times F_h \exp[-i\alpha_h] = F_h^2 \quad (6)$$

When the intensity of the scattered x-rays is measured, the phases associated with the structure factors are therefore lost. The key issue in crystallographic structure determinations is the restoration of the phase information.

Figure 1 shows the Fourier reconstruction of a few atoms distributed along an axis based upon structure factors. Notice that the summation of the first $h = 10$ structure factors provides an approximate image of the atoms. On the left of the figure, the amplitude and phase of each structure factor are indicated. It should be clear that without the phase information, the distribution of electron density could not be determined.

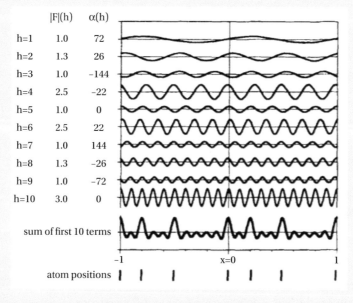

	|F|(h)	α(h)
h=1	1.0	72
h=2	1.3	26
h=3	1.0	−144
h=4	2.5	−22
h=5	1.0	0
h=6	2.5	22
h=7	1.0	144
h=8	1.3	−26
h=9	1.0	−72
h=10	3.0	0

sum of first 10 terms

−1   x=0   1

atom positions

Figure 1 (From Blow D [2002] *Outline of Crystallography for Biologists*. Oxford University Press. With permission.)

phases, which are reapplied to the observed amplitudes, and an improved map is calculated. The modeling of the structure into the map is then improved by correcting errors, such as wrongly connecting the densities with the chain, and extended (to previously un-modeled parts of the structure), and the process is repeated until the result converges and changes are very small. Refinement is both useful and dangerous, because errors in assigning and building the initial structures can propagate during refinement; indeed, it has even happened that incorrect structures have been reported with the amino acid sequence modeled backwards! But in most cases, refinement works well, particularly if the resolution of the data set is high (~2 Å) (**Box 9.9**).

9.5 CRYSTALLOGRAPHIC MODELS OF PROTEINS ARE DERIVED FROM ELECTRON DENSITY MAPS

As critical scientists, we need to know how much we can rely on models derived from crystallographic data. There are several keys to assessing the result, but the two most useful are the resolution (Box 9.9) and the *R* factors. We have discussed resolution at several points along our way in this book, and the basic idea is that the higher the resolution, the more we know with confidence: at high resolution, atoms can be reliably modeled; at medium resolution, side chains can be identified with confidence; at low resolution, secondary structure elements are clear; and at very low resolution, the shapes of molecules are seen. In **Figure 9.11**, a tryptophan sidechain is seen as fitted to good maps at four different resolutions.

There are two useful residual R factors: the crystallographic and free *R* factors. These parameters measure the extent to which the data are accounted for by the structural model. Both *R* factors have the same form:

$$R = \frac{\sum \left| |F_{obs}| - |F_{calc}| \right|}{\sum |F_{obs}|}$$

In this equation, the F_{obs} are the observed amplitudes (square roots of the measured intensities), and the F_{calc} are the amplitudes that are calculated from the model. A smaller *R* means a better agreement between the model and the data. The *R* will not be zero for a number of reasons, including experimental errors in measurement of the intensities and errors in the model. For the crystallographic *R* factor (R_{cryst}), the sums run over all reflections except a small, random subset that is set aside and not used in refining the structure. The free *R* factor, *R*-free, is calculated from the subset, with the idea that it is an independent assessment unbiased by the refinement process. *R*-free will always be greater than R_{cryst}, because the model is not fitted to the intensities used in *R*-free, but the two should be close for a good model. These analytical parameters, properly used, give a quantitative confidence measure—how much should we trust that the model is correct? Of course, it is easier to fit the data at low resolution, where there are far fewer data to account for, but it is harder at high resolution. The strengths and weaknesses of MP crystal

Figure 9.11 Electron density maps and structures at different resolutions. A skeleton model of a tryptophan side chain is shown compared with the density at different resolutions. At 1.0 Å, the atoms can be resolved. This is true "atomic resolution" and is seldom obtained for biological macromolecules. At 2.5 Å, generally considered to be a good structure in crystallography, one can clearly identify and position the Trp in the map, giving assurance that the atom locations are known. At 3 Å, currently considered a good level in cryo-EM, identity and orientation are less certain, and at 4.0 Å, a more usual cryo-EM structure, the fit and identity are still less clear. Refer back to Figure 9.5 and study the density map in (A), asking how well you think the side chains are defined and positioned at 3.4 Å. (With permission from Philip R. Evans Laboratory of Molecular Biology, Cambridge, UK.)

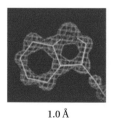

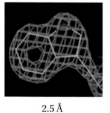

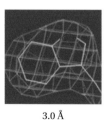

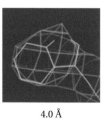

| 1.0 Å | 2.5 Å | 3.0 Å | 4.0 Å |

BOX 9.9   FOURIER ANALYSIS OF A GAUSSIAN FUNCTION TO ILLUSTRATE ASPECTS OF FOURIER RECONSTRUCTION, THERMAL DISORDER, AND RESOLUTION

The Gaussian function is useful for describing atomic scattering density in both protein crystals and lipid bilayers (Chapter 2). A Gaussian representation of a collection of atoms or group of atoms (e.g., an amino acid) located at z_0 on the z-axis, is given by

$$\rho(z) = \frac{b}{A\sqrt{\pi}} \exp\left\{-\left[\frac{z-z_0}{A}\right]^2\right\} \quad (1)$$

In this equation, b is the scattering length of the group and A is the $1/e$ half-width of the Gaussian. A collection of Gaussians with half-widths of A located at z_0 and repeating at intervals of d along the z-axis has the familiar appearance shown in **Figure 1**]. The value of b does not matter for present purposes; let $b = 1$. The term $1/(A\sqrt{\pi})$ simply normalizes the density so that the integral over z is equal to b. As discussed in Box 9.1, constructive interference from a large array with many atoms (i.e., Gaussians) will occur only at angles satisfying the Bragg condition, $1/\lambda = h/d$. We can compute the expected structure factors from the inverse FT (Box 2.5):

$$F(h) = 2b \exp\left[-\left(\pi A \frac{h}{d}\right)^2\right] \cos\left(2\pi z_0 \frac{h}{d}\right) \quad (2)$$

Note that the FT of a Gaussian is another Gaussian.

An important question is how the width of the Gaussian affects the number and amplitudes of the structure factors. The structure factors for Gaussians with half-widths of $A = 1$ Å and $A = 3$ Å distributed along the z-axis at $d = 50$ Å intervals reveal an important aspect of crystallography (**Figure 2**): very narrow features produce many more significant structure factors than broader features. One can imagine that a very "stiff" structure with tightly held atoms will produce more diffraction spots than a structure with loosely held atoms. In the former case, for example, the atoms undergo very limited thermal motion, whereas in the latter case, the atoms have much more thermal motion. Even though the atoms might occupy the same average positions in the lattice, atoms with more thermal motion produce fewer structure factors. This is the basis for so-called B factors in crystallography; thermal vibrations lead to broader distributions of the atoms and consequently, fewer diffraction orders.

What are the consequences of these thermal motions for Fourier reconstructions? In **Figure 3**, Gaussians have been reconstructed by FT using the first nine ($h = 1 \ldots 9$) structure factors. Reconstruction with nine structure

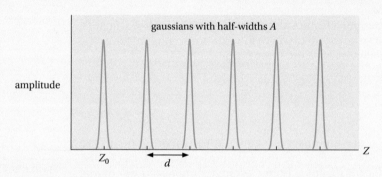

Figure 1 A set of Gaussians distributed uniformly along the z-axis.

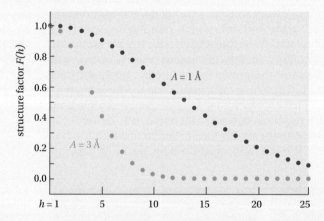

Figure 2 The potential number of structure factors produced in a diffraction experiment varies inversely with feature width. A narrow Gaussian width results in more structure factors than a broader one. (From Weiner MC, White SH [1991] *Biophys J* 59: 162–173.)

(Continued)

BOX 9.9   FOURIER ANALYSIS OF A GAUSSIAN FUNCTION TO ILLUSTRATE ASPECTS OF FOURIER RECONSTRUCTION, THERMAL DISORDER, AND RESOLUTION (CONTINUED)

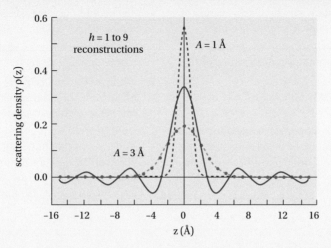

Figure 3 The number of structure factors used in the Fourier reconstruction of a Gaussian peak affects the reconstructed peak appearance. The broad $A = 3$ Å peak (dashed blue line) can be accurately reproduced using nine orders (blue data points), whereas for the $A = 1$ Å peak (dashed red curve), nine orders (solid red line) are not sufficient; the peak is under-resolved and is accompanied by oscillating tails (Fourier noise). (From Weiner MC, White SH [1991] *Biophys J* 59: 162–173.)

factors reproduces very well the 3 Å Gaussian but not the 1 Å Gaussian. Notice the oscillations at the edges of the 1 Å Gaussian, which are characteristic of under-resolved features. In crystallography, these oscillations produce so-called Gaussian noise due to under-resolution, which makes accurate reconstruction of the unit cell difficult. Accurate reconstruction of the very narrow 1 Å distribution would require more than 25 diffraction orders.

Can we estimate the number of diffraction orders required for the accurate determination? The canonical resolution r of a structure is generally defined as $r = d/h_{max}$ where d is a unit cell dimension and h_{max} is the maximum number of observed diffraction spots in a particular direction. For the example in Figure 3, with $d = 50$ Å and $h_{max} = 9$, the resolution is about 5 Å, corresponding roughly to the half-width of the $A = 3$ Å Gaussian. How many diffraction orders would be necessary to reconstruct the narrow

Gaussian in Figure 3? Taking the half-width of the narrower Gaussian as about 2 Å, the formula for the canonical resolution tells us that the h_{max} required is at least 25, consistent with Figure 2.

Considering a diffraction experiment on a crystal with a lattice dimension of d, the resolution is estimated from the highest-order diffraction spots seen on the detector. For the previous example, to resolve a narrow Gaussian of width $A = 1$ Å, we would need to see spots corresponding to $d/h_{max} \approx 2$ Å. For the $A = 3$ Å Gaussian, we need to see diffracted intensities out to only about 5 Å. And if we see diffraction only to 5 Å, we can be certain of not being able to resolve any atomic-level features. As a reference point, the first structure of bacteriorhodopsin (Figure 1.27) was a 7 Å structure, which allowed the helices with diameters of about 10 Å to be resolved, but nothing else.

structures are presented in **Box 9.10**. A major weakness of crystallographic structures is that the lipid/detergent environment of the MP is not seen except under special circumstances (**Box 9.11**).

9.6 ELECTRON CRYSTALLOGRAPHY IS USEFUL FOR DIFFRACTION FROM SMALL CRYSTALS

Crystal structures of proteins can also be obtained, though not yet commonly, using electron microscopy. The principles are the same as for x-ray crystallography. There are two big advantages of electrons, however. First, structures of MPs can be obtained using 2-D crystals following the methods pioneered by Richard Henderson and Nigel Unwin in their determination of the structure of

BOX 9.10   MEMBRANE PROTEIN CRYSTALLOGRAPHY: STRENGTHS AND WEAKNESSES

Strengths

- Models can be built that show the chemical structure of a membrane protein or complex in terms of atoms.

- The detail is enough to show water molecules and bound ions in many cases, aiding understanding of mechanism.

- Additional features, such as prosthetic groups and bound lipids, are seen.

- As with soluble proteins, it is possible to infer functional mechanisms by solving variations in the structure using mutants, bound ligands, and other perturbations (such as light).

Weaknesses

- The membrane protein is, in most (but not all) cases, removed from a membrane bilayer environment.

- Crystals must be obtained and are usually studied at low temperature.

- While perturbations by the crystal contacts are thought to be small, they may nonetheless be significant in some cases; for example, crystallization may select a subset of conformations.

- As with cryo-EM, the result is usually an atomic model based on fitting a density map that is at lower resolution than the atomic structure, so it may have errors arising from incomplete constraints.

---

bacteriorhodopsin (Figure 1.27), and second, data can be obtained from very small crystals.

A major problem with crystallography of MPs is often the small sizes of the crystals that can be obtained. Although modern synchrotron x-ray sources have surmounted this hurdle to a large extent, electron crystallography offers major advantages because of the high-quality diffraction patterns obtainable from microscopic crystals. An electron beam interacts with molecules $\sim 10^6$ times more strongly than an x-ray beam does, so it can produce diffraction data from extremely small crystals (up to six orders of magnitude smaller in volume than those typically used for x-ray crystallography). In the field of view shown in **Figure 9.12A**, circled crystals of a soluble protein are the kind used for x-ray studies, while the arrows point to crystals that can be diffracted using electrons. The quality of diffraction pattern from the microcrystals is shown in **Figure 9.12B**.

While the method has not yet been applied to MPs, it seems likely that it will be, because microcrystals are much more easily obtained than large crystals. Until recently, the study of such microcrystals was thought to be the province of the pulsed x-ray sources, but it now appears that better data can be obtained using the EM. It remains true, however, that the pulsed x-ray sources can view crystals at room temperature, so there may still be a use for such data if the technical hurdles can be overcome.

(A)

(B)

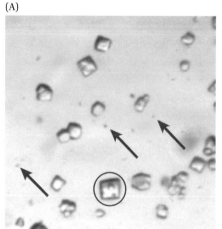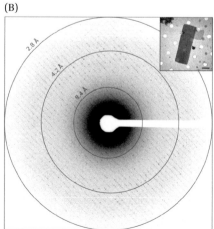

Figure 9.12 Electron diffraction can be used to examine microcrystals and may be useful for membrane proteins in the future. (A) A field of crystals with the size used in x-ray studies circled and arrows indicating the size for electron diffraction. (B) An electron diffraction pattern. (A, From Shi D et al. [2013] *eLife* 2: e01345. B, From Nannenga BA et al. [2014] *eLife* 3: 303600.)

BOX 9.11   LIPIDS AND DETERGENTS IN PROTEIN CRYSTALS CAN BE VISUALIZED USING X-RAY AND NEUTRON DIFFRACTION

The function and structure of MPs depend, of course, upon the structure and properties of the surrounding lipid bilayer. Lipids tightly bound to an MP can be—and often are—seen in high-resolution structures (**Figure 1**). But, the fluid lipid bilayer surrounding the protein is generally not revealed in crystallographic structures. This may seem odd, but it really isn't, for two reasons. First, except for type I crystals (**Figure 2A**), the crystal lattice lacks the lamellar organization of membranes. Second, the detergents or

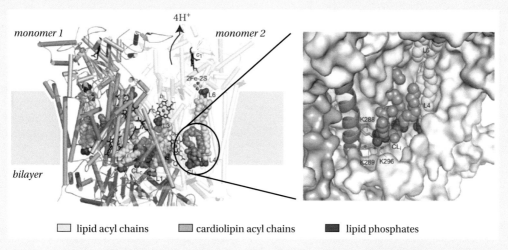

Figure 1 Specifically bound lipids are important in membrane protein function. A critically placed cardiolipin molecule in cytochrome bc₁ (Complex III; PDB: 1P84, 3CX5) stabilizes the protein for efficient build-up of the proton gradient across the mitochondrial inner membrane. (From Wenz T, Hielscher R, Hellwig P et al. [2009] *Biochimica et Biophysica Acta (BBA) Bioenergetics* 1787: 609–615. With permission from Elsevier.)

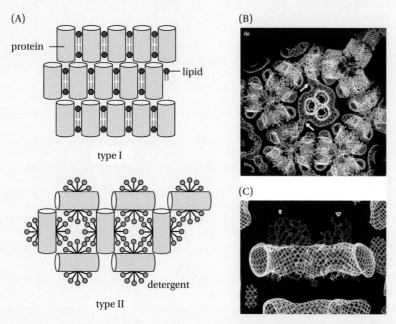

Figure 2 Membrane protein crystal lattices. (A) Two types of crystals encountered in membrane protein crystallography. Type I crystals have lamellar arrangements of the of the proteins, approximating a stack of membranes; the lipid/detergent mixtures form a pseudo bilayer containing the proteins. Type II crystals lack the lamellar organization. (B) The arrangement of OmpF trimeric porins (orange) in a type II crystal obtained by solubilizing OmpF in β-octyl glucoside (βOG) detergent (green). The detergent is not seen in x-ray structures, because it forms thermally disordered bands around the protein. (C) If the detergent is deuterated, it can be revealed by low-resolution neutron diffraction crystallography using the same methodology as for visualizing the double bonds of fluid lipid bilayers (Figure 2.18). (From Pebay-Peyroula E, Garavito RM, Rosenbusch JP et al. [1995] *Structure* 10: 1051–1059. With permission from Elsevier.)

(Continued)

BOX 9.11   LIPIDS AND DETERGENTS IN PROTEIN CRYSTALS CAN BE VISUALIZED USING X-RAY AND NEUTRON DIFFRACTION (CONTINUED)

detergent-lipid mixtures used for crystallization are highly thermally disordered, which precludes atomic-level images.

If special efforts are made, however, thermally disordered lipids or detergents can be visualized in crystal structures in two ways. First, just as thermally disordered double bonds can be visualized in fluid lipid bilayers using lamellar neutron diffraction (Figure 2.19), deuterated detergents in MP crystals can be seen using low-resolution neutron crystallography. The organization of the detergent around the protein is found by superimposing the high-resolution x-ray crystal structure of the protein on the low-resolution neutron crystal structure (**Figure 2B**). Looking along one of the crystal axes in the superimposed images shows that the detergent forms a belt around the protein, thus acting as stand-in for the missing lipid bilayer (**Figure 2C**).

Second, if the membrane protein crystals are of type I, low-resolution images of lipids and detergents around the protein can sometimes be obtained by x-ray diffraction (**Figure 3**). Certain orientations of the crystals in the x-ray beam give diffraction spots at low angles (panel A) that are similar to those seen in lamellar x-ray diffraction (Figure 1.18A) due to scattering from the electron-dense phosphates of the lipids. As in lamellar diffraction from bilayers, thermal motion prevents atomic-level resolution. Nevertheless, the thermal envelopes of the phosphates of lipids co-crystallized with the protein are resolved (panel B), which reveals the location of the bilayer (or its detergent-lipid stand-in) surrounding the protein. The envelopes are very similar to those built from molecular dynamics simulations of the protein in a lipid bilayer (panel C).

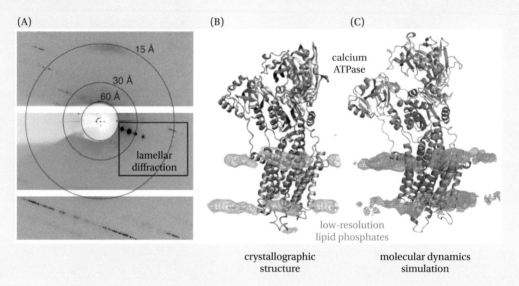

Figure 3 Small-angle lamellar diffraction peaks from type I crystals of calcium ATPase can be used to find the thermal envelopes of lipid-headgroup phosphates. (A) Diffraction pattern from calcium ATPase crystals obtained by solubilizing the protein in detergent-phospholipid mixtures. The crystal has been oriented so that the low-angle diffraction is equivalent to that from oriented lipid bilayer arrays (Figure 1.18A). (B) Superposition of the high-resolution crystal structure of the ATPase (PDB: 2YFY) on the low-resolution structure of the thermal envelopes of the lipid phosphate groups. (C) A model for the phosphate thermal envelopes can be constructed from molecular dynamics simulations of the protein in a lipid bilayer. (Sonntag Y et al. [2011] *Nature Communications* 2: Article 304. With permission from Springer Nature.)

KEY CONCEPTS

- The experimental result from crystallography or cryo-EM is a density map from which a chemical model is built and refined. It is not usually a determination of the "atomic resolution structure," as is often said, and involves the use of other information, such as the side chain structures and backbone constraints of a polypeptide.

- The resolution of the data is a key to understanding how much it can tell us. The information content is related to the volume of reciprocal space, so, for example, a structure at 2.0 Å resolution has, in principle, eight times as much information as a structure at 4.0 Å, not counting errors in measurement and model building.

- In addition to the resolution, the R factors give measures of how much one can rely on a model. These are, however, imperfect for a variety of reasons beyond the scope of this book.

- It should be kept in mind that the protein or complex has to be isolated, stabilized, and studied at low temperature, as the structure or conformation may be altered by these factors.

- It is, nonetheless, remarkable how much insight can be gained on the structures and chemical mechanisms of membrane proteins using current structural methods.

FURTHER READING

Blow, D. (2002) *Outline of Crystallography for Biologists*, 1st edition, Oxford University Press, USA; (June 20, 2002) (ISBN-10:0198510519).

Rhodes, G. (2006) *Crystallography Made Crystal Clear: A Guide for Users of Macromolecular Models*, 3rd edition, Academic Press, San Diego; (March 2, 2006) (ISBN-10:0125870736) Web.

Moore, P.B. (2012) *Visualizing the Invisible*, Oxford University Press, New York.

Lattman, E.E., and Loll, P.J. (2008) *Protein Crystallography: A Concise Guide*, Johns Hopkins University Press, Baltimore.

Harburn, G., Taylor, C.A, and Wellbury, T.R. (1975) *Atlas of Optical Transforms*, Cornell University Press, Ithaca, N.

Chiu, W., and Downing, K.H. (2017) Cryo Electron Microscopy: Exciting advances in CryoEM herald a new era in structural biology. A special issue of *Current Opinion in Structural Biololgy*. See *Cur Opin Struc Biol* 46:iv–viii.

Cheng et al. (2015) A primer to single-particle cryo-electron microscopy. *Cell* 161:438–449.

Nogales, E., and Scheres, S.H.W. (2015) Cryo-EM: A unique tool for the visualization of macromolecular complexity. *Molecular Cell* 58:677–689.

EXERCISES

1. For each of the following two-dimensional objects, sketch the diffraction pattern you would expect.

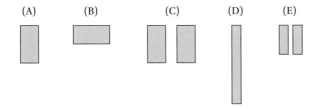

(A) (B) (C) (D) (E)

2. Given that the wavelength of a 200 keV electron is ~0.025 Å, why is it that electron microscopy of fields of single particles does not give scattering density maps with much better resolution than ~2 Å (at best)? You should identify at least four reasons in your answer.

3. Rapid freezing of samples for electron microscopy is a key to the study of many specimens. One benefit is that the hydrogen bonds in the ice are trapped before microcrystals are formed, resulting in vitreous ice with no regular structure.

   a. Why might ice crystals be undesirable? (Hint: consider the effect of relatively slow freezing and thawing on biological structures.)

   b. How do the timescales of different kinds of molecular motions compare with the freezing rate?

c. Given that an ensemble of molecules at room temperature will have a diversity of conformations at different length scales, how does the freezing inform your answer to question 2?

4. Considering your answers to (3), do you expect all of the unit cells in a crystal to be identical? What effects might differences between unit cells have on the resolution and accuracy of models from crystallographic data?

5. Convolution is an operation that produces an array of objects by combining the array with a single object. To test your understanding, sketch the result of a convolution for each of the following pairs.

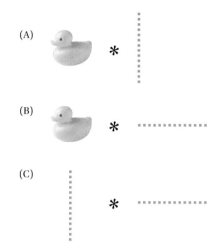

6. Match the following diffraction patterns with the diffracting objects. Explain each match.

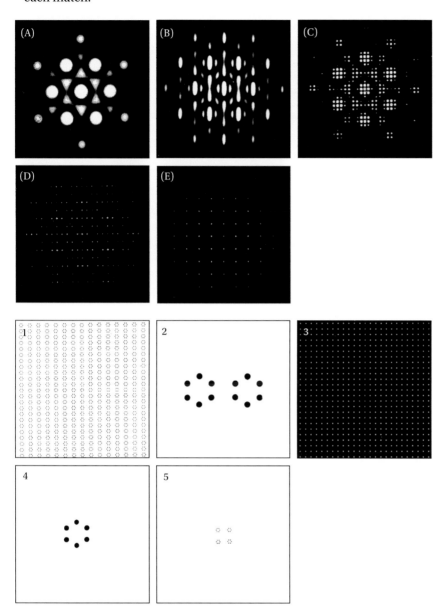

7. Crystallography is a better way to get chemical models for relatively small (~<100 kDa) macromolecular structures, and cryo-EM is emerging as a stronger approach for much larger (~>1000 kDa) entities. Name and explain at least two reasons why.

Small-Molecule Channels

10

The lives of cells depend upon exquisite control over every aspect of their interactions with the environment. Necessarily, these interactions require ions, molecules, and signals to traverse membranes. Some cell activities require fast responses; others do not—for example, uptake of metabolic precursors can be measured on the timescale of biosynthesis (minutes, for macromolecules), while responses to some signals must be fast (milliseconds for nerve signals). An important principle of control is the use of natural high-energy barriers that prevent a process of some kind from reaching equilibrium quickly. These barriers are then selectively lowered with the help of proteins to allow processes to occur in a controlled fashion. This barrier-control approach is nowhere more obvious than in the movement of ions across cell membranes to produce and use electrochemical gradients. The lipid bilayer of the cell membrane is an excellent barrier; neither ions nor water can cross with ease, allowing control of their movements by opening and closing protein channels across the membrane or by using pumps to create gradients of concentration.

In this chapter and the following one, we examine the controlled flow of small molecules and ions down their concentration gradients. In later chapters, we examine the slower processes by which ATP energy is used to create and use concentration gradients, and the coupling of concentration gradients to use an energetically downhill movement to drive an uphill movement of a different species. We begin with the simplest cases of how the membrane barrier can be overcome and how the barrier can be breached while maintaining selectivity for one or a few small-molecule species. How can water be allowed to pass while small ions cannot? How can emergency needs, such as osmotic shock, be coped with? Answers are provided by examining α-helix-based channels, and pores in plasma membranes involved in the transmembrane movement of small molecules such as water and glycerol are the primary subject of this chapter. But we also consider the β-barrel-based channels and pores found in the outer membranes of Gram-negative bacteria and mitochondria.

What are the structural features of proteinaceous channels that lead to selective solute permeation? Can the permeability of such channels be controlled, and how? We described in Chapter 1 how studies of water movements across cells, particularly protoplasts, provided key insights into the existence and properties of cell membranes. We noted that water can cross pure lipid bilayers, driven by the high concentration (~55 M) of water. Does this mean that water channels are not necessary? Applying the principle of barrier control, without any data whatsoever, one has to conclude that cells must control water transport using channels. What is the evidence that water channels exist? We begin this chapter by examining the critical experiments that demonstrated the existence of water channels.

DOI: 10.1201/9780429341328-11

10.1 WATER CAN CROSS MEMBRANES EITHER BY DIFFUSION THROUGH THE LIPID BILAYER OR THROUGH CHANNELS

10.1.1 Water Diffuses through Lipid Bilayers but with a High Activation Energy

In the mid-20th century, little was known about protein structure or the genetic code, and as we noted in Chapter 1, even the existence of lipid bilayers as the structural fabric of membranes was controversial. Nevertheless, in the virtual absence of structural data, physicochemical measurements led inexorably toward the idea that water and ions pass through pores embedded in a thin lipid membrane. A key advance, due to Henry Eyring (1901–1981), was a mathematical description of the diffusion of solutes through membranes based upon Transition-State Theory, which posits that chemical reactions occur by the formation of an activated (high-energy) transition-state complex that is in quasi-equilibrium with the reactants. From the activated state, the transition-state complex can decay into reaction products; the job of enzymes is to lower the activation state barrier and speed up the reaction (**Figure 10.1A**). The rather abstract reaction coordinate can be thought of as the progress of a reaction along a series of structural and chemical steps. Over time, the concentration of products will balance the energies on either side of the barrier, but the barrier itself remains for both the forward and reverse reactions. When applied to diffusion through membranes, the high-energy state is the energy required to surmount obstacles to diffusion (such as the displacement of neighboring molecules by an amount λ), and the local reaction coordinate is the distance between the barriers.

Diffusion of a water molecule across a lipid membrane can be described by rate constants for entering (k_{sm}) or leaving (k_{ms}) the membrane at the membrane–water interface (Figure 10.1B). Within the membrane, the rate constant moving forward or backward a distance λ between collisions with membrane

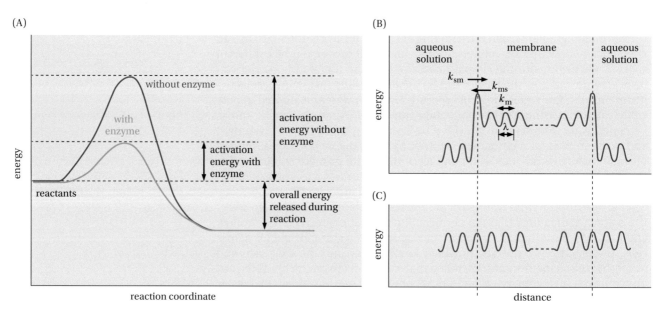

Figure 10.1 Eyring Transition-State idea. (A) Reactants in a chemical reaction form a high-energy transition-state complex represented by an activation energy E_{act}. Enzymes can lower this barrier and thus, speed up the reactions. As the reaction proceeds, the net free energy goes to zero when equilibrium is reached, but the role of the enzyme in the forward and reverse reactions persists. (B) Diffusion across a membrane can be depicted as a diffusing molecule jumping over a series of barriers separated by a distance λ. In the depiction here, the main barrier is solution/membrane interface. (C) Energy barrier representation for the diffusion of the solute through a water-filled pore. (B, C, From Price HD, Thompson TE [1957] *J Mol Biol* 41: 259–277. With permission from Elsevier.)

molecules is k_m. The distance λ is typically 4–5 Å. Eyring and his colleagues showed that the permeability P of the membrane (see **Box 1.1**) is given by

$$\frac{1}{P} = \frac{2}{k_{sm}\lambda} + \frac{m}{k_m\lambda(k_{sm}/k_{ms})} \tag{10.1}$$

where m is the number of steps of length λ required to cross the membrane. The ratio k_{sm}/k_{ms} defines the solution: membrane partition coefficient of water, defined as K_w (Box 1.2). The first term expresses the resistance to membrane entry at the interface. The solubility of water in the lipid membrane is low, so that $k_{sm} << k_{ms}$. Assuming that $k_{ms} \geq k_m$, as in the situation shown in **Figure 10.2C**, Equation 10.1 reduces to

$$P = k_m K_w \lambda / m \tag{10.2}$$

The diffusion coefficient D_m of water in the membrane is given by $k_m\lambda^2$, and the membrane thickness d is $m\lambda$, so that

$$P \simeq \frac{D_m K_w}{d} \tag{10.3}$$

When d is measured in cm, the permeability P has units of cm s^{-1} because the diffusion coefficient D_m has units of cm^2 s^{-1} and K_w is dimensionless. This equation has been validated experimentally by measuring the water permeability of vesicles made from lipids with different hydrocarbon chain lengths, resulting in different hydrocarbon layer thicknesses (see Chapter 2). Light scattering is related to vesicle size, so that osmotic swelling or shrinking resulting from

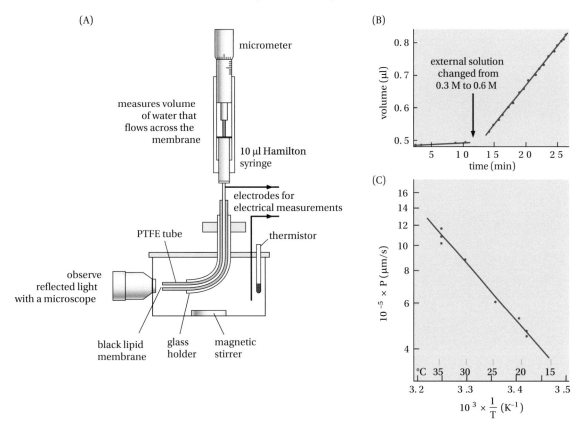

Figure 10.2 Measuring water diffusion through black lipid membranes (Figure 1.19). (A) Experimental set-up. (B) Changing the salt solution on one side of the membrane creates an osmotic gradient, resulting in a net flux of water across the membrane. The micrometer syringe removes or adds fluid to keep the membrane flat, thus measuring the rate of water flow, which permits calculation of the membrane's water permeability P. (C) The activation energy of transport can be determined from the temperature dependence of permeability. (A, R Fettiplace and DA Haydon (1980) *Physiological Reviews* 60:510–550. B, From Price HD, Thompson TE [1957] *J Mol Biol* 41: 259–277. With permission from Elsevier. C, From Petersen DC [1983] *Biochim Biophys Acta (BBA) Biomembranes* 734: 201–209. With permission from Elsevier.)

Table 10.1 Permeability of Various Cells to Water

Cell	Permeability (μm s^{-1})
Human red blood cells	53
Amoeba (*Chaos chaos*)	0
Amoeba proteus	0
Frog ovarian cell	1
Zebra fish ovarian cell	1
Xenopus egg	1
Salmon egg	2
Ehrlich ascites tumor cells	45
Valonia (an alga)	24

changing the osmolarity of vesicle environments versus their contents can be observed. The rate of light scattering change as vesicles swell or shrink can then give values for water permeability, which range from $P = 0.25$ to 50 μm s^{-1} for phosphatidylcholines with chains ranging from 24 to 14 carbons.

One of the important features of black lipid membranes (Figure 1.9) is that their permeabilities can be readily measured using osmotic gradients (Figure 10.2) or tritiated water (tritium gives the measurement reported as radioactivity moving across the bilayer, assuming the ^{3}H isotope is a close mimic of ^{1}H). Depending upon lipid composition, P ranges from 20 to 75 μm s^{-1}. For single biological cells, the reported permeability (**Table 10.1**) is generally smaller than observed for black lipid membranes but more varied. The low values are likely due to various protective layers that hinder diffusion. But, perhaps, the black lipid membranes are exceptionally permeable to water? That does not appear to be the case, because the black film water permeabilities agree with those of large unilamellar vesicles.

Comparisons of the values of P alone do not provide useful information about the existence of water-conducting pores in membranes. For example, one membrane might have a large number of small pores and another a small number of large pores, with the result that the permeabilities are the same. The critical parameter is not permeability but rather, its temperature dependence, which describes how easily a molecule can cross the membrane. The rate constant k_m in Equation 10.2 depends upon the height of the activation barrier E_k and temperature according to the Arrhenius equation, $k_m = a \cdot \exp(-E_k/RT)$, where a is a constant. Thus, the temperature dependence of the permeability can be described by

$$P = A\exp\left(-E_{act}/RT\right) \tag{10.4}$$

E_{act} is readily determined by plotting $\ln P$ against $(1/T)$, as shown in Figure 10.2C. Values of E_{act} have been determined for black lipid membranes formed from many different lipids. The activation energies are 10–12 kcal mol^{-1} for black lipid membranes and large unilamellar vesicles alike. These values are considerably higher than the activation energy of 4.2 kcal mol^{-1} for diffusion of water molecules through water (self-diffusion). To understand the significance of these numbers, think about how large a rise in temperature would be required for the permeability of a lipid membrane to be the same as for water. You'll see that a three-fold temperature rise would be required, meaning that the bilayer permeability will be the same as an equivalent thickness of water at about 900 K!

10.1.2 Water Diffuses through Pores in Erythrocyte Membranes

Physiologists argued for years in the mid-20th century about the existence and nature of membrane water pores, particularly in erythrocytes. One of the great biophysicists of the day, Arthur K. Solomon (1912–2002), carried out defining experiments in 1970, which proved beyond doubt that most water passing across red cell membranes passes through channels. His measurements of the temperature dependence of erythrocyte permeability revealed the activation energy to be 4.9 kcal mol^{-1}. Completely inconsistent with passage through a bilayer, this number is close to the activation energy of the self-diffusion of water. The inescapable conclusion was that there are pores in the membrane that present little resistance to water diffusion. The nature of the pores remained a mystery until 1992. Further evidence for the existence of pores came from studies by Robert Macey, who showed that the water permeability of erythrocytes could be reversibly inhibited by an agent that acts on proteins, $HgCl_2$. Taken together, these studies suggested that water transport occurs through a protein channel that has free sulfhydryl groups that are accessible to mercury. The channel is now known to be an aquaporin. Peter Agre shared the 2005 Nobel Prize in Chemistry for discovering and characterizing aquaporins. Although, as we have seen, water can move across lipid bilayers, aquaporin channels evolved to allow the selective movement of water across membranes at higher rates that facilitate biological function. That is, aquaporins allow water movement while excluding the movement of metabolically important solutes, especially protons and other ions whose transport must be closely regulated.

10.2 WATER AND OTHER SMALL MOLECULES MOVE ACROSS MEMBRANES USING CHANNELS: AQUAPORINS AND AQUAGLYCEROPORINS

The first water channel, AQP1, was identified during a search for the identity of the Rh blood group antigens. One of the Rh antigens was isolated from radiolabeled red blood cell membranes using hydroxylapatite chromatography and was found to be about 32 kDa. The protein did not stain well with Coomassie stain, which targets amino groups. When silver reagent was used to detect the protein instead, a protein of 28 kDa was also detected. This protein was missed in all the early red cell protein studies, even though there are about 100,000 molecules of AQP1 in each red cell, because of the insensitivity of staining by Coomassie blue. It was thought that the 28 kDa protein might be a fragment of the 32 kDa protein, but it was found that an antibody to the 28 kDa protein did not bind to the 32 kDa protein, calling into question the possibility of the 28 kDa protein being a fragment. Further work on the protein showed that it had physical properties similar to those of a channel. The 28 kDa protein was thus named CHIP28 (Channel-like Integral Protein 28 kDa).

The cDNA that encodes CHIP28 was cloned from an erythroid library, enabling a dramatic experiment. *Xenopus laevis* oocytes are largely impermeable to water (Table 10.1), which protects them against osmotic effects when the frog lays them in dilute water environments. When oocytes expressing CHIP28 were placed into dilute solution (70 milliosmolar instead of 200 milliosmolar), they rapidly swelled and exploded, while the control oocytes swelled negligibly (**Figure 10.3**). Later experiments showed that the pure protein, placed in lipid vesicles, dramatically increased their permeability to water while not increasing proton permeability (**Figure 10.4**). These experiments and others established the identity of CHIP28 as the long-sought water channel, later renamed Aquaporin-1.

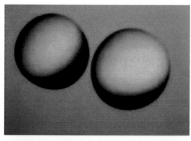

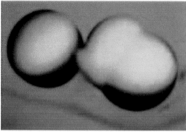

Figure 10.3 The upper panel shows a pair of oocytes in isotonic solution, and the lower panel shows the result of placing them in dilute solution. The oocyte expressing AQP1, on the right, rapidly took up water and exploded, whereas the control oocyte showed a resistance to lysis because it has very limited water permeability. (Photo from Peter Agre's Nobel Prize lecture 2003. With permission.)

10.2.1 Aquaporins and Aquaglyceroporins Facilitate the Passage of Water and a Few Small Molecules, Such as Glycerol, across Membranes

Aquaporins are ancient membrane-channel proteins that are present at all levels of life, from bacteria to mammals. Most aquaporins are selectively permeated by water, although some family members are also permeated by other small, uncharged molecules such as glycerol or urea. The subgroup that is permeable to water, glycerol, and a few other small molecules is named the aquaglyceroporins. Membrane permeability to glycerol has been known for a century, and water permeability for some time as well, but the nature of the transport system was obscure until Peter Agre's discoveries. While not so abundant as ion channel genes, there are 13 aquaporins and aquaglyceroporins in the human genome with a range of important functions (**Figure 10.5**).

Why are there so many? Aquaporins in lung, kidney, red cells, and the cornea, for example, have different targeting and regulation, so there is a need for different genes. There is an increasing interest in the aquaporins in disease mechanisms and as drug targets; e.g., transgenic mice lacking AQP4 show involvement of AQP4 in cerebral water balance, astrocyte migration, and neural signal transduction. AQP4-null mice have reduced brain swelling and improved neurological outcome in models of (cellular) cytotoxic cerebral oedema, including water intoxication, focal cerebral ischemia, and bacterial meningitis.

10.2.2 The Structure of an Aquaglyceroporin Reveals Tetramers with a Transport Pore in Each Subunit

The family members form tetramers, and, initially, it was thought that the channel must be formed at the center of the tetramer. But, the channels are actually found in each monomer. Recent hydrogen exchange measurements suggest that the role of the tetramer may be to stabilize the structures of the monomers, helping to define the channel structures. All members of the family have very similar structures, so that a general view is informative. The basic tetramer and monomer structures, solved by the group of Robert Stroud, are seen in **Figure 10.6**.

The structure reveals a common design for the creation of a channel: a bundle of helices surrounds the channel, separating it from the lipid bilayer. The lipid-exposed faces of the helices have hydrophobicity distributions that roughly match the profile of the bilayer, allowing a stable, sequestered interior channel to span the membrane. The properties of the channel can then be tailored to the needs for transporting water, glycerol, or other species. What are the properties

Figure 10.4 Proteoliposomes reconstituted with CHIP28 become very permeable to water. (From Zeidel ML, Ambudkar SV, Smith BL [1992] *Biochemistry* 31: 7436–7440. With permission from American Chemical Society.)

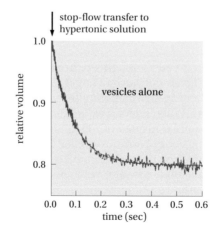

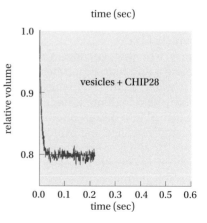

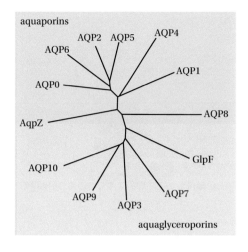

aquaporin	permeability	tissue distribution	subcellular distribution*
AQP0	water (low)	lens	plasma membrane
AQP1	water (high)	red blood cell, kidney, lung, vascular endothelium, brain, eye	plasma membrane
AQP2	water (high)	kidney, vas deferens	apical plasma membrane, intracellular vesicles
AQP3	water (high), glycerol (high), urea (moderate)	kidney, skin, lung, eye, colon	basolateral plasma membrane
AQP4	water (high)	brain, muscle, kidney, lung, stomach, small intestine	basolateral plasma membrane
AQP5	water (high)	salivary gland, lacrimal gland, sweat gland, lung, cornea	apical plasma membrane
AQP6	water (low), anions (NO^{3-}>Cl^-)	kidney	intracellular vesicles
AQP7	water (high), glycerol (high) urea (high) arsenite	adipose tissue, kidney, testis	plasma membrane
AQP8†	water (high)	testis, kidney, liver, pancreas, small intestine, colon	plasma membrane, intracellular vesicles
AQP9	water (low), glycerol (high), urea (high), arsenite	liver, leukocytes, brain, testis	plasma membrane
AQP10	water (low), glycerol (high), urea (high)	small intestine	intracellular vesicles

* homologues that are present primarily in either the apical or basolateral membrane are noted as residing in one of these membranes, whereas homologues that are present in both of these membranes are described as having a plasma-membrane distribution.
†AQP8 might be permeated by water and urea. AQP, aquaporin

Figure 10.5 Aquaporins and aquaglyceroporins found in the human genome. Eight aquaporins and five aquaglyceroporins have been identified in the human genome on the basis of sequence homology. (From King LS et al. [2004] *Nature Rev Mol Cell Biol* [2004] 5: 687–698. With permission from Springer Nature.)

of the channel? To answer that question, we focus on an aquaglyceroporin, because we can see the transported glycerol molecules in the channel structure.

10.2.3 Sequence Analysis of the Family Reveals Key Features of Aquaglyceroporins

The most prominent conserved sequence features of the aquaglyceroporins are the two Asn-Pro-Ala (NPA) motifs near the N-termini of the two short helices (M3 and M7), which were postulated to be part of the pore even before structural data were available. The overall sequence similarity is in the range of 25–40% and is highest near the region of the pore (**Figure 10.7**). The structure appears to have evolved by gene duplication, with the two halves each having three transmembrane segments (TMs) and a shorter hydrophobic sequence. A remarkable aspect of the organization is that the two halves have opposite topologies across the membrane.

Figure 10.6 General structure of a tetrameric aquaglyceroporin (PDB: 1FX8). The structure is a tetramer with four channels, one in each subunit. (A) View of the tetramer from the extracellular surface. Glycerol molecules are shown as stick models. (B) A monomer is shown, emphasizing the conserved NPA motifs that play a key role in selectivity (purple) and marking the gene duplication in blue and green. Glycerol molecules are shown in spacefill, occupying positions in the channel. As in the membrane region of the K+ channels, short helices and extended chains are seen, but here, the two short helices meet at the NPA motifs in the middle of the membrane. The extended chains are sequestered from the bilayer hydrophobic region.

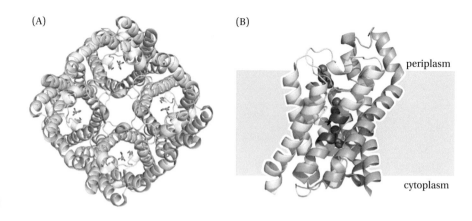

(A)   (B)

periplasm

cytoplasm

10.2.4 The Charge Distribution in GlpF Is in Accord with the "Positive-Inside Rule"

The tetramer of the *E. coli* channel, GlpF, shows a dramatic charge difference between the surfaces that face the aqueous sides of the membrane, with a negative face at the periplasmic side and a positively charged face toward the cytoplasm. This difference is in accord with the "positive-inside rule" that influences the topology of TMs generated during biosynthesis, as suggested by von Heijne and his colleagues. Views of the electrostatics are seen in **Figure 10.8**. Because the translocated molecules, principally water and glycerol, are uncharged, this charge distribution is likely to reflect constraints other than interactions with substrates, and the "rule" may provide an explanation. See Chapter 6 for a full discussion of this interesting point.

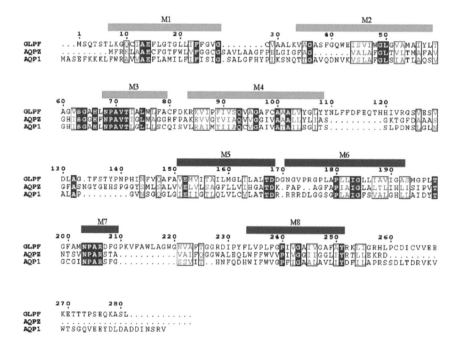

Figure 10.7 Sequence alignment of aquaglyceroporins reveal key helical and conserved structural features. (From Stroud RM, Savage D, Miercke LJ et al. [2003] *FEBS Lett* 555: 79–84. With permission from John Wiley and Sons.)

periplasmic face

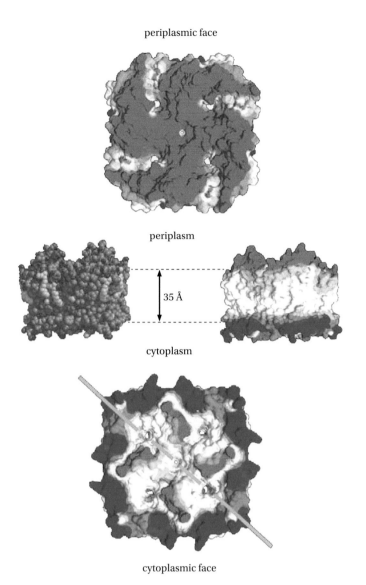

periplasm

35 Å

cytoplasm

cytoplasmic face

Figure 10.8 Electrostatics of the GlpF aquaglyceroporin (PDF: 1FX8). The periplasmic surface is strongly negative (red), while the cytoplasmic surface is strongly positive (blue). The cross-sections in the middle panel give another view of the transmembrane distribution, with the presumed location of the lipid bilayer indicated. (From Stroud et al. [2003] *Curr Opin Struct Biol* 13: 424–431. With permission from Elsevier.)

10.2.5 An Ancient Gene Duplication Resulted in the Present Structure

Aquaporins arose in evolution before the divergence of the three kingdoms of life, perhaps two billion years ago. That the homologies persist (*E. coli* has both an aquaporin and an aquaglyceroporin) suggests that effective molecular designs evolved early and that they need all of their parts. An example of the functions of the different sidechains in an aquaglyceroporin, GlpF, is shown in **Figure 10.9**. Two of the themes (see later) seen in the selectivity filters of K+ channels are found: short helices that are partly sequestered from lipid, and stabilizing extended chains that contribute carbonyl groups to the channel. The extended chains provide polar oxygens that accept H-bonds from water or glycerol molecules, much as the carbonyls in the ion channels contribute replacements for the water of hydration.

10.3 AQUAPORIN STRUCTURE REVEALS STRUCTURE–FUNCTION RELATIONSHIPS

How do the molecular features of GlpF combine to facilitate glycerol permeability? The GlpF channel provides a pathway that fits the dimensions of water

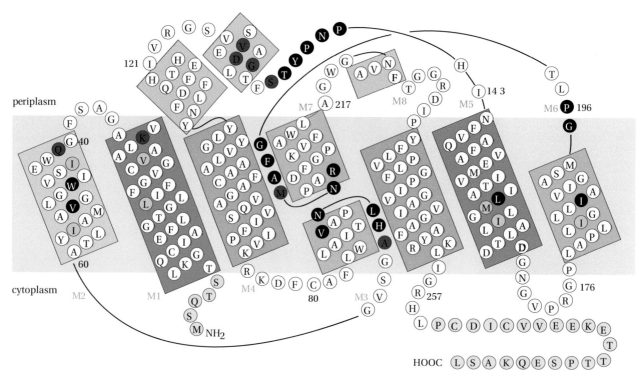

Figure 10.9 The structure of the aquaglyceroporin GlpF shows the two-fold relationship derived from the early gene duplication. Black residues interact with the glycerol molecules observed within the channel. Red residues contribute carbonyls to the central channel. Purple side chains contribute hydrocarbon surfaces to the channel. The extended chains that precede M3 and M7 contribute a line of eight carbonyls within the 28 Å-long channel that accept hydrogen bonds from the permeant molecule. (From Stroud et al. [2003] *Curr Opin Struct Biol* 13: 424–431. With permission from Elsevier.)

or glycerol and that presents a polar environment for the -OH groups and a non-polar surface for the backbone of glycerol. The glycerol molecule has distinctly differing polar and non-polar surfaces (**Figure 10.10**), and the channel likewise has two distinct surfaces: hydrophobic and hydrophilic. This division can be seen in the cross-section views of **Figure 10.11A**, where a space-filling monomer has been opened to reveal the distinction. Because glycerol molecules are found in the channel in the structure, it is possible to examine closely the interactions that stabilize the molecules in the channel; it is apparent that there is a close matching of non-polar features of the glycerol (mainly the three-carbon backbone and C-H hydrogens) with non-polar sidechains in the channel. Correspondingly, there is a matching of the polar features of the three -OH groups with polar features of sidechains and backbone carbonyl groups in the extended chains. A continuous strip of polar features lines one side of the channel, and a continuous strip of non-polar features lines the opposite side, creating an asymmetric slide for the glycerol to ride on and explaining the facile permeation. Thus, we can account for the transport properties based on the chemistry and geometry of the channel.

Figure 10.10 Structure and chemistry of glycerol. The molecule has non-polar methylene groups on the upper side and polar hydroxyl groups on the lower side.

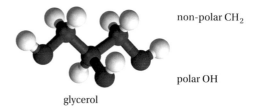

non-polar CH_2

polar OH

glycerol

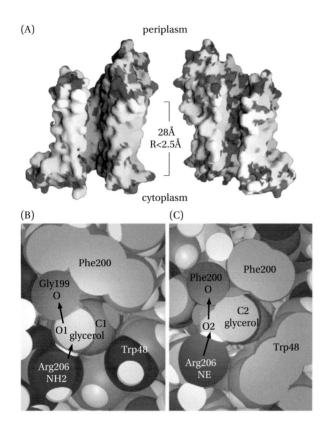

(A)   periplasm

28Å
R<2.5Å

cytoplasm

(B)

(C)

Phe200

Gly199
O

O1   C1
       glycerol

Arg206
NH2

Trp48

Phe200   Phe200
O

           C2
O2   glycerol

Arg206
NE

Trp48

Figure 10.11 Glycerol selectivity of the GlpF channel. (A) Two faces of the GlpF channel. Opening the channel into two halves, it is seen that the surfaces differ in polarity: Red = negative charge, Blue = positive charge, Yellow = Sulfur, White = non-polar. The non-polar surface cannot substitute for waters of hydration, so that ions cannot pass easily through the channel, because they cannot be easily dehydrated. (B) Two cross-sections of a glycerol molecule in the channel. The views show carbon 1 and carbon 2 of the glycerol. Note that the polar OH groups are accommodated by positive charges on arginine 206 interacting with the OH oxygen and a backbone carbonyl oxygen accepting H-bonds from the OH hydrogen, while relatively non-polar aromatic groups interact with the CH_2 side of the glycerol. (From Stroud et al. [2003] *Curr Opin Struct Biol* 13: 424–431. With permission from Elsevier.)

10.3.1 Ions Are Blocked from the Channel by Their Hydration

As discussed later, when an ion enters the potassium channel selectivity filter, it is dehydrated by substituting carbonyl oxygens for the water oxygens. In the Na^+ channel, water is positioned along with carbonyl groups to substitute for hydration by bulk water. In the GlpF channel, one side has no carbonyl groups exposed and no other groups with adequate polarity to substitute for waters, so an ion will not be sufficiently dehydrated to pass through the channel (Figure 10.11B). Thus, the problem of excluding ions is solved at the same time as facilitating the passage of glycerol. But, a problem remains: the channel must resist the translocation of protons, which can account for a significant part of the transmembrane potential in prokaryotes and mitochondria—if the proton gradient is abolished, ATP synthesis and many key transport events will also be abolished, and the cell will die. How does the channel prevent proton transport?

10.3.2 Ideas for the Block of Proton Transport Are under Investigation

The reason that there is a concern about proton translocation is that water allows proton movements via the Grotthuss "hop-turn" mechanism, first proposed in 1806 (!) by Theodor Grotthuss (1785–1822) to explain the conductivity of ice, which is greater than that of water. In ice, the water molecules are well aligned in chains, and the mechanism allows protons to move with ease. The idea is illustrated in **Figure 10.12**.

If chains of water molecules can occupy the channel, as they do, why can't protons move freely across the membrane? There are two current ideas under discussion based on computational studies: (1) Interruption of the chain of water molecule hydrogen bonds and (2) charge blockade. The idea of an interruption of the chain of water molecule hydrogen bonds arises from molecular dynamics simulations that suggest the immobilization of a water at the center

Figure 10.12 The Grotthuss mechanism of proton motion in water. (I) The first step is a hop of protons in a cooperative motion along a chain of water molecules. This motion is fast and effectively moves the charge from one end of the chain to the other. (II) In a second step, the waters turn to reset the chain to its original state.

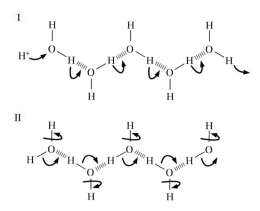

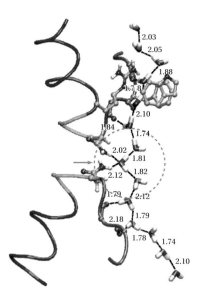

Figure 10.13 Simulation shows a stabilized water that breaks the chain. Illustrated is the suggested chain of waters, showing an immobilized water near the center of the membrane (arrow). This water fixes the orientation of the hydrogen bonding of water chains up or down in the channel, so that they cannot turn to allow proton translocation via the Grotthuss mechanism. (From Stroud et al. [2003] *Curr Opin Struct Biol* 13: 424–431. With permission from Elsevier.)

of the membrane such that it cannot turn. Immobilization results in the water establishing directions for the chains of water hydrogen bonds on either side of the key water in opposite directions. Thus, the Grotthuss mechanism cannot work because the required turning of waters cannot occur (Figure 10.13).

Charge blockade has been suggested as an alternative by a different set of simulation results arguing that the barrier is not attributable to the orientation of the water molecules across the channel but rather, to the electrostatic penalty for moving the proton charge to the center of the channel, such that the main reason for the high barrier is the loss of the generalized solvation upon moving the proton charge from the bulk to the center of the channel, where asparagines create a region of positive charge. While these ideas still await clear experimental tests of their validity, two chemically plausible mechanisms for the important property of preventing proton leakage are in hand. Aquaglyceroporins have evolved to prevent passive proton permeation. Yet, the facilitation of proton permeation across membranes is important in some biological processes, such as the activation of influenza virus, as discussed in **Box 10.1**.

10.4 GATING AND CONTROL OF AQUAPORIN FUNCTION

As with any catalytic process, whether it is a metabolic enzyme catalyzing a reaction in metabolism or a channel catalyzing the transit of a molecule across a membrane, the enhancement of the rate of a process creates the possibility of controlling it. By altering the rates of reactions in a metabolic pathway, the levels of metabolites are regulated; likewise, by altering the rates of transport across membranes, changes in the concentrations of molecules or ions across a membrane are controlled. In the discussion of small-molecule channels, we have focused on the task of catalyzing the selective passage of small molecules across a membrane. In the next chapter, we will add gating and control as major themes, but here, we discuss a simple case of controlling an aquaporin channel.

10.4.1 Plant Aquaporins Reveal Gating Mechanisms for Control

Vascular plants rely on water in a variety of ways that are not used by animals or microorganisms. Their aerial parts absorb CO_2 for photosynthesis and release water by transpiration, a process that is controlled by stomata, microscopic pores located in the aerial parts. Transpiration is enabled by a flow of water that moves in a stream from the roots to the spaces under the stomata, from whence it evaporates (**Figure 10.14**). The stream of water (sap) drives the flow of nutrients from the roots throughout the plant. Water uptake by the roots and its control are required in order to maintain water physiology in the whole plant. The presence of strong cell

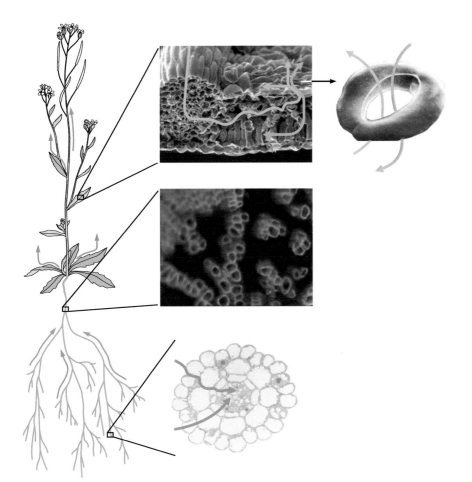

Figure 10.14 Water movements in a plant. Plants use both short- and long-distance water transport. Release of water from the leaves creates a stream of water from the roots to the shoots. Water from the soil flows radially in the root before reaching the xylem vessels. which conduct the flow to the shoots. In leaves, water reaches chambers under the stomata, through which it diffuses out of the plant as a gas. (From Maurel C, Boursiac Y, Luu D-T et al. [2006] *Amer Physiol Soc* 95: 1321–1358. With permission from The American Physiological Society.)

walls allows the maintenance of intracellular hydrostatic pressure, as much as several atmospheres, which contributes to the erect shape of the plant (as we observe when we forget to water, and our houseplants droop). So, the key roles of water in a plant require that it be moved across membranes and controlled, functions that involve aquaporins at cell surfaces and in cellular compartments.

Genetic studies of plants have expanded rapidly in recent years, with many reports of whole genome sequences. Analysis of these sequences have revealed

BOX 10.1 INFLUENZA VIRUS M2 CHANNEL

When the influenza virus infects a cell, it is taken up by endocytosis. The endosome pH is lowered by the proton pumps that are included in the endosome membrane (these normally pump protons out of the cell), and the virus is triggered by the low pH to fuse with the membrane and deliver its nucleic acid–containing capsid into the cytoplasm, where it proceeds to infect the cell. The virus must have proton channels so that the pH change can be effective inside the virus, where it dissociates the capsid from the nucleic acid.

The M2 protein of influenza viruses forms tetrameric proton channels that open to allow entry of the protons into the virus. In a later stage of virus replication, the M2 protein has an additional function: to maintain the high pH of the trans-Golgi network during viral maturation. The M2 protein is targeted by the amantadine class of antiviral drugs. Resistance to these drugs has emerged, so the structure

of the drug–channel complex is of interest as a basis for further drug design.

There was a controversy over the binding of the amantadine, because solution nuclear magnetic resonance (NMR) and crystallography gave different answers, but it is now clear that the primary binding site at both low and high pH is in the channel, acting as a plug in the central cavity. The crystal structure of the transmembrane region is shown in **Figure 1**, and the functional assignments of parts of the structure are from mutational analysis.

Solid-state NMR was used to gain further insights into the changes induced by pH, and the behavior of the channel in the presence of amantadine is consistent with the gating region identified by mutagenesis. Ongoing computational efforts are aimed at understanding the proton translocation pathway. **Figure 2** shows the high- and low-pH structures.

(Continued)

BOX 10.1 INFLUENZA VIRUS M2 CHANNEL (CONTINUED)

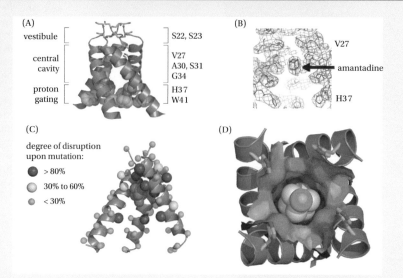

Figure 1 Crystal structure of the transmembrane region of the M2 channel, with bound amantadine, at low pH (PDB 3C9J). (A) The critical residues identified by site-directed mutagenesis line the pore of the M2 channel. Gly34, His37, and Trp41 are shown as space-filling spheres. The carbon atoms of His37 are colored tan. The sidechains of the other critical residues are shown as sticks. (B) An electron density map showing electron density in the amantadine-binding region. (C) Positions of Cys mutations that disrupt the channel blocking by amantadine are shown by red balls (>80% disruption), yellow balls (30% to 80%), and green balls (no significant disruption) in full-length M2. Amantadine is shown in magenta. For clarity, the front helix was removed. (D) Structure showing amantadine (nitrogen in cyan and carbon in white) inside the binding site along with the surface associated with Val27 (red surface), Ala30 (green), Ser31 (blue), and Gly34 (orange). (From Stouffer AL et al [2008] *Nature* 451:596–599. With permission from Springer Nature.)

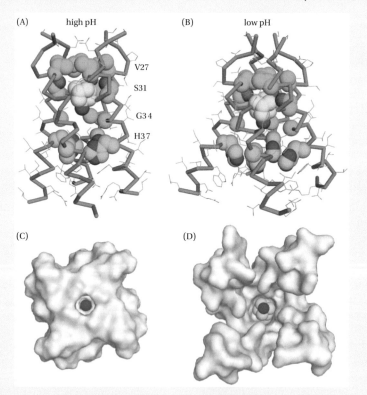

Figure 2 Comparison of the high-pH NMR structure of M2 in bilayers with bound amantadine (PDB 2KQT) with the low-pH crystal structure. (A) Side view of the high-pH NMR structure shows that amantadine is enclosed by Val27 at the top and His37 at the bottom. (B) Side view of the low-pH crystal structure, showing that the helices are splayed far apart near the C-terminus. (C) View of the high-pH structure at the C-terminal end, showing that amantadine is well sequestered. (D) View of low-pH structure at the C-terminal end. This structure reveals that the drug is more solvent accessible. (From Cady SD et al. [2010] *Nature* 463:689–692. With permission from Springer Nature.)

many aquaporins in higher plants, with more than 30 isoforms identified so far. Many plants are highly polyploid (many copies of each chromosome); the soybean genome encodes 66 aquaporin homologs, for example. While animal and bacterial aquaporins are found to be divided into the water channel (aquaporin) and aquaglyceroporin groups, most plant subfamilies are aquaporins. Studies of some of the different genes have been enabled by expression in frog oocytes, yeast, and lipid vesicle systems, revealing that different channels are specialized for transport not only of water but also of urea, H_2O_2, CO_2, hydrated ions, and ammonia, among other species (see **Box 10.2** on gas transport).

Control of water channels to cope with the needs of the plant water system is an evident need. A sessile life form encounters challenges through seasonal temperature changes, drought, and flood as well as changes in acidity, ion concentrations, etc. It is no surprise that there are mechanisms to respond to these challenges, and roles have been found for H^+, cations, and phosphorylation to gate or modulate aquaporin activity. A unifying model of pore gating has been proposed for the channels across plasma membranes, based on structural data (**Figure 10.15**). In the model, the channel closes in response to low pH by protonation of a perfectly conserved His residue in loop D (His193 in Figure 10.15). At acidic pH, the histidine is charged and interacts with Asp28, Glu31, and loop B (Ser115) to stabilize loop D in a conformation that blocks the pore with a leucine sidechain—a simple mechanism!

Water transport can also be inhibited by divalent cations, which can be explained by the direct binding of the cation (yellow sphere), which in turn, mediates H-bonding between the NH_2 terminus (Gly30 and Asp31) and loop D

BOX 10.2   GAS TRANSPORT THROUGH AQUAPORINS

Aquaporin (AQP1) is often highly expressed in cells where water transport does not seem essential. Walter Boron's laboratory asked whether there might be another function for the channels: gas transport. They used *Xenopus* oöcytes, either expressing AQP1 or not, and injected each of them with carbonic anhydrase (CA), so that when CO_2 enters the oocyte, it is converted to bicarbonate ion and a proton. By measuring the pH using electrodes inserted into the oöcyte, the transport of CO_2 can be followed. In **Figure 1**, the changes of pH resulting from pulses of CO_2 are shown on the left. The first two pulses are followed by the addition of an inhibitor of the AQP1 channel, ETX, and then a third pulse is applied. After each pulse, there is a drop in pH followed by recovery after the CO_2 is removed. When the channel is blocked, the gas entry is slowed, as is the recovery when CO_2 is removed. The rates are shown in the bar graph. In subsequent work, physiological roles for the transport were found, for example, in the proximal tubule of the kidney.

Why might gas permeation across a membrane need assistance? It was presumed that gas permeation was facile, but there are several reasons why channels are needed. A gas molecule has to penetrate the lipid headgroup layer. Then, it will be trapped in the non-polar region of the bilayer, which measurements have shown to be a preferred environment for many gases—nitrous oxide (NO), for example, has a partition coefficient of about 5, favoring lipid over water. So, free diffusion is slow. Some gas molecules, such as NO, are found to exert important functional influences on neighboring cells in short times, so channels are needed.

There is an active area of research aimed at understanding the transport of gases, small molecules, and their control. Among the gases of interest are CO_2, O_2, NH_3, and NO. Interestingly, the different aquaporins show selectivity for different gases, suggesting specific roles in different tissues. **Figure 10.5**, from a review, summarizes the state of human AQP transport properties for water, glycerol, urea, and ammonia as of 2015.

(Continued)

BOX 10.2   GAS TRANSPORT THROUGH AQUAPORINS (CONTINUED)

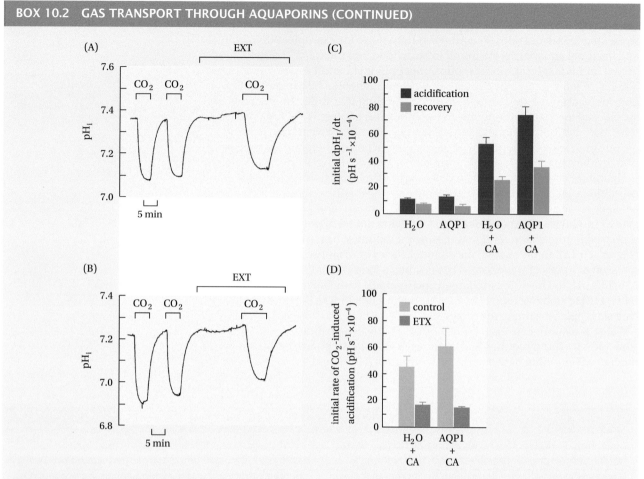

Figure 1 Permeation of CO_2 into *Xenopus* oocytes through AQP-1. (From Nakhoul NL et al. [1998] *Am J Physiol* 274 (*Cell Physiol* 43): C543–548.)

through loop B, thereby stabilizing the closed pore conformation. Although this mechanism was revealed by Cd^{++} binding in the crystal structure of SoPIP2;1, the binding site may, rather, be occupied by Ca^{++} *in vivo*. Further, covalent modification can modulate function. The studies in oocytes suggest a role for several cytosol-exposed phosphorylation sites in controlling water transport. For example, phosphorylation of a conserved serine residue in loop B would destabilize the loop D–loop B anchor, thereby favoring the open-pore conformation. Studies of the plant proteins have given insights into a simple set of gating and control mechanisms. In the next chapter, we will examine a range of gating events that endow animals with an amazing range of capabilities, including the senses and brain function.

10.5 AMMONIA CHANNELS HAVE MANY ROLES ACROSS PHYLA BUT SIMILAR STRUCTURES

Ammonia, which is toxic in mammals, is used as the key nitrogen source in bacteria and is a useful nitrogen source in plants. While ammonia can diffuse across membranes in its uncharged form, the rates are not sufficient for most uses, so

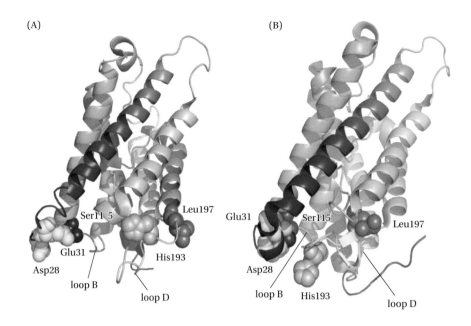

Figure 10.15 Gating a plant aquaporin. A spinach aquaporin structure was solved in open (A: PDB code 2B5F) and closed (B: PDB code 1Z98) conformations. In the open conformation (A), the His193 (orange) is not protonated, and loop D is distal from the other cytoplasmic loop (loop B). In the closed conformation (B), the protonation of His193 allows interaction with an acidic residue (Asp28, in yellow) of the NH_2 terminus. The conformational change of loop D moves the Leu 197 side chain (in magenta) into the cytosolic pore mouth. Binding of divalent cations (Cd^{++} in the atomic structure, in red) also involves Asp28 and an adjacent acidic residue (Glu31, in yellow). In this model, phosphorylation of loop B, at Ser115 (in purple), would disrupt this network of interactions and unlock loop D to allow the open conformation.

channels are required. The diversity of roles, dependent on the different metabolic schemes across phyla, is satisfied by a set of homologous structures that use themes that we have already encountered, including an ancestral internal anti-parallel duplication, the positive-inside rule, and oligomeric association of subunits, with a channel in each. As an example, we consider AmtB, a channel from *E. coli* that serves to provide nitrogen for the bacterial metabolism. Its structure was reported at very high resolution (1.3 Å) by Robert Stroud and his colleagues (**Figure 10.16**); there are a number of structures of other ammonia channels as well. The structure has a vestibule on each side of the membrane, formed by the divergence of the array of helices. The vestibules contain a number of water molecules, but none are seen in the 20 Å hydrophobic channel, indicating that they are mobile.

Ammonia can exist as either NH_4^+ or NH_3, with a pK_a of 9.25 in aqueous solution. However, placing the ion in a hydrophobic environment changes the pK to favor the neutral form, which is the transported form. The channel also can serve as a path for CO_2 exchange, but the main purpose is nitrogen access—in *E. coli*, nitrogen starvation induces the gene for expression of the channel. Faster regulation is provided by a trimeric cytoplasmic protein, GlnK, that can respond to cytoplasmic nitrogen levels by stoppering/

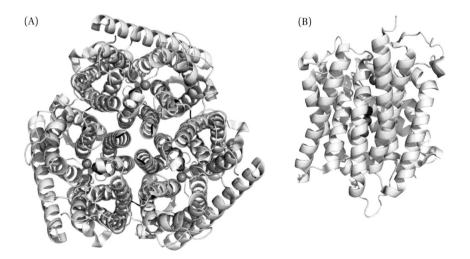

Figure 10.16 Crystal structure of an ammonia channel, AmtB (PDB: 1U7G). The trimeric arrangement of channels is seen in (A), where each monomer has a channel surrounded by 11 transmembrane helices. Ten of the helices have a quasi-twofold in the plane of the membrane, emphasizing an ancient gene duplication like those found in aquaporin and other cases. By crystallizing in ammonium sulfate, it was possible to partially immobilize three NH_3 (magenta spheres) and an ammonium ion (orange) (20% occupancy), marking the channel. (B) shows a view down the pseudo-twofold of a monomer.

BOX 10.3 REGULATION OF AN AMMONIA CHANNEL

Nitrogen is, as we know, an essential element for building the molecules of life, and its capture by fixation of the N_2 in the atmosphere builds the precious supply that is carefully managed by organisms. In *E. coli*, the key uptake is the flow of NH_3 through the ammonia channel, AmtB, as discussed in the text. But, free ammonium ion levels can be toxic, as we know from the need to exclude it from the blood by transport in the kidneys. Free ammonium is also toxic in the cytoplasm of *E. coli*, but it is also essential—so, a need for careful regulation is created. Here, we explore an example of channel regulation in the context of ammonia uptake.

Long-term regulation of the AmtB ammonia channel is accomplished via gene regulation, and shorter-term regulation is mediated by a protein that is linked to its expression: GlnK. GlnK is regulated by nitrogen-linked metabolic signals that result in modification by uradylylation/deuradylylation, controlling the inhibitory binding of GlnK to the channel.

The channel regulation is remarkably simple in its concept: a trimer of GlnK binds to the trimer of AmtB, inserting a protrusion that simply blocks the channel by inserting an Arg side chain. In **Figure 1**, we see the close trimer-to-trimer fit of the GlnK to AmtB:

Figure 2 shows how the loop inserts the Arg into the channel. The energy of association must be poised so that uridylation can dissociate the GlnK from the channel. Presumably, the trimeric association creates cooperativity so that the control can be finely managed—an insight, perhaps, into a reason for the oligomeric association of channels.

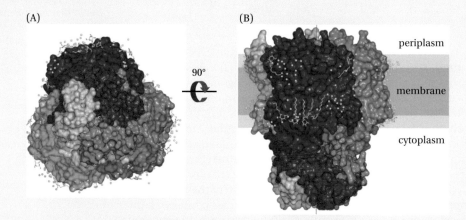

Figure 1 Ammonia channel inhibition by a complex with a matching trimer of GlnK (PDB 2NS1). The two trimers have closely matching structures that facilitate the insertion of loops into the channels of the subunits, each blocking a channel with an Arg side chain. (From Gruswitz F et al. [2007] *PNAS* 104:42–47. With permission from Robert Stroud.)

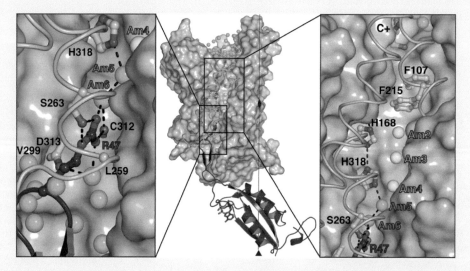

Figure 2 Channel inhibition and ammonia conductance pathway. Blockade of the ammonia conduction pathway by R47 of GlnK (red) is shown in context (center) and in detail (left). The open channel detail is shown on the right. The surface of AmtB (blue) is shown with residues 211–270 as a tubular backbone trace to allow visualization of the channel and the inserted Arg loop. (From Gruswitz F et al. [2007] *PNAS* 104:42–47. With permission from Robert Stroud.)

opening the inner opening of the channel by the insertion/removal of an Arg (see **Box 10.3**).

Interestingly, the channel recruits NH_4^+ and then removes H^+ to give a neutral NH_3 that readily moves through a hydrophobic channel (**Figure 10.17**). The attractive potential for NH_4^+ appears to arise from the influences of π orbitals in two adjacent Phe and one Trp near the outside opening of the channel. A tight constriction may facilitate proton removal, allowing the uncharged NH_3 to pass through the ~20 Å-long hydrophobic channel. This mechanism has support from experiments using AmtB reconstituted into liposomes. Once inside the *E. coli* cell, NH_3 is a substrate for glutamine synthetase in Gln synthesis, from whence it is used in intermediary metabolism. The metabolism of ammonia as a nitrogen source depletes its cytoplasmic concentration, favoring by mass action the transport through a passive channel.

10.6 SPECIALIZED MECHANOSENSITIVE CHANNELS RELIEVE OSMOTIC STRESS

10.6.1 Mechanosensitive Channels Respond to Mechanical Forces by Opening to Allow Transmembrane Flow of Osmolytes

Animal cells live in a closely regulated extracellular environment maintained by kidneys, lungs, and other organs. In fact, the aquaporins play a key role in this regulation (Figure 10.5). Bacteria, on the other hand, live in an often-changing environment. *E. coli* can survive even when challenged with distilled water. The strong cell wall provides containment to suppress lysis, but there is an emergency measure held in reserve for countering extreme osmotic stress, such as immersion in hypotonic solutions. The countermeasure consists largely of mechanosensitive channels that open when the lateral tension of the membrane is increased by swelling. Although the specific types of mechanosensitive receptors found in bacteria and many plants are not found in animals, all organisms have uses for the ability to sense mechanical forces on membranes in responding both to the external environment and to internal events. Signals ranging from osmotic stress in microorganisms to the senses of hearing and touch in vertebrates are sensed by triggering flow through mechanosensitive channels that can serve, for example, as ion flux signals that alter the membrane potential, or as sensors of osmotic stress by allowing bulk flow of small molecules and water.

The osmotic channel proteins sense deformations of the lipid bilayer through coupling of their conformations to the bilayer profile, responding to changes in the hydrophobic thickness, lateral pressure, and curvature. The best-understood examples are in microorganisms, which must contend with osmotic stresses that inevitably arise from changes in the environment. When a bacterium suddenly finds itself in a dilute solution, osmotic forces cause water to rush into the organism. Were it not for the opening of large, non-selective pores that allow water and solute molecules to equilibrate transiently across the membrane, the cell membrane would rupture. Recent structural, functional, and spectroscopic studies have provided insights into the molecular basis of the lipid–protein interactions that drive channel opening. There are four types of mechanosensitive channels (Msc) in *E. coli* that provide, collectively, graded responses to increasing stress (**Figure 10.18**). The channels range from the Mini channels (MscM) that open readily with small conductances to the Large MscL channels that open only when the membrane is about to break.

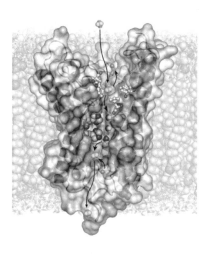

Figure 10.17 Mechanistic features of an ammonia channel, AmtB. Part of the structural model is removed to reveal features of the channel. The NH_4^+ is shown in green and NH_3 in purple. Orange spheres are protons. The exterior vestibule, with many bound waters and a net negative charge, attracts NH4+. The tight entry and surrounding π electrons strip a proton to give NH_3, which then passes through the hydrophobic channel. It is reprotonated to give NH_4^+ in the cytoplasm. (From Khademi S, Stroud RM [2006] *Physiology* 21: 419–429. With permission from The American Physiological Society.)

Figure 10.18 Four types of mechanosensitive channels (Msc) found in *E. coli*. Collectively, the channels provide graded responses to increasing stress. The MscM Mini channels respond with small conductance increases at low membrane tensions, whereas the MscL Large channels respond with very large conductance increases only under extreme stress; they are the last line of defense.

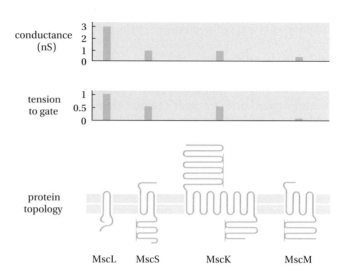

conductance (nS)

tension to gate

protein topology

MscL   MscS   MscK   MscM

10.6.2 Mechanosensitive Channels Couple Lateral Membrane Tension to Channel Area

When the lateral tension on a membrane increases, several interfaces are perturbed. Taking a simple view, these are the lipid–lipid, lipid–protein, and protein–protein (including internal) interactions (**Figure 10.19A**). In each of these cases, the result is a tendency to increase membrane area, and the resulting structural changes will depend on the relative strengths of the interactions. The first two are strongly stabilized by the hydrophobic effect, as discussed in Chapters 2 and 3, but the protein internal interactions are more complex and variable. Some proteins will be stable and although stressed, will not change their structure significantly. Others will be distorted and functionally altered (and therefore are "mechanosensitive"). Still others will be the weakest links, changing their shape considerably as the tension mounts. By taking advantage of this last class, biology has created channels that are opened by the shape-response to stress, changing their area in the membrane and performing the functions of signaling and osmotic adaptation, among others.

We can think about the energetics of opening a channel in terms of changes in the surface tension σ of the membrane (Figure 10.19B). Recalling that surface tension with units of force per unit length is equivalent to surface free energy per unit area (Box 1.3), we can write the free energy change

Figure 10.19 Force distribution in a membrane when tension σ is applied. (A) The force is felt by lipid–lipid interactions (LL), lipid–protein interactions (LP), and internal protein or protein subunit interactions (PP). (B) A schematic of a pore opening with an increase of membrane surface tension. (B) (From Sukharev et al. [1997] *Ann Rev Physiol* 59 :633–657. With permission from Springer Nature.)

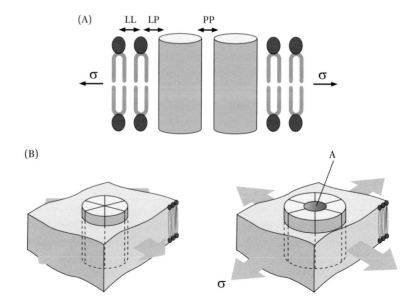

(A)   LL   LP   PP

σ   σ

(B)   A

σ

associated with channel opening as $\Delta G = \Delta G^0 - \sigma(A_{open} - A_{closed})$, where $\Delta G^0 = -RT\ln(n_{open}/n_{closed})$ is the standard free energy for channel opening when $\sigma = 0$. When $\Delta G = 0$, $\sigma_{1/2} = \Delta G^0/\Delta A$, and half of the channels are open. The different types of channels have different sensitivities to surface tension (Figure 10.18). The most sensitive channels will have the largest area changes for a given tension change. An example is helpful. Consider a channel with $\Delta G^0 \approx 10RT \approx 24$ kJ mol^{-1} (i.e., with a stability 10 times the average thermal energy). The free energy associated with a single channel will be 4×10^{-20} J or 4 pN nm. For $\Delta A = 10$ nm^2, a single channel would require a tension of 4 mN m^{-1} to for it to be open half of the time. Because the rupture tension for membrane patches studied with patch clamps is typically 20–30 mN m^{-1}, this channel would open long before the rupture point of the membrane. The MscM, MscS, and MscL channels have characteristic opening tensions ($\sigma_{1/2}$) of ~1, 5, and 10 mN m^{-1}, so they would open in order as the tension increases toward the rupture point, producing a graded, increasing sequence of transmembrane fluxes to respond to osmotic tension.

10.6.3 McsL Opens in a Squashed Iris Motion

How is the area increase coupled to the opening of a channel? The MscL channel has five subunits, each of which has two TM segments (**Figure 10.20**), and the pentamer is stable in a variety of detergents, as might be expected, since the closed form of the channel must be cooperative and have strong internal interactions that make it resistant to opening right up to the point of membrane rupture, or there would be wasteful losses from random openings of the membrane. The determination of the crystal structure of MscL from *Mycobacterium tuberculosis* (Tb-MscL), initially at 3.5 Å resolution, has provided an important scaffold to integrate a variety of experimental information. The structure (Figure 10.20), most likely in a closed conformation, shows how the TM1 helices interact to form a tightly packed bundle that narrows to a hydrophobic constriction, which is thought to function as the channel gate. Structures of MscL in gradations of the open state show that the area of the channel can increase dramatically in response to stress.

A variety of functional and biophysical studies have defined properties of MscL, including the use of patch-clamping methods pioneered by Sakmann and Neher (Chapter 11). But before bacteria can be patch-clamped, giant spheroplasts must be prepared (**Box 10.4**). Patch clamping gives measures of the responses of MscL to osmotic stress, first revealing that it opens at high stress. Other experiments using the channel reconstituted into bilayer vesicles large enough to clamp revealed that changing the bilayer thickness influences opening but does not trigger it. But, changing the lateral pressure profile by additives that modulate the area of the bilayer, such as lysophosphatidylcholine, was found to open the channel spontaneously. If we think about the lateral stretching of the bilayer, we know it to be limited; when lipid bilayer vesicles containing fluorescent dyes are stretched by osmotic forces, they begin to leak when the area is increased by 10%. Consistent with this observation, we know that the area per lipid tends to be conserved (see Chapter 2). Thus, bilayers, although fluid, are not very stretchable. On the other hand, force changes can be large in the context of a limited area change; it is just a matter of what the force constant is for the expansion. So, it seems consistent to think of the bilayer, while liquid, as difficult to expand, and that the force exerted on the MscL pentamer is great over small distances. The pentamer just has to have a smaller internal force constant than the bilayer, and it becomes the weakest link.

The result of the lateral force is conformation changes that lead to channel opening. Models for the change, which have been established using electron paramagnetic resonance (EPR) and other measurements, suggest that the helices tilt relative to each other to open a large space at the center of the oligomer. Strong support for this idea is found in the structure of a truncated

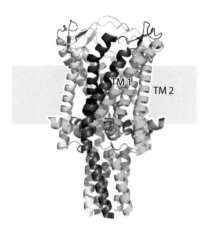

Figure 10.20 The structure of the closed pentameric mechanosensitive channel MscL from *M. tuberculosis* (PDB 2OAR).

Because the dimensions of a bacterial cell are considerably smaller than the opening of a patch pipette, patch clamping bacteria was one of the ultimate technical challenges in electrophysiology. To overcome this problem, Hans-Jürgen Ruthe and Julius Adler devised a "giant spheroplast" preparation by taking advantage of the antibiotic cephalexin (an inhibitor of septation during cell division) and the enzyme lysozyme (which attacks the protective cell walls of bacteria). In the presence of cephalexin, *E. coli* cells cannot septate and grow into long filaments (of up to 150 μm in length). Their subsequent digestion with lysozyme and EDTA promotes the relaxation of these filaments into spheres that are up to 10 μm in diameter, which are amenable to patch-clamp studies (**Figure 1**).

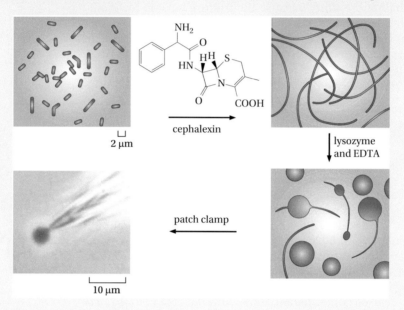

Figure 1   (From Perozo E [2005] Nature Rev. Mol Cell Biol 7: 109–119. With permission from Springer Nature.)

MscL from *Staphylococcus aureus*, which is tetrameric. Comparison of the structure with that of *M. tuberculosis* McsL (**Figure 10.21**) shows a difference in which pairs of helices from adjacent subunits remain as rigid bodies, tilting from the MscL native state to a form with the *S. aureus* McsL thinner across the bilayer and occupying more area. This idea is consistent with a body of evidence that the inter-subunit interactions are strong at this helix–helix interface. Each of these forms is thought to be closed, because the hydrophobic pores are not large enough to permit water or ion movement. But, the mutant structure is in the direction of opening a still larger pore, which has been modeled as being on the pathway toward opening a 22 Å pore, large enough to be the open state.

10.6.4 McsS Also Gates by Helix Rearrangement

The structures of the closed and open states of McsS have been solved at moderate resolution (**Figure 10.22**). The McsS from *E. coli* is a heptamer of subunits, each of which has three TMs. A structure of the open form of the channel was obtained at 3.4 Å, while a structure of a point-mutant at 3.45 Å appears to be locked in the closed form. Together, the two structures provide a view of the conformational changes involved in channel opening (**Figure 10.23**). As with the model of opening McsL, there is a tilting and mutual reorientation of the TMs to open a channel. Models derived from EPR measurements are similar but not identical. As with MscL, forces in the bilayer cause the channel to open but at a lower force for MscS, providing a graded response to osmotic stress, depending on the size of the stress. One might imagine that it is an advantage to

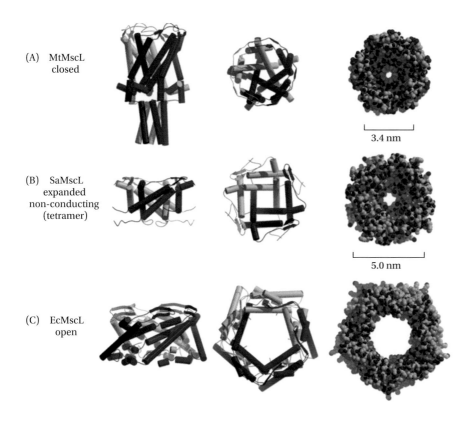

(A) MtMscL
closed

3.4 nm

(B) SaMscL
expanded
non-conducting
(tetramer)

5.0 nm

(C) EcMscL
open

Figure 10.21 Opening of the Mechanosensitive Channel of Large Conductance (MscL): (A) Crystal structure of *M. tuberculosis* MscL in the closed state. (B) Crystal structure of a tetrameric fragment of *S. aureus* MscL in an expanded, non-conducting state. (C) Model of the open state of *E. coli* MscL. The left-hand row shows side views. Chain traces of the subunits in each oligomeric channel are depicted in different colors; α-helices and β-sheets are shown as cylinders and arrows, respectively. The cytoplasmic region is positioned at the bottom. The middle row shows chain traces of the transmembrane region viewed down the membrane normal. Pay attention to the very large changes in TM-helix tilt as the channel opens. Space-filling representations of the structures are shown in the right-hand row. Crudely, the cross-sections of the membrane-spanning region of the channel may be approximated as a regular polygon. The approximate areas of the channels, from top to bottom, are 20, 25, and 43 nm². The change in area as the channel opens is remarkably large. (From Haswell ES, Phillips R, Rees DC [2011] *Structure* 19: 1356–1359. With permission from Elsevier.)

open smaller channels if they are sufficient, allowing less of the leakage of cell contents and ions than the larger MscL would provide. The "portals" observed in the MscS structure might act as filters to keep large cytoplasmic objects, such as proteins, transport vesicles, or ribosomes, away from the open channel, limiting the flow to ions and small molecules.

In the mechanosensitive channels, we see that there is a set of ways to open channels in response to membrane stress, giving a graded series of responses. Other sensors of bilayer properties include the recently discovered (2010) shape sensor, Piezo, which detects changes in bilayer curvature (see **Box 10.5**).

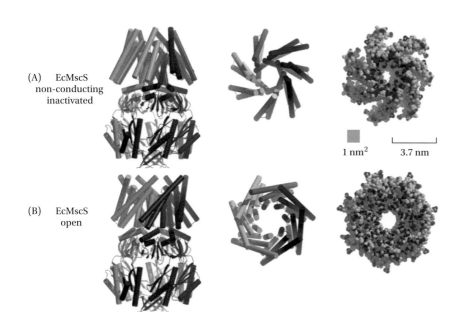

(A) EcMscS
non-conducting
inactivated

1 nm²   3.7 nm

(B) EcMscS
open

Figure 10.22 Opening of the *E. coli* McsS channel is suggested from two crystal structures, an inactivated, closed form (PDB: 4HW9) and an open form (PDB 2VV5). The TM regions are at the tops of the structures at the left. Note the smaller changes in tilts of the TM helices compared with MscL (Figure 10.21). (From Haswell ES, Phillips R, Rees DC [2011] *Structure* 19: 1356–1359. With permission from Elsevier.)

Figure 10.23 Helix motions in the opening of MscS. (A) Superposition of the crystal structures of MscS in the open (PDB 2VV5 in grey, TM1-TM2 in one subunit in green) and closed (PDB 2OAU in blue, TM1-TM2 in one subunit in orange) states. Note the portals that may limit the size of cytoplasmic components that can reach the open channel. Also note that the cytoplasmic domain remains unchanged as the channel opens. (B) Same as in (A), but for a single subunit. TM3 and the cytoplasmic domain remain stationary, while TM1 and TM2 move as a unit.

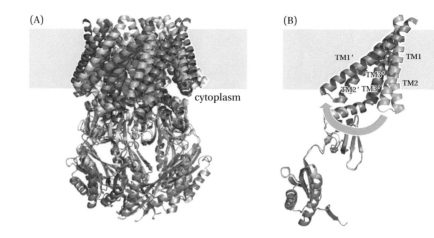

BOX 10.5 VERTEBRATE PIEZO CHANNELS SENSE CHANGES IN MEMBRANE CURVATURE

The Piezo family of cation-selective mechanosensitive ion channels was discovered in 2010 and has been implicated in a wide range of physiological processes, such as hearing, regulation of blood pressure, and breathing. Ardem Patapoutian was awarded the Nobel Prize in Medicine or Physiology 2021 for his discovery of the Piezo1 channel. Piezo channels are very large, homotrimeric structures, with each subunit composed of ~2500 amino acids. Cryo-electron microscopy (cryo-EM) studies reveal a surprising structure, with a central channel domain surrounded by three very long, curved arms,

each composed of 36 TMs (**Figure 1**). The membrane-embedded arms are at a ~60° angle relative to the trimeric symmetry axis, rather than ~90° as we would expect for a typical membrane protein. The obvious implication is that Piezo induces strong curvature on the surrounding membrane, essentially forming a "dimple" with the channel domain in the middle. Increased tension in the membrane surrounding the dimple will presumably cause the dimple to flatten and the curved arms to bend away from the channel domain. The arms on Piezo may thus act as springs that resist flattening, with the

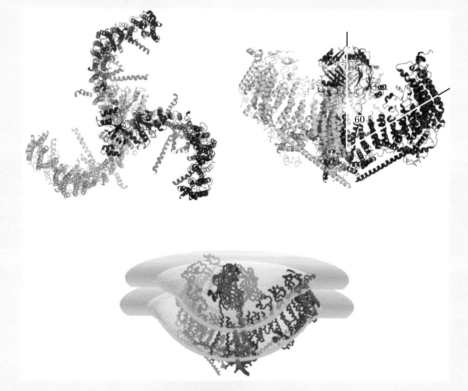

Figure 1 The Piezo channel (PDB: 6B3R), seen from above (top left) and from the side (top right). As shown in the bottom panel, the protein will distort the membrane into a dimple-like shape that may be flattened out by an increase in membrane tension. (From Guo and MacKinnon [2017] *eLife* 6: e33660.)

(Continued)

energy being stored as an increase in the angle between the arms and the central channel domain. The detailed mechanics of gate opening in the central channel domain are not clear, but the model predicts that the conformational change brought about by the increased membrane tension will be focused on the region that connects the arms to the ion channel in the center of the protein, i.e., on the gate region.

An interesting aspect of this model for Piezo gating is that the dimple can be flattened out without requiring that the ion channel in the center of Piezo opens very much. When the dimple flattens, the area increase in the surrounding membrane is equal to the increase in the area of the dimple as projected onto the membrane plane. This area increase can be quite large—total flattening of a semi-spherical dimple of a size corresponding to the Piezo trimer (~400 nm^2) would lead to an area increase of more than 100 nm^2—but since the energy is largely taken up by the conformational change caused by the arms pivoting relative to the channel, does not in principle require the ion channel in the center of Piezo to expand at all. Indeed, electrophysiological measurements suggest that the open Piezo channel is cation-selective and has a diameter of only 4–5 Å, while the open pore in MscL is non-selective, has 100-fold higher conductance, has a diameter of more than 20 Å, and is associated with an area expansion of only ~10 nm^2.

10.6.5 Non-specific Porins Allow Solute Passage Based Mostly on Molecular Weight

The most abundant porins in the outer membranes of *E. coli* are OmpF, OmpC, and PhoE. Each is a homotrimer composed of 3 barrels with 16 strands (Figure 10.24A). With effective pore diameters of about 8 Å due to a mid-barrel constriction, they will allow passage of solutes that are smaller than 600 Da by simple diffusion, although there is a dramatic decrease in permeability to solutes with molecular weights greater than 350 Da, as shown by permeability measurements made on vesicles with incorporated porins (Figure 10.24B). Because they allow all solutes below the 600 kDa cutoff to pass through, these porins are referred to as general diffusion porins. Of course, even between molecular weights 150 and 350 Da, there is a three-orders-of-magnitude range of permeabilities, meaning that the relative magnitudes of the fluxes of permeable molecules can vary considerably.

Small solutes such as ions, amino acids, and small sugars find their way into the periplasm through these general porins with reasonable ease. Although not strongly selective, a closer look at the permeability data reveals that the general diffusion porins are somewhat sensitive to charge. Notice in Figure 10.24B that PhoE is more permeable to molecules carrying a negative charge than is OmpF. The explanation is that there are differences in charged residues at the mid-barrel constriction and of Loop 3 of PhoE compared with OmpF (Figure 10.24C). These charges affect, of course, the passage of ions.

10.6.6 Metabolically Important Sugars Use Specialized Porins for Gaining Access to the Periplasm of Gram-Negative Bacteria

Both maltose and sucrose, each with a molecular weight of 342 Da, can pass through general diffusion porins (Figure 10.24B) and into the periplasm of *E. coli*, but not as efficiently as many other solutes. Because of their importance as energy sources, specialized porins have evolved for managing these two sugars. Furthermore, these porins are highly selective. Reconstitution of LamB (maltoporin) into liposomes, for example, makes the vesicles highly permeable to maltose and maltodextrins but not to sucrose (Figure 10.25A). Another specific porin, ScrY from *Salmonella typhimurium*, is highly selective for sucrose. Both porins, LamB and ScrY, form homotrimers from monomers with 18 strands. How do they differ from 16-stranded OmpC and PhoE? Their principal features, besides a pore, are belts of aromatic residues that spiral down the insides of the porins forming a "greasy slide" and in parallel, a string of charged residues referred to as the "polar track" (Figure 10.25B,C). This arrangement is very

BOX 10.6   GAP JUNCTIONS

In many tissues, there is a need for molecules to pass from the cytoplasm of one cell to the cytoplasm of an adjacent cell. This is particularly true in cardiac tissue, in which gap junctions allow rapid electrical communication between cardiac cells, allowing the heart, in effect, to behave as one large cell (functional syncytium). To accomplish communications between cells, one structure that has evolved is the gap junction, an oligomer of six transmembrane connexin subunits, each with four TMs (the name comes from early EM views showing a gap between cell membranes with the junction between them). The hexamer crosses the membrane of one cell and binds to an identical hexamer in the adjacent cell, forming a 12-subunit channel between the cells. The channel allows the transfer of a wide variety of solutes of different sizes, including ions, metabolites, nucleotides, peptides, and secondary messengers. Gap junction channels function in many biologically important processes, including cardiac development, fertility, the immune system, and electrical signaling in the nervous system. The structure of one of the human connexins, Cx26, has been revealed by x-ray crystallography (3.5 Å resolution; **Figure 1**) and electron microscopy (**Figure 2**).

The x-ray map shows the two membrane-spanning structures and the arrangement of the four transmembrane helices of each of the six connexin molecules forming them. Each hexamer features a positively charged cytoplasmic entrance,

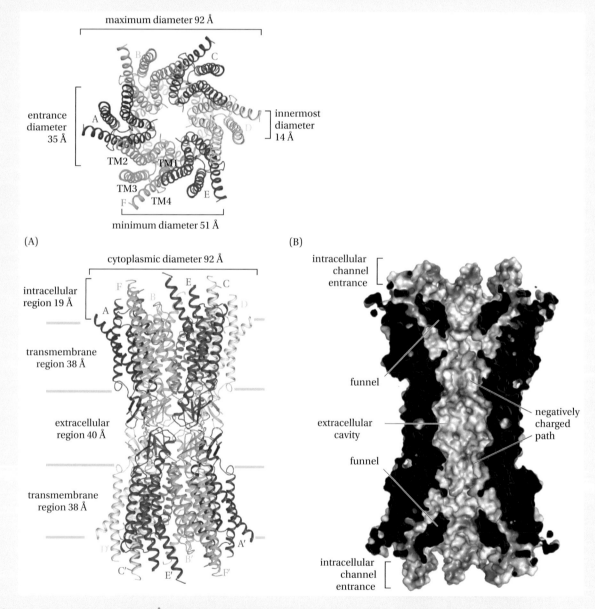

Figure 1 X-ray structure at 3.5 Å resolution of the connexin 26 (Cx26) gap junction (PDB 2ZW3). (A) Ribbon representation of Cx26, top and side views. (B) Cut-away space-filling representation of Cx26 showing the passageway between cells. (From Maeda S et al. [2009]. *Nature* 458:597–602. With permission from Springer Nature.)

(Continued)

BOX 10.6 GAP JUNCTIONS (CONTINUED)

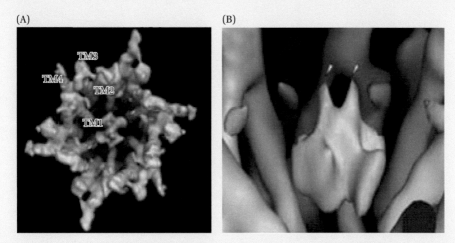

Figure 2 EM reconstruction from two-dimensional crystals of a gap junction. (A) View along the channel axis. (B) View perpendicular to the channel axis. These images suggest a plug structure that may be related to permeability regulation. (From Fujiyoshi Y [2011] *Microscopy* 60: 5149–4159. With permission from Oxford University Press.)

a funnel, a negatively charged transmembrane pathway, and an extracellular cavity between the connected cells. The pore is narrowed at the funnel, which is formed by the six amino-terminal helices lining the wall of the channel, which thus determines the molecular size restriction at the channel entrance.

The Cx26 gap junction conductivity is modulated by cellular substances, but the mechanisms of ligand action are largely unknown. It is known that protonated aminosulfonates, notably taurine, directly and reversibly inhibit homomeric and heteromeric channels that contain Cx26 but not homomeric Cx32 channels, indicating specificity, but mechanistic understanding awaits further work. Interestingly, while the crystal structure shows no block in the channel (Box Figure 1), studies of two-dimensional crystals using the EM reveal a plug structure that may be the key to regulation (Figure 2).

BOX 10.7 PORINS AND THE VERSATILE β-BARREL

All outer membrane (OM) proteins of Gram-negative bacteria are β-barrels, because as noted in Chapter 6, the Sec apparatus must treat them as secreted proteins so that they will be exported to the periplasm for folding and assembly into the outer membrane. The features that cause export by SecYEG translocon are short strings of 8–10 amino acids alternating between hydrophobic and polar residues that form the barrel strands. The strands are simply not long enough or hydrophobic enough to be selected for insertion into the cytoplasmic membrane. Despite these seeming structural limitations, OM proteins have evolved to satisfy various needs of bacteria ranging from simple anchors to selective pores and enzymes (**Table 1** and Box **Figure 1**). The most common function of porins, as their generic name suggests, is to act as pores in the outer membrane to allow movement of metabolites between the periplasmic space and the extracellular fluid.

The β-barrel structural motif, consisting of antiparallel β-sheets folded into a cylindrical shape, is not unique to outer membrane proteins. It is also found in all sorts of soluble proteins, such as the retinol binding protein (**Figure 2A**). All β-barrels adhere to structural principles summarized in Figure 2B,C. The strands are never parallel to the barrel axis but rather, wrap spirally around the axis in a tilted fashion (panel B). Barrels are characterized mathematically by the number of strands n and the so-called shear number S (panel C). The shear number can be understood by unrolling the barrel to form a flat β-sheet. Because the strands are tilted but must be parallel to each other, S describes the resulting off-set of the registry between strands. An important feature of the motif is that, moving along a strand, the sidechains alternate between facing into the barrel and facing away from it. In OM proteins, the less polar residues face outward, with more polar residues facing inward. But in soluble barrels, the pattern is reversed to achieve protein stability and solubility.

(Continued)

BOX 10.7 PORINS AND THE VERSATILE β-BARREL (CONTINUED)

Table 1 Structural and Functional Features of Prototype Outer Membrane Proteins from *E. coli*

Protein family	Small β-barrel membrane anchors	Small β-barrel membrane anchors	Membrane-integral enzymes	General (non-specific) porins	Substrate-specific porins	TonB-dependent receptors
Protein	OmpA	OmpX	PldA (OMPLA)	OmpF	LamB	FhuA
Function	Links OM to peptidoglycan	Neutralization of host defense mechanisms	Phospholipid hydrolysis	Diffusion pore for ions and small molecules	Maltose and maltodextrin uptake	Takes up iron–siderophore complexes
Oligomeric state	Monomer	Monomer	Monomer/dimer	Homotrimer	Homotrimer	Monomer
Domain structure	Two co-linear domains	One domain	One domain	One domain	One domain	Two interconnected domains
Size (# residues)	171	148	269	340	421	714
PDB code	1BXW	1QJ8	1QD5	2OMF	1MAL	1BY3, 2FCP
Resolution (Å)	2.5	1.9	2.4	2.4	2.6	2.5
Number of transmembrane β-strands, *n*	8	8	12	16	18	22
Shear number, *S*	10	8	16	20	22	24

(Table simplified from Koebnik et al. (2000) *Mol Microbiol* **37**:239–253)

(Continued)

BOX 10.7 PORINS AND THE VERSATILE β-BARREL (CONTINUED)

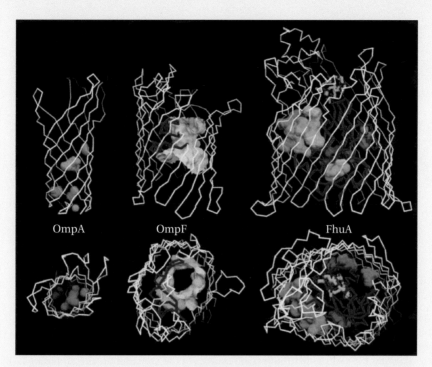

OmpA   OmpF   FhuA

Figure 1 Porins serve many functions in the outer membranes of bacteria, summarized in Table 10.1. The functions range from simple anchors (OmpA; PDB 1QJP) to sophisticated receptors (FhuA; PDB 1BY5). All β-barrels, whether as soluble or membrane proteins, adhere to structural principles described by the number of strands and the shear number (Box 10.7, Figure 2). (From Koebnik et al. [2000] *Mol Microbiol* 37: 259–253. With permission from John Wiley and Sons.)

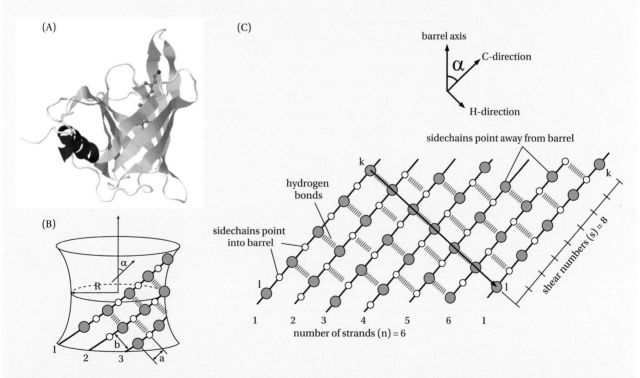

Figure 2 Parameterization of the β-barrel motif. (A) The β-barrel is a common structural motif in soluble as well as outer membrane proteins, as illustrated by the structure of the retinol binding protein. (B) Strands are never parallel to the barrel axis. Rather, they are tilted at an angle α relative to the axis so as to form a spiral wrap. (C) The tilt is characterized by the shear number (S). In this view, the β-sheet wrap has been laid out flat. The shear number is measured by counting the number strands along a normal to the starting strand and ending on the starting strand. (B, C, From Murzin AG et al [1994] *J Mol Biol* 236:1369–1381. With permission from Elsevier.)

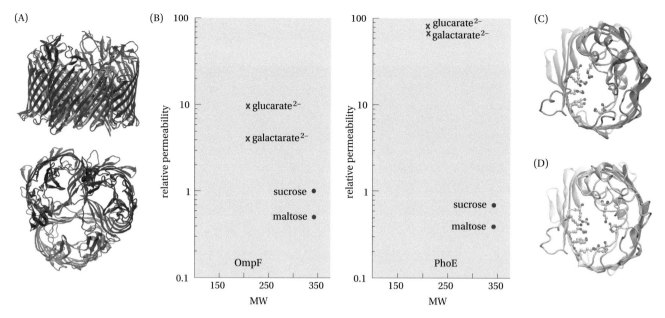

Figure 10.24 Structures and properties of general diffusion porins. (A) General diffusion porins are "triple-barreled" homotrimeric β-barrels with 16 transmembrane strands composing each barrel. (B) Relative permeabilities of OmpF and PhoE porins reconstituted into liposomes. Relative permeabilities were determined by osmotic-swelling measurements. Overall, permeabilities are determined by the molecular weight of the solute used. But, notice that PhoE is more permeable to anionic solutes than is OmpF, because the amino acid composition of the mid-barrel constriction and loop 3 differ. Notice that these general diffusion porins are permeable to sucrose and maltose (solid red circles). (C, D) Pore structures of OmpF (PDB 1OPF) and PhoE (PDB 1PHO), respectively, showing the charge-distribution differences in the constriction site and loop 3. (B, Redrawn and adapted from Nikaido H, Rosenberg EY [1983] *Journal of Bacteriology* 153: 241–252.)

similar to the scheme found in aquaglyceroporins (Figure 10.11). The mechanism of permeability is believed to involve sliding of the sugars down the greasy slide while charged residues of the polar track interact with sugar hydroxyls by hydrogen bonding to achieve selectivity. Because of the polar track, one can well imagine that these specific porins have differences in selectivity.

The structures of LamB and ScrY are nearly superimposable but show important differences at three positions in the mid-barrel restriction site. Compared with LamB, an Arg protruding into the channel lumen is replaced by an Asn that now points toward the channel wall; a Tyr that constricts the channel is replaced by an Asp; and an Asp is replaced by a Phe. If these substitutions are made in LamB by site-directed mutagenesis, then the LamB mutant becomes permeable to sucrose.

10.6.7 Why Are Beta-Barrels not Used in the Plasma Membrane Instead of Helical Bundles?

This interesting question can only be addressed in a speculative way at present, but it may be that barrels are relatively rigid and hard to vary and so, cannot accommodate as wide a range of functions as helical bundles do. Also, it appears that the helical proteins follow evolutionary paths that involve gains of function through gene duplication events, such as that giving the structure of aquaporins, and it seems hard to accommodate such domain events with a barrel structure. Further, the complex oligomeric structures we will come to in subsequent chapters would be difficult to duplicate, and the helix conformational changes and interactions needed for function might not readily occur in the context of a rigid barrel. While barrels exhibit some versatility (see Box 10.6), mutual interactions between barrels are limited, whereas helical proteins can be built one helix at a time to form large functional units such as those found in electron transport or photosynthesis. Thus, barrels are limited by the nature of their design—a rigid wall. Another perspective is that the porins must cross the plasma membrane and insert into the outer membrane of bacteria after synthesis, so a different pathway and set of principles is used (see Chapter 6).

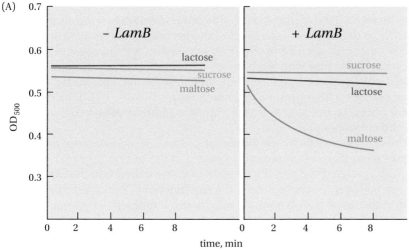

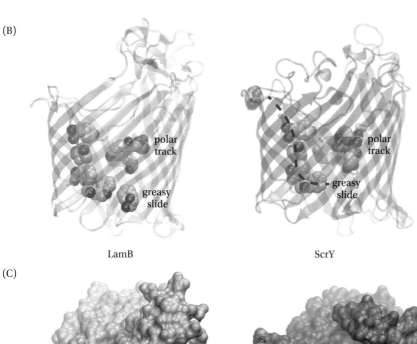

Figure 10.25 Specific porins for sugar transport. (A) Liposome-swelling experiments show that LamB is selective for maltose but not sucrose. In the left-hand panel, permeabilities (represented as changes in the optical densities of liposome solutions) are about the same in the absence of LamB. In the presence of LamB, the vesicles become selectively permeable to maltose. (B) Structures of LamB (PDB 1MPQ) and ScrY (PDB 1AOT) showing the so-called "greasy slide" and polar track. Sugars are envisioned to slide down the greasy slide, while polar residues form hydrogen bonds with sugar hydroxyl groups to gain selectivity. (C) Space-filling images of the LamB and ScrY pores. Notice the differences in size and shape of the pores. (A, From Luckey M, Nikaido H. [1980] *PNAS* 77: 167–171. With permission from M. Luckey.)

10.7 β-BARRELS IN OUTER MEMBRANES REVEAL DIVERSE STRATEGIES FOR SOLUTE PERMEATION

While the plasma membrane is the barrier that defines the cytoplasm, there are additional barriers that have evolved as useful to define functional spaces. In Gram-negative bacteria, an additional "outer" membrane creates a peripheral space just outside the plasma membrane, which defines the periplasm between the two membranes. The outer membrane controls the flow of materials to define the composition of the periplasm that is used to facilitate various functions, such as nutrient accumulation, for the organism. During evolution, organisms with such a pair of membranes are thought to have infected eukaryotic cells and subsequently evolved into specialized cytoplasmic organelles: chloroplasts and mitochondria (Figure 1.5). Barrier control must apply to the outer as well as the inner membrane of bacteria, mitochondria, and chloroplasts.

In the structures of the mechanosensitive channels, we have seen that a large aqueous channel can be transiently opened in the bilayer by an arrangement of helices. A related use is in the creation of stable helical bundles that provide aqueous pathways to connect cells, the gap junctions (**Box 10.6**). The proteins found in the outer membranes of bacteria use a different structural theme to create channels: the β-barrel, and those barrels that form channels are called *porins* (**Box 10.7**). Perhaps as a consequence of a different biosynthetic pathway being needed to create them in the absence of direct contact with the cytoplasm (see Chapter 6), the barrel proteins define transmembrane pathways by walling off the lipid bilayer with a beta-sheet and using internal loops of polypeptide to confer selectivity and control (**Figures 10.24** and **10.25**). Given how β-barrels are formed, it is easy to create different polypeptide structures inside them, where the folding constraints are not much different from those facing a soluble protein except that the volume is limited and the ends of the loops are constrained to be bound to the barrel strands.

Thus, despite Gram-negative bacteria being limited to β-barrels in their outer membranes, OM proteins have nevertheless evolved to serve numerous sophisticated biological functions by arranging their interior loops. The principal activity of interest in the present discussion is passive solute transport through pores, which is obviously necessary for the uptake and discharge of metabolites. The pores in some cases are simple, permitting the transit of molecules based upon size. In other cases, pores act selectively to admit only molecules with particular chemical properties. The major OM porin of mitochondria is even voltage dependent. Some Gram-positive bacteria, lacking outer membranes, have evolved uses for β-barrels in the form of secreted toxins that are structurally elegant despite their malevolent purpose (Chapter 6).

KEY CONCEPTS

- Channels create pathways across membranes by creating barriers between their interiors and the lipid bilayer so that the bilayer properties are not seen by a molecule moving in the pathway. The pathway across the barrier can be a set of helices or a β-barrel.

- Reentrant structures, including short helices and extended polypeptide chains, are often used inside the barriers to provide greater flexibility in channel designs. These structures are more readily formed in the fully or partly sequestered environments inside the barriers.

- The water channel structures create relatively non-polar surfaces that face the lipid bilayer and more polar interiors for the channels by choice of side chains, so the helices tend to be amphipathic and the barrels use alternating non-polar/polar amino acids along each strand of the barrel.

- The channel surfaces avoid trapping solutes in the channel by avoiding large energy differences for positions along the channel.

- Water and glycerol channels avoid H^+ translocation either by preventing the Grotthuss hop-turn or by creating an electrostatic barrier to hydrated protons (e.g., H_3O^+). While there are competing theories on the details (since the theories depend on molecular modeling, it remains to find experiments to resolve the issues), it is gratifying that there are ideas that work to explain the resistance to proton flow through water and glycerol channels. If protons were allowed, the membrane potential would collapse, and the cell would die.

- Plant aquaporins can be gated in response to environmental challenges, such as acidity.

- Ammonia transport relies upon hydrophobic channels that pass uncharged NH_3 molecules, rejecting water and NH_4^+. Protons are stripped off the NH_4^+ at the entry of the channel, and transport is driven by mass action through depletion of the ammonia on one side of the membrane.

- Mechanosensitive channels can protect against osmotic shock by opening in response to stress, allowing the passage of osmolytes.

- β–barrels, found in the outer membranes of bacteria, mitochondria, and chloroplasts, can provide pathways for a variety of small molecules, such as sugars and amino acids, and can be less specific than channels found in plasma membranes, since they are secondary barriers defining antechambers outside the primary plasma membranes, but may be more specific for efficiency with needed metabolites.

FURTHER READING

Stroud, R.M., Miercke, L.J., O'Connell, J., Khademi, S., Lee, J.K., Remis, J., Harries, W., Robles, Y., and Akhavan, D. (2003) Glycerol facilitator GlpF and the associated aquaporin family of channels. *Curr Opin Struct Biol* 13:424-431.

Haswell, E.S., Phillips, R., and Rees, D.C. (2011) Mechanosensitive channels: What can they do and how do they do it? *Structure* 19:1356-1369.

Kefauver, J.M., Ward, A.B., and Patapoutian, A. (2020) Discoveries in structure and physiology of mechanically activated ion channels. *Nature* 587:567-576.

Preston, G.M., Carroll, T.P., Guggino, W.B., and Agre, P. (1992) Appearance of water channels in Xenopus oocytes expressing red cell CHIP28 protein. *Science* 256:385-387.

Maurel, C., Boursiac, Y., Luu, D.-T., Santoni, V., Shahzad, Z., and Verdoucq, L. (2015) Aquaporins in plants. *Physiological Reviews* 95:1321-1358.

Khademi, S., and Stroud, R.M. (2006) The Amt/MEP/Rh family: Structure of AmtB and the mechanism of Ammonia Gas Conduction. *Physiology* 21:419-429.

Törnroth-Horsefield, S., Hedfalk, K., Fischer, G., Lindkvist-Petersson, K., and Neutze, R. (2010) Structural insights into eukaryotic aquaporin regulation. *FEBS Letters* 584:2580-2588.

Slutsky, J.S. (2016) Outer membrane protein design. *Curr Opin Struct Biol.* 45:45-52.

KEY LITERATURE

Paganelli, C.V., and Solomon, A.K. (1957) The rate of exchange of tritiated water across the human red cell membrane. *J Gen Physiol.* 41:259-277.

Price, H.D., and Thompson, T.E. (1969) Properties of liquid bilayer membranes separating two aqueous phases: Temperature dependence of water permeability. *J Mol Biol.* 41:443-457.

Agre, P. (2004) Nobel Lecture. Aquaporin water channels. *Biosci Rep.* 24:127-163.

Stroud, R.M.1, Miercke, L.J., O'Connell, J., Khademi, S., Lee, J.K., Remis, J., Harries, W., Robles, Y., and Akhavan, D. (2003) Glycerol facilitator GlpF and the associated aquaporin family of channels. *Curr Opin Struct Biol.* 13:424-431.

King, L.S.1, Kozono, D., and Agre, P. (2004) From structure to disease: The evolving tale of aquaporin biology. *Nat Rev Mol Cell Biol* 5:687-698.

Maurel, C., Boursiac, Y., Luu, D.T., Santoni, V., Shahzad, Z., and Verdoucq, L. (2015) Aquaporins in Plants. *Physiol Rev.* 95:1321-1358.

Khademi, S.1, and Stroud, R.M. (2006) The Amt/MEP/Rh family: structure of AmtB and the mechanism of ammonia gas conduction. *Physiology (Bethesda).* 21:419-429.

Haswell, E.S.1, Phillips, R., and Rees, D.C. (2011) Mechanosensitive channels: What can they do and how do they do it? *Structure.* 19:1356-1369.

Perozo, E. (2006) Gating prokaryotic mechanosensitive channels. *Nat Rev Mol Cell Biol.* 7:109-119.

Wang, W., Black, S.S., Edwards, M.D., Miller, S., Morrison, E.L., Bartlett, W., Dong, C., Naismith, J.H., and Booth, I.R. (2008) The structure of an open form of an *E.coli* mechanosensitive channel at 3.45Å resolution. *Science.* 321:1179-83.

Schulz, G.E. (1996) Porins: General to specific, native to engineered passive pores. *Curr Opin Struct Biol.* 6:485-490.

Dumas, F., Koebnik, R., Winterhalter, M., and Van Gelder, P. (2000) Sugar transport through maltoporin of *Escherichia coli.* Role of polar tracks. *J Biol Chem.* 275:19747-19751.

EXERCISES

1. There are several ideas about roles for aquaporins in cancer. Find and read a recent review.

2. Given the observed binding of glycerol, suggest a basis for the interaction of water in the channel. Then, look up an article where molecular modeling has been used to describe possible interactions.

3. Control of water transport might not be needed in some cases. Contrast the erythrocyte with plant root channels, and propose a reason why regulation may be much less important in the erythrocyte.

4. Regulation is important because ammonia is both needed and toxic. Explore and understand why ammonia can be toxic in the cytoplasm of *E. coli*.

5. Autotransporters are β-barrel channels that combine with protein passengers to expose them as folded domains on the outside surfaces of bacteria. Read a paper on the function of these remarkable channels.

6. Why are proteins more likely to have a wide range of responses to changes in lateral membrane tension than lipids?

7. Why are the structures of native mechanosensitive channels found to be in the closed state?

Ion Channels

11

How do cells recognize and respond to their environments? Single cells are able to seek food by chemotaxis, respond to threats such as osmotic stress, and outgrow the competition by being efficient with available resources, to note only a few of the many responses. Cells in multicellular organisms, on the other hand, must work in concert over long distances with other cells and tissues to respond to environmental challenges in a coordinated manner. The nervous systems of higher organisms allow high-speed communication between diverse cells and tissues by means of electrical and chemical signaling. It is exciting that we are beginning to understand in chemical detail these processes in living cells. Without fast signaling, we would not be able to move or think as we do. Ion channels dominate fast signaling. They do this through exquisite control of ion flows across membrane barriers. The lipid bilayer of cell membranes is an excellent barrier; neither ions nor water can cross this thin hydrocarbon barrier with ease (Chapters 1 and 2). As a result, it is possible for proteins embedded in the bilayer to regulate the passive flow of ions, water, and other molecules across the membrane.

The need for barrier control becomes obvious when considering resting quiescent cells. Except for acidophiles, all organisms have a resting transmembrane potential that is negative inside (cytoplasm) with respect to the extracellular space. Furthermore, relative to the extracellular space, the intracellular concentration of K^+ is high inside, while the Na^+ concentration is low. We will learn in Chapter 12 how Na^+/K^+-ATPases produce these ion gradients. Barrier control occurs by means of transmembrane ion-selective channels that allow cells to control the passive movements of ions down their electrochemical gradients by controlling the number of channels per unit area and the ion conductivity of the individual channels. The principles of how quiescent cells maintain constant transmembrane ionic and voltage gradients are summarized in **Box 11.1**. These principles are the starting point for understanding fast signaling via the propagation of action potentials in nerve and muscle cells. We focus primarily on the α-helix-based plasma membrane channels and pores involved in the movement of ions down their electrochemical potential gradients. But, we also look briefly at a gated β-barrel-based channel found in the outer membrane of mitochondria. As in prior chapters, we examine a few examples to explore the principles. To set the stage, we review key steps in the history of neurobiology and ion channels.

DOI: 10.1201/9780429341328-12

BOX 11.1   HOW THE SQUID AXON GETS ITS TRANSMEMBRANE POTENTIAL

We have learned that all cells have resting transmembrane potentials V_{rest} across their cytoplasmic membranes. Except for acidophilic microorganisms, the potential V_{rest} is always negative inside relative to outside. The other universal feature of non-acidophiles is high cytoplasmic concentrations of K^+ and low concentrations of Na^+. These concentrations give Nernst potentials V_K and V_{Na} that are, relative to the outside, negative for K^+ and positive for Na^+ (see Chapter 0). These concentration gradients are due to active inward transport of K^+ and active outward transport of Na^+ by Na^+,K^+-ATPase pumps (Chapter 12). Because both V_{rest} and V_K have negative values, one suspects that somehow K^+ must play a major role in setting the resting membrane potential. But how?

The silent hero is negatively charged organic anions A^- that serve as counterions for K^+. Exactly what molecules comprise A^- is rather fuzzy, but basically, any molecule bearing a negative charge counts as an organic ion as long as it is membrane impermeable. We can understand the origin of the resting membrane potential using the giant axon of squid as an example (see Box 11.2). Squid have no kidneys; the extracellular ion concentrations are the same as for seawater. Because the interior of the axon must be isotonic with respect to seawater, the intracellular concentrations of ions are much higher than in vertebrates (about fourfold).

Figure 1A shows the intracellular and extracellular ion concentrations for squid axons. There must be charge

(A)   squid axon membrane
electrochemical gradients

intracellular	extracellular (sea water)	
$[K^+]_i = 400$ mM	$[K^+]_o = 10$ mM	$V_K = -93$ mV
$[Na^+]_i = 50$ mM	$[Na^+]_o = 460$ mM	$V_{Na} = +56$ mV
$[Cl^-]_i = 70$ mM	$[Cl^-]_o = 540$ mM	$V_{Cl} = V_{rest}$
$[A^-]_i = 345$ mM	$[A^-]_o = 0$ mM	

membrane potential: V_{rest} 0 -50 mV

squid axon membrane

(B)   permeable only to K^+

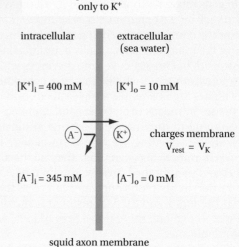

intracellular	extracellular (sea water)
$[K^+]_i = 400$ mM	$[K^+]_o = 10$ mM
	charges membrane $V_{rest} = V_K$
$[A^-]_i = 345$ mM	$[A^-]_o = 0$ mM

squid axon membrane

(C)   permeable to K^+
and somewhat to Na^+

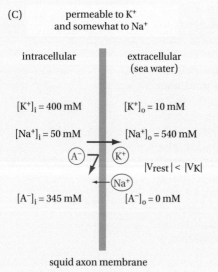

intracellular	extracellular (sea water)
$[K^+]_i = 400$ mM	$[K^+]_o = 10$ mM
$[Na^+]_i = 50$ mM	$[Na^+]_o = 540$ mM
	$|V_{rest}| < |V_K|$
$[A^-]_i = 345$ mM	$[A^-]_o = 0$ mM

squid axon membrane

Figure 1 Determinants of the resting membrane potential. (A) Squid axon ion concentrations and Nernst potentials. (B) If the membrane were permeable only to K^+, then the resting membrane potential would be equal to the potassium equilibrium potentials. Because the membrane is impermeable to the organic ions, there is no Nernst potential associated with it. (C) The membrane is permeable to both Na^+ and K^+, but in the resting state, it is much more permeable to K^+ than Na^+. This means that the resting membrane potential is dominated by the K^+ equilibrium potential.

(Continued)

BOX 11.1   HOW THE SQUID AXON GETS ITS TRANSMEMBRANE POTENTIAL (CONTINUED)

neutrality inside and outside the axon, but only the concentrations of ions relevant to the resting membrane potential are shown. The resting potential V_{rest} of squid axons is about −50 mV (see Box 11.2, **Figure 2**). The panel also lists the equilibrium potentials for the principal ions. Why is a value for A⁻ not listed? Because the membrane is impermeable to A⁻, the Nernst equation does not apply; it can only be used when the membrane is permeable to an ion.

To understand the origin of V_{rest}, assume for a moment that the membrane is not permeable to any ion. Then, imagine that at some instant, the membrane is made permeable only to K⁺ (Figure 1B). There will immediately be a net movement of K⁺ outward across the membrane because of the high internal K⁺ concentration. A⁻ wants to follow K⁺ to maintain electroneutrality, but it can't. But, what it can do is hang out close to the membrane to be as close as possible to the leaked K⁺ so that electroneutrality is maintained across the membrane. This separation of charges creates a potential that is negative inside the cell. But, how big will the potential be? There can be a net flow of K⁺ only as long as it is not in electrochemical equilibrium. The flow will stop when the Nernst equation is satisfied, that is, the membrane potential is equal to the potassium equilibrium potential. In this hypothetical scenario, this would mean that $V_{rest} = V_K =$ −93 mV. The actual value of V_{rest} is about −50 mV. Why?

Suppose we now make the membrane permeable to Na⁺, which has a high concentration outside. Na⁺ will then flow into the cell, replacing some of the lost K⁺ and thereby reducing (making less negative) the membrane potential,

i.e., $|V_{rest}|$ will be smaller than $|V_K|$ (Figure 1C). The actual value of V_{rest} thus depends upon the relative ease with which K⁺ and Na⁺ can cross the membrane. That is, the value depends upon the relative values of the membrane conductances G_K and G_{Na}. Given the conductances, can we calculate V_{rest}? Yes! The simplest way is to consider the resting membrane potential as the conductance-weighted average of the K⁺ and Na⁺ equilibrium (Nernst) potentials:

$$V_{rest} = \frac{G_K V_K + G_{Na} V_{Na}}{G_K + G_{Na}} = \frac{V_K + \dfrac{G_{Na}}{G_K} V_{Na}}{1 + \dfrac{G_{Na}}{G_K}} \qquad (1)$$

This equation shows that under resting conditions, when the net flux of ions from all causes is 0, the membrane potential is determined by the ion equilibrium potentials and the conductance of Na⁺ relative to K⁺. The resting membrane potential for the squid axon membrane indicates that $G_{Na}/G_K \approx 0.4$. If this ratio changes, then the membrane potential will also change. But, what about Cl⁻? How does it fit in? As discussed further later, because Cl⁻ is not actively transported, it is constrained to be in equilibrium with the membrane potential as long as the cell is in a quiescent resting state. That is, $V_{Cl} = V_{rest}$, which sets the internal concentration of Cl⁻.

Several points should be kept in mind. First, the amount of charge that moves across the membrane in Figure 1C is minuscule compared with the bulk ion concentrations; typically, only picomoles of charge are separated. The amount of charge Q separated can be calculated by treating the membrane as a capacitor with capacitance C and using the formula $Q = CV$. This holds true always, so that during an action potential, only very small amounts of charge need move across the membrane to change the membrane potential.

Second, we have ignored completely the existence of active transport except to note that it maintains the ion gradients. In the resting steady state, V_{rest} is constant, as are the concentrations of ions inside the axon (and outside, too, determined by the ionic composition of the seawater), meaning that there is no net flux of any ion across the membrane; if there were, the membrane potential would change. For a real axon, all of the ion fluxes from all sources must balance to zero, as summarized in the cartoon in Figure 2. Everything must be perfectly balanced and regulated in the steady state. Because $V_K \neq V_{rest} \neq V_{Na}$, there must be leakage of Na⁺ and K⁺ down their electrochemical gradients. This means that active transport must create opposing fluxes of Na⁺ and K⁺ that precisely balance the leak fluxes. If other ions are actively transported as well, the same rules apply: active transport must precisely balance leakage down electrochemical gradients. For any ion that is not actively transported, the ion must be distributed at equilibrium relative to the resting membrane potential, as in the case of Cl⁻.

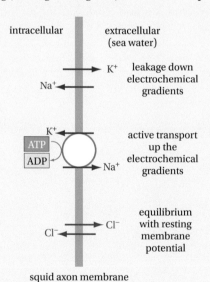

intracellular

extracellular (sea water)

K⁺ → leakage down electrochemical gradients

Na⁺ ←

ATP / ADP

K⁺ → active transport up the electrochemical gradients

→ Na⁺

Cl⁻ → equilibrium with resting membrane potential

squid axon membrane

Figure 2 In the resting axon, active transport of Na⁺ and K⁺ against their electrochemical gradients precisely balances the leakage of the ions down their electrochemical gradients. Because Cl⁻ is not actively transported, it must be distributed at equilibrium with respect to the resting membrane potential determined by the relative permeabilities of Na⁺ and K⁺.

11.1 IONIC CURRENTS ACROSS MEMBRANES UNDERLIE NERVE ACTION POTENTIALS

11.1.1 Measurements of Macroscopic Electrical Properties of Nerves Reveal the Essential Features of Action Potentials

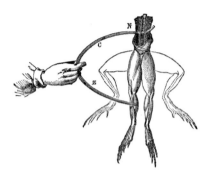

Figure 11.1 One of Luigi Galvani's experiments on "bioelectricity." The metal instrument (C) in contact with the muscle (Z) and the spinal cord (N) stimulates motor neurons in the spinal cord due to electrode potentials between metal and electrolyte.

After Luigi Galvani (1737–1798) discovered that metal dissecting instruments caused the legs of dissected frogs to contract (**Figure 11.1**), biologists and physicists became intensely interested in "bioelectricity." Galvani suggested that the contraction was generated by "electrical fluids" carried to muscles from the nervous system. But Alessandro Volta (1745–1827) believed, correctly, that the muscle excitation was due to the metal instruments generating electric currents, which led him to invent the first battery.

In Galvani's experiments, motor neurons in the frog spinal cord were being stimulated, which caused electrical signals (nerve impulses or action potentials (APs)) to travel down the nerve axon and stimulate muscles through nerve–muscle chemical synapses (**Figure 11.2A**). More than a century later, advances in technology were coupled with a model system to find the chemical basis of the nerve function. Squid, it was found in 1909, have a giant axon (up to 1 mm in diameter, but typically about 0.5 mm) that controls part of the water jet propulsion system. In the 1930s, it was realized that the large axon size made it easy to insert electrodes for measuring and controlling transmembrane electrical changes.

Inserting electrolyte-filled micropipettes (~1 μm tip diameter) connected to a sensitive voltmeter at various positions along the axon revealed that the AP travels along the axon at constant velocity and amplitude (Figure 11.2B and C). The AP shown in Panel C is a famous image recorded by A.L. Hodgkin (1914–1998) and A.F. Huxley (1917–2012) in 1939. Although it was known that the AP was due to ions crossing the axon membrane in response to changes in the membrane's ion conductivity, Hodgkin and Huxley, using an intracellular

Figure 11.2 Nerves and action potentials (APs). (A) A motor neuron that conducts APs from the nerve's cell body to the terminal branches of the axon. (B) A schematic experiment for measuring the shape and conduction velocity of an AP using electrodes inserted at intervals along the axon. (C) Each electrode in panel B records the same voltage changes across the membrane as a function of time as the AP passes by. The AP propagates at a constant velocity along the nerve with a constant amplitude. (From Huxley AF, Hodgkin AL [1945] *J Physiol* 104: 176–195. With permission from John Wiley and Sons.)

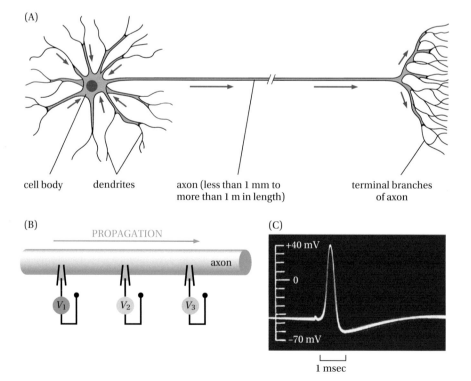

electrode for the first time to measure the electrical potential difference across the axon membrane, showed that the transmembrane potential changed transiently from negative to positive (inside relative to outside). This was the first step toward understanding in quantitative and molecular detail the time and voltage dependence of the underlying ion conductances. This magnificent achievement, published as four papers in *Journal of Physiology* in 1952, was accomplished by Hodgkin and Huxley using the voltage-clamp method introduced by Kenneth S. Cole (1900–1984).

The voltage clamp (**Box 11.2**) is an experimental method used to measure ion currents through the membranes of excitable cells, such as neurons, while holding the membrane voltage at a set level. A basic voltage clamp will measure the membrane potential iteratively and then change it to a desired value by adding the necessary current. This response "clamps" the cell membrane at a desired constant voltage, allowing the voltage clamp to record what currents are delivered. Because the currents applied to the cell must be equal (and opposite in charge) to the current going across the cell membrane at the set voltage, the recorded currents indicate how the cell reacts to changes in membrane potential. The voltage clamp allows the membrane voltage to be manipulated independently of the ionic currents, allowing current–voltage relationships of membrane channels to be studied. Cole had elegantly demonstrated with Howard J. Curtis (1906–1972) huge increases in the ionic conductance of squid axons during the AP. Curtis and Cole published their work in the same year (1939) that Hodgkin and Huxley reported their transmembrane measurements of the squid axon (Figure 11.2C). The voltage clamp allowed the action potential to be dissected in detail, which revealed the existence of two kinds of voltage-dependent channels: sodium channels responsible for the rising phase of the AP and potassium channels responsible for recovery. The necessity for two kinds of channels was hinted at by the refractory period that follows each AP **Figure 11.3**.

The discovery of how voltage-gated ion channels work is a marvelous scientific tale, which we unfold in this chapter. The most fundamental questions are (1) why is the membrane potential voltage and time dependent and (2) what is the molecular basis for these dependences? An important conclusion of Hodgkin and Huxley was that in order for conductances to be voltage dependent, there must be charged molecules within the membrane that move in response to changes in the electric field. They hunted for evidence of the movements in their voltage-clamp recordings but were unable to detect them with confidence, largely because the electrical recording technology did not permit it. Years later, it became possible to measure these "gating charges" directly (Box 11.3).

11.1.2 Excitability of Membranes Is Revealed by I–V Curves

A feature of excitable membranes that intrigued electrophysiologists for many years was the phenomenon of "negative resistance." If one applies voltages (V) of different values across a simple resistor of conductance G (G equals $1/R$, where R is the resistance), the current (I) flowing through the resistor increases linearly and proportionally to voltage following Ohm's law: $I = GV$. Furthermore, changing the polarity of the voltage changes the direction of current flow. A simple resistor is an example of a passive electrical device, defined by a linear relation between current flow and applied voltage; an example is shown in **Figure 11.4**, marked as $I_{passive}$. One can imagine that membranes that are not excitable would behave passively over physiological voltage ranges. But, what about excitable axons? How do they behave? Look at the membrane current in Figure 11.4 and think about what the membrane current does as the clamp voltage is raised from very negative membrane potentials to very positive potentials. Because the transmembrane potential varies with time during the AP, we must measure the current response to an applied voltage at a specific time after stimulating the axon. If we use a time

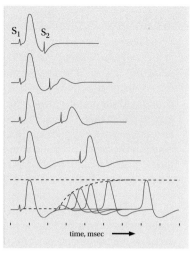

time, msec ⟶

Figure 11.3 Nerve stimulations reveal a refractory period. One of the early mysteries of action potentials was the limit on the rate of firing of action potentials illustrated here. S_1 and S_2 are two stimuli separated in time. If the time is too short, another action potential cannot be initiated (top frame). But as the interval between stimuli is increased, action potentials are possible, but with reduced amplitude. If the interval is sufficiently long, full-blown action potentials are again possible. The time interval during which a second action potential cannot be fired is called the refractory period, and it ensures that the nerve conduction is directional, since the backward direction will be kinetically blocked. Hodgkin and Huxley's voltage-clamp experiments (Box 11.2) showed that the refractory period is due to differences in the voltage and time dependence of the opening and closing of Na$^+$ and K$^+$ ion channels. (From Katz B [1966] *Nerve, Muscle, and Synapses.* McGraw-Hill. With permission.)

BOX 11.2   VOLTAGE-CLAMP MEASUREMENTS OF SQUID GIANT AXONS

Squid have evolved to escape predators by jetting away at high speeds. The jet is controlled by the large stellate nerve running the length of the mantle cavity (**Figure 1A**). Experiments show that the velocity at which squid axons conduct APs is proportional to the square root of the diameter, meaning that the larger the diameter of the nerve, the faster the jet velocity and escape speed. A squid's survival depends on speed; the squid with the largest axons survive best—serious evolutionary pressure for large axons! The giant axons of squid are typically about 0.5 mm in diameter, which is large enough for electrophysiologists to insert electrodes down the central axis of the axon (Figure 1B). This made it possible for Hodgkin and Huxley to make the first accurate measurements of an AP (Figure 11.2).

Besides its short duration of about 1 ms, two features of the AP stand out. First, before the axon is stimulated by passing a brief depolarizing current across the membrane (the "blip" in the recording just before the AP), the resting membrane potential, as for all vertebrate cells, is negative inside relative to outside. This had been surmised early on

for axons, but accurate values were not possible without intracellular recordings. The resting potential is only a bit less negative than predicted by the Nernst equation (Equation 0.30) from the inside (high) and outside (low) concentrations of potassium ions (**Table 1**), implying that the axon in the resting state is selectively permeable mostly to potassium ions (Box 11.1). Second, and this is the breakthrough finding of Hodgkin and Huxley, at the peak of the AP, the potential reverses, becoming positive inside relative to the outside, which implied that the axon was no longer dominated by high K⁺ conductance. Despite the importance of this observation, Hodgkin and Huxley did not comment on it in their 1939 paper in *Nature*.

Because the concentration of sodium ions is low in the axon interior and high in the external seawater (squid have no kidneys; their blood is isotonic with seawater) (Table 1), Hodgkin and Huxley surmised later that the major ion involved at the peak of the AP was sodium, because the equilibrium potential for Na⁺ is positive inside relative to outside. An elegant experiment conducted by Hodgkin,

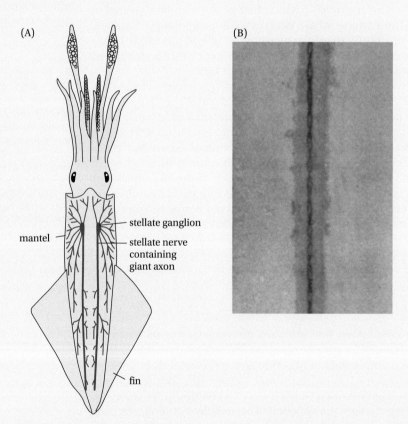

(A)

(B)

mantel — stellate ganglion

stellate nerve containing giant axon

fin

Figure 1 Giant axon of squid. (A) A squid and the nerve fibers that control jet propulsion. Depending on species and age, the diameter of the giant stellate axon is typically about 0.5 mm in diameter—large enough to insert electrodes down the center axis. (B) Microphotograph of a giant axon with inserted voltage-clamp electrode. The diameter of the axon is about 0.6 mm. (A, Figure from Thomas J. Herbert. B, From Hodgkin AI, Huxley AF, Katz B [1952] *J Physiol* 116: 424–428. With permission of John Wiley and Sons.)

(Continued)

BOX 11.2   VOLTAGE-CLAMP MEASUREMENTS OF SQUID GIANT AXONS (CONTINUED)

Huxley, and Bernard Katz (1911–2003) showed that indeed, the positive potential of the AP was due to sodium ions (**Figure 2**). The ionic basis of the action potential thus began to emerge: In the resting state, the conductance (G) of the axon is high for K^+ and low for Na^+ ($G_K \gg G_{Na}$), while at the peak of the AP, the reverse is true ($G_{Na} \gg G_K$). Because the AP is transient and arises when a depolarizing external current is applied, G_{Na} and G_K must depend on membrane potential (V_m) and time (t). To sort out the voltage and time dependence of the conductances, Hodgkin and Huxley carried out their classic voltage-clamp experiments on squid axons, which yielded quantitative empirical descriptions of $G_K(V_m,t)$ and $G_{Na}(V_m,t)$. The voltage-clamp experiments revolutionized electrophysiology and earned

Hodgkin and Huxley the 1963 Nobel Prize in Medicine or Physiology.

The principle of the voltage clamp is shown in **Figure 3A**. Command signals to the control amplifier cause the amplifier to apply currents very quickly—about 5 μs, which is much faster than the time course of an AP—in order to

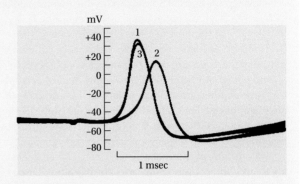

Figure 2 Effect on the squid axon action potential of diluting the bathing seawater by 50% isotonically. Trace 1, AP in normal seawater; trace 2, 50% seawater; trace 3, restoration of normal seawater. Three observations can be made: the amplitude of the AP is reduced when [Na^+] is reduced, the AP rises to its maximum more slowly, and the AP is broader (slower). These results show clearly that the AP depends strongly on [Na^+], implying that the flow of Na^+ across the axon reverses the membrane polarity transiently. (From Hodgkin AL, Katz B [1949] *J Physiol* 108: 37–77. With permission from John Wiley and Sons.)

Table 1 Approximate Ionic Composition of Seawater and Squid Axon Interior

Ion	Seawater	Axon interior
Na^+	460	50
K^+	10	400
Cl^-	540	70
A^-	—	345

A^-: organic anions. Concentrations are mM.

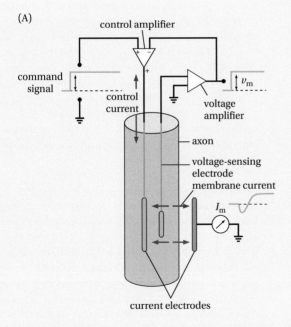

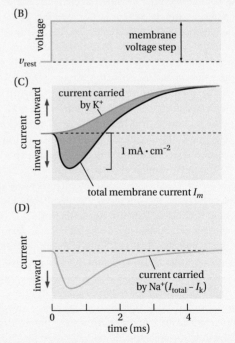

Figure 3 Measurement of the currents that underlie the squid axon action potential by means of the voltage-clamp apparatus (Panel A). As described further in the text, the apparatus keeps the membrane at a constant, selected potential (panel B) and measures the ion currents that flow in response (panels C and D). (Figure from Thomas J. Herbert.)

(Continued)

BOX 11.2 VOLTAGE-CLAMP MEASUREMENTS OF SQUID GIANT AXONS (CONTINUED)

change the transmembrane potential quickly to a selected value V_m. Changes in V_m cause G_{Na} and G_K to change. Consequently, an ionic current I_m begins to flow across the membrane. This current starts to change the membrane potential, but the fast-acting control amplifier applies counter currents I_{cc} to keep the membrane potential fixed at the selected V_m. The current flowing across due to ions is given by $I_m = -I_{cc}$. A typical membrane current recording is shown in Figure 3C in response to a step change in voltage (panel B). Typically, one observes an inward flow of positive ions followed by an outward flow. I_m is the sum of the sodium I_{Na} and potassium I_K currents. To separate the two currents from one another, ion concentrations and clamp voltages are manipulated. For example, if the external sodium ion concentration is changed so that the sodium equilibrium potential is equal to the chosen clamp voltage, then only a potassium current will be recorded (Figure 3C). After restoring the sodium concentration to its normal value, subtraction of the measured potassium ion current from the total current yields the sodium current (Figure 3D). In experiments such as these, one is really interested in the changes of membrane conductance. But those are easy to determine

from Ohm's law, because the membrane potential is held constant at chosen values.

Analysis of hundreds of measurements of the sort illustrated in Figure 3 yield what we are really interested in, namely, the conductance changes of the axon membrane that cause the voltage changes observed during an AP. These are summarized in **Figure 4**.

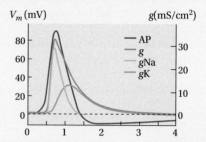

Figure 4 An action potential and the accompanying currents of sodium and potassium ions determined from voltage-clamp experiments. Here, the AP has been shifted vertically so that the $V_m = 0$ corresponds to the resting membrane potential. The total conductance $g = g_{Na} + g_K$.

BOX 11.3 GATING CURRENTS REVEAL THE NUMBER OF CHARGES NEEDED FOR CHANNEL OPENING

Voltage-gated channels are fundamental elements in the functioning of cells and the key to the nerve transmission events that enable you to read this text. Their salient property is that they open or close in response to changes in the transmembrane voltage, and to do this, they must have a way to sense the voltage. To sense the voltage, charges are driven to change their exposure from one side to the other across the high resistance barrier of the membrane, and these charges are referred to as "gating charges," since they regulate the gating of the channel. Measuring the motion and number of gating charges is a very challenging experimental task, and the invention of patch clamping together with advances in electronics provided a great improvement in such measurements since they were first made by Clay Armstrong and Francisco "Pancho" Bezanilla in 1973.

Gating charges can be measured when a voltage clamp is applied to the membrane (Box 11.2). In the example shown in **Figure 1**, the membrane protein contains two positive charges (blue in panel A; negative countercharges are red). When the membrane voltage is reversed (panel B), these charges move from inside to outside the membrane, crossing the entire electric field. In order to keep the membrane potential constant, the voltage-clamp circuit therefore must remove two negative charges from inside the membrane capacitor (small dotted circles below "protein") and deliver

two negative charges to the external side. The current recorded will therefore echo the outward movement of two positive charges. If the two charges of the sensor traverse only half the field, then only one external charge will be required to move. Thus, experimentally, the charge transferred is the product of the magnitude of the moving charge times the fraction of the field it traverses. Panel C compares the voltage dependence of the gating charge $Q(V_m)$ and the probability of channel opening $P_o(V_m)$. Because there are several sensors per conduction pore, the two curves differ in shape.

The principle of measuring gating currents is straightforward but technically demanding because of their small size and masking by ionic currents and capacitive charging currents. Armstrong and Bezanilla, using giant axons of squid (Box 11.2), suppressed the ionic currents by replacing K^+ with impermeant Cs^+ and adding tetrodotoxin to block Na^+ currents. The capacitive charging currents that, precede and follow each voltage clamp (panel D) were removed by summing the currents of exactly matched pulses of opposite sign. Original published data of Armstrong and Bezanilla (**Figure 2**) show the time course of the gating current compared with the early inward Na^+ current. Notice that the gating current precedes the ionic current, as expected from the fact that the movement of the voltage sensors

(Continued)

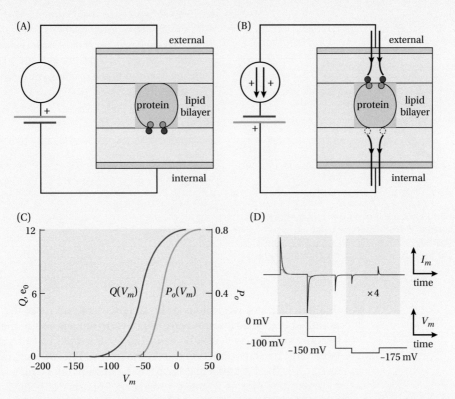

Figure 1 Measuring gating currents. For illustrative purposes, consider that the membrane protein contains two positive charges (panel A). When the membrane potential is reversed (panel B), these charges move from inside to outside the membrane. The voltage dependence of the gating charge (the $Q(V_m)$ curve) has a sigmoid shape (panel C), which differs from the $P_o(V_m)$ that reports the probability P_o of opening because there are several sensors per conduction pore. In experimental conditions, gating currents (I_g) must be extracted from ionic currents and from the linear capacitive current from the discharge of stored ionic potential across the membrane (I_c). To separate I_g from I_c (panel D), the current produced by a subtracting pulse, which generates only linear current, is subtracted from the current produced by a test pulse, which generates both linear (black) and gating current (red). If the gating charge is low, or there are only a few molecules, or the movement is slow, I_g is undetectable. (Bezanilla F [2008] *Nature Reviews Molecular Cell Biology* 9: 232–332. With permission from Springer Nature.)

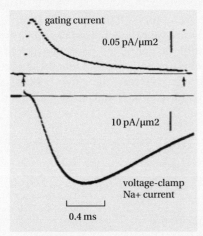

Figure 2 Comparison of the time course of gating and ion currents. Notice that the gating-current peak occurs well before the voltage-clamp current and is much smaller. (From Armstrong CM, Bezanilla F [1973] *Nature* 242: 459–461. With permission from Springer Nature.)

must occur before the channels open. Also notice that the gating-current amplitude is about 200 times smaller than the ionic-current amplitude, which explains why Hodgkin and Huxley were unable to detect it with the electronic techniques then available. Similar approaches can be used for measuring the gating charges associated with the opening of K+ channels.

A logical question that arises from gating-current measurements is the amount of charge per channel that moves across the membrane during channel opening. How can we measure it? One way, used by the MacKinnon laboratory, is to express, in effect, different numbers of K+ channels in frog oöcytes and to measure the gating currents. The number of channels is counted by binding a radiolabeled potassium channel–specific toxin to the oöcytes. The gating charges moved per channel across the membrane are calculated from plots of total gating charge against number of channels, as shown in the plot in **Figure 3**. Do 13.6 positive charges per channel make sense? Recognizing that there

(Continued)

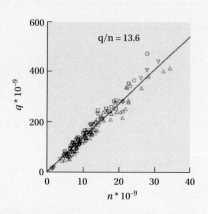

are four voltage sensors per channel (Figures 11.12, 11.25), this number corresponds to about 3.4 charges per sensor. This agrees well with the four or five arginines per sensor, because there are also negative charges present, potentially reducing the net amount of charge moved.

Figure 3 The gating charges moved per channel across the membrane calculated from plots of total gating charge against number of channels. (From Aggarwal SK, MacKinnon R [1996] *Neuron* 16: 1169–1177. With permission from Elsevier.)

of about 0.25 ms, we will be at the steeply rising phase of the AP. As the clamp potential is raised closer to zero, the net current across the membrane does not become smaller, as would be expected for a simple resistor; rather, it becomes larger and reverses direction due to the inward flow of Na+ (noted in Figure 11.4 as $I_{excitable}$). The I–V curve in this case has a negative slope near the origin, referred to in earlier times as negative resistance. This sort of behavior is often seen in active electric components such as transistors. The measurement of I–V curves has proven to be a useful way of identifying excitable membranes.

11.1.3 I–V Curves and Chemical Agents Show That Ions Flow across Membranes through Channels

The controversy of the 1960s and 1970s about the architecture of cellular membranes (Chapter 1) extended to mechanisms of membrane ion permeability as well. The question was: do ions penetrate membranes in a diffuse general way, or do they pass through structured channels? If they pass through channels, do they share a common channel, or are there specialized channels for each type of ion? Studies of membrane antibiotic peptides—particularly gramicidin and alamethicin—in black lipid membranes had revealed that these small peptides form pores that allow selective transmembrane movement of ions and importantly, revealed single-channel conductance behavior for the first time. These

Figure 11.4 Measurements of current (*I* in picoamps, pA) versus applied voltage (*V*) yield *I–V* curves. If conductance is not voltage dependent, then linear *I–V* curves result ($I_{passive}$), but if conductance is voltage dependent, as for squid axons, then an *I–V* curve with a region of negative slope is obtained ($I_{excitable}$).

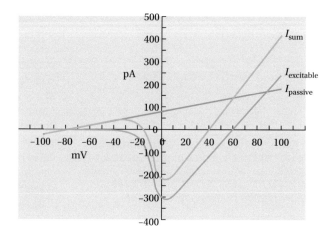

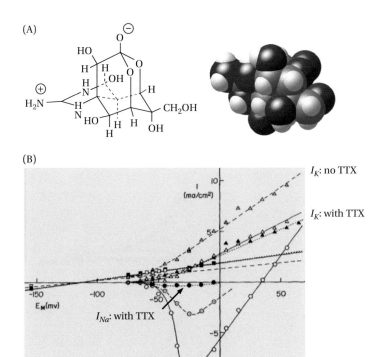

(A)

(B)

I_K: no TTX

I_K: with TTX

I_{Na}: with TTX

I_{Na}: no TTX

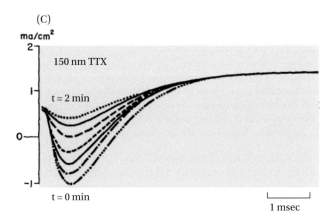

(C)

ma/cm²

150 nm TTX

t = 2 min

t = 0 min

1 msec

Figure 11.5 Tetrodotoxin (TTX) selectively blocks sodium conductance in lobster and squid axons. (A) TTX structure. (B) TTX causes loss of Na^+ current in lobster axons, because it knocks down only the negative-slope region, leaving the K^+ current unaffected. (C) Voltage-clamp records from squid axons show most clearly that TTX inhibits Na^+ currents without affecting K^+ currents. (B, From Narahashi T, Moore JW, Scott WR [1964] *J Gen Physiol* 47: 965–974. With permission from Rockefeller University Press. C, From Moore JW, Blaustein MP, Anderson NC et al. [1967] *J Gen Physiol* 50: 1401–1411. With permission from Rockefeller University Press.)

studies helped propel investigations of channels in excitable tissues. The voltage-clamp methodology of the 1950s had by this time been extended to other species and cell types, such as lobster axons and frog myelinated nerves. Three sets of studies secured the case for ion-specific channels.

John Moore and Toshio Narahashi examined puffer fish tetrodotoxin (TTX, **Figure 11.5A**), which is one of the deadliest neurotoxins known (nM concentrations are lethal). They hypothesized that it inhibited APs. Moore voltage-clamped lobster axons and applied TTX. Based upon the *I–V* curves (Figure 11.5B), he concluded that TTX selectively blocks the sodium conductance in lobster axons. Subsequent voltage-clamp measurements on squid axons with Narahashi provided the most direct and dramatic proof that TTX selectively inhibits Na^+ conductance (Figure 11.5C). The results were not consistent with either diffuse models or the passage of ions through a single all-encompassing channel. Another important finding was that TTX was active only if applied externally; perfusion of TTX into the axon interior had little effect. This observation suggested that the channel has "sidedness."

The case for ion channels was strengthened further by Clay Armstrong, who discovered that quaternary ammonium (QA) compounds, such as tetraethylammonium (TEA), block potassium currents when perfused into the interior of squid axons (**Figure 11.6**). These soluble, positively charged compounds with different hydrophobicities were found to inactivate K^+ currents to different

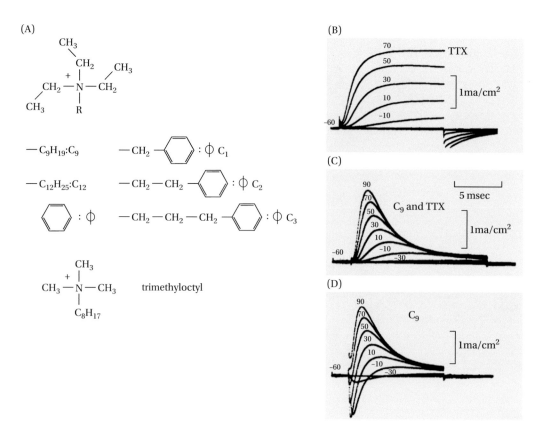

Figure 11.6 Quaternary ammonium (QA) compounds block potassium currents in squid axons when perfused into the axon interior. (A) Quaternary nitrogen compounds that block K+ currents. (B) Voltage-clamp records from a squid axon at different levels of TTX. In this case, only outward K+ currents are observed. (C) The infusion of the C_9 QA causes inactivation in a manner suggesting that, like K+, the ion is swept into a channel. But being too large, and perhaps by interacting hydrophobically with the channel mouth, it becomes lodged in the mouth, thus blocking passage to K+ ions. (D) The C_9 QA causes inactivation in the absence of TTX, showing that C9 does not interfere with the Na+ conductance. Importantly, these data show that the QAs do not interfere strongly with the voltage sensitivity of opening. That is, gating and blocking are separate phenomena. (From Armstrong CM [1971] *J Gen Physiol* 58: 413–437. With permission from Rockefeller University Press.)

extents. Importantly, although the QAs caused inactivation, they did not interfere very much with the voltage sensitivity of the potassium conductance. After extensive studies, Armstrong made a strong case for channels and started to give them structure. He noted in his 1971 paper:

> The impermeability of QA ions and many aspects of their action on G_K are easily explained if one imagines that the outer part of a K+ channel of a squid axon is a gramicidin-like tunnel with a large mouth at its inner end. The mouth is capable of accepting a hydrated K+ ion or a QA ion, but the tunnel can accept only dehydrated K+ ions, and QA ions are too large to enter. If this is so, it may be that two factors are involved in determining the selectivity of a K+ channel. The first would be the ease with which an ion enters the mouth of a channel, and the second factor would be the ease with which it enters the tunnel.

As will become apparent, this description of K+ channels was broadly accurate.

In a third milestone study, Bertil Hille examined the selectivity of sodium conductance, reasoning that if the sodium permeability is due to channels with defined structures, then the channel-pore geometry could be defined by studying the permeability of Na+ channels to organic cations. Nodes of Ranvier are periodic gaps in the insulating myelin sheath wrapping the axons of certain

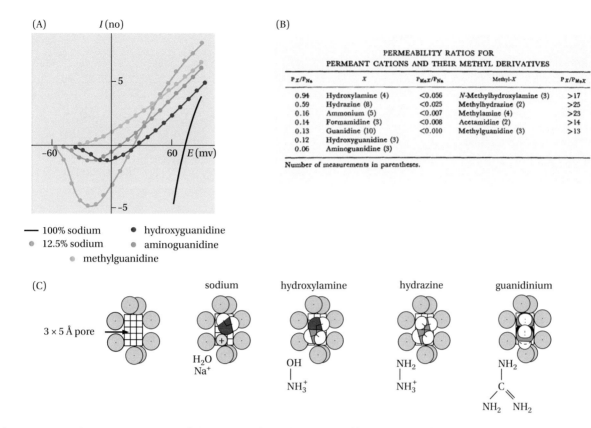

Figure 11.7 Defining the geometry of the sodium channel selectivity filter through relative permeability measurements. (A) *I–V* curves for various sodium ion substitutes. (B) Summary of relative permeabilities of organic cation replacements for Na$^+$. (C) Surmised structure of the Na$^+$ channel sodium selectivity filter based on the shapes of organic cations. These data revealed that sodium channels have a fixed geometry. (From Hille B [1971] *J Gen Physiol* 58: 599–619. With permission from Rockefeller University Press.)

neurons, where many ion channels are concentrated. Rather than squid axons, Hille used voltage-clamped nodes of Ranvier from frog nerves. He examined a wide range of organic substitutes for Na$^+$ using measurements of *I–V* curves and voltage-clamp current measurements. His findings, summarized in **Figure 11.7**, not only strongly supported the idea of a structured channel but suggested geometry and properties of the Na$^+$ selectivity filter (Figure 11.7C). An important property besides geometry, Hille suggested, was the hydrogen bond–forming ability of both the channel and the cation. The 3 × 5 Å pore, he proposed, would admit partially hydrated Na$^+$ ions. These ideas of Armstrong and Hille were prescient in the light of what we now know about the structure of voltage-gated ion channels.

11.1.4 Noise in Voltage-clamp Experiments Suggests Stochastic Opening and Closing of Ion Channels

Conductance measurements of gramicidin single channels in black lipid membranes (Figure 1.19), the discovery of the ability of TTX and QAs to block G_{Na} and G_K selectively, and the characterization of size-selective permeability of G_{Na} and G_K made it seem likely that ions cross membranes through ion-selective channels whose probability of opening and closing depends on the transmembrane potential in a time-dependent manner. Suppose that each channel has an inherent conductance when open and that opening and closing kinetics depend upon channel structure and voltage. What are the implications for the currents measured during a voltage-clamp experiment? The currents should have characteristic noise signatures due to the stochastic opening and closing of channels. The noise signature should be related to the frequency of opening and closing and the amplitude of the open-channel current. Experiments

Figure 11.8 Noise signatures from squid axons reveal the existence of single Na⁺ and K⁺ channels. Broadly, two types of noise are seen. $1/f$ noise is seen in all electrical measurements, including current flow through a simple passive resistor. Shot noise, on the other hand, arises from short-duration current pulses such as those arising from off-on fluctuations of ion channels. These two types of noise can be computed and parameterized. The sum of the two types of noise (solid lines) fit the experimental curves well enough to yield information on the shot noise attributed to channel opening/closing. (From Conti F, Hille B, Neumcke B et al. [1975] *J Physiol* 248: 45–82. With permission from John Wiley and Sons.)

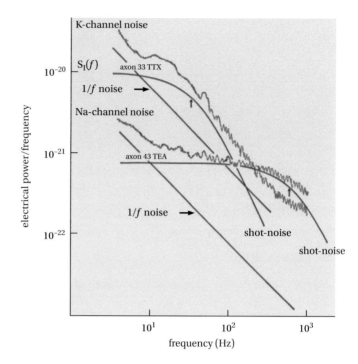

reported by Franco Conti, Louis De Felice, and Enzo Wanke in 1975 revealed exactly such behavior. Furthermore, the noise signature could be modified using TTX and QAs (**Figure 11.8**). These experiments and those of other investigators firmly established the existence of voltage-dependent ion-selective channels. Theoretical analyses of the noise signatures suggested that the conductances of single open Na⁺ and K⁺ channels were roughly $5–10 \times 10^{-12}$ siemens (5–10 pS).

11.1.5 Patch Clamping Reveals Individual Ion Channels

Hodgkin and Huxley's analysis of their voltage-clamp records allowed them to write empirical expressions for the voltage and time dependence of the Na⁺ and K⁺ conductances. Given what was known about membrane and protein structure in the 1950s, they could say very little about the structure of the conductance pathways other than that "sodium movement depends on the distribution of charged particles" in the membrane. A fundamental question was whether the shapes of the conductance curves reflected the properties of each of the molecular-level conductance events or was due to ensembles of channels of fixed conductance that individually open and close stochastically. The noise measurements showed definitively that underlying the macroscopic conductances were ensembles of single channels. But, could an experimental method be developed that would allow the individual channels to be observed?

Erwin Neher and Bert Sakmann accomplished this revolutionary step in 1978 by inventing the patch clamp (**Figure 11.9A**). It so revolutionized ion channel research that Neher and Sakmann were awarded the 1991 Nobel Prize in Physiology or Medicine. The basic principle is to make glass micropipettes with tip diameters of about 10 μm. The tips of the patch pipettes must be carefully fire polished to make them very smooth. After the pipettes are filled with an electrolyte solution, they are manipulated mechanically on to the surface of the cell, and then a slight suction is applied, which seals the pipette to the cell membrane. For this process to work, the cells must be treated with collagenase to remove any connective tissue adhering to the cells. If it is done properly, the leak resistance between patch and cell surface is ~10¹⁰ ohms, while the resistance of the pipette interior is ~10⁷ ohms. This means that the only current entering the pipette is that which passes through the membrane itself. As shown in Figure 11.9A,

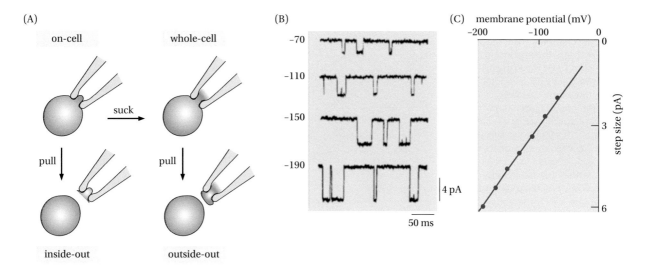

Figure 11.9 Patch-clamp recordings of single ion channels. (A) Preparation of patches. A polished pipette tip makes a seal with a membrane. If the pipette is pulled from the cell, it removes a tiny patch of the membrane with the inside surface of the membrane now facing out. If suction is used to break the membrane to create an opening, events in the whole cell can be observed. If the whole-cell pipette is pulled away from the cell, the membrane may reseal to give a patch with an outside-out topology. (B) Patch-clamp recordings of the acetylcholine receptor show single channels opening and closing, and that they are uniform in conductivity. (C) Each open channel acts like a resistor, i.e., current is proportional to voltage. (B, From Hamill OP, Marty A, Neher E et al. (1981) *Pflügers Arch* 391: 85–100. https://doi.org/10.1007/BF00656997. With permission from Springer Nature.)

there are several ways the pipettes can be used, the most common being pulling a patch of membrane away from the cell, either inside-out or outside-in. This ability makes it possible to examine the effects of pharmacological agents applied to the inside or outside of cells.

The very first single-channel recordings were made from acetylcholine (ACh) receptors in denervated frog pectoris muscles (Figure 11.9B). Square-wave current pulses were recorded as channels open and close. The duration and frequency of the openings are stochastic, as is expected for single channels. Think about that. The patch clamp allowed observations of single channels doing their thing! Although openings and closings of the channels are stochastic, the amplitudes of the currents depend linearly upon the applied potential (Figure 11.9C); once open, the channel behaves as a simple passive resistive element. For the ACh receptors shown in the figure, the conductance is about 30×10^{-12} siemens (30 pS). Single-channel conductances of Na^+ and K^+ channels are typically about 20 pS. However, there are many, many different types of channels, and each has a characteristic conductance that depends upon molecular structure.

Can single-channel recordings of Na^+ and K^+ currents be reconciled with macroscopic voltage-clamp recordings? The two types of recordings can indeed be reconciled through voltage-clamp measurements on single cells and membrane patches (**Figure 11.10**). When a step voltage is applied to a patch containing voltage-gated Na^+ channels, the channels open shortly after the step and then close again due to spontaneous inactivation. A sum of the single-channel currents in time gives exactly the form expected from macroscopic voltage-clamp measurements (Figure 11.10A). K^+ channels under similar conditions open more slowly and do not show spontaneous inactivation, but the ensemble sum gives exactly the expected macroscopic K^+ current (Figure 11.10B).

11.1.6 Cloning and Expression of Sodium Channels Provide the First Glimpse of Channel Structure

Another revolutionary change in ion channel physiology and structure was initiated by Shosaku Numa (1929–1992) who determined the sequence of

Figure 11.10 Single-channel recordings from Na⁺ and K⁺ channels. (A) Na⁺ channels from mouse toe-muscle fibers in response to a voltage step (top) vary (traces). When all the traces are averaged together, the average exactly matches the macroscopic data for opening and spontaneous closing of Na⁺ channels. (B) Similarly, averaging of single K⁺ channel signals from a squid axon reveals slower opening and no spontaneous inactivation, as in the macroscopic observation that they close only when the voltage step is reversed.

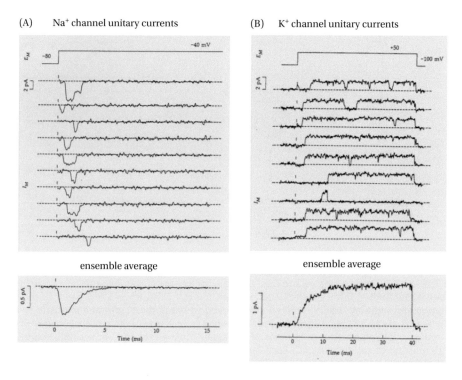

a sodium channel from the electric eel *Electrophorus electricus* by molecular cloning methods. A hydropathy-plot analysis revealed a protein with four internal repeats, assumed to be the principal structural components of the channel (**Figure 11.11A**). Subsequent studies showed that sodium channels could also be cloned from rat brains. Numa's analysis of the brain and *Electrophorus* sodium channel sequences suggested that each internal repeat contained six transmembrane helices (Figure 11.11B), which was a rather daring proposal because transmembrane (TM) helix 4 is highly positively charged due to the presence of several arginine residues. Numa proposed that helix 4 must be involved in sensing transmembrane potential and channel gating. To prove that he had really cloned sodium channels, he injected sodium channel mRNA into frog eggs, which then expressed the protein and inserted it into the egg plasma membrane. *I–V* curves (Figure 11.11C) and voltage-clamp measurements (Figure 11.11E) suggested the presence of sodium channels, which are not naturally present in frog eggs. The crowning experiment was the demonstration that the channels could be inhibited by TTX (Figure 11.11D).

We learn several things from these and the voltage-clamp experiments (see Figure 11.3). First, during the AP, there is initially an inward flow of Na⁺ that moves the membrane potential toward positive values, followed by an outward flow of K⁺ that moves the membrane potential back toward negative values. In the resting state, the membrane potential is close to the negative K⁺ equilibrium potential, whereas at the height of the APs, the membrane potential is close to the positive Na⁺ equilibrium potential. Second, and very importantly, over the full range of the APs, the net passive movement of K⁺ is *always* outward, while the net passive movement of Na⁺ is always *inward*; what Na⁺ takes away, K⁺ gives back. Third, the increase in Na⁺ current (conductance) is transient; without any intervention, it decreases spontaneously. This decrease is called sodium inactivation. Fourth, the K⁺ conductance remains high as long as the membrane is kept depolarized; there is no potassium-conductance inactivation (this is not true for all potassium channels, however). Fifth, the voltage-dependent changes in K⁺ conductance are slower than the changes for Na⁺ channels. For this reason, the squid axon K⁺ channels are referred to as delayed rectifiers (a rectifier in electric circuit theory is a conductor whose conductance is voltage dependent; see Figure 11.4). The refractory period, during which the Na⁺ channel cannot open again, prevents the nerve impulse from traveling backward—the only

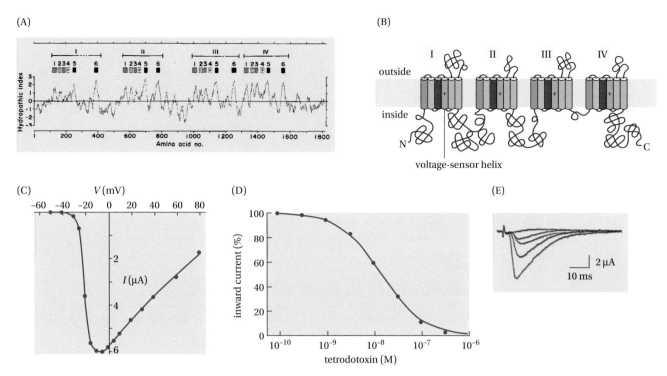

Figure 11.11 Cloning and expression of sodium channels. (A) Hydropathy plot of a sodium channel cloned from *Electrophorus electricus* shows the existence of four internal repeats. (B) Proposed secondary structure of the internal repeats of a sodium channel cloned from rat brains. Helix 4, containing 4 or more arginine residues, was suggested to be involved in sensing transmembrane potentials. (C) *I–V* curve for sodium channels expressed in frog eggs showing the expected behavior of voltage-sensitive channels (c.f., Figure 11.4). (D) The inward sodium current of the expressed sodium channels in frog eggs is sensitive to TTX, proving that the expressed protein is that of a sodium channel. (E) Inward sodium currents determined by voltage clamping are inhibited by TTX. (A, From Noda M et al. [1984] *Nature* 312: 121–127. With permission from Springer Nature. B, D, From Noda M et al. [1986] *Nature* 322: 826–828. With permission from Springer Nature.)

adjacent channels that can be stimulated by a given channel are those that are at their resting potential.

11.1.7 Sodium, Potassium, and Calcium Channels Share Common Structural Motifs and Are Found in All Kingdoms of Life

The general approach of Numa and his colleagues has become a standard method for identifying ion channels in a wide variety of organisms. Indeed, sodium and potassium channels are found in all kingdoms of life (**Figure 11.12**). Sequence analyses and structural studies (see later) show that an ion-selective tetrameric pore domain is common to all the channels (Figure 11.12A). A domain monomer is composed of two TM helices connected by an amino acid sequence that forms the lining of the pore when four of the monomers assemble into a tetramer to form the pore, which can be highly selective for sodium, potassium, or calcium, although some channels show only general selectivity for all cations. Voltage-sensitive channels are defined additionally by voltage-sensor domains (VSDs) with four TM helices, the fourth of which contains four or more arginine residues. As Numa suggested, the fourth helix is responsible for sensing transmembrane voltage. But, the helix works in concert with the other three helices to form the VSD, which can act on the pore domain to control its opening and closing. Some channels, lacking VSDs, are controlled by chemical ligands. In other cases, the VSD itself can act as a channel for protons. The voltage-sensor domain can also work in concert with enzymes to make them voltage sensitive. Evolution seems to have tried every possible combination of pore and sensor domains (Figure 11.12A). The importance of ion

(A)

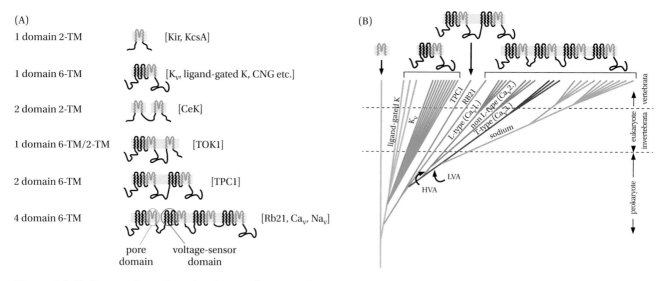

1 domain 2-TM   [Kir, KcsA]

1 domain 6-TM   [K_v, ligand-gated K, CNG etc.]

2 domain 2-TM   [CeK]

1 domain 6-TM/2-TM   [TOK1]

2 domain 6-TM   [TPC1]

4 domain 6-TM   [Rb21, Ca_v, Na_v]

pore   voltage-sensor
domain   domain

(B)

Figure 11.12 Structural motifs and evolution of cationic channels. (A) Structural motifs showing the variety of ways that cation-selective channels can be constructed from pore domains (light blue) and voltage-sensor domains (black and red). (B) A scheme for the evolution of cation channels. (From Anderson APV, Greenberg RM [2001] *Comp Biochem Physiology B* 129: 17–28. With permission from Elsevier.)

channels to human life is evident from the existence of more than 400 genes for ion channels in the human genome, many (perhaps all) of which have disease phenotypes when they fail to function properly.

These and other elegant experiments defined the tasks that an ion channel must accomplish, and they are summarized in **Box 11.4**. For most of the 20th century, physiologists struggled to understand how a protein channel might accomplish this set of objectives. As we have seen, progress was slow (but steady) until the determination of the KcsA crystal structure provided the first chemical view of channels. The amazing progress in ion channel structural biology since the KcsA structure has led to a revolution in our understanding of channels at the chemical level.

11.2 POTASSIUM CHANNELS ARE HIGHLY SELECTIVE AND PERMIT DIFFUSION-LIMITED ION TRANSPORT

Potassium channels are the most common type of ion channel and are found in most cells, where they control many cell functions. In excitable cells, such as neurons, they shape APs and set the resting membrane potential. They also regulate cellular processes such as the secretion of hormones (e.g., insulin release from pancreatic beta cells). An important example involves the transmission of signals along nerve axons. As we have seen, sodium channels open in response to voltage changes, and potassium channels subsequently open in response to the potential changes resulting from the sodium ion movement, allowing potassium ions to move across the membrane at nearly their diffusion rate in water and restoring the potential to its resting state after the nerve impulse has passed (Box 11.3, Figure 4).

X-ray crystallography has revolutionized many fields by providing models that describe proteins at the level of chemistry, enabling interpretation of function in atomic mechanistic terms. Because of the difficulty of obtaining suitable crystals, applications to membrane proteins have lagged behind studies of soluble proteins, but the pace has been accelerating; many structures are now in hand, and advances in electron microscopy have further

BOX 11.4   ION CHANNEL REQUIREMENTS

As discussed in Chapter 2, the barrier for an ion to cross a membrane lipid bilayer is large, and more so for positive than for negative ions. It is interesting to speculate that the use of positive ions for many signaling events might have arisen from the greater barrier that the bilayer presents for a positive charge versus a negative charge to cross spontaneously. Despite the great variety of cation channels, they share many structural features and common functions that are adapted to specific requirements evolution has imposed on them.

1. Ion Dehydration

Ions in solution are hydrated by tightly bound shells of water molecules; the free energies of hydration are inversely related to diameter, because the radial electric field at the surfaces of small ions is greater compared with larger ions. The amount of energy required to strip all the waters from any ion is immense. This high cost means either that ions pass through channels partially hydrated or that the channel structure can satisfy the hydration requirement without incurring a high energy penalty. For some channels, particularly the K^+-selective ones, the channel dehydrates the ion as part of the selectivity process, substituting elements of protein structure for the hydration water, although others, such as Na^+ and Ca^{++} channels, operate by moving hydrated ions.

Ion	Radius (Å)	Dehydration energy (kcal/mole)
Li^+	0.69	−113
Na^+	1.02	−87
K^+	1.38	−70
Rb^+	1.49	−66
Cs^+	1.70	−60
Ca^{++}	1.00	−359

(Data from Marcus, Y. [1991] *J. Chem. Soc. Faraday Trans.* 87: 2995–2999.)

2. Select a Specific Ion Type

Channels often distinguish kinds of ions, for example, allowing potassium through the channel while excluding sodium. This is accomplished by a "selectivity filter" in the channel structure.

3. Lower the Energy Barrier

The barrier for an ion to cross a bilayer can be very high (tens of kilocalories per mole), so the protein must act to lower the barrier in order to facilitate transport on useful timescales. The barrier-lowering function of channels arises from their atomic structure.

4. Allow High Ion Flux

Channels must allow the rapid passage of large numbers of ions per second (flux) while maintaining selectivity. This is a challenging trick, which is related to dehydration and barrier lowering.

5. Control the Opening and Closing of the Channel

Barrier control is an essential feature of ion channels. For ion channels, barrier control means control of channel opening and closing, called gating, in response to a stimulus. A gate is simply an energy barrier that prevents flow through the channel in the closed state and allows ions to flow in the open state. There are three basic kinds of gating stimuli that do this: binding a ligand, subjecting the channel to a mechanical force, or changing the voltage across the membrane; hence the names "ligand-gated," "mechanosensitive," and "voltage-gated" channels.

---

accelerated the pace of discovery. The study of channels has derived a great leap in understanding from the crystallographic work, but as with most areas of science, a gain in understanding raises new and different questions for further work.

Seldom has so much been learned from a single crystal structure as from the first example of a potassium channel transmembrane pore domain determined at 3.2 Å resolution by the laboratory of Roderick MacKinnon, who shared the 2005 Nobel Prize in Chemistry for his work on channel structures. The breakthrough structure reported by MacKinnon and his colleagues was that of the membrane-embedded part of a pH-gated potassium channel from the Gram-positive actinobacterium *Streptomyces lividans* (KcsA, **Figure 11.13**). Many potassium channels are ligand-gated; for KcsA, the gating ligands are protons. The pore of the channel was readily identified by the presence of three potassium ions, two in the selectivity filter and

(A)

(B)

(C)

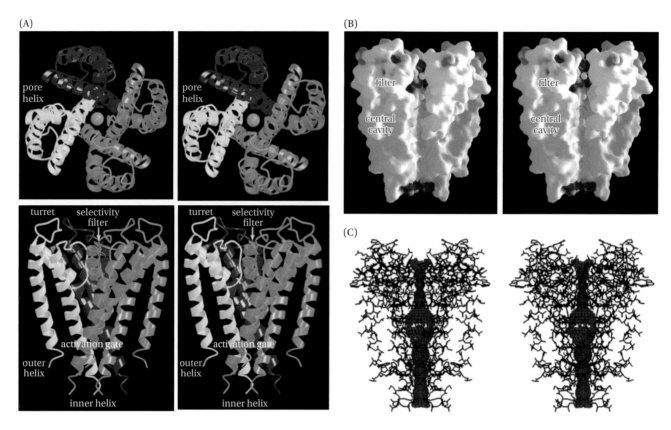

Figure 11.13 Ribbon and space-filling views of the transmembrane pore domain of the tetrameric KcsA potassium channel (PDB 1BL8). All images are cross-eye stereo pairs. (A) Channel viewed from outside the cell (top) and along the membrane plane (bottom). The green sphere in the upper panel shows the location of potassium ions. (B) Cut-away space-filling view of KcsA showing two potassium ions (green) in the selectivity filter and one in the channel's central cavity, corresponding to the midplane of the membrane. Blue regions are positively charged, and red regions are negatively charged. Hydrophobic side chains are indicated in yellow. (C) The entire internal pore of the channel shown as a red surface. Below the selectivity filter, the pore expands, creating a central cavity or vestibule. QA ions can enter the vestibule but not the pore, thus blocking the passage of ions through the selectivity filter. This pore-vestibule structural scheme was proposed by Clay Armstrong long before the structure was determined by MacKinnon. (Doyle et al. [1998] *Science* 280: 69–77. With permission from AAAS.)

one in the central cavity (Figure 11.13B). Because the channel is not voltage sensitive, the structure did not reveal how voltage-gated channels function. However, it did reveal, in concert with site-directed mutagenesis studies, that channel opening/closing must be determined by the conformation of the inner helices that form the so-called activation gate (Figure 11.13A), which was closed in the KcsA structure.

The structure showed that the channel pore is formed by the association of four identical subunits, each of which has two inner TM helices forming a conical structure, a short "pore helix" inserted in the wide end of the cone, and an extended polypeptide associated with the short helix (Figure 11.13). How do the features of this structure accomplish the channel functions?

11.2.1 Dehydration and Ion Selectivity Result from the Properties of the Selectivity Filter

Dehydration and ion selectivity result from the properties of the selectivity filter, formed by the four short, extended polypeptides associated with the pore helices (**Figure 11.14**). The selectivity filter was first identified in the sequence by mutational analysis, determining which residues in the amino acid sequence caused loss of selectivity when mutated. A five–amino acid sequence, typically TVGYG, was found to be a nearly universal feature of potassium-selective channels (**Figure 11.15**). The 3.2 Å structure revealed two potassium ions in complex with four extended chains of the filter that form the narrow selectivity filter of the pore. The remarkable feature of the filter is that it is not the side-chains that face the ions, but the extended backbones!

11.2.2 Cyclic Peptide Ionophores hint at the Selectivity Mechanism of Potassium Channels

Many investigators had speculated over the years that potassium selectivity was due to interactions with carbonyl groups that served as stand-ins for water oxygens. The speculations were based on the selectivity of the natural cyclic ionophores valinomycin and nonactin, which bind K^+ in cages formed by carbonyl groups (**Figure 11.16A**). Because the outer rims of these iono-phores are formed by non-polar side chains, the ionophore can solubilize ions in non-polar environments, allowing them to be transported across black lipid membranes by the application of electric fields. These iono-phores, which are uncharged without bound ions, have a very strong prefer-ence for K^+ compared with Na^+.

In crystallographic structures of valinomycin and nonactin, K^+ is coor-dinated by the carbonyl groups acting as substitutes for water oxygens. For valinomycin, the coordination is by six carbonyls in an octahedral configura-tion, whereas for nonactin, the coordination number is eight. Why are these ionophores so selective for K^+? Molecular dynamics simulations of Li^+, Na^+, and K^+ in water show how the ions are hydrated, i.e., how they are coordi-nated by water. The number of coordinating waters is probabilistic (Figure 11.16B); there is not a fixed number but rather, a range of numbers, meaning that ions will exist in solution in an ensemble of coordination states. Most sodium ions will be coordinated with five or six waters. The larger K^+, on the other hand, is more forgiving; the ensemble will have potassium ions with six, seven, or eight coordinating waters. The energetics of moving Na^+ or K^+ from water into carbonyl coordination in the ionophore are complicated to model and compute, but experimentally, valinomycin and nonactin clearly prefer to coordinate K^+ rather than Na^+, aided, perhaps, by the larger radius and more forgiving coordination requirements of K^+.

Considering the huge hydration energies of ions (Box 11.4), one might think that the energetic difference between an ionophore binding K^+ and binding Na^+ might be equally huge. How big must the difference be to explain the selec-tivity? The answer is provided by the Boltzmann function (Equation 0.29). Valinomycin has a selectivity of 10^4 for K^+ over Na^+, meaning that the probability of Na^+ being caged relative to K^+ is 10^{-4}, corresponding to an unfavorable bind-ing free energy of about +7 kcal mol^{-1} for Na^+ compared with K^+. This is a rather small number compared with the hydration energies of the ions. Precise num-bers for selectivity are difficult to obtain computationally, because one must compute small differences between large numbers. What about the selectivity of potassium channels? Selectivity for K^+ over Na^+ can be as high as 10^3, which translates into an unfavorable binding free energy of +4 kcal mol^{-1} for Na^+ com-pared with K^+.

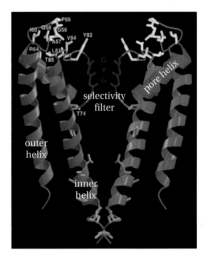

Figure 11.14 Structure of the KcsA selectivity filter. Two subunits have been removed to reveal the selectiv-ity filter backbone carbonyl groups (red) that substitute for water oxygens. The white residues interact with toxins that can block access to the filter from the outside. The orange residues affect interactions of tetraethylammonium from inside the pore. Site-directed mutagenesis revealed the essential features of the channel before the structure was determined. Replacement of the green residues with cysteines allowed cysteine-reactive reagents to bind to the channel whether open or closed, whereas replacement of the purple residues allowed binding only when the channel was open. (Doyle et al. (1998) *Science* 280: 69–77. With permission from AAAS.)

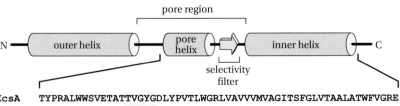

```
KcsA    TYPRALWWSVETATTVGYGDLYPVTLWGRLVAVVVMVAGITSFGLVTAALATWFVGRE
KCNA1   SIPDAFWWAVVSMTTVGYGDMYPVTIGGKIVGSLCAIAGVLTIALPVPVIVSNFNYFY
KCNSK   SIPASFWFVLVTMTTVGYGDLVPLSPFGKVVGGMCAMIGVLTLALPVPIIVANFKHFY
```

Figure 11.15 Sequence alignment of several ligand-gated potassium channels that shows the high con-servation of the GYG motif of the potassium selectivity filter. KcsA is from the bacterium *Streptomyces lividans*, KCNA1 is a human K^+ chan-nel, and KCNSK is from the worm *Caenorhabditis elegans*. (Doyle et al. (1998) *Science* 280: 69–77. With permission from AAAS.)

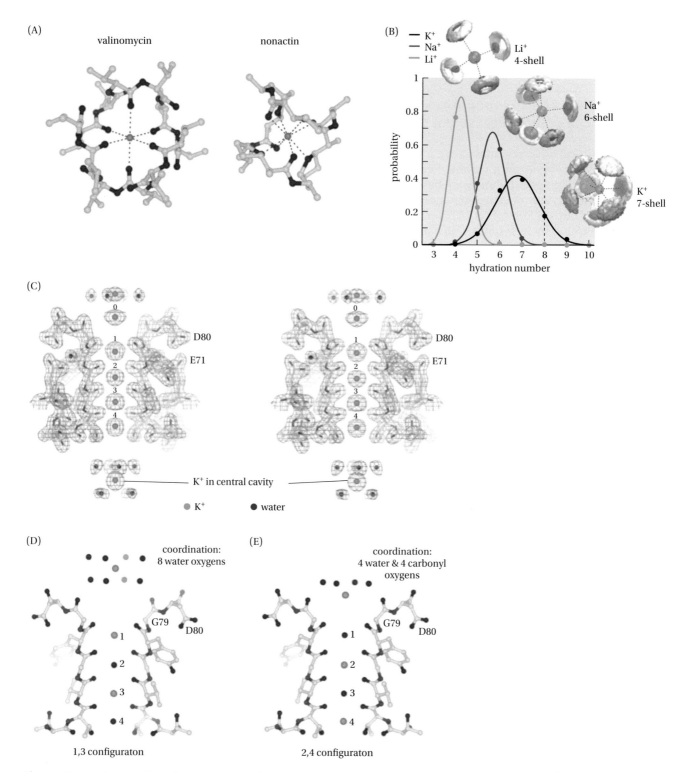

Figure 11.16 Coordination of potassium ions by carbonyl groups in ionophores and the KcsA selectivity filter. (A) Structures of the ionophores valinomycin and nonactin with bound potassium ions. In valinomycin, K+ is coordinated by six carbonyls, whereas in nonactin, it is coordinated by eight. (B) Hydration shells of ions apparently do not involve a fixed number of waters. Rather, based on molecular dynamics simulations, there are ensembles of ions with different numbers of waters. (C) Locations and oxygen coordination of K+ observed in high-resolution structures of KcsA. Not all the sites in the filter are occupied 100% of the time; typically, the occupancy for each site is about 50%. Sites not occupied by K+ are occupied by water. (D) Locations of ions and waters for channels in the 1,3-configuration. (E) Locations of ions and waters for channels in the 2,4-configuration. (A, From Morais-Cabral JH et al. [2001] *Nature* 414: 37–42. With permission from Springer Nature. B, From Bostick DL, Brooks CL III [2007] *PNAS* 104: 9260–9265. With permission from CL Brooks III. C–E, From Zhou Y et al. [2001] *Nature* 414: 43–48. With permission from Springer Nature.)

11.2.3 The Carbonyl Groups of the Filter Determine Selectivity

The 3.2 Å resolution of the first KcsA structure was too low to obtain definitive information about the coordination of K+ by the filter's carbonyl groups. A subsequent high-resolution (2.0 Å) structure revealed in marvelous detail where and how K+ ions are coordinated by carbonyl and water oxygens (Figure 11.16C). Look first at the ions located in the filter (numbered 0 through 4). The stereo view of panel C might suggest that the filter has four bound K+ ions. But, that is not correct. Any particular filter in the crystallographic structure really has only two ions, located either at positions 1 and 3 or at positions 2 and 4. What we see in the figure is an average structure with each position occupied about 50% of the time. For channels in the 1,3-configuration, waters occupy the 2,4-positions (Figure 11.16D). For channels in the 2,4-configuration, waters occupy the 1,3-positions (Figure 11.16E). At the extracellular entrance to the channel (labeled 0), there are two positions for K+ (two partial-occupancy ions are seen). Ions at these locations are poised to enter or leave the channel. Finally, in the central cavity, there is a single K+ surrounded by eight waters in one of the coordination configurations expected from the simulation studies (Figure 11.16B). This ion is poised to enter or leave the channel. The key idea that emerges from the structure is that potassium ions pass through the channel by exchanging the water-oxygen partial charges for the negative partial charges of the filter's carbonyl oxygens. Noting that in the figure only two KcsA monomers of the homo-tetramer are shown for clarity, throughout the ion path, each ion is always coordinated by eight oxygens from either water or backbone carbonyls.

The structure suggests that the filter may be rigidly organized to allow a "snug fit" with the backbone carbonyls. But, this view must take into account the motions arising from thermal energy. As expected, molecular dynamics simulations suggest that the carbonyls have dynamic motion. When free energies are calculated from molecular dynamics trajectories taking into account the electrostatic properties of dynamic carbonyl groups, energy differences of about the right magnitude are seen (**Figure 11.17**).

11.2.4 The Energy Barrier for Ion Passage Is Lowered by Helix-end Polarity and a Water Chamber

We have seen how the structure accounts for dehydration and selectivity of the channel, but how is the barrier lowered for an ion to cross the membrane? The constrained water found in the middle of the central cavity provides a clue that leads to an important principle: the protein creates a water-filled chamber for an ion in the middle of the membrane and at the same time, a favorable electrostatic environment.

α-helices have delocalized partial charges at their ends as a result of the systematic orientation of the backbone peptide bonds, with about half a negative charge at the C-terminal end and half a positive charge at the N-terminal end of a helix. These partial charges are delocalized, with relatively large Born radii. This effect is sometimes referred to as a "helix dipole," but, in reality, it is seen as two separate regions of charge at the distances found within proteins. The four short pore helices (Figure 11.14) point their C-terminal ends toward the central cavity so that a significant region of negative charge is created, providing a favorable environment for a positively charged ion (**Figure 11.18**). The cavity formed by the channel tetramer provides space for water molecules to hydrate potassium ions, substituting hydration for the carbonyl groups in the selectivity pore and thus allowing movement of ions into or out of the pore. By providing a

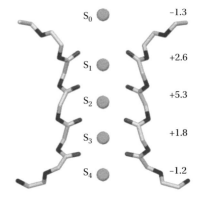

S_0 −1.3

S_1 +2.6

S_2 +5.3

S_3 +1.8

S_4 −1.2

Figure 11.17 Filter selectivity. Unfavorable free energies for replacing K+ by Na+ calculated from molecular dynamics simulations. The free energy costs (in kcal mol^{-1}) of the replacements for each position along the filter are shown on the right-hand side of the figure. (From Noskov SY et al. [2004] *Nature* 431: 830–834.)

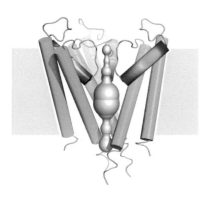

Figure 11.18 Partial charges at the ends of oriented pore helices and a central hydration chamber overcome the membrane barrier for entry of potassium ions. Three of the four pore helices are shown, with their negatively charged ends (red) pointed toward the central chamber. These charges can stabilize a potassium ion in the center of the channel prior to entering the filter, where it is dehydrated for passage through the channel.

favorable environment at the center of the membrane, the barrier to passing an ion across is greatly lowered.

11.2.5 High Ion Flux Is Facilitated by Cooperative Ion Movement

The remaining property of the KcsA transmembrane region that was understood from the structure is how the channel manages to allow the passage of ions at nearly their diffusion limit. Under physiological conditions, about 10^8 ions flow through a channel each second! The positions of ions and waters in the two configurations provide the answer (Figure 11.16D, E). In the structure, as discussed earlier, the five positions, 0 through 4, are not fully occupied, and it is clear from steric and electrostatic considerations that potassium ions do not fit in all five positions at the same time: only configurations in which the cations are separated by one water molecule are allowed. So, there are groupings of occupied sites that are allowed, and ion movement through the channel takes place by hopping transitions of the K^+ between stable multi-ion configurations (**Figure 11.19**). Without ion flow, two states (1,3 and 2,4) are possible that are energetically equivalent, based on the equal occupancy of the positions in the structure. The K^+ ions in these states are coordinated by eight oxygens (as in nonactin). During ion flow, however, the situation changes, because a third K^+ must enter the filter (configuration in middle of Figure 11.19), which changes the coordination of K^+ to six oxygens (as in valinomycin). The three-ion state was not seen in the structure, and it is hypothesized that it is somewhat higher in energy than the doubly occupied states as a result of electrostatic repulsion between ions and also as a result of the necessary dehydration and rehydration of entering and exiting ions; thus, the three-ion state can be viewed as a transition state for the throughput cycle.

Figure 11.19 Coordination and motion of channel K^+ ions. The selectivity filter is depicted as five sets of four in-plane oxygen atoms (the top is outside the cell), with K^+ ions and water molecules shown as green and red spheres, respectively. K^+ ions undergo coordination by eight oxygen atoms when in the 1,3 and 2,4 configurations. Movement would involve octahedral coordination by six oxygen atoms, two provided by the intervening water molecules. (From Morais-Cabral JH. et al. [2001] *Nature* 414: 37–42. With permission from Springer Nature.)

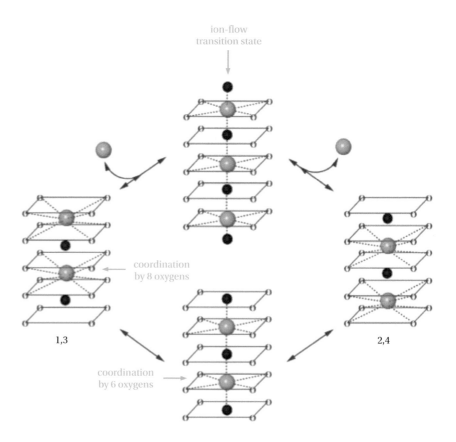

11.3 LIGANDS CAN CONTROL THE OPENING AND CLOSING OF CHANNELS

How are the fluxes through channels controlled? Two of the major themes in all of biochemistry are acceleration and control of chemical processes. Enzymes, channels, other proteins, and folded RNA chains (ribozymes) facilitate the rates of reactions, while allosteric, modification, competition, and gating events control them. In the case of channel gating, there has been an astonishing growth in our understanding of control by ligand binding, mechanical stress, and voltage changes. How ligands can control gating of potassium channels is a useful starting point. To illustrate some general principles, we place emphasis on the ligand-controlled gating of MthK channels because of the insights it provides into voltage-gated channels. But there is a wide variety of ligand-gated channels, such as transient receptor potential (TRP) channels, which are involved in a large number of sensory functions (**Box 11.5**). Another ligand-gated channel of exceptional interest is the nicotinic ACh receptor channel involved in important processes such as linking nerve impulses to muscular contraction (**Box 11.6**).

11.3.1 Gating Energy Determines Regulatory Specificity

The energy difference between the open and closed states of the gate in the absence of ligand describes the extent of regulation, in the sense that the spontaneous distribution of open and closed states is set by this energy. The extent to which the channel is subject to spontaneous opening is given by $\Delta G = -RT \ln([\text{open}]/[\text{closed}])$ at equilibrium with no ligand stimulus. The binding of the triggering ligand must provide this energy to open the channel. Biological systems tend not to set limits that they do not need, so channels that must be tightly regulated (such as the ion channels in the nervous system) will usually have large gating energies, and less critical channels will have lower ones.

11.3.2 Gating by Ligand Binding Can Pull a Channel Open by Lateral Forces

Soon after the seminal discovery of the structure of the KcsA transmembrane region, the structure of a second potassium channel was solved by MacKinnon and his colleagues, the calcium-gated channel MthK from *Methanobacterium thermoautotrophicum*. Calcium is often used by cells to control important processes, because it has a very limited ability to cross a bilayer (it is both small and doubly charged) and so can be maintained at low concentration in the cytoplasm by means of sequestering machineries. Opening a channel can let Ca^{++} ions into the cell, and they in turn can activate other processes—here, the opening of a K^+ channel. The KcsA structure was assumed to be in a closed state because the opening at the activation gate formed by the inner helices (Figure 11.14A) was found to be too small for hydrated-ion passage. Recall that the KcsA structure consists only of the transmembrane part of the proton-gated channel, the rest having been removed to allow crystallization. In the case of MthK, the structure is of the entire channel, including the cytoplasmic domains that control the gate. The general scheme for ligand gating of ion channels involves motions of the cytoplasmic domain that open the channel in response to ligands (**Figure 11.20A**). For the MthK channel, the cytoplasmic domain consists of eight so-called RCK (Regulators of K^+ conductance) domains that form a gating ring (Figure 11.20B).

Like KcsA, the channel is a tetramer, and the upper membrane-embedded pore domain seen in Figure 11.20B is closely homologous to the KcsA pore structure. The space-filling structure of MthK (Figure 11.20C) shows that the gating

BOX 11.5 TRANSIENT RECEPTOR POTENTIAL (TRP) CHANNELS

TRP channels contribute to a broad spectrum of sensory functions, such as sensing heat, cold, irritant agents, inflammatory molecules, pH, osmolality changes, and UVB radiation. In humans, there are TRP channels from seven related families. Their name derives from the discovery of a *Drosophila* mutant that had a vision defect characterized by a TRP. Subsequent investigations revealed that the visual defect is caused by a gene encoding a voltage-gated cation channel. TRP channels are under intensive study because of their role in pain sensation and are only moderately gated by voltage, as it turned out. The classic

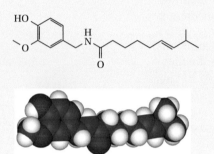

Figure 1 Structure of capsaicin. Note that it is hydrophobic and can readily cross a membrane.

TRP channel example is the temperature receptor, which is also gated by capsaicin (**Figure 1**), the principal ingredient of hot chili peppers. The capsaicin receptor (also called the vanilloid receptor 1 or TRPV1) selectively activates sensory neurons, yielding a burning sensation. Capsaicin or high temperature activates TRPV1 channels to conduct Na^+ and Ca^{++}. The channel activity is modulated by lipids or low pH (**Figure 2**). Interestingly, capsaicin acts by penetrating the membrane and binding on the cytoplasmic surface of the receptor, explaining the delay in heat sensation when it is ingested as well as the difficulty of removing it. David Julius received the Nobel Prize in in Medicine or Physiology in 2021 for his discovery of TRPV1.

TRPV1 has six transmembrane domains and a short, pore-forming hydrophobic stretch between the fifth and sixth TM domains, which form a pore domain like those of many other ion channels (**Figure 3**). The channel is activated by heat (T > 43 °C), acids, and various lipids as well as by capsaicin. Similarly to MthK channels, TRP channels have a N-terminal cytoplasmic domain responsible for ligand gating. TRP channels were hard to purify and study, but the field has been revolutionized by single-particle cryo-electron microscopy studies of the TRPV1 channel. The current (as of this writing) status is illustrated in **Figure 4**.

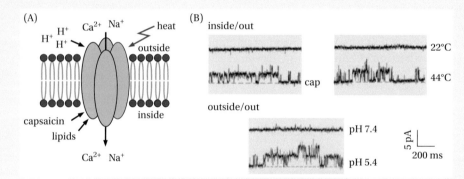

Figure 2 Channel activity of the TRPV1 receptor. (A) Cartoon representation of a tetrameric channel summarizing its cation selectivity for Na^+ and Ca^{++} and its sensitivity to various ligands. (B) Single-channel recordings that demonstrate the activation of TRPV1 by capsaicin (cap, 100 nM) or temperature (44 °C). The channel activity is modulated by lipids and extracellular protons. (From Tominaga M, Tominaga T [2005] *Pflügers Arch Eur J Physiol* 451: 143–150. With permission from Springer Nature.)

(Continued)

ring is connected to the pore domain by four polypeptide strands. Comparing the KcsA and MthK structures is informative (**Figure 11.21**). While the selectivity pore and the central chamber are similar, the bottom of the MthK channel is open, as if it had been pulled apart by the strands of polypeptide connected to the gating domains. The MthK structure apparently represents a channel in the open state, supporting the idea that inner helices serve as activation gates (Figure 11.21A). In the open state, the helices bend outward at Gly83, corresponding to Gly99 in KcsA, suggesting a hinge-like motion. Glycines are often found at points of flexibility in proteins because their lack of side chains allows more conformational freedom. Because this gating-hinge glycine is highly conserved phylogenetically (Figure 11.21B), it appears to be central to gating in all

BOX 11.5   TRANSIENT RECEPTOR POTENTIAL (TRP) CHANNELS (CONTINUED)

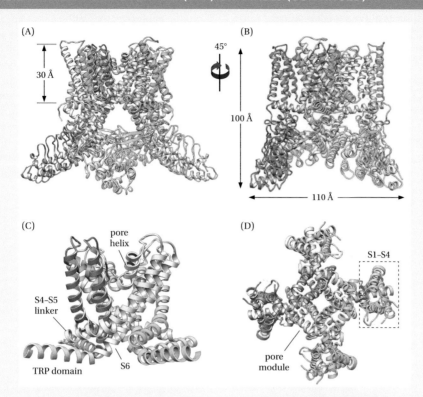

Figure 3 TRPV1 atomic model with each of the four identical subunits color-coded, showing views from two sides (A and B; PDB 3J5P). The approximate location of the ~30 Å-thick membrane is shown in (A). (C) Ribbon diagram focusing in on side view of S5-P-S6 pore with TRP domains. (D) Bottom view of model, focusing on the transmembrane core. Note that the S1–S4 helices flank and interact with the S5–P–S6 pore helices from adjacent subunit, facilitating cooperativity. (From Liao M et al. [2013] *Nature* 504: 197–112. With permission from Springer Nature.)

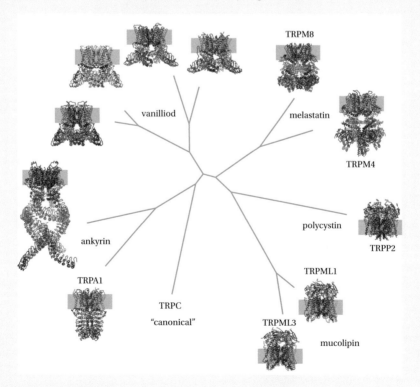

Figure 4 Phylogenetic similarity of TRP channels with well-resolved EM maps. In each image, the grey rectangle indicates the approximate position of the membrane.

BOX 11.6   NICOTINIC ACETYLCHOLINE RECEPTOR (AChR)

The ACh receptor connects nerve signals to muscle contraction. When an AP reaches a nerve/muscle synapse, it transmits the nerve signal by opening a large cation channel in the muscle endplate, triggering the release of Ca^{++} from the sarcoplasmic reticulum and stimulating the contraction process. The response is mediated by the release of ACh from synaptic vesicles that fuse with the membrane of a motor neuron's presynaptic terminal. The ACh diffuses across the synaptic cleft to bind with and activate the ACh receptors on the postsynaptic muscle membrane. The channel is typically comprised of two α-subunits and single copies of the β, γ, and δ subunits, but the subunit composition varies among different tissues. The highly homologous subunits, each with four TM helices, form a fivefold-symmetric pore (**Figure 1A**). A proton-activated channel from the bacterium *Erwinia chrysanthemi* has a very similar structure (Figure 1B) and provides a good model system for understanding ion permeation mechanisms. ACh receptors are members of the Cys-loop receptors, named for the characteristic Cys-containing loop formed by a disulfide bridge that joins the extracellular β-sheet-rich receptor domain to the helical TM domain (**Figure 2**). This loop is important for coupling ACh binding to channel gating.

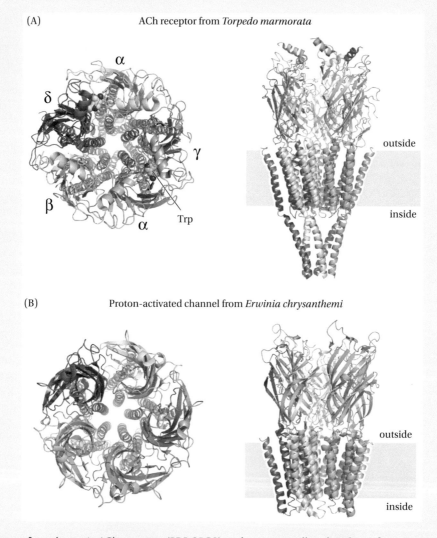

(A)   ACh receptor from *Torpedo marmorata*

(B)   Proton-activated channel from *Erwinia chrysanthemi*

Figure 1 Structures of a eukaryotic ACh receptor (PDB 2BG9) and a structurally related one from Gram-negative bacteria (PDB 3EHZ). (A) The structure, seen at 4 Å resolution by cryo-electron microscopy, shows that the α-subunits, which bind the acetylcholine, have slightly different organization than the other subunits. A tryptophan, shown in spacefill, marks the ACh binding site. The vestibule, formed by the five subunits, has a negatively charged interior that favors the concentration of cations for passage through the channel. (B) The overall structure of the proton-activated channel from *Erwinia chrysanthemi* is remarkably similar to that of the ACh receptor. However, this channel is a homo-pentamer. (A, From Unwin N [2005] *J Mol Biol* 346: 967–989. With permission from Elsevier.)

(Continued)

BOX 11.6   NICOTINIC ACETYLCHOLINE RECEPTOR (AChR) (CONTINUED)

ACh binds to the α-subunits, which opens the channel. The channels lack the high selectivity of K^+ and Na^+ channels; rather, they admit both ions (cation selective). The influx of cations causes local depolarization of the muscle postsynaptic membrane and consequently, initiation of muscle APs that propagate along individual muscle fibers in the manner of nerve APs. These APs trigger muscle contraction by depolarizing the muscle T-tubules, which initiate calcium release from the fiber's endoplasmic reticulum (called the sarcoplasmic reticulum in muscle), and the Ca^{++} ions activate contraction.

Clever kinetic trapping experiments, illustrated in **Figure 3**, have suggested how the gating works. When ACh binds to the α-subunits, it initiates rotations of the protein chains on opposite sides of the entrance to the membrane-spanning pore. These rotations are communicated to the pore-lining α-helices and open the gate—a constricting hydrophobic girdle at the middle of the membrane—by breaking it apart. The movements are small and involve energetically favorable displacements parallel to the membrane plane.

The channel-opening event at the synapse, and the experiment to capture this event *in vitro*, are shown schematically in Figure 3. In the experiment (panel A), ACh-containing droplets are sprayed onto an aqueous film containing isolated postsynaptic membranes. Rapid freezing traps the channels in the open state. Ferritin marker particles, included in the spray solution, allow the electron microscope to reveal the regions of the film where the droplets have impinged and spread, and so identify the membranes containing open channels. Using this method, Nigel Unwin was able to visualize the open structure and could thus compare it with the closed structure. Binding of ACh to the ligand-binding domains initiates rotational movements in the α-subunits (arrows, Figure 3B and Figure 3C) that are communicated to the inner helices shaping the pore. The rotation opens a narrow region of the pore by repositioning bulky side chains.

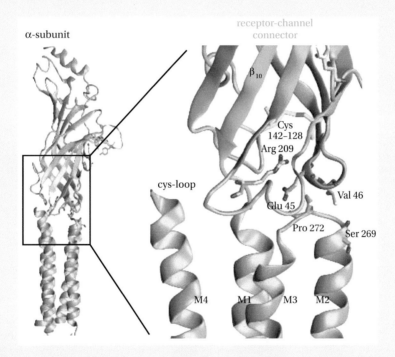

Figure 2 Defining features of Cys-loop receptors (PDB 2BG9). The Cys-loop is stabilized by a disulfide bridge between Cys142 and Cys128 (yellow). (Sine SM, Engel AG [2006] *Nature* 440–455. Used with permission of Springer Nature.)

(Continued)

potassium channels, suggesting that a closed channel can be opened by applying a lateral force at the C-termini of the inner helices. In the case of MthK, the gating ring would supply a radial-outward force to open the gate.

A plausible mechanism proposed by McKinnon is that the cytoplasmic gating ring responds to Ca^{++} binding by a conformational change that pulls on the channel via the polypeptide links, separating the lower part of the channel subunits and opening it to ion flow. Lowering of the Ca^{++} concentration would

BOX 11.6   NICOTINIC ACETYLCHOLINE RECEPTOR (AChR) (CONTINUED)

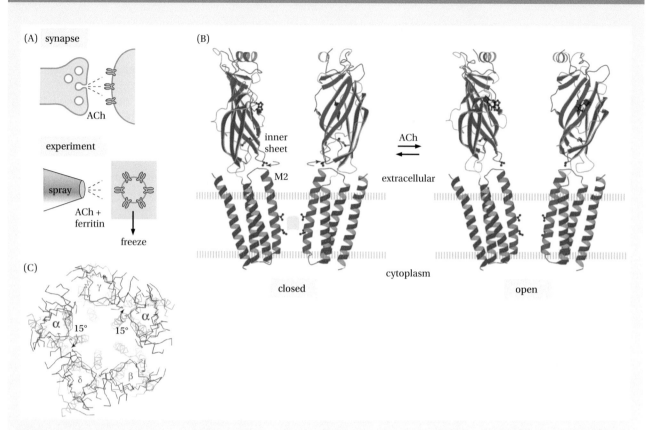

Figure 3 The mechanism of ligand gating of the nicotinic ACh receptor by acetylcholine. (A) Summary of an experiment designed to capture ACh receptors in the open state. In the experiment (panel A), ACh-containing droplets are sprayed onto an aqueous film containing isolated postsynaptic membranes. Rapid freezing traps the channels in the open state. Ferritin marker particles, included in the spray solution, allow the electron microscope to reveal the regions of the film where the droplets have impinged and spread, and so identify the membranes containing open channels. (B) Views along the membrane plane of the ACh receptor in the closed and open states. (C) View normal to the membrane showing the rotations of the ACh-receptor α-subunits that lead to channel opening and the influx of cations that depolarize the muscle membrane. (From Unwin N [2003] *FEBS Lett* 555: 91–95. With permission from John Wiley and Sons.)

reverse the process. In this mechanism, the stability of the closed channel must be sufficient to hold it closed so as to prevent spontaneous opening and consequent leakage of K⁺ ions. The energy required to open the channel must come from the Ca⁺⁺ binding energy, which must be sufficient to overcome the intrinsic channel interactions that hold the channel closed. Channel activity would thus be controlled by Ca⁺⁺ levels in the cytoplasm, a common mechanism. The control is likely cooperative because there are eight RCK domains operating as four structural pairs. The binding of Ca⁺⁺ to the first pair may cause a conformational change that affects the neighboring pairs cooperatively. Cooperative ligand gating is a simple and direct way to use ligand binding energy to pull the channel open.

An important consequence of channel opening becomes apparent when a simple electrostatic model is applied to the closed KcsA channel and the open MthK channel (**Figure 11.22**). The open-channel conformation favors ion conduction by creating a region of high electrostatic field strength that can concentrate ions near the filter entrance, increasing the probability of ions occupying the conducting positions and thereby facilitating high flux. Conversely, the

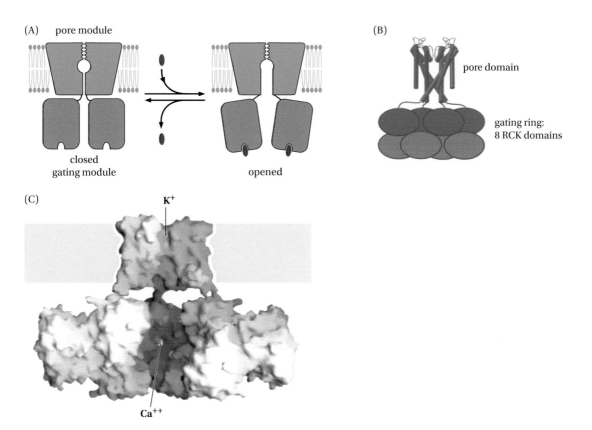

Figure 11.20 Ligand-controlled gating of ion channels. (A) The general scheme for gating includes a pore module and gating module, which responds to the ligand. Structural changes are transmitted mechanically to the pore domain to cause it to open or close. (B) The pore domain of the MthK channel (PDB: 1LNQ) is controlled by a gating ring containing eight RCK (Regulation of Conductance K⁺) domains. (C) Surface representation of the MthK channel showing K⁺ ions in the selectivity filter of the pore module and Ca⁺⁺ bound to the gating ring, which has twisted the pore domain into an open conformation. (From Jiang Y et al. [2002] *Nature* 417: 515–522. With permission from Springer Nature.)

lower field strength of the closed channel will produce a lower local ion concentration.

Recent studies have revealed the structure of intact KcsA at moderate resolution, and differences suggest a mechanism for the opening of the channel by protons. The cytoplasmic domain responsible for controlling channel opening in response to pH changes is a long four-helix bundle (**Figure 11.23**). Although the pH-sensing domain is very different from the Ca⁺⁺ binding domains of MthK, the opening of the gate in the membrane is similar in the sense that the domain conformation must change and exert a radial force on the channel domain, causing it to open or close. Although the exact structural changes remain to be

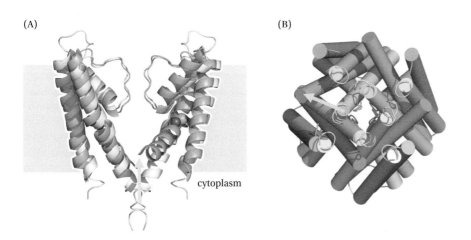

Figure 11.21 Comparisons of the MthK and KcsA channels. (A) Overlays of structures of the closed KcsA pore domain (PDB 1BL8, blue) and the open MthK pore domain (PDB 3LDC, green) viewed along the membrane plane. Only two of the four subunits are shown. The yellow arrow indicates the movement of the inner transmembrane helices during opening. A comparison of the two structures allows identification of a Gly residue (red) that serves as a gating hinge of the activation gate. (B) A view from inside the cell, with helices represented as cylinders. Sequence alignments show that the Gly gating hinge is highly conserved, implying that all gated K⁺ channels are controlled by inward/outward movements of the inner helices at the hinge. In the case of MthK, the movements are powered by the gating ring (Figure 11.20A). In voltage-gated channels, the movements are powered by the voltage-sensor domains (Figure 11.25).

Figure 11.22 Simplified electrostatic model of the closed and open states of a K+ channel. Using a simple electrostatic model with a uniform dielectric for the membrane and the protein, the change in the distribution of the electrostatic field when the channel opens is shown. Note that the field lines are much closer together, meaning that the field gradient is stronger. A steeper gradient means an increased force on ions to drive them across the membrane through the channel. (From Jiang Y et al. [2002] *Nature* 417: 515–522. With permission from Springer Nature.)

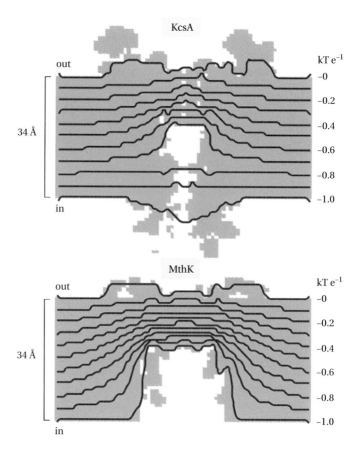

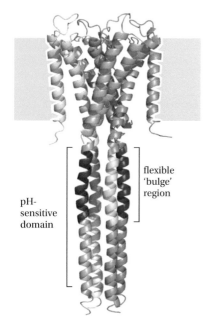

Figure 11.23 Structure of the full-length KcsA pH-sensitive channel (PDB 3EFF). The bulge region, unlike the channel, has twofold symmetry and high flexibility. The bulge region plays a key role in controlling channel opening.

elucidated, there is a flexible bulge region immediately below the channel that is believed to be the key to channel gating.

11.4 CHANGES IN TRANSMEMBRANE VOLTAGE CAN GATE CHANNELS RAPIDLY BY CHANGING CHARGE EXPOSURE ACROSS THE MEMBRANE BARRIER

11.4.1 Voltage Gating Is Needed for Fast Responses

Ligand gating is a relatively slow process because a series of diffusion events must occur—for example, in MthK, Ca++ ions must enter the cell, diffuse to the gating domain, induce the conformational change, and open the channel, which would be too slow a sequence of events for nerve signals to be responsive on a millisecond timescale. Instead, as we discussed earlier, voltage changes are used to open and close ion channels rapidly. The studies of voltage gating by physiologists over a large number of years resulted in a set of constraining parameters that must be components of the gating mechanism (Section 11.1). As in the case of ligand gating, a series of crystal structures have given a mechanistic idea of the process. The basic process utilizes movements within a voltage-sensing domain to open and close the channel. All gated channels so far observed structurally are two-module systems, one module being the pore domain with a flexible activation gate and the other a gating module, which in the case of the MthK channels, is the homo-octameric gating ring. In voltage-gated channels, the gating ring is replaced by voltage-sensor domains.

11.4.2 Voltage Gating Builds on the Theme of K⁺ Channels, with Added Domains for Voltage Sensing

A number of voltage-gated sodium and potassium channel tetramers are built from a basic unit with six TM helices. Mutagenesis and physiology experiments have provided a functional sequence map, which indicates that the first four helices, usually called S1–S4, form the voltage-sensing domain, with positive charges on S4 controlling the opening of the channel (**Figure 11.24**). The two remaining helices form a channel structure very much like that of KcsA, with two helices (S5 and S6) separated by a loop that forms the pore helix and selectivity filter. Rather than being formed as tetramers, some voltage-gated channels are single polypeptides with fourfold repeats that accomplish the same organization (Figure 11.12). It seems logical that the voltage sensor should be across the membrane, where the voltage difference is found, while the ligand sensor (see earlier) is in the space next to the bilayer, where the ligands are found in solution. The voltage-sensing domain is independent of the rest of the structure, as indicated by the properties of chimeric channels where the two domains have been spliced together from different genes and yet, the gating is accomplished. The independence of the voltage-sensing and pore domains (Figure 11.24) is emphasized by the discovery that some proton channels are essentially single voltage sensors. Independence, judged by evolutionary recycling, is further emphasized by the discovery of voltage-activated enzymes that are regulated by a single voltage-sensing domain.

11.4.3 Determination of the Structure of Voltage-Gated Ion Channels Presents Many Challenges

The first structure of a voltage-gated potassium channel was obtained from crystals of a Kv channel (voltage-gated potassium channel) isolated from the archaeon *Aeropyrum pernix*, referred to as the KvAP channel. The structure, determined by MacKinnon and colleagues, provided important clues to Kv channel gating despite distortions in the all-important voltage-sensor domain (VSD) caused by the detergent that produced type II crystals (Chapter 9, Box 11). Subsequent studies of mammalian Kv channels that produced type I crystals yielded accurate structural models and also showed the disposition of phospholipid acyl chains surrounding the channel (**Figure 11.25A**). The structures showed that Kv channels have their native structure only when they are embedded in a lipid bilayer. A structure of the full KvAP channel is shown in Figure 11.25B.

The KvAP channel serves here as a general structural map for understanding Kv architecture. The pore domain, comprised of TM helices S5 and S6 and the pore helix, is very similar to that of the KcsA channel. The VSD is comprised of TM helices S1-S4. The key part of the VSD is the S3b-S4 paddle, which carries four charged arginines (Figure 11.25C). This membrane-embedded paddle moves up and down in response to changes in the transmembrane electric field to produce a torque on the inner-helix activation gate that opens and closes the

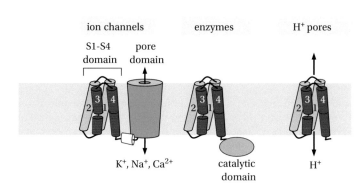

Figure 11.24 Voltage-sensor domains of voltage-gated K⁺ channels (far left) have been repurposed by evolution for regulating voltage-dependent enzymes (center) and to serve independently as proton channels (right). (From Krepkly D et al. [2009] *Nature* 462: 473–479. With permission from Springer Nature.)

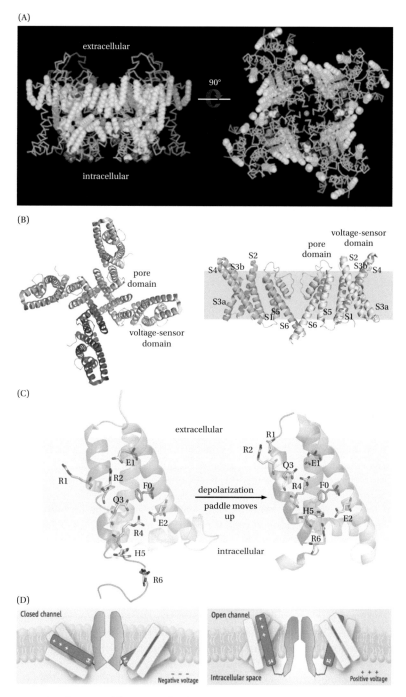

Figure 11.25 Structural features of voltage-gated potassium channels. (A) Structure of a chimeric Kv channel in which the S3b–S4 paddle domain (panel C) of the Kv2.1 channel replaces the native paddle of the Kv1.2 channel (PDB 2R9R). The structure also shows the hydrocarbon chains of the phospholipids (yellow) that co-crystallized with the channel. The phospholipids helped the channel crystallize in a bilayer-like Type I form (Chapter 9, Box 11) that allowed the voltage sensor to adopt a native configuration. (B) Structure of the full KvAP voltage-gated channel (PDB 6UWM). In the in-plane image (right), the purple and blue subunits in front and back have been removed to afford a view of the selectivity filter. (C) Structure of the S2–S3–S4 part of the VSD from the mammalian KCNQ1 potassium channel in an intermediate (left; PDB 6MIE) and activated (right; PDB 5VMS) conformation. The arginines (R1–R4) in the S3b–S4 paddle are responsible for sensing changes in the electric field across the membrane, which drives them to move past the "aromatic plug" residue F0 in S2. Acidic residues E1 and E2 in S2 interact with the positively charged arginines as they move. (D) An early cartoon suggesting how movements of the VSD in the transmembrane electric field might gate the channel. (From Long SB et al. [2007] *Nature* 450: 376–382. With permission from Springer Nature.)

channel (Figure 11.25D). Phospholipid phosphates interact strongly with the S4 arginines, consistent with the observation that gating cannot occur when channels are reconstituted into black lipid membranes formed from lipids lacking phosphates.

11.4.4 Channel Structures Lead to Mechanistic Ideas about Voltage Sensing

The importance of the design features of channels is revealed by the fact that a chimeric potassium channel consisting of a rat Kv1.2 K$^+$ channel in which the voltage sensor S3b-S4 paddle was replaced by the paddle from the rat Kv2.1 K$^+$ channel is functional. A comparison of the selectivity filters of the chimera and KcsA reveals high structural conservation (**Figure 11.26**). Thus, the design is very precise in evolutionary terms, constraining the structural solutions to a narrow range. A consequence is that we learn a great deal from a few examples.

Electrophysiological experiments established that the voltage sensor uses the positive charges located on the S4 helix to measure the transmembrane voltage, effectively using the energy of the potential to drive a conformation change that moves the exposure of the charges from one side of the membrane barrier to the other, from high to low potential energy (Figure 11.25D). The conformational change therefore couples the electrical potential across the membrane to a mechanical movement, which is used to control an ion channel. A striking feature of the S4 helix is that it has a 3_{10} helical structure, placing the α-carbons of the arginines bearing charged groups in a line on one side of the helix. The exact mechanism of the channel opening has been an active subject of debate in recent years, and because it has not been possible to impose a membrane potential on the molecules in a crystal, a model for the mechanism has been inferred from a body of molecular dynamics simulations and biochemical and structural evidence, including changes in chemical modification of groups when the channel is open or closed.

When there is no voltage across the membrane, the channel is open, so it is expected that the crystal structure of a native channel will represent this state (the "active state"). Increasing the negative charge on the inside of the cell (and the positive charge on the outside) forces the cooperative movement of the positively charged arginine or lysine groups inward, closing the channel. If the biological function is to open the channel when the voltage decreases, as is the case in nerve axons, then the voltage would be used to keep the channel closed (the "resting state"), and a reduction in voltage below a critical threshold would open the channel. A variety of measurements are consistent with an S4 displacement of about 15 Å and a large reorganization of the voltage-sensor structure associated with gating. Without the structure of a closed conformation, the chemical details of the mechanism remain unknown, and different plausible models have been proposed. One idea, from the MacKinnon group, is that the VSD "paddles" are positioned inside the membrane when the channel is closed, and that the paddles move a large distance across the membrane from inside to outside when the channel opens. KvAP channels were reconstituted into planar lipid membranes and studied using monoclonal Fab fragments, a voltage-sensor toxin, and avidin binding to tethered biotin, and the results suggest that the

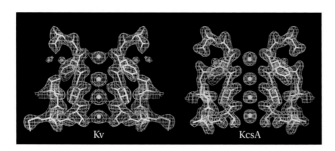

Figure 11.26 Structural comparison of the Kv1.2/2.1 selectivity filter with that of the KcsA filter reveals high structural conservation. The filters are nearly indistinguishable at the small error levels of these high-resolution structures, emphasizing the way that evolution continues to use solutions to functional problems over again in different contexts. (From Long SB et al. [2007] *Nature* 450: 376–382. With permission from Springer Nature.)

voltage-sensor paddles operate somewhat like hydrophobic cations attached to levers, enabling the membrane electric field to open and close the pore.

On the other hand, extensive molecular modeling by a number of groups suggests that the conformational change occurring during activation is a helical screw–sliding helix mechanism in which the S4 segment retains its helical conformation and reorients as it moves, principally along its long axis. The gating charges are not directly exposed to the lipid hydrocarbon, and the S3–S4 helix-turn-helix does not move as a highly concerted structural motif across the membrane during voltage gating as proposed in the paddle model. Rather, salt bridges involving the gating residues play an important role, as proposed by Clay Armstrong in 1981, and the S4 helix moves across a "focused electric field" in which the spatial variation in the transmembrane potential affecting the gating charges of the VSD is concentrated over a narrow region that is considerably thinner than the full bilayer membrane (like the electrostatics in KcsA channel; Figure 11.22). This second mechanism is also suggested by structural work on sodium channels (see later).

Considerable energy is available for the gating mechanism, and recent calculations give about 14 kcal mol^{-1} for opening a related channel (the Shaker channel, a much-studied example from *Drosophila*) by moving charges across the membrane potential, so the inactive channel can remain tightly closed by a factor of 10^5 over the open state to resist any false signaling.

11.4.5 Lipid Composition and Mechanical Distortions Can Affect Channel Function

One aspect of the structure of the Kv1.2/2.1 chimeric potassium channel is particularly significant because it reveals the close association of the channel with lipids (Figure 11.25A). The fact that these lipid regions are seen in the crystal structure means that they occupy preferred locations that are well defined, contributing to the structure. From what we have learned about lipid–protein interactions, we expect the bound lipids, by inference, to contribute to the stability of the protein, which raises the possibility that channel properties are coupled to the composition and mechanical properties of the membrane.

The dependence on composition is supported by studies of the voltage dependence of the chimeric channel reconstituted into oöcytes and black lipid membranes with different lipid compositions (**Figure 11.27A**). The measurements show changes in voltage dependence and thus, the potential importance

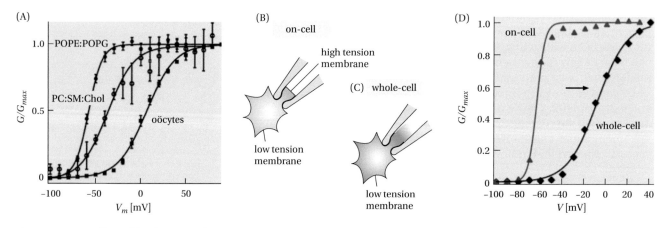

Figure 11.27 Effect of lipid and mechanical stress on the voltage dependence of a Kv1.2/2.1 chimera. (A) The voltage dependence of the Kv channel depends strongly on lipid composition. (B) Tension can be applied to a whole cell by sucking the cell into the recording pipette without rupturing the membrane (called an on-cell recording; Figure 11.9). (C) If suction is applied to rupture the membrane so as to leave the cell attached (whole-cell recording), tension on the cell is relieved. (D) On-cell recordings show a much steeper dependence on voltage due to membrane stress. (A,B, From Schmidt D, MacKinnon R [2008] *PNAS* 195: 19276–19281. With permission from R. MacKinnon. C,D, From Schmidt D, Marmol del J, MacKinnon R [2012] *PNAS* 109:10352–10357. With permission from R. MacKinnon.)

of the lipid composition of the membrane surrounding the channels. We should not be surprised, then, if we find that the voltage dependence also responds to membrane stress. The effects of stress in live cells can be observed by comparing channel voltage dependence in cells stressed by suction and in relaxed cells (Figure 11.27B). Membrane stress dramatically increases the steepness of the voltage dependence, suggesting that K^+ channels might play a role in mechanosensation.

11.5 VOLTAGE-GATED NA⁺ CHANNEL MECHANISM AND HYDRATED-ION SELECTION

Voltage-gated sodium channels (Na_Vs or Navs) allow Na^+ ions to cross membranes, opening channels in response to changes in the membrane electrical potential; they are essential for producing APs in nerve fibers (Figure 11.10, Figure 11.11). Prokaryotic sodium channel crystal structures have provided detailed views of sodium channels, which have been interpreted to suggest functions for structural features in human sodium channels. Since the first structure of a voltage-dependent sodium channel was reported in mid-2011 (from *Arcobacter butzleri,* hence named NavAb) by William Catterall and his colleagues, a number of structures have been reported, notably from the laboratories of Catterall and Bonnie Wallace. Here, we focus on key features revealed by the structural studies. The Nav AP cycle in vertebrates has three principal states: (1) held in the closed, resting state by the membrane potential, (2) opened by changing the membrane potential, and (3) spontaneously inactivated after opening (see Figure 11.10A). The structural data are mainly informative about the closed and open states.

Vertebrate Na^+ channels are composed of approximately 2000 amino acid residues in four linked homologous domains, whereas bacterial Na^+ channels are homo-tetramers, but the overall organizations are similar (see Figure 11.12). Prokaryotic channels have been used in a number of structure studies, often with mutational strategies to stabilize them. Recent work has revealed structures for the full-length proteins, including the cytoplasmic domains (**Figure 11.28**). The overall architecture is very similar to that of Kv channels, but significant differences are seen in the selectivity filter structure and possibly, a somewhat different gating mechanism.

(A)

(B)

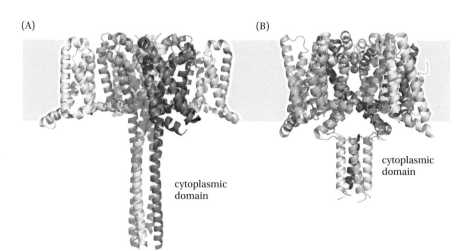

cytoplasmic domain

cytoplasmic domain

Figure 11.28 Voltage-activated sodium (Na_V) channels. Recent work has given structures for full-length Na_V channels, including their C-terminal cytoplasmic domains. (A) Na_VAb/FY channel in a double mutant form that locks the channel in a closed state (PDB 5VB2). (B) The native NavMs channel in the open state (PDB 5HVX). Compare with KcsA (Figure 11.24).

11.5.1 The Na$_V$ Channel Is Similar to the KcsA Channel but Selects Hydrated Na$^+$ Ions

The Na$_v$ channel region is formed as a tetramer of subunits, each with two TMs (S5 and S6) but with two pore helices. A space-filling cross section reveals the selectivity filter, a central chamber that can accommodate a *hydrated* Na$^+$ (**Figure 11.29**). The Catterall lab found through clever mutational changes that Na$_v$Ab channels could be converted into Ca^{2+} channels, and further, that these mutants could yield better structures, including a closed form to compare with the Na$_v$Ms (meaning Na$_v$ from *Magnetococcus marinus*) from the Wallace lab—Figure 11.29 compares the pore domains of open and closed channels, showing that *hydrated* Na+ ions can pass through the open but not the closed channel.

If we compare the details of the Na$^+$ and K$^+$ selectivity filters, several features are strikingly different. First, the Na$^+$ channel relies on an intensely negative electrostatic field created by four Glu side chains at the opening. Second, no Na$^+$ ions are stably bound in the filter, and third, water molecules can be accommodated in the filter to hydrate the ion near the ring of Glu$^-$ charges. The intense negative charge will attract cations and facilitate their passage through the channel (**Figure 11.30**). Perhaps, the intense charge at the surface of the smaller Na$^+$ ion cannot be accommodated by the same strategy of using carbonyl groups that is used in the K$^+$ case. Structural comparisons give images of the gates in the closed, open, and inactivated states of the Na$_v$ channel cycle. Although investigation of the states of the Na$_v$Ab channel has required the use of mutants and truncations that may alter some of the details, the resulting structures have given a plausible view of the functional cycle of the gate, as shown in **Figure 11.31**. It is interesting that the fourfold symmetry is broken in the inactivated state. The open state of the truncated Na$_v$Ab/1-126 is sufficiently large to allow access by local anesthetic and antiarrhythmic drugs, which bind to sites in the central cavity.

11.5.2 The Structure and Environment of the Gating Helix Suggest a Mechanism for Na$_v$

While the paddle mechanism suggested for voltage gating in the K$_v$ structure seems plausible, a different set of ideas emerges from the Na$_v$ structures and is supported by a body of biochemical and biophysical evidence. The domains of the Na$_v$ tetramer are interleaved (**Figure 11.32A**); the yellow S1–S4 domain (the VSD) has a light blue S5–S6 domain between it and the yellow S5–S6 domain. This interleaving, or domain swapping, causes a strong coupling between the motions of one subunit and the next, giving the overall structure a high cooperativity, meaning that the channel will open or close over a narrow voltage range, so that it can exist in a poised state, ready to open. As in the proposed mechanisms for K$_v$2.1, the exposure of the S4 helix is changed from one side of the membrane to the other, and the suggested motion resembles the second idea of a rotation of the VSD and a relatively independent motion of the S4 helix with its four arginines (also in a 3$_{10}$ helix). A possible sequence of events is the

Figure 11.29 (A) Open and (B) closed Na$_v$ and Ca$_v$ pores. The pore dimensions as calculated by the program MOLE are shown. Clever mutagenesis converted Na$_v$Ab into a Ca^{2+} channel, which gave a better structure for the closed form.

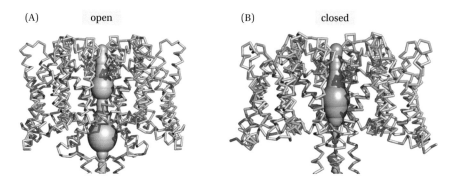

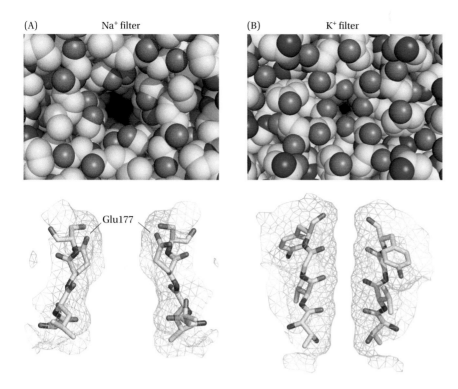

(A) Na⁺ filter (B) K⁺ filter

Glu177

Figure 11.30 Comparison of (A) Na⁺ and (B) K⁺ selectivity filters as seen from the top (upper panels) and side (lower panels; only two of the four subunits shown) of the channel. Note that the Na⁺ filter is significantly larger than the K⁺ filter, even though the K⁺ ion is ~1.4 times as large. The explanation is that the Na⁺ ion is hydrated when in the filter, while the K⁺ is not. Note the negatively charged Glu177 residues and the absence of regularly spaced backbone carbonyl groups in the Na⁺ filter.

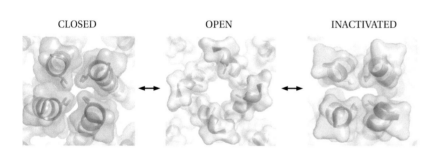

CLOSED   OPEN   INACTIVATED

Figure 11.31 View of the gate in the Na_vAb functional cycle. Close-up view of the activation gate at the level of residue 217. The view is as if one were looking at the permeation pathway toward the outside of the cell. Na_vAb/FY (locked closed by the mutation) is shown in grey, Na_vAb/1–226 (lacking 40 residues of the C-terminal domain) is shown in wheat, and Na_vAb/WT is shown in green—with each model containing cartoon helices, a transparent surface representation, and a stick representation of the side chain at position 217. (From Lenaeus MJ et al. [2017] *PNAS* 114: E5031–E3060. With permission.)

following (Figure 11.32B). First, the VSD and the S4–S5 linker move as a unit. Second, the closed-to-open pore transition is mediated by a single molecular hinge at the base of S5 (indicated by an asterisk in the figure). Third, rotation of the VSD and the S4–S5 linker, operating as a structural unit, pulls the S5–S6 helices outward to open the pore. Concerted pore opening is believed to occur as a result of an S4–S5 from one subunit forcing the neighboring subunits to undergo similar motions.

11.6 A VOLTAGE-GATED PORIN IN THE OUTER MEMBRANE OF MITOCHONDRIA CONTROLS EXCHANGE OF METABOLITES WITH THE CYTOPLASM OF EUKARYOTES

Mitochondria evolved from a proteobacterium taken up by an ancient eukaryotic precursor cell, causing mitochondria to bear a likeness to Gram-negative bacteria, including β-barrel proteins in the outer membrane. But unlike bacteria, which face a highly variable environment, mitochondria

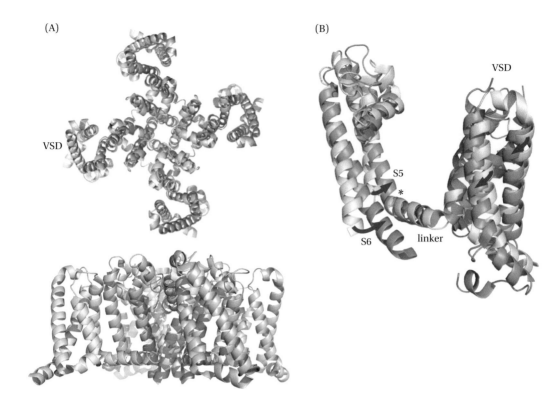

Figure 11.32 Overview of transmembrane structure of Na$_v$Ab (PDB 3RVY) and a model for voltage-gated pore opening based on superposition of K$_v$2.1 (PDB 3LUT) and Na$_v$Ab structures. (A) Note the interleaving of subunits, seen as the intrusion of the light blue channel subunit between the channel and sensor domains of the yellow subunit. (B) Single subunits are shown of Na$_v$Ab (closed) in yellow and K$_v$2.1 (open) in green. Voltage-induced motion of the voltage-sensing module (right) exerts force on the pore helices through the S4–S5 linkers, pulling open the bottom of the channel, where the gate is located. A hinge is indicated by the asterisk.

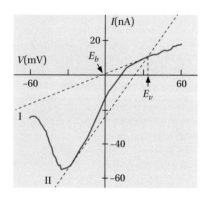

Figure 11.33 The voltage dependence of the VDAC channel reconstituted into planar lipid membranes. (From Schein SJ et al. [1976] *J Membr Biol* 30: 99–120. With permission from Springer Nature.)

live in a stable environment, and the relaxation of several constraints has allowed more evolutionary divergence. Rather than protecting the mitochondrion from harsh environments, the outer membrane has become essentially a "platform" for passive ion and metabolite exchange. The most abundant porin is VDAC (voltage-dependent anion channel). In bacteria, the outer membrane porin protein family of β-barrels usually have an even number of β-strands; in contrast, VDAC and Tom40 (Chapter 6) have an uneven number of 19 β-strands but a similar molecular architecture. The voltage gating of VDAC was discovered long before the importance of the channel in mitochondrial physiology was fully appreciated (**Figure 11.33**). The gating, which causes a shift between weak anionic and cationic selectivity, is important for controlling metabolite exchange between the cytoplasm and mitochondrial inner membrane space in the face of changes of chemical and electrical gradients.

A high-resolution structure of a VDAC (**Figure 11.34**) gives an impression of the electrostatics that are important in selectivity and gating. Particularly interesting is the N-terminal helix of 26 amino acids within the barrel (Figure 11.34B), which is believed to be responsible for gating the channel. Evidence for gating comes from site-directed mutagenesis studies of the helix in which mutations change the voltage dependence. The importance of the gating helix is more apparent in surface representations of the pore that include electrostatic potential surfaces (Figure 11.33C). The helix partly occludes the channel, causing the channel passageway to be more electropositive than when absent. Although details are lacking, it is clear that movements of the helix can be expected to change the electrostatics and selectivity of VDAC.

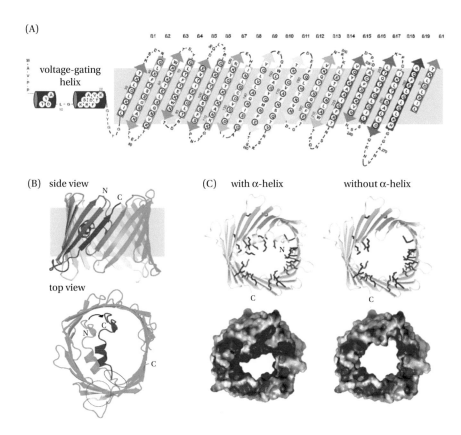

(A)

voltage-gating helix

(B) side view

(C) with α-helix   without α-helix

top view

Figure 11.34 The structure of the mouse VDAC channel (PDB 3EMN). (A) Channel topology showing 19 transmembrane segments. (B) Structure of VDAC. Note the short N-terminal helix enclosed by the barrel, which believed to be the gating helix. (C) Electrostatic surfaces show the changes in potential in the presence and absence of the gating helix (blue—positive, red—negative). (From Ujwal R, Cascio D, Colletier J-P et al. [2008] *PNAS* 105:17742–17747. With permission from Jeff Abramson.)

11.7 CONCLUSION

Fast signaling is needed for an organism to be responsive to its external world and to coordinate many of its internal activities, and ion channels are the key devices that are used for speedy responsiveness. From structural analyses (**Box 11.7**) and a large body of biochemical and physiological data, we now have a plausible chemical view of how channels manage to be selective, to facilitate the transit of ions across a membrane, and to control the opening and closing of the channels by ligands, mechanical forces, and voltage. Some of the details are likely to be revised as science proceeds, but it seems unlikely that the current view is seriously in error.

KEY CONCEPTS

- Cells in multicellular organisms must signal each other to team up in tissues, culminating in conscious thought. Even single-celled organisms must have rapid responsiveness to changes in their environments. In each case, different external cues require a range of response times, from slow (e.g., changes in food molecules, differentiation signals) to fast (osmotic stress, nerve signals). Ion channels are the key to rapid signaling, amplifying a signal by the transmembrane passage of many ions while maintaining specificity for the types of ion.

- Channels have a modular construction, usually with a tetrameric channel domain and separate sensor domains for ligands or voltage gating.

- Ion channels select a specific ion type by precisely tuning the size and dynamics of a narrow part of the channel, the selectivity filter. Channels often distinguish kinds of ions, for example, allowing potassium through a channel while excluding sodium. The selectivity filter in the channel structure exploits differences in ion size and surface charge to favor the passage of one ion over another. The dynamics of the filter play a role in combining selectivity with high flux.

- Ion channels lower the barrier for the passage of ions across a membrane by sequestering a central aqueous cavity from the lipid and by orienting the negative ends of short helices that help to counter the charge on hydrated ions in the cavity. The barrier for an ion

BOX 11.7   ION CHANNEL FAMILIES

Properties of the three classes of ion channels of known three-dimensional structure (Figure 1).

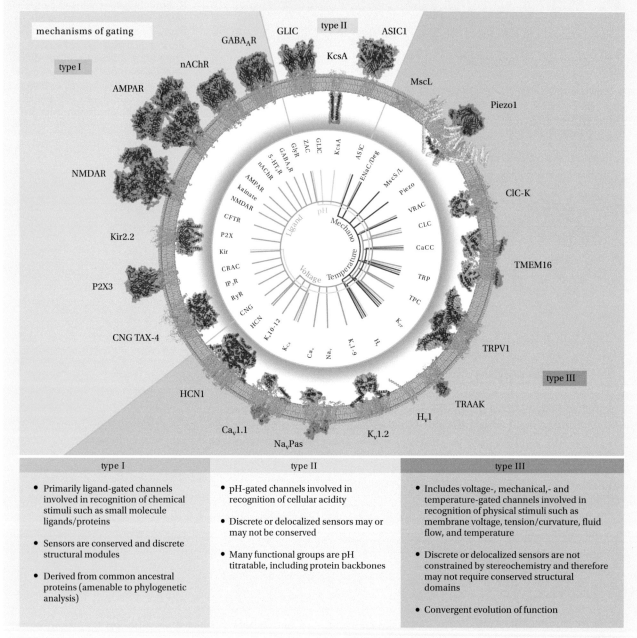

type I	type II	type III
• Primarily ligand-gated channels involved in recognition of chemical stimuli such as small molecule ligands/proteins	• pH-gated channels involved in recognition of cellular acidity	• Includes voltage-, mechanical,- and temperature-gated channels involved in recognition of physical stimuli such as membrane voltage, tension/curvature, fluid flow, and temperature
• Sensors are conserved and discrete structural modules	• Discrete or delocalized sensors may or may not be conserved	• Discrete or delocalized sensors are not constrained by stereochemistry and therefore may not require conserved structural domains
• Derived from common ancestral proteins (amenable to phylogenetic analysis)	• Many functional groups are pH titratable, including protein backbones	• Convergent evolution of function

Figure 1 Properties of the three classes of ion channels. (From Goldschen-Ohm, Chanda [2017] *Cell* 170: 214.e1. With permission.)

to cross a bilayer can be very high (tens of kilocalories per mole), so the protein must create a lowering of the barrier to facilitate transport in useful timescales. We have seen how the barrier can be lowered using these features of the channel structure.

• Many channels allow the rapid passage of many ions while maintaining selectivity. This is a challenging trick, which is accomplished by the movement of a stack of ions and water molecules in a cooperative single file through the selectivity filter via a set of closely spaced isoenergetic positions that cannot all be occupied at the same time.

• Ion channels control the opening and closing of the channel by positioning gates that can be opened or closed by voltage, mechanical force, or ligand binding. For ion channels to be used as responsive signaling devices, there must be a way to open or close a "gate" in the channel in response to a stimulus. A gate is simply an energy barrier that prevents flow through the channel in the closed state and allows ions or molecules to flow in the open state, and must have a closed-state stability energy large enough to oppose the energy difference created by the chemical potential of the transported ion or molecule. Opening or closing is usually in response to a stimulus. There are three basic kinds of gating stimulus that do this: binding a ligand, subjecting the channel to a mechanical force, and changing the voltage across the membrane; hence the names "ligand gated," "mechanosensitive," and "voltage gated". Structural studies have revealed how examples of these gates function.

- Gated channels rely on domain architectures that are used repeatedly over evolutionary time with modifications for specific functions.

FURTHER READING

Armstrong, C.M., and Hille, B. (1998) Voltage-gated ion channels and electrical excitability. *Neuron* 20:371–380.

Hille, B., Dickson, E.J., Kruse, M., Vivas, O., and Suhb, B.-C. (2015) Phosphoinositides regulate ion channels. *Biochim Biophys Acta* 1851:844–856.

Sula, A., and Wallace, B.A. (2017) Interpreting the functional role of a novel interaction motif in prokaryotic sodium channels. *J Gen Physiol* 149:613–622.

Hille, B. (2001) *Ion Channels of Excitable Membranes*, Sinauer Associates, Sunderland, Mass.

Becker, T., and Wagner, R. (2018) Mitochondrial outer membrane channels: Emerging diversity in transport processes. *Bioessays* 40:e1800013.

Zaydman, M.A., Silva, J.R., and Cui, J. (2012) Ion channel associated diseases: Overview of molecular mechanisms. *Chem Rev* 112:6319–6333.

Changeux, J.P., and Christopoulis, A. (2017) Allosteric modulation as a unifying mechanism for receptor function and regulation. *Diabetes Obes Metab* 19 Suppl 1:4–21.

Catterall, W.A., and Swanson (2015) Structural basis for pharmacology of voltage-gated sodium and calcium Channels. *Mol Pharm* 88:141–150.

KEY LITERATURE

Hodgkin, L., and Huxley, A.F. (1945) Resting and action potentials in single nerve fibers. *J. Physiol.* 104:176–195.

Cole, K.S., and Curtis, H.J. (1939) Electric impedance of the Squid Giant Axon during activity. *J Gen Physiol.* 22:649–670.

Noda, M., Shimizu, S., Tanabe, T., Takai, T., Kayano, T., Ikeda, T., Takahashi, H., Nakayama, H., Kanaoka, Y., Minamino, N. et al. (1984) Primary structure of Electrophorus electricus sodium channel deduced from cDNA sequence. *Nature.* 312:121–127.

Anderson, P.A., and Greenberg, R.M. (2001) Phylogeny of ion channels: Clues to structure and function. *Comp Biochem Physiol B Biochem Mol Biol.* 129:17–28.

Armstrong, C.M. (1971) Interaction of tetraethylammonium ion derivatives with the potassium channels of giant axons. *J Gen Physiol.* 58:413–437.

Doyle, D.A., Cabral, J.M., and Pfuetzner, R.A. et al. (1998) The structure of the Potassium channel: Molecular basis of K⁺ Conduction and selectivity. *Science* 280:69–77.

Long, S.B., Tao, X., Campbell, E.B., and MacKinnon, R. (2007) Atomic structure of a voltage-dependent K⁺ channel in a lipid membrane-like environment. *Nature* 450:376–382.

Payandeh, J., Scheuer, T., Zheng, N., and Catterall, W.A. (2011) The crystal structure of a voltage-gated sodium channel. *Nature* 475:353–358.

Noskov, S.Y., Bernèche, S., and Roux, B. (2004) Control of ion selectivity in potassium channels by electrostatic and dynamic properties of carbonyl ligands. *Nature.* 431:830–834.

Jiang, Y., Lee, A., Chen, J., Cadene, M., Chait, B.T., and MacKinnon, R. (2002) Crystal structure and mechanism of a calcium-gated potassium channel. *Nature.* 417:515–522.

Krepkiy, D., Mihailescu, M., Freites, J.A., Schow, E.V., Worcester, D.L., Gawrisch, K., Tobias, D.J., White, S.H., and Swartz, K.J. (2009) Structure and hydration of membranes embedded with voltage-sensing domains. *Nature.* 462:473–439.

Jiang, Y., Ruta, V., Chen, J., Lee, A., and MacKinnon, R. (2003) The principle of gating charge movement in a voltage-dependent K+ channel. *Nature.* 423:42–48.

Vargas, E., Yarov-Yarovy, V., Khalili, F., Catterall, W.A., Klein, M.L., Tarek, M., Lindahl, E., Schulten, K., Perozo, E., Bezanilla, E., and Roux, B. (2012) An emerging consensus on voltage-dependent gating from computational modeling and molecular dynamics simulations. *J Gen Physiol.* 140:587–594.

Schein, S.J., Colombini, M., and Finkelstein, A. (1976) Reconstitution in planar lipid bilayers of a voltage-dependent anion-selective channel obtained from paramecium mitochondria. *J Membr Biol.* 30:99–120.

Ujwal, R., Cascio, D., Colletier, J.P., Farnham, S., Zhang, J., Toro, L., Ping, P., and Abramson, J. (2008) The crystal structure of mouse VDAC1 at 2.3 A resolution reveals mechanistic insights into metabolite gating. *Proc Natl Acad Sci U S A.* 105:17742–17747.

Liao, M., Can, E., Julius, D., and Cheng, Y. (2013) Structure of the TRPV1 ion channel determined by electron cryo-microscopy. *Nature* 504:107–112.

Madej, M.G., and Ziegler, C.M. (2018) Dawning of a new era in TRP channel structural biology by cryo-electron microscopy. *Pflugers Archiv-European Journal of Physiology* 470:213–225.

EXERCISES

1. Using a diagram of the structure of the Ca-gated MthK K^+ channel, discuss each of its tasks in one sentence, including a labeled circle on the structure to indicate where the task is accomplished.

   Dehydration/Hydration of the ion
   Selection of the ion
   Lowering the energy barrier
   Allowing high flux
   Opening the gate

2. The known structures of membrane proteins show that (True/False):

   — the only structural motifs allowed within the hydrophobic slab of the bilayer are helices and beta barrels.
   — lipids mainly interact with membrane proteins via hydrogen bonds.
   — ion and sugar translocation pathways follow the lipid–protein interface.
   — hydrogen bonding of the main chain is a dominant energy term in dictating peptide folding in a bilayer.

3. Using a diagram, explain the APs in Figure 11.3 using the observations in Figure 11.10.

4. By sequence changes it is possible to convert a sodium channel into a calcium channel, as was done in the studies by the Catterall group. Why might this change be more easily accomplished for the conversion of Na^+ to Ca^{++} than for Na^+ to K^+?

5. Given the function of the S4–S5 link in voltage-gated ion channels, what would you expect the basis of the structural design of the linker to be? Compare the linker in Na_v with the linker connecting the Ca-binding domains to the K^+ channel in MthK.

6. An Na^+ channel refractory state is needed for the progression of an AP, but there is no refractory state for the K^+ channel in an axon. The channel gating regions of K^+ and Na^+ channels each have approximate fourfold symmetry, but the Na^+ channel has a refractory state that lacks this symmetry. Which property of the two kinds of channel may facilitate this asymmetry, seen in Figure 11.32?

7. The gating of a K^+ channel involves the bending of one of the TM helices at a glycine. Why would glycine be used for this position? How might this explain the common occurrence of glycines at the ends of helices?

Primary Transporters
Transport against Electrical and Chemical Gradients

12

As we learned in Chapter 11, the lipid bilayer–based permeability barrier to ions and polar molecules allows the interiors of cells to have chemical compositions that differ from their surroundings, enabling such diverse functions as synaptic signaling, nutrient uptake, and energy interconversions. Pores and channels control the flow of ions and other molecules across membrane barriers, utilizing electrochemical gradients. In nerve axons, for example, the intracellular concentration of potassium ions is very high relative to the extracellular space, whereas the intracellular concentration of sodium ions is relatively very low. Voltage-dependent ion channels tightly control the flow of sodium and potassium ions down their electrochemical gradients to produce the resting membrane potential (Box 11.1) and the action potential (Box 11.2).

Physiologists have long recognized that electrochemical gradients can only exist due to the expenditure of cellular metabolic energy. An early question was exactly how metabolic energy was tapped to maintain the gradient. Another question was whether the expenditure of metabolic energy is involved directly in the production of action potentials. One idea was that the rising phase of the action potential was due to the passive influx of sodium ions, while the falling recovery phase was due to active pumping of sodium ions out of the cell. Alan Hodgkin and Richard Keynes tested the idea using the squid axon. Without understanding at the time exactly why, they knew that dinitrophenol (DNP) interrupted cell metabolism and should consequently inhibit the active transport of ions across the membrane. They showed that while DNP caused a gradual decline (minutes to hours) in the amplitudes of the resting and action potentials, it had no effect on the millisecond time scale of the action potential (Figure 12.1A). This was clear evidence that metabolic energy expenditure is not required to produce the action potential; even metabolically inactive axons could produce action potentials. But, what is the nature of the active transport process? By measuring K^+ and Na^+ fluxes using radioactive tracers, they discovered that active transport of the two ions is coupled via some kind of active exchange process (Figure 12.1B): Na^+ ions were actively "pumped" out in exchange for K^+ pumped in.

We now know that DNP is an agent that uncouples electron transport in mitochondria from ATP synthesis. When mitochondria are poisoned with DNP, ATP is not produced. This observation means that the source of energy for active Na^+/K^+ transport is the energy stored in high-energy compounds such as ATP (see Chapter 14). How does active transport of ions work? Which membrane proteins are involved? Which ions can be actively transported? Can substances other than ions be actively transported using ATP as the energy source? Is ATP the only energy source for active transport?

Creating and maintaining transmembrane compositional differences of ions and small molecules are the jobs of different categories of transport proteins. Major categories are: electron transporters that use metabolic or light energy to excite electrons, which then expend their energy in a controlled manner to

DOI: 10.1201/9780429341328-13

Figure 12.1 Role of active ion transport in the production of the action potential and the electrochemical gradient. (A) Addition of the metabolic inhibitor dinitrophenol (DNP) has little effect on the squid nerve action potential on the millisecond time scale of the action potential. (B) The active transport of Na⁺ and K⁺ ions is coupled. The efflux of Na⁺ is reduced when the external K⁺ concentration is reduced or during long incubation with DNP present. Experiments such as these showed that the active transport of K⁺ and Na⁺ were coupled, likely in a single process. (From Hodgkin AL, Keynes RD [1955] *J Physiol* 128: 28–60. With permission from John Wiley and Sons.)

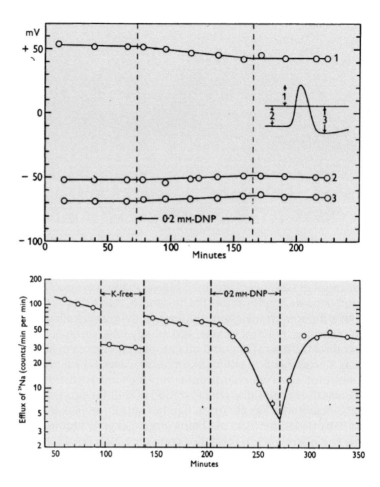

create proton-based electrochemical gradients for the production of ATP; primary transporters that use ATP directly to drive transport processes; secondary transporters that use the energy stored in electrochemical gradients to move other molecules and ions uphill; and group translocators that couple translocation of a substrate to its chemical modification, resulting in release of a modified substrate at the opposite side of the membrane. This chapter is about primary transporters that couple transmembrane pumping of ions and other solutes directly to the hydrolysis of ATP. Secondary transporters are discussed in Chapter 13, and the synthesis of ATP via electron transporters in discussed in Chapter 14; group translocators are briefly mentioned in Chapter 13.

How are ion and solute pumping coupled to ATP consumption in primary transporters? We first examine the family of P-type ATPases, and in particular the Na⁺/K⁺-ATPase, responsible for establishing the transmembrane electrochemical gradient of axons and other cells, and the Ca²⁺-ATPase that maintains strong Ca²⁺ gradients across the muscle sarcoplasmic reticulum as an essential feature of the control of muscle contraction. We then consider another very large family of proteins, the so-called **ATP binding cassette (ABC) transporters**, which use ATP for pumping all sorts of molecules across cell membranes. Finally, we consider an interesting family of ATP-driven transporters, the energy-coupling factor (ECF) transporters, in which the substrate-binding subunit appears to "topple" in the membrane in response to ATP hydrolysis.

12.1 P-TYPE ATPASES

P-type ATPases are found throughout the living world. P-type ATPases actively transport a variety of ions—and even lipids—across membranes. The name is

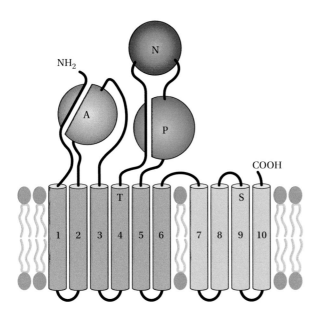

Figure 12.2 Schematic structure of P-type ATPases. The actuator (A) domain is comprised of the N-terminal tail and the first cytoplasmic loop. The nucleotide-binding (N) and phosphorylation (P) domains are inserted into the intracellular loop between TM4 and TM5. The first six TM helices form the transport (T) domain. The support (S) domain is comprised of the last four TM helices. (Palmgren MG, Nissel P [2011] *Annual Review of Biophysics* 40: 243–266. With permission from Maike Bublitz.)

meant to reflect a key mechanistic feature, namely, that a conserved Asp residue is phosphorylated during the reaction cycle. All P-type ATPases have five domains that cooperate during the reaction cycle: the cytoplasmic actuator (A), nucleotide-binding (N), and phosphorylation (P) domains, and the two membrane-embedded transport (T) and class-specific support (S) domains (**Figure 12.2**). In many cases, there is also a sixth regulatory (R) domain attached to the N- or C-terminus (or both) of the protein. We will focus the discussion on two P-type ATPases: the historically important Na^+/K^+-ATPase and the currently best understood P-type ATPase, the Ca^{2+} ATPase.

12.1.1 A P-Type ATPase Is Responsible for Na^+/K^+ Active Transport

Following the discovery of the importance of the coupled active transport of Na^+ and K^+ by Hodgkin and Keynes, Jens Skou in Denmark embarked on biochemical studies of active transport in isolated crab nerves that led him to discover and characterize membrane-embedded Na^+/K^+-ATPases. This discovery opened a new era in biochemistry and earned him a share of the 1997 Nobel Prize in Chemistry. His experiment was elegantly simple: measure the production of phosphate liberated from ATP as a function of the concentrations of NaCl and KCl bathing the nerve preparation (**Figure 12.3**). Because magnesium ions

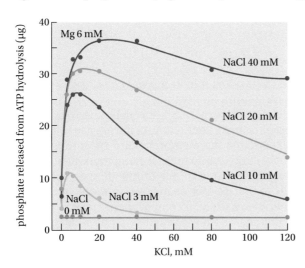

Figure 12.3 Discovery of the sodium-potassium ATPase. Experiments by Jens Skou using crab nerves revealed for the first time the biochemistry of the Na^+,K^+-ATPase. The data show that the enzyme activity, judged by the production of PO_4^{-3} from the splitting of ATP, depends upon both Na^+ and K^+ ions, as expected for a Na^+/K^+ ion pump that catalyzes the reaction $Na^+(in) + K^+(out) + ATP <=> ADP + P_i + Na^+(out) + K^+(in)$. (From Skou JC [1957] *Biochimica et Biophysica Acta* 23: 394–401. With permission from Elsevier.)

were found to be necessary for ATPase activity, the ATPase became known as the Na^+-K^+-Mg^{2+}-ATPase.

Skou's initial discovery raised important questions such as where, how, and when ATP is split into $ADP + P_i$ during ion pumping, how the protein is arranged in the membrane, and where K^+ and Na^+ are located during the pump cycle. Skou and other distinguished scientists, including Ian Glynn, Robert Post, and R. Wayne Albers, worked on these problems. Biochemical experiments revealed that a critical feature of ion pumping is the phosphorylation of the ATPase during the pump cycle and that this phosphorylation step is required to change the pump from being Na^+ selective to K^+ selective. Post and Albers worked out the consecutive steps of the pump cycle, and Glynn showed that the chemical reactions in the cycle are reversible. That is, reversal of the normal ion concentrations across a membrane will lead to the production of ATP from ADP and P_i. This is a result of the phosphorylation of the ATPase, which can keep the high-energy phosphate localized spatially within the enzyme. Another important result of the investigations of Post, Albers, and others was that the Na^+ and K^+ ions must be occluded within the membrane at some point in the pump cycle so that they cannot simultaneously access the aqueous phases on either side of the membrane. The first three-dimensional structures of P-type ATPases, determined in the laboratories of Chikashi Toyoshima in Japan and Poul Nissen in Denmark, revealed the structural basis for the P-type ion pumps at the atomic level.

In all P-type pumps, there is an inward-facing state called E1 that is selective for the ion to be pumped out (e.g., Na^+) and an outward-facing state called E2 that is selective for the ion to be pumped in the opposite direction (e.g., K^+) (**Figure 12.4A**). Each of these states can convert to an occluded state with either ion trapped within the membrane-embedded domain. Phosphorylation/dephosphorylation activates a structural switch that makes the protein cycle between the E1 and E2 states. The first Na^+,K^+-ATPase structure (Figure 12.4B and Figure 12.4C) revealed the enzyme in the $E2P_i$ state (boxed in orange, Figure 12.4A).

These biochemical features are similar for all P-type ATPases. They all undergo a phosphorylation/dephosphorylation cycle that involves an invariant Asp residue located in the P-domain, quite far away from the ion-transporting T-domain (Figure 12.4B). Phosphorylation and dephosphorylation of this Asp residue drives large conformational changes in the cytoplasmic part of the transporter that are transmitted to the membrane domain, thereby driving alternating access of the substrate-binding sites to the two sides of the membrane. Motions of the nucleotide-binding N-domain and the actuator A-domain underlie the conformational changes of the ATPase that define the pump cycle structurally. Incidentally, one of the oldest drugs known to physicians used to treat heart ailments, digitalis, exerts its effects by partially inhibiting the cardiac Na^+,K^+-ATPase (**Box 12.1**).

12.1.2 The Complete Pump Cycle of the Ca^{2+}-ATPase Is Understood Structurally

The Ca^{2+} ATPase plays a key role in muscle cells. Muscle contraction is initiated when a motor neuron innervating the muscle cell releases acetylcholine into the synaptic cleft (**Figure 12.5**). Acetylcholine triggers the activation of acetylcholine receptors in the membrane of the muscle cell, leading to ion influx into the cell, depolarizing the membrane and triggering an action potential. The action potential spreads into the cell along transverse T-tubules, activating voltage-dependent Ca^{2+} channels (dihydropyridine receptors) that are located in close proximity to Ca^{2+}-release channels (ryanodine receptors) in the sarcoplasmic reticulum, an intracellular compartment that is part of the endoplasmic reticulum (ER). This causes a massive efflux of Ca^{2+} ions from the sarcoplasmic reticulum to the cytoplasm, where the sudden rise in Ca^{2+} initiates contraction of the actin-myosin filaments. To end the contractile phase, Ca^{2+} must be

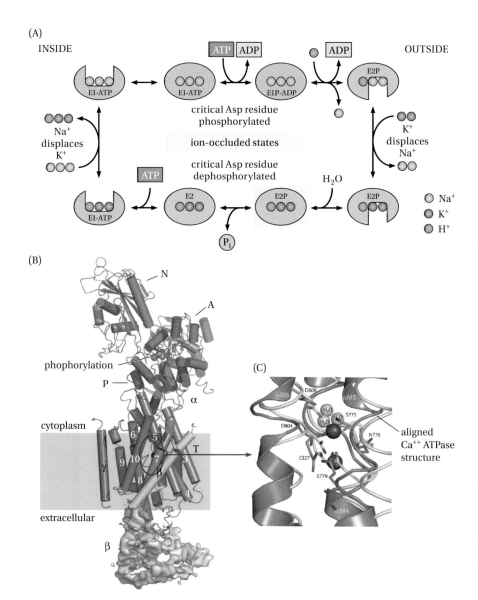

(A)

INSIDE

OUTSIDE

critical Asp residue
phosphorylated

ion-occluded states

critical Asp residue
dephosphorylated

Na⁺ displaces K⁺

K⁺ displaces Na⁺

○ Na⁺
○ K⁺
○ H⁺

(B)

N

A

phophorylation

P

α

cytoplasm

T

γ   β

extracellular

β

(C)

aligned Ca⁺⁺ ATPase structure

Figure 12.4 The function and structure of the Na^+,K^+-ATPase. The defining feature of the ATPase is the phosphorylation of the enzyme during the pump cycle; hence, the classification of the ATPase as P-type. (A) The Post-Albers pump cycle involves two states, the E1 inward-facing state and E2 outward-facing state. In the E1-ATP state, the enzyme with bound ATP favors the binding of Na^+ over K^+, which causes a conformational change and occlusion of the Na^+ within the membrane (E1-ATP). This conformational change leads to the splitting of ATP to yield a high-energy phosphorylated state (E1P-ADP). Release of the ADP leads to the E2P phosphorylated state, which favors the release of Na^+ and the binding of K^+. Upon dephosphorylation, K^+ becomes occluded (E2-P_i state). Loss of P_i (E2 state) and the binding of ATP returns the enzyme to the E1-ATP state, initiating another pump cycle. (B) The structure of Na^+,K^+-ATPase in the E2P_i state. K^+ is occluded in the center of the membrane in the T-domain, with no exit paths into an aqueous phase. (C) Close-up view of the two bound K^+ ions (in red). The image also shows the structural similarity of T-domains of the Na^+,K^+- and the Ca^{2+}-ATPase (in yellow). (A, From Morth JP et al. [2011] *Nature* 12: 60–70. With permission from Springer Nature. B, C, From Morth JP et al. [2007] *Nature* 450: 1043–1049. With permission from Springer Nature.)

pumped back into the sarcoplasmic reticulum. This is where the Ca^{2+}-ATPase (**Figure 12.6A**) enters the game, using ATP to drive the pump. In order to quickly clear out the Ca^{2+}, the concentration of Ca^{2+}-ATPase in the sarcoplasmic reticulum membrane is very high.

The different conformational states during the reaction cycle of the Ca^{2+}-ATPase have been visualized by x-ray crystallography, using various ATP analogs and inhibitors to trap the protein in different states (Figure 12.6B). Starting from the inward-open E1·2Ca^{2+} state with two Ca^{2+} ions bound in the T-domain, ATP binding induces a transition to the E1·ATP state in which the ATP bound to the N-domain is held close to the conserved Asp residue in the P-domain. A further slight rearrangement in the P-domain creates a binding site for Mg^{2+} near the conserved Asp residue. The Mg^{2+} ion neutralizes the negative charges on the Asp and the terminal phosphate group on the ATP molecule, allowing the transfer of the phosphate to the Asp residue to form the ~E1P state.

Phosphate transfer from ATP to the Asp residue breaks the ATP-mediated interaction between the N- and P-domains. The P-domain bends forward, inducing a stretching of the linker between the transmembrane helix M3 and the A-domain, which in turn causes the A-domain to rotate by almost 90°. This conformational change destroys the high-affinity of the Ca^{2+} binding site, opens an exit channel in the T-domain towards the extra-cytoplasmic compartment,

BOX 12.1   DIGITALIS

Perhaps the most well-known cardiac drug—and certainly the oldest—is digitalis. Digitalis is prepared from the foxglove plant (*Digitalis lanata*). Nowadays, one uses digoxin, Box **Figure 1**, a pure compound extracted from the plant.

Digoxin and related compounds inhibit the Na⁺,K⁺-ATPase in the cardiac muscle, leading to an increase in intracellular Na^+ and a decrease in the Na^+ gradient across the cell membrane. This in turn lowers the driving force for the Na^+,Ca^{2+} antiporter (see Chapter 13 for a discussion of antiporters) and results in increased intracellular Ca^{2+} concentrations in the sarcoplasmic reticulum. When the muscle is stimulated, more Ca^{2+} can thus be released from the sarcoplasmic reticulum, leading to stronger muscle contraction.

The scientific discovery of digitalis goes back to 1775, when an English country doctor and botanist, William Withering, had a sick patient who became much better after receiving a potion from a gypsy woman. Withering searched throughout Shropshire for the woman. He finally found her and enticed her to part with her secret formula, which contained—among other things—a preparation of the purple foxglove. After experimenting with different kinds of preparations, he found that dried, powdered leaves gave the best results, although one had to be very careful when administering the drug because of the fine line between concentrations that confer beneficial effects on the heart and toxic effects.

Withering is also known for his treatise on the plants of Britain (Box **Figure 2**), a two-volume tome with the impressive 130+ word title: *A Botanical Arrangement of All the Vegetables Naturally Growing in Great Britain. With Descriptions of the Genera and Species, According to the System of the Celebrated Linnaeus. Being an Attempt to Render Them Familiar to Those who are Unacquainted with the Learned Languages. Under Each Species are Added, the Most Remarkable Varieties, the Natural Places of Growth, the Duration, the Time of Flowering, the Peculiarities of Structure, the Common English Names; the Names of Gerard, Parkinson, Ray and Bauhine. The Uses as Medicines, Or as Poisons; as Food for Men, for Brutes, and for Insects. With Their Application in Oeconomy and in the Arts. With an Easy Introduction to the Study of Botany. Shewing the Method of Investigating Plants, and Directions how to Dry and Preserve Specimens. The Whole Illustrated by Copper Plates and a Copious Glossary.* Coming up with catchy titles was a very different business in Withering's day!

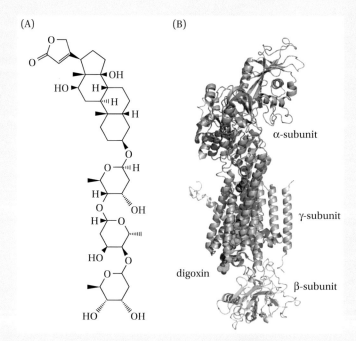

(A)   (B)

Figure 1 Digoxin interacts with the Na⁺,K⁺-ATPase (PDB 4RET). (A) Chemical structure of digoxin. (B) Digoxin (green, in spacefill) binds to the Na⁺,K⁺-ATPase in the entrance path of K⁺ and Mg²⁺ to inhibit the ATPase.

(Continued)

and positions the phosphatase site in the A-domain close to the phosphorylated Asp residue in the P-domain. The protein is now in the outward-open E2P state, in which the Ca^{2+} ions are free to diffuse away. For the Ca^{2+}-ATPase, protons serve as counter-ions, replacing the two Ca^{2+} ions from the extra-cytoplasmic side. Binding of H^+ causes an additional, small movement of the A-domain,

BOX 12.1   DIGITALIS (CONTINUED)

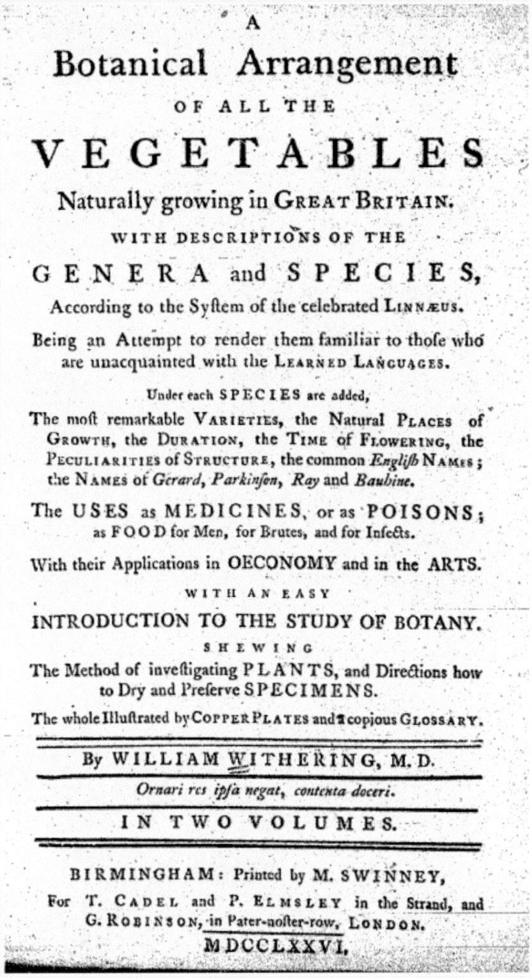

A

Botanical Arrangement

OF ALL THE

VEGETABLES

Naturally growing in GREAT BRITAIN.

WITH DESCRIPTIONS OF THE

GENERA and SPECIES,

According to the System of the celebrated LINNÆUS.

Being an Attempt to render them familiar to those who
are unacquainted with the LEARNED LANGUAGES.

Under each SPECIES are added,

The most remarkable VARIETIES, the Natural PLACES of
GROWTH, the DURATION, the TIME of FLOWERING, the
PECULIARITIES of STRUCTURE, the common *English* NAMES;
the NAMES of *Gerard, Parkinson, Ray* and *Bauhine*.

The USES as MEDICINES, or as POISONS;
as FOOD for Men, for Brutes, and for Insects.

With their Applications in OECONOMY and in the ARTS.

WITH AN EASY

INTRODUCTION TO THE STUDY OF BOTANY.

SHEWING

The Method of investigating PLANTS, and Directions how
to Dry and Preserve SPECIMENS.

The whole Illustrated by COPPER PLATES and a copious GLOSSARY.

By WILLIAM WITHERING, M.D.

Ornari res ipsa negat, contenta doceri.

IN TWO VOLUMES.

BIRMINGHAM: Printed by M. SWINNEY,
For T. CADEL and P. ELMSLEY in the Strand, and
G. ROBINSON, in Pater-noster-row, LONDON.
MDCCLXXVI.

Figure 2 Title page of William Withering's treatise on the plants of Britain. (With permission from Bibliothèque nationale de France.)

aligning the phosphatase site with the phosphorylated Asp residue, which can now be dephosphorylated. Finally, the A-domain rotates back, the P-domain bends upwards, the N-domain binds ATP, and the T-domain changes back to the inward-open state with high affinity for Ca^{2+}, returning the pump to the $E1 \cdot 2Ca^{2+}$ state. Representative structures of the Ca-free E2 and the $E1 \cdot 2Ca^{2+}$ states are shown in **Box 12.2**.

12.1.3 P-type ATPases Are Structurally Similar but Serve Many Different Transport Functions

The P-type ATPases have obvious structural similarities (**Figure 12.7**); all are believed to function much as the Ca^{2+}-ATPase, differing only in details such as the number of transmembrane helices in their T- and S-domains and their association with regulatory proteins. Based on sequence similarity, P-type ATPases have been classified into five subfamilies (P1–P5). Ca^{2+}-ATPase and Na^+,K^+-ATPase both belong to the P2 subfamily, as does the gastric H^+,K^+-ATPase that acidifies the gastric juice in our stomach to begin digestion of our food and to kill harmful organisms.

Figure 12.5 Schematic view of the regulation of the Ca^{2+} concentration within a muscle fiber. Muscle contraction is initiated by an action potential spreading along the surface of the muscle cell, causing release of Ca^{2+} from the sarcoplasmic reticulum (SR) via the dihydropyridine (DHP) and ryanodine (Ry) receptors. The dramatic rise in cytoplasmic Ca^{2+} triggers contraction of the actin-myosin filaments. Relaxation of the filaments ensues when the Ca^{2+}-ATPase pumps Ca^{2+} back into the SR in exchange for protons.

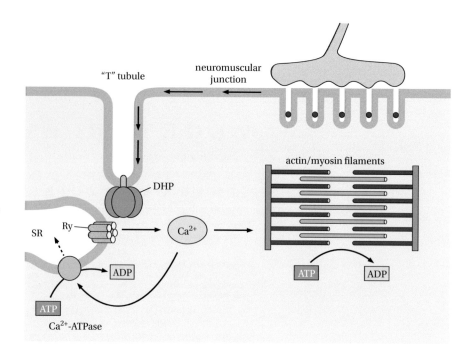

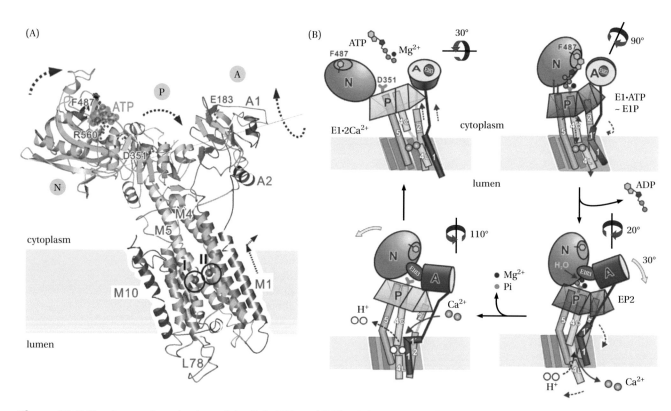

Figure 12.6 Structure and mechanism of the Ca^{2+}-ATPase. (A) The Ca^{2+}-ATPase in the $E1 \cdot 2Ca^{2+}$ state, viewed parallel to the membrane plane (PDB 1SU4). The ribbon diagram is colored using a spectrum that changes gradually from the amino terminus (blue) to the carboxy terminus (red). The purple spheres (circled) within the T-domain represent bound Ca^{2+}. The protein has three cytoplasmic domains (A, N, and P). The α-helices in the A-domain and in the T-domain (M1–M10) are indicated. ATP, shown in transparent space-filling format, is docked to the N-domain, and the protein is ready to move into the $E1 \cdot ATP$ state. Shown in ball-and-stick format are key residues involved in ATP binding (E183, F487, and R560) and the phosphorylation site D351. The rotation axis of the A-domain is indicated by the thin orange line. (B) Cartoon of the structural changes of the Ca^{2+}-ATPase during the reaction cycle, based on crystal structures in seven different states (see text). (From Toyoshima C [2009] *Biochim Biophys Acta <BBA> Molecular Cell Research* 1793: 941–946. With permission from Elsevier.)

BOX 12.2   THE E2 AND E1 STATES OF Ca²⁺ ATPASE

Representative structures of sarcoplasmic reticulum Ca²⁺-ATPase (SERCA) in the Ca-free $[H_n]E2$:ATP state and the $[Ca_2]E1$:ATP state are shown in Box **Figure 1**. Close-up views of the cytoplasmic domains of the two states are shown in Box **Figure 2**. An inactive state with the small transmembrane peptide inhibitor sarcolipin (magenta) bound to the membrane domain is also shown. Sarcolipin is a regulatory protein found in skeletal muscle SERCA that regulates contractility and non-shivering heat generation. In cardiac SERCA, these roles are fulfilled by the protein phospholamban.

In the $[H_n]E2$:ATP state, H⁺ ions have replaced the Ca²⁺ ions in the T-domain, and the critical Asp³⁵¹ in the P-domain (in spacefill) that becomes phosphorylated during the reaction cycle is far from the ATP molecule. In the $[Ca_2]E1$:ATP state, two Ca²⁺ ions (red spheres) have entered the T-domain from the cytosolic side, displacing the H⁺ ions. The N- and P-domains are closer together, bringing Asp³⁵¹ close to the ATP and priming it for the phosphotransfer reaction. Note how the A- and T-domains move in concert between the two states. The single-spanning transmembrane peptide sarcolipin (magenta in Box Figure 1)

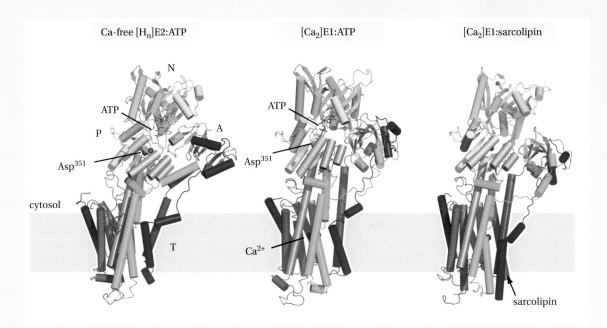

Figure 1 Structures of the sarcoplasmic reticulum calcium ATPase (SERCA) in three different states (PDB 2C8K, 1T5S, 4H1W).

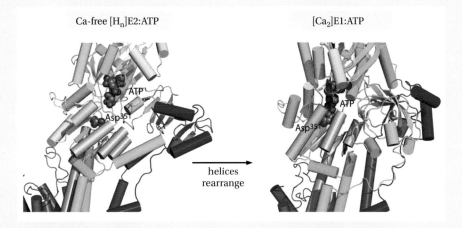

Figure 2 Close-up views of the Asp³⁵¹ region in SERCA in the Ca-free $[H_n]E2$:ATP and $[Ca_2]E1$:ATP states. Non-hydrolyzable ATP analogs were used in the crystallization experiments in order to trap these intermediates.

(Continued)

BOX 12.2 THE E2 AND E1 STATES OF CA²⁺ ATPASE (CONTINUED)

[Ca₂]E1:sarcolipin   [Ca₂]E1:ATP

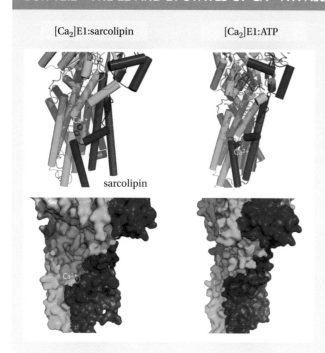

sarcolipin

traps the protein in a [Ca₂]E1-like state but with some differences, mainly in the T-domain (Box **Figure 3**). In all the figures, proteins are colored from the N terminus (blue) to the C terminus (red).

Figure 3 Sarcolipin (red helix) affects the accessibility of Ca²⁺ to the calcium binding sites.

P1-ATPases are involved in K^+ transport in bacteria, although it is unclear what, if any, ion they pump. The P1 group also includes heavy-metal pumps that can extrude ions such as Cu^+, Ag^+, Cu^{2+}, Zn^{2+}, Co^{2+}, Pb^{2+}, and Cd^{2+} from both prokaryotic and eukaryotic cells.

P3-ATPases pump protons out of plant and fungal cells, creating steep electrochemical gradients that can drive secondary active transporters (see Chapter 13).

P4-ATPases constitute a large gene family in most eukaryotes (there are 14 genes encoding P4-ATPases in the human genome). Some of these proteins are phospholipid flippases that help create and maintain phospholipid asymmetry in biological membranes (**Box 12.3**), while others are involved in vesicular fission reactions in the exo- and endocytic transport pathways. It appears that when large phospholipids are flipped across the membrane, the polar headgroup is transported along a "canyon" in the T-domain, while the lipid tails stick out into the surrounding membrane. Another family of proteins, referred to as lipid scramblases, are also involved in lipid flipping. These proteins are Ca^{2+}-activated but structurally unrelated to the P-type ATPases.

P5-ATPases so far have only one known function, namely to dislocate a transmembrane helix from a mitochondrial tail-anchored protein that has become mislocalized to the ER membrane. In humans, the P5-ATPases are found in the ER and lysosomal membranes.

12.2 THE TRANSPORT PRINCIPLES OF ABC TRANSPORTERS DIFFER FROM THOSE OF THE P-TYPE ATPASES

As do the P-type ATPases, ABC transporters use readily available ATP, the cell's main "energy currency," to drive uphill transport of many different substrates. Under normal cellular conditions, the hydrolysis of 1 mole of ATP to ADP and

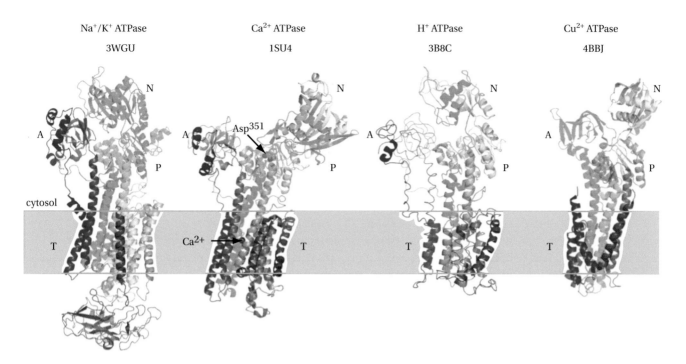

Na⁺/K⁺ ATPase   Ca²⁺ ATPase   H⁺ ATPase   Cu²⁺ ATPase

Figure 12.7 The P-type ATPases are grossly similar structurally. The polypeptide chains are colored from N-terminus (blue) to C-terminus (red). Although there are many differences in structural details, all of them have A-, N-, P-, and T-domains. The binding sites for the transported ions are located in the T-domain. For the Ca²⁺ ATPase, the locations of the two bound Ca²⁺ ions in the T-domain (red spheres) and the Asp residue in the P-domain (in space filling), which becomes phosphorylated during the reaction cycle, are highlighted. The PDB codes of the structures are indicated in the figure.

phosphate liberates about 50–70 kJ mol⁻¹ (12.17 kcal mol⁻¹) of free energy (see Chapter 14). In theory, this is enough to transport 1 mole of an uncharged substrate against a 10^6-fold concentration gradient, although in practice, various losses lead to a somewhat lower gradient. The trick that primary transporters must perform is to couple ATP hydrolysis to a transport event in such a way that most of the energy is converted into useful work to create ion or other solute gradients. This requires a tight coupling between the binding of substrates and the binding and hydrolysis of ATP.

In the highly simplified "alternating access" reaction scheme shown in **Figure 12.8**, the relaxed state of an ABC transporter in the absence of both substrate and ATP is assumed to face outwards. Binding of substrate stabilizes a conformation with increased affinity for ATP. ATP is hydrolyzed, shifting the conformational equilibrium of the transporter to an inward-open, ADP-bound state. ADP dissociates from the transporter (the concentration of ADP inside the cell is low), leaving the transporter in a state with low affinity for the substrate, which then also dissociates. Finally, the empty transporter returns to its most stable outward-facing conformation. The net result is that one ATP has been hydrolyzed to ADP (an energetically favorable reaction), and one substrate molecule has been transported from a low-concentration to a high-concentration compartment (an energetically unfavorable reaction).

ABC transporters are found in all organisms. Their most common function is to couple ATP hydrolysis to the uphill transport of small molecules, either into cells (e.g., ions or enzyme co-factors) or out of cells (e.g., toxic drug-like compounds or various kinds of peptides). For instance, tumors can become resistant to chemotherapy through selection for cells that express high levels of ABC exporters, and antigen-presenting cells in the immune system use a peptide exporter in the ER to translocate virus-derived peptides into the secretory pathway for their ultimate display on the cell surface.

All ABC transporters are composed of two membrane domains and two cytoplasmic ATP binding domains. The different domains usually correspond to

BOX 12.3   PHOSPHOLIPID TRANSPORT: FLIPPASES AND SCRAMBLASES

P4-ATPases carry out ATP-driven transport of phospholipids from one membrane monolayer to the other, i.e., they flip lipids across the membrane and hence are called flippases. One example is flippases that flip phosphatidylserine (PS) from the outer to the inner surface of the plasma membrane in eukaryotic cells; an important activity since the presence of PS on the outer surface of cells triggers their destruction by phagocytic cells.

The overall structural features of the P4-ATPases are similar to the Na+,K+-ATPase and the Ca2+ ATPase (Box **Figure 1**), but in contrast to the latter, the P4-ATPases translocate big lipid substrates rather than small ions across the membrane. The structure shows a phospholipid substrate bound to the flippase with its headgroup buried in the T-domain and the acyl chains sticking out into the membrane. This suggests that the head group is translocated across the membrane inside the flippase, while the acyl chains remain embedded in the surrounding lipid bilayer during the flipping reaction.

Lipids can also be flipped across the membrane by ABC flippases, i.e., flippases that belong to the ABC transporter superfamily. The PglK ABC flippase transports lipid-linked oligosaccharides across the membrane as a critical step in the pathway that allows cells to glycosylate Asn-residues in secreted proteins. A typical substrate is shown in Box **Figure 2A**. The structure of an outward-occluded state of PlgK is shown in Box Figure 2B, C. Biochemical experiments indicate that only outward-open states are involved in flipping of the lipid substrate. The model shown in Box **Figure 3** posits that the end of the polyprenyl tail initiates translocation by binding in a pocket underneath the two short α-helices at the top of the fully outward-open transporter, and that the pyrophosphate moiety that connects the long undecaprenyl chain to the oligosaccharide enters the cavity, pulling the oligosaccharide in behind it. ATP hydrolysis then squeezes the cavity to a smaller volume, pushing the large lipid headgroup to the trans-side of the membrane. The hydrophobic lipid tail of the substrate remains in the lipid bilayer at all times.

In contrast to flippases, scramblases catalyze the equilibration of lipid species across a membrane bilayer and do not require an energy input. The first structure of a scramblase (from the fungus *Nectria haematococca*) was determined in 2014 (Box **Figure 4**). The protein is a dimer. At the extreme end of each subunit is a canyon that extends across the entire width of the surrounding membrane, with many polar residues facing into the canyon. An attractive hypothesis is that lipid headgroups can move back and forth across the membrane via this hydrophilic canyon, while the lipid tails remain embedded in the surrounding bilayer. Molecular dynamics simulations support this hypothesis and suggest that lipid headgroups can pass through the canyon with ease: the energy barrier has been estimated to be on the order of only 1 kcal mol^{-1}.

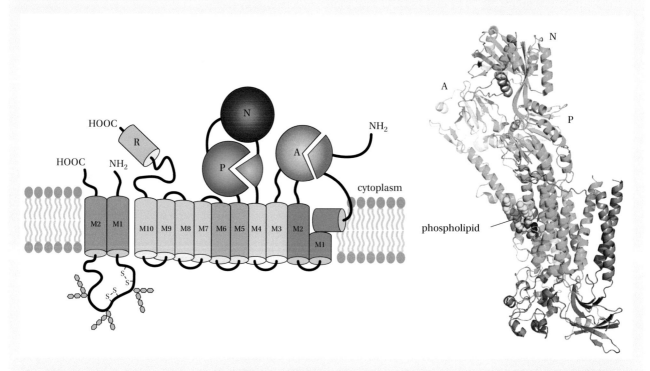

Figure 1 Topology and structure of a P4-ATPase. In addition to the canonical A-, N-, P-, and T-domains, there is a regulatory R-domain and an associated β subunit (CDC50; pink) that is necessary for maturation and intracellular transport of the protein. The structure shows the enzyme trapped in the E1P$_i$-state with a bound phospholipid (PL: headgroup—red, acyl chains—yellow; PDB 6K7M).

(Continued)

BOX 12.3   PHOSPHOLIPID TRANSPORT: FLIPPASES AND SCRAMBLASES (CONTINUED)

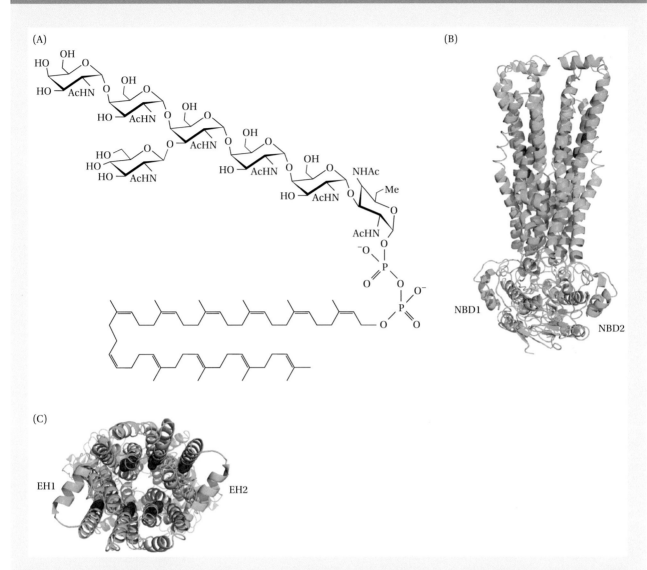

Figure 2 The PglK ABC transporter. (A) Structure of the lipid-linked oligosaccharide substrate. (B) The outward-occluded state of PglK (PDB 5C73). Note the long slit in the membrane domain, through which the pyrophosphate and oligosaccharide chain are thought to enter the central cavity. Arginine residues critical for substrate binding lining the cavity are shown in red. The ATPase domains NBD1 and NBD2 are indicated. (C) Top view. Two short α-helices at the extracellular end (EH1, EH2) below which the end of the polyprenyl chain may bind are indicated. (A, From Perez C, Köhler M, Janser D et al. [2015] *Nature* 524: 433–438.)

(Continued)

different polypeptides but are sometimes fused together. The ABC importers generally have an additional substrate-binding receptor that delivers the substrate to the membrane domain. In the ATP-bound state, the two ATP binding domains form a tight dimer with two ATPs bound between them. The ATP binding domains are well conserved in evolution and show strong sequence similarity across all ABC transporters, but the extant membrane domains appear to have evolved from at least three independent origins. ABC transporters are therefore classified into three distinct superfamilies (aptly named ABC1, ABC2, and ABC3) based on the kind of membrane domain they possess.

BOX 12.3   PHOSPHOLIPID TRANSPORT: FLIPPASES AND SCRAMBLASES (CONTINUED)

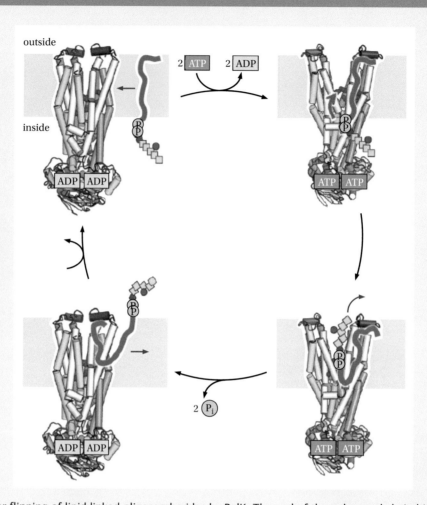

Figure 3 Model for flipping of lipid-linked oligosaccharides by PglK. The end of the polyprenyl chain binds underneath helices EH1 and EH2 (in red), while the pyrophosphate moiety (circled Ps) enters the central cavity, followed by the oligosaccharide chain (yellow). ATP hydrolysis induces a tightening of the cavity, squeezing the headgroup to the trans-side of the membrane (much as when one squeezes a toothpaste tube). (From Perez C, Köhler M, Janser D et al. [2015] *Nature* 524: 433–438.)

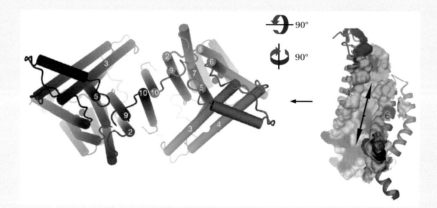

Figure 4 A lipid scramblase, seen from above (left) and within the plane of the membrane looking in the direction of the arrow (right). As seen in the right-hand panel, the polar canyon running across the membrane is lined with polar residues. The hypothetical path for passage of lipid headgroups across the membrane is indicated by the double arrow. (From Brunner JD et al. [2014] *Nature* 516: 207–212. With permission from Springer Nature.)

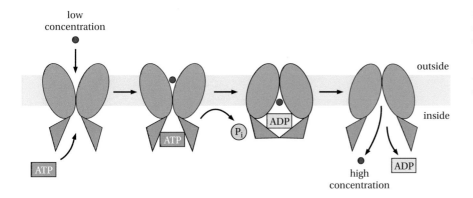

Figure 12.8 Basic "alternating access" reaction scheme for an ATP-driven transporter of the ABC transporter superfamily. The substrate is in red. ATP hydrolysis switches the transporter from an outside-facing to an inside-facing conformation.

12.2.1 ABC Importers: The Maltose Transporter

A particularly well-understood ABC importer is the inner membrane maltose transporter from *E. coli*. Maltose first gains access to the periplasmic space between the outer and inner membranes via the outer membrane trimeric porin LamB (**Figure 12.9**). The channel through each subunit of LamB is lined by aromatic residues on one side and hydrogen-bond donors and acceptors on the opposite side. This allows maltose to move along the "greasy slide" formed by the aromatic residues while satisfying the polarity of its hydroxyl groups by hydrogen bonding to polar residues (see Figure 10.25).

Once in the periplasm, maltose binds with high affinity to the soluble maltose-binding protein MalE. The favorable energy of binding to MalE reduces the periplasmic concentration of free maltose, trapping it and causing accumulation of the sugar by mass action. MalE closes up around the bound maltose, taking up a conformation that can bind to the maltose transporter, an inner membrane protein composed of two membrane subunits, MalF and MalG, and a dimer of the cytoplasmic nucleotide-binding domain (NBD) protein MalK. The inactive transporter exists mainly in an inward-facing, ATP-free form (**Figure 12.10A**). Interaction with maltose-loaded MalE stabilizes a pre-translocation intermediate with high affinity for ATP (Figure 12.10B). ATP binding to MalK leads to further structural rearrangements, resulting in an outward-facing conformation in which MalE has been forced open (Figure 12.10C), thereby allowing maltose to move out of MalE and

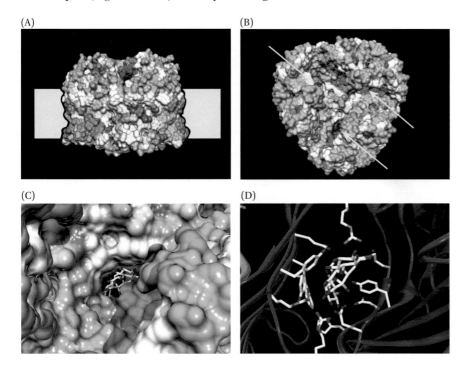

Figure 12.9 LamB (PDB 1MPM) is a trimeric porin that allows maltose to diffuse across the outer membrane. (A) Side view. Hydrophobic, lipid-facing residues are shown in orange. The membrane is indicated by the grey strip. (B) Top view. Note the three narrow channels (arrows), each with a bound maltose molecule. (C) Close-up of one channel with a bound maltose. (D) Same as in (C). Note the aromatic residues lining the left-hand wall of the channel and the polar residues lining the right-hand wall.

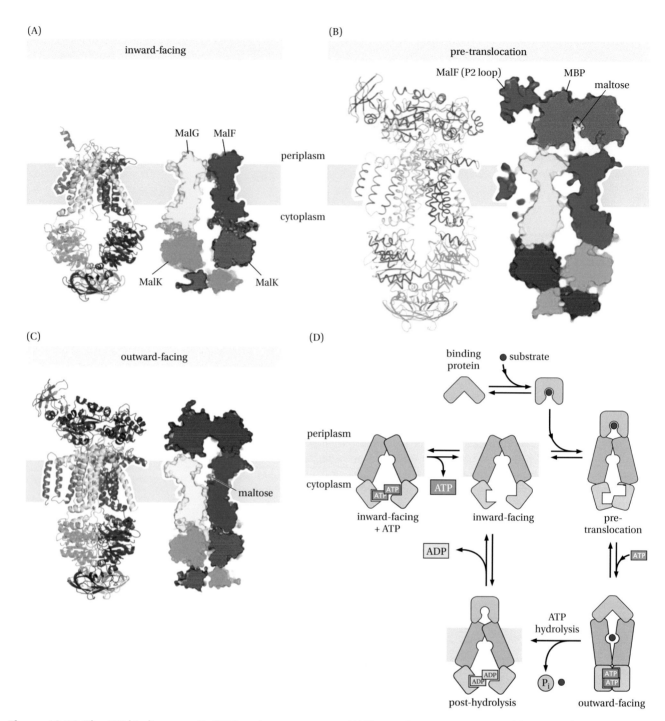

(A) inward-facing

MalG   MalF

periplasm

cytoplasm

MalK   MalK

(B) pre-translocation

MalF (P2 loop)   MBP

maltose

(C) outward-facing

maltose

(D)

binding protein   ● substrate

periplasm

cytoplasm

ATP   ATP

inward-facing + ATP   inward-facing   pre-translocation

ADP   ATP

ATP hydrolysis

ADP   ATP ATP

P_i ●

post-hydrolysis   outward-facing

Figure 12.10 The ATP binding cassette (ABC) maltose transporter. (A) The resting state is an inward-facing conformation that binds neither the periplasmic maltose-binding protein MalE nor ATP. The two ATP binding domains in the MalK dimer are far apart. (B) In the pre-translocation state, maltose-loaded MalE binds to the periplasmic face of the transporter. This forces the transporter to close off towards the cytoplasm and brings the two ATP binding domains closer together. (C) ATP binding in the interface between the two MalK subunits converts the transporter to an outward-facing state, thereby forcing the bound MalE to open up and release maltose into the transporter. Finally, ATP hydrolysis allows the transporter to return to the original inward-facing resting state. (D) Summary of the transport cycle. (A, C, From Khare D, Oldham ML, Orelle C et al. [2009] *Molecular Cell* 33: 528–536. With permission. B, D, From Oldham ML, Chen J [2011] *Science* 332: 1202–1205. With permission from The American Association for the Advancement of Science.)

into a binding pocket inside the transporter. Finally, ATP hydrolysis breaks the interaction between the two MalK NBDs, and the transporter returns to its inward-facing resting state, releasing maltose-free MalE to the periplasm and maltose to the cytoplasm. While there is not yet complete clarity, one way to understand the energetics is that the binding energy of ATP drives the change in conformation and

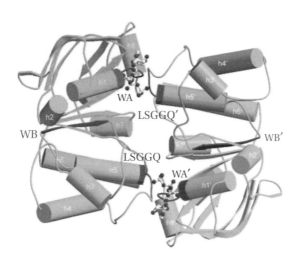

Figure 12.11 MalK nucleotide-binding domains (NBDs) in the ATP-bound state. The two NBDs come together to form two complete ATP binding sites, each involving residues from both NBDs. The ATP molecules are shown as stick models. (From Davidson AL, Chen J [2004] *Annu Rev Biochem* 73: 241–268. With permission from Amy Davidson.)

liberation of maltose from MalE, while its cleavage to ADP+P$_i$ allows the relaxation of MalFGK to the inward state with dissociation of the ADP and maltose to the inside. The whole cycle is shown schematically in Figure 12.10D.

In the maltose transporter, the energy supplied by ATP binding and hydrolysis is used not only to drive the transitions between the inward- and outward-facing states but also to make MalE release its rather tightly bound substrate (the K_D for binding maltose is around 1 μM). Indeed, the whole system, including the LamB porin, seems to be tuned to work at maximal velocity for maltose concentrations in the μM range; for higher concentrations, the amount of LamB in the outer membrane becomes rate-limiting.

A key feature of the MalK ATPase is that each NBD contains only a part of the two ATP binding sites present in the dimer (**Figure 12.11**). The active ATPase is thus formed only when the two NBDs simultaneously bind two ATPs, using the binding energy to drive the interface into a configuration in which both active sites are fully formed and simultaneously stabilizing the outward-facing conformation of the membrane subunits. The use of this mechanism may explain why the ABC transporters are dimers. Inhibition of the maltose transporter and related sugar transporters plays an important role in carbon catabolite repression, an important regulatory process in bacteria (**Box 12.4**).

12.2.2 ABC Exporters: P-glycoprotein

While ABC importers such as the maltose transporter have been found only in prokaryotes and archaea, ABC exporters are found in all domains of life. The human genome encodes 48 different ABC exporters, including P-glycoprotein (also called multidrug resistance protein 1 or MDR1), multidrug resistance-associated protein 1 (MRP1), and breast cancer resistance protein (BCRP). As their names suggest, these particular ABC transporters have been implicated in processes that render tumor cells resistant to chemotherapy by continuously pumping the drugs out of the cells, thereby lowering their concentration inside the cell. ABC exporters often have very broad substrate specificity and hence, are able to pump out many different kinds of drugs. While it is usual to think of the process as a pumping of molecules out of the cell, the actual event is to move relatively hydrophobic molecules, such as drugs, from the inner to the outer monolayer of the membrane bilayer, from where equilibration with the cytoplasm or external environment takes place.

Like the ABC importers, the exporters are built from two membrane domains and two cytosolic NBDs; one membrane domain and one NBD are often fused into a single polypeptide chain, and the complete transporter is thus either a homo- or a hetero-dimer. The NBDs are highly homologous between transporters. The membrane domains are more divergent, although they generally conform to a topology with six transmembrane helices in each monomer.

BOX 12.4   INHIBITION OF THE MALTOSE TRANSPORTER IN CARBON CATABOLITE REPRESSION

Structural studies of the maltose transporter have illuminated a central regulatory process in bacteria: carbon catabolite repression. When bacteria grow in media containing a preferred carbon source such as glucose, they shut down the activity and/or synthesis of enzymes required to metabolize other, less preferred sugars. They also block certain sugar transporters, including the maltose transporter. In *E. coli*, the key regulatory protein is the enzyme EIIA^{Glc}. EIIA^{Glc} can be reversibly phosphorylated depending on the level of glucose in the cell, and in its un-phosphorylated form, it binds to and inhibits a variety of transporters.

The structure of the MalF-MalG-MalK-EIIA^{Glc} complex is shown in Box **Figure 1**. Two EIIA^{Glc} molecules are wedged between the two MalK subunits, thereby stabilizing the transporter in the inward-facing resting state, preventing the binding of maltose-loaded MalE and formation of the pre-translocation intermediate. The ATP-induced transition to the outward-open state is hence blocked, and the transporter remains inactive as long as glucose is present.

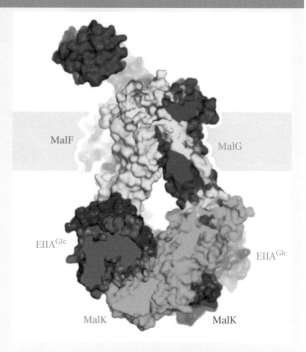

Figure 1 Two copies of the regulatory enzyme EIIA^{Glc} (purple) bind between the two MalK subunits, trapping the transporter in an inward-facing conformation (PDB 4JBW).

The first structure of an ABC exporter was of a bacterial protein called Sav1866, solved in 2006. Overall, the architecture is reminiscent of the MalF-MalG-MalK ABC importer (note that there is no equivalent of the maltose-binding protein MalE in this case), with a dimeric membrane domain connected to two NBDs (**Figure 12.12A**). However, the membrane domain has no structural similarity to the membrane domain in the maltose transporter and is divided into two "wings," each with six transmembrane helices. Two of the transmembrane helices in each monomer have been swapped into the other monomer, such that each wing is composed of helices 1–2 from one subunit and helices 3–6 from the other. The two NBDs are closely apposed with two ADPs bound between them. The membrane domain is outward-open with a large cavity that can fit various substrates (Figure 12.12B). Swapping of structural elements between subunits is also evident at the level of the interaction interface between the membrane domain and the NBDs. The interaction is mediated by coupling helices in the membrane domain that fit into cavities at the top of the NBDs, forming ball-and-socket joints. The coupling helices from one subunit mainly contact the NBD from the other subunit (Figure 12.12C).

Structures of the eukaryotic P-glycoprotein, both from mouse and from the worm *Caenorhabditis elegans*, are shown in **Figure 12.13**. In both, the NBDs are far apart, and the membrane domain is in an inside-open conformation. Portals between the wings in the membrane domain provide direct access from the cytosolic leaflet of the membrane into the substrate-binding cavity. This presumably allows hydrophobic substrates to partition directly from the lipid phase into the transporter, a key aspect of its function in drug resistance. Co-crystallization experiments with the mouse P-glycoprotein show that two different substrates can bind simultaneously in the transporter and hence, presumably, be co-transported (Figure 12.13B).

Comparing the inward-open conformations of the P-glycoprotein structures with the partially inward-open conformation of an ABC transporter from

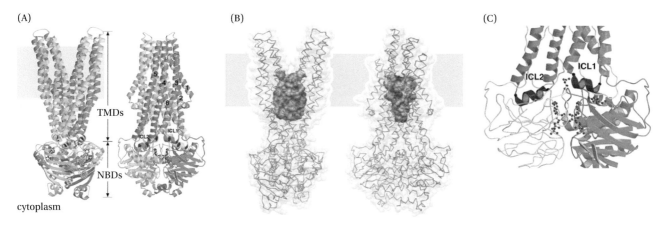

Figure 12.12 Structure of the bacterial Sav1866 ABC exporter involved in multidrug resistance. (A) Structure in ribbon format showing two views parallel to the membrane. The right-hand view is rotated 90° about the membrane normal. The transmembrane domains (TMDs) and the NBD domains are indicated. (B) The same structures as in panel A but shown in wire-frame format with a transparent surface representation superimposed. The molecular surface of the central cavity is indicated (blue: positive charges, red: negative charges). (C) Close-up view of the NBD domains. The coupling helices are in red. (A–C, From Dawson RJP et al. [2006] *Nature* 443: 180–184. With permission from Springer Nature.)

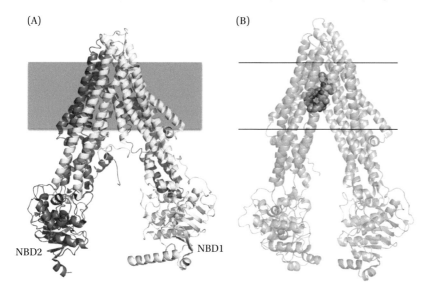

Figure 12.13 P-glycoprotein transporter from mouse (A; PDB 3G5U) and *C. elegans* (B; PDB 3G61). In (B), two different substrates are bound simultaneously in the central chamber.

Thermotoga and the outward-open conformation of Sav1866 (**Figure 12.14**), it is clear that binding of ATP and substrate to the inward-open state stabilizes a conformation where the NBDs are brought together to form the two intact ATP binding sites, closing off the cytosolic opening into the substrate-binding cavity in the membrane domain while simultaneously opening it towards the extracellular side. As the two bundles of transmembrane helices that define the inward-open state come together, the wings seen in the outward-open conformation separate along an axis that is perpendicular to the axis along which the NBDs come together, as seen in Figure 12.14.

Some ABC exporters have only one canonical ATP binding site in the NBDs, the second site being mutated to be catalytically inactive. This raises the possibility that one ATP remains bound to the non-canonical site throughout the transport cycle, which is then driven by the hydrolysis of just one ATP per transport event. The reaction scheme may thus be different for ABC exporters that have one or two canonical ATP binding sites (**Figure 12.15**).

Among all the ABC transporters, there is one odd man out: the Cystic Fibrosis Transmembrane Conductance Regulator (CFTR). CFTR is an ATP-regulated chloride channel, not an ATP-driven transporter, but otherwise conforms to the basic ABC transporter architecture (**Box 12.5**).

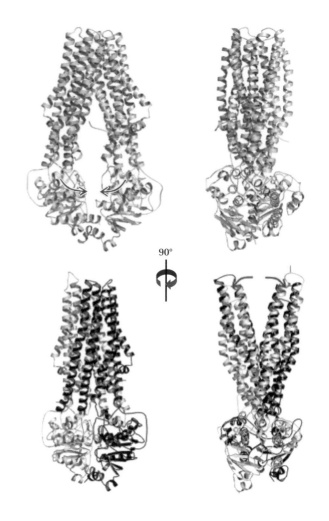

Figure 12.14 Comparison of inward-open (top) and outward-open (bottom) states of two ABC transporters. Note that the cytoplasmic and extracellular substrate-binding cavities open up along perpendicular axes. Top: ABC transporter from *Thermotoga maritima* in partially inward-open state. Bottom: Sav 1866 in outward-open state. (From Hohl M et al. [2012] *Nature Struct Mol Biol* 19: 395–402. With permission from Springer Nature.)

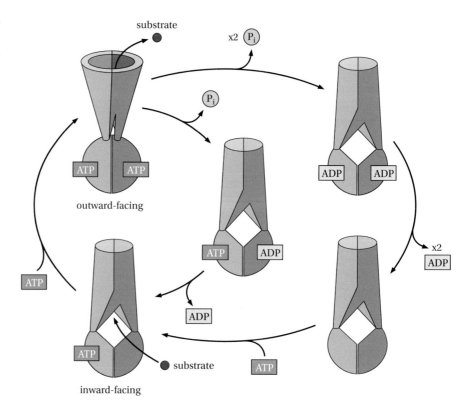

Figure 12.15 Reaction schemes for ABC transporters with one or two catalytically active ATP binding sites. The substrate is represented by the red dot. (From Hohl M et al. [2012] *Nature Struct Mol Biol* 19: 395–402. With permission from Springer Nature.)

BOX 12.5 THE CYSTIC FIBROSIS TRANSMEMBRANE CONDUCTANCE REGULATOR

Cystic fibrosis (CF) is a debilitating, genetically inherited disease affecting approximately 1 person out of 3000 among people of Northern European ancestry. The most severe effects are on the lungs, which accumulate a viscous mucus that makes breathing difficult and opens up the body to infections. CF is caused by mutations in the gene encoding the CFTR; both the cDNA and the gene encoding CFTR were sequenced in 1989, revealing that CFTR belongs to the family of ABC transporters and that the most common disease-causing mutation is a deletion of a phenylalanine residue in position 508 (ΔF508), found in some two-thirds of all CF patients.

Despite belonging to the ABC transporter family, CFTR is an ATP-regulated chloride channel, not an ATP-driven transporter. Unlike *bona fide* ABC transporters, CFTR also has a regulatory R-domain that must be phosphorylated for the channel to open. Further, only one of the two ATPase sites in the CFTR NBDs is catalytically active.

Despite intense work in many laboratories, it was not until 2016 that the structure of CFTR (from zebrafish) was determined thanks to the developments in cryo-electron microscopy (EM) (see Chapter 9). The structure—which represents the ATP-free, unphosphorylated closed state of the channel—shows a typical ABC transporter fold (Box Figure 1A). The unphosphorylated R domain is not well ordered in the structure but can be seen inserted between the two NBDs such that it prevents them coming together to activate the ATPase and open the channel. The ion-conduction pathway is funnel-shaped, closed at the extracellular end, and lined with positively charged residues that presumably contribute to the anion selectivity.

Phosphorylation of the R-domain displaces it from the NBD domain interface and allows NBD dimerization driven by ATP binding, as shown by a structure obtained for the phosphorylated, ATP-bound conformation (Box Figure 1B). In this particular structure the ion-conduction pathway is caught in an almost fully open state, and only a small loop displacement at the very top would be necessary to fully open the channel.

The critical F508 residue is located in NBD1 and is inserted into a hydrophobic pocket at the end of TMH11 (Box Figure 2). In vivo, the ΔF508 mutation renders CFTR highly unstable, and very little protein reaches the cell surface.

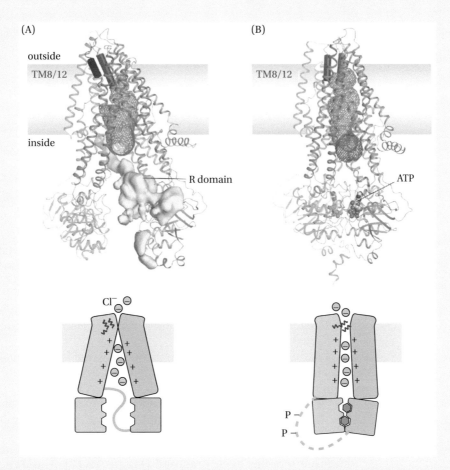

Figure 1 Cryo-EM structures of the unphosphorylated, ATP-free (A) and phosphorylated, ATP-bound (B) conformations of zebrafish CFTR (PDB 5W81), whose sequence is 55% identical to human CFTR. The grey mesh indicates the ion-conduction pore. The R-domain is sufficiently ordered to show visible density only in the ATP-free state. (From Zhang Z, Liu F, Chen J [2017] *Cell* 170: 483–491. With permission from Elsevier.)

(Continued)

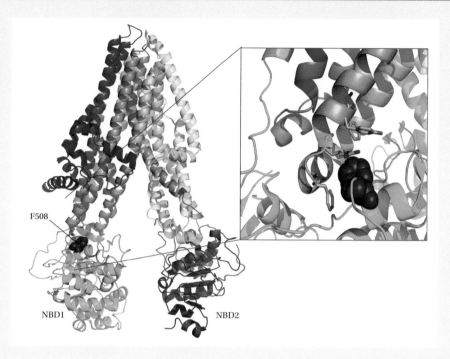

Figure 2 The ΔF508 mutation is a deletion of residue F508 (in purple) that removes an important hydrophobic interaction between NBD1 and TMH11. This causes the protein to become unstable and leads to its rapid degradation in the cell (PDB 5TSI).

12.3 ENERGY-COUPLING FACTOR TRANSPORTERS COUPLE ATP HYDROLYSIS TO "TOPPLING" OF THE SUBSTRATE-BINDING SUBUNIT

Energy-coupling factor (ECF) transporters are prevalent in prokaryotes, where they transport, e.g., water-soluble vitamins or transition metal ions. They generally consist of two modules: a substrate-binding membrane subunit (called the S-component) and an ECF module that is composed of a membrane subunit (called the T-component) and two ATPase subunits (the A-components). In some ECF transporters, a single ECF module can interact with a number of different S-components and hence, drive transport of a number of different substrates.

A key to the function of ECF transporters came from the first x-ray structures of an isolated S-component protein and a full ECF complex (**Figure 12.16**). Quite unexpectedly, the structures suggested that the S-component protein rotates or "topples over" in the plane of the membrane by almost 90° when it binds to the ECF component, thereby moving the substrate-binding site from the periplasmic to the cytoplasmic side of the membrane.

Based on biochemical studies, a model for ATP-driven transport by ECF transporters has been proposed (**Figure 12.17**). The ATP-bound "closed" form of the ECF module has little affinity for the substrate-free S-component but binds tightly to the substrate-loaded form. Binding of the S-component triggers ATP hydrolysis in the two closely apposed ATPase domains,

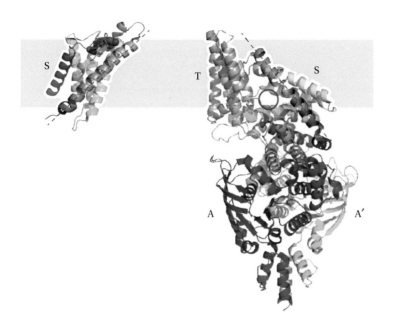

Figure 12.16 Energy-coupling factor (ECF) transporter. The S-component on the left (PDB 3RLB) undergoes a large rotation in the plane of the membrane when it binds to the ECF module; notice the change in the position of dashed line. Bound substrate in the isolated S-component (left) is shown in red space-filling form; the empty sub-strate-binding site (red circle) faces the cytoplasm in the full complex (right; PDB 4HUQ).

which then detach from each other while at the same time toppling the S-component across the membrane. This "open" form of the complex has low affinity for substrate, which is released to the cytoplasm. Re-binding of ATP to the ATPase domains resets the system for a new round of transport. This "release-and-catch" mechanism, where the S-component cycles on and off the ECF module, neatly explains how a single ECF module can serve many different S-components and thereby sustain transport of a range of substrates.

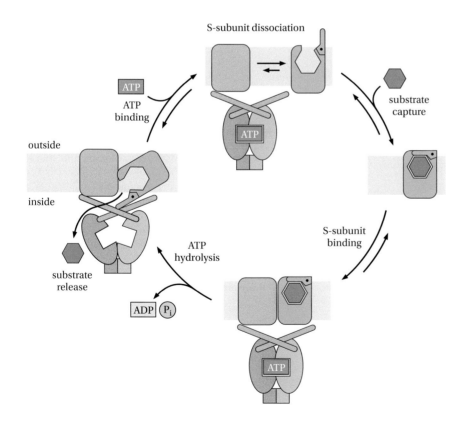

Figure 12.17 Proposed reaction scheme for ECF transporters. Only the substrate-bound form of the S-component has affinity for the ATP-bound form of the ECF module. ATP hydrolysis triggers a rotation of the S-component in the membrane, mediated by two coupling helices (in yellow) that slide against each other. (From Karpowich NK et al. [2015] *Nature Struct Mol Biol* 22: 565–571. With permission from Springer Nature.)

KEY CONCEPTS

- Primary transporters use ATP hydrolysis to directly drive transport processes.

- P-type ATPases couple ATP-dependent phosphorylation/dephosphorylation of an invariant Asp residue in the cytosolic P-domain to translocation of substrate across the membrane-embedded T-domain.

- P-type ATPases transport many different substrates, such as protons, mono- and divalent ions, phospholipids, and transmembrane helices.

- ABC transporters are composed of two cytosolic NBDs and a membrane-embedded transport domain. ATP

binding and hydrolysis involve elements from both NBDs acting in concert.

- ABC transporters commonly transport ions or small molecules into cells, or small peptides or toxic compounds out of cells.

- Up-regulation of clinically important ABC transporters can make cells resistant to, e.g., anti-cancer drugs used in chemotherapy.

- In ECF transporters, a single energy-coupling module can interact with, and drive transport through, many different substrate-biding S-components.

FURTHER READING

Andersen, J.P., Vestergaard, A.L., Mikkelsen, S.A., Mogensen, L.S., Chalat, M., and Molday, R.S. (2016) P4-ATPases as phospholipid flippases–Structure, function, and enigmas. *Front Physiol* 7:275.

Boos, W., and Shuman, H. (1998) Maltose/maltodextrin system of *Escherichia coli* : Transport, metabolism, and regulation. *Microbiol. Mol. Biol. Rev.* 62:204–229.

Chen, J. (2013) Molecular mechanism of the *Escherichia coli* maltose transporter. *Curr. Opin. Struct. Biol.* 23:492–498.

Glynn, I.M. (2002) A hundred years of sodium pumping. *Annu. Rev. Physiol.* 64:1–18.

Locher, K.P. (2016) Mechanistic diversity in ATP binding cassette (ABC) transporters. *Nature Struct. Mol. Biol.* 23:487–493.

Morth, J.P., Pedersen, B.P., Buch-Pedersen, MJ., Andersen, J.P., Vilsen, B., Palmgren, M.J., and Nissen, P. (2011) A structural overview of the plasma membrane Na^+,K^+-ATPase and H^+-ATPase ion pumps. *Nature Rev. Mol. Cell Biol.* 12:60–70.

Palmgren, M., and Nissen, P. (2011) P-type ATPases. *Annu. Rev. Biophys.* 40:243–266.

Slotboom, D.J. (2014) Structural and mechanistic insights into prokaryotic energy-coupling factor transporters. *Nature Rev. Microbiol.* 12:79–87.

KEY LITERATURE

Brunner, J.D., Lim, N.K., Schenck, S., Duerst, A., and Dutzler, R. (2014) X-ray structure of a calcium-activated TMEM16 lipid scramblase. *Nature* 516:207–212.

Dawson, R.J., and Locher, K.P. (2006) Structure of a bacterial multidrug ABC transporter. *Nature* 443:180–185.

Dutzler, R., Wang, Y-F., Rizkallah, P.J., Rosenbusch, J.P., and Schirmer, T. (1996) Crystal structures of various maltooligosaccharides bound to maltoporin reveal a specific sugar translocation pathway. *Structure* 4:127–134.

Lykke-Møller Sørensen, T., Vuust Møller, J., and Nissen, P. (2004) Phosphoryl transfer and calcium ion occlusion in the calcium pump. *Science* 304:1672–1675.

Oldham, M.L., Khare, D., Quiocho, F.A., Davidson, A.L., and Chen, J. (2007) Crystal structure of a catalytic intermediate of the maltose transporter. *Nature* 450:515–521.

Perez, C., Gerber, S., Boilevin, J., Bucher, M., Darbre, T., Aebi, M., Reymond, J.L., and Locher, K.P. (2015) Structure and mechanism of an active lipid-linked oligosaccharide flippase. *Nature* 524:433–438.

Riordan J.R., Rommens, J.M., Kerem, B., Alon, N., Rozmahel, R., Grzelczak, Z., Zielenski, J., Lok, S., Plavsic, N., Chou, J.L., Drumm, M.L., Iannuzzi, M.C., Collins, F.S., and Tsui, L.-C. (1989) Identification of the cystic fibrosis gene: Cloning and characterization of complementary DNA. *Science* 245:1066–1073.

Skou, J.C. (1957) The influence of some cations on an adenosie triphosphatase from peripheral nerves. *Biochim. Biophys. acta* 23:394–401.

Toyoshima, C., Nakasako, M., Nomura, H., and Ogawa, H. (2000) Crystal structure of the calcium pump of sarcoplasmic reticulum at 2.6 Å resolution. *Nature* 405:647–655.

Zhang, Z., and Chen, J. (2016) Atomic structure of the cystic fibrosis transmembrane conductance regulator. *Cell* 167:1586–1597.

EXERCISES

1. Figure 12.4 illustrates the states of the Na^+,K^+-ATPase. It is clear that the energy of ATP cleavage is a major driver of the cycle, and that the pumping of Na^+ and K^+ are uphill in concentration. But, there is another ion that is

moved across the membrane. What is it? Is it moved up or down its concentration gradient? Might its movement contribute to the energy of the transport of the other ions?

2. Given that muscle contraction is driven by the opening of Ca^{2+} channels and relaxation is via lowering of the cytoplasmic Ca^{2+} concentration by the Ca^{2+}-ATPase, do you expect a difference in the concentration of the channel and the ATPase in the sarcoplasmic reticulum membrane? Why? What implications are there for the rates of triggering contraction and its relaxation?

3. Why is the binding of maltose to MalE so tight (KD ~ 1 μM)? Compare the energy of binding of maltose to MalE with the typical energy available from the cleavage of ATP. Speculate on the possible difference—how does the reaction go?

4. What do the considerations in (3) tell us about the transporters that use only one ATP, as in Figure 12.15? How much of a concentration gradient might they support?

5. Examine the role of H^+ in the mechanism of the Ca^{2+} pump (Figure 12.6). Is its net flow with or against the H^+ gradient? Again, might it have a favorable or an unfavorable role in the energetics? What other role does it play?

6. Some ABC transporters transport drugs and toxins away from the cytoplasm and into contact with the exterior milieu. It seems unreasonable to suppose that evolution would have anticipated a need for drug resistance, so what happened? There must be toxic molecules that are drug-like in the environment. What does "drug-like" mean? Think of a major source of active molecules in the environment (hint: plants do not have immune systems).

Secondary Transport

13

Life depends on the ability of cells to create, maintain, and exploit compositional differences between their insides and the outside world. Creation, maintenance, and exploitation often involve pumps that can use a source of energy to transport a species uphill against its concentration gradient. In secondary transport, the driving energy source can be the favorable downhill flow of an ion across the membrane, coupling a favorable free energy to drive an unfavorable one. The term "secondary" is used because the energy to form the driving ion gradient is provided by "primary" transporters that are connected to energy sources from outside the cell, for example from light energy or the metabolism of nutrients. In this chapter, we discuss examples of how secondary membrane transporters accomplish such feats.

Since the early 20th century, ideas of ionic and osmotic flows across membranes have fascinated physiologists, who sought explanations for phenomena such as the osmotic shrinkage and swelling of cells that had been observed over previous centuries (Chapter 1). Early ideas included the notion that pumps might act as carriers, shuttling between the two surfaces of a membrane, but this idea faded as data accumulated. Long before any structures of membrane proteins were known, clever ideas of how secondary transporters might work were proposed independently by Peter Mitchell (1920–1992) in 1957, Clifford Patlak (1926–2014) in 1957, Oleg Jardetzky (1929–2016) in 1966, and George A. Vidaver (1930–1986) in 1966. These models are now generally referred to as "alternating access" models. The main concepts are that a pump must have at least two forms: one open to one side of the membrane and another open to the other side. In addition, it must have a binding site for the pumped molecule that has different affinities in the two conformations, and a way to couple energy from some source to drive the process. In later developments, it was recognized that there may be "occluded states" that are intermediates not open to either side (see **Box 13.1**). The evolution of the alternating-access concept is shown schematically in **Figure 13.1**. These prophetic ideas have a remarkable correspondence to contemporary views of how secondary transporters appear and work, although they are (forgivably) incomplete in many respects. But, keep in mind that in early days, scientists were arguing about whether the lipid bilayer was the basic membrane fabric and how proteins might be situated within the bilayer (Chapter 1). Since the beginning of the 21st century, though, the structures of a large number of secondary transporters have been determined. These structures and biochemical measurements (**Box 13.2**) have revealed three main structural themes for alternating access (**Figure 13.2**). These form the basis for our discussion of the molecular principles of alternating-access transport.

The menagerie of secondary transporters is traditionally subdivided using special names. A transporter that moves the driving ions in the same direction as the substrate molecule being transported is called a co-transporter or a symporter. If the motion of the driving ions across the membrane is opposite to the

DOI: 10.1201/9780429341328-14

BOX 13.1 THERMODYNAMICS OF SECONDARY TRANSPORT: AN EXAMPLE

Because there are variations in the ways that secondary transporters function in detail, including the uses of alternative ion gradients in addition to Na$^+$, the following should be viewed as a discussion of how the steps in one kind of example, sodium symporters, might be viewed from the perspective of energetics. Several examples are discussed in the chapter. An advantage of the sodium-driven structures is that we can see how the driving ion(s) are bound; and in a few examples, we can understand the chemistry of interactions with the binding of the transported substrate.

A tricky point is that the energy diagrams we typically draw for reactions describe initial conditions and do not reflect the situation as a transporter approaches a steady state. Note that each of the transport steps is reversible and that their concentrations adjust over time, so that the free energy changes become small. It is better to think of the transporters as operating in a steady state with flow through the system. It is the *net flow* of sodium ions down their electrochemical gradient that drives the transport of the substrate. We must thus distinguish equilibrium of the protein among its various states from the ion/substrate distribution across the membrane. Neither ion nor substrate can be at equilibrium across the membrane under normal physiological conditions; otherwise, transport would not occur. For discussion, we postulate a two-gate structure, but the points would apply to any of the mechanisms we discuss in this chapter.

1. **Interconversion between inward-open, outward-open, and occluded states** involves a set of conformations that include, in this discussion, an intermediate occluded state in which both gates are closed (**Figure 1**). If both gates were open, leakage would result, as confirmed by structural evidence that occluded states exist. Occlusion may be managed with gates, as shown, or by a continuous seal that surrounds the ion–substrate region during alternate access, or by a slide that moves the ion and substrate through a seal in an elevator mechanism. It may be that the energy difference between the inward and outward states is small, but the conformational changes and topological steps during transport are large, implying significant energy barriers or complex pathways, and likely to be slow. The occluded state will be less stable than either the outward- or the inward-facing state and will have a low occupancy.

2. **The empty transporter** is a good place to start our discussion. Under this hypothetical condition, which can be studied *in vitro*, there will be an equilibrium between inward- and outward-facing states, with a small occupancy of the occluded state. One might expect a bias toward an outward-facing orientation, but in fact, such a bias is not needed and does not always occur; rather, under physiological conditions, the bias is provided by binding of the driving ion from the outside.

3. **Substrates bind from the outside** with an ion-binding event that helps stabilize the outward-open state, facilitating the binding of the transported substrate. Ion binding is favored by the high concentration of Na$^+$ on the outside of the membrane. The sodium-bound complex is at lower energy than the empty transporter; otherwise, the empty state would be favored. Thus, the occupancy of the sodium-bound outward state is higher than that of the outward-empty state. With the Na$^+$ bound, the binding of substrate becomes energetically favored. One might ask why the substrate might not bind first, but in the physiological state, the ion is abundant and the substrate much less so on a molar basis, and the ion binding must ensure the abundance of the outward-facing conformation relative to the inward, empty conformation. The ternary transporter–Na$^+$–substrate complex can either reverse, releasing the substrate, or go through the next steps. With substrate and Na$^+$ both bound, the conformational energy of the transporter favors the inward-facing state. But in that state, the unbinding of the Na$^+$ is favored because of the relatively low concentration of Na$^+$ inside the cell, and the substrate binding is compromised, favoring release. Thus, the net transmembrane flow of Na$^+$ and substrate is driven by the release of the Na$^+$ on the inside of the membrane, where its concentration is low, although all states are reversible. A key concept is that the binding of the ion and that of the substrate are coupled.

4. **The outside-open complex conversion to the inside-open complex.** When the substrate binds, there is a low energy barrier to closing the outside gate rapidly. Then, the outside gate locks as the conformation changes slowly to inside-closed, again with a set of large energy barriers, reflected in a long time-course. This step is essentially between two occluded states with both gates closed. Then, the inside gate opens, completing the topological transformation. Although the system moves toward equilibrium among the various states, the Na$^+$ and substrate are not at equilibrium, because there must be a net flow of ion and substrate across the membrane. One can imagine that for experiments *in vitro*, one could arrange the concentrations of substrate and ion so that the pump runs backward, and such observations have been reported for other pumps such as the Na$^+$/K$^+$ ATPase.

(Continued)

BOX 13.1   THERMODYNAMICS OF SECONDARY TRANSPORT: AN EXAMPLE (CONTINUED)

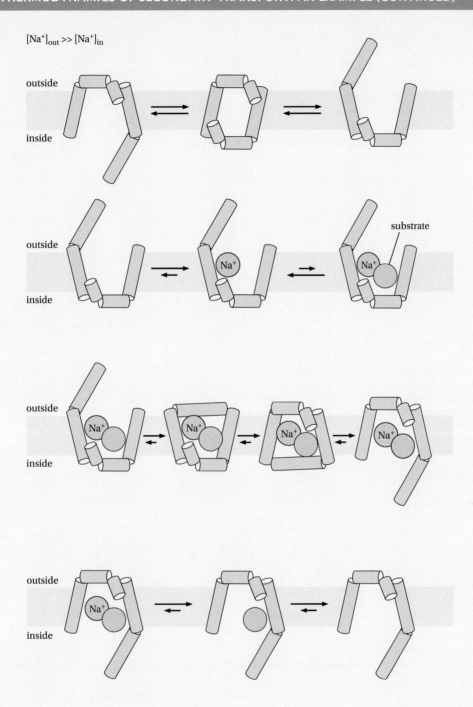

Figure 1

5. **Ion and substrate dissociation happen sequentially**, returning the empty transporter to an equilibration of the inward-open and outward-open states. The ion is destabilized once the inward gate is open because it is now in a region of low Na$^+$ concentration inside the cell. When Na$^+$ dissociates, its contribution to the binding of the substrate is lost, so substrate dissociation is favored even though it is at a higher concentration than on the outside. Each of these events is relatively fast. While the release of the ion may be favored as the first step in some cases, it need not be the case in all mechanisms, because the structure of the inward-open state is not superimposable with the

(Continued)

BOX 13.1   THERMODYNAMICS OF SECONDARY TRANSPORT: AN EXAMPLE (CONTINUED)

outward-open state, and the shift of the structure to the inward-open state may alter the relative binding affinities. But the coupling of Na⁺ and substrate binding must remain. Again, the binding of Na⁺ at the high concentration outside and its release at the low concentration inside drives the net flow of substrate through the cycle, and the changes in the substrate-binding affinity induced by the ion binding couple the Na⁺ flow to substrate transport (**Figure 2**).

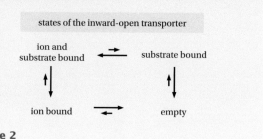

Figure 2

Figure 13.1 Origins of the alternating-access idea. (A) In this early scheme, referred to as a gate-type mechanism, the substrate molecule enters the transport molecule via a gate G (G_1 state open on the left side). After a series of energy-driven transformations to state G_4, the substrate is released on the opposite side of the membrane. (B) Alternating-access scheme proposed by Peter Mitchell that visualized the alternating access as changes in the molecular structure of the transporter. (C) As we gained knowledge of membrane structure (Chapter 1), the ideas of panels A and C morphed into the modern alternating-access cartoon. (A, Patlak CS (1957) *Bul Math Biophys* 19:209–235. B, Mitchell P [1957] *Nature* 180:134–136. With permission from Springer Nature. C, Figure reproduced with permission from Singer SJ [1974] *Annu Rev Biochem* 43: 805–833. Copyright 1974 Annual Reviews.)

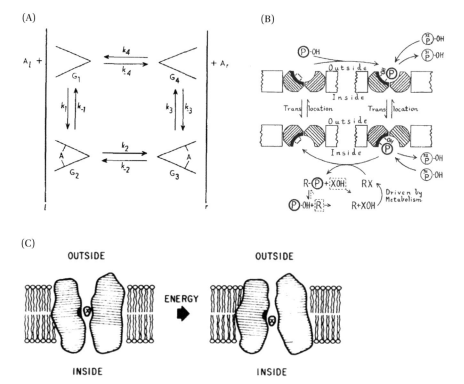

direction of motion of the substrate, it is called an antiporter. If the substrate molecule is simply moved down its concentration gradient but is assisted in crossing the membrane barrier, the transporter is a facilitator.

We begin by examining the requirements for secondary transport function established by the early investigators. We then examine current structural views of how these requirements are met, using several examples. Not all of the steps are fully revealed structurally in the present state of understanding, but there has been remarkable progress in recent years. A set of structural intermediates is likely to be found separating the various parts of the transport pathway and may vary in individual cases, but the energy differences between the chemical potentials of the transported molecules in the compartments on the sides of the membrane are the key energies in the process. Thus, the energies of intermediate states cannot be too high or low; else, the process would be too slow physiologically. This situation is similar to the situation in pores and channels (Chapter 11), where the pathway cannot have energy traps or barriers that abrogate function.

BOX 13.2   ISOLATED MEMBRANE VESICLES ALLOW MEASUREMENT OF ION-DRIVEN TRANSPORT

Using methods pioneered by H. Ronald Kaback in the 1960s, membranes began to be studied by preparing sealed, empty vesicles from cells. When the vesicles were prepared from bacteria, one knew that the membranes were the plasma membranes and that they had little content from the cytoplasm. A typical preparation was to expose bacteria to agents that disrupt the cell wall, such as digestion by lysozyme in the presence of EDTA (to remove ions that stabilized the cell wall), followed by exposure to a low-osmolarity environment that ruptured the membranes by osmotic shock. Subsequent washing by cycles of centrifugation and resuspension gave membrane preparations like those shown in **Figure 1**.

These preparations, and similar ones from mammalian cells, allowed the study of transport across the membranes. Because the transporters were components of the isolated membrane vesicles, the environment could be readily changed, and materials would be trapped inside if they were transported.

An excellent example is stoichiometry of ion/γ-amino butyric acid (GABA) transport into synaptic vesicles. To obtain the stoichiometry, it is assumed that GABA is translocated in its zwitterionic form. The transport stoichiometry can then be written as n_{Na+}:m_{Cl-}:GABA. At equilibrium, the chemical potentials can be written:

$$n\Delta\mu_{Na+} + m\Delta\mu_{Cl-} + \Delta\mu_{GABA} = 0$$

Upon rearrangement and using the Nernst equation (Chapter 0),

$$\ln\frac{[GABA]_{in}}{[GABA]_{out}} = n\ln\frac{[Na^+]_{out}}{[Na^+]_{in}} + m\ln\frac{[Cl^-]_{out}}{[Cl^-]_{in}} - (n-m)\frac{F}{RT}\Delta\psi$$

The stoichiometry of transport can be determined when both the chloride ion gradient and the membrane potential are kept constant, and the sodium ion gradient is varied by plotting the logarithm of the GABA concentration gradient (internal/external) versus the logarithm of the sodium ion gradient (external/internal), which should give a straight line. Under these conditions, the value of the slope of the line gives the stoichiometry of n. Similarly, it is possible to obtain the value of m. The data in **Figure 2** reveal an excellent fit, giving the stoichiometry $n = 1.5$. Similar experiments with chloride ions yield $m = 0.5$. One can thus conclude that two Na^+ and one Cl^- are transported for each GABA molecule transported into the synaptic vesicles, and that the movements of GABA, Na^+, and Cl^- are coupled.

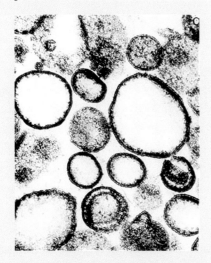

Figure 1 Membrane vesicles from *E. coli*. Thin sections of embedded vesicles stained with metal ions were prepared by Dr. Samuel Silverstein of the Rockefeller University and observed in the electron microscope. Closed, empty vesicles are seen. Image provided by H. Ron Kaback.

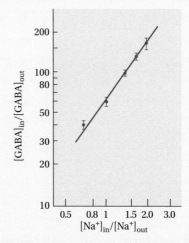

Figure 2 Relationship between the sodium gradient and GABA gradient in synaptic membrane vesicles isolated from rat brains. (From Radian R, Kanner BI [1983] Biochemistry 22: 1236–1241. With permission from American Chemical Society.)

Nonetheless, some of the intermediates are, or can be made (by mutations, inhibitor binding, etc.), sufficiently stable for structural studies and have contributed important snapshots of how these remarkable molecules work.

The main questions we seek to answer are: How does a transporter selectively bind driving ions and a substrate? How do these binding events lead to conformation changes that close access to one side of the membrane while opening access to the other without leaks? How are the driving ions and substrate

Figure 13.2 Cartoon schemes for alternating-access transport. The schemes recognize that the pumps have at least two domains that move with respect to one another to enable alternating access. The bold black lines indicate contacts between domains. The bold grey lines represent barriers to substrate movement. The transported substrate is represented by red spheres. The simplest pumps have two structurally distinct domains within the structure (e.g., sugar transporters such as lactose permease). The more complex ones are multi-subunit proteins in which the subunits move relative to one another (e.g., amino acid transporters). In some cases, there may be occluded states along the pathway, not open to either side (see Figure 13.3 and Box 13.1). (Figure reproduced with permission from Drew D, Boudker O [2016] *Annu Rev Biochem* 85: 543–572. Copyright 2016 Annual Reviews.)

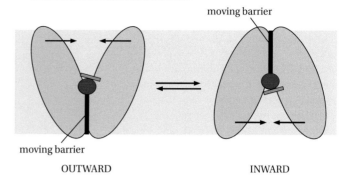

(A) rocker switch (moving barrier)

moving barrier

moving barrier

OUTWARD   INWARD

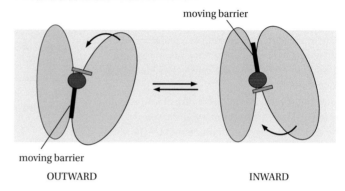

(B) rocking bundle (moving barrier)

moving barrier

moving barrier

OUTWARD   INWARD

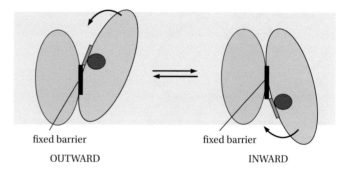

(C) elevator (fixed barrier)

fixed barrier   fixed barrier

OUTWARD   INWARD

released? Fundamentally, how can the processes be interpreted using chemical and thermodynamic concepts?

13.1 REQUIREMENTS FOR SECONDARY TRANSPORTERS

We can use concepts of structure and energy to describe a logical sequence of events that frame the process of driving an energetically uphill transport of a secondary species, an ion or another substrate, using the downhill energy flow of a primary species, usually an ion. A focus of the chapter is connecting transporter structure to function, but it is useful to begin by considering in general terms the requirements for a symporter. Lactose is a sugar used as an energy source by *E. coli*, which uses a transporter, LacY, to accumulate it to fuel growth and replication. The sequence of transport events is illustrated in **Figure 13.3**.

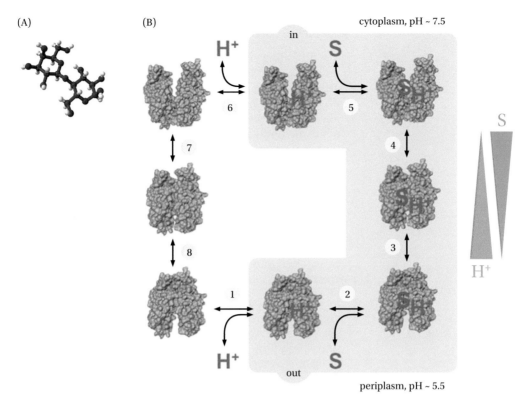

Figure 13.3 Structural cartoon based on lactose transport by LacY to illustrate how coupled secondary transport works. (A) Lactose, the natural substrate (S) for LacY. (B) For the example of an ion-coupled symporter, based upon LacY, the cycle starts with an empty, outward-facing transporter that binds the substrate and coupled ion (blue and orange, respectively) from the extracellular solution, and the combined binding energies drive the closure of the gate. The transporter undergoes a conformational transition to an inward-facing state with the outward gate locked shut, and the opening of the intracellular gate is now possible. The release of the ion to its region of low concentration destabilizes the substrate. Exchange or counterflow involves only steps 2–5 (grey shaded area). Release of the substrate into the cytoplasm yields an empty inward-facing state, the gate closes, and the transporter equilibrates with the outward-facing state. (From Kaback HR [2015] *PNAS* 112: 1259–1264. With permission.)

The directionality of transport is due to the H$^+$ electrochemical potential. In thinking about how such molecules work, we must keep in mind the dynamics of the molecules involved—ever-present thermal energy causes molecular motions that facilitate the steps in both directions, emphasizing that the fixed drawings (or crystal structures) we use are abstractions that represent averages at different points in the cycle. A more detailed discussion of the thermodynamic transport steps is in Box 13.1, but as an introduction to the basic steps of transport, we use our current understanding of LacY from *E. coli* (Figure 13.3) based upon the pioneering work of H. Ronald Kaback (1936–2019), which has led to a mass of LacY data unmatched for any other transporter.

1.   *A low-energy form of the transporter is open to the external face of a membrane, where it sequentially binds driving ions and a substrate (steps 1 and 2 in the figure)*

There must be a low-energy, stable form of the transporter that is set to accept binding of the transported species. In general, the opening will permit access of the primary ion and secondary species as single molecular events, allowing them to find specific binding sites in the interior of the transporter. The binding is ordered, with the primary H$^+$ ion binding typically as the first step, which helps stabilize the outward-open transporter to lower the K_m for binding the substrate. Because the primary ion is present at high concentration, the task is easily done. In the example of LeuT, discussed later, two sodium ions bind and help to form the binding site for leucine, the transported molecule.

2. *The substrate-binding energy causes the transporter to close access to the external face (step 3)*

All binding sites for the driving species, usually an ion, and the secondary species must be occupied. Generally, the binding of the secondary species is weak unless the primary driver is bound but strong with the driving ion present. A part of the combined binding energy shifts the equilibrium to favor a structural change that closes access to the external face of the membrane to form an occluded state that is closed to both sides of the membrane.

3. *The transporter changes conformation from outward to inward facing (step 4)*

In a conformation change that also is favored by energy from the binding of the driving ion and secondary species, the transporter rearranges its structure across the membrane via occluded states to an inward-facing form that opens to the inside. These steps (3 and 4), which require considerable structural rearrangements, can be very slow—seconds in some cases—but examples also exist where they are fast.

4. *Access to the inside is created by an opening in the transporter*

Opening should be thought of as a separate step because inward-facing, occluded states are seen. The observation of such intermediates is enabled by the very slow kinetics of the conformational changes. The slow steps imply large energy barriers that are being overcome (see **Box 13.3**).

5. *The ligands dissociate sequentially (steps 5 and 6)*

The conformation change that reverses the direction of the transporter from outward to inward relaxes the binding of the transported driving ion and secondary species, allowing them to dissociate. The dissociation of the driving ion is facilitated by conformation changes that destabilize the binding sites and by its low concentration on the inside; Arg302 comes into proximity with protonated Glu325, facilitating deprotonation; and the departure of the substrate helps to destabilize the binding of the secondary species, even if it dissociates first (see Box 13.1; bear in mind that these are equilibria). As with the binding events, a specific sequence of release events may be employed. In the example of LeuT (see later), release of the two sodium ions destabilizes the leucine binding site, and the leucine is delivered into the cytoplasm, still as a consequence of the binding energy in the fully occupied, outward-facing conformation. The key point is that while the transporter can rebind both the ion and substrate and reverse to the outside, the low ion concentration favors a path via the empty transporter.

6. *The empty transporter closes its opening to the inside and changes conformation to the outward-facing form, which is stabilized by binding a driving ion (steps 7 and 8)*

When the transporter is empty, its most stable state is, once again, an equilibrium with the outward-facing form in 1, which is captured and stabilized by the binding of ions from the outside. To reach this state, it must close the opening to the inside, reverse the conformation change in 3, and reopen to allow the ion binding from the outside: altogether, a slow process. In some examples, such as LacY, the switch to the outside form is driven by another gradient: the proton electrochemical gradient difference between the inside and the outside of the cell. If the secondary substrate is not available, the equilibrium gives a mixture of inward- and outward-facing empty transporters. Thus, while each step in the cycle is reversible, the direction is biased by the electrochemical gradient(s) of the driving ion(s), resulting in net movements of the ion(s) and the transported species.

To examine current understanding of these events, we focus on well known cases. We begin with two examples based upon the major facilitator superfamily

BOX 13.3   RATES OF CHEMICAL REACTIONS

How is the rate of a process related to thermal energy and other factors? Although reaction rate ideas were developed for reactions between molecules in solution, it turns out that the same concepts can be applied to conformational changes. In 1889, the Swedish chemist Svante Arrhenius (1859–1927) proposed a simple expression for the rate of a chemical reaction:

$$k = Ae^{\frac{-E_a}{RT}}$$

The equation expresses how the rate of the reaction, k, depends on the activation energy E_a, the absolute temperature, T, and a "pre-exponential factor," A. R is the gas constant. It is remarkable how useful the concept has been for over a hundred years. Many approaches have been used to improve and refine it, but the basic ideas stand for understanding membrane protein function if we keep the following thoughts in mind:

k: The number of collisions resulting in a chemical reaction per second.

T: Most biology is conducted at approximately constant temperature, so the T becomes a constant in the equation. This is almost a shame from the point of view of the theory, because part of its triumph is its prediction of the temperature dependence of chemical reactions. In biochemical experiments, T is sometimes varied as a parameter to obtain thermodynamic values, however.

R: If, as we have mostly been doing, we use a chemical view and expressions in molar units, the gas constant is used, and RT is the molar energy. An alternative is the physics point of view, in which single molecules are studied, in which case the Boltzmann constant is used, and $k_B T$ is the energy per molecule.

E_a: Arrhenius' concept of an activation energy was a stroke of genius, but we now generally use the Gibbs activation free energy, $\Delta G^{\ddagger}$, which allows us to separate the enthalpic and entropic components, $\Delta H^{\ddagger}$ and $T\Delta S^{\ddagger}$. Essentially, the activation free energy tells us that a slow reaction (small k) is likely to have a large energy barrier.

A: If we take a simple view, k is the number of collisions that result in a reaction per second, A is the total number of collisions (leading to a reaction or not) per second, and $\exp(E_a/RT)$ is the probability that any given collision will result in a reaction. A has the same units as k: s^{-1} if the reaction is first order.

A point of view: Although the expression is highly useful, the activation energy and the rate constant are not simply related to an activation free energy and collisions at the molecular level. Biological molecules are generally asymmetric and also have internal energies arising from thermal motion. In a collision between biological molecules, the collision angle, the relative translational energy, and the internal (particularly vibrational) energy will all contribute to the chance that the collision will produce a reaction. So, a way to speed up a reaction is to orient reacting molecules, hold them near each other, and provide a conducive chemical environment—essentially, what enzymes often do.

(LacY and FucP) that exemplify the rocker-switch scheme (Figure 13.2A). We then consider the sodium ion–driven transporter LeuT (a rocking-bundle transporter; Figure 13.2B) that is a leucine symporter. To complete our discussion of alternating-access transporters, we consider the bacterial glutamate transporter Glt$_{Ph}$, which follows the "elevator" scheme (Figure 13.2C). We conclude with two other secondary transporters: a member of the mitochondrial carrier family (the ADP/ATP carrier) and the erythrocyte glucose facilitator Glut1.

13.2 ROCKER-SWITCH TRANSPORTERS: MAJOR FACILITATOR SUPERFAMILY (MFS) SYMPORTERS

13.2.1 The MFS Family Is Large

About a quarter of all known membrane transport proteins in prokaryotes belong to the MFS, the largest and most diverse superfamily of secondary active transporters. The MFS is classified as having 58 distinct families, with about 15,000 sequenced members so far. These transport a wide range of substrates and include antiporters, symporters, or uniporters.

While exceptions exist, most MFS members have 12 transmembrane helices (TMs), with a similarity of sequence between the first 6 and second 6 helices, and each of these sets of 6 helices has an internal inverted repeat, giving a 3+3 antiparallel topology. Crystal structures show that this homology is real, with an apparent gene duplication in ancient times that produced a twofold axis that became pseudo-twofold as evolution diversified the sequence of each half independently. But in contrast to the transporters LeuT and Glt$_{Ph}$, discussed later, there is an even number of helices in each 6-helix half, so the 6+6 pseudo-twofold is perpendicular to the plane of the membrane rather than parallel as in LeuT and its relatives. At the time of this writing, crystal structures have been reported for the lactose transporter (LacY), discussed earlier; a glycerol phosphate antiporter (GlpT), which uses an inorganic phosphate gradient; a d-xylose:H$^+$ symporter (XylE); the fucose symporter (FucP), which uses H$^+$; and the multidrug transporter, EmrD, which expels amphipathic molecules from the *E. coli* cytoplasm. Both the LacY and GlpT structures are inward-open, while EmrD is occluded. By comparison, the fucose-proton symporter structure is in an outward-open conformation, but its inward-open structure is not known. Closely related structures are available for XylE. The available structures emphasize the general conservation of fold and architecture, consisting of two domains with a transport pore between them, and are consistent with a rocker-switch model with occluded states. It is to be expected that many new structures will be found in this very active field, so more details will emerge with time, but enough is known to examine some principles.

13.2.2 Structures of the Proton-Lactose Symporter, LacY, and the Proton-Fucose Symporter, FucP, Can Be Combined to Give a View of MFS Function

The lactose transporter was discovered as a part of the famous *lac* operon that François Jacob and Jacques Monod used in their pioneering studies to understand the control of gene expression, and it has been extensively studied ever since. The transporter uses the proton gradient across the membrane, in which proton concentrations are high outside and low inside, to drive the accumulation of lactose, a nutrient sugar, by *E. coli* (Figure 13.3A). Because it is so dynamic in its conformation, structural studies historically relied on indirect methods to test ideas of its molecular organization (see **Box 13.4**). Many years of effort with inhibitors and mutations finally resulted in a crystal structure at 2.9 Å resolution. A 3.52 Å view is shown in **Figure 13.4**. Panel A shows the structure with a lactose homolog bound, and a twofold pseudo-symmetry, which had been missed in the previous studies, is immediately apparent. Panel B shows the structure in cross-eyed stereo. The locations of mutants known to influence substrate and proton binding are shown in panel C. A single sugar-binding site is observed at the apex of the cavity near the approximate middle of the molecule.

The structure confirms many previous findings obtained by site-directed mutagenesis in conjunction with biochemical and biophysical studies. Fewer than ten amino acyl side chains are irreplaceable with respect to transport activity, an observation that is in agreement with many cases in which a large number and variety of amino acid changes are compatible with function in enzymes. Such a plasticity of substitutions also tells us that it is expected that the duplication giving rise to the two-domain structure will be obscured by sequence changes in evolutionary time, since many independent mutations will be tolerated in the domains. The sequences are so different in the LacY structure that the duplication was only seen when the structure was solved.

The relationship of the two domains and the positioning of the substrate between them bear a remarkable resemblance to the early ideas of Mitchell and Jardetzky (Figure 13.1); an obvious hypothesis is that the domains rock against each other to alternate access between the inside and outside of the cell, just as they suggested long ago. The structure of the fucose symporter, FucP, gives more

BOX 13.4 HYDROGEN–DEUTERIUM EXCHANGE TO MEASURE DYNAMICS OR SOLVENT EXPOSURE

Hydrogen–deuterium exchange (also called H–D or H/D exchange) is a reaction that replaces a covalently bound hydrogen atom with its deuterium isotope, or vice versa. A common application is in studying solvent access or dynamics of proteins, and such studies of membrane proteins have given some useful insights. Usually, the examined protons are the backbone amide hydrogens, which are stabilized by hydrogen bonding in secondary structures, such as the TMs of a membrane protein. The experiment typically involves changing the deuteration of the aqueous environment at time zero, then stopping the exchange by changing the pH, and detecting the results by nuclear magnetic resonance (NMR), infrared spectroscopy, mass spectrometry, or neutron scattering. The exchange rates are interpreted as resulting from solvent exposure, which can be simply topological, as in the restriction of the water contact of a single TM by immersion in a bilayer, or dynamic, as in variation of solvent exposure during conformational changes.

Because the exchange reaction can be catalyzed by acids or bases, the rate is strongly influenced by pH; backbone amide hydrogens exchange very slowly at low pH, so an experiment can be conducted at neutral pH and quenched as a function of time by a pH drop. For detection by NMR, the pH may be moved to around 4.0–4.5. For detection by mass spectrometry, the pH is dropped to the minimum of the exchange curve, pH 2.6. In the most basic experiment, the reaction is allowed to take place for a set time before it is quenched.

Hydrogen and deuterium differ greatly in mass, magnetic properties, and scattering length for neutrons; hence the use of NMR, infrared spectroscopy, mass spectrometry, and neutron scattering for detection. If NMR assignments have been done, a heteronuclear single quantum coherence (HSQC) experiment can follow the exponential process of exchange at specific sites. Mass spectrometry requires less material and is very useful for regions of a protein analyzed after protease digestion. Infrared detection gives overviews of structures and relies on the changes in vibrations resulting from the mass changes. Neutron scattering is much more demanding, but some useful data have been obtained, for example, in studies of protein crystals where specific sites can be viewed.

The case of LacY is an instructive example. Although x-ray crystal structures of membrane proteins are essential, transport mechanisms are dynamic processes, and a study of H/D exchange reveals the probable reason why LacY and related proteins are so hard to crystalize. H/D exchange of ~85% of the backbone amide protons in LacY is observed by infrared within 10–15 min at ~20 °C with the protein either embedded in detergent micelles or reconstituted into liposomes, with 100% exchange at elevated, non-denaturing temperatures (Box **Figure 1**).

In contrast, with Photosystem II at room temperature, only ~20% exchange is seen after 18 h!

The dramatic exchange of LacY clearly indicates that this very hydrophobic protein (~65% unequivocally hydrophobic residues) is in a highly dynamic state, with solvent exposure of its inner regions. By contrast, proteins that do not use large conformation changes, such as the photosynthetic proteins and bacteriorhodopsin, have large amounts of their structures protected from the aqueous environment. It is no surprise that many of the early crystal structures were photosynthetic proteins! Another "rigid" protein is the KcsA potassium channel (Figure 11.26), which undergoes slow and incomplete H–D exchange compared with almost complete H–D exchange in LacY under comparable experimental conditions (**Figure 2**).

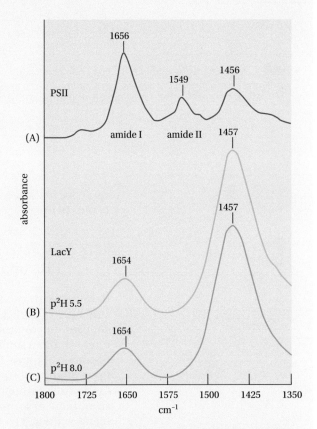

Figure 1 Infrared spectra of ^{2}H-exchanged proteins in β-dodecyl maltoside micelles. The intensity of the amide II band is strongly affected by H–D exchange, decreasing in intensity as H is replaced by D. For photosystem II (A), the amide II band persists (~20% exchange after 18 h) during H–D, whereas for LacY (B, C), the band disappears quickly (~85% exchange after 15 min). (From Patzlaff JS, Moeller JA, Barry BA et al. [1998] *Biochemistry* 37: 15363–15375.)

(Continued)

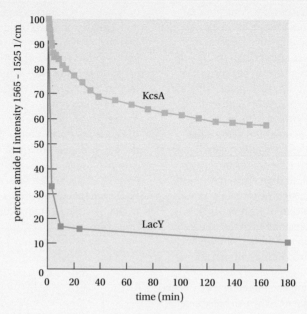

Figure 2 LacY undergoes rapid H–D exchange compared with KcsA. The percent amide II intensity declines to only 75% 20 min after exchange, whereas the LacY amide II intensity declines to 15% in the same time period. (From leCoutre J et al. [1998] *PNAS* 95: 6114–6117. With permission from J leCoutre.)

credence to these mechanistic ideas because it shows the opposite orientations of the opening.

13.2.3 Structures of the Substrate-Free LacY at Different PHs Imply an Induced-Fit Mechanism

Because the transport of lactose is driven by the downhill movement of protons, structures of the ligand-free transporter were solved at pH 6.5 and pH 5.6 to find changes due to protonation. The sugar-binding site in both ligand-free structures exhibits a similar configuration, which differs from the ligand-bound configuration (**Figure 13.5A,B**), strongly indicating that a rearrangement occurs upon binding of sugar. It appears that the side chains interacting directly with the hydroxyl groups of the sugar are not in position for binding and that the sugar induces formation of the binding site, mainly through sidechain movements without a global conformational change. Comparison of the unliganded structures at the two pH values shows a change in the position of one of the glutamic acid side chains that can be interpreted as a proton transfer step, linking the motion of a proton to the process of transport. Unfortunately, the current resolution does not permit the determination of the location of water molecules that may be involved as well.

Nonetheless, a detailed set of events is postulated whereby protonation of Glu325 increases the affinity for the sugar (Figure 13.5C), advancing understanding of the initial step in coupling the transport of lactose to a proton movement. However, it is still not clear in this model how sugar binding leads to the global conformational change between the inward- and outward-facing conformations. An idea of how this might work is provided by studies of the fucose transporter (next subsection).

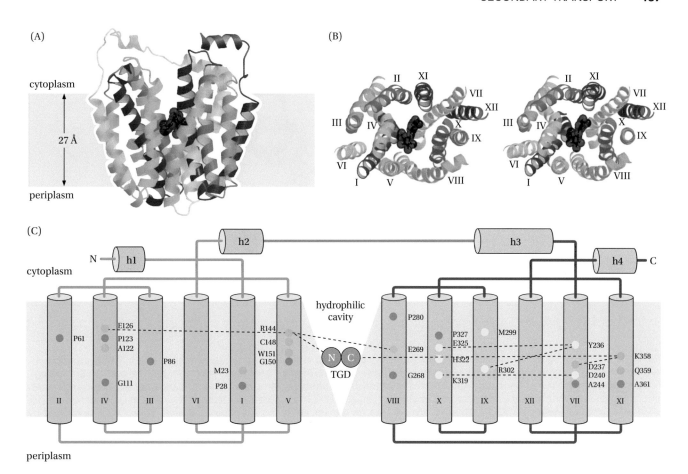

Figure 13.4 Structure of the lactose permease. The figures are based on the C154G mutant structure with bound β-d-galactopyra nosyl-1-thio-β-d-galactopyranoside (TDG). (A) Ribbon representation of LacY viewed parallel to the membrane. The 12 transmembrane helices are progressively colored from the N-terminus (N) in purple to the C-terminus (C) in pink with TDG represented by black spheres. (B) Stereo view of the ribbon representation of LacY viewed along the membrane normal from the cytoplasmic side. For clarity, the loops between helices have been omitted. The color scheme is the same as in (A), and the 12 transmembrane helices are labeled with roman numerals. (C) Secondary structure schematic of LacY. The N- and C-terminal portions of the enzyme are colored in blue and red, respectively. Residues at the kinks in the transmembrane helices are marked with purple circles; residues marked with green and yellow circles are involved in substrate binding and proton translocation, respectively; while residue E269, colored aqua, is involved in both substrate binding and H+ translocation. The large hydrophilic cavity is designated by a blue triangle, and TDG is depicted as two black circles. (From Kaback HR [2005] *Comptes Rendu Biol* 328:557–567.)

13.2.4 The Fucose Symporter, FucP, Is Outward, Open, and Supports a Rocker-Switch Alternating-Access Mechanism

L-fucose is a major constituent of N-linked glycans on the cell surfaces of many organisms and can serve as the sole carbon source for some bacteria. In *E. coli*, the uptake of L-fucose is mediated by FucP, a fucose/H+ symporter, another member of the MFS superfamily. Its structure is known at moderate (3.1 Å) resolution (**Figure 13.6**), and remarkably, the protein is found in a conformation that is outward-open, perhaps because a molecule of detergent is bound in the sugar-binding site. Thus, it might give important mechanistic insights if it can be paired with members of the family that are known in the inward-open state, such as LacY. Such comparisons are sometimes very useful in gaining understanding of mechanisms (see **Box 13.5**). A comparison of the sequences reveals only limited sequence homology, as might be expected given the sequence divergence and insensitivity to mutation found between the domains of LacY. Given that we have reasonably good structures with strong similarities, how do we compare them? One way is to compare the six TM domains with each other at the structural level using a calculation of the root mean square distances (rmsd) between α-carbon positions when the domains are aligned to minimize

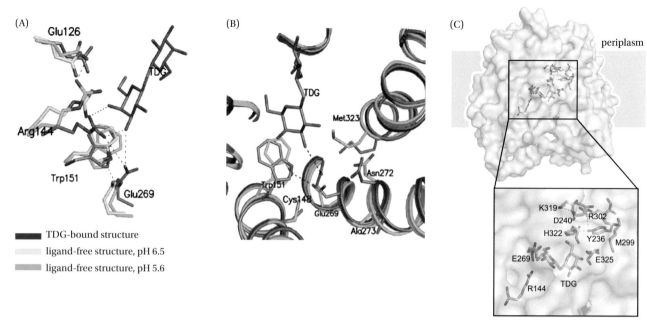

Figure 13.5 LacY sugar-binding site rearranges in the absence and presence of sugar and changes in pH. (A) Superimposed cytoplasmic views of the sugar-binding sites show changes with pH (compare green and yellow structures) and sugar binding (blue structure). (B) View of the critical Glu269 in a relatively hydrophobic environment that influences its pK_a. Rearrangement of the active site in the absence and presence of sugar is shown. TDG is a high-affinity lactose homolog, β-D-galactopyranosyl -1-thio-β-D-galactopyranoside. (C) Residues involved in proton translocation and coupling. Hydrogen bonds are represented by dashed black lines in all panels. (A, B, From Kaback HR, Iwata So, Verner G et al. [2005] *EMBO J* 328: 557–567. With permission from John Wiley and Sons.)

Figure 13.6 Structures of FucP and LacY give a view supporting alternating access. The structures of FucP and LacY are compared by superimposing their six TM domains. It is found that they are remarkably similar, differing in the complete structures by a rotation of relatively rigid domains that is very like the idea of alternating access proposed by Mitchell and Jardetzky decades earlier. (From Dang S et al. [2010] *Nature* 467: 734–738. With permission from Springer Nature.)

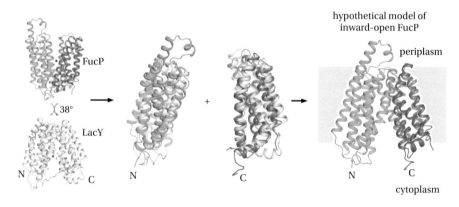

this parameter (**Box 13.6**). This allows one to build a model for FucP in the inward-facing state, but more detail will be needed to gain chemical insight into the mechanism.

13.3 ROCKING-BUNDLE TRANSPORTERS: NEUROTRANSMITTER SODIUM SYMPORTERS

When nerve transmission crosses a synapse, it does so by releasing neurotransmitters from the transmitting neuron into the synaptic space, where they bind to receptors in the receiving neuron that open ion channels (see discussion in Chapter 11). Once the signal has been transmitted, the transmitting neuron

BOX 13.5 HOMOLOGS AND MUTANTS ARE OFTEN THE KEYS TO PROGRESS WITH MEMBRANE PROTEIN STRUCTURES

Because humans regulate their body temperature, their proteins are often relatively unstable—they denature at temperatures slightly above 37 °C. A general principle is that proteins are only as stable as they have to be in order to function at the growth temperature, so random mutations that tend to destabilize the structure are not selected against until the denaturation temperature is just above the growth temperature, resulting in a minimal stability. Because these proteins are relatively unstable, they are subject to denaturation when they are in less favorable environments, such as the detergents used to prepare them in solution.

On the other hand, bacterial and archaeal proteins must survive more extreme thermal environments and so are often more stable. Thus, it has proven advantageous to seek homologs of human proteins for study, and the key examples of structures that are discussed in this chapter, LeuT, Glt$_{Ph}$, and XylE, are cases in point. The related human proteins of great interest are the neurotransmitter sodium symporters (NSS) and excitatory amino acid transporters (EAAT) families of neurotransmitter uptake transporters, and the glucose transporter GLUT1. While they have differing sets of ions that drive them, they are sufficiently similar to inform us of mechanistic ideas (**Figure 1**)

An alternative to finding naturally stable versions of membrane proteins is the rapidly emerging technique of finding stabilizing mutations. This approach has been applied to the study of G-protein coupled receptors, which are of great importance as regulators of cell activity and as the targets for many drugs. As with the MFS superfamily, it was a challenge, probably because of conformational degeneracy, to obtain crystals for study. By screening a

large number of mutants for stability and ligand affinity, it is now possible to find versions of a GPCR that has enhanced stability and ligand binding. An example is shown in Box **Figure 2**.

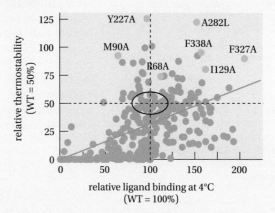

Figure 2 Point mutations are useful for stabilizing membrane proteins, such as the β$_1$ adrenergic receptor. The idea illustrated here is to find mutants with high thermal stability that have the same ligand binding as the wild-type protein. The proteins (318 of them!) were solubilized in a detergent and the stability determined by calorimetry (Chapter 1). Ligand binding was determined after the denatured samples were returned to 4 °C. The dashed curves intersect at the thermostability and ligand binding constant of the wild-type protein. The ellipse shows the approximate experimental uncertainty of the measurements. Another variable, not shown here, is how well the protein can be expressed. A particularly stable mutant was formed by making multiple mutations (green points). (From Tate CG [2012] *Trends Biochem Sci* 37:343–352. With permission from Elsevier.)

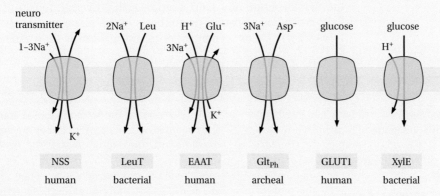

Figure 1 Archaeal and bacterial homologs of human symporters. Although the transport stoichiometries may differ, the basic mechanisms are similar enough to gain understanding of the human proteins. NSS, neurotransmitter sodium symporter. EAAT, Excitatory amino acid transporters.

must reclaim the neurotransmitter molecules for further use, shutting down the signal. The reuptake is accomplished by sodium-driven symporters for serotonin, noradrenaline, glycine, dopamine, and γ-aminobutyric acid, which comprise the NSS superfamily. The transporters are found in neurons and glial cells, as shown in **Figure 13.7**. When these symporters do not function

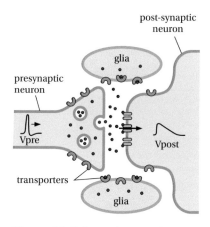

Figure 13.7 Synaptic reuptake. After a nerve signal is transmitted across a synapse, Na⁺-driven symporters (blue) in the neuron and glial cells recover the transmitter molecules.

well, serious health issues can arise, including Parkinson's disease, depression, schizophrenia, and epilepsy. A number of drugs act to modulate the NSS, including cocaine, antidepressants, and serotonin reuptake inhibitors (SSRIs) (**Box 13.7**). Because of their medical and societal importance, they have been the subject of extensive studies. Great advances have been made using bacterial homologs—the leucine symporter, LeuT, and the glutamate transporter, Glt$_{Ph}$—resulting in mechanistic insights that show how many of the tasks mentioned are accomplished.

Structures representing the outward-facing closed state and an inhibitor-bound state have been obtained for LeuT. The use of mutations and conformation-stabilizing antibodies has enabled structures corresponding to the outward-open and inward-open states also to be determined. Inhibition by biological molecules, such as Trp, which locks LeuT in the open-out conformation, has been useful in crystallographic studies. The SSRIs work similarly by binding molecules to active sites and locking transporters, such as the serotonin transporter, thereby modulating synaptic transmission.

13.3.1 Key Topology Theme of LeuT Is an Inverted Repeat

Many transport proteins have inverted repeat structures that arise from gene duplication events for sequences with an odd number of TM helices (see Figure 6.29). LeuT has two principal domains that appear to have originated in this way, each with five TMs (**Figure 13.8**). The inverted repeat gives a framework that might naturally lead to the alternating-access model by producing an interface with two points of stability in the interaction of the two domains, and perhaps this is a feature in some cases. But evolution has taken LeuT in a more elaborate direction, as shown by analysis of its structural intermediates. The changes seen in the crystal structures reveal a scaffold domain, which remains relatively unchanged, and a core domain, where motions of parts of helices result in the formation of the intermediates.

13.3.2 The LeuT Structure in the Outward-Facing, Closed State Reveals Binding Sites for Na+ and Leucine

The structure, determined at 1.6 Å resolution, is shown in **Figure 13.9**. In this, the first structure of a LeuT homolog, the inverted repeat of helices 1–5 and 6–10 is seen as a pseudo-twofold axis in the plane, and so the repeated structures are similar, as seen by superimposing them as shown in Figure 13.9D. This

---

BOX 13.6   COMPARING CRYSTAL STRUCTURES

So, given that we have structures of LacY and FucP that may be related to different states of transport, how do we compare them? One way is to compare the six TM domains with each other at the structural level using a calculation of the rmsd between alpha carbon positions when the domains are aligned to minimize this parameter. Each X is the distance between an alpha carbon in one structure and the corresponding alpha carbon in the other structure.

$$rmsd = \sqrt{\frac{1}{n}\left(X_1^2 + X_2^2 + X_3^2 + \cdots + X_n^2\right)}$$

By using the backbone α-carbon positions, the effects of the sequence divergences are lessened. This approach is in common use for comparing structures, and if it is

applied to compare the N-terminal 6TM domain with the C-terminal 6TM domain of FucP, they are found to have the same structures within about 2.97 Å rmsd, revealing their pseudo-twofold relationship. Applying the analysis to compare the N-terminal domains of FucP and LacY gives an rmsd of 2.86 Å, and comparing the C-terminal domains gives 2.90 Å. The comparisons are shown in Figure 13.6. Thus, there is great similarity of the domains in the inward and outward forms of these proteins, and a rigid body rotation of the domains relative to each other by about 40 degrees converts one form to the other. Unlike the LeuT case, where the alternating access involves complex internal motions in the subunits, the Jardetzky idea of a rocking motion between two rigid subunits appears to apply rather well in this interpretation!

BOX 13.7   SIDE EFFECTS OF DRUGS THAT INHIBIT NEUROTRANSMITTERS

One class of drugs is aimed at inhibiting reuptake of neurotransmitters, thereby affecting the activities of selected classes of neurons. There are roughly 10^{11} neurons in an adult human brain and on average ~5000 synapses per neuron. These staggering numbers underscore the enormous complexity of processing that is allowing you to read and interpret this text at this moment while at the same time maintaining the functions of your body. The synapses operate when action potential signals arriving at one neuron, the presynaptic neuron, cause the release of neurotransmitters that bind to receptors on another neuron, the postsynaptic neuron, causing it to fire an action potential and thus receive the message. Then, the neurotransmitters must be pumped from the synaptic space to clear the synapse so that it can fire again.

In addition to the neurons, there is a roughly equal number of glial cells that function to support the neurons, holding them in place, supplying nutrients, and helping to keep the synapses from firing adjacent synapses when they are flooded with neurotransmitters. Both the neurons and the glial cells have pumps to scavenge the neurotransmitters. These pumps are co-transporters that use the Na^+ gradient to drive the uptake of neurotransmitters from the synaptic space, using mechanisms like those discussed in this chapter.

Many agents are known that block reuptake and influence activity, such as the SSRIs used to treat depression. One adverse consequence is that the transmitters may leak to trigger adjacent synapses, as shown schematically here (**Figure 1**).

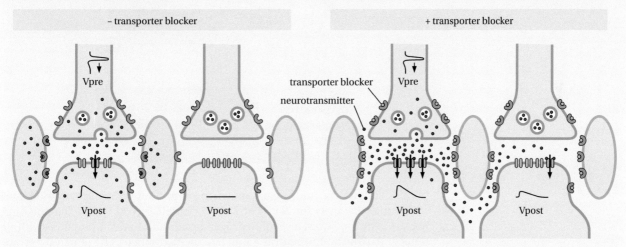

Figure 1 The origin of some side effects of re-uptake inhibitors is shown schematically. (A) In a normal synapse, transmitters are removed from the synapse by pumps such as LeuT. (B) If the pumps are inhibited, transmitters accumulate and can leak into other synapses, triggering them.

structural inverted repeat was not anticipated on the basis of sequence analysis and came as a surprise when the structure was determined, as has been the case for other transporters, such as LacY, the lactose permease. The binding sites for two Na^+ ions and for leucine are found halfway across the membrane in a site that is devoid of water and where the regular structure of the helices is interrupted to contribute to the binding sites. Conserved amino acids are found at the helix interfaces and at the binding sites for Na^+ and leucine. Helices 1 and 6 are discontinuous in the middle of the structure.

13.3.3 Ion- and Substrate-Binding Sites Use Principles Seen in Ion Channels

Some of the principles discussed in Chapter 11 are evident in the secondary transporter structures. Main-chain atoms and the partial charges at helix ends participate in ion and substrate binding, a theme we have already encountered in the organization of the potassium channels. In KcsA, for example, the partial negative charges at the ends of the short helices help stabilize ions in the central cavity, and main-chain carbonyl oxygen atoms form the selectivity filter where they substitute for hydration water oxygens coordinated to K^+ ions. In LeuT,

Figure 13.8 Topology of LeuT and homologs. Schematics of the TM topologies of LeuT in (A) and BetP in (B, C) domain structures shown in comparison with other members of the family. These transporters share the core structural arrangement of LeuT-like secondary transporters, containing an apparent five–TM helix inverted-topology repeat (blue and red). Driving ions are indicated using green circles, and the transported species is shown as a green triangle. The grey helices are individually added elements. (Khafizov K, Staritzbichler R, Stamm M et al. [2010] *Biochemistry* 50: 10702–10713. With permission.)

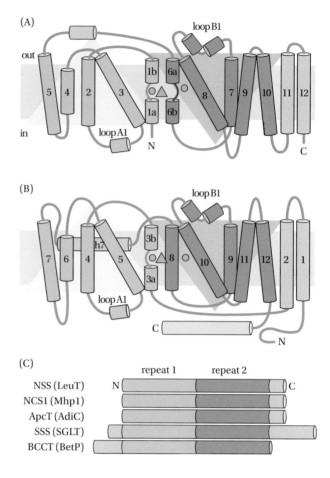

binding sites for Na$^+$ and Leu are shown in Figure 13.9, **Figure 13.10**, and **Figure 13.11**; the selectivity for Na$^+$ over K$^+$ is very high (~1000:1). The leucine binding site is formed from a set of hydrogen bonds and ionic interactions with the polar end of the amino acid, as might be expected, since COO$^-$ and NH$_2^+$ groups must be accommodated in a water-free environment. The carboxyl group interacts with partial charges on the backbone and side chains, but the side chain contacts are particularly interesting, since they both help to neutralize charge and allow positioning of the carboxyl and amino groups at helix ends, where the helical charges also provide an electrostatic contribution. The hydrophobic side chain of the leucine fits into a non-polar pocket with good van der Waals contacts, as shown in Figure 13.10C. Thus, a set of ionic and hydrophobic interactions give a chemical fit to the leucine and stabilize it in an interior location.

13.3.4 LeuT Is Active in Crystals Used in Structure Determination

To check on the activity of the transporter studied in the crystals, LeuT was reconstituted into liposomes, and the transport of Leu, K$^+$, Na$^+$, and Cl$^-$ was followed. Na$^+$ and Leu were translocated, and transport was found to be independent of Cl$^-$ and K$^+$ (**Box 13.8**), although a Cl$^-$ is found in the structure (Figure 13.9C). The two sodium-binding sites, Na1 and Na2, seen in the structure are shown in Figure 13.11. The ions have key roles in stabilizing the partially unwound structures of TM1 and TM6 as well as the bound leucine molecule. Octahedral coordination of Na1 is provided by the leucine carboxy oxygen, as noted earlier, by a mixture of backbone and side chain carbonyl oxygens, and by a Thr sidechain oxygen. Na2 is positioned between the TM1 unwound region and TM8, about 7.0 Å from Na1 and 5.9 Å from the α-carbon of bound leucine. Trigonal bi-pyramidal coordination to Na2 is achieved by carbonyl oxygens and the hydroxyl

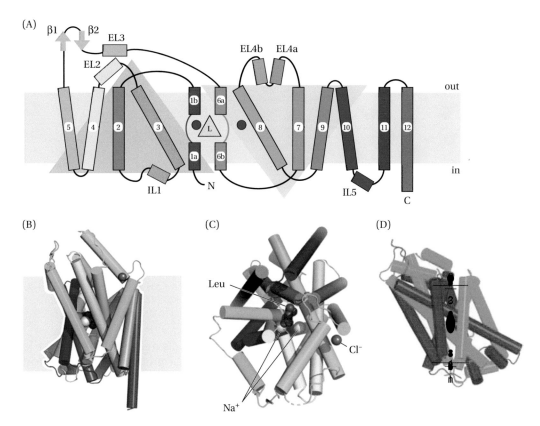

Figure 13.9 Structure of a LeuT secondary transporter homolog (PDB 2A65). (A) Topology; compare with Figure 13.8. (B) A view along the membrane plane. The structure reveals one bound L-leucine, two sodium ions, and one chloride ion, which are shown as CPK models in magenta, grey, and purple, respectively. (C) View as seen from the extracellular side of the membrane. (D) Pseudo-symmetry in the molecule is indicated as a pseudo-twofold axis through the molecule, relating TM1–TM5 (red) and TM6–TM10 (green). The rotation axis is depicted as a black ellipsoid. (From Yamashita A, Singh SK, Kawate T et al. [2005] *Nature* 437: 215-223. With permission from Springer Nature.)

oxygens from Thr354 and Ser355. Importantly, coordination for both Na1 and Na2 is accomplished by partial charges from the protein, with the carboxy group of the bound leucine being the only ligand bearing a formal charge. It appears that the discrimination between Na⁺ and K⁺ may be explained by the sizes of the ions—the K⁺ is simply too large to fit into these binding sites. There may also be

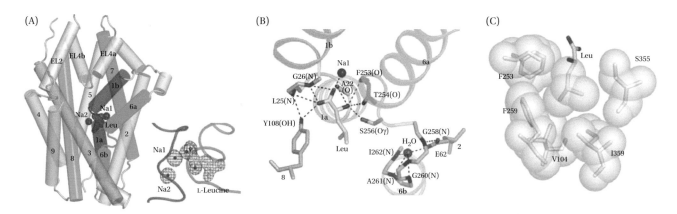

Figure 13.10 Binding of leucine in LeuT$_{Aa}$. (A) Structure of LeuT with contoured electron densities of the sodium ions and leucine. The inset shows an amplified view of the electron densities of leucine and sodium ions. TM10–TM12 have been omitted to reveal the binding sites. (B) This close-up of the LeuT structure shows hydrogen bonds and ionic interactions in the leucine binding pocket, which are indicated as dashed lines. (C) Space-filling representations show the hydrophobic interactions between the leucine and LeuT$_{Aa}$. (From Yamashita A, Singh SK, Kawate T et al. [2005] *Nature* 437: 215-223. With permission from Springer Nature.)

(A)

(B)

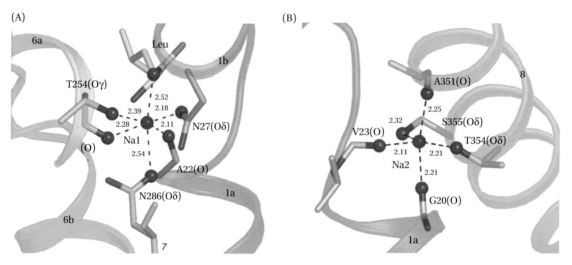

Figure 13.11 Na⁺-binding sites in LeuT Na⁺-binding sites are highly specific, favoring Na⁺ over K⁺ by 1000:1. (A) Na1 is specifically bound and forms part of the binding site for Leu (shown in yellow). (B) Na2 is also specifically bound. Note that a number of backbone carbonyls are involved—a theme often found in ion binding. Distances are in Å. (From Yamashita A, Singh SK, Kawate T et al. [2005] *Nature* 437: 215–223. With permission from Springer Nature.)

an electrostatic difference, since the surface charge density on the K⁺ is lower due to its larger radius. The structuring of the protein by interaction with the sodium ions further supports the idea that the leucine must bind after the sodium is in place. Thus, this driving ion does double duty, helping to form the binding site for the transported leucine. If sodium must be bound before the substrate leucine can bind, removing the sodium will destabilize the leucine binding. The removal of sodium in the inward-open state would facilitate the release of the substrate leucine in the low-sodium environment to which the leucine is delivered, giving a structural understanding of how the concentration difference of sodium across the membrane is used to translocate leucine.

BOX 13.8 MEASURING TRANSPORTER ACTIVITY USING RECONSTITUTED SYSTEMS

While the crystal structures that dominate the chapter are essential in providing mechanistic ideas of channel functions, the functions themselves must be carefully measured so that there is clarity on what needs to be explained. Many methods have been developed in recent times, including single molecule measurements using black lipid membranes (Figure 1.19). But perhaps the most useful and accessible measurements are based on reconstitution of pure transporters into lipid vesicles. Generally, one finds a system for expressing the transporter molecule, in bacteria, insect cells, mammalian cells, or cell-free systems, subsequently obtaining the molecule in a detergent-stabilized environment. Then, the molecule is mixed with lipids, also in detergent, and the detergent is removed (by dialysis, for example), resulting in the formation of vesicles with the transporters across their bilayers if all goes well (which it often doesn't).

In the example shown in **Figure 1**, LeuT reconstituted into liposomes was studied. The His-tagged protein was produced in *E. coli*; cell membranes were isolated and solubilized with 40 mM dodecylmaltoside (DDM). The LeuT (in detergent) was purified on a metal ion affinity column and then digested with thrombin to remove the His tag. A mixture of total *E. coli* lipid extract and egg lecithin solubilized in detergent was mixed with the purified LeuT, and the detergent was removed by incubation with polystyrene beads that bind detergent (Biobeads™), resulting in liposomes containing LeuT.

The resulting liposomes were incubated with tritium-labeled leucine in the presence of different ions added outside the vesicles to create ion gradients across the liposomal membranes, and the resulting transport of leucine was measured. The transport assay was initiated by diluting liposomes into buffers containing either 100 mM NaCl, 100 mM sodium gluconate, or 100 mM KCl. The liposomes were captured on filters, washed, and counted to measure Leu uptake. The data, in Figure 1 show that uptake is stimulated by Na⁺ but not by K⁺, and that transport does not require Cl⁻ in this case (it is required by some other transporters).

(Continued)

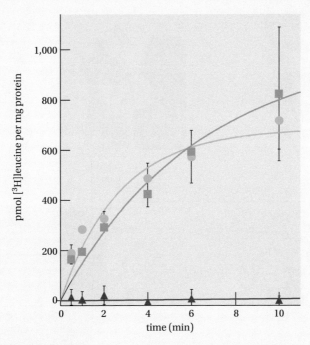

Figure 1 [^{3}H]Leucine transport in LeuT is driven by Na$^+$ and not K$^+$ in liposomes. The filled circles (green, NaCl), squares (blue, sodium gluconate), and triangles (red, KCl) show the uptake of labeled leucine in the presence of various ion gradients. Transport occurs in the presence of Na$^+$ regardless of anion type (Cl$^-$ or gluconate$^-$), whereas transport does not occur in the presence of KCl. (From Yamashita A et al. [2005] *Nature* 437: 215–223. With permission from Springer Nature.)

13.3.5 Structures of the Outward-Open and Inward-Open States Have Been Determined

The structure of LeuT discussed earlier is the outward-facing, occluded state (Figure 13.10 and Figure 13.11). Through a structure- and function-based set of mutations, coupled with antibody binding, it has been possible to obtain other intermediates for crystallographic studies (**Figure 13.12, Figure 13.13**). The objective is to adjust the free energy of the solubilized transporter so that key intermediates are the lowest-energy structures. Of course, there is always the problem of whether these changes and binding events might alter the results in significant ways, but functional assays and other data provide at least some reassurance. Another concern is that these structures are at a lower resolution (~3 Å) than the original outward-open case (1.6 Å). The fact that the results make sense adds to confidence (but carries the danger that one is biased toward what one is looking for).

13.3.6 The Substrate-Free, Outward-Open State Reveals the Chemistry of the Outward-Open to Outward-Closed Transition

The first step in the cycle of transport happens when both the sodium ions and the leucine substrate are bound, closing the gate. Crystals of substrate-free

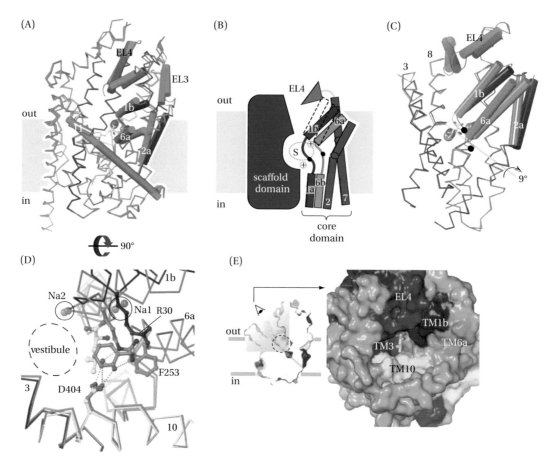

Figure 13.12 The outward-open state of LeuT, with bound Na⁺ but no leucine. (A) Shows the superposition of the outward-open (colored) and leucine-bound outward-occluded (grey) conformations so that the changes can be seen. (B) Using a schematic helps to show the motion when leucine binds (S) and the outside is occluded. (C) By choosing the right view, it is seen that a rotation about an axis (yellow) describes the conformational change associated with opening/closing to the outside. (D) Opening the gate requires rupture of gate interactions (grey dashed lines) in the outward-open structure. Two water molecules that bridge Arg30 and Asp404 in the outward-occluded state are shown as red spheres. (E) View of vestibule. (From Krishnamurthy H et al. [2012] *Nature* 481: 469–471. With permission from Springer Nature.)

LeuT were obtained by starting with a natural variant, K288A, adding a Tyr108-to-Phe mutant in TM3, and binding a conformation-specific antibody fragment (Fab). Care was taken to purify the protein so that no remaining free leucine was present. The structure is shown in Figure 13.12. Comparison of the structure with that of the outward-facing occluded structure previously solved (Figure 13.9) shows that the gate action involves a hinged movement of two helical segments that are parts of helices 1 and 6; Figure 13.12A. The hinges are near the locations of the binding sites for the two sodium ions and the leucine substrate, indicated schematically in Figure 13.12B and in the structures in Figure 13.12C. By zooming in, the interactions that are disrupted to open the gate (or formed to close it) are seen in Figure 13.12D to involve bridging water molecules and other hydrogen bonds.

13.3.7 The Inward-Open Structure of LeuT Allows the Structural Conformation Change and Internal Gating to Be Understood

With the determination of the crystal structure of the inward-open state, comparisons can be made that reveal both the inward gating and the key conformational changes involved in the transport event. These are wonderfully clarifying and reveal some key ideas that will likely be applicable to many transporters. A

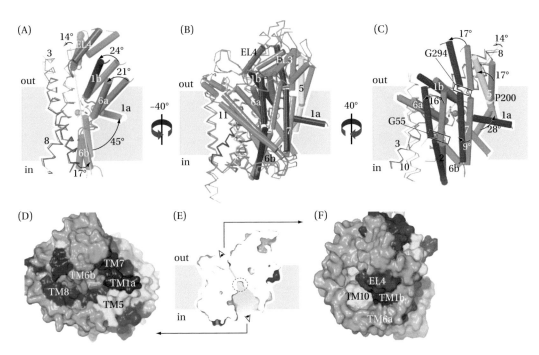

Figure 13.13 Superposition of outward-occluded and inward-open structures of LeuT shows large movements of domains and helices. The overall changes are shown in (B) and are divided into two parts, (A) and (C), for clarity, with the outward, occluded structure in grey and the inward, open structure in colors. There are large movements of TMs between the outward-occluded and inward-open conformations, as shown in (A) and (C). The axis of rotation of the core domain with the exclusion of TM1a is depicted in yellow. (D) Surface representation of the inward-open structure looking "up" into the binding site from the intracellular side, as depicted in (E). (F) Surface representation of the inward-open structure showing the elements forming the extracellular thick gate. It is helpful to compare the images in (E) and (F) with Figure 13.12E. (From Krishnamurthy H et al. [2012] *Nature* 481: 469–471. With permission from Springer Nature.)

major idea is that local hinge movements of transmembrane helices are energetically driven by the formation and disruption of substrate- and sodium-binding sites, which are translated through nearly rigid body movements of groups of helices and loops into opening and closing of the gates (Figure 13.13). This finding is reminiscent of the motions of the gate in a potassium channel, where bends in the channel helices move half-helix segments to open the gate (Figure 11.21). An important question is how the local substrate- and ion-binding events drive conformation and gating of LeuT during the transport cycle.

A comparison of the outward-occluded structure with the inward-open structure shows that the original rocking-domains idea proposed by Jardetzky is vaguely correct but that the real motions are more complex. The comparison of the inward and outward conformations is shown in Figure 13.13. The anti-parallel pseudo-twofold symmetric relationship of TMs 1–5 and 6–10, together with the four-helix bundle organization of TMs 1, 2, 6, and 7, might suggest a rocking-bundle mechanism of transport with relative motion about a rotation axis parallel to the membrane and intersecting the substrate-binding site. However, detailed comparison reveals that only part of the core moves as a unit and that the motion does not have strict twofold symmetry. Further, the gates and occluded states are features not clearly recognized in the earlier, simple ideas.

13.3.8 Combining the Three Structures of LeuT Gives a View of the Sequence of Events That Lead to Symport

The three structural views are shown schematically in **Figure 13.14** and can be tied together to give a "movie" of the transport process (**Figure 13.15**). The use of hydrogen bonds, including main chains, the partial charges on helix ends,

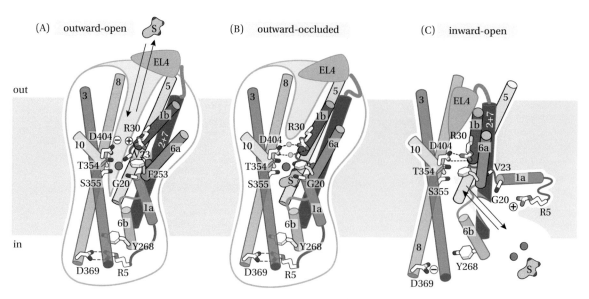

Figure 13.14 A schematic view helps to show overall changes in the LeuT transport mechanism. (A), (B), and (C) show structural elements and gating residues associated with the transition from the outward-open (A) to the outward-occluded state (B) and the inward-open state (C). At present, there is no crystal structure for an inward-occluded state. (From Krishnamurthy H et al. [2012] *Nature* 481: 469–471. With permission from Springer Nature.)

electrostatic interactions stabilizing charges in the ions and substrate, and the waters are all found by inference from the positions of the atoms in the models. Changes in these interactions are clearly documented as components of the structural changes that result from binding of the two sodium ions and the substrate leucine and that result in the release of ions and substrate when the low sodium concentration inside the membrane dissociates the ions.

While the resulting model is satisfying in most respects, one must note that the inward-facing occluded structure is not yet documented and that two of the structures are at relatively low resolution. A next goal would be to follow the energies of all the interactions to move the understanding to a higher level of chemistry, but such understanding is still generally in the future for almost all protein–ligand interactions.

13.4 ELEVATOR-LIKE TRANSPORTERS: STRUCTURAL STUDIES OF A GLUTAMATE TRANSPORTER, GLT$_{PH}$, REVEAL SIMILAR PRINCIPLES

At about the same time as the work on the LeuT transporter emerged, structures of Glt$_{Ph}$, an archaeal glutamate transporter homolog from *Pyrococcus horikoshii* that transports aspartate, also appeared. These help us flesh out a picture of different ways that the alternating-access transport can work in detail while exploiting the same basic principles. The glutamate transporter is a homolog of a different branch of the NSS family and has a trimeric structure in which each monomer is a symporter that moves three Na$^+$ downhill into the cytoplasm to drive the transport of one molecule of aspartic acid in the same direction. The structure of Glt$_{Ph}$ is shown in **Figure 13.16**. Like LeuT, the architecture is derived from inverted repeats, but it is rather different in having two of them (compare with Figure 13.4).

Crystal structures of Glt$_{Ph}$ are presently available for the outward-open, outward-closed, and inward-closed forms. Figure 13.15 is a schematic that compares the basic steps inferred from the Glt$_{Ph}$ and LeuT structures, bearing in

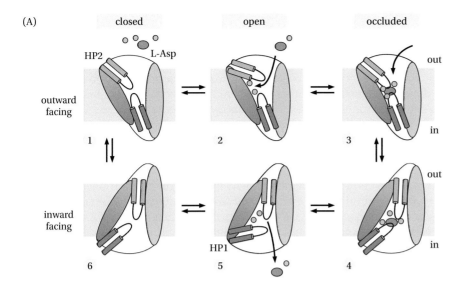

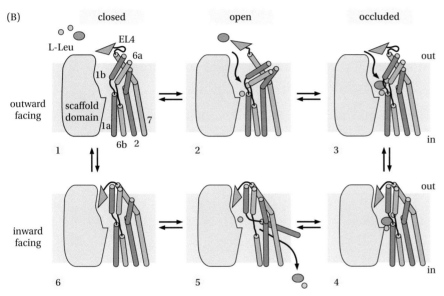

Figure 13.15 Schematic views show that the transport mechanisms in Glt$_{Ph}$ (panel A) and LeuT (panel B) use similar principles. (1) External gate movements allow sodium binding to stabilize outward-facing open states in the absence of other ligands (2). Na$^+$ binding (blue dots) creates stable binding sites for substrates and additional ion(s), and their binding energy closes the structures to form outward-occluded states (3). Large, slow structural rearrangements lead to inward-facing, occluded states (4). These large motions are somewhat different in Glt$_{Ph}$ and LeuT. A sodium is released in response to the low concentration inside the membrane, leading to the opening of intracellular gates (5) and de-stabilization of the binding and consequent release of substrate and additional ion(s), resulting in inward, unliganded states. (6), the intracellular gates close and the conformations can return to the outward-facing unliganded states. (From Focke PJ, Wang X, Larsson HP [2013] *Structure* 21: 694–705. With permission from Elsevier.)

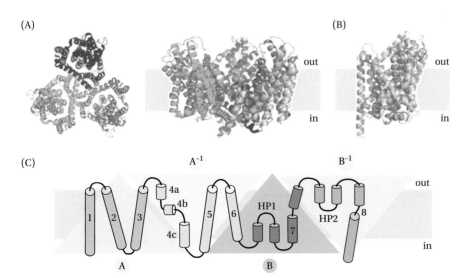

Figure 13.16 Structural features of the glutamate transporter, Glt$_{Ph}$. (A) Glt$_{Ph}$ assembles to form a bowl-shaped trimer, in which each subunit is a transporter. Left, extracellular view. Right, view parallel to the membrane. Individual monomers are colored wheat, blue, and green. (B) LeuT monomer, viewed parallel to the membrane, for comparison with one of the subunits. (C) Inverted repeats are seen, as shown in this schematic view. (From Focke PJ, Wang X, Larsson HP [2013] *Structure* 21: 694–705. With permission from Elsevier.)

mind that the inward-closed state in LeuT and the inward-open state in Glt_{Ph} were not structurally determined as a basis for this scheme. While the details of the motions in the proteins are not the same, the principles of binding, gating, and conformation change are similar. One notable difference, though, is that in Glt_{Ph}, the Asp substrate is translocated across the membrane by an elevator-like up-down movement of a transport domain relative to the membrane, while in LeuT, the substrate-binding site remains more or less stationary.

Interestingly, a crystal structure for Glt_{Ph} has been found in which the trimer is asymmetric, with one transporter in an intermediate and two in an inward-facing conformation. This observation raised the question of whether the motion of the trimer units may be interdependent. But by using Förster resonance energy transfer (FRET) measurements to follow motions of the subunits in single trimers, it was found that the motions of the subunits are independent of each other during transport (**Figure 13.17**). Further, in the absence of Asp, the transporter cycles either between inward and intermediate or between outward and intermediate states, with dynamics that are similar to those during transport. Thus, the energy differences between the conformations appear to be small. An important implication is that the intermediate state, which is occluded to both sides, may prevent transport of Na^+ that would degrade the membrane potential in the absence of Asp transport by preventing a monomer from switching directly between outward- and inward-facing conformations.

13.4.1 FRET studies of Glt_{Ph} reveal the dynamic nature of protein function

We can surmise many important functional principles from transporter crystal structures, but unlike the ion channels whose dynamics are readily observed through single-channel recordings (Chapter 11), observation of facilitator dynamics is more challenging. An important approach is the use of Förster Resonance Energy transfer (**Box 13.9**) between donor and acceptor dyes

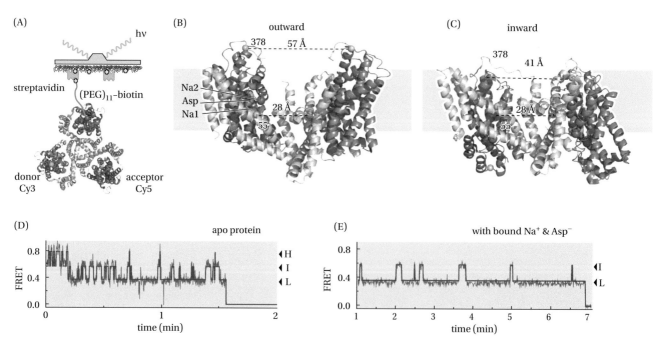

Figure 13.17 Determination of Glt_{Ph} dynamics using FRET labels. (A) The protein is tethered to an optical surface in a total internal reflectance microscope that allows FRET changes to be observed during structural transitions of the protein. (B) The outward-facing form of the protein showing the positions and distances between the FRET donors and acceptors. (C) Distances between donors and acceptors in the outward-facing state. (D) FRET signals observed for labels at N378C for the apo-protein. (E) FRET signals observed with Na^+ and Asp^- bound to the protein. (From Akyuz N et al. [2013] *Nature* 502: 114–118. With permission from Springer Nature.)

BOX 13.9 FÖRSTER RESONANCE ENERGY TRANSFER (FRET)

In FRET, light energy emitted by one fluorophore (donor) at a particular wavelength is absorbed by a second nearby fluorophore (acceptor), which fluoresces at a longer wavelength. The extent of energy transfer from the donor to the acceptor is strongly dependent on the distance between the two, typically tens of Å. FRET has become a widely used biophysical method for examining molecular function at the molecular and cellular levels. For membrane proteins, the method can be used to study changes in the distance between two parts of a molecule that have been labeled with a donor fluorophore and an acceptor fluorophore, generally by attaching them to Cys residues engineered into the protein by site-directed mutagenesis.

Fluorescent molecules fluoresce when exposed to ultraviolet light that raises the molecule to an excited electronic state by the absorption of a photon with energy $E_{ex} = hc/\lambda_{ex}$, where h is Planck's constant, c the speed of light, and λ_{ex} the wavelength of the exciting light. The energy states available depend upon various vibrational and other states of the molecule. Consequently, the excited states are represented schematically in a "Jablonski diagram" as a stack of quantized energy states (**Figure 1**, panel A). The excited molecule generally relaxes non-radiatively to a lower energy state (E_{em}), from which decay to the ground state occurs by emission of a photon of wavelength λ_{em}. The values of λ_{ex} and λ_{em}

depend upon the molecular structure of the fluorophore. If there is a nearby acceptor fluorophore whose $E_{ex}(A)$ overlaps the $E_{em}(D)$ of the donor, the donor photon can excite the acceptor, which subsequently fluoresces at wavelength $\lambda_{em}(A)$. Excitation and emission generally occur over a range of wavelengths (panel B), which can be determined using a fluorescence spectrometer that allows one to measure the emission intensity at a particular (λ_{em}) wavelength while scanning through a range of exciting wavelengths (excitation spectrum). By fixing the exciting wavelength (λ_{ex}) and scanning the fluorophore emission, the range of emitted wavelengths (emission spectrum) can be determined. FRET can occur when the donor emission spectrum overlaps the acceptor excitation spectrum (called spectral overlap), shown schematically in panel B.

The efficiency of energy transfer (E_{FRET}) depends very strongly upon the distance R between the donor and the acceptor (panel C). In the equation, R_0 is a parameter characteristic of the fluorophores and their relative orientations. The general assumption is that the fluorophores explore all angles in space. We generally ignore numerous technical and theoretical issues and use FRET measurements in an empirical way. A more or less typical plot of E_{FRET} against the separation distance R between the fluorescence labels on the molecule is shown in panel D.

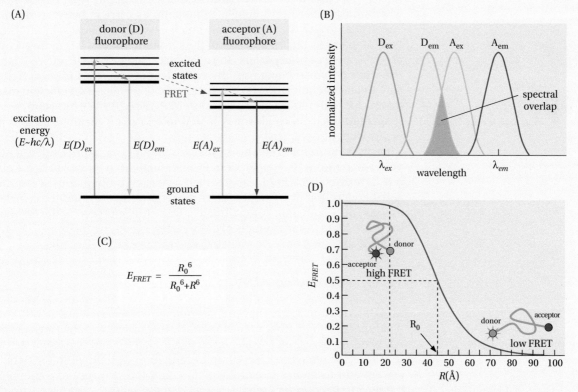

Figure 1 Basic principles that underlie measurements of changes in distance between donor and acceptor fluorophores attached to a molecule. The method is named in honor of the German physical chemist Theodor Förster (1910–1974). (D, From Sahoo H [2011] *J Photochem Photobiol C: Photochem Rev* 12: 20–30. With permission from Elsevier.)

covalently attached to the molecule via Cys residues engineered into the molecule at strategic locations. An example is shown in Figure 13.17 for Glt$_{Ph}$. To observe FRET, the labeled molecules are tethered by a polyethylene glycol (PEG) linker to an optical surface in a total internal reflectance microscope. Unlike most other facilitators, the structures of outward-open and inward-open Glt$_{Ph}$ structures are known (Figure 13.17B,C), which allows selection of dye-binding sites that report on changes in protein conformation in the presence of substrates and inhibitors. The dynamics of Glt$_{Ph}$ under different conditions are apparent in Figure 13.17D and E; the dynamics of the apo and the Na$^+$,Asp$^-$-bound states are quite different. Noteworthy is the slowness of the transitions between states, which for the apo form, occur with a frequency of only about 0.5 s^{-1}. One can conclude from these data that the apo form samples both outward- and inward-facing states and does so relatively rapidly compared with the substrate-loaded form, whose transition rates decrease dramatically to 0.02 s^{-1}. The great slowness of the transitions in the laboratory likely reflects the fact that the protein is from a hyper-thermophile (Ph = *Pyrococcus horikoshii*, which grows optimally at 98 °C).

13.5 THE ADP/ATP TRANSPORTER FROM MITOCHONDRIA

13.5.1 Mitochondrial Membranes Transport Key Metabolites, Including ATP and ADP

Mitochondrial membranes transport many metabolites that are essential for eukaryotic metabolism. Specific transport through the inner mitochondrial membrane is achieved by transporters that belong to another large family, the mitochondrial carrier family (MCF). The family was named at a time when transporters were thought to function as carriers between the outside and inside surfaces of a membrane, much as valinomycin acts in carrying ions (see Chapter 11), but on the basis of the subsequent work, it is more accurate to describe them as transporters. The efficient exchange of ADP and ATP is essential, since a human being uses the equivalent of their own mass of ATP every day! The regeneration of ATP in mitochondria thus needs an efficient machine that is able to import ADP and to export ATP: the ADP/ATP transporter.

Adenine nucleotides also play a vital role in plant metabolism and physiology. Heterotrophic cells regenerate most of the ATP in their mitochondria, whereas autotrophic cells also possess chloroplasts, representing a second powerhouse for ATP regeneration. Even though the synthesis of these nucleotides is restricted to a few locations, their use is nearly ubiquitous across the cell, and therefore, highly efficient systems are required to transport these molecules into and out of different compartments. These transport events are emphasized in **Figure 13.18**. As an example of these carrier events, we focus on a case where detailed structural data are available, the bovine ATP/ADP carrier.

13.5.2 The Bovine ADP/ATP Transporter

All ADP/ATP transporters have a characteristic consensus sequence, RRRMMM, that is not found in other mitochondrial transporters, and they are specifically inhibited by bongkrekic acids, atractyloside (ATR), and carboxyatractyloside (CATR). As is often the case, complex formation with an inhibitor stabilizes a protein and facilitates structural studies, and the structure of a monomer of the bovine transporter (see later) was determined in a complex with CATR. Atractylosides are lethal poisons produced by a Mediterranean thistle known to the ancient Egyptians and described by the Greek physician Dioscorides (~40–90 AD) as a medicinal herb. The two families of inhibitors provide tools that have been useful in studies of the carrier.

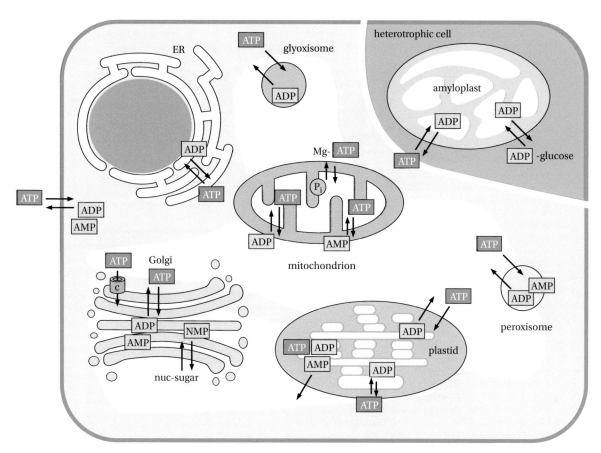

Figure 13.18 There are many adenine nucleotide transporters in plant cell organelle membranes. The passage of adenine nucleotides in organelles is mediated by specialized carrier proteins (which are now referred to as group transporters). Many of these exchange ATP with ADP or AMP, some are transporters driven by ions, and one is a channel.

The transporter has six TMs, and the amino and carboxy termini are oriented toward the intermembrane space. The helices are in three pairs, which arose from gene duplication events of the kind we have seen in other cases, including all the transporters discussed in this chapter. Each pair of helices has a straight TM and a kinked TM (shown schematically in **Figure 13.19A**). Kinking of the TMs is seen here and in all the structures in the preceding figures—it is generally found that about half of the TMs in polytopic membrane proteins are bent or kinked. The stoichiometry of exchange of metabolites is one to one, and the adenosine phosphates are transported without magnesium bound. As might be anticipated, the carrier appears to exist in two conformations involved in the translocation mechanism: one is stabilized by the atractylosides and the second by bongkrekic acid. It is thought that the functional transporter unit is a dimer, as inferred from the observed dimeric state of the CATR-inhibited transporter, which has one CATR per dimer.

13.5.3 The Structure of a Monomer of the Bovine ATP/ADP Transporter Has Been Determined

The structure is at a resolution of 2.2 Å and is in a complex with CATR, but no clear dimer is seen, and the ratio of the monomer to the inhibitor is one to one (Figure 13.19). Six α-helices form a compact transmembrane domain, which, at the surface toward the space between inner and outer mitochondrial membranes, reveals a deep depression surrounded by a basket of helices (**Figure 13.20**). At its bottom, a hexapeptide carrying the signature of nucleotide carriers (RRRMMM) is located. The cavity traverses the lipid bilayer hydrophobic

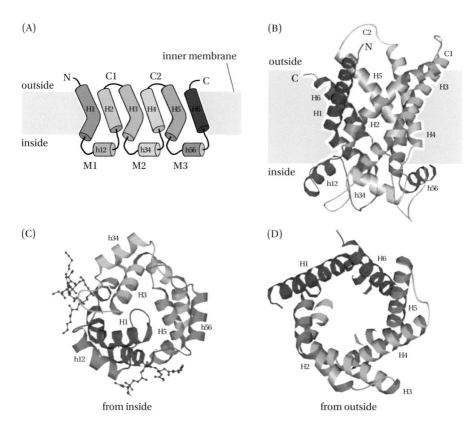

Figure 13.19 The structure of the ATP/ADP transporter (PDB 1OKC) reveals the use of a repeated motif. (A) A schematic diagram of the carrier secondary structure shows the repeated motif. Odd-numbered helices are kinked by the presence of prolines (P27, P132, P229). Inside and outside refer to the matrix and the intermembrane spaces, respectively. (B) A ribbon diagram of a side view shows how the motifs are arranged in three dimensions. (C) A view from the inside includes two immobilized cardiolipins represented in black (D) Outside view. (From Pebay-Peyroula E et al. [2003] *Nature* 426: 39–44. With permission from Springer Nature.)

Figure 13.20 (A) A cross-section surface of the ATP/ADP transporter shows a cone-shaped cavity. The characteristic RRRMMM hexapeptide is found at the bottom of the cavity. The conical pit is open to the outside, and the RRR sequence spans the closed part of the carrier. The highlighted arginines are thought to provide counter charges for the ATP/ADP phosphates. (B) A close-up view of the RRRMMM motif in the same orientation. (From Pebay-Peyroula E et al. [2003] *Nature* 426: 39–44. With permission from Springer Nature.)

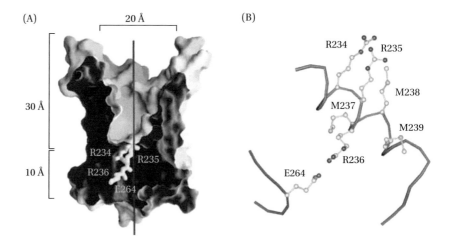

region. The structure, together with earlier biochemical results, suggests that transported substrates bind to the bottom of the cavity and that translocation results from a transient transition from a "pit" to a "channel" conformation, but as in the case of the MFS family, the absence of additional structures leaves the matter open.

13.5.4 Is the Transporter a Monomer or a Dimer?

Given that the monomer defines an apparent channel, the need for a dimer is not obvious, but biochemical studies continue to reveal monomers and dimers, depending on conditions. In a later crystallographic study, protein–protein interactions were seen, mediated by immobilized cardiolipins and emphasizing the critical role of compositional variables. A dependence on cardiolipin for activity upon reconstitution had been previously reported. Note the cardiolipins seen in Figure 13.19C. One speculative idea under study is that the functional unit may be a pair of transporter molecules linked by cardiolipin with one member of the pair open to the inside of the membrane and the other open to the outside, and that the coupling would reverse the conformations during transport.

13.6 UNIPORTERS (AKA FACILITATORS) FACILITATE TRANSPORT WITHOUT CHEMICAL GRADIENTS

While the main subject of this chapter is the use of coupled gradients to drive transport uphill energetically, there are biologically important examples where the task is simply to allow specific molecules to move across membranes down their concentration gradients. While channels work well for ions, specificity for small molecules is limited, and except for the aquaglyceroporins (Chapter 10), passive transporters use mechanisms more related to the secondary transporters. Transport can then be biased by mass action if the transported species is modified after transport, so that its concentration is constantly lowered by metabolism—a different kind of driving force. An example is the transporter for glucose that allows facile motion of the sugar across the erythrocyte membrane, GLUT1. While glucose can permeate a membrane by itself, it does so too slowly for adequate function. The transporter lowers the barrier by providing a different, specific pathway.

13.6.1 The Erythrocyte Glucose Transporter Resembles the MFS Proteins LacY and GlpT.

Before there was a three-dimensional structure of GLUT1, the results from a number of studies using homology modeling, cysteine scanning, and chemical modification showed that its topology, substrate binding, and sequence features are consistent with a structure resembling LacY. GLUT1 comprises 492 amino acids; is hydrophobic; contains a single, exofacial N-linked glycosylation site; and is predominantly α-helical. Hydropathy analysis, scanning glycosylation mutagenesis, proteolysis, antibody binding, and covalent modification studies indicated that GLUT1 contains intracellular N- and C-termini and 12 transmembrane α-helices. As an example of these kinds of data, cleavage studies are shown in **Figure 13.21**. Amphipathic α-helices were proposed to form an aqueous translocation pathway for glucose transport across the plasma membrane using the alternating-access mechanism. The mechanism was thought to be similar to the transport of lactose by LacY but with no coupling to a concentration gradient except that of glucose itself.

When the structure of GLUT1 was determined, it revealed that the inferences from indirect methods were substantially correct, showing how progress can be made even in the absence of a crystal structure. **Figure 13.22** shows the structure in the inward-open state, with an aqueous cavity for entry/exit of a glucose molecule to the inside of the erythrocyte. Of course, much chemical detail is seen that was not part of the earlier views, including the presence (and possible mechanistic roles) of surface helices, locations of mutations that cause diseases, and the details of sugar binding.

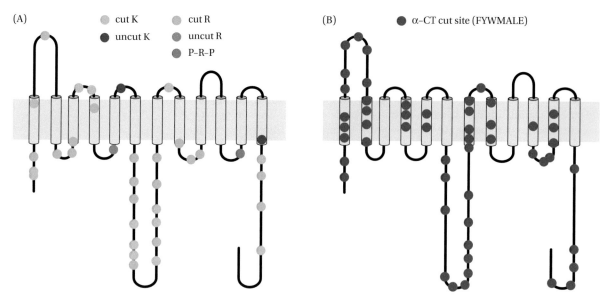

Figure 13.21 Proteolytic cleavage has been used to map the topology of the erythrocyte glucose uniporter, GLUT1. GLUT1 was digested with trypsin or α-chymotrypsin, and the cleavage sites were determined by identification of the fragment using reverse phase high-performance liquid chromatography (HPLC)-electrospray ionization (ESI)-tandem mass spectrometry (MS/MS). (A) GLUT1 contains 35 potential trypsin cleavage sites, of which 32 are cleaved by trypsin. The idea is that cleavage requires access from the aqueous environment, so the solvent-exposed regions are defined. (B) GLUT1 contains 197 potential α-chymotrypsin (α-CT) cleavage sites (Phe, Tyr, Trp, Leu, Met, Ala, and Glu). The 52 detected cleavage sites are indicated by brown circles. Potential α-CT cleavage sites are present in all TM domains. (From Blodgett DM, Graybill C, Carruthers A [2008] *J Biol Chem* 283: 36416–36424. With permission from Elsevier.)

Figure 13.22 The structure of the glucose transporter GLUT1 (PDB 4PYP) resembles LacY and FucP. The structure is in the inward, open conformation. Views are shown in the membrane plane and from the intracellular side, schematically and in electrostatic surface views. The cavity is also marked. Crystallization was facilitated by using a mutant, E329Q, that locks the transporter and by preventing glycosylation by mutation N45T. As expected, there is an aqueous cavity open to the inside and a clearly homologous structure to the LacY and FucP transporters. Comparisons with other structures, such as the bacterial antiporter XylE, support the kind of alternating-access mechanism shown in Figure 13.6. (From Deng D et al. [2014] *Nature* 510: 121–125. With permission from Springer Nature.)

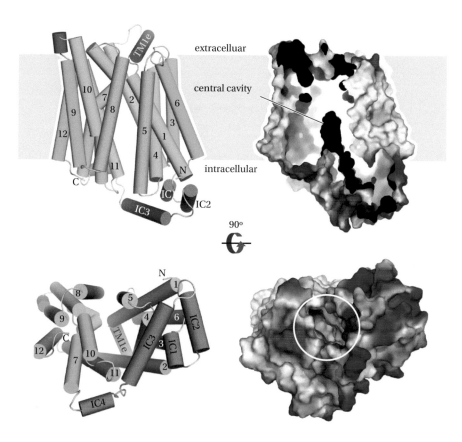

Glucose transport is driven by metabolism in the erythrocyte. The reason why erythrocytes need glucose is that they are metabolically active and feed the glucose into the glycolytic pathway. The first step in glycolysis is the phosphorylation of glucose by hexokinase. Once this modification has happened, the internal concentration of unmodified glucose is lowered, and mass action causes the net migration of more glucose molecules into the cell. Thus, the net transport of glucose is driven by metabolism.

13.7 CONCLUSION

The interpretation of a vast number of physiological, biochemical, and biophysical studies accumulated over decades has been greatly aided by recent crystal structures. Before these recent findings, the central mystery was to understand how an ion gradient could be coupled to drive the accumulation of a small molecule across a membrane against its concentration gradient. This mystery has now been solved to a good first approximation.

A part of the answer that should not have been a surprise is the coupling of the binding of driving ions to the binding of the substrate. We have seen such events before in biochemistry, for example in the kinases. Hexokinase transfers the terminal phosphate from ATP to an -OH group on glucose. But why, it was asked, does it not transfer the phosphate to an -OH on water if glucose is absent? The answer is that the catalytic groups of the enzyme are only positioned for catalysis through a conformational change driven by binding energy when both ATP and glucose are bound. Similarly, binding driving ion(s) to the transporter helps create the binding site for the substrate, and the conformation change to face the other side of the membrane is only favored when both are present.

The complex formed on the outside (transporter + driving ion(s) + substrate) equilibrates with the inward, open state. Ion(s) are released, since they are now exposed to a low cytoplasmic concentration, which destabilizes the substrate binding, and the now empty transporter relaxes to face outward, where it is stabilized by binding ions in the absence of substrate. There are variations and embellishments, such as the use of a K^+ gradient to push the transporter toward the outside or the use of an H^+ ion, but the basic story persists. Remarkably, it is only from the series of structural studies since about 2002 that such perspectives have been gained, following many decades of indirect experiments, intense curiosity, and imaginative speculation.

KEY CONCEPTS

- As in previous examples, such as ion channels, the co-transporters surround internal cavities that allow bound ions and substrates to cross the bilayer. The cavities may be deep, moving the transport events near the inner surface of the membrane, or partway across.

- Co-transport commonly proceeds via a set of identifiable states: outward-facing-open-empty; outward-facing–with bound ions and substrate; occluded transition; inward-facing–open with bound ions and substrate; inward-open-empty; occluded transition; and back to outward-open-empty.

- Binding the abundant ions to the outward-open-empty state shifts the equilibrium of the inward-open-empty state toward the outward-open-empty state.

- Some themes that recur are the uses of helix ends and backbone carbonyl groups to counter charges while creating ion-binding sites. Of course, side chain groups are also used.

- Ions are used to stabilize the binding of the transported substrate on the outside of the membrane, where the ion concentration is high.

- Conformation changes, via occluded states, close the outside-facing conformation to an inside-facing conformation, which opens to form the inside-open state, where the low ion concentration favors the dissociation of the ions and the substrate.

- The inward-empty transporter equilibrates with the outward-open-empty state via an occluded intermediate.

- Similar conformation changes can be used to facilitate the specific translocation of substrates to equilibrate them across a membrane in the absence of a driving ion gradient. Net transport can be driven by mass action when metabolism depletes the substrate concentration inside the cell.

FURTHER READING

Focke, P.J., Wang, X., and Larsson, H.P. (2013) Neurotransmitter Transporters: Structure meets function. *Structure* 21:694-705.

Forrest, L.R., Krämer, R., and Ziegler, C. (2011) The structural basis of secondary active transport mechanisms. *Biochim. Biophys. Acta* 1807:167-188.

Krishnamurthy, H., Piscitelli, C.L., and Gouaux, E. (2009) Unlocking the molecular secrets of sodium-coupled transporters. *Nature* 459:347-355.

Kaback, H.R. (2015) A chemiosmotic mechanism of symport. *Proc Natl Acad Sci USA* 112:1259-1264.

Yan, N. (2015) Structural biology of the major facilitator superfamily of transporters. *Ann. Rev. Biophys.* 44:257-283.

Quistgaard, E.M., Löw, C., Guettou, F., and Nordlund, P. (2016) Understanding transport by the major facilitator superfamily (MFS): Structures pave the way. *Nature Reviews Mol. Cell Biol* 17:123-132.

Drew, D., and Boudker, O. (2016) Shared molecular mechanisms of membrane transporters. *Ann. Rev. Biochem.* 85:543-572.

Blodgett, D.M., Graybill, C., and Carruthers, A. (2008) Analysis of glucose transporter topology and structural dynamics. *J Biol Chem* 283:36416-36424.

KEY LITERATURE

Study of membranes as isolated vesicles (these preparations allowed the finding that transport can be driven by ions):

Kaback, H.R. (1974) Transport studies in bacterial membrane vesicles. *Science.* 186: 882-892.

Radian, R., and Kanner, B.I. (1983) Stoichiometry of sodium- and chloride-coupled gamma-aminobutyric acid transport by synaptic membrane vesicles isolated from rat brain. *Biochemistry* 22:1236-1241.

early ideas of coupled transport (these are interesting, as they anticipated much that was later found)

Mitchell (1957) A general theory of membrane transport from studies of bacteria. *Nature* 180:134-136.

Patlak, C.S. (1957) Contributions to the theory of active transport. II. The gate type non-carrier mechanism and generalizations concerning tracer flow, efficiency, and measurement of energy expenditure. *Bull. Math. Biophys.* 19:209-235.

Jardetzky (1966) Simple allosteric model for membrane pumps. *Nature.* 211:969-970.

Vidaver, G.A. (1966) Inhibition of parallel flux and augmentation of counter flux shown by transport models not involving a mobile carrier. *J. Theor. Biol.* 10:301-306.

Dang, S., Sun, L., Huang, Y., Lu, F., Liu, Y., Gong, H., Wang, J., and Yan, N. (2010) Structure of a fucose transporter in an outward-open conformation. *Nature* 467:734-738.

Focke, P.J., Wang, X., and Larsson, H.P. (2013) Neurotransmitter transporters: Structure meets function. *Structure* 21:694-705.

Yamashita, A., Singh, S.K., Kawate, T., Jin, Y., and Gouaux, E. (2005) Crystal structure of a bacterial homologue of Na/Cl dependent Neurotransmitter transporters". *Nature* 437:215-223.

Krishnamurthy, H., and Gouaux, E. (2012) X-ray structures of LeuT in substrate-free outward-open and apt inward-open states. *Nature* 481:769-774.

Verdon, G., and Boudker, O. (2012) Crystal structure of an asymmetric trimer of a bacterial glutamate transporter homolog. *Nat Struct Mol Biol.* 19:355-357.

Akyuz, N., Altman, R.B., Blanchard, S.C., and Boudker, O. (2013) Transport dynamics in a glutamate transporter homologue. *Nature* 502:114-118.

Pebay-peyroula, E., Dahout-Gonzalez, C., Kahn, R., Trezeguet, V., Lauquin, G.J., and Brandolin, G. (2003) Structure of mitochondrial ADP/ATP carrier in complex with carboxyatractyloside. *Nature* 426:39-44.

Deng, D., Xu, C., Sun, P., Wu, J., Yan, C., Hu, M., and Yan, N. (2014) Crystal structure of the human glucose transporter GLUT1. *Nature* 510:121-125.

EXERCISES

1. Using the hydrogen–deuterium exchange approach discussed in Box 13.4, you make a measurement on the exchange of the KcsA transmembrane domain. Do you expect that the exchange will be like that of LacY or Photosystem II? Why?

2. In LacY, the binding of a proton facilitates binding of the sugar. Why is it not the other way around, with the sugar binding first and then the proton? Identify a biological advantage for the ordered binding, imagining two

strains of bacteria in competition, with one strain binding sugar first and the other protons.

3. Disulfide cross-linking has been used to identify points of close approach in many cases. But, it has been misleading in studies to find the structure of LacY. Suggest a reason for the problem.

4. Mass action plays a role in several of the steps in transport. Which of the steps in Figure 13.3 would likely be driven by mass action?

5. Transport is a relatively slow process compared with channel gating. Compare the timescale of Figure 13.17 with gating timescales in Chapter 11. Suggest some reasons why.

6. Given the Mitchell–Patlak–Jardetzky alternating-access ideas, the finding of inverted sequence repeats may not be a complete surprise. But, speculate why the independent evolution of the repeated structures might be an advantage for the transport mechanism.

7. In this chapter, we focus on symporters, which move ions in the same direction as the substrate being transported against its concentration gradient. Speculate on the steps that an antiporter might use to move the substrate in the opposite direction to the ion flow driving its accumulation.

Bioenergetics

14

The organized structures and functions of a cell would decay to nothing under the influences of entropy were it not for the flow of energy. All cellular processes, including the transport processes discussed in Chapters 12 and 13, require some kind of energy input. Energy is typically supplied to these processes in the form of compounds such as ATP that are kept far above their equilibrium concentrations or tapped from actively maintained ion concentration gradients across cellular membranes. Ultimately, photosynthesis is the source of energy for most life on earth, solar energy is used to produce reduced forms of carbon, and respiration elaborates this energy in much of biology by re-oxidizing carbon to drive ATP synthesis and build up ion gradients across membranes. Certain bacteria and archaea that live in, e.g., thermal vents on the ocean floor can use inorganic sources of reducing power such as sulfur, molecular hydrogen, or ferrous iron, but these chemotrophs will not concern us further here.

Respiration and photosynthesis stand out as two of the most impressive products of molecular evolution. The respiratory chain and the photosynthetic apparatus (**Figure 14.1**) weave together elements of photochemistry, electrochemistry, and the making and breaking of covalent bonds with purely physical phenomena such as charge separation across a membrane and mechanical movement of protein parts, forming macromolecular complexes of exquisite functional sophistication.

Plants, green algae, and cyanobacteria harvest the high energy of the incident solar light (which has a black-body temperature of around 6000 K) to convert water and carbon dioxide into free oxygen and carbohydrates. Aerobic organisms, such as ourselves, convert these resources back to water and carbon dioxide while extracting the energy needed to run all our cellular processes. As do all irreversible processes, photosynthesis and respiration also produce heat, which in due time is radiated out from the earth to the ultimate heat sink: empty space (well, not quite empty but filled with the cosmic background radiation at a temperature of 2.7 K). Life as we know it is possible because the earth is located between a high-temperature heat source (the sun) and a low-temperature heat sink (space). It is the flow of energy that allows ordered systems such as living organisms to exist.

Bioenergetics as a scientific field of enquiry dates back at least to the early 18th century, when Stephen Hales realized the importance of air and sunlight for plant growth. Joseph Priestly discovered in 1772 that plants when kept under sunlight in an airtight glass jar can "refresh" the air after a flame (or a live mouse) has "consumed" it; this was the first observation that plants take up CO_2 and release O_2. The term itself dates back to the late 1950s and came into general use with the book *Bioenergetics* published by Albert L. Lehninger in 1963. The defining moment, however, came with the publication of Peter Mitchell's chemiosmotic theory in 1961, which laid the foundation for our present understanding

DOI: 10.1201/9780429341328-15

CHLOROPLAST   MITOCHONDRION

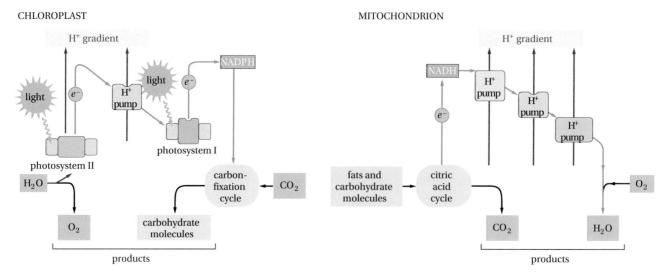

Figure 14.1 Photosystems I and II in plant chloroplasts convert the energy in incident solar photons into a proton gradient across the thylakoid membrane while producing O_2 and carbohydrates from H_2O and CO_2 (left). The respiratory chain in animal mitochondria re-oxidizes carbohydrate to CO_2 and H_2O while conserving energy in the form of a proton gradient across the mitochondrial inner membrane (right). (From Alberts B, Johnson A, Lewis JH et al. [2014] *Molecular Biology of the Cell*, 6th Edition. Garland Science, New York. With permission from W. W. Norton.)

of both respiration and photosynthesis (**Figure 14.2**). Given the general ignorance of membrane structure in the early 1960s (see Chapter 1), it is perhaps not surprising that Mitchell's theory was met with much skepticism. But as knowledge of membranes matured, Mitchell eventually prevailed. He was awarded the Nobel Prize in Chemistry in 1978.

14.1 THE CHEMIOSMOTIC THEORY IS THE FOUNDATION OF BIOENERGETICS

Pre-Mitchell, a rough outline of the mitochondrial respiratory chain had been established: electron transfer from a reduced substrate (e.g., NADH or succinate) to O_2 was somehow coupled to the production of ATP from ADP+P_i. What was not understood was how the two processes were coupled, and years had been spent looking for an elusive high-energy chemical intermediate that would transfer the energy siphoned off from the electron transport chain to the ATP synthase enzyme. Mitchell's insight—which appears not to have sprung from any one critical experiment but rather, was arrived at by reasoning—was that such a chemical intermediate was not needed. Instead, he proposed that the energy is stored as a proton gradient across the mitochondrial inner membrane, which in turn drives a proton-translocating ATP synthase. The enzymes that catalyze the oxidation-reduction (redox) reactions in the electron transport chain would thus be proton pumps that pump protons out of the mitochondrial matrix. This view turned bioenergetics from a field focused on redox enzymes that catalyze covalent bond formation to one focused on transporters that mediate the transfer of electrons and protons across a lipid bilayer.

Our contemporary view of mitochondrial energy conversion and the respiratory chain is depicted in **Figure 14.3** and **Figure 14.4**. As electrons flow through the respiratory chain from NADH to O_2, protons are translocated by three protein complexes that act sequentially: NADH dehydrogenase (complex I), cytochrome *c* reductase (complex III), and cytochrome *c* oxidase (complex IV). Electrons can also be fed into the respiratory chain from succinate via succinate dehydrogenase (complex II; a component in the citric acid cycle), but complex II does not translocate protons.

Figure 14.2 A facsimile of the first page of Mitchell's massive review of the chemiosmotic theory, published by his own private publishing house in 1966.

CHEMIOSMOTIC COUPLING IN OXIDATIVE
AND PHOTOSYNTHETIC PHOSPHORYLATION

By PETER MITCHELL

Glynn Research Laboratories,
Bodmin, Cornwall, England

CONTENTS

All four complexes harbor redox-sensitive cofactors that contain either Fe or Cu ions; these cofactors must be held in precise spatial relations relative to each other by a protein scaffold in order to form "molecular wires" that allow electrons to pass through each complex. Electrons are carried between the complexes by either ubiquinone—(UQ)—a redox carrier that moves within the lipid bilayer—or cytochrome c, a small soluble redox protein located in the inter-membrane space. **Box 14.1** contains a collection of cofactors that are used in redox reactions in the respiratory and photosynthetic complexes.

14.1.1 Redox Potentials Provide a Quantitative Foundation for Bioenergetics

At their typical concentrations in the mitochondrial matrix, NADH is a strong reductant and O_2 a strong oxidant. There is thus a large drop in free energy as electrons are transferred from NADH to O_2, which is almost fully converted into free energy stored in the proton gradient across the inner membrane. To put the energies on a quantitative footing, it is convenient to use redox potentials

Figure 14.3 Overview of mitochondrial energy conversion. CO_2 and reduced NADH are produced in the citric acid cycle. NADH feeds electrons into the respiratory chain, ultimately reducing O_2 to H_2O and maintaining the H^+ gradient across the inner membrane, which in turn drives ATP synthesis. (From Alberts B, Johnson A, Lewis JH et al. [2014] *Molecular Biology of the Cell*, 6th Edition. Garland Science, New York. With permission from W. W. Norton.)

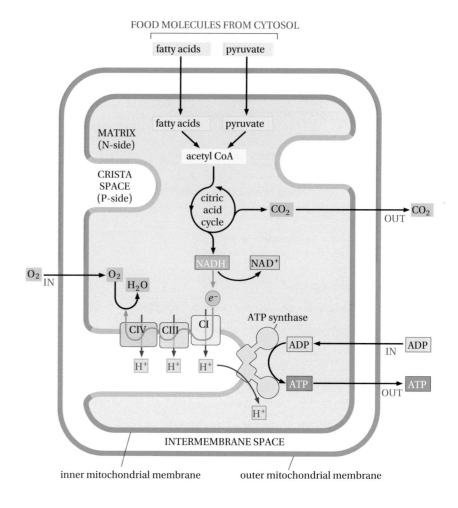

Figure 14.4 The mitochondrial electron transport chain is composed of complexes I–IV: NADH dehydrogenase, succinate dehydrogenase, cytochrome *c* reductase, and cytochrome *c* oxidase. Only succinate dehydrogenase does not translocate protons. (From Alberts B, Johnson A, Lewis JH et al. [2014] *Molecular Biology of the Cell*, 6th Edition. Garland Science, New York. With permission from W. W. Norton.)

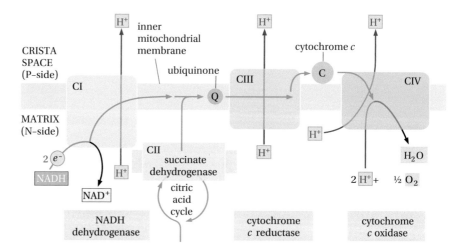

(measured in mV) rather than the Gibbs free energy (measured in kJ mol^{-1} or kcal mol^{-1}), although one can readily convert between the two, as shown in **Box 14.2**. Redox potentials have the added advantage that they help us see the close analogy between electron transfer chains and standard electric circuits.

Redox potentials can be used to describe the equilibrium $A_{ox} \rightleftharpoons A_{red}$ between oxidized (A_{ox}) and reduced (A_{red}) forms of an ion or a molecule. To cite a few examples:

$$NAD^+ + 2e^- + H^+ \rightleftharpoons NADH$$

BOX 14.1 REDOX COFACTORS

A number of cofactors are used to transport electrons in the respiratory and photosynthetic protein complexes. Here is a collection of the most common ones.

Nicotinamide adenine dinucleotide (NAD^+/NADH) & nicotinamide adenine dinucleotide phosphate ($NADP^+$/NADPH)

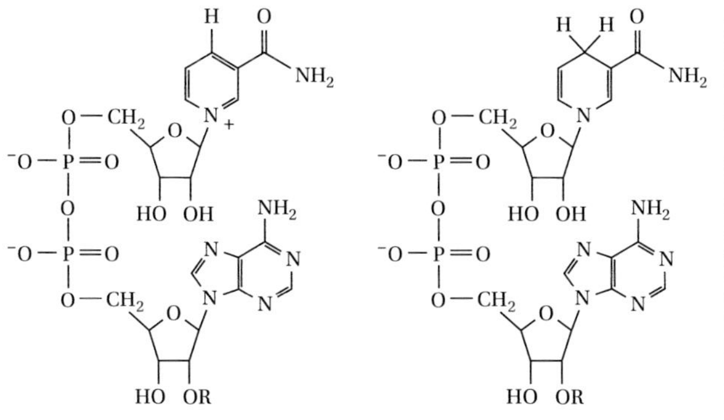

NAD^+ : R = H
$NAD(P)^+$: R = PO_3^{2-}

NADH : R = H
NAD(P)H : R = PO_3^{2-}

Flavine mononucleotide (FMN) & flavin adenine dinucleotide (FAD)

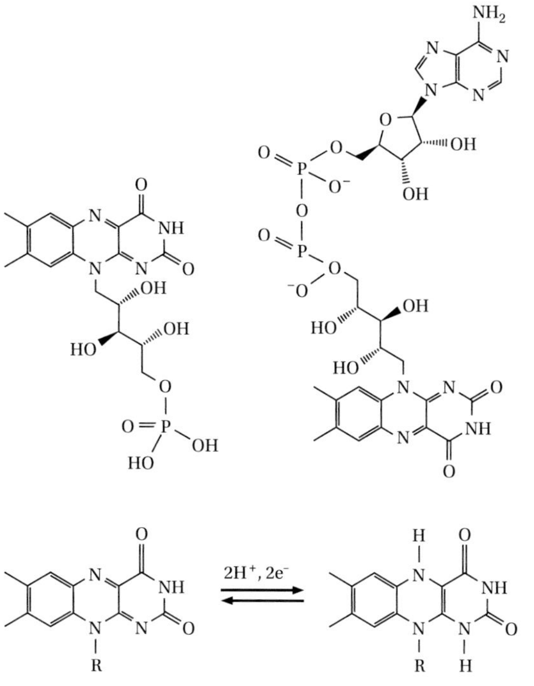

Ubiquinone (UQ)/Ubiquinol (UQH_2)

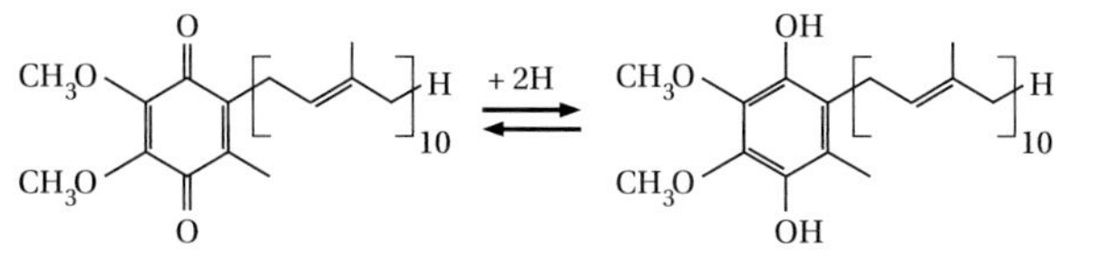

Heme groups

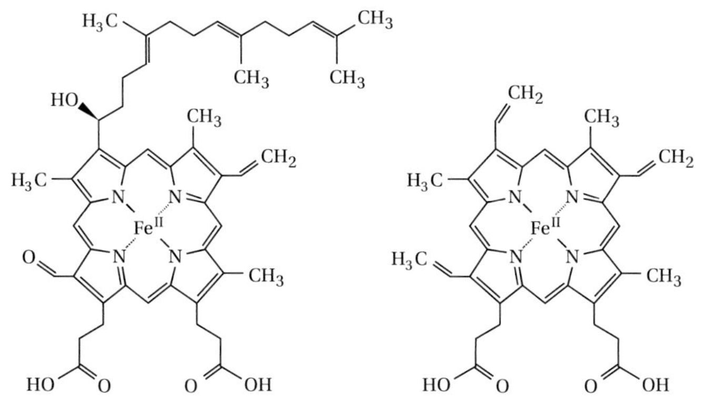

heme a

heme b

Iron-sulfur (Fe-S) clusters

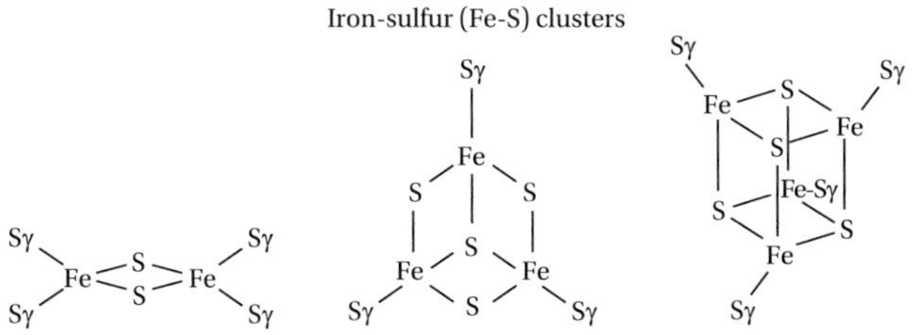

2Fe–2 S

3 Fe–4 S

4 Fe–4 S

Sγ ligands are usually from Cys but can also be N atoms from His

Figure 1

BOX 14.2   SOME CONVENTIONS AND CONVERSIONS

The side of the membrane to which protons are pumped is called the P-side (positive side); the other side is called the N-side (negative side). In mitochondria, the intermembrane (crista) space is the P-side and the matrix is the N-side; in thylakoids, the lumen is the P-side and the stroma is the N-side; and in bacteria, the periplasm is the P-side and the cytoplasm is the N-side.

ΔpH is defined as $pH(P\text{-side}) - pH(N\text{-side})$

$\Delta\psi$ is defined as $\psi(P\text{-side}) - \psi(N\text{-side})$

$1\,eV = 1.602 \times 10^{-19}\,J = 96.5\,kJ\,mol^{-1} = 23.1\,kcal\,mol^{-1}$

BOX 14.3   MEASURING REDOX POTENTIALS

To measure the standard redox potential E^0 of a redox couple $A_{ox} \rightleftharpoons A_{red}$, one connects a beaker of a pH 0 solution in which $[A_{ox}] = [A_{red}] = 1$ M to a beaker containing a Pt electrode around which one bubbles H_2 gas at a pressure of 1 atm immersed in a pH 0 solution(**Figure 1**). The whole system is kept at T = 25 °C. Under these conditions, the potential of the Pt reference electrode is by definition 0. The beakers are connected through a voltmeter by an electric wire, and the circuit is completed by a salt bridge (a tube containing concentrated solution of KCl). If electrons flow from A_{red} to H^+ (i.e., if there is a net gain in free energy upon oxidizing A_{red} to A_{ox} while simultaneously reducing H^+ to H_2), then by definition, E^0 for the redox pair formed by A_{ox}/A_{red} is negative and equal to the measured voltage.

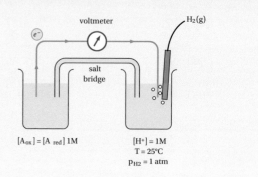

Figure 1

$$\text{fumarate} + 2e^- + 2H^+ \rightleftharpoons \text{succinate}$$

$$UQ + 2e^- + 2H^+ \rightleftharpoons UQH_2$$

$$\text{Cytochrome c}(Fe^{3+}) + e^- \rightleftharpoons \text{Cytochrome c}(Fe^{2+})$$

In electrochemistry, standard redox potentials (E^0) are defined relative to the standard hydrogen electrode (**Box 14.3**). If, under the conditions defined in the Box, electrons flow from the reduced component A_{red} to H^+ (i.e., if there is a net gain in free energy upon oxidizing A_{red} to A_{ox} and simultaneously reducing H^+ to H_2), then, by definition, E^0 for the redox pair formed by A_{ox}/A_{red} is negative and equal to the measured voltage.

It is important to keep in mind that E^0 values refer only to the situation when $[A_{ox}] = [A_{red}] = 1$ M at pH 0. When these conditions do not hold (as is generally the case, of course), the actual redox potential E at pH 0 for a redox couple where n electrons are transferred between the oxidized and reduced states ($A_{ox} + ne^- \rightleftharpoons A_{red}$) is given by the relationship

$$E = E^0 + \frac{2.3RT}{nF}\log_{10}\left(\frac{[A_{ox}]}{[A_{red}]}\right) \tag{14.1}$$

where R is the gas constant, F is Faraday's constant (the total electric charge in 1 mol of electrons), and the factor 2.3 is because we use $\log_{10}$ rather than natural logarithms (note that pH is measured on a $\log_{10}$ scale, and we therefore use $\log_{10}$ for all concentrations). Obviously, $E = E^0$ if $[A_{ox}] = [A_{red}]$. Compare this with the Nernst equation (see Equation 0.30).

However, as biochemists, we are rarely interested in reactions proceeding at pH 0; but rather, we are concerned with aqueous environments around pH 7, where the chemistry of life mainly occurs. This becomes important if protons are involved in the redox reaction:

$$A_{ox} + ne^- + mH^+ \leftrightharpoons A_{red}$$

In this case, the standard redox potential at a pH different from zero is more negative than E^0 by m/n 2.3 RT/F [mV per pH unit]. At $T = 37\,°C$, if $m = n$, this is equal to -61 mV per pH unit, and if $m = 1$, $n = 2$ (e.g., for the redox couple $NAD^+/NADH$) to -30 mV per pH unit (always relative to the standard hydrogen electrode at pH 0).

For our purposes, it is convenient to tally redox potentials for relevant redox couples at pH 7 and with $[A_{ox}]$ and $[A_{red}]$ equal to their approximate concentrations in healthy, respiring mitochondria. In analogy with Equation 14.1, the actual redox potential for a given redox couple at pH 7 is given by

$$E_{mito,pH7} = E^{0'} + \frac{2.3RT}{nF}\log_{10}\left(\frac{[A_{ox}]}{[A_{red}]}\right) \tag{14.2}$$

where $E^{0'}$ is the standard redox potential of the redox couple at pH 7 (also called the midpoint potential, because this is the potential at pH 7 where $[A_{ox}] = [A_{red}]$). Some examples are given in **Table 14.1**.

Electron transfer between two redox couples with actual redox potentials E^a and E^b corresponds to a change in the Gibbs free energy G:

$$\Delta G = -nF\left(E^a - E^b\right) \tag{14.3}$$

where n is the number of electrons being transferred. As an example, in mitochondria, a transfer of 2 mol electrons from NADH to O_2 (over a redox span of 1.07 V; Table 14.1) corresponds to $\Delta G = -207$ kJ mol^{-1} (1 kJ mol^{-1} = 0.24 kcal mol^{-1}). Note that if the electron transfer is "downhill" across a membrane (i.e., from the N-side to the P-side; see Box 14.2) with a membrane potential $\Delta\Psi$, the reduction in G caused by charge transfer across $\Delta\Psi$ must be added to Equation 14.3:

$$\Delta G = -nF\left(E^a - E^b\right) + \Delta\psi \tag{14.4}$$

Table 14.1 Standard Redox Potentials at pH 7 (E^0) and Actual Redox Potentials In Healthy, Respiring Mitochondria ($E_{mito,pH\,7}$) Measured Relative to the Standard Hydrogen Electrode (at pH 0). Adapted from Nicholls, D.G and Ferguson, S.J., Bioenergetics, 4th edition [2013] Academic Press.

Redox couple	Number of e⁻ (n)	Number of H⁺ (m)	$E^{0'}$ (mV)	$[A_{ox}]/[A_{red}]$ in mitochondria	$E_{mito,pH\,7}$ (mV)
H⁺/0.5H₂ (1 atm)	1	1	−420		
NAD⁺/NADH	2	1	−320	10	−290
NADP⁺/NADPH	2	1	−320	0.01	−380
Glutathione, ox/red	2	2	−172	0.01	−240
Fumarate/ succinate	2	2	+30		
UQ/UQH₂	2	2	+4		
Cytochrome c, ox/red	1	0	+220		
O₂ (1 atm)/2H₂O (55 M)	4	4	+820		+780 (0.2 atm O₂)

Obviously, for "uphill" transfer from the P- to the N-side, $\Delta\Psi$ must be subtracted instead. If the two redox couples are inside a membrane of thickness D and separated by a distance d along the membrane normal, the drop in the electrical membrane potential between them is $\approx\Delta\Psi(d/D)$.

The central concept in the chemiosmotic theory is that energy from the electron transfer chain is stored as a proton gradient across the mitochondrial inner membrane, which is then used to synthesize ATP. Both ΔpH and $\Delta\Psi$ contribute to the Gibbs free energy of the electrochemical gradient:

$$\Delta G = -F\Delta\psi + 2.3RT\Delta pH \qquad (14.5)$$

This can also be expressed in terms of mV rather than kJ mol^{-1} and is then called the protonmotive force (pmf or Δp; note the conventional change of sign):

$$\Delta p(mV) = -\frac{\Delta G}{F} = \Delta\psi - 2.3\frac{RT}{F} = \Delta\psi - 61\Delta pH \ \left(\text{for } T = 310K\right) \qquad (14.6)$$

In respiring mitochondria, $\Delta p \approx 180$ mV, mostly contributed by $\Delta\Psi$ (≈ 150 mV). ΔpH is only ≈ 0.5 pH units. Thylakoids exposed to light also have $\Delta p \approx 180$ mV, but here, the major part comes from a ΔpH of 2–3 pH units, with only a small contribution from $\Delta\Psi$.

14.1.2 Quantum Mechanical Tunneling Allows Long-Range Transfer of Electrons

Electron transfer between redox pairs can happen over rather long distances, up to around 20 Å. This is because electrons, having very low mass, can "tunnel" through energy barriers. Tunneling is a quantum mechanical effect that reflects the wave function description of an electron: even if an electron is trapped in an energy well, there is some probability of finding it outside the well, and this probability decreases exponentially with the distance from the well.

Rudolph A. Marcus (Nobel Prize in Chemistry 1992) worked out the implications of quantum mechanical tunneling for electron transport and calculated the Gibbs activation energy (ΔG^*) for the transition state for electron transport between a donor and an acceptor redox center:

$$\Delta G^* = \frac{\left(\Delta G^0 + \lambda_0\right)^2}{4\lambda_0} \qquad (14.7)$$

where ΔG^0 is the standard free energy difference between the donor and acceptor and λ_0 is the so-called reorganization energy. When an electron tunnels to the acceptor, the change in charge will cause the donor, receptor, and surrounding water molecules to reorganize their electronic structures. The reorganization energy essentially stems from the fact that the solvent molecules that surround the donor–acceptor pair relax to the new equilibrium configuration imposed by the altered charge distribution between the donor and the acceptor on a much longer time scale than that of the electron transfer itself; hence, right after the transfer, they will be in a non-optimal high-energy configuration that then relaxes to one of lower energy.

Finally, Marcus derived the rate of electron transfer:

$$k = ce^{-\beta r} \cdot e^{-\Delta G^*/RT} \qquad (14.8)$$

where β is a function of the environment between the donor and acceptor (water or protein, say), r is the distance between the donor and the acceptor, and c is a constant to be determined experimentally. Using Marcus' theory and experimentally determined values for λ_0, β, and c, one finds that electron tunneling can proceed on the millisecond time scale (the relevant time scale for fast biochemical reactions, since it is the time scale for large molecular motions) for donor–acceptor distances ≤20 Å. Longer distances can be covered if donors and acceptors are organized in chains with each individual pair being ≤20 Å apart.

14.2 ELECTRONS MOVE ALONG A CHAIN OF PROTEIN COMPLEXES

With the basic ideas of the energetics in place, we now see how nature manages to capture the energy from electron transport to move protons across membranes. The process in mitochondria is organized in four protein complexes, aptly called complex I, II, III, and IV. A schematic picture of the redox potentials along the mitochondrial electron transport chain is shown in **Figure 14.5**, and the corresponding proton stoichiometries in **Figure 14.6**. The changes in redox potential across complexes I and III are more or less compensated by the increases in free energy due to proton translocation—i.e., these complexes work close to equilibrium—whereas the drop in free energy across complex IV is larger than what is regained in the proton gradient, rendering the overall reaction essentially irreversible. The redox potential of the succinate/fumarate couple and complex II are both near zero mV, and electrons donated by succinate feed into the electron transport chain at the level of UQ. As shown in Figure 14.6, as 2 e$^-$ flow from NADH to O$_2$ via complex I, UQ, complex III, cytochrome c, and complex IV, 10 H$^+$ are translocated from the N-side to the P-side of the inner membrane: 4 by complex I, 2 via UQ, 2 in complex III, and 2 in complex IV. Figure 14.6 also shows that only 6 H$^+$ are translocated to the P-side when 2 e$^-$ flow from succinate to O$_2$.

Structures at atomic or near-atomic resolution are available for complexes I–IV and ATP synthase, from both bacterial and mitochondrial sources, allowing us to follow the pathways of electron transfer back and forth across the membrane in three dimensions. The precise mechanisms of proton translocation in the different complexes are not entirely clear, however, and are intensely discussed in the field.

14.2.1 Complex I (NADH Dehydrogenase)

Structures of complex I from bacteria, yeast, and cow show that the enzymes have a similar overall architecture but with a large number of accessory

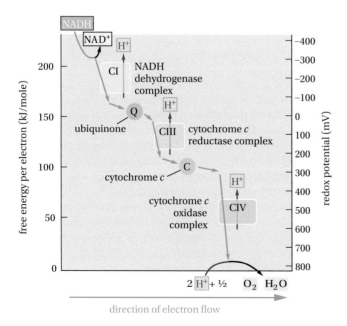

Figure 14.5 Redox potentials along the mitochondrial electron transport chain. The energy released during the downhill transport of electrons is largely conserved in the uphill transport of protons, except for complex IV, which catalyzes an essentially irreversible reaction. (From Alberts B, Johnson A, Lewis JH et al. [2014] *Molecular Biology of the Cell*, 6th Edition. Garland Science, New York. With permission from W. W. Norton.)

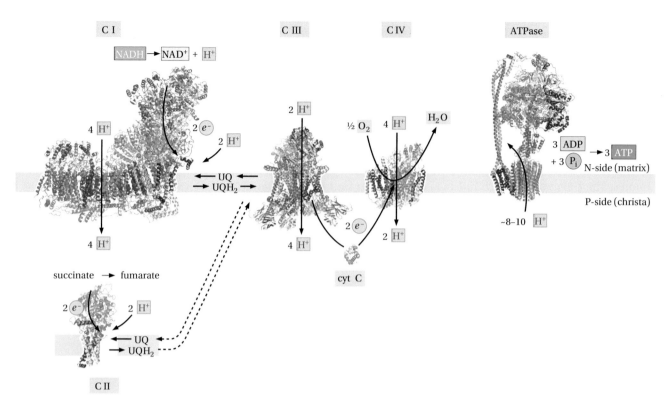

Figure 14.6 Proton stoichiometries in the mitochondrial electron transport chain. Ten H+ are translocated from the N-side to the P-side per two e− flowing from NADH to O₂. Six H+ are translocated from the N-side to the P-side per two e− donated by succinate.

subunits in the eukaryotic enzymes. Thus, while complex I from the thermophilic bacterium *Thermus thermophilus* has 14 subunits, the bovine version has 45 (**Figure 14.7**). The "supernumerary" subunits in the mammalian enzyme are wrapped around the core subunits but leave the region around the NADH binding site exposed, as in the bacterial enzyme. Many of the supernumerary subunits are essential for the assembly and stability of the complex (**Box 14.4**).

The NADH binding site is located ≈50 Å above the membrane (on the N-side), near the tip of the peripheral domain, and is connected to the membrane domain by a flavin mononucleotide (FMN) and an electron conduction "wire" of seven FeS clusters (**Figure 14.8**). The edge-to-edge distances of neighboring clusters are all ≲15 Å, which means that electrons can tunnel between them. The redox potentials of the FMN and the FeS clusters are close to that of the NAD+/NADH couple, except for the cluster closest to the membrane—called cluster N2—which has a redox potential close to that of the UQ/UQH₂ couple (see Table 14.1).

The UQ binding site is at the end of a ≈30 Å-long cavity that extends from the membrane surface to within ≈15 Å of the N2 cluster (**Figure 14.9**), again within tunneling distance. UQ therefore has to move out of the membrane and into the cavity in order to pick up the two electrons. Because UQ²⁻ has low affinity for the binding site, it is released back into the membrane. Presumably, the two protons needed to form UQH₂ are added from the N-side once UQ²⁻ leaves the cavity.

How is electron transfer coupled to proton translocation? Notably, three of the membrane subunits show structural similarities to Na+/H+ antiporters (see Chapter 13), and a fourth H+ translocation pathway is very likely formed between the subunits closest to the peripheral domain. A striking feature of the membrane domain is a string of polar and charged residues halfway across the membrane that extends throughout the entire membrane domain (Figure 14.8). The currently favored model suggests that the two negative charges that appear on UQ when electrons are transferred from the N2 cluster create an electrostatic force that is transmitted through the string of charged residues, resulting in proton pumping.

BOX 14.4   SMALL SINGLE-SPAN MEMBRANE PROTEINS ABOUND IN EUKARYOTIC RESPIRATORY AND PHOTOSYNTHETIC COMPLEXES

In bacteria, the respiratory and photosynthetic complexes are composed of a basic set of essential core subunits. In eukaryotic complexes, the core subunits are surrounded by a plethora of additional small subunits, many of which have only a single transmembrane helix (Box **Figure 1**). In general, only the core subunits are encoded in the mitochondrial and chloroplast genomes, and the additional subunits have to be imported into the organelles.

Why do the eukaryotic complexes have so many extra subunits? Gene knock-out studies in human cell lines have shown that the majority of the additional "supernumerary" subunits in mitochondrial complex I are strictly required for the assembly of a functional complex, and that loss of a subunit leads to destabilization of the surrounding part of the complex.

And, why are so many of the extra subunits single-span transmembrane proteins? We simply do not know; perhaps it is easier to import single-span membrane proteins than multi-span membrane proteins—small single-span proteins would not be expected to be able to interact cotranslationally with the signal recognition particle (see Chapters 5 and 6) and hence, may run less of a risk of being mis-targeted to the endoplasmic reticulum than larger, multi-spanning proteins.

In addition to the supernumerary subunits, the respiratory complexes also require a number of other proteins called assembly factors for proper complex formation; these factors interact only transiently with other subunits and do not end up in the final complex. Human complex I needs more than ten assembly factors. Assembling big complexes like these clearly involves quite complicated processes under strict genetic control, and we do not understand them in molecular detail.

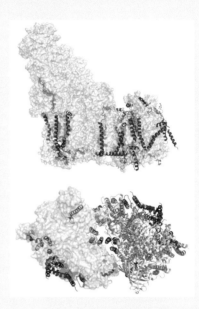

Figure 1 Small, single-span subunits (in red) surrounding mitochondrial complex I (top; PDB 5LC5) and the photosystem II core dimer (bottom; PDB 5KAF).

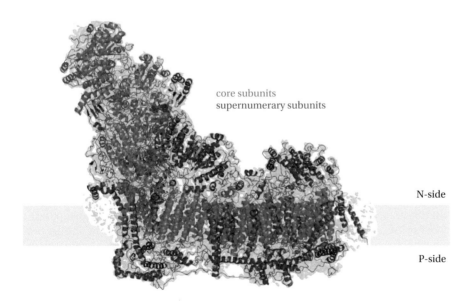

core subunits
supernumerary subunits

N-side

P-side

Figure 14.7 Bovine complex I (PDB 5LDW). The 14 core subunits are shown in blue and 30 additional, "supernumerary" subunits are in red. (From Zhu Y et al. [2016] *Nature* 536: 354–358. With permission from Springer Nature.)

14.2.2 Complex II (Succinate Dehydrogenase)

Complex II feeds electrons from succinate—an intermediate in the citric acid cycle—into the respiratory chain at the level of UQ. The structure of mitochondrial complex II is shown in **Figure 14.10**. Two electrons are delivered from succinate to a FAD (see Box 14.1) and then pass down a wire of three FeS clusters

Figure 14.8 Bovine complex I (PDB 5LC5). Electrons are transported from NADH, via FMN and a "wire" of FeS centers, to UQ. A string of charged residues appears to connect the UQ binding cavity to the proton-pumping subunits in the membrane domain, possibly coupling reduction of UQ to proton pumping.

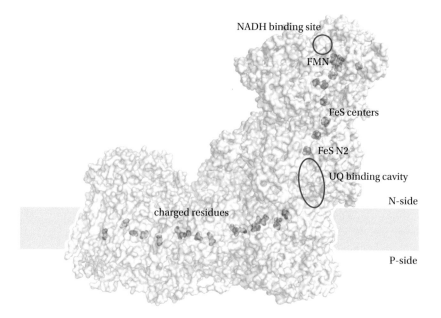

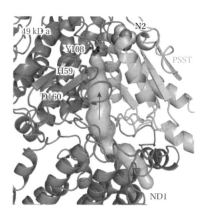

Figure 14.9 The UQ binding cavity (wheat) opens up near the base of the peripheral domain and extends ~30 Å toward the N2 FeS cluster. (From Zhu Y et al. [2016] *Nature* 536: 354–358. With permission from Springer Nature.)

to UQ. The role of the heme group is unclear, and it is not on the direct pathway to UQ. Interestingly, the second FeS cluster has a rather negative $E^{0\prime}$ of −260 mV and hence, constitutes a barrier to electron transfer, but because the edge-to-edge distances between this cluster and its neighboring clusters are small, electrons *en route* to UQ can still pass at a sufficiently high rate. The likely binding site for UQ is near the interface between the peripheral domain and the membrane domain (**Figure 14.11**), and reduced UQ^{2-} therefore picks up the two protons from the N-side of the membrane (Figure 14.6).

14.2.3 Complex III (Cytochrome c Reductase)

The next enzyme in the chain, complex III, transfers $2\,e^-$ from UQH_2 to the small, soluble protein cytochrome *c*, which is located in the intermembrane space, i.e., on the P-side of the membrane. In the process, two protons are picked up by UQH_2 from the N-side and delivered to the P-side (Figure 14.6). Cytochrome *c* contains a heme group that can take up electrons and carry them from complex III to complex IV. The surface charges on cytochrome *c* are organized to orient the protein to make loose complexes with complexes III and IV. Electron transfer through complex III is a little more involved than for the other enzymes in the respiratory chain, and the mechanism has been given a special name: the Q-cycle. The Q-cycle proceeds in two steps (**Figure 14.12**).

There are two binding sites for UQH_2 in complex III, one near the P-side of the membrane (hence called the Q_P site) and one near the N-side (called—you guessed it—the Q_N site). In the first step, UQH_2 binds to the Q_P-site, releases its two H^+, and delivers a first e^- to an FeS cluster in a subunit called the Rieske protein (discovered by John Rieske in 1964). The e^- tunnels to a heme group in the cytochrome c_1 subunit and eventually to a heme group in the soluble shuttle protein cytochrome *c*. The radical $UQ^{\bullet-}$ (where the signifies an unpaired electron) is now in the Q_P site, and it loses its unpaired e^- to a heme called b_L, from where the e^- rapidly moves on to a second heme, b_H, both present in the cytochrome *b* subunit of complex III. The loss of the unpaired, second e^- leads to the formation of UQ in the Q_P site, and because UQ is soluble in the lipid membrane, it can equilibrate between the Q_P site, the membrane, and the Q_N site. Once a UQ binds in the Q_N site, it can be reduced back to $UQ^{\bullet-}$ by the neighboring heme b_H.

The second step is basically a repeat of step 1, except that $UQ^{\bullet-}$ is now present in the Q_N site from the beginning. Hence, when the e^- appears on heme b_H, it will reduce $UQ^{\bullet-}$ to UQ^{2-}, which then picks up two H^+ from the N-side and is

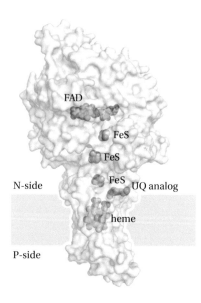

Figure 14.10 Complex II from pig mitochondria (PDB 1ZOY). It has one FAD, three FeS clusters, and one heme group. A UQ analog is also present in the structure.

released into the membrane as UQH_2. A full turn of the Q-cycle thus produces two molecules of reduced cytochrome c (each picks up one e$^-$) and transfers four H$^+$ from the N-side to the P-side (remember that the UQH_2 that takes part in step 1 picked up its two H$^+$ from the N-side of complex I or complex II).

The structure of complex III/cytochrome c from yeast mitochondria is shown in **Figure 14.13A**. As discussed earlier, a key aspect of complex III is the bifurcation in the electron transfer pathway in both steps in the Q-cycle, with one e$^-$ going to heme c_1 and one to heme b_L. This is explained by structural studies of the enzyme, which show that the globular part of the Rieske protein, including the 2Fe-2S center, can move by about 20 Å between a position close to the Q_P site and a position close to heme c_1 in the cytochrome c_1 subunit. When the Rieske protein is in its oxidized (Fe^{3+}) form, the affinity for cytochrome c_1 is low, and it can pick up an e$^-$ from the Q_P site, whereas the opposite is true when it is in its reduced (Fe^{2+}) form (Figure 14.13B, C). Therefore, once the first e$^-$ from UQH_2 has been transferred to the FeS cluster, the Rieske protein will move away from the Q_P site, leaving no option for the second e$^-$ than to tunnel to heme b_L.

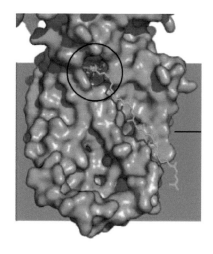

Figure 14.11 The membrane domain of complex II with UQ modeled into the putative binding site. N-side is up. (From Sun F, Huo X, Zhai Y et al. [2005] *Cell* 121: 1043–1057. With permission from Elsevier.)

14.2.4 Complex IV (Cytochrome c Oxidase)

Complex IV, the last complex in the electron transport chain, accepts 2 e$^-$ from cytochrome c, reduces 1/2 O_2 to H_2O (the 2 H$^+$ are taken up from the N-side), and pumps 2 H$^+$ from the N-side to the P-side of the membrane (Figure 14.6). In mitochondrial complex IV, the electrons are passed from cytochrome c, first to a binuclear Cu center (Cu$_A$) and then via heme a to the heme a_3-Cu$_B$ catalytic site, where oxygen is reduced (**Figure 14.14**).

The key questions concerning complex IV are how the reduction of oxygen is catalyzed and how this is coupled to proton pumping. Neither question is fully resolved, but plausible mechanisms have been proposed. It is known that O_2 initially binds to the reduced Fe^{2+} ion in heme a_3 and that one O atom is eventually transferred to the reduced Cu$_B^{1+}$. Of the four electrons required to split the O_2 molecule, two are thought to come from heme a_3 (with Fe^{2+} oxidized to Fe^{4+}), one from Cu$_B$ (which is oxidized to Cu^{2+}), and one from a nearby tyrosine side chain (which forms a Tyr$^{\cdot}$ radical). Four e$^-$ transferred sequentially from incoming cytochrome c molecules and four H$^+$ taken up from the N-side then return heme a_3, Cu$_B$, and the Tyr residue to their original states, leading to the release of 2 H_2O. Two possible proton pathways (called the D- and K-channels, after a critical Asp and Lys residue, respectively) that lead from the N-side of the enzyme to the vicinity of the heme a_3/Cu$_B$ site have been identified in the

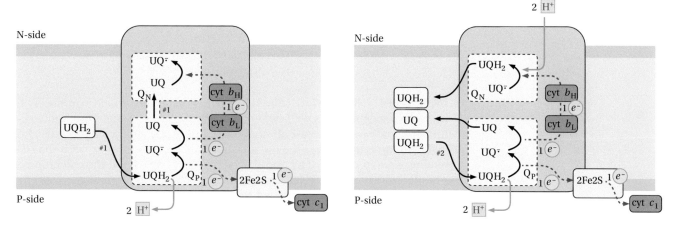

Figure 14.12 The Q-cycle describes electron transport through complex III. In the first step (left), a UQH_2 molecule binds in the Q_P site and delivers one e$^-$ via the Rieske protein to cytochrome c_1, and one e$^-$ via heme b_L and heme b_H to a UQ bound in the Q_N site, producing the radical UQ$^{\cdot-}$. In the second step (right), another UQH_2 molecule binds in the Q_P site and delivers two e$^-$, eventually resulting in the formation of UQH_2 in the Q_N site. (From Nicholls DG, Ferguson SJ [2014] *Bioenergetics 4*, Academic Press, Elsevier, New York.)

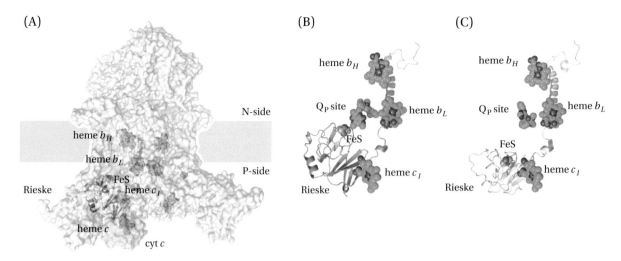

Figure 14.13 Principal features of complex III. (A) Dimeric mitochondrial complex III with bound cytochrome *c* (shaded in green) and the Rieske protein marked in purple (PDB 3CX5). The electron transport chain contains heme *c*, heme c_1, the Rieske FeS center, heme b_L, and heme b_H. (B and C) The Rieske protein acts as a mechanical switch, either picking up an e⁻ from UQH₂ in the Q_P site and transporting it to heme c_1 (B) or leaving no option for the e⁻ but to move to heme b_L (C). The approximate location of the Q_P site is marked by two co-crystallized inhibitors (stigmatellin and azoxystrobin; PDB 1PP9, 3L71).

Figure 14.14 Complex IV from bovine mitochondria (PDB 1OCC). Protons enter from the N-side through the D- and K-channels, while electrons are delivered from the P-side by cytochrome *c* to Cu_A. O₂ reduction takes place between the Fe ion in heme a_3 and Cu_B. O₂ is soluble in lipid and enters complex IV from the membrane.

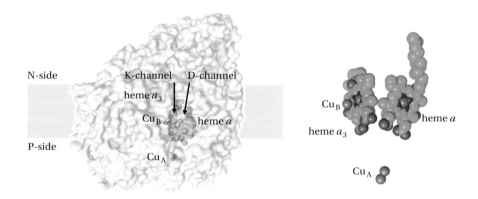

structure. Of the 8 H⁺ that are taken up from the N-side for each O₂, 4 (the "scalar" protons) end up in the 2 H₂O molecules, and 4 (the "vectorial" protons) are pumped across to the P-side. It is not clear which of these protons pass via the D- and K-channels, and there is not yet consensus on how proton pumping is driven by the O₂ reduction process.

14.2.5 The Mitochondrial Electron Transport Chain Components Form Supercomplexes

Using gentle solubilization conditions, it is possible to isolate complexes I, III, and IV together as a well-defined supercomplex. A model based on electron cryomicroscopy (cryo-EM) analysis of individual supercomplex particles is shown in **Figure 14.15**. The organization of the supercomplex, with the UQ binding sites in complexes I and III being ~13 nm apart and the cytochrome *c* binding sites on complexes III and IV being ~10 nm apart, is suggestive of a geometry that has been optimized to minimize the distances that need to be travelled by UQ and cytochrome *c*, so-called substrate channeling. The idea of substrate channeling in the respiratory chain has been challenged, however, and the functional significance of supercomplex formation is debated.

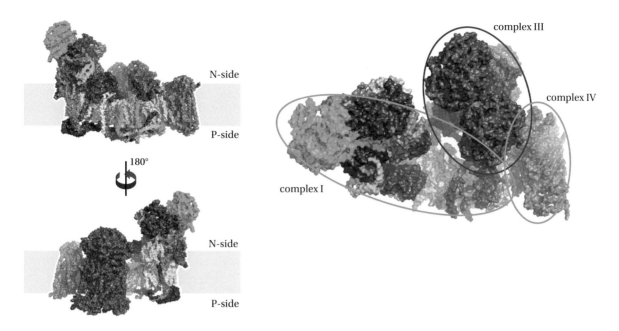

Figure 14.15 Side and top views of the electron transport chain supercomplex from bovine mitochondria (PDB 5J4Z).

14.3 ATP SYNTHASE PRODUCES ATP USING THE PROTON GRADIENT

Given the capture of energy in the form of a transmembrane proton gradient, living systems can use the gradient to drive other processes, such as secondary transport, flagellar rotation, and ATP synthesis. By far the most useful form of energy is ATP, the energy currency of cells (**Box 14.5**), and the conversion of the proton gradient to ATP is accomplished by the ATP synthase. ATP is continually hydrolyzed to ADP and P_i in myriad processes throughout the human body. ATP is then re-synthesized in the mitochondria at the staggering rate of ~60 kg of ATP per day! The ATP synthase converts electrochemical energy to mechanical energy to chemical energy in an amazing series of intertwined events. In respiring mitochondria, the main use of the proton motive force (Δp) built up by the electron transport chain is to drive ATP synthesis by the ATP synthase (although under other conditions, the ATP synthase can work in reverse, using ATP hydrolysis to build up a Δp). A classic experiment demonstrating that Δp can drive ATP synthesis is described in **Box 14.6**.

14.3.1 ATP Synthase Is a Rotary Enzyme

The ATP synthase is one of only a small number of enzymes in nature that use a rotary motion to drive a chemical reaction; in this sense, it is a "molecular machine" *par excellence*. A structural model of the *E. coli* ATP synthase is shown in **Figure 14.16**. The key to understanding the function of the ATP synthase is the Δp-driven proton flow through the membrane domain of the enzyme (the F_o part), which induces a rotation in the c-ring relative to the a-subunit (much as water flowing through a water mill makes the waterwheel turn relative to its casing). The rotation of the c-ring is transmitted to the attached central stalk, which rotates in a cavity in the middle of the catalytic domain (the F_1 part). The catalytic domain is anchored to the a-subunit in the membrane by the long stator that runs down its side, ensuring that the central stalk will rotate relative to the catalytic domain. The rotary movement of the central stalk was inferred from biochemical and biophysical experiments and beautifully confirmed by single-molecule experiments (**Box 14.7**). In 1997, John Walker and Paul Boyer were awarded the Nobel Prize in Chemistry for their work on ATP synthase.

BOX 14.5 HOW CAN THE CELL EXTRACT ENERGY FROM ATP HYDROLYSIS?

We often say that ATP is a "high-energy molecule." What we mean by that is not that ATP in and of itself can supply energy but rather that, thanks to the respiratory chain and ATP synthase, cells displace the concentrations of ATP, ADP, and P_i far away from their equilibrium values.

Any system that is away from equilibrium can perform work: as explained in Chapter 0 (E-words), heat transferred from a hot to a cold body can perform work, and as explained in Chapter 13, an ion gradient across a membrane can be used to drive the transport of other ions or small molecules against their concentration gradient. Indeed, ATP synthase itself is a macromolecule that is driven by the pH gradient across the inner mitochondrial membrane created by the respiratory chain.

All reversible chemical reactions are characterized by their equilibrium constant K. For the reaction ATP $\rightleftharpoons$ ADP + P_i at equilibrium:

$$K = \frac{[ADP]_{eq}[P_i]_{eq}}{[ATP]_{eq}} = 10^5 \, M$$

For instance, if we plug in typical cellular values for ADP and P_i ([ADP] = 10 mM and [P_i] = 1 mM), we get [ATP]

$= 10^{-10}$ M, or [ATP]/[ADP] $= 10^{-8}$. This is the equilibrium value of the [ATP]/[ADP] ratio at the given P_i concentration. However, thanks to mitochondrial respiration and ATP synthase, the [ATP]/[ADP] ratio in the cytoplasm can be as high as 10^3, meaning that the [ATP]/[ADP] ratio is much higher than its equilibrium value. This displacement from equilibrium means that work can be extracted if the conversion of ATP to ADP + P_i is somehow coupled to a molecular "machine" of some kind, such as the ATP-driven transporters discussed in Chapter 12. Conversely, to synthesize ATP from ADP and P_i at such high [ATP]/[ADP] ratios, the ATPase needs an energy input of 50–60 kJ mol^{-1}, which it extracts from the pH gradient across the membrane.

To be more quantitative, from thermodynamics, we know that the free energy of the reaction ATP $\rightarrow$ ADP + P_i is

$$\Delta G = \Delta G^0 + RT \ln \frac{[ADP][P_i]}{[ATP]} = -RT \ln K + RT \ln \frac{[ADP][P_i]}{[ATP]}$$

where [ATP], [ADP], and [P_i] are the actual concentrations in the cell. At T = 298 K, and with the earlier values ($K = 10^5$ M, [ATP]/[ADP] $= 10^3$, [P_i] $= 10^{-3}$ M), we get $\Delta G = -63$ kJ mol^{-1} ($=-15$ kcal mol^{-1}).

BOX 14.6 A DEMONSTRATION THAT A PROTON GRADIENT CAN DRIVE ATP SYNTHESIS

In 1974, Efiam Racker and Walther Stoeckenius conducted an experiment that remains the most important test of the Mitchell chemiosmotic hypothesis: that a proton gradient can cause a membrane-bound complex from mitochondria to synthesize ATP. Bacteriorhodopsin, discovered by Stoeckenius, uses light energy to move protons across a membrane (Box 14.8). As shown in Figure 1, molecules of oriented bacteriorhodopsin were put into lipid vesicles

that contained the ATP synthase from mitochondria [1], and ADP and P_i were added to the solution [2]. When in the dark, no ATP was made [3]; however, in the light, the bacteriorhodopsin used light energy to pump protons into the vesicle, creating a proton gradient, and ATP was produced [4]! This remarkable experiment definitively settled the debates that had been raging for and against the Mitchell hypothesis.

1 ATP synthase and bacteriorhodopsin were incorporated into membrane vesicles

2 ADP and P_i were added on the outside of the vesicles

3a one sample was kept in the dark: no ATP was made

3b one sample was exposed to light: ATP was made

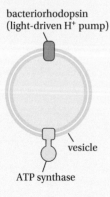
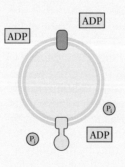

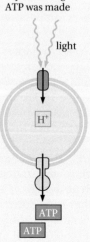

Figure 1

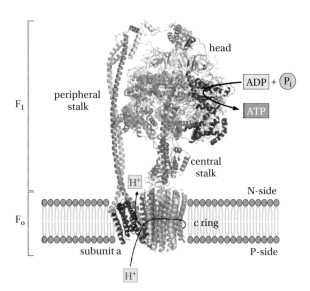

In yeast mitochondria, the c-ring is composed of ten copies of the c-subunit, which in turn, is composed of two transmembrane α helices. The α helices are arranged in two concentric rings (**Figure 14.17**). The narrow space in the middle of the c-ring is thought to house a few lipid molecules. The catalytic domain is composed of three α and three β subunits, arranged in a circular α-β-α-β-α-β structure. There are six nucleotide binding sites, but only three are catalytic (**Figure 14.18A**). The central stalk is somewhat bent and also has a protrusion right underneath the catalytic domain (Figure 14.18B). When the stalk rotates inside the catalytic domain, the bent "elbow" and the protrusion will push and pull on the α and β subunits as they pass by. Rotation of the central stalk will therefore force conformational changes in the α and β subunits.

14.3.2 Mechanism of ATP Synthesis

How do the protons make the c-ring rotate relative to the a-subunit, and how can rotation of the central stalk relative to the catalytic domain catalyze the reaction ADP + P_i → ATP? The current thinking is that a critical Arg residue located near the middle of the membrane in the a-subunit plays a key role, together with a likewise critical Glu residue in the middle of each c-subunit in the c-ring (**Figure 14.19**). The c-ring is in contact with two highly tilted transmembrane helical hairpins in the a-subunit, with the Arg residue neutralizing the Glu residue in one of the c-subunits (**Figure 14.20**); let's call this particular subunit c*. The helical hairpins essentially act as a barrier in the middle of the membrane, allowing protons to access the c-ring through a half-channel that leads from the P-side to the vicinity of the Glu residue in subunit c*. On the N-side of the barrier, another half-channel leads from the vicinity of the Glu residue in subunit c*-1.

The Δp drives an H^+ into the P-side half-channel, where it protonates the Glu residue in subunit c*, breaking the Arg-Glu salt bridge. This Glu residue is now uncharged and can move away from the a-subunit into the membrane. As the whole c-ring rotates by $360/n$ degrees (where n is the number of c-subunits in the ring), the protonated Glu residue from subunit c*-1 in the c-ring will move close to the Arg residue in the a-subunit, lose its H^+ to the N-side proton half-channel, and form a new salt bridge to the Arg residue. In this way, n protons will pass through the membrane for each full turn of the c-ring.

Biochemical analysis of the ADP + P_i ⇋ ATP reaction in the catalytic domain has shown that the rotation of the central stalk induces transitions of the catalytically active nucleotide binding sites between three different conformations: open (O), loose (L), and tight (T) (**Figure 14.21**). The O state has very low affinity for nucleotide and hence is empty. The L state has loosely bound ADP and

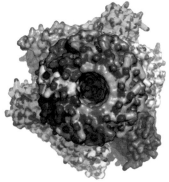

Figure 14.17 Yeast mitochondrial ATP synthase seen from the c-ring end (PDB 2WPD). The narrow central space (most clearly visible in the space-filling model at the bottom) is likely filled with a few lipid molecules. Note that the a-subunit and the stator are not shown.

BOX 14.7 VISUALIZATION OF THE ROTATION OF THE CENTRAL STALK IN ATP SYNTHASE BY SINGLE-MOLECULE MICROSCOPY

In an experiment conducted in 1997, Kinosita and co-workers attached purified F_1 ATP synthase to a coverslip (using an engineered His-tag on the β subunits that binds strongly to Ni-NTA on the coverslip) and connected a ~1 μm-long, fluorescently labeled actin filament to the end of the γ subunit by a streptavidin-biotin linker (biotin is a small organic molecule that binds with exceptionally high affinity to the streptavidin protein). The actin filament was then observed under a fluorescence microscope. Upon addition of ATP, the actin filament was seen to rotate in an anti-clockwise direction. ATP hydrolysis by the αβ subunits drove rotation of the γ subunit and dragged the actin filament in a circle, directly demonstrating the rotary nature of ATP synthase (Figure 1).

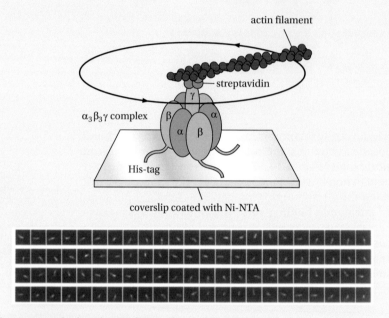

Figure 1 Top: Actin filament attached to the γ subunit of a surface-attached F_1 ATPase. Bottom: Addition of ATP induces rotation of the fluorescently labeled actin filament, as seen in the time-lapse photographs. (From Noji H et al. [1997] *Nature* 386: 299–301. With permission from Springer Nature.)

Figure 14.18 The catalytic domain and the central stalk (in dark blue) in the yeast enzyme, as seen from the top (A) and from the side (B). Bound nucleotides are in red.

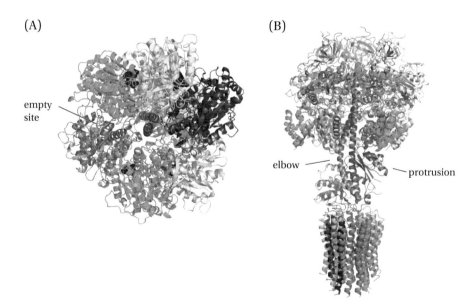

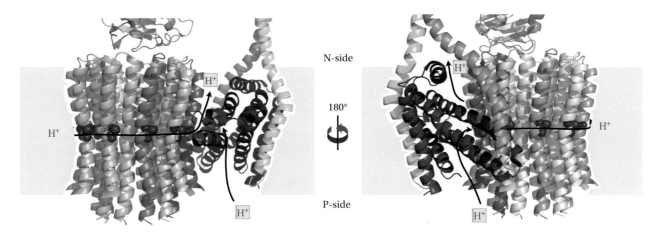

Figure 14.19 The membrane domain of the *E. coli* ATP synthase, with the critical Glu residues in the c-ring in red. Protons enter from the P-side between two highly tilted helical hairpins in subunit a, are carried around the c-ring as it rotates, and leave to the N-side through a cavity on the upper side of the tilted helices.

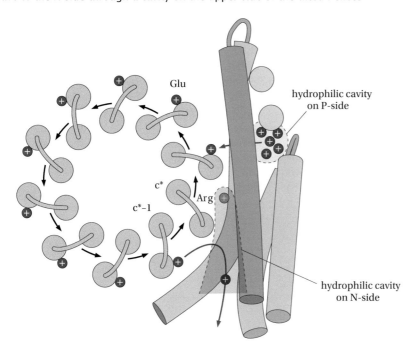

Figure 14.20 Path of protons through the membrane domain of *E. coli* ATP synthase. A proton enters through the P-side half-channel and protonates the Glu residue in subunit c*, allowing it to move away from the Arg in the a-subunit and into the lipid membrane. As a result of this small rotation in the c-ring, the protonated Glu residue in subunit c*-1 moves into contact with the Arg residue in the a-subunit and loses its proton through the N-side half-channel. (From Allegretti M et al. [2015] *Nature* 521: 237–240. With permission from Springer Nature.)

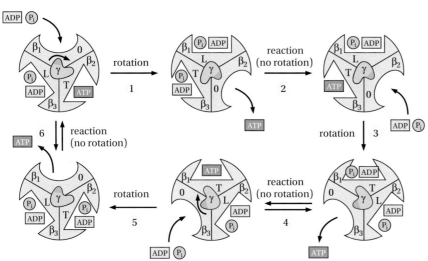

Figure 14.21 Phosphorylation of ADP to ATP. The three catalytic nucleotide binding sites cycle between the O, L, and T states, driven by the rotation of the central stalk.

Pi, while the T state binds ATP with very high affinity (the dissociation constant may be as low as 10^{-12} M). The unfavorable ΔG for the formation of ATP from ADP and P_i in the mitochondrial matrix (+40 kJ mol^{-1}) is compensated by the very high binding affinity of ATP in the T state. The main work done by the ATPase during catalysis is expended in the T→O transition, when the torque on the central stalk pushes the nucleotide binding site from the low-energy T state to the high-energy O state, releasing ATP in the process.

14.3.3 The H⁺/ATP Ratio Varies between Different Organisms

One full rotation of the c-ring catalyzes the synthesis of three ATP molecules. The corresponding number of protons that flow through the c-ring is equal to n, the number of c-subunits in the c-ring. For the yeast enzyme described earlier, n = 10. The ATP synthase in mammalian mitochondria has n = 8, whereas the ATP synthase in plant thylakoids has n = 14. The higher the Δp, the higher is the free energy gained per proton flowing across the membrane, and the fewer c-subunits are needed in the c-ring to compensate for the ΔG required for the synthesis of one ATP. Conversely, for a given Δp, the ATP/ADP ratio can be displaced further from its equilibrium value by increasing n. It is possibly significant that the number of c-subunits is not a multiple of three, i.e., there is a symmetry mismatch between the c-ring and the threefold symmetric catalytic domain, which may facilitate smooth turning of the c-ring relative to the catalytic domain.

If the supply of oxygen or substrates to the respiratory chain is compromised, the Δp is reduced, and the pH in the matrix (the N-side) drops. If this happens, the ATPase must be inhibited lest it starts to turn backwards and hydrolyze ATP. The small inhibitor protein IF₁ inhibits the cycle of conformational changes in the catalytic subunit that are necessary for ATP hydrolysis. When activated by low pH on the N-side, IF₁ inserts itself into the cleft formed between an α and a β subunit (**Figure 14.22**), thereby preventing further conformational changes and blocking the enzyme. Essentially, and befitting a "mechanical" enzyme such as the ATPase, inhibition by IF₁ is based on the simple mechanical principle "throw a monkey wrench into the works."

Figure 14.22 Inhibition of the catalytic domain. The small inhibitory protein IF1 (red) binds between an α and a β subunit, jamming the catalytic domain (PDB 3ZIA). The central stalk (blue) is highlighted for reference.

14.3.4 ATP Synthase Dimers Are Organized into Rows in Mitochondria

Cryo-EM studies of intact mitochondria and isolated mitochondrial membranes show that ATP synthase forms dimers that assemble into long rows

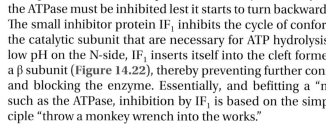

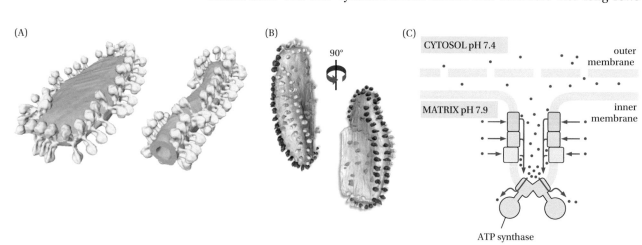

Figure 14.23 Structural organization of mitochondrial cristae. (A) Cristae membranes showing rows of ATP synthase dimers. (B) Complex I (green) is more randomly dispersed. (C) Model for how protons may flow within a crista. (A, B, From Davies KM et al. [2011] *PNAS* 108: 14121–14126. With permission from Werner Kühlbrandt. C, From Alberts B, Johnson A, Lewis JH et al. [2014] *Molecular Biology of the Cell*, 6th Edition. Garland Science, New York. With permission from W. W. Norton.)

along highly curved ridges on the cristae membranes, while complex I has a more random distribution (**Figure 14.23**). A speculative idea is that protons pumped into the confined cristae space by the electron transport chain diffuse along the membrane surface towards the ATP synthase, maintaining a "local" proton current within the cristae.

14.4 PHOTOSYNTHESIS HARVESTS LIGHT TO PRODUCE OXYGEN

If respiration is what keeps us animals (and most bacteria) ticking, it is photosynthesis that is the ultimate life-giver. We might like to think—anthropocentric as we are—that the human brain is the pinnacle of evolution. But on the scale of biology as a whole, photosynthesis, with its mastery of light-driven redox chemistry, surely cannot be trumped (**Figure 14.24**). Thanks to the very high surface temperature of the sun, about 6000 K, sunlight has a very high energy content when it hits the earth (which is at around 300 K). The energy of a photon with frequency ν is $h\nu$ (where h is Planck's constant $=6.6 \times 10^{-34}$ Js). Multiplying by Avogadro's number (N) and converting from frequency to wavelength (λ, in nm), the Gibbs free energy of one mol of photons is $Nhc/\lambda = 1.2 \times 10^5/\lambda$ kJ mol^{-1} (where c is the speed of light). Visible light has $\lambda = 400$–700 nm, and the corresponding energies available in 1 mol of photons are 300–170 kJ mol^{-1}. Or, expressed as a potential difference, $\Delta p = -\Delta G/F = -3.1$ to -1.7 V. With sunlight, it should thus be possible to shift redox potentials by more than 1 V and provide very strong driving forces for electron transport. Photosynthesis is a very efficient way to capture this energy. A much simpler but exceedingly well-understood system for capturing light energy, the light-driven proton pump bacteriorhodopsin, is described in Box 14.8.

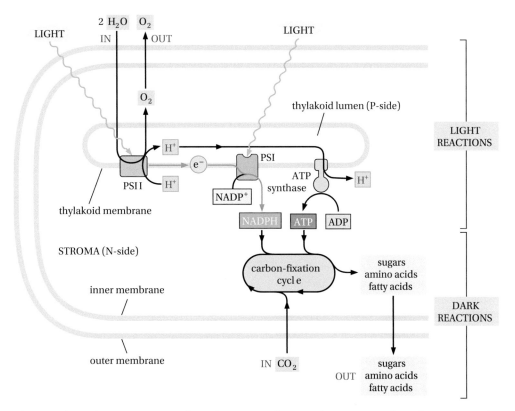

Figure 14.24 Photosynthesis harnesses energy from the sun. In photosynthesis, energy from the sun is used to produce O_2 and complex biomolecules—sugars, amino acids, and fatty acids—from H_2O and CO_2, with NADPH and ATP serving as intermediates. (From Alberts B, Johnson A, Lewis JH et al. [2014] *Molecular Biology of the Cell*, 6th Edition. Garland Science, New York. With permission from W. W. Norton.)

BOX 14.8 BACTERIORHODOPSIN AND RELATED PROTEINS USE LIGHT ENERGY FOR ION PUMPING AND SIGNALING

In addition to the use of light for photosynthesis in plants, evolution has found alternative mechanisms for capturing light energy. A set of simple proteins in the archaea use retinal to capture visible light photons, converting the energy into ion gradients across membranes or gating channels, or guiding taxis as light sensors. Each of these proteins has seven transmembrane helices with similar structures adapted to a range of functions, including a proton pump (bacteriorhodopsin), a chloride pump (halorhodopsin), and a sodium pump (KR2). Perhaps the most interesting of these from the standpoint of bioenergetics is the proton pump, since the proton gradient it generates can drive ATP synthesis directly and thus provide a general source of energy for an organism, *Halobacterium salinarium,* that lives in high-salt environments. By locating the bacteriorhodopsin molecule in a vesicle membrane with an ATP synthase, light-driven ATP synthesis provided a key test of the Mitchell chemiosmotic theory (see Box 14.6).

Its simplicity has motivated ongoing research attention to the bacteriorhodopsin molecule as a pioneer structure, as a model for protein folding studies, and as a transducer of light energy into electrochemical energy. Here, we discuss the energy conversion aspect. A photon of visible light is captured by a retinal group that is bonded as a protonated Schiff base to a lysine on helix G, the seventh helix in the protein. The retinal has a set of conjugated double bonds that create an "electron in a box" with a resonant frequency in the visible range. Absorption of a photon causes a directional motion of protons along a pathway, causing the movement of a single proton across the membrane. The path is shown in a portion of the structure in **Figure 1**.

The functional cycle of bacteriorhodopsin is described by a sequence of steps, originally resolved by kinetic analysis as changes in the optical spectrum of the retinal and further studied by other methods. The steps in the photocycle have been assigned letters K through O for the excited states plus BR, the resting state. For this discussion, we consider a simplified version of the photocycle, shown in **Figure 2**. The local structure of the Schiff base in the resting state is known at very high resolution, as shown in **Figure 3**.

When a photon is absorbed (the J state), the *trans*-retinal isomerizes in an ultrafast step from *trans* to *cis* at the 13 position, thereby pushing on the surrounding protein and causing a strained rearrangement of the H-bonds in the Schiff base region (the K state). The protein cannot move as fast as the retinal isomerization and responds via structural changes to relax the forces, leading to the L intermediate, which positions the Schiff base proton for transfer to the unprotonated carboxyl group on Asp[85]. Relaxation of the K intermediate, with further structural

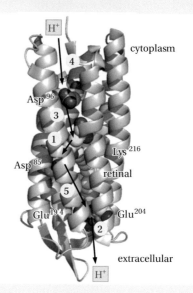

Figure 1 Proton pathway in bacteriorhodopsin (PDB 4XXJ). The numbers indicate the temporal order of proton movements that follow the conformational change driven by the absorption of light by the retinal: (1) Schiff base transfer to Asp[85], (2) H[+] release to the extracellular side, (3) Asp[96] re-protonation of the Schiff base, (4) proton uptake from the cytoplasm, and (5) Asp[85] re-protonation of the H[+] release group. (From Kandori H [2015] *Frontiers Mol Biosci* 2: 52.)

changes in the protein, leads to the formation of the L intermediate, which then transfers the proton to the carboxyl group of Asp[85], forming the M intermediate. When the M intermediate is formed, the key proton transfer has occurred, and the remaining steps are to re-protonate the Schiff base from a different carboxyl, Asp[96], forming the N intermediate. The proton transfers to Asp[85] and from Asp[96] are thought to define the unidirectionality of proton pumping from the cytoplasm to the extracellular side of the membrane.

The bacteriorhodopsin crystal structure reveals an asymmetric pattern of hydration, with seven internal water molecules in the extracellular path from the Schiff base but only two positioned in the pathway from the cytoplasm to the Schiff base. The water molecules provide a hydrogen-bonding network for fast exit of a proton from Asp[96], but the cytoplasmic side may be inaccessible in the dark and only opens when a photon is absorbed. Thus, a part of the asymmetry in the pump is generated by the alternation of proton access between extracellular (EC) and cytoplasmic (CP) sides, as indicated in the photo cycle figure (Figure 2). The conformational change is mainly due to a relatively large motion of helix F (the sixth helix in the sequence), and would be a slow step between N and BR states. So, a change of access between M and N is

(Continued)

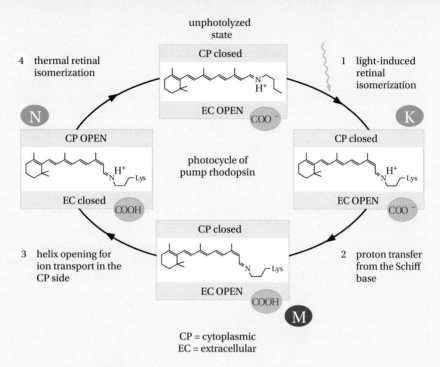

Figure 2 Simplified operating cycle of bacteriorhodopsin as a light-driven proton pump. The resting state is at the top, with the retinal in its all-*trans* isomer. Absorption of a photon by the retinal drives a series of steps, resulting in the net transfer of a proton across the membrane. The full set of steps is called the "photocycle" (see text). (From Kandori H [2015] *Frontiers Mol Biosci* 2: 52.)

accompanied by the transfer of the proton to Asp[96]. In the relaxation from N via O to BR, the retinal isomerizes to all-*trans*, and the access changes back to a state open to the extracellular side.

An important feature of the mechanism is that each step is slower than the one before, most likely because larger-scale motions are involved. We see that the asymmetrical pumping may arise from the kinetic hierarchy!

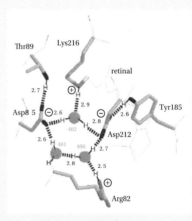

Figure 3 Schiff base region of bacteriorhodopsin in the resting state, shown at high (1.55 Å) resolution (PDB 1C3W). The proton is on the base, and the key charged carboxyl groups of Asp[85] and Asp[212] are shown along with three water molecules. (From Kandori H [2015] *Frontiers Mol Biosci* 2: 52.)

In plants, algae, and cyanobacteria, there are two related photosystems acting in tandem: photosystem II (PSII) and photosystem I (PSI) (**Figure 14.25**). Between PSII and PSI, electrons flow through plastoquinone (PQ, similar to UQ), the cytochrome b_6f complex (similar to mitochondrial complex III), and plastocyanin (a soluble, Cu-containing protein). Other photosynthetic bacteria have only one photosystem. Large parts of photosynthetic proteins are often rigid, since the positioning of the prosthetic groups involved in electron transfer is key, and since nothing needs to move much for this part of their functions. Because rigid membrane proteins are often easier to crystallize than more flexible ones, and because bacterial homologs are smaller and easier to handle than eukaryotic ones, many early crystal structures of membrane proteins were photosynthetic complexes of bacterial origin. In fact, the first high-resolution crystal structure of an integral membrane protein, published in 1985, was of the photosynthetic reaction center from the bacterium *Rhodopseudomonas viridis* (see Figure 1.28). Here, however, we will focus on photosynthesis in plants.

14.4.1 Light-Harvesting Complex II

While many of the chemical steps carried out by the photosystems are similar to those we have already seen in the mitochondrial electron transport chain, a major difference is the addition of various light-harvesting devices in

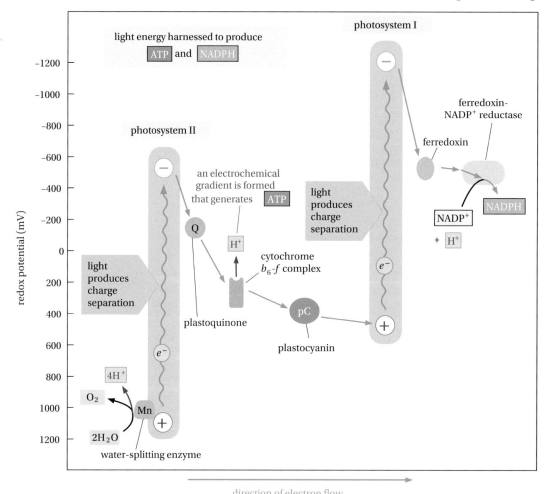

Figure 14.25 The electron transport chain through PSII and PSI, also called the Z-scheme. For every two e⁻ extracted from the water-splitting reaction on the P-side (i.e., the thylakoid lumen), six H⁺ are released to the P-side. The two e⁻ are ultimately used to reduce NADP⁺ to NADPH on the N-side (i.e., in the stroma), and the protons drive an ATP synthase located in the thylakoid membrane, producing ATP in the stroma. (From Alberts B, Johnson A, Lewis JH et al. [2014] *Molecular Biology of the Cell*, 6th Edition. Garland Science, New York. With permission from W. W. Norton.)

(A)

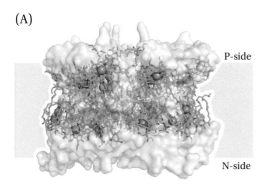

P-side

N-side

(B)

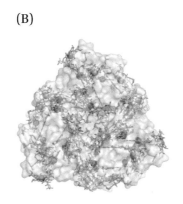

Figure 14.26 Organization of the light-harvesting complex II. Pea LHCII (PDB 2BHW) as seen in the plane of the membrane (A) and from above (B). The prosthetic groups are shown as sticks, with Mg atoms in orange.

photosynthesis that capture photons. In plants, the two major light-harvesting complexes are called LHCI and LHCII and are mainly associated with PSI and PSII, respectively. LHCII is composed of a trimeric protein scaffold that holds a large number of prosthetic groups—no fewer than 52 chlorophylls and 12 carotenoids—in a fixed geometry (**Figure 14.26**). Incident photons excite electrons in the highly conjugated bond systems of the prosthetic groups. The energy stored in an excited state is quickly delocalized over the entire LHCII complex, either by Förster Resonance Energy Transfer (FRET; named after the German scientist Theodor Förster. See Box 13.9) or by a mechanism known as exciton coupling. FRET is most efficient over distances around 20 Å or so, while exciton coupling requires shorter distances. These processes are very fast, occurring on the fs to ps time scale.

14.4.2 Photosystem II

In the intact thylakoid, four LHCII complexes, together with additional light-harvesting complexes called CP24, CP26, and CP29, encircle two PSII complexes (**Figure 14.27**), providing a large number of photon absorbers that can feed energy into PSII. The key to the redox chemistry of PSII is the transfer of the energy absorbed by the surrounding light-harvesting complexes to a pair of chlorophylls (called P_{680} because they absorb light at 680 nm) that are situated in the center of the complex (**Figure 14.28**). The electron that is excited by the light quantum is transferred on the ps time scale from P_{680} (creating the charged form P_{680}^+), via a nearby chlorophyll and a pheophytin, to a plastoquinone (PQ_A) on the stromal side (N-side) of the thylakoid membrane (see Figure 14.25). It then crosses over via an Fe atom to a second plastoquinone (PQ_B) with a redox potential of ~0 mV located ~20 Å away from PQ_A. Similarly to UQ in mitochondrial complex III, PQ_B can pick up two e^- in sequel to form PQ_B^{2-}, which after picking up 2 H^+ from the N-side, is released into the thylakoid membrane as PQH_2.

The P_{680}^+/P_{680} couple has an extremely high redox potential of about +1.3 V. In air at pH 7, the $O_2/2\ H_2O$ redox couple has a redox potential of +820 mV (Table 14.1), meaning that P_{680}^+ can in principle extract electrons from water to produce O_2 and protons. This is in fact what happens in PSII, in one of the most intriguing biochemical reactions in nature. This water-splitting reaction is carried out by the oxygen-evolving complex (OEC), a metal-oxygen cluster of composition Mn_4CaO_5 (**Figure 14.29**). The OEC is too distant from P_{680}^+ for direct electron tunneling to be possible, so the electrons travel via a strategically located Tyr residue (called Tyr_Z; Figure 14.28) that loses an e^- to P_{680}^+, which is rapidly replaced from the OEC. The four Mn atoms in the OEC can each switch between the Mn^{3+} and Mn^{4+} states, and the OEC can thus release four e^- in sequel to P_{680}^+ before it has to be "reset." These four e^- are taken from two molecules of H_2O, ultimately releasing one molecule of O_2 and four H^+ to the P-side (i.e., the thylakoid lumen). Crystallographic studies using a femtosecond-pulse free electron laser (FEL) and two-flash illumination to convert PSII to the S3 state in which three e^- have been transferred from the OEC, together with quantum

(A)

(B)

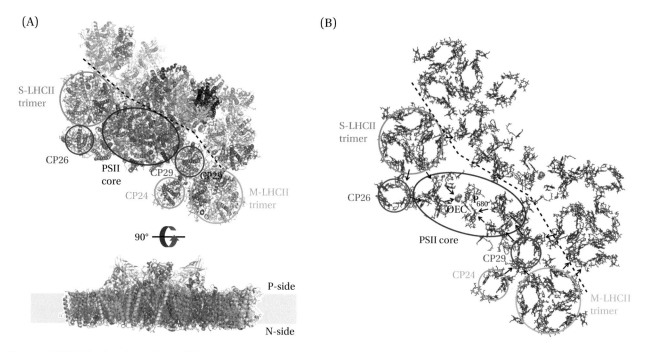

Figure 14.27 The low-light form of the LHCII–PSII supercomplex (PDB: 5XNM) from spinach (A: top and side views). During high-light conditions, the M-LHCII trimer and the CP24 light-harvesting protein dissociate from PSII. Prosthetic groups are shown in stick representation. (B) shows potential energy transfer pathways (arrows) from M- and S-LHCII to P_{680} (chlorophyll a in red, chlorophyll b in blue).

Figure 14.28 Structure of the electron transport chain in PSII (PDB 5KAF). Electrons pass from P_{680} down the right-hand path and then cross over from PQ_A to PQ_B via the central Fe. Electrons from the OEC reach P680$^+$ via Tyr$_Z$.

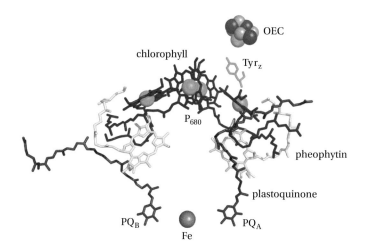

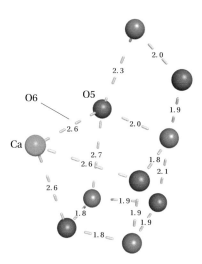

Figure 14.29 Structural relationships of the OEC (Ca, green; Mn, purple; O, red). Distances are given in Å. O_2 likely is formed from O5 and an O atom (O6) from a water that inserts itself between O5 and the Ca atom in the S3 state. (PDB 5WS6)

mechanical calculations, support a reaction scheme where the O_2 molecule is produced from oxygen O5 in the OEC together with an additional oxygen atom (O6) from a water molecule that inserts between O5 and the Ca atom.

14.4.3 The Cytochrome b$_6$f Complex and Plastocyanin

The cytochrome b_6f complex is situated between PSII and PSI in the photosynthetic electron transport chain (see Figure 14.25). It accepts electrons from PQH$_2$ generated by PSII and transfers them to plastocyanin, a small Cu-containing protein located in the thylakoid lumen. The cytochrome b$_6$f complex plays a role in photosynthesis similar to that of complex III in mitochondria and works according to the Q-cycle mechanism described earlier for the latter.

14.4.4 Photosystem I

PSI works similarly to PSII except that the electrons are delivered from plastocyanin rather than water. In addition, the LHCI antennae are arranged differently than the antennae of LHCII: although the LHCI and LHCII monomers are similar in structure, LHCI monomers form dimers rather than trimers and are packed along one face of PSI (**Figure 14.30**). After light activation, the electron passes from the P_{700} chlorophyll dimer (**Figure 14.31**), via either of two electron transfer chains composed of two chlorophylls, a phylloquinone, and three FeS centers, and is finally delivered to the small, water-soluble protein ferredoxin located in the stroma. After the loss of its excited electron, P_{700}^+ is reduced back to P_{700} by the luminal plastocyanin. Ferredoxin has a markedly negative redox potential of $-530\,mV$ and serves as the electron donor for a number of reactions, such as the reduction of $NADP^+$ to NADPH. Overall, the light-driven transfer of two electrons from H_2O to NADP+ results in six H+ being released to the P-side (lumen) of the thylakoid membrane, four H+ being removed from the N-side (stroma), and two e^- being translocated from the P-side to the N-side. The protons drive ATP synthesis by the thylakoid ATP synthase. In turn, ATP, together with NADPH, delivers the energy and reducing equivalents needed for CO_2 fixation in the so-called Calvin cycle.

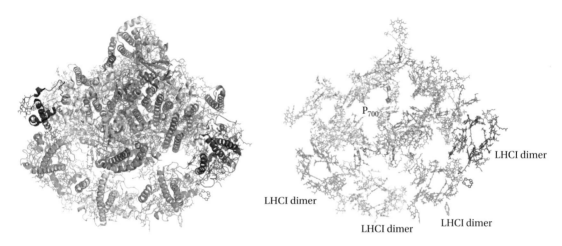

Figure 14.30 PSI from pea (PDB 4Y28), seen perpendicular to the membrane from the P-side. The four LHCI dimers (green, yellow, cyan, purple) cover the bottom side of the complex. The right-hand image shows only the prosthetic groups (mainly chlorophylls).

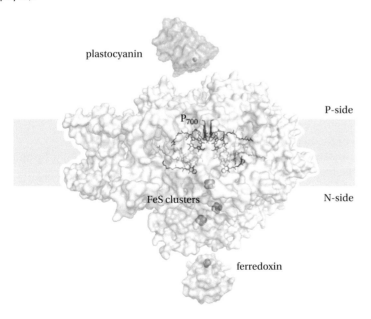

Figure 14.31 The electron transfer chain in PSI (PDB 4Y28). The light-excited e^- passes from P_{700} to ferredoxin (PDB 3B2F) and is replaced by an e^- from plastocyanin (PDB 9PCY).

Figure 14.32 Electron micrograph of a chloroplast from tobacco leaves. CW: cell wall, EM: envelope membranes, S: stroma, GT: stacked grana thylakoids, ST: unstacked stroma thylakoids. Arrowheads show that thylakoids make connections with the inner envelope membrane of the chloroplast. (Staehlin LA and Paolillo DJ [2020] *Photosynthesis Research* 145: 237–258.)

Figure 14.33 A model for how a phosphorylated LHCII trimer (PDB 2BHW) is bound to PSI (PDB 2O01), increasing its capacity to absorb light by augmenting the LHCI antennae system. (From Amunts A et al. [2007] *Nature* 447: 58–63. With permission from Springer Nature.)

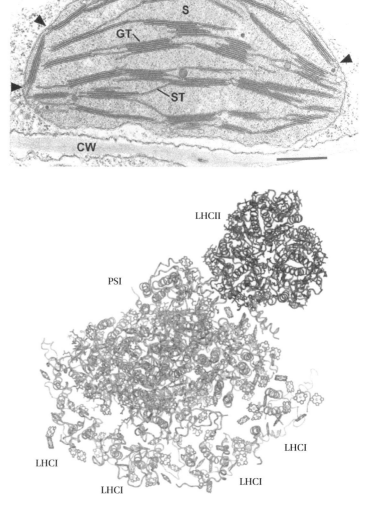

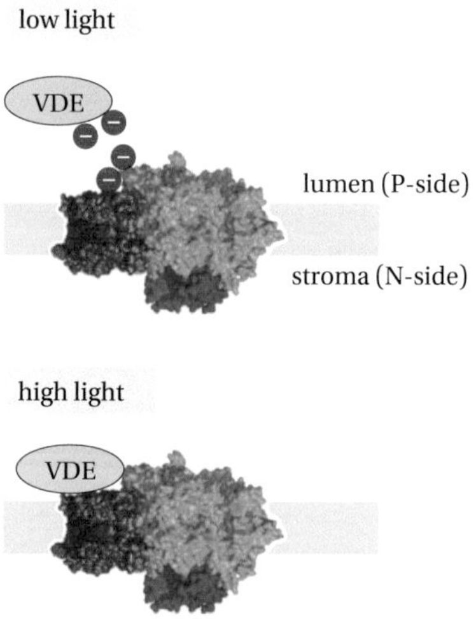

Figure 14.34 Model for photoprotection by NPQ. Acidification of the lumen during high-light conditions reduces the negative surface charge on VDE and PSI, allowing the enzyme to deliver the protective carotenoid zeaxanthin into PSI. (From Mazor Y, Borovikova A, Nelson N [2015] *eLife* 4: e07433.)

PSI and PSII are located in different regions of the thylakoid membrane: PSII in tightly stacked grana lamellae and PSI in unstacked stroma lamellae (**Figure 14.32**). To balance the activities of PSII and PSI, LHCII monomers can be phosphorylated by a kinase that is regulated by the balance between reduced (PQH_2) and oxidized (PQ) plastoquinone. Phosphorylated LHCII disengages from PSII, exits the grana stack, and associates with PSI in the unstacked lamellae (**Figure 14.33**). This gives rise to a negative regulatory feedback loop: an increase in PSII over PSI activity leads to a higher production of PQH_2 and a migration of LHCII from PSII to PSI, reducing the PSII antenna system and its light-harvesting capability while promoting light absorption by PSI. Thus, the activities of the two photosystems can be kept in balance.

Plants must adapt to changing light conditions and especially, be able to protect themselves against high intensities of light. In PSI, one protective mechanism, called non-photochemical quenching (NPQ), is to decrease the efficiency of energy transfer from the antennae to the core of the complex. This is achieved by the pH-dependent synthesis of a special carotenoid, zeaxanthin, that binds to LHCII and quenches excited chlorophyll triplet states. The responsible enzyme, violaxanthin deepoxidase (VDE), has an acidic domain that is repelled by the overall negatively charged lumenal surface of LHCII. Under high light intensity, however, the luminal pH will drop, and many of the acidic residues on LHCII will be titrated. VDE can now bind and deliver zeaxanthin into PSI (**Figure 14.34**), helping to dissipate some of the absorbed energy as heat.

KEY CONCEPTS

- Photosynthesis uses the energy in sunlight to produce O_2 and carbohydrates from H_2O and CO_2.

- Respiration is "photosynthesis in reverse": it reacts carbohydrates with O_2 and converts the extracted energy into ATP.

- Both photosynthesis and respiration use transmembrane proton gradients as an intermediary form of energy storage.

- The central photosynthetic and respiratory complexes are large membrane-embedded proteins that scaffold precisely shaped chains of redox centers, making it possible for electrons to flow from primary electron donors with low redox potential (e.g., P680, NADH) to acceptors with high redox potential (e.g. plastocyanin, O_2) coupled to proton translocation across the membrane.

- The respiratory chain in mammalian mitochondria is composed of complexes I–IV and ATP synthase.

- Complex I (NADH dehydrogenase) transfers electrons from NADH to UQ, and translocates four protons per two electrons.

- Complex II (succinate dehydrogenase) transfers electrons from succinate to UQ. It does not translocate protons.

- Complex III (cytochrome c reductase) transfers electrons from UQH_2 to cytochrome c, and translocates four protons per two electrons.

- Complex IV (cytochrome c oxidase) transfers electrons from cytochrome c to O_2, and translocates two protons per two electrons.

- ATP synthase uses the proton gradient across the inner mitochondrial membrane (maintained by complexes I–IV) to make ATP. It is a rotary enzyme, where the rotation of the rotor part against the stator part is driven by the proton current.

- The photosynthetic apparatus in higher plants is located in the thylakoid membrane and is composed of photosystems I and II and cytochrome b_6f.

- Photosystem II uses light energy to extract electrons from water, producing O_2. The electrons are transferred to the cytochrome b_6f complex via PQH_2.

- Cytochrome b_6f transfers electrons to photosystem I via plastocyanin.

- Photosystem I uses light energy to extract electrons from plastocyanin and transfers them down an electron transport chain to the final electron acceptor, $NADP^+$.

- Thylakoid ATP synthase converts the proton gradient created by the photosystems into ATP.

- NADPH and ATP are used in the Calvin cycle to fix CO_2.

FURTHER READING

Kandori, H. (2015) Ion-pumping microbial rhodopsins. *Frontiers Mol Biosci* 2:52.

Nicholls, D.G., and Ferguson, S.J. (2013) *Bioenergetics*, Vol. 4, Academic Press, London.

KEY LITERATURE

Allegretti, M., Klusch, N., Mills, D.J., Vonck, J., Kühlbrandt, W., and Davies, K.M. (2015) Horizontal membrane-intrinsic α-helices in the stator a-subunit of an F-type ATP synthase. *Nature* 521:237–240.

Björkholm, P., Harish, A., Hagström, E., Ernst, A.M., and Andersson, S.G. (2015) Mitochondrial genomes are retained by selective constraints on protein targeting. *Proc Natl Acad Sci USA* 112:10154–10161.

Blaza, J.N., Serreli, R., Jones, A.J.Y., Mohammed, K., and Hirst, J. (2014) Kinetic evidence against partitioning of the ubiquinone pool and the catalytic relevance of respiratory-chain supercomplexes. *Proc Natl Acad Sci USA* 111:15735–15740.

Deisenhofer, J., Epp, O., Miki, K., Huber, R., and Michel, H. (1985) Structure of the protein subunits in the photosynthetic reaction centre of Rhodopseudomonas viridis at 3Å resolution. *Nature* 318:618–624.

Iwata, S., Lee, J.W., Okada, K., Lee, J.K., Iwata, M., Rasmussen, B., Link, T.A., Ramaswamy, S., and Jap, B.K. (1998) Complete structure of the 11-subunit bovine mitochondrial cytochrome bc1 complex. *Science* 281:64–71

Kampjut, D., and Sazanov, L.A. (2020) The coupling mechanism of mammalian respiratory complex I. *Science* 370:547.

Kern, J. et al. (2018) Structures of the intermediates of Kok's photosynthetic water oxidation clock. *Nature* 563: 421–425.

Letts, J.A., Fiedorczuk, K., and Sazanov, L.A. (2016) The architecture of respiratory supercomplexes. *Nature* 537:644–648.

Mazor, Y., Borovikova, A., and Nelson, N. (2015) The structure of plant photosystem I super-complex at 2.8 Å resolution. *eLife* 4:e07433.

Mitchell, P. (1966) Chemiosmotic coupling in oxidative and photosynthetic phosphorylation. Reprinted in: *Biochim Biophys Acta.* 2011 180712(?):1507–1538 [check reference].

Noji, H., Yasuda, R., Yoshida, M., and Kinosita, K. Jr. (1997) Direct observation of the rotation of F_1-ATPase. *Nature* 386:299–302.

Racker, E., and Stoeckenius, W. (1974) Reconstitution of purple membrane vesicles catalyzing light-driven proton uptake and adenosine triphosphate formation. *J Biol Chem* 249:662–663.

Stroud, D.A., Surgenor, E.E., Formosa, L.E., Reljic, B., Frazier, A.E., Dibley, M.G., Osellame , L.D., Stait, T., Beilharz, T.H., Thorburn, D.R., Salim, A., and Ryan, M.T. (2016) Accessory subunits are integral for assembly and function of human mitochondrial complex I. *Nature.* 538:123–126.

Tsukihara, T., Aoyama, H., Yamashita, E., Tomizaki, T., Yamaguchi, H., Shinzawa-Itoh, K., Nakashima, R., Yaono, R., and Yoshikawa, S. (1996) The whole structure of the 13-subunit oxidized cytochrome c oxidase at 2.8 A. *Science* 272:1136–1144.

Vinothkumar, K.R., Zhu, J., and Hirst, J. (2014) Architecture of mammalian respiratory complex I. *Nature* 515:80–84.

Su, X., Ma, J., Wei, X., Cao, P., Zhu, D., Chang, W., Liu, Z., Zhang, X., and Li, M. (2017) Structure and assembly mechanism of plant C2S2M2-type PSII-LHCII supercomplex. *Science* 357:815–820.

Suga, M. et al. (2017) Light-induced structural changes and the site of O=O bond formation in PSII caught by XFEL. *Nature* 543:131–135.

EXERCISES

1. What is the "electrochemical potential"? What do "electro" and "chemical" refer to?

2. Why are positive ions rather than negative ions used for maintaining an electrochemical potential across the mitochondrial membrane?

3. What is the role of the antenna (light-harvesting complex) in photosynthetic organisms? How does it improve the overall energy yield of photosynthesis?

4. What are the key structural elements and functional principles of a redox-driven membrane-bound proton pump?

5. What is the advantage of using two photosystems in plants (as compared with a single photosystem)? Hint: Consider the energy needed to transfer an electron from H_2O to $NADP^+$ (Figure 14.25). Compare this energy with that available from "red/blue photons" assuming a yield of about 50%.

6. What are the factors determining the rate of an electron transfer reaction in a protein?

7. What structural characteristics of the ATP synthase determine the number of protons required to synthesize an ATP molecule? What is the $[H^+]/[ATP]$ ratio for the (a) mammalian, (b) *Saccharomyces cerevisiae*, and (c) plant thylakoid ATP synthase?

8. Formation of ATP from ADP and P_i requires 40 kJ mol^{-1}. Assuming $\Delta pH = 0.5$ (higher proton concentration on the outside of the membrane) and $\Delta\psi = 150$ mV (positive on the outside), what is the minimum number of charges (H^+) transferred across the membrane (via the ATP synthase) to synthesize 1 mol of ATP?

9. Assuming an $[NAD^+]/[NADP]$ ratio of unity and 0.2 atm O_2 (pH 7), what is the maximum number of protons that can be translocated in the respiratory chain per two electrons transferred from NADH to O_2 for $\Delta p = 180$ mV? How does the number change for $[NAD^+]/[NADP] = 10$?

Information Transfer: Signaling in Cells

15

All cells in an organism have the same genome, yet they develop into different organs and become specialized for different functions during development. Even single-cell organisms must be able to react to changes in their environment. How can cells react to what goes on in their surroundings? How are they triggered by extracellular or intracellular signaling molecules to activate intracellular regulatory circuits that move them down specific developmental pathways or guide their migration? In short, how can information be transferred from the outside of the cell to its inside?

In both single-cell and multicellular organisms, information transfer in most cases involves transmembrane (TM) receptors located at the cell surface (**Figure 15.1**). The classical view is that binding of an extracellular ligand traps the receptor in an "active" state—a conformation that allows intracellular effector proteins to bind to cytosolic parts of the receptor and to be activated in turn. The activated effector proteins then diffuse into the cell, triggering cellular responses such as the opening of ion channels or activation of gene transcription. Certain small signaling molecules do not need receptors on the cell surface but can diffuse across the cell membrane into the cell and trigger soluble nuclear receptors that bind directly to DNA. Still other signals arise from the cytoplasm and interact directly with cytoplasmic domains of kinases.

In this chapter, we focus on four classes of cell-surface receptors: two-component receptor histidine kinases (RHKs), receptor tyrosine kinases (RTKs), G protein–coupled receptors (GPCRs), and receptors that connect the cell cytoskeleton to neighboring cells and to the extracellular matrix (cadherins and integrins), although other types, such as ion channel–linked receptors, death receptors, and cytokine receptors, also exist.

The critical questions we address concern the mechanisms that transmembrane receptors use to transfer information across the membrane, the structures of different types of receptors, and the regulation of information transfer. We will see that cadherins, integrins, and transmembrane subunits in RHKs and RTKs generally have only one or two transmembrane helices and often form dimers or higher oligomers. In contrast, GPCRs generally have seven transmembrane helices and may be monomers or dimers. All types of receptors can exist in multiple low-energy conformations, and in most cases, the binding of ligand stabilizes the active form of the receptor.

15.1 BACTERIA SENSE THEIR ENVIRONMENT BY TWO-COMPONENT SIGNALING SYSTEMS

Bacteria live in unpredictable and often rapidly changing environments. Consequently, they must be able to respond quickly to changing circumstances

DOI: 10.1201/9780429341328-16

to obtain food, seek light, escape environmental stresses (e.g., extremes in pH, temperature, or osmolarity), or communicate among themselves. The minimum requirements for a system that can trigger a cellular response include some kind of sensor to detect a change in the concentration of a ligand (usually a small molecule), a means of converting this binding event into a chemical reaction inside the cell, and a mechanism to translate the chemical reaction into a physiological response.

Two-component systems are remarkably versatile systems built from a few basic building blocks that can be combined to construct an essentially unlimited number of regulatory pathways. They are called two-component systems because during signaling, a phosphate group is passed between two proteins: an RHK for sensing the environmental signal and a response regulator (RR) for triggering the physiological response (see **Box 15.1** for a primer on phosphorylation). Two-component signaling is the most common signaling system in bacteria. Although relatively uncommon, two-component signaling is also found in some eukaryotes.

15.1.1 General Design of Two-Component Systems

The prototypical two-component system is composed of a receptor histidine kinase (often a homodimeric transmembrane protein) and a cytoplasmic response regulator protein (**Figure 15.2**). For a few particularly well-studied cases, it is known that binding of ligand to the extracellular sensor domain in

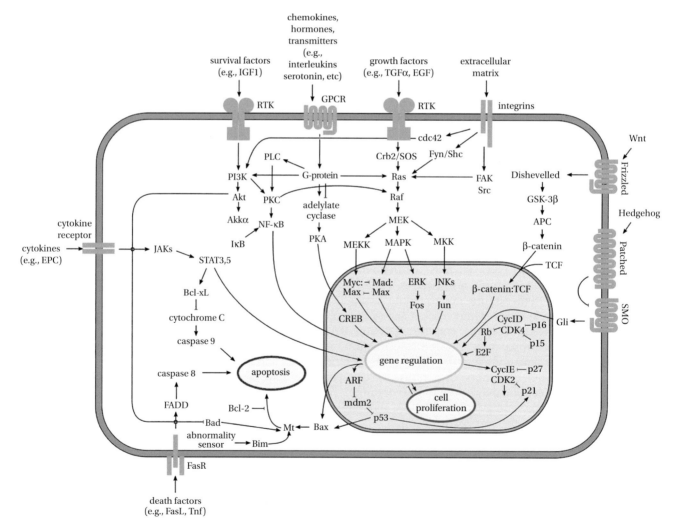

Figure 15.1 A few of the signal transduction pathways in eukaryotes. The pathways are numerous and complex, and interact with one another. In this chapter, we emphasize receptor tyrosine kinases (RTKs), G protein–coupled receptors (GPCRs), and integrins.

BOX 15.1 PHOSPHORYLATION OF AMINO ACID SIDE CHAINS IS CRUCIAL FOR SIGNALING

Reversible phosphorylation of amino acid side chains is essential for intracellular signaling in all branches of life. Many different side chains can be phosphorylated, including serine, threonine, tyrosine, histidine, aspartate, arginine, and lysine. The basic scheme for phosphorylation/dephosphorylation is shown in **Figure 1**, using serine as an example. In bacteria, the most common signaling system is two-component signaling that involves phosphorylation of histidine and aspartate. Although uncommon, two-component signaling is also found in some eukaryotes. On the other hand, RTKs are most commonly used in eukaryotic signaling systems, but they also occur to a limited extent in prokaryotes.

Why is phosphorylation an effective signaling method? First, at physiological pH, each phosphate group carries ~1.5 negative charges. The consequent change in charge state can modify the conformation of the phosphorylated protein or affect the binding of ligands. Second, the attachment of a phosphate can affect recognition of the protein by other proteins. The regulation of signaling pathways by phosphorylation/dephosphorylation allows exquisite control of signaling, because the relative activities of kinases and phosphatases can determine at any moment the activity of the target protein.

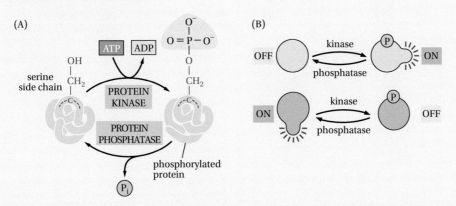

Figure 1 Schematic of protein phosphorylation. (A) Phosphorylation involves the transfer of a phosphate group from ATP catalyzed by a kinase. In dephosphorylation, a phosphatase catalyzes the removal of the phosphate group. (B) For some proteins, the phosphorylated state is the active one, while for others, it is the de-phosphorylated state. (From Alberts B, Johnson A, Lewis JH et al. [2014] *Molecular Biology of the Cell*, 6th Edition. Garland Science, New York. With permission from W. W. Norton.)

the kinase triggers a conformational change that is transmitted through the transmembrane domain. The conformational change activates the intracellular catalytic ATP binding domain (CA; Figure 15.2), which autophosphorylates a His residue in the "dimerization and histidine phosphotransfer" domain (DHp; Figure 15.2). In other examples, the signal is cytoplasmic and directly alters the conformation of the DHp domain, stimulating autophosphorylation.

The RR is generally composed of a receiver domain (RD) and an output domain. The RD can interact with the DHp domain to catalyze the transfer of the phosphate from the His to an Asp residue in the RD, thereby activating the output domain.

An example of the general structural organization of dimeric RHKs is the PhoQ histidine kinase involved in cationic antimicrobial peptide resistance of *E. coli* (**Figure 15.3**). The mechanism of signal transfer across the membrane in this case may involve signal-induced rearrangements of the four transmembrane helices.

In some two-component systems, the RD is part of the histidine kinase itself (Figure 15.2A). In such cases, the phosphorelay system has an extra histidine phosphotransferase (HPt) that couples to a second, cytoplasmic RR, providing an additional point of regulation. Sometimes, the HPt domain is a separate protein.

15.1.2 Phosphoryl Transfer Is Carried out by the Kinase Core

The DHp and CA domains together form a dimeric kinase core in which the CA domain can phosphorylate the DHp domain (**Figure 15.4A**) in some cases by a

(A)

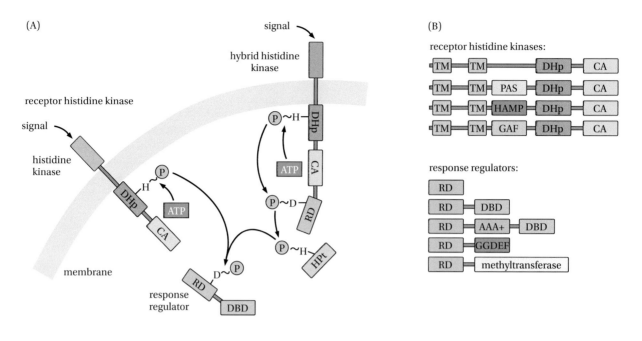

CA: catalytic ATP binding domain
DHp: dimerization and histidine phosphotransfer domain
RR: response regulators
RD: receiver domain
DBD: DNA-binding domain
HPt: histidine phosphotransferase

Figure 15.2 Two-component signaling systems. (A) The response regulator can either be a separate protein (left) or part of the histidine kinase (right). The RHK (receptor histidine kinase) is in most cases a homodimer. (B) Small regulatory domains (PAS, HAMP, GAF) are sometimes inserted between the transmembrane (TM) and DHp domains in the RHK. In the RR, the receiver domain can be coupled to many different output domains. (Figure reproduced with permission from Capra EJ, Laub MT [2012] Annu Rev Microbiol 66: 325–347. Copyright 2012 Annual Reviews.)

trans-phosphorylation reaction whereby the CA domain in one of the subunits phosphorylates the His residue in the DHp domain of the other subunit. It is not entirely clear how the conformational change in the sensor domain triggers phosphorylation of the DHp domain by the CA domain. One model is that the signal is transmitted down the central coiled-coil in the DHp dimer, disturbing the packing between the DHp and CA domains such that the CA domain can seek out and phosphorylate the nearby His residue, while another model posits that a relatively unstructured region in the four-helix bundle flanking the critical histidine in the DHp domain becomes more ordered, stimulating autophosphorylation.

15.1.3 Phosphorylation of the Receiver Domain Activates the Output Domain

Efficient phosphoryl transfer from the His residue in the DHp domain of the histidine kinase to the Asp residue in the RD in the response regulator requires participation of active-site residues from both proteins. The conformational change upon phosphorylation of the RD is subtle and restricted mainly to two "switch" residues—a conserved Ser/Thr and a conserved Phe/Tyr—located near the active site (Figure 15.4B). The Phe/Tyr residue is known to interact with residues in at least some output domains. In this way, phosphorylation of the conserved Asp residue can be coupled to the attached output domain.

While the DHp, CA, and RD domains that are directly involved in the phosphate transfer reaction are well conserved in two-component signaling systems, the number of different sensor and output domains is vast. The sensor domains

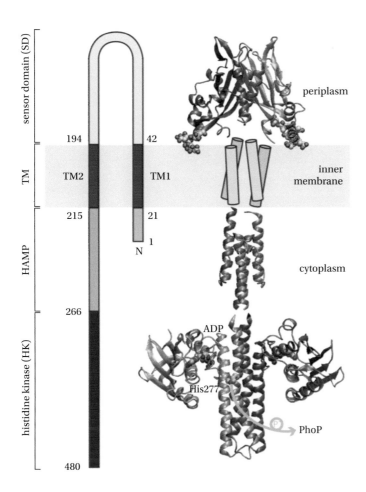

Figure 15.3 Organization of the PhoQ dimeric RHK. Crystallographic structures of the individual domains, except for the TM domains, are shown. Proceeding from the periplasm, the PDB codes are 3BQ8, 2L7H, and 2C2A. (Adapted from Lemmin T, Soto CS, Clinthorne G et al. [2013] *PLOS Compu Biol* 9: e1002878.)

of the classical chemoreceptors have two transmembrane helices flanking a periplasmic domain (see below) and interact with a soluble histidine kinase. There are unrelated sensor domains with up to at least 20 transmembrane helices, and some RHKs have cytoplasmic sensor domains. There is a large variety of output domains, including DNA binding domains that can activate or repress gene transcription; enzymatic domains involved in, for example, regulation of cyclic diguanylate synthesis, methylesterases, or phosphatases; and domains that bind to other proteins to regulate their function. A single species of bacteria can contain two-component systems built from more than 100 different sensor and output domains. *E. coli* devotes about 5% of its membrane proteins to two-component signaling.

15.1.4 Bacteria Seek Nutrients by Chemotaxis Using Two-Component Signaling

Many bacteria, including *E. coli*, can actively seek out favorable environments—those containing higher concentrations of a useful amino acid, say—or move away from unfavorable ones. This process, called chemotaxis, depends on specialized systems that sense concentration gradients of chemical attractants or repellants in the surroundings and tune the action of flagellar motors to generate an appropriate swimming behavior. A very simple experiment by Wilhelm Pfeffer (1845–1920) demonstrated that bacteria accumulate in and around glass capillaries containing chemicals of metabolic importance, such as galactose (**Figure 15.5A**). An important question was whether the attraction was related to the metabolic processing of the chemical by the bacteria. Julius Adler at the University of Wisconsin demonstrated that *E. coli* are attracted directly

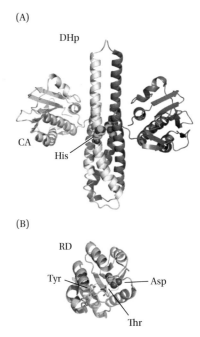

Figure 15.4 A kinase core and a receiver domain. (A) Dimeric kinase core of the PhoQ histidine kinase from *Thermotoga maritima* (PDB: 2C2A) showing the catalytic CA domain and the phosphorylatable histidine. (B) CheY receiver domain (PDB 1F4V) with the phosphorylatable Asp and switch residues Thr and Tyr shown. (Figure reproduced with permission from Gao R, Stock AM [2009] *Annu Rev Microbiol* 63: 133–154. Copyright 2009 Annual Reviews.)

Figure 15.5 Bacteria are attracted to metabolically important molecules. (A) A glass capillary filled with galactose attracts motile *E. coli*. The strength of the attraction can be quantitated by counting the number of bacteria found inside the capillary after 1 h. (B) The number of bacteria found in the capillary as a function of the galactose concentration. The decline at high concentrations is due to rapid diffusion of galactose into the growth medium. The increase in numbers of bacteria is the same for bacteria unable to metabolize galactose (galactose⁻) as for wild-type *E. coli* (galactose⁺). This demonstrates that *E. coli* respond directly to attractants rather than indirectly as a result of galactose metabolism. (From Adler J [1969] *Science* 166: 1588–1597. With permission from the American Association for the Advancement of Science.)

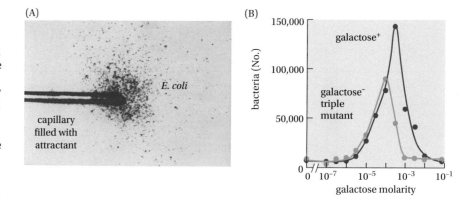

Figure 15.6 *E. coli* are motile because of the motion of flagella driven by molecular motors, which are among the most complex membrane protein assemblies in biology. (A) A micrograph of a flagellated bacterium and the three-dimensional trajectory of a swimming *E. coli* projected onto the surfaces of a cube (labeled 1, 2, 3), showing how a biased random walk can produce net motion up a chemoattractant gradient. If you imagine folding the labeled squares along the dashed lines to form a cube, the trajectory would be seen in three dimensions. If the left and upper panels of each figure are folded out of the page along the dashed lines, the projections appear in proper orientation on three adjacent faces of a cube. (B) Depictions of the molecular motor that drives the motion of the flagella. The flagellar motor switches between anti-clockwise and clockwise rotation in response to the binding of phosphorylated CheY. (C) Structure of a flagellar motor determined using electron cryotomography. (A, From Berg HC, Brown DA [1972] *Nature* 239: 500–504. With permission from Springer Nature. B, Porter SL et al. [2011] *Nature Rev Microbiol* 9: 153–165. With permission from Springer Nature. C, DeRosier D [2006] *Curr Biol* 16: R928–R930. With permission from Elsevier.)

to chemoreceptors—a term coined by Adler—rather than indirectly through a metabolic product (Figure 15.5B).

E. coli and other motile bacteria swim toward attractants (or away from irritants) by switching the actions of flagella as the bacteria sample chemical gradients. Flagella are long filaments that extend from the surface of the bacterium (**Figure 15.6A**). Flagella are attached at their base to a rotary motor that is driven by the proton gradient across the inner membrane (Figure 15.6B). The motor rotates either clockwise or anti-clockwise depending on whether the phosphorylated form of the response regulator CheY is bound to a cytoplasmic switching device in the motor. When rotated anti-clockwise, the helical flagella assemble cooperatively into a long trailing bundle that propels the cell forward (the "run" phase), while clockwise rotation leads to a loss of the cooperative packing of the helices in the bundle, causing them to fly apart and allowing the

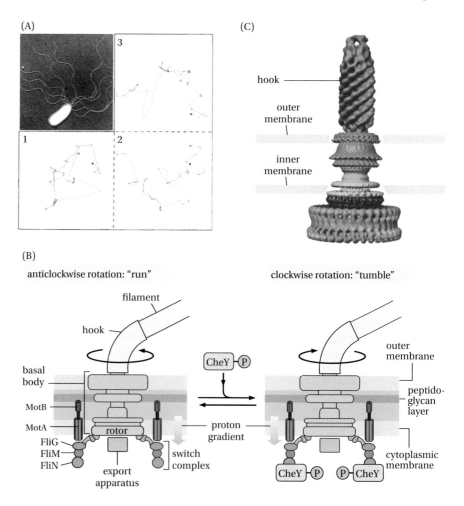

bacterium to tumble randomly. During the run phase, the cell can swim at rates of a few tens of μm per second in aqueous solution.

Imagine that the cell finds itself in an environment where there is a gradient of an attractant, such as the amino acid aspartic acid. The difference in attractant concentration between the two ends of the cell is too small to be sensed directly. Instead, the cell monitors the change in attractant concentration as it moves through the medium and uses this information to bias the time it spends running versus tumbling. The cell biases the motor toward anti-clockwise rotation (the run mode: the helical flagella work together as a bundle) when it finds itself in an environment with high levels of attractant, and toward clockwise rotation (tumble mode: the helical flagella clash to disassemble the bundle) when the attractant levels are low. This gives rise to a biased random walk up the concentration gradient: if the cell happens to swim toward a lower concentration of attractant, the run will be short. After the ensuing tumble, the cell will set off in a new random direction; if it happens to swim toward a higher concentration of attractant, the run will be longer. The net effect is that the cell wanders up the concentration gradient using a biased random walk (**Figure 15.7A**).

There are many assays that can be used to study chemotaxis, including tracking the motions of individual cells on a microscope slide (Figure 15.6A). A simpler chemotaxis assay is shown in Figure 15.7B, which takes advantage of the ability of E. coli to move in swarms in dilute agar medium. E. coli has two different chemoreceptors for sensing Ser and Asp concentration gradients that feed into the CheA/CheY two-component system. Cells were placed at two foci on a Petri dish containing growth medium enriched in both Ser and Asp, and then allowed to grow and swarm for a number of hours. The initial swarm consumes all the Ser as it moves away from the inoculation point, leaving Asp for the second swarm. This leads to the formation of two concentric rings of cells, where the cells in the outer ring live on serine and those in the inner ring on aspartate. This can only happen if the bacteria prefer one chemical over another. Growth assays of this sort can be used to identify cells that have genetic mutations that perturb chemotaxis. All the components of the two-component systems in model bacteria, such as E. coli, were originally identified through genetic

(A)

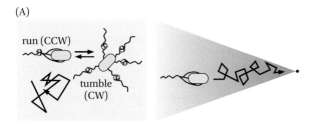

run (CCW)

tumble (CW)

(B)   very soft agar-agar plate containing serine and aspartate

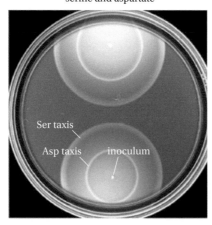

Ser taxis

Asp taxis   inoculum

Figure 15.7 Chemotaxis by *E. coli*. (A) Biased random walk up a concentration gradient is based on switching between straight swimming and tumbling triggered by counterclockwise and clockwise rotations of the motor, respectively. (B) A common chemotaxis assay invented by Julius Adler that takes advantage of the ability of *E. coli* to swarm in weakly polymerized agar gels, here containing Ser (which the bugs love to eat) and Asp (which they eat as a second choice). The first swarm phase consumes all the Ser, leaving the Asp to induce the second swarm. Courtesy of Parkinson Lab, University of Utah.

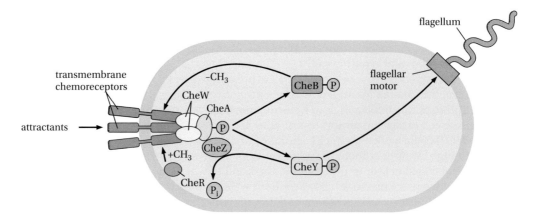

Figure 15.8 Schematic describing the control of motility through the CheA/CheY two-component signaling system. See text for details. (Porter SL et al. [2011] *Nature Rev Microbiol* 9: 153–165. With permission from Springer Nature.)

screens based upon various growth assays, such as the swarming motility assay illustrated in Figure 15.7B.

The key response regulator controlling running versus tumbling is CheY. Its phosphorylation state, and therefore its ability to bind to the flagellar motor and induce the tumbling motion, is controlled by the RHK system shown in **Figure 15.8.** In the case of the aspartate-sensing system, the histidine kinase CheA is a separate subunit and binds to the transmembrane receptor via an adaptor protein called CheW. CheA is active when aspartate is not bound to the periplasmic sensor domain of the chemoreceptor. Thus, when the aspartate concentration is low, CheA autophosphorylates its acceptor His residue, the phosphoryl group is transferred to the Asp residue in the response regulator CheY, and phosphorylated CheY binds to the flagellar motor, inducing clockwise rotation and tumbling of the cell. If the aspartate concentration increases, less phospho-CheY is produced, biasing the flagellar motor toward anti-clockwise rotation and inducing the cell to perform longer runs. To shorten the response time of the system, a phosphatase, CheZ, continually removes phosphate from CheY.

The aspartate two-component system has an additional built-in device that allows it to reset its baseline activity to reflect the average concentration of attractant in the surroundings. This is achieved by reversible methylation of glutamate side chains in the cytoplasmic part of the sensor. The higher the number of methylated glutamates, the higher is the activity of the CheA histidine kinase. The methyl groups are added by the methyltransferase CheR and are removed by the methylesterase CheB. Feedback control is established through CheB, which is active only if it has been phosphorylated by the CheA kinase. In the case of a sudden increase in the overall concentration of aspartate, the CheA activity drops and less phospho-CheB is produced, resulting in an increase in the number of methylated glutamates on the sensor. This brings the activity of CheA back up again, effectively resetting the system to work at the new set-point.

The classical cartoon representation of the whole aspartate chemoreceptor and its histidine kinase, and a contemporary molecular model of the related serine chemoreceptor, are shown in **Figure 15.9.** Seeing these structures, an obvious question is how the binding of a chemoattractant to the periplasmic sensor domain can affect the activity of the CheA kinase located at the base of the long cytoplasmic coiled-coil domain. Electron cryomicroscopy (cryo-EM) and coarse-grained molecular dynamics simulations suggest that the active unit is a trimer of dimers (**Figure 15.10**). Signaling is hypothesized to occur through an expansion and contraction of the trimeric bundle, but a detailed molecular understanding will require higher structural resolution.

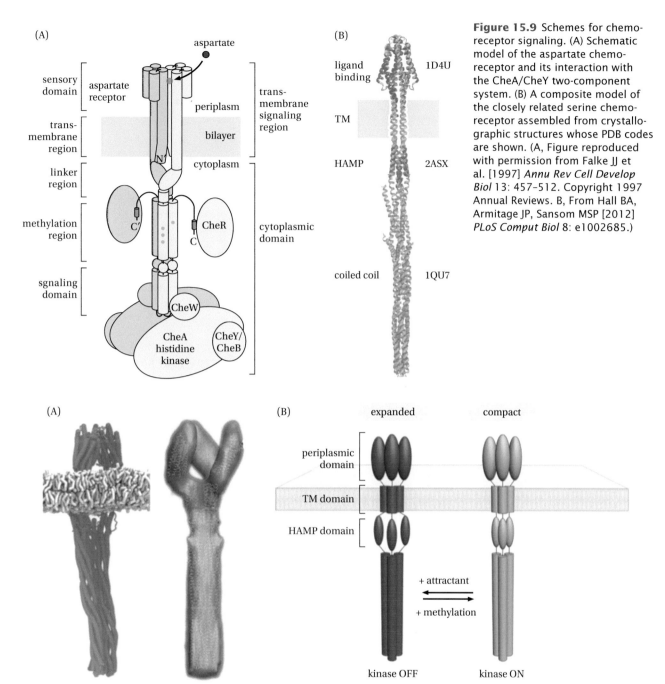

Figure 15.9 Schemes for chemo-receptor signaling. (A) Schematic model of the aspartate chemo-receptor and its interaction with the CheA/CheY two-component system. (B) A composite model of the closely related serine chemo-receptor assembled from crystallo-graphic structures whose PDB codes are shown. (A, Figure reproduced with permission from Falke JJ et al. [1997] *Annu Rev Cell Develop Biol* 13: 457–512. Copyright 1997 Annual Reviews. B, From Hall BA, Armitage JP, Sansom MSP [2012] *PLoS Comput Biol* 8: e1002685.)

Figure 15.10 Trimer-of-dimers model for the serine chemoreceptor. (A) Cryo-EM image (right) and coarse-grained molecular dynamics simulation (left). (B) One hypothesis for how the trimer of dimers might transmit a signal. The key idea is that attrac-tant binding causes a slight expansion of the trimeric bundle of the kinase. (From Hall BA, Armitage JP, Sansom MSP [2012] *PLoS Comput Biol* 8: e1002685.)

15.1.5 Bacteria Respond to Osmotic Stress by Regulating the Expression of OmpF and OmpC Porins in the Outer Membrane

Another classic example of two-component signaling is the response of *E. coli* to changes in the osmolarity of their bathing solution (**Figure 15.11**). One of the protection mechanisms is to change the relative amounts of OmpF and OmpC porins (Box 10.7) in the outer membrane; the larger pore size of the OmpF porin is thought to facilitate nutrient uptake under poor, low-osmolarity conditions, while the smaller OmpC pore may prevent uptake of toxic bile salts

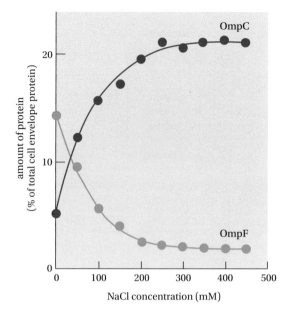

Figure 15.11 *E. coli* changes the relative amounts of the outer membrane porins OmpF and OmpC in response to changes in the osmolarity of the surrounding medium. (Modified from Alphen WV, Lugtenberg B [1977] *J Bacteriol* 131:623–630.)

present under high-osmolarity growth conditions, such as those encountered in the human gut. Osmolarity changes are detected by the inner membrane RHK EnvZ, which has two TM helices, similar to the chemoreceptors discussed earlier. The relative amounts of OmpF and OmpC are varied by changing the amounts of proteins produced, i.e., by controlling the expression of the *ompF* and *ompC* genes by OmpR, a two-domain cytoplasmic response regulator (RR).

A frequent structural motif found in transcription factors is the helix-turn-helix (H-T-H) motif, which has a recognition helix that binds in the major groove of the DNA (**Figure 15.12**). In the case of OmpR, the output domain is a so-called winged H-T-H DNA binding domain, and the receiver domain houses the critical aspartate that can be phosphorylated to produce phospho-OmpR (OmpR-P). The *ompF* gene has four OmpR-P biding sites (F1–F4) in its promoter region, and the *ompC* gene has three (C1–C3). The affinity for OmpR-P decreases in the order F1, C1 > F2, F3 > C2 > C3, F4. At low levels of OmpR-P (i.e., at low osmolarity). only the F1–F3 and C1 sites are occupied by OmpR-P, leading to strong induction of *ompF* transcription but only weak induction of *OmpC*. At high levels of OmpR-P (i.e., under high osmolarity), OmpR-P binds also to the C2, C3, and F4 sites, increasing transcription of the *ompC* gene while reducing expression of *ompF*, because binding of OmpR-P to the F4 sites blocks transcription. The relatively complex osmolarity-dependent regulation of OmpC and OmpF thus depends not only on the EnvZ-OmpR two-component system but also on the specific design of the *ompC* and *ompF* promoter regions.

15.2 EUKARYOTIC CELLS COORDINATE THEIR INTERACTIONS USING RECEPTOR TYROSINE KINASES

Compared with multicellular organisms, such as mammals, the signaling needs of bacteria are quite rudimentary, involving primarily, as we have seen, the search for nutrients and protection against environmental stress. Two-component signaling systems are quite sufficient for the needs of bacteria but not, apparently, for eukaryotes and archaea, where they are rarely found. Two-component signaling systems are never found in mammals, where one finds a more elaborate range of signaling mechanisms, including RTKs. These kinases coordinate cell proliferation and differentiation, cell-cycle control, and

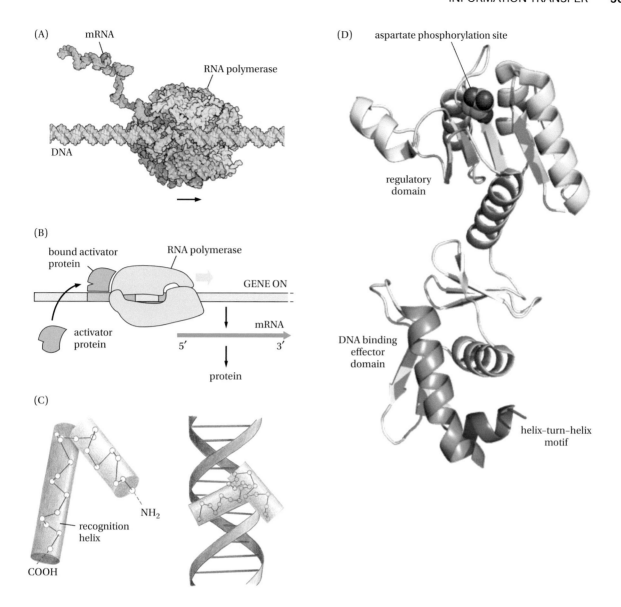

Figure 15.12 Regulation of gene expression by response regulators in two-component signaling systems. (A) RNA polymerase recognizes a promoter site on the DNA and produces mRNA from the gene. (B) Many promoters, although specific for the gene of interest, are weak but can be enhanced by the binding of one or more activator proteins. (C) Promoter sites are recognized by specific DNA binding proteins. There are many different protein structural motifs that recognize and regulate gene expression, but a very common one is the helix-turn-helix motif. (D) Two-component signaling response regulators have two domains: the DNA binding (effector) domain and a phosphorylatable regulator domain that essentially controls the effectiveness of the response regulator. The structure shown is that of DrrD (PDB 1KGS), which is homologous to OmpR, which regulates the expression of the *ompF* and *ompC* genes. (A, From David S. Goodsell and the RCSB PDB. B, C, From Alberts B, Johnson A, Lewis JH et al. [2014] *Molecular Biology of the Cell*, 6th Edition. Garland Science, New York. With permission from W. W. Norton.)

migration, and are thus essential for the development of mammals and other eukaryotes that have differentiated tissues (gut, heart, muscle, etc.). A possible explanation for the absence of two-component signaling systems in mammals is that two-component systems do not readily allow the interplay of inputs by the different signals that are useful in mammals (note the complex web of interactions in Figure 15.1).

As in the case of bacteria, the environmental signals are carried by chemical compounds. Those produced within organisms are referred to generically as growth factors. This term arose in the 1950s when Rita Levi-Montalcini (1909–2012) discovered that a chemical compound secreted by mouse tumor cells promoted neurite outgrowths in chicken embryos. She and Stanley Cohen (1922–2020) subsequently showed the compound to be a protein, which they

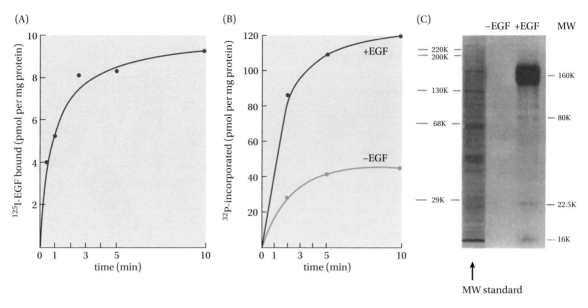

Figure 15.13 The discovery of the epidermal growth factor (EGF) receptor and its phosphorylation in the presence of EGF. (A) ^{125}I-labeled EGF binds to the membranes of A-431 human epidermoid carcinoma cells. (B) Membrane proteins are phosphorylated in the presence of EGF. (C) A membrane protein with molecular weight of 150–170 kDa (the EGF receptor) is heavily phosphorylated when membranes are incubated with [γ-^{32}P]ATP in the presence of EGF. The use of radioactive protein allows visualization by autoradiography. (From Carpenter G et al. [1978] *Nature* 276: 409–410. With permission from Springer Nature.)

called nerve growth factor (NGF). Following the idea that mammalian cells generally communicate chemically, Cohen discovered epidermal growth factor (EGF), which stimulated the proliferation of epithelial cells.

The discovery of growth factors revolutionized thinking about developmental biology and earned Levi-Montalcini and Cohen the 1986 Nobel Prize in Physiology or Medicine. A key step in the discovery of RTKs was the elegant experiments by Cohen and his colleagues in 1978 (**Figure 15.13**). These experiments showed definitively that EGF bound to membrane proteins (Figure 15.13A), which became phosphorylated in the presence of EGF (Figure 15.13B). Extraction of membrane proteins with detergents in the presence and absence of EGF revealed the EGF receptor (EGFR) to be a 170-kDa protein (Figure 15.13C).

As a result of these pioneering discoveries, we now know that many crucial signaling events in mammals are mediated by RTKs, which are single-span plasma membrane receptors. All RTKs are composed of an extracellular ligand-binding domain, a single transmembrane helix, and an intracellular tyrosine kinase domain, but the ligand-binding domains can differ greatly between different receptors. Examples of RTKs are shown in **Figure 15.14**.

15.2.1 Ligand Binding Induces Receptor Dimerization or Dimer Reorganization

RTKs work on the principle that the binding of ligand to the extracellular domain induces receptor dimerization or rearrangement of a pre-existing dimer, which in turn leads to autophosphorylation and activation of the intracellular tyrosine kinase domain. The phosphorylated active receptor then recruits various intracellular signaling proteins that bind to phosphotyrosines in the receptor and in turn trigger intracellular signaling pathways. The various signaling proteins are apparent in Figure 15.13C as bands of lower molecular weight that appear in the presence of EGF.

Structural studies of ligand-binding domains have uncovered at least four different ways that a ligand can induce, enhance, or modify receptor dimerization (**Figure 15.15**). The importance of interactions involving the transmembrane helices in RTK activation appears to differ between receptors.

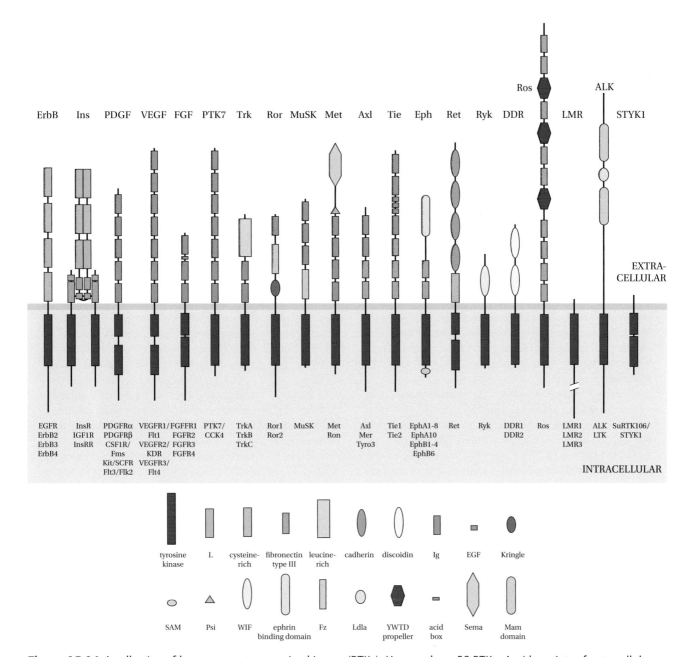

Figure 15.14 A collection of human receptor tyrosine kinases (RTKs). Humans have 58 RTKs. A wide variety of extracellular ligand-binding domains are combined with an intracellular tyrosine kinase domain. (From Lemmon MA, Schlessinger J [2010] *Cell* 141: 1117–1134. With permission from Elsevier.)

Some have GXXXG motifs (see Chapter 6) that might be important for the dimerization process, while in others, the sequence seems to matter less. A dimeric ligand can directly mediate dimerization in several ways: the two ligands can bind between two receptors (panel A) or provide part of the dimer interface (panel B), or two monomeric ligands can bind separately to a dimeric receptor (panel C). Finally, two monomeric ligands can each bind to one receptor molecule, with the entire dimer interface provided by the receptor itself (panel D). In case A, dimerization is mediated directly and uniquely by the ligand, while in cases B–D, binding of ligand induces a conformational change in the receptor that leads to dimerization and activation. An example of case D is the epidermal growth factor receptor (EGFR), described in Spread 15.1.

SPREAD 15.1 THE EPIDERMAL GROWTH FACTOR RECEPTOR

EGFR is historically the most important receptor tyrosine kinase. It was the first one to be cloned and the first one that was clearly implicated in cancer. There are four members of the human EGFR family (EGFR itself—also called Erb1—and ErbB2–4), and they play essential roles in, e.g., epithelial development, cardiac development, and neurogenesis. For these reasons, they are important drug targets.

There are two main types of Erb ligands: agonists such as EGF and transforming growth factor α (TGFα) that bind Erb1, and the neuregulins that bind Erb3 and Erb4 (**Figure 1**). Some ligands are bispecific and can bind both Erb1 and Erb3-4. Erb2 appears not to have a soluble ligand and may act by forming heterodimers with other Erb family members.

The four Erb family members share the same structure, with an extracellular part composed of four domains of which domains I and III are ligand-binding, a single transmembrane helix, a short intracellular juxtamembrane domain, a tyrosine kinase domain, and a C-terminal regulatory domain containing the tyrosine residues that become autophosphorylated during activation of the receptor.

Composite models of EGFR in the inactive monomeric state, an inactive dimeric state, and an activated dimeric state, based on crystal structures of the extracellular ligand-binding part and the intracellular kinase domain, are shown in **Figure 2**. In the monomeric state, domains I and III in the extracellular part are far apart, and there are extensive interactions tethering domains II to domain IV (panel A). In the active dimeric state, one EGF molecule is bound to each monomer between domains I and III, stabilizing a conformation in which domain II is no longer tethered to domain IV but rather, has swung out to form a large dimer interface with domain II in the other monomer. At the same time, the tip of domain IV, which is connected to the transmembrane helix, engages the tip of domain IV in the other monomer. It has been estimated that there is only a modest energy difference of ~2 kcal mol⁻¹ between the tethered and untethered forms of the unliganded monomer, and it may be that binding of EGF to the monomer provides enough binding energy to shift the equilibrium to the untethered conformation, thereby promoting dimerization.

Studies using a range of techniques, from nuclear magnetic resonance (NMR) and mutagenesis experiments to molecular dynamics simulations, suggest that the EGFR transmembrane helix can exist in three different states of comparable stability: one monomeric and two different dimeric states, only one of which is compatible with the structure of the activated kinase domain. In the inactive dimeric state, a GXXXG-type motif mediates interaction between the cytoplasmic ends of the two transmembrane helices (Figure 2B). This is incompatible with the formation of a structural element composed of two short antiparallel α-helices in the juxtamembrane domain, which is critical for activation of the dimeric kinase domain. In the active dimeric state, the two transmembrane helices instead interact via two overlapping GXXXG motifs located near the extracellular ends of the helices (**Figure 3**). The intracellular ends of the helices are splayed apart in this state, allowing the antiparallel α-helical element in the juxtamembrane domain to form and inducing the active state of the dimeric kinase domain.

The tyrosine kinase domain, finally, is in an inactive conformation, in which the so-called activation loop displaces the C-helix away from the active site in both the monomeric and inactive dimeric states of the receptor (**Figure 4**). This prevents the formation of a critical ion pair between a glutamate residue on the C-helix and a lysine residue in the active site, rendering the kinase inactive. In the active, asymmetric dimeric structure, the activation loop in one of the two kinase domains (the "receiver" domain) adopts a more extended conformation, allowing the C-helix to approach the active site and restoring the Glu-Lys ion pair that is required to coordinate the α- and β-phosphates of bound ATP. This activates the kinase, which proceeds to trans-phosphorylate tyrosines in the C-terminal regulatory region of the other kinase domain. In contrast to other receptor tyrosine kinases, the activation loop in the Erb family members is not displaced by being phosphorylated but as a result of allosteric interactions in the asymmetric dimer.

Finally, the second kinase domain is also activated, leading to autophosphorylation of both regulatory regions. The whole activation process is summarized in **Figure 5**.

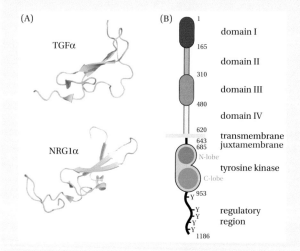

Figure 1 (A) Erb ligands. TGFα (PDB 1HRE) activates Erb1 (EGFR), and NRG1α (PDB 1MOX) activates Erb3 and Erb4. (B) The domain structure of EGFR. (Figure reproduced with permission from Ferguson KM [2008] *Annu. Rev. Biophys.* 37: 353–373. Copyright 2008 Annual Reviews.)

(Continued)

SPREAD 15.1   THE EPIDERMAL GROWTH FACTOR RECEPTOR (CONTINUED)

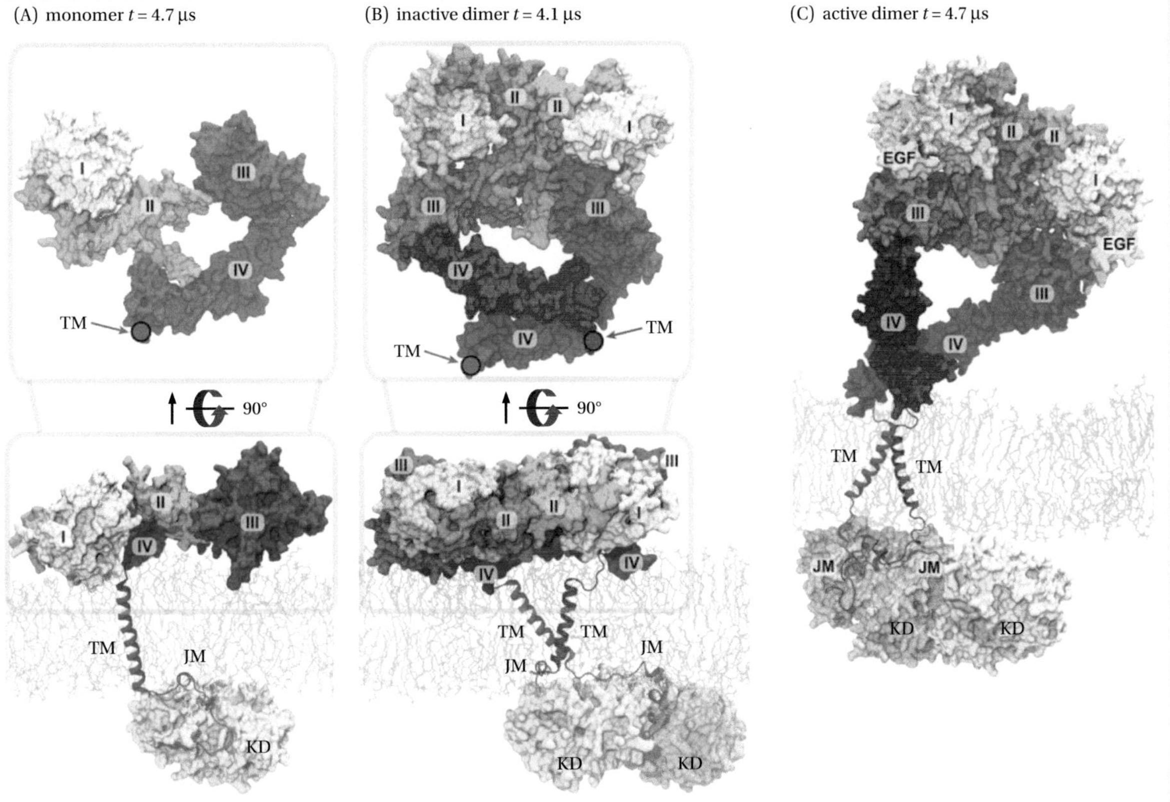

(A) monomer $t = 4.7\ \mu s$ (B) inactive dimer $t = 4.1\ \mu s$ (C) active dimer $t = 4.7\ \mu s$

Figure 2 The extracellular ligand-binding domains in EGFR are modeled as lying flat on the membrane in the inactive monomeric and dimeric states (panels A and B), and in a more upright position in the active dimeric state (panel C). (From Arkhipov A, Shan Y, Das R et al. [2013] *Cell* 152: 557–569. With permission from Elsevier.)

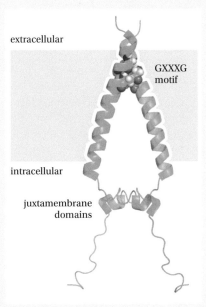

Figure 3 NMR structure of the transmembrane and juxtamembrane regions of the EGFR (PDB 2M20). The TM helices interact via GXXXG motifs in their extracellular halves. The two juxtamembrane domains form two short interacting α-helices. (From Endres NF, Das R, Smith AW et al. [2013] *Cell* 152: 543–556. With permission from Elsevier.)

(Continued)

SPREAD 15.1 THE EPIDERMAL GROWTH FACTOR RECEPTOR (CONTINUED)

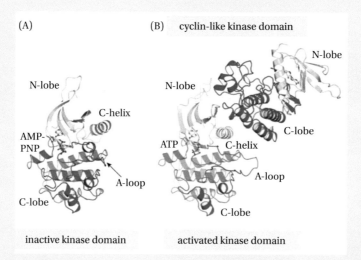

Figure 4 Activation of the kinase domain. (A) Inactive monomeric conformation (PDB 2GS7), where the activation loop (magenta) prevents the catalytically important C-helix (yellow) from reaching the nucleotide in the active site. (B) Active dimeric conformation (PDB 2GS6), in which the activation loop in the "receiver" kinase domain adopts an extended structure that allows the C-helix access to the nucleotide. (Figure reproduced with permission from Ferguson KM [2008] *Annu Rev Biophys* 37: 353–373. Copyright 2008 Annual Reviews.)

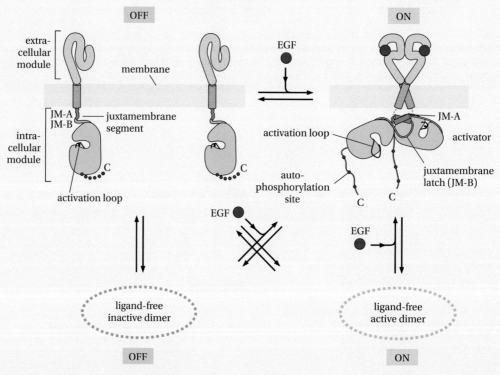

Figure 5 Model for activation of EGFR. Note the structure of the transmembrane and juxtamembrane regions, and the asymmetric dimer formed by the two kinase domains in the activated state. (From Endres NF, Das R, Smith AW et al. [2013] *Cell* 152: 543–556. With permission from Elsevier.)

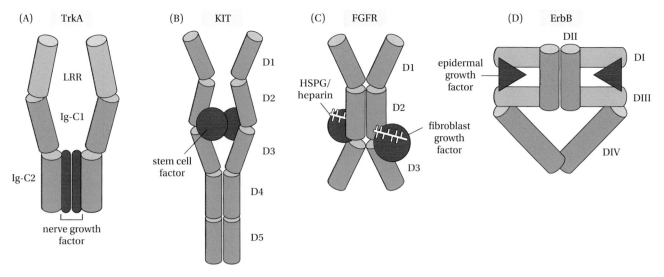

Figure 15.15 Ligand binding can stabilize RTK dimers in at least four different ways to induce signaling. The ligand, shown in red, may provide the entire dimer interface (A) or part of the interface (B), may bind to accessory molecules such as heparin (C), or may induce dimerization indirectly by altering the conformation of the monomer (D). Abbreviations: TrkA, Tropomyosin receptor kinase A; KIT, derived from the gene name encoding a tyrosine kinase; FGFR, fibroblast growth factor receptor; ERB, an epidermal growth factor. (From Lemmon MA, Schlessinger J [2010] *Cell* 141: 1117–1134. With permission from Elsevier.)

15.2.2 Phosphorylation of the "Activation Loop" Activates the Kinase

Regardless of the mechanism whereby binding of ligand triggers or alters dimerization of the extracellular domain, formation or reorganization of the dimer brings the two cytosolic tyrosine kinase domains into active contact. The key event in the activation of the kinase is typically that an "activation loop," which otherwise inhibits the kinase, is removed from the vicinity of the active site (**Figure 15.16**). Activation is usually induced by trans-phosphorylation of a tyrosine residue in the activation loop by the neighboring kinase in the dimer. A simple way by which phosphorylation can be achieved is that the activation loop in each kinase domain may spontaneously dissociate from the active site with some frequency and thereby allow one of the two kinases in the dimer to phosphorylate the activation loop in the other. Once phosphorylation of the activation loops has taken place, the now-activated kinase domain can phosphorylate other tyrosines in the receptor to create the multiple binding sites required to recruit cytosolic signaling proteins. But, there are many different physiological schemes for activating receptor tyrosine kinases. In some RTKs, the key tyrosine that is phosphorylated in the activation step is located in a different part of the cytosolic domain, and in the case of EGFR, phosphorylation is not even required, as described in Spread 15.1.

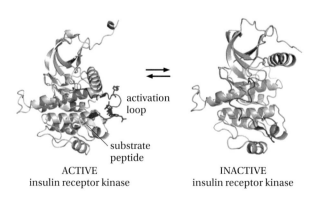

ACTIVE
insulin receptor kinase

INACTIVE
insulin receptor kinase

Figure 15.16 Phosphorylation of an "activation loop" can control receptor kinase activity (PDB 1IR3, 1IRK).

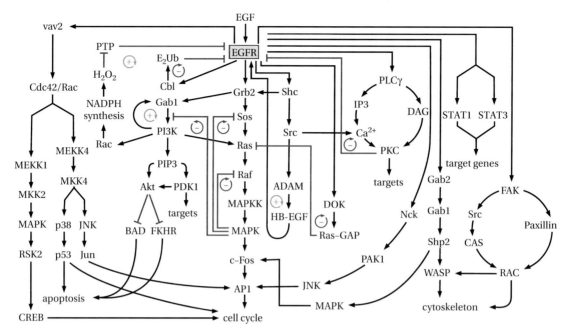

Figure 15.17 The intricacies of the EGFR signaling network. Stimulatory signals are shown in black, inhibitory signals in red. Abbreviations, such as MEKK1, indicate various proteins involved in the signaling pathways. Their definitions are not important for appreciating the complexity of the signaling network. (From Lemmon MA, Schlessinger J [2010] *Cell* 141: 1117–1134. With permission from Elsevier.)

In multicellular organisms, many complex processes, such as cell division and proliferation, are closely coordinated through several signaling networks. The complexity of even a single network, such as that associated with the EGF receptor, boggles the mind (**Figure 15.17**). Presumably, there is a network of some kind for each of the 58 human RTKs. But considering that these networks must coordinate their activities, the full complexity is almost unimaginable. The complexity, however, allows efficient integration of sets of input signals required by multicellular organisms.

15.3 G PROTEIN RECEPTORS TRANSMIT SIGNALS ACROSS THE PLASMA MEMBRANE IN RESPONSE TO HORMONES AND OTHER COMPOUNDS

G protein-coupled receptors—GPCRs—have been studied biochemically for more than 50 years and by pharmacologists for nearly a hundred years (see **Box 15.2**); yet, it is only recently that we have attained a detailed chemical understanding of how they function. GPCRs transmit signals across the plasma membrane of cells in response to the binding of extracellular ligands such as hormones, odorants, and drugs. In the case of the GPCR rhodopsin, the light-sensitive pigment in our eyes, the activating agent is a photon that is absorbed by the cofactor retinal lodged inside the receptor. GPCRs are common drug targets; ~30% of the drugs currently on the market target GPCRs (**Figure 15.18**).

Interestingly, the family of GTP-binding G proteins that are activated by GPCRs was identified before the receptors themselves were shown to exist as separate entities. The first ligand-binding GPCRs (including the β_2 adrenergic receptor; see later) were identified in the 1970s using radioactive ligands and could later be purified by affinity chromatography (Box 15.2). Eventually, an entire signaling pathway could be reconstituted in lipid vesicles from purified

BOX 15.2   RECEPTORS AND THE DISCOVERY OF THE GPCR PROTEIN FAMILY

John Newport Langley (1852–1925) arguably was the first scientist to suggest explicitly the existence of cell-surface receptors. He studied the action of the hormone adrenaline and drugs such as nicotine and curare on skeletal muscle, and concluded that these chemicals likely bind to what he called "receptive substances" on the surface of muscle cells. In 1905, he wrote:

> So we may suppose that in all cells two constituents at least are to be distinguished, a chief substance, which is concerned with the chief function of the cell as contraction and secretion, and receptive substances which are acted upon by chemical bodies and in certain cases by nervous stimuli. The receptive substance affects or is capable of affecting the metabolism of the chief substance.

The idea that there are cell-surface receptors was long controversial, however, and it was not until 1970 that the existence of the first ligand-binding GPCR was demonstrated by Robert Lefkowitz, who used radioactive adrenocorticotropic hormone (ACTH) to show that specific binding sites for the hormone were present in membranes prepared from the adrenal gland (**Figure 1**). Rhodopsin, which in contrast to the ligand-binding GPCRs can be obtained in large quantities from natural sources (e.g., cows' eyes), was purified by Ruth Hubbard in 1954, and its amino acid sequence was determined by the Russian biochemist Yuri Ovchinnikov (1934–1988) in 1982, but it was only with the cloning of the first ligand-binding GPCR in 1986 that the similarity between ligand-binding GPCRs and rhodopsin was discovered (see Figure 15.23). The first x-ray structure of rhodopsin was determined by Palczewski et al. in 2000, and that of a ligand-binding GPCR (the β_2 adrenergic receptor), both with and without bound agonists and in complex with G proteins, by Brian Kobilka's laboratory, starting in 2007.

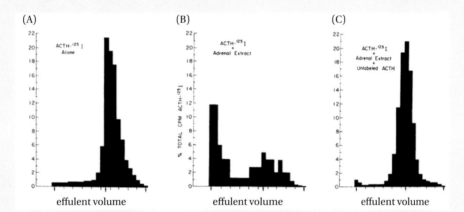

Figure 1 An early study showing the presence of binding sites (i.e., receptors) for adrenocorticotropic hormone (ACTH) using solubilized membranes prepared from mice adrenal tumors. In panel A, radioactive ^{125}I-labeled ACTH in buffer was run on a gel filtration column, fractions were collected, and the radioactivity of each fraction from the column was determined in a scintillation counter. Panel B shows the same experiment, except that ^{125}I-ACTH was incubated with solubilized membranes before the gel filtration. A fraction of ^{125}I-ACTH binds to the receptor and now runs in the void volume. In panel C, non-radioactive ACTH was added in excess to ^{125}I-ACTH and solubilized membranes before gel filtration; the non-radioactive ACTH displaces ^{125}I-ACTH from the receptor, and there is no radioactivity in the void volume. (From Lefkowitz RJ, Roth J, Pricer W et al. [1970] *PNAS* 65: 745–752. With permission from RJ Lefkowitz.)

β_2 adrenergic receptor, purified G protein, and purified effector protein (adenylate cyclase, an enzyme that produces the second messenger cyclic AMP) in the mid-1980s.

Similar biochemical work on rhodopsin was done in parallel to, and independently of, the work on the β_2 adrenergic receptor, but it was not until the gene encoding the β_2 adrenergic receptor was cloned and sequenced in 1986 and could be compared with the sequence of rhodopsin that it became clear that the ligand-activated GPCRs and the light-activated rhodopsins have a common architecture, with seven TM helices (**Figure 15.19**), and signal through G proteins in a similar way.

Attesting to their importance, there are about 800 different GPCRs encoded in the human genome (**Figure 15.20**). Based on sequence similarity, they have been

Figure 15.18 G protein–coupled receptors (GPCRs) are the targets for about 30% of all drugs currently on the market. Other membrane proteins such as ion channels, small-molecule transporters, and integrins account for another 15%. (From Hopkins AL, Groom CR [2001] *Nature Rev Drug Discovery* 1: 727–730. With permission from Springer Nature.)

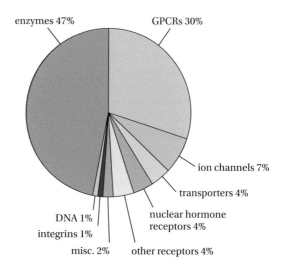

Figure 15.19 The original topology model of rhodopsin published by Y.A. Ovchinnikov in 1982; the amino acid sequence was determined by peptide sequencing, an impressive accomplishment at the time. The cytoplasmic side is up. (From Ovchinnikov YuA [1982] *FEBS Lett* 148: 179–191. With permission from John Wiley and Sons.)

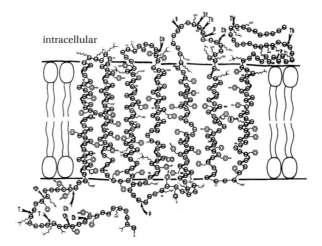

classified into five superfamilies: rhodopsin-like receptors, adhesion receptors, secretin receptors, glutamate receptors, and the Frizzled/TAS2 receptors (there are also other classification schemes in use). The rhodopsin-like receptors constitute the largest superfamily, which is itself subdivided into the α, β, γ, and δ branches. The β_2 adrenergic receptor and the rhodopsins belong to the α class, while the olfactory receptors (which are about half of the human GPCRs) are found in the δ class. GPCRs are ubiquitous in the animal kingdom, as illustrated in **Figure 15.21**.

15.3.1 GPCRs Bind a Vast Variety of Ligands

Because they have a significant externally exposed area with many possible clefts and structural variations among the helices, GPCRs can bind a vast variety of both natural and synthetic ligands, ranging in size from small molecules to oligopeptides. Most GPCRs have a basal activity in the absence of ligand (**Figure 15.22**). Full or partial agonists activate the receptor, while inverse agonists reduce the activity of the receptor. Neutral antagonists do not in themselves alter the basal activity but prevent agonists from binding to and activating the receptor.

A cascade of more than 600 3-D structures of GPCRs—beginning with bovine rhodopsin in 2000 and the β_2 adrenergic receptor in 2007 (**Figure 15.23**)—have revolutionized our understanding of GPCR function. As surmised from the early sequence analyses, the three-dimensional structures of the rhodopsin and the β_2 adrenergic receptor are similar (Figure 15.23C). In addition, the structures of a number of GPCRs with bound ligand are now known. Most ligands are bound deep within the protein (**Box 15.3**). In some cases (e.g., the β_2 adrenergic

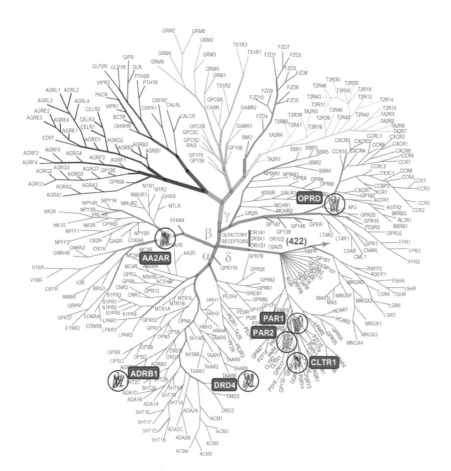

Figure 15.20 The family tree of human GPCRs includes more than 800 protein species. A few receptors of known structure are highlighted (AA2AR: adenosine receptor A2a, OPRD: δ-type opioid receptor, PAR1: proteinase-activated receptor 1, PAR2: proteinase-activated receptor 2, ADRB1: β-1 adrenergic receptor, DRD4: dopamine D4 receptor, CLTR1: Cysteinyl leukotriene receptor 1). The database GPCRdb (https://gpcrdb.org) collects a lot of information on GPCRs. (Adapted from Stevens RC et al. [2013] *Nature* 12: 25–34. With permission from Springer Nature.)

receptor), the binding pocket is open toward the extracellular space; in others (e.g., rhodopsin) it is closed off by extracellular loops that form a lid over the ligand when it binds.

15.3.2 Signaling Begins with Stabilization of an Active Conformation of the GPCR Induced by Agonist Binding

In classical G protein signaling (**Figure 15.24**), signaling is initiated by an agonist that binds to the extracellular binding pocket of its cognate GPCR (Box 15.3), thereby stabilizing a conformation of the receptor with a high affinity for the GDP-bound form of a cytosolic hetero-trimeric G protein composed of α, β, and γ subunits. Association with the receptor weakens the affinity of the G protein for GDP, which is released. The nucleotide-free form of the G protein then binds GTP (which is at a much higher concentration than GDP in the cytosol), which in turn leads to the dissociation of the α subunit and the βγ dimer from each other and from the receptor. In the case of the β_2 adrenergic receptor, the GTP-loaded α subunit activates adenylate cyclase to produce the intracellular second messenger cyclic AMP (cAMP), while the βγ dimer regulates Ca^{2+} channels in the membrane. Finally, GTP is hydrolyzed on the α subunit to GDP and phosphate, allowing α to reassemble with βγ in preparation for another round of signaling.

The ligand-bound form of the receptor can activate multiple G proteins, and a single GTP-loaded α subunit can promote the formation of a large number of cAMP molecules. The signal initiated by a single ligand molecule is thus strongly amplified and can have major effects on the state of the target cell.

Initiating a signal through a GPCR is only one side of the coin, however. Continued exposure to ligand eventually induces desensitization, a process

Figure 15.21 GPCRs are found in high numbers throughout the eukaryotic kingdom. The number of GPCRs belonging to different sequence families in each organism is indicated. (From Nordström KJV, Sällman Almén M [2011] *Mol Biol Evol* 28: 2471–2480. With permission from Oxford University Press.)

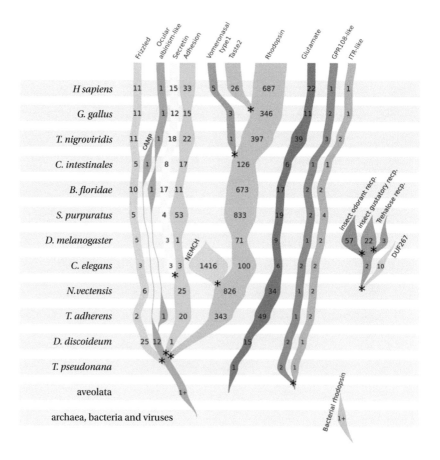

Figure 15.22 Pharmacological classification of receptor ligands. Basal or constitutive activity is the physiological activity in the absence of a ligand. Agonists bind to and activate the receptor, eliciting a physiological response higher than the basal activity. Partial agonists activate a receptor but with a smaller maximal response compared with a full agonist. Neutral antagonists bind to receptors without elevating the response above the basal level. Inverse agonists lower activity compared with the basal response. (From Tate CG [2012] *Trends Biochem Sci* 37: 343–352.)

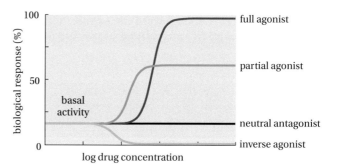

through which the receptor becomes less sensitive to ligand or is removed altogether from the cell surface by internalization. Desensitization is initiated within seconds of receptor activation by GPCR kinases that phosphorylate sites in the cytosolic part of the receptor (**Figure 15.25**). A second protein, arrestin (vertebrates have four different arrestin subtypes), then binds to the phosphorylated receptor to prevent further interactions with G proteins and to promote the receptor's internalization and ultimate degradation via the cell's endocytic machinery. Arrestins can also serve to initiate intracellular signaling pathways independently of the G proteins.

15.3.3 β_2 Adrenergic Receptor and Bound G Proteins Change Conformation in Response to Ligand Binding

The structure and function of rhodopsin (see later) and the β_2 adrenergic receptor are the best understood among the GPCRs. Understanding the β_2 adrenergic receptor is important, because this receptor underlies many physiological

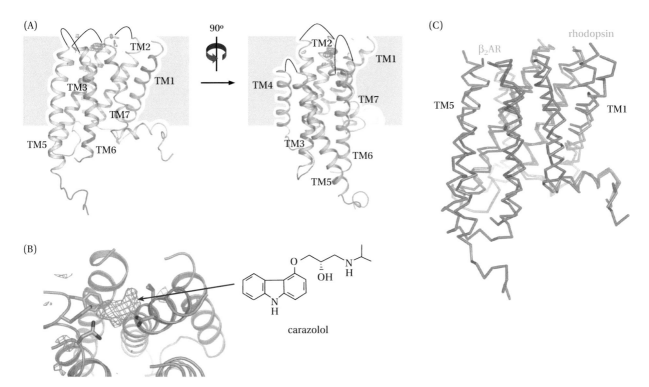

Figure 15.23 The structures of rhodopsin and β_2 adrenergic receptor (β_2AR) are very similar, as expected from sequence comparisons. (A) The structure of β_2 adrenergic receptor (PDB 2R4R), which was the first structure determined of a GPCR other than rhodopsin. The grey bands represent the approximate boundaries of the lipid bilayer. (B) The location of carazolol (a beta-blocker, green mesh), which is a high-affinity antagonist of the receptor. (C) Structural comparison of the α-carbon backbone structures of β_2AR and rhodopsin (PDB 1GZM). (From Rasmussen S, Choi HJ, Rosenbaum D et al. [2007] *Nature* 450: 383–387. With permission from Springer Nature.)

functions controlled by the sympathetic nervous system in response to stimulation by adrenaline (epinephrine), an important one being the relaxation of smooth muscle causing vasodilation. Robert Lefkowitz and Brian Kobilka were awarded the Nobel Prize in Chemistry in 2012 for their work on the β adrenergic receptor.

Structures of the β_2 receptor have been determined with both agonists and antagonists bound to the protein. Interestingly, binding of agonist alone is not sufficient to stabilize the receptor in its active conformation; instead, the active state is stable only when both agonist and G protein are bound to the protein simultaneously.

The structure of the active, ligand- and G protein–bound β_2 receptor is shown in **Figure 15.26**, and the intricate work required to obtain this first-of-its-kind structure is described in Box 15.4. The main interaction between the receptor and the G protein involves the C-terminal α helix from the α subunit of the G protein (G_α), which penetrates into a crevice in the cytoplasmic part of the receptor. The crevice is not seen in the inactive receptor and is created by an extension and slight movement of TM5 and a large outward movement of the cytoplasmic part of TM6 when the receptor switches between inactive and active conformations (**Figure 15.27**). In contrast, the inactive and active conformations are very similar in the extracellular half of the receptor.

The G protein also undergoes a major structural change when it binds to the receptor (**Figure 15.28**). In particular, the α-helical (AH) domain of the α subunit is displaced from its position in the free G protein. This perturbs the nucleotide binding pocket and leads to a decrease in its affinity for GDP. The C-terminal helix of the α subunit also shifts in position as it lodges itself in the receptor.

How does agonist binding shift the equilibrium between the different receptor conformations toward the active one that can bind the G protein? A possible mechanism, based on a comparison of the structures of an inactive receptor

BOX 15.3 GPCRS HAVE CHARACTERISTIC BINDING POCKETS FOR EACH LIGAND

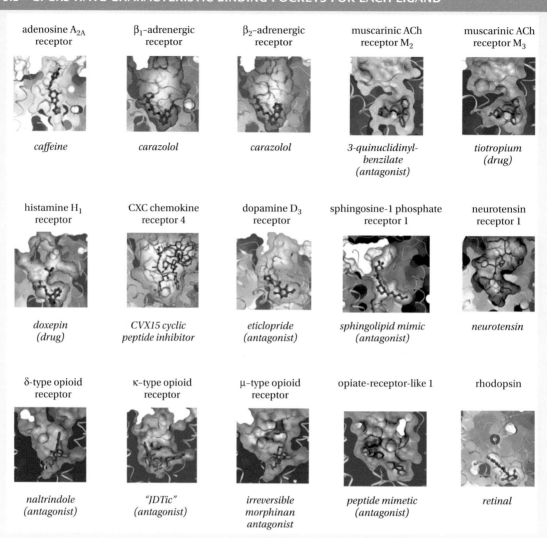

Adapted from Stevens et al. [2013] *Nature Rev Drug Discov* 12:25-34. With permission from Springer Nature.

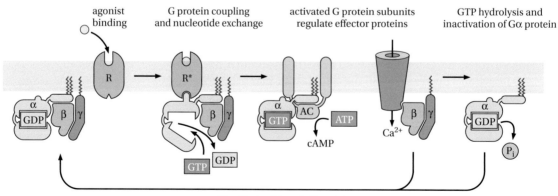

Figure 15.24 Activation of G proteins by a GPCR. Binding of agonist stabilizes a conformation of the cytoplasmic part of the GPCR to which the G protein can bind. This leads to release of bound GDP from the G protein, which is replaced by GTP. GTP binding induces disassociation of the G protein α and βγ subunits from the receptor and activation of downstream targets such as adenylate cyclase and ion channels. Hydrolysis of bound GTP in the α subunit resets the system. (From Rasmussen S, DeVree B, Zou Y et al. [2011] *Nature* 477: 549–555. With permission from Springer Nature.)

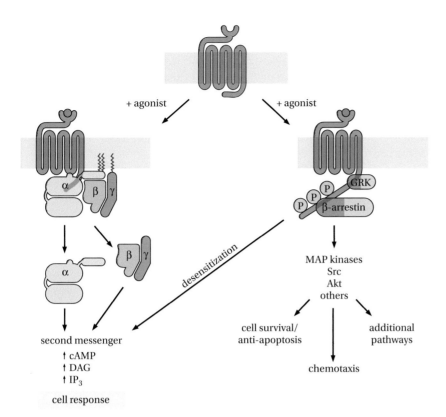

Figure 15.25 Activated GPCRs are phosphorylated by GPCR kinases (GRKs). Binding of arrestin to the phosphorylated receptor leads to desensitization. Arrestin can also initiate intracellular signaling in its own right. (From Lefkowitz RJ [2007] *Acta Physiologica* 190: 9–19. With permission from John Wiley and Sons.)

with an inverse agonist in the binding pocket and an agonist-bound, active form stabilized by a nanobody (a single-chain camelid antibody), is shown in **Figure 15.29**. The key observation is that Ser203 and Ser207 in TM5 form multiple hydrogen bonds to the agonist but not to the inverse agonist. The formation of these hydrogen bonds leads to an inward shift in the position of this part of TM5 by 1.5–2 Å, which in turn is only possible if Ile121 shifts upwards to allow Pro211 in TM5 and Phe282 in TM6 to come closer together. The repositioning of Phe282 into the space vacated by Ile121 is coupled to a rotation of TM6 that finally causes the outward movement of its cytoplasmic end required to form the crevice into which the Gα C-terminal helix binds. These very subtle differences between the inactive and active states of the receptor imply that the two states are nearly equally stable, which is also reflected in the fact that the β_2 receptor has a fairly large basal activity.

15.3.4 Rhodopsin's Photocycle Underlies Vision

Rhodopsin, one of the best-studied GPCRs, is a key molecule in vision. It is the low-light detector in our retina and therefore, must combine high sensitivity and low background activity. A single rod cell in the dark produces spontaneous signals at the remarkably low rate of about once every 100 seconds. This extraordinarily low rate is due to the thermal stability of rhodopsin's 11-*cis* retinal, whose activation energy barrier for spontaneous conversion to the activated all-*trans* form exceeds 40 kcal mol⁻¹, meaning that an individual rhodopsin molecule would only be activated once every 500 years (!) if kept in the dark at room temperature. Although this energy barrier is high for thermal excitation, it is modest for photons in the visible spectrum, which have energies equivalent to 60–80 kcal mol⁻¹ depending upon wavelength.

Different versions of rhodopsin are found in the rod and cone cells of the retina (**Figure 15.30**), which are specialized nerve cells that form chemical synapses with so-called bipolar cells. The rod cells are much more numerous and light-sensitive than the cones but are not sensitive to color. The cones serve as

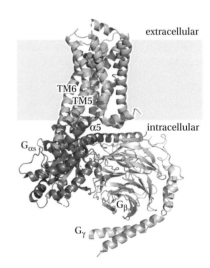

Figure 15.26 Structure of a β_2 adrenergic receptor–G protein complex (PDB 3SN6). The receptor is in green, $G_{\alpha s}$ in blue, G_β in light orange, and G_γ in light blue. The agonist is in spacefill.

BOX 15.4 DETERMINING THE FIRST STRUCTURE OF A GPCR–G PROTEIN COMPLEX

A landmark achievement in the GPCR field was the determination of a three-dimensional structure of a GPCR–G protein complex by Brian Kobilka and his collaborators in 2011. The work behind this structure provides a nice illustration of the difficulties often encountered when one attempts to crystallize membrane proteins and the various kinds of biochemical tricks that researchers have come up with to increase the chances of success.

The basic procedure that was eventually successful is shown in **Figure 1**. The first problem that had to be solved was to produce sufficient amounts of β_2AR and G protein. This was achieved using insect cells. The proteins then each had to be solubilized and purified. This required much optimization to find the best detergent and the

proper purification protocols. To stabilize the β_2AR against denaturation, a suitable high-affinity agonist was used. When mixing β_2AR and G protein, it was realized that any GTP or GDP remaining in the preparations destabilizes the receptor–G protein complex; therefore, the enzyme apyrase had to be added in order to hydrolyze all the nucleotides.

The initial solubilization was done in DDM, but this detergent did not yield good crystals. Instead, more than 50 different detergents were tested until a good one was found (it happened to be a newly developed amphiphile).

Electron microscopy imaging of the solubilized β_2AR–G protein complex showed large micelles in which most of the protein would be inaccessible for making crystal contacts. In order to increase the mass of protein located outside the

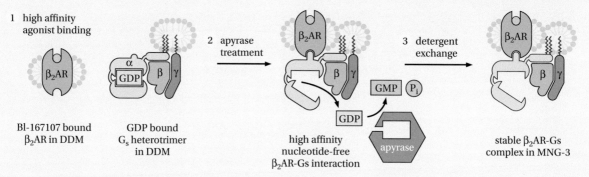

Figure 1 Preparation of a stable receptor–G protein complex suitable for x-ray crystallography. (From Rasmussen SGF et al. [2011] *Nature* 477: 549–555. With permission from Springer Nature.)

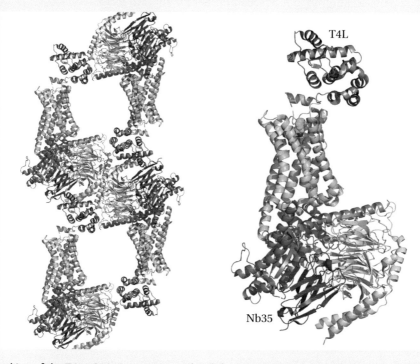

Figure 2 Crystal packing of the T4L–β2AR–G protein complex (left; PDB 3SN6) and the structure of a single complex (right). Note the T4 lysozyme (T4L in orange) and the nanobody (Nb35 in red). The agonist BI-167107 is shown in spacefill.

(Continued)

micelles, monoclonal antibodies were raised against the protein and added to the purified complex, but none gave sufficiently good crystals.

As an alternative to antibodies, T4 lysozyme, a small water-soluble protein, was cloned into the floppy N-terminal tail of β_2AR at a number of different positions. All these constructs were analyzed by electron microscopy to find one with a stable orientation of T4L relative to receptor.

The full T4L–β_2AR–G protein complex was also analyzed by electron microscopy, and it was seen that the Gα subunit

was not held in a stable position relative to the rest of the complex. As a remedy, single-chain antibodies (so-called nanobodies) against the whole complex were raised in llamas, and one nanobody was found that maintained the complex in a stable state.

Finally, sufficiently good crystals could be grown in lipid cubic phase, and the structure could be solved after hundreds of these crystals had been subjected to x-ray diffraction measurements at a synchrotron beamline (**Figure 2**).

our red, green, and blue light detectors, but only at higher illumination levels; we cannot perceive color in a dark forest. A specialized part of the rod cells, the rod outer segment (ROS), contains tightly stacked membrane structures called disk membranes that are stuffed full of rhodopsin molecules: in human rod cells, one ROS contains some 10^3 disks, and each disk contains some 10^5 rhodopsins. Thanks to the essentially unlimited availability of cows' eyes from slaughterhouses, bovine rhodopsin was first purified in 1954 by Ruth Hubbard, who was the first woman to hold a tenured professorship in biology at Harvard University.

The light-absorbing moiety in rhodopsin is 11-*cis* retinal (vitamin A), which is attached to Lys296 on helix TM7 through a protonated Schiff base ($=NH^+-$) (**Figure 15.31A**). Upon absorption of a photon, the rhodopsin-bound retinal undergoes a photo-induced isomerization to the all-*trans* isomer, putting strain into the system that is eventually relieved by conformational changes in the protein that can be detected spectroscopically as the photocycle (Figure 15.31B). The most important state in the photocycle is the Meta II state, which binds and activates the G protein (transducin). Finally, the retinal is released from the protein (the empty form of the protein is called opsin), reduced by short-chain alcohol dehydrogenases to the alcohol, and then converted back enzymatically to 11-*cis* retinal that can recombine with opsin to complete the cycle. The 11-*cis* retinal isomer acts as an inverse agonist, lowering the intrinsic activity of rhodopsin compared with opsin. The protein is strongly structured

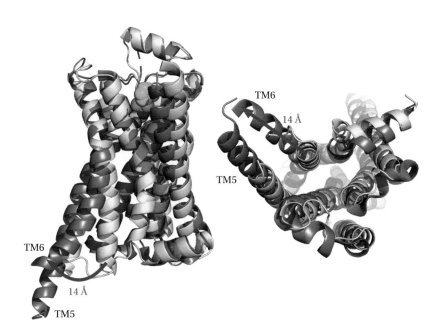

Figure 15.27 Overlay of inactive (yellow; PDB 2RH1) and active (blue; PDB 3SN6) conformations of the β_2 adrenergic receptor. The cytoplasmic end of TM6 moves 14 Å between the two structures (red arrow), opening up a crevice for G protein binding. The inactive form has a bound inhibitor, carazolol (in spacefill).

Figure 15.28 Overlay of the $G_{\alpha s}$ subunit from the β_2 adrenergic receptor/G protein (green/blue) complex (PDB 3SN6) and from an isolated $G_{\alpha s}$ subunit (grey; PDB 1AZT) with a bound GTP analog. Note how the $G_{\alpha s}$AH domain swings up in the receptor-bound form.

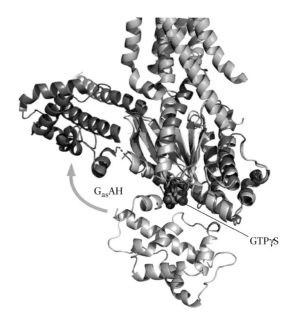

to prevent isomerization of retinal, which happens readily and spontaneously in free retinal, requiring the large energy from an absorbed photon to provide the strain leading to a signal. Thus, the background is very low, and the photon detection can be very sensitive.

The high barrier to thermal activation, noted earlier, is essential because of the extraordinary sensitivity of the cyclic-nucleotide cascade initiated by light activation of rhodopsin (**Figure 15.32A**). A single rhodopsin activated by a single photon is sufficient to cause a change in the membrane conduction of the rod cell as a result of the cascade (Figure 15.32B). The amplification due to the cascade is impressive: a single photoactivated rhodopsin catalyzes the activation of ~50 transducin molecules, each of which can stimulate a single phosphodiesterase that can hydrolyze 10^3 cGMP molecules per second to GMP. Thus, a single activated rhodopsin can cause the hydrolysis of ~10^5 molecules of cGMP per second. The loss of cGMP through the hydrolysis cascade causes cGMP-gated channels that conduct Na^+ and Ca^{2+} to close; these closures cause hyperpolarization of the rod by increasing G_K/G_{Na} (see Box 11.2), closure of Ca^{2+}

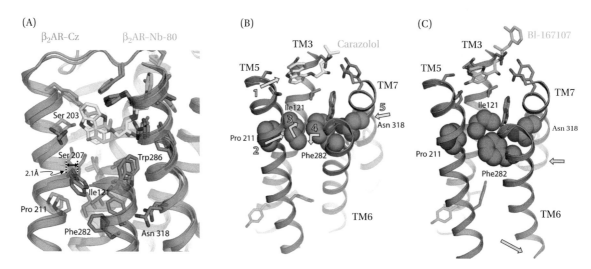

Figure 15.29 Conformational changes around the ligand-binding pocket of the β_2 adrenergic receptor. (A) Overlay of an inactive, inverse agonist–bound structure (blue; PDB 2RH1) and an active, nanobody-stabilized and agonist-bound (orange; PDB 3POG) structure. (B) Packing interactions stabilizing the inactive state. (C) The packing of Ile121 and Pro211 is destabilized by the inward movement of TM5 during activation. (From Rasmussen S, Choi H-J, Fung JJ et al. [2011] *Nature* 469: 175–180. With permission from Springer Nature.)

(A)

(B)

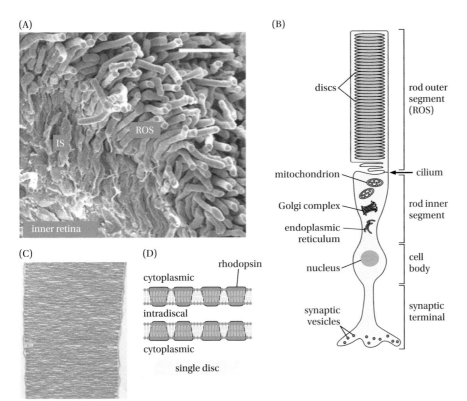

(C)

(D)

Figure 15.30 Organization of the retina. (A) Rod cells in a mouse retina visualized by scanning electron microscopy. (B) Schematic representation of a rod cell. (C) Disk membranes visualized by electron microscopy. (D) Packing of rhodopsin in the disk membrane. (A,C, Liang Y, Fotiadis D, Maeda T et al. [2004] *J Biol Chem* 279: 48189–48196. Published under CC BY 4.0. B, D, Palczewski K [2011] *Annual Review of Biochemistry* 75: 743–767. With permission.)

channels at the synapse, and consequently, a reduction of neurotransmitter release to the bipolar cells.

Thanks to the environment created by the protein around the retinal, the quantum yield of 11-*cis* retinal is as high as 0.67; i.e., a photon impinging on the molecule has a 67% probability of triggering the retinal 11-*cis* to all-*trans* transition. The high sensitivity is partially caused by the orientation of the retinal by the rhodopsin plus the disks, keeping the excitation dipole perpendicular to the path of the light. Another example of clever chemistry by evolution!

The structure of rhodopsin is known in the dark, inactive state, and a structure of opsin with the G_α C-terminal helix from transducin bound to the protein has also been determined (**Figure 15.33**). Just as seen for the β_2 adrenergic receptor, activation leads to an outward movement of TMH5 and 6, opening a crevice for binding of the G_α C-terminal helix.

15.3.5 Crystal Structures Reveal General Principles of GPCR Activation

From what we know based on the crystal structures and sequence comparisons, the vast superfamily of GPCRs seems to operate on a few, rather simple general principles (**Figure 15.34**). The extracellular half of the receptors constitutes the ligand-binding module, while the intracellular half is the signaling module. The ligand-binding module has adapted over evolutionary times to bind many different ligands; consequently, it is characterized by high sequence diversity and many differently shaped binding pockets. On the other hand, the binding pockets are embedded within a rather static structural framework, which undergoes only small conformational changes between inactive and active conformations of the receptor.

As one might expect, the signaling module is more conserved between different receptors and has evolved to undergo larger conformational changes during activation. The interactions with G proteins, GPCR kinases, and arrestins are probably quite similar in all cases, with the outward movement of TM5 and 6 creating a crevice for G_α binding that provides a common mechanistic

(A)

(B)

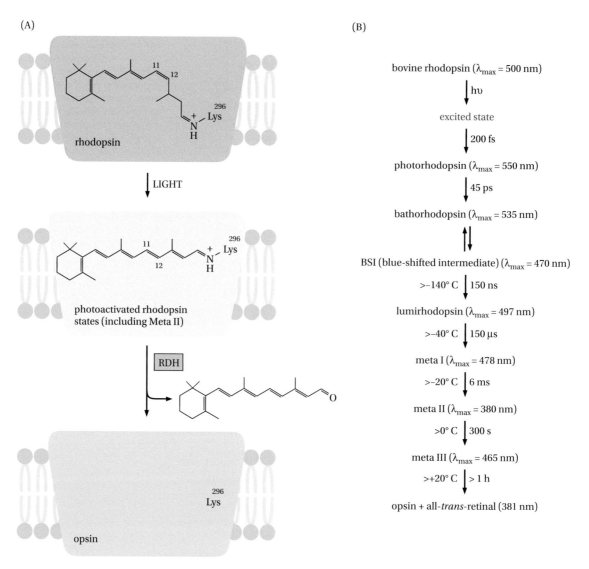

bovine rhodopsin (λ_{max} = 500 nm)

↓ hυ

excited state

↓ 200 fs

photorhodopsin (λ_{max} = 550 nm)

↓ 45 ps

bathorhodopsin (λ_{max} = 535 nm)

⇅

BSI (blue-shifted intermediate) (λ_{max} = 470 nm)

>-140° C ↓ 150 ns

lumirhodopsin (λ_{max} = 497 nm)

>-40° C ↓ 150 µs

meta I (λ_{max} = 478 nm)

>-20° C ↓ 6 ms

meta II (λ_{max} = 380 nm)

>0° C ↓ 300 s

meta III (λ_{max} = 465 nm)

>+20° C ↓ >1 h

opsin + all-*trans*-retinal (381 nm)

Figure 15.31 Rhodopsin chemical changes during excitation. (A) Conformational changes in the retinal. (B) Photocycle intermediate states defined by optical spectroscopy. (A, Palczewski K [2011] *Annual Review of Biochemistry* 75: 743–767. With permission. B, Shichida Y, Imai H [1998] *Cell Mol Life Sci* 54: 1299–1215. With permission from Springer.)

principle. Because there are ~20 different G proteins expressed in human cells, and different receptors bind different G proteins with different affinities, there must be selectivity also in the intracellular half of the receptor, albeit more limited than in the extracellular half.

Different GPCRs have different basal activities in the absence of ligand, and the relative stabilities between inactive and active states in the presence of ligand or G protein also differ between receptors. A conceptual energy landscape that describes the conformational transitions between inactive states (R), various possible metastable intermediate states (R′, R″), and one or more active states (R*) serves to illustrate this point (**Figure 15.35**). The receptor alone has some small probability to occupy the active state(s) depending on the energy difference ΔE between R and R*. In the presence of agonist, ΔE is reduced. For the β_2 receptor, E(R) is smaller than E(R*), and the conformation of the receptor–agonist complex is similar to that of the receptor alone. For rhodopsin, E(R*) < E(R), meaning that the state with an open binding crevice for the G_α C-terminal peptide is the most stable one when the retinal is in the activated all-*trans* conformation. Finally, with both agonist and G protein present, E(R*) <<

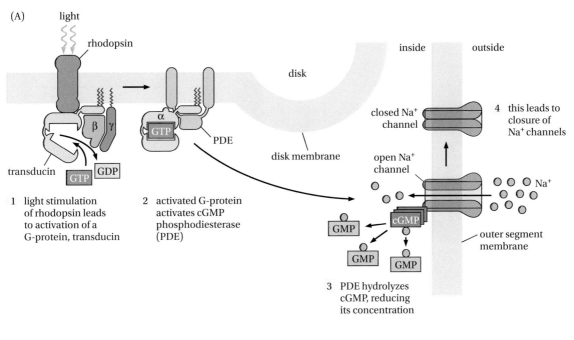

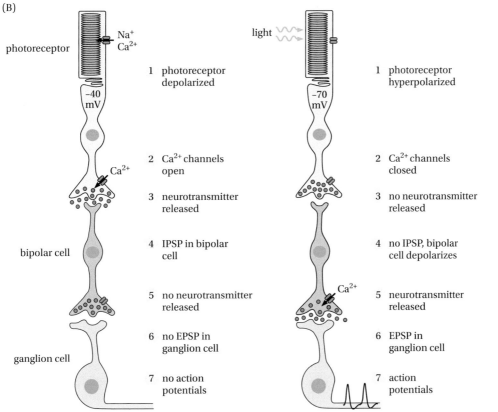

EPSP: excitatory postsynaptic potential

IPSP: inhibitory postsynaptic potential

Figure 15.32 Signaling by photoactivation of rhodopsin in ROS membranes. (A) Schematic of signaling via the cyclic GMP (cGMP) cascade that leads to closing of cGMP-gated channels. These closures cause the resting membrane potential to become more negative (hyperpolarized). (B) Schematic representation of signaling between the photoreceptor, intermediary bipolar cells, and ganglion cells that transmit signals to higher levels of the brain for interpretation.

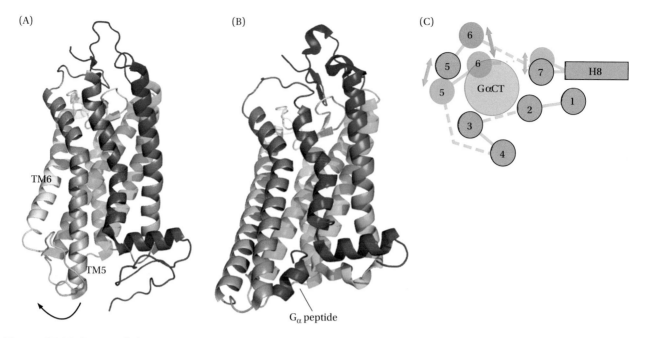

Figure 15.33 Structural changes that occur during activation of a GPCR. Crystal structures of inactive dark-state rhodopsin (panel A; PDB 1U19) and opsin bound to the Gα C-terminal peptide from transducin (panel B; PDB 3DQB). The schematic view from the cytoplasmic side in panel C shows how the movement of TM5 and TM6 creates a binding site for the Gα peptide (GαCT). (Figure reproduced with permission from Smith SO [2010] *Annu Rev Biophys* 39: 309–328. Copyright 2010 Annual Reviews.)

E(R), and the active agonist–receptor–G protein complex is the most stable species. Binding of an inverse agonist, in contrast, makes E(R) << E(R*) even in the presence of G protein and therefore, reduces the basal activity of the receptor.

As in the tyrosine kinase receptors, signals from GPCRs are integrated physiologically to give a broad range of responses, and the integration occurs both within the cell (Figure 15.1) and at higher organizational levels. The retina is a good example (Figure 15.32).

15.4 CADHERINS AND INTEGRINS MEDIATE MECHANICAL INTERACTIONS WITH NEIGHBORING CELLS AND THE EXTRACELLULAR MATRIX

In multicellular organisms, every cell must be able to interact with its neighbors, both directly by physical contact and indirectly by signaling molecules such as hormones. Cells comprising tissues are physically connected and organized in different ways depending upon tissue function and architecture. Epithelial and connective tissues, found together most prominently in the gut, represent opposite extremes (**Figure 15.36A**): the former is organized by strong cell–cell contacts, while in the latter, individual cells are dispersed in the extracellular matrix (ECM), composed of a variety of polymers, including proteoglycans (e.g., heparin sulfate), polysaccharides (e.g., hyaluronic acid), and proteins (e.g., collagen, elastin, and fibronectin). Depending on its precise composition, the ECM can help build tissue as different as teeth, bone, tendons, the cornea, or the basal lamina in epithelia.

Different kinds of cell-surface receptors mediate the two kinds of interactions and allow the cell to respond to mechanical forces from the surrounding tissue.

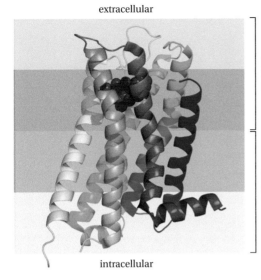

extracellular

ligand binding
higher sequence diversity
smaller conformational changes

signalling
lower sequence diversity
larger conformational changes

intracellular

Figure 15.34 Distribution of sequence diversity and conformational changes in GPCRs (PDB 3PBL). The outward-facing, ligand-binding part is characterized by high sequence diversity (to allow binding of many different ligands) and small conformational changes upon ligand binding. The inward-facing, G protein–binding part is more conserved and undergoes larger conformational changes upon ligand binding.

The most important class of receptor for cell–cell adhesion is the cadherins, while for cell–ECM interactions, it is the integrins; both are single-span plasma membrane proteins that attach, via adaptor proteins, to the intracellular actin or intermediate cytoskeletal filament networks and to extracellular partner proteins on neighboring cells or in the ECM (Figure 15.36B).

15.4.1 Cadherins Mediate Cell-Cell Contacts

In a classic experiment performed in 1955, Philip Townes and Johannes Holtfreter showed that disaggregated cells isolated from an early amphibian embryo could reassemble spontaneously into a structure that resembled the original embryo (**Figure 15.37**). We now know that this is because cell type–specific cadherin molecules (and other cell-surface receptors) interact among themselves to make cells of the same type adhere to each other. Cadherin-mediated cell–cell interactions are important not only for maintaining tissue integrity but also during embryonic development, where groups of cells can turn on and off specific cadherins to steer the continuous remodeling of the growing embryo.

The human genome contains ~120 cadherin superfamily members. Cadherins are characterized by an extracellular domain that contains two or more extracellular cadherin (EC) repeats. As a rule, cadherins mediate homophilic cell–cell interactions; i.e., only cadherins of the same type bind to each other. Cadherins bind to each other via their N-terminal EC repeats. Two binding modes have been seen by x-ray crystallography (**Figure 15.38**): S-dimers, where a Trp residue in each of the two EC domains inserts into a pocket in the other, and X-dimers, where EC domain linker segments interact between the two cadherins. Ca^{2+} can bind in the linker segments; in the absence of Ca^{2+}, the structure becomes floppy, and the binding affinity is decreased.

Cadherin dimers are only weakly bound, with dissociation constants in the micromolar range. However, the dimers can form large and very stable clusters called adherens junctions, containing thousands of cadherin molecules. On the intracellular side, cadherins interact with the cytoskeleton—mainly the actin filaments—via adaptor proteins such as the catenins (**Figure 15.39**). Adherens junctions can respond to forces acting on them from neighboring cells by increasing their size, anchoring to more actin filaments, and hence, balancing the forces across the junction; an example of mechanotransduction.

receptor
receptor + agonist
receptor + agonist + G protein

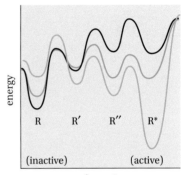

energy

R   R′   R″   R*

(inactive)   (active)

conformation

Figure 15.35 A schematic "energy landscape" for the β_2 adrenergic receptor. The simultaneous binding of agonist and G protein is required to stabilize the active state of the receptor. (From Rosenbaum DM et al. [2011] *Nature* 469: 236–240. With permission from Springer Nature.)

(A)

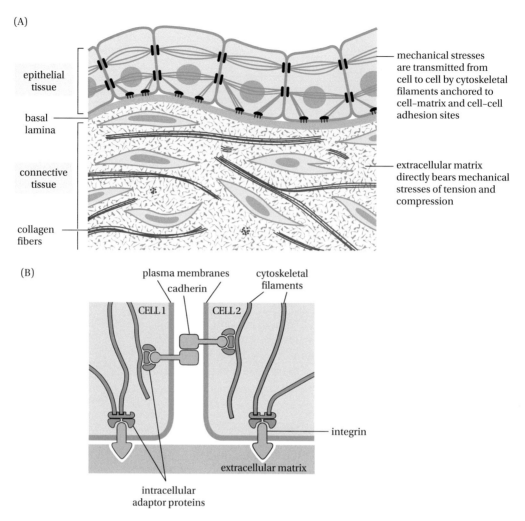

epithelial tissue

basal lamina

connective tissue

collagen fibers

mechanical stresses are transmitted from cell to cell by cytoskeletal filaments anchored to cell–matrix and cell–cell adhesion sites

extracellular matrix directly bears mechanical stresses of tension and compression

(B)

plasma membranes

cadherin

cytoskeletal filaments

CELL 1

CELL 2

integrin

extracellular matrix

intracellular adaptor proteins

Figure 15.36 Stabilization of animal tissues. (A) Tissues are stabilized by cell–cell contacts and cell–matrix interactions. (B) The main mediators of cell–cell and cell–matrix contacts are the cadherins and the integrins, respectively. (From Alberts B, Johnson A, Lewis JH et al. [2014] *Molecular Biology of the Cell*, 6th Edition. Garland Science, New York. With permission from W. W. Norton.)

15.4.2 Integrins Mediate Two-Way Signaling between Cells and the Surrounding Extracellular Matrix

Integrins were first identified through experiments such as those described in **Figures 15.40** and **15.41**, which were performed by Alan Horwitz and collaborators. Single immature muscle cells (myoblasts) cultured in a plastic Petri dish generally adhere to and spread out upon the bottom of the dish (Figure 15.40A), but if monoclonal antibodies against cell-surface receptors are added to the culture dish, the cells detach and become spheroidal (Figure 15.40B). Molecules on the outer surface of a cell can cause rearrangement of the cell's interior structure. This experiment was a critical step toward identifying a particularly important class of cell-surface receptors, called integrins, that mediate direct interactions between cells and the surrounding extracellular matrix and are involved in, e.g., regulation of the cell cycle and the intracellular cytoskeleton.

An important clue to the identity of the surface receptors came from electron micrographs of cultured hamster embryo fibroblasts that showed intracellular microfilaments connecting to extracellular fibronectins at the cell membrane

(Figure 15.41). The fibronectin filaments appeared to be continuations of the microfilaments, as though the filaments just passed through the membrane. However, fibronectin antibodies bound selectively only to the outer fibronectins, meaning that there must be an intervening protein in the cell membrane that somehow connected the two types of filaments. Using the antibodies that caused myoblast cells to detach from culture-dish surfaces and round up (Figure 15.40), the transmembrane integrin proteins were eventually purified, cloned, and sequenced.

Binding of extracellular ligands to integrins is critical for promoting growth of certain cell types that need to be anchored to a substratum. Integrins can also stimulate receptor tyrosine kinases and thereby influence a host of intracellular signaling pathways.

15.4.3 Two-Way Signaling Mediated by Integrins Allows Cells to Shape the Extracellular Matrix, and Vice Versa

Compared with other cell-surface receptors, integrins have the interesting property that they can be controlled by inside-out activation while carrying out outside-in signaling; that is, they are bidirectional signaling devices that help integrate the extra- and intracellular environments. The mechanistic basis for bidirectional signaling is that binding of intracellular ligands to integrins can modulate the affinity for extracellular ligands, and vice versa.

Integrins are heterodimers composed of an α and a β subunit, and attach to ECM proteins such as fibronectin on the extracellular side and to actin filaments (or sometimes to intermediate filaments) via adaptor proteins such as talin and vinculin on the cytoplasmic side (**Figure 15.42**). Both the α and β subunits are single-span plasma membrane proteins with short intracellular tails and large extracellular parts composed of multiple domains. In vertebrates, 18 different types of α subunit and 8 different types of β subunit can combine into at least 24 different αβ heterodimers.

Inside-out activation of integrins proceeds through a series of large conformational transitions in the extracellular part that depend on the binding of intracellular signaling molecules to the cytoplasmic tails and the TM helices. The relative stabilities of the different conformational states of the αβ heterodimer depend on which intra- and extracellular ligands are bound, and the equilibrium between active and inactive conformations shifts in response to the different binding events.

The two TM helices in the α and β subunits dimerize in the absence of ligands, but the dimer is easily disrupted by binding of intracellular ligands to the cytoplasmic tails. The TM helix dimer is held together by two sets of interactions: the outer and inner membrane clasps (**Figure 15.43**). Intracellular ligands such as talin can insert between the cytoplasmic ends of the TM helices and break up the clasps, shifting the equilibrium toward conformations where the two TM helices no longer interact.

The ligand-binding domain in the extracellular part is located in the headpiece at the tip of the αβ dimer (**Figure 15.44**). The headpiece can adopt either closed conformations, which cannot bind ligand, or open, ligand-binding conformations. The open conformation of the headpiece can only form if the two TM helices do not interact, and the closed form only if the TM helices do interact. In inside-out activation, binding of a cytoplasmic effector protein such as talin to the cytoplasmic end of the integrin stabilizes the open conformation, increasing the affinity for extracellular ligands. Conversely, in outside-in signaling, the presence of an extracellular ligand that binds to the open form of the headpiece will favor conformations where the TM helices do not interact and the cytoplasmic tails are far apart, facilitating binding to cytoplasmic effector proteins (Figure 15.44).

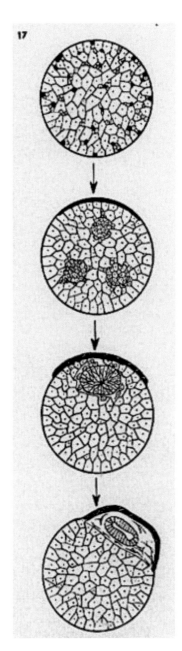

Figure 15.37 Resilience and specificity of cell–cell contacts. In this experiment from 1955, an amphibian embryo was disassembled into its constituent mesoderm, neural plate, and epidermal cells, and the cells were remixed. Over time, the cells sorted themselves into a structure similar to the original embryo. (From Holtfreter J, Townes PL [1955] *J Exp Zool* 128: 53–120. With permission from John Wiley and Sons.)

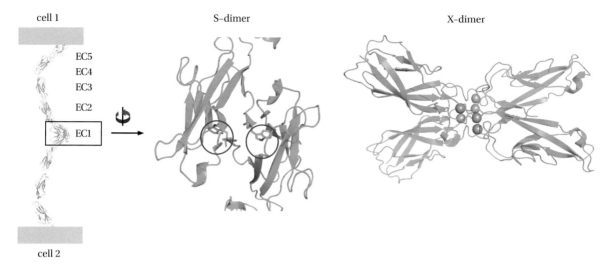

Figure 15.38 E-cadherin dimerizes via its most N-terminal EC repeats into S-dimers (PDB 3Q2V) and X-dimers (PDB 3NLH). Red circles indicate Trp residues that mediate S-type dimerization. Ca²⁺ ions in the linker regions between the EC repeats are shown as grey balls. (From Vae Priest A, Shafraz O, Sivasankar S [2017] *Exp Cell Res* 358: 10–13. With permission from Elsevier.)

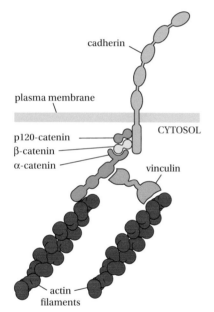

Figure 15.39 Cadherins bind to actin filaments via adaptor proteins such as catenins and vinculins.

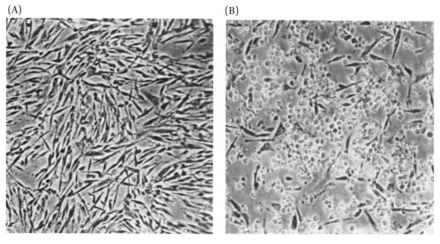

Figure 15.40 Receptors of a then (1982) unknown type affect the shapes of cultured myoblast cells. Monoclonal antibodies against the receptors were applied to cultured cells that naturally attach to the culture dish. (A) In the absence of antibodies, the cells are elongated. (B) In the presence of antibodies, the cells detach from the dish and round up. (From Neff NT, Lowrey C, Decker C et al. [1962] *J Cell Biol* 95: 654–555. With permission from Rockefeller University Press.)

Figure 15.41 Electron micrograph of hamster embryo fibroblasts showing the continuity between intracellular microfilaments and extracellular filaments of fibronectin. (From Singer II [1979] *Cell* 16: 675–685. With permission from Elsevier.)

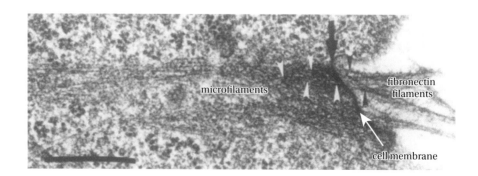

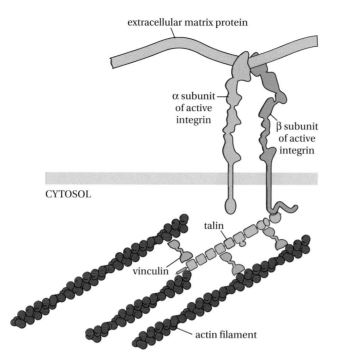

extracellular matrix protein

α subunit
of active
integrin

β subunit
of active
integrin

CYTOSOL

talin

vinculin

actin filament

Figure 15.42 Integrins connect the intracellular cytoskeleton with the extracellular matrix.

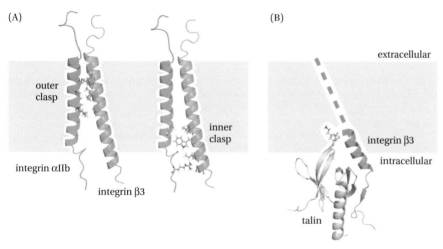

(A)

outer
clasp

integrin αIIb

integrin β3

(B)

inner
clasp

extracellular

integrin β3

intracellular

talin

Figure 15.43 Structures and interactions of the integrin αIIb–β3 transmembrane domains. (A) NMR structure (PDB 2K9J). The outer and inner "clasps" are indicated. (B) Talin (PDB 2H7E) binds to integrin β and breaks the α–β interaction.

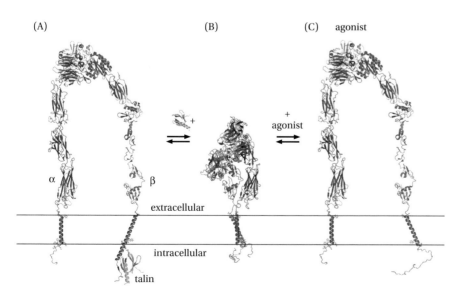

(A)

(B)

(C)   agonist

α

β

extracellular

intracellular

talin

+

+
agonist

Figure 15.44 Functional states of integrins. (A) Model for the inside-out activated state. Talin stabilizes the open form of the integrin, increasing its affinity for extracellular ligands. (B) Resting state, with low affinity for both intra- and extracellular ligands. (C) Outside-in activated state. Binding of an extracellular ligand stabilizes a state where the cytosolic tails are free to bind intracellular effector proteins. (From Lau T-L, Kim C, Ginsberg MH et al. [2009] *EMBO J* 28: 1351–1361. With permission from John Wiley and Sons.)

KEY CONCEPTS

- Bacteria use two-component receptor histidine kinases (RHKs) to sense and respond to environmental cues, such as concentration gradients of metabolically important substrates.

- Mammals mainly use receptor tyrosine kinases (RTKs) and G protein–coupled receptors (GPCRs) to react to extracellular signaling molecules.

- In response to ligand binding, signal transmission by two-component and tyrosine kinase receptors is mediated by dimers and higher aggregates that can assemble or rearrange in response to signals.

- In two-component systems, the RHK autophosphorylates in response to a signal and then passes the phosphate group to a cytoplasmic response regulator that mediates the physiological response.

- RTKs also autophosphorylate in response to external signals, in turn activating other downstream effector proteins in the cytosol. Mammalian cells have large numbers of RTKs that form complex,

interacting networks for controlling physiological processes.

- Ligand binding to GPCRs stabilizes a conformation of the receptor that can bind intracellular G proteins, triggering the exchange of bound GDP for GTP in the G protein, which then dissociates from the receptor and activates downstream effector proteins such as enzymes or ion channels. Rhodopsin is stimulated by light absorption by retinal but uses the same G protein pathway concepts.

- Cadherins and integrins are dimeric proteins that mediate attachment between cells, and between cells and the extracellular matrix, respectively. On the intracellular side, they attach to cytoskeletal actin and intermediate filaments.

- Integrins can mediate outside-in signaling such that changes in the extracellular matrix can modify, e.g., the cytoskeleton. Similarly, integrins can also mediate inside-out signaling, such that changes in the cytoskeleton can modify the organization of the ECM.

FURTHER READING

Hazelbauer, G.L. (2012) Bacterial chemotaxis: The early years of molecular studies. *Annu Rev Microbiol* 66:285–303.

Porter, S.L., Wadhams, G.H., and Armitage, J.P. (2011) Signal processing in complex chemotaxis pathways. *Nature Rev Microbiol* 9:153–165.

Capra, E.J., and Laub, M.T. (2012) Evolution of two-component signal transduction systems. *Annu. Rev. Microbiol.* 66:325–347.

Kearns, D.B. (2010) A field guide to bacterial swarming motility. *Nature Rev Microbiol* 8:634–644.

Lemmon, M.A., and Schlessinger, J. (2010) Cell signaling by receptor tyrosine kinases. *Cell* 141:1117–1134.

Gschwind, A., Fischer, O.M., and Ullrich, A. (2004) The discovery of receptor tyrosine kinases: Targets for cancer therapy. *Nature Rev Cancer* 4:361–370.

Broughton, S.E., Hercus, T.R., Lopez, A.F., and Parker, M.W. (2012) Cytokine receptor activation at the cell surface. *Current Opinion in Structural Biology* 22:350–359.

Wang, X., Lupardus, P., LaPorte, S.L., and Garcia, K.C. (2009) Structural biology of shared Cytokine receptors. *Annu. Rev. Immunol.* 27:29–60.

Smith, S.O. (2010) Structure and activation of the visual pigment Rhodopsin. *Annu. Rev. Biophys.* 39:309–328.

Katritch, V., Cherezov, V., and Stevens, R.C. (2013) Structure-function of the G Protein-Coupled receptor superfamily. *Annu. Rev. Pharmacol. Toxicol.* 53:531–556.

Thal, D.M., Glukhova, A., Sexton, P.M., and Christopoulos, A. (2018) Structural insights into G-protein coupled receptor allostery. *Nature* 559:45–53.

Gul, I.S., Hulpiau, P., Sayes, Y., and van Roy, F. (2017) Evolution and diversity of cadherins and catenins. *Exp. Cell Res.* 358:3–9.

Priest, A.V., Shafraz, O., and Sivasankar, S. (2017) Biophysical basis of cadherin mediated cell-cell adhesion. *Exp. Cell Res.* 358:10–13

Luo, B.-H., Carman, C.V., and Springer, T.A. (2007) Structural basis of integrin regulation and signaling. *Annu. Rev. Immunol.* 25:619–647.

Abram, C.L., and Lowell, C.A. (2009) The Ins and Outs of Leukocyte integrin signaling. *Annu. Rev. Immunol.* 27:339–362.

Kim, C., Ye, F., and Ginsberg, M.H. (2011) Regulation of Integrin activation. *Annu. Rev. Cell Dev. Biol.* 27:321–345.

Wolfenson H., Lavelin I., and Geiger, B. (2013) Dynamic regulation of the structure and functions of integrin adhesions. *Developmental Cell* 24:447–456.

KEY LITERATURE

G Protein Coupled Receptors, Including Rhodopsin

Langley, J.N. (1901) Observation on the physiological action of extracts of the supra-renal bodies. *J Physiol (London)* 17:231–256.

De Lean, A., Stadel, J.M., and Lefkowitz, R.L. (1980) A ternary complex model explains the agonist-specific binding properties and the adenylate cyclase-coupled beta-adrenergic receptor. *J Biol Chem* 255:7108–7117.

Ovchinnikov, Yu. A. (1982) Rhodopsin and bacteriorhodopsin: structure-function relationships. *FEBS Lett.* 148:179-191.

Dixon, R.A., Kobilka, B.K., Strader, D.J., Benovic, J.L., Dohlman, H.G., Frielle, T., Bolanowski, M.A., Bennet, C.D., Rands, E., Diehl, R.E., Mumford, R.A., Slater, E.E., Sigal, I.S., Caron, M.G., Lefkowitz, R.J., and Strader, C.D. (1986) Cloning of the gene and cDNA for mammalian beta-adrenergic receptor: primary structure and membrane topology. *Nature* 321:75-79.

Schertler, G.F., Villa, C., and Hendersson, R. (1993) Projection structure of rhodopsin. *Nature* 362:770-772.

Palczewski, K., Kumasaka, T., Hori, T., Behnke, C.A., Motoshima, H., Fox, B.A., Le Trong, I., Teller, D.C., Okada, T.,

Stenkamp, R.E., Yamamoto, M., and Miyano, M. (2000) Crystal structure of rhodopsin: A G-protein coupled receptor. *Science* 289:739-745.

Scheerer, P., Park, J.H., Hildebrand, P.W., Kim, Y.J., Krauss, N., Choe, H.W., Hofmann, K.P., Ernst, O.P. (2008) Crystal structure of opsin in its G-protein-interacting conformation. *Nature* 455:497-502.

Rasmussen, S.G., DeVree, B.T., Zou, Y., Kruse, A.C., Chung, K.Y., Kobilka, T.S., Thian, F.S., Chae, P.S., Pardon, E., Calinski, D., Mathiesen, J.M., Shah, S.T., Lyons, J.A., Caffrey, M., Gellman, S.H., Steyaert, J., Skiniotis, G., Weis, W.I., Sunahara, R.K., and Kobilka, B.K. (2011) Crystal structure of the human β_2 adrenergic receptor-Gs protein complex. *Nature* 477:549-555.

Two-Component Signaling Systems

Van Alphen, W., and Lugtenbery, B. (1977) Influence of osmolarity of the growth medium and on the outer membrane protein pattern of *Escherichia coli*. *J. Bacteriology* 131:623-630.

Hall, M.N., and Silhavy, T.J. (1981) The *ompB* locus and the regulation of the major outer porin proteins of *Escherichia coli* K12. *J. Mol. Biol.* 146:23-43.

Ferris, H.U., Dunin-Horkawicz, S., Hornig, N., Hulko, M., Martin, J., Schultz, J.E., Zeth, K., Lupas, A.N., and Coles, M. (2012) Mechanism of regulation of receptor histidine kinases. *Structure* 20:56-66.

Cowan, W., Schirmer, T., Rummel, G., Steiert, M., Ghosh, R., Pauptit, R.A., Jansonius, J.N., and Rosenbusch, J.P. (1992)

Crystal structures explain functional properties of two *E. coli* porins. *Nature* 358:727-733.

Wang, L.C., Morgan, L.K., Godakumbura, P., Kenney, L.J., and Anand, G.S. (2012) The inner membrane histidine kinase EnvZ senses osmolality via helix-coil transitions in the cytoplasm. *EMBO J.* 31:2648-2659.

Forst, S., Comeau, D., Norioka, S., and Inouye, M. (1987) Localization and membrane topology of EnvZ, a protein involved in osmoregulation of OmpF and OmpC in *Escherichia coli*. *J Biological Chemistry* 262:16433-16438.

Yoshida, T., Qin, L., Egger, L.A., and Inouye, M. (2006) Transcription regulation of *ompF* and *ompC* by a Single Transcription Factor, OmpR. *J. Biol. Chem.* 281:17114-17123.

Bacterial Motility

Adler, J. (1966) Chemotaxis in bacteria. *Science* 153:708-716.

Adler, J. (1969) Chemoreceptors in bacteria. *Science* 166:1588-1597.

Wolfe, A.J., and Berg, H.C. (1989) Migration of bacteria in semisolid agar. *Proc Natl Acad Sci USA* 86:6973-6977.

Berg, H.C., and Brown, D.A. (1972) Chemotaxis in *Escherichia coli* analyzed by three-dimensional tracking. *Nature* 239:500-504.

Berg, H.C., and Anderson, R.A. (1973) Bacteria swim by rotating their flagellar filaments. *Nature* 245:380-82

Armstrong, J.B., Adler, J., and Dahl, M.M. (1967) Nonchemotactic mutants of *Escherichia coli*. *J Bacteriol* 93:390-98

Armstrong, J.B., and Adler, J. (1967) Genetics of motility in *Escherichia coli*: complementation of paralysed mutants. *Genetics* 56:363-373.

Silverman, M., and Simon, M. (1974) Flagellar rotation and the mechanism of bacterial motility. *Nature* 249:73-74.

Receptor Tyrosine Kinases

Levi-Montalcini, R. (1952) Effects of mouse tumor transplantation on the nervous system. *Ann NY Acad Sci* 55:330-340.

Carpenter, G., Lembach, K.J., Morrison, M.M., and Cohen, S. (1975) Characterization of the binding of 125I-labeled epidermal growth factor to human fibroblasts. *J Biol Chem* 250:4297-4304.

Carpenter, G., King, L. Jr, and Cohen, S. (1978) Epidermal growth factor stimulate phosphorylation in membrane preparations *in vitro*. *Nature* 276:409-410.

Ushiro, H., and Cohen, S. (1980) Identification of phosphotyrosine as a product of epidermal growth factor-activated protein kinase in A-431 cell membranes. *J Biol Chem* 255:8363-8365.

Ullrich, Coussens, L., Hayflick, J.S., Dull, T.J., Gray, A., Tam, A.W., Lee, J., Yarden, Y., Libermann, T.A., Schlessinger, J.,

Downwardt, J., Mayest, E.L.V., Whittlet, N., Waterfieldt, M.D., and Seeburg, P.H. (1984) Human epidermal growth factor receptor cDNA sequence and aberrant expression of the amplified gene in A431 epidermoid carcinoma cells. Nature 309:418-425.

Downward, Yardent, Y., Mayes, E., Scrace, G., Totty, N., Stockwell, P., Ullrich, A., Scblessingert, J., and Waterfield, M.D. (1984) Close similarity of epidermal growth factor receptor and v-erb-B oncogene protein sequences. *Nature* 307:521-527.

Privalesky, M.L., Ralston, R., and Bishop, J.M. (1984) The membrane glycoprotein encoded by the retroviral oncogene v-erb-B is structurally related to tyrosine-specific protein kinases. *Proc Natl Acad Sci USA* 81:704-707.

Cahderins and Integrins

Townes, P.L., and Holtfreter, J. (1955) Directed movements and selective adhesion of embryonic amphibian cells. *J. Exp. Zool.* 128:53–120.

Hynes, R.O., and Destree, A.T. (1978) Relationships between fibronection (LETS protein) and actin. *Cell* 15:875–886.

Geiger, B. (1979) A 130K protein from chicken gizzard: Its localization at the termini of microfilament bundles in cultured chicken cells. *Cell* 18:193–205.

Singer, I.I. (1979) The fibronexus: A transmembrane association of fibronectin-containing fibers and bundles of 5 nm microfilaments in hamster and human fibroblasts. *Cell* 16:675–685.

Neff, N.T., Lowrey, C., Decker, C., Tovar, A., Damsky, C., Buck, C., and Horwitz, A.F. (1982) A monoclonal antibody detachs embryonic skeletal muscle from extracellular matrices. *J Cell Biol* 95:654–666.

Knudsen, K.A., Horwitz, A.F., and Buck, C.A. (1985) A monoclonal antibody identifies a glycoprotein complex involved in cell-substratum adhesion. *Exp Cell Res* 157:218–226.

Tamkun, J.W., DeSimone, D.W., Fonda, D., Patel, R.S., Buck, C., Horwitz, A.F., and Hynes, R.O. (1986) Structure of integrin, a glycoprotein involved in the transmembrane linkage between fibronectin and actin. *Cell* 46:271–282.

Lau, T.-L., Kim, C., Ginsberg, M.H., and Ulmer, T.S. (2009) The structure of the integrin αIIbβ3 transmembrane complex explains integrin transmembrane signaling. *EMBO J* 28:1351–1361.

EXERCISES

1. Bearing in mind the fluidity of the lipid bilayer, why does it make sense that oligomers would be involved in signaling by RTK receptors? (Hint: how could a monomer get a specific signal across?)

2. There are ~60 receptor tyrosine kinases and ~ 800 G protein–coupled receptors in the human genome. Compare the diversity of signals activating these two classes, including molecular sizes, polarities, and light. Propose an explanation for the difference in the abundance of these receptors.

3. Give two plausible reasons why GPCRs are favorable targets for drugs.

4. What is the characteristic morphology of a flagellum that enables bundle formation for counterclockwise rotation but causes disassembly for clockwise formation?

Electrostatics Appendix

I FROM THE HUMBLE CAPACITOR: PRINCIPLES OF ELECTROSTATICS AND THE POISSON EQUATION

Descriptions of the electrostatic properties of charged membranes begin with the Poisson equation that connects electric potentials to spatially distributed charges. We derive it informally here. The Poisson equation is a consequence of fundamental principles of electrostatics embodied in the first of James Clerk Maxwell's (1831–1879) four equations, which summarize compactly all of the principles of electricity and magnetism. The principles of the first equation, however, represent the work of other scientists, particularly Charles-Augustin de Coulomb (1736–1806) and Johann Carl Friedrich Gauss (1777–1855). We introduced in Chapter 2 the idea of using electrical capacitance to describe the properties of lipid bilayers, but we did not explain the origin of the equation for capacitance (Equation 2.5). We do that here as a way of reviewing the basic principles of electrostatics and as a starting point for deriving the Poisson equation.

Coulomb's famous inverse-square law describes mathematically the force between two charges (Q and q; **Figure 1A**) in vacuum separated by a distance r:

$$F = k\frac{Qq}{r^2} = \frac{1}{4\pi\varepsilon_0}\frac{Qq}{r^2} \tag{1}$$

The constant $k = 1/4\pi\varepsilon_0$ indicates the use of Standard International (SI) units, which allow all physical units to be connected to fundamental standards of length, time, mass, and the speed of light. The so-called permittivity of free space, $\varepsilon_0 \approx 8.854 \times 10^{-12}$ Farads meter^{-1}, is a scale factor on the dielectric constant of a material. When considering electric fields in a dielectric medium with dielectric constant ε, ε_0 is replaced by $\varepsilon\varepsilon_0$. For a vacuum, $\varepsilon = 1$. The electric field (force per unit charge), which is the centerpiece of electrostatics, is defined from Equation 1 as F/q (**Figure 1B**):

$$E = \frac{Q}{4\pi\varepsilon_0 r^2} \tag{2}$$

F and E are vectors that indicate the directions of the Coulomb force and electric field in space. But to simplify the presentation, we consider point charges or flat surfaces that do not require the use of vector algebra. We represent schematically the intensity of electric fields surrounding a charge by using radial lines, called flux lines (Figure 1B); the greater the field intensity, the larger the number of lines emanating from the charge. A revolutionary advance in electrostatic theory was made by Gauss when he formalized this idea. He defined the electrical flux Φ as equal to $E \cdot S$, where S is the area of an imaginary surface surrounding the charge through which the flux lines pass. To demonstrate this idea, in Panel A, an imaginary sphere of radius r (dashed line), surrounding and centered on Q, has been drawn. The flux through the sphere's surface of area $S = 4\pi r^2$ is $E \cdot 4\pi r^2$. Multiplying both sides of Equation 2 by the sphere's area yields the result that $E \cdot S = Q/\varepsilon_0$. Ah ha! This suggests that if one integrates $E \cdot S$ over a closed surface—any closed surface regardless of shape—one can determine the

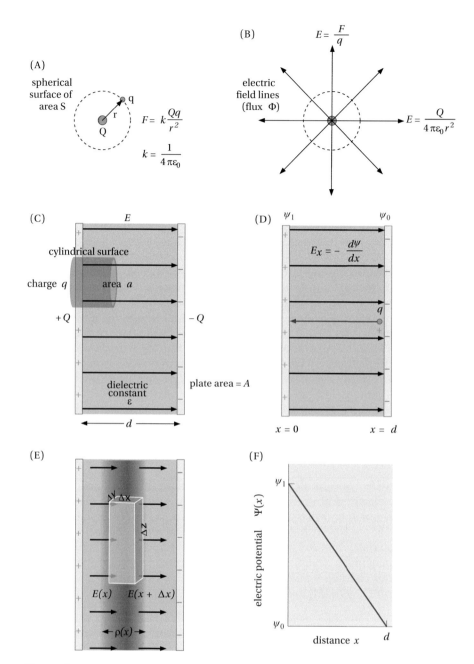

Figure 1

amount of charge contained within the surface. This is Gauss's law (Maxwell's First Equation), which is written formally as

$$\oint_S E \cdot dS = \frac{Q}{\varepsilon_0} \qquad (3)$$

The integral symbol with the superimposed circle indicates that the integration is over the entire surface S. Importantly, the surface must be closed; i.e., the charge must be completely surrounded; there can be no gaps. The neat thing is that the electric field emanating from the surface depends only on the charge contained within the surface. So, if one draws a closed surface of any shape around any charge distribution, even in the presence of neighboring charges, the associated electrical field is determined solely by Equation 3!

Gauss's law makes it almost trivial to calculate the intensity of the uniform electric field of a charged parallel-plate capacitor. As illustrated in **Figure 1C**, construct a right-cylindrical Gauss surface (called a Gauss "pillbox") that includes the surface charges q contained within the box. Because the electric field is normal to the capacitor plates and parallel to the sides of the pillbox, the electric flux passes through only the end of the pillbox (area = a) (electric fields are always 0 inside a conductor, so we need consider only the end of the box jutting into the space between the plates). Applying Gauss's law, $Ea = q/\varepsilon_0$ (or $q/\varepsilon\varepsilon_0$ if there is a dielectric medium between the plates). Dividing by a, the electric field between the plates is

$$E = q/a\varepsilon_0 = \sigma/\varepsilon_0 \tag{4}$$

and $\sigma = q/a$ is the surface charge density on the plate.

The field is uniform and independent of distance away from the plate. The electric potential $\psi(x)$ between the plates, however, does depend on distance (**Figure 1D**). The electric potential is useful because it provides a convenient way of describing the work W involved in moving a charge in the field. To move the positive charge q from $x = d$ to $x = 0$ (Panel D), we must apply a force $F = -qE$ (negative sign because our applied force must be opposite to the field direction). The work done is given by $w = Fd = -qEd$. Compared with measurements of electric field strength, measurements of electric potential are easily done. We therefore introduce the electric potential $\psi(x)$, defined so that the work of moving charges in an electric field can be computed from differences in potential:

$$w = q\Delta\psi = q\left[\psi(0) - \psi(d)\right]$$

To do this, we need a relationship between E and ψ. That is easy: expend work w to move the charge q an infinitesimal distance dx through the electric field so that the potential changes an infinitesimally small amount $d\psi$. Energy must be conserved, which means that $w = qd\psi = -qEdx$. Dividing by q yields the important result that

$$E_x = -\frac{d\psi}{dx} \tag{5}$$

The subscript on E has been introduced to remind us that for more general cases, we must use vector calculus. The quantity $d\psi/dx$ is referred to as the gradient of the electric potential. In the case of our parallel-plate capacitor, the electric field is constant (Equation 4), and so is the gradient (Equation 5). This is summarized in **Figure 1F**. The voltage V across the capacitor is given by $V = \psi_1 - \psi_0$. Absolute potentials do not exist; all potentials are measured relative to some reference value, which can be chosen arbitrarily. We can thus choose ψ_0 as the reference potential and set it equal to 0, in which case $\psi_1 = V$. Due to the constant field (or gradient of the electric potential), we can also say that $E/d = V$. The formula for capacitance (Equation 2.5) is now easily derived using the definition of capacitance (Equation 2.4) and Equation 4, assuming that the dielectric constant $\varepsilon > 1$.

The last step toward deriving the Poisson equation is to convert the integral form of Gauss's law (Equation 3) into the differential form. Shown in Figure 1E is a parallel-plate capacitor that contains, in addition to the dielectric medium, a region of additional fixed charge described by the charge density $\rho(x)$. For illustrative purposes, we have arranged the density so that it varies only along the x-axis; i.e., for any value of x, there is no variation along y or z. To figure out how this additional charge affects the electric field within the capacitor, construct a rectangular Gauss surface with dimensions Δx, Δy, and Δz. The field entering the surface from the left is $E(x)$, and because of the charge contained in the box, the field exits with a value $E(x + \Delta x)$. The net electric flux through the box's surface will be $E(x + \Delta x) \cdot \Delta y\Delta z - E(z) \cdot \Delta y\Delta z$ and must equal the charge contained within the box. If the box is so thin that $\rho(x)$ is essentially constant over the distance Δx, the charge inside the thin box will be $\rho(x)\Delta x\Delta y\Delta z$. By Gauss's law,

$$\left[E(x+\Delta x)-E(X)\right]\Delta y\Delta z = \frac{\rho(x)\Delta x\Delta y\Delta z}{\varepsilon\varepsilon_0}$$

or

$$\frac{\left[E(x+\Delta x)-E(x)\right]}{\Delta x} = \frac{\rho(x)}{\varepsilon\varepsilon_0} \tag{6}$$

As is the usual practice in differential calculus, we take the limit of the left-hand side as Δx approaches 0:

$$\lim_{\Delta x \to 0}\frac{\left[E(x+\Delta x)-E(x)\right]}{\Delta x} = \frac{dE(x)}{dx}$$

This yields from Equation 6 the differential form of Gauss's law:

$$\frac{dE(x)}{dx} = \frac{\rho(x)}{\varepsilon\varepsilon_0} \tag{7}$$

We can now take advantage of the relationship between the electric field and the field gradient (Equation 5) to arrive at Poisson's equation:

$$\frac{d^2\psi(x)}{dx^2} = -\frac{\rho(x)}{\varepsilon\varepsilon_0} \tag{8}$$

II Determination of the Distribution of Ions around Membranes Using the Poisson–Boltzmann Equation: Gouy–Chapman Theory

We have shown that the charge distribution in a capacitor containing a fixed charge distribution $\rho(x)$ is related to the electric potential through the Poisson equation (Equation 8). The equation can be used to describe how a charged membrane affects the distributions of ions and other charged molecules in its vicinity. Unlike the example of a fixed distribution of charge, the ions in solution are mobile and subject to thermal motion. In the absence of the charged membrane, the ions in solution are uniformly distributed in a way that satisfies electroneutrality; i.e., in any volume of space, there are equal numbers of negative and positive charges. But if a charged membrane is present, the situation changes. All biological membranes are negatively charged. This means that positive ions will be attracted to and negative ions repelled from the membrane, which must lead to a non-uniform distribution in which the concentration of cations is higher in the vicinity of the membrane and the concentration of anions is lower (Figure 2). This non-uniform distribution is referred to as a diffuse double layer.

Several questions arise. How strong is the electric potential at the surface of the membrane? How does the potential depend on membrane surface charge density? How does the surface potential change with distance and the bulk ion concentration? To answer these questions, the Poisson equation must be modified to account for the thermal motion of ions and the effect of the electric potential on that motion. This can be done using the Boltzmann function and the principles underlying the Nernst equation (Equation 0.30). Let's consider a symmetrical salt solution with a concentration of N moles of salt per liter; symmetrical means that the cations and anions have the same valence z. Very far away from the membrane where $\psi \approx 0$, the number of cationic and anionic charges must always be equal, so that $n^+(x) = n^-(x) = N$, where $n^+(x)$ and $n^-(x)$ are the concentrations of cations and anions, respectively. The concentrations

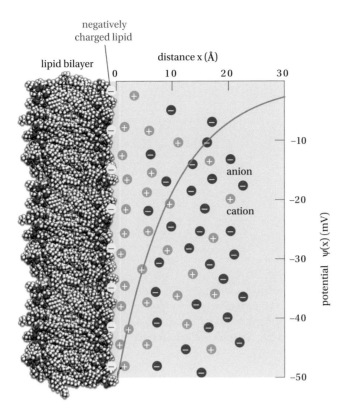

negatively
charged lipid

lipid bilayer

distance x (Å)

anion

cation

potential ψ(x) (mV)

Figure 2

are not equal close to the membrane, where the concentrations are described by the Boltzmann equations:

$$n^+(x) = N \exp\left[\frac{-F\psi(x)}{RT}\right] \qquad (9a)$$

$$n^-(x) = N \exp\left[\frac{+F\psi(x)}{RT}\right] \qquad (9b)$$

The charge density in the vicinity of the membrane will then be given by

$$\rho(x) = zF\left[n^+(x) - n^-(x)\right] \qquad (10)$$

The Poisson–Boltzmann equation is obtained by combining Equations 8–10 and recalling the definition of the hyperbolic sine function, which is $\sinh(x) = (e^x - e^{-x})/2$:

$$\frac{d^2\psi(x)}{dx^2} = \frac{2zFN}{\varepsilon\varepsilon_0} \sinh\left(\frac{zF\psi(x)}{RT}\right) \qquad (11)$$

This rather innocent-looking second-order differential equation has a very cumbersome-looking solution after setting the required boundary conditions. The potential at the membrane surface is defined as ψ_0, but very far away from the membrane, the potential must be very small. Besides these obvious conditions, there is another. As x increases without bound and $\psi(x)$ approaches 0, the curve becomes flat, described as having a slope of 0. The boundary conditions are thus

$$\psi(x) = \psi_0 \text{ for } x = 0; d\psi(x)/dx \text{ and } \psi(x) \to 0 \text{ as } x \to \infty \qquad (12)$$

With these boundary conditions, the solution to Equation 4 is

$$\psi(x) = \frac{2RT}{zF} \ln\left(\frac{1 + \alpha \exp(\kappa x)}{1 - \alpha \exp(\kappa x)}\right) \tag{13}$$

where

$$\alpha = \frac{\exp(zF\psi_0/2RT) - 1}{\exp(zF\psi_0/2RT) + 1} \tag{14}$$

and

$$\kappa = \left(\frac{2z^2 F^2 N}{\varepsilon\varepsilon_0}\right)^{\frac{1}{2}} \tag{15}$$

The constant κ is useful. Its inverse, $1/\kappa$, defines a distance referred to as the Debye length, which is a measure of the decay of $\psi(x)$ with distance away from the membrane. Happily, Equation 13 can be simplified if ψ_0 is not large, which is the usual case in cells. When that is true, $\exp(zF\psi_0/2RT) \approx 1 + zF\psi_0/2RT$, and Equation 6 becomes

$$\psi(x) = \psi_0 \exp(-\kappa x) \tag{16}$$

meaning that $\psi(x)$ falls to ψ_0/e at $x = 1/\kappa$. It is handy to have a mathematical relation between charge density σ and ψ_0. Electroneutrality demands that $\sigma = -\int_0^\infty \rho(x)dx$. By using the boundary conditions of Equation 12 and carrying out the first integration of Equation 4, the Gouy equation is obtained:

$$\sigma = (8N\varepsilon\varepsilon_0 RT)^{1/2} \sinh\left(\frac{zF\psi_0}{2RT}\right) \tag{17}$$

Index